HANDBOOK OF
CHARACTER RECOGNITION
AND
DOCUMENT IMAGE ANALYSIS

HANDBOOK OF CHARACTER RECOGNITION AND DOCUMENT IMAGE ANALYSIS

Editors

H. Bunke
Institut für Informatik und Angewandte Mathematik
Universität Bern

P. S. P. Wang
College of Computer Science, Northeastern University

World Scientific
Singapore • New Jersey • London • Hong Kong

Published by

World Scientific Publishing Co. Pte. Ltd.

P O Box 128, Farrer Road, Singapore 912805

USA office: Suite 1B, 1060 Main Street, River Edge, NJ 07661

UK office: 57 Shelton Street, Covent Garden, London WC2H 9HE

British Library Cataloguing-in-Publication Data
A catalogue record for this book is available from the British Library.

HANDBOOK OF CHARACTER RECOGNITION AND DOCUMENT IMAGE ANALYSIS

ISBN 981-02-2270-X

Printed in Singapore by Uto-Print

CONTENTS

CHARACTER RECOGNITION

SPECIAL APPLICATIONS AND SYSTEMS

BASIC METHODOLOGY

Handbook of Character Recognition and Document Image Analysis, pp. 1–47
Eds. H. Bunke and P. S. P. Wang

CHAPTER 1

IMAGE PROCESSING METHODS FOR DOCUMENT IMAGE ANALYSIS

THIEN M. HA
Institut für Informatik und angewandte Mathematik
Universität Bern
Neubrückstrasse 10, CH-3012 Berne
Switzerland

and

H. BUNKE
Institut für Informatik und angewandte Mathematik
Universität Bern
Neubrückstrasse 10, CH-3012 Berne
Switzerland

This chapter describes image processing methods for document image analysis. The methods are grouped into four categories, namely, image acquisition, image transformation, image segmentation, and feature extraction. In image acquisition, we describe the process of converting a document into its numerical representation, including image coding as a means to reduce the storage requirement. Image transformation addresses image-to-image operations, which comprise a large spectrum of techniques ranging from geometrical correction, filtering and figure-background separation to boundary detection and thinning. In image segmentation, we describe four popular techniques, namely, connected component labeling, X-Y-tree decomposition, run-length smearing, and Hough transform. Finally, a number of feature extraction methods, which constitute the basis of image classification, are presented.

Keywords: Image acquisition; Image transformation; Segmentation; Feature extraction; Spatial sampling; Quantization; Image coding; Geometrical correction; Filtering; Figure-background separation; Boundary detection; Thinning; Statistical features.

1. Introduction

Document image analysis, as its name indicates, is a subfield of image analysis. This implies that, on the one hand, it inherits the more general techniques of image analysis, and on the other, it can serve as a platform for testing various image analysis methodologies. Furthermore, with time, document image analysis has also acquired its own techniques, specifically designed for its needs. The aim of this chapter is to introduce the reader to the basic analysis techniques that have been proposed in literature or are currently used in various commercial systems. However, the main concern is to provide a tutorial rather than a description of the most up-to-date techniques, which are subjects of the other chapters of the handbook.

This chapter comprises four main sections, namely, image acquisition, image transformation, image segmentation, and an introduction to feature extraction. Image acquisition (Sec. 2) describes the process of converting a document into its numerical representation, which is suitable for subsequent processing on a digital computer. Image transformation (Sec. 3) addresses image-to-image operations, which comprise a large spectrum of techniques ranging from geometrical correction and filtering to line detection and representation. Image transformation can be applied globally to the whole image or locally to some specific areas, in which case a prior segmentation (Sec. 4) is required. Section 5 gives an introduction to feature extraction, which is the basis of image classification (see chapter 2 of this book).

2. Image Acquisition

In order to process a document on a digital computer, the very first step consists in converting the document into a numerical representation. The conversion process is physically accomplished by a *digitiser*, which can either be a scanner or a camera. There exists another type of digitiser, called "tablet digitiser", which acquires in real-time the position of the pen tip as the user writes and thus does not operate on a document. This section addresses the conversion process from a formal point of view. That is, we consider the document image as an analog two-dimensional signal that has first to be *spatially sampled.* Then each sample value is *quantized* into one of a finite number of gray-levels. The amount of data thus obtained may be very large in some applications where high quality is required. Many *coding* techniques are available to reduce the amount of data to be stored or processed. Sections 2.1 to 2.3 address the spatial sampling problem, quantization, and image coding with emphasis on the case of binary images, respectively.

2.1. *Sampling*

In this section we consider the spatial sampling of a document image. The document image is considered as a function $i(x, y)$ of two independent spatial continuous variables x and y, i.e., $i(x, y)$ represents the gray-level of the document image at the position $(x, y) \in \mathcal{D} \subset R^2$, where $\mathcal{D}$ corresponds to the spatial extent of the document. The spatial sampling is defined as the process of evaluating $i(x, y)$ at discrete positions, being usually the cross points of a square grid. The process results in a sampled function of the form $i_s(k \cdot \Delta x, l \cdot \Delta y); k, l \in N$. Δx and Δy are the *spatial sampling periods* in x- and y-direction, respectively. In the following we will discuss two intertwined aspects of spatial sampling, namely, how to obtain the value of i_s at a given position from $i(x, y)$, and the choice of Δx and Δy. In general, it can be said that the former is a well understood problem in the context of sampling theory, whereas the latter is highly application-dependent.

At the heart of the sampling process are the conditions under which $i(x, y)$ can be reconstructed from $i_s(k \cdot \Delta x, l \cdot \Delta y)$ exactly, or at least to a good degree of

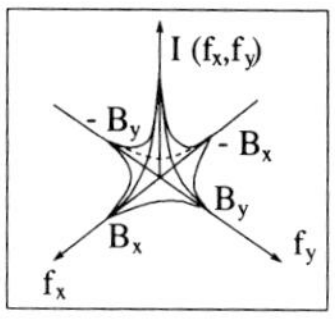

(a)

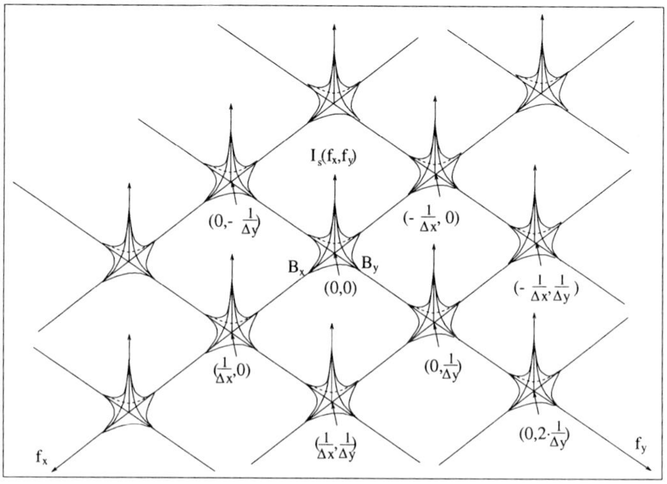

(b)

Fig. 1. Illustration of the Sampling Theorem. (a) Fourier transform of an image $I(f_x, f_y)$ and (b) Fourier transform of its sampled version $I_s(f_x, f_y)$.

approximation. The answer is provided by Shannon's Sampling Theorem [1, 2, 3] given below.

Theorem 1. If the Fourier transform of $i(x, y)$, $\mathcal{F}\{i(x, y)\} = I(f_x, f_y)$, is zero for $|f_x| > B_x$ and $|f_y| > B_y$, then it is possible to choose Δx and Δy sufficiently small that $i(x, y)$ can be exactly reconstructed from $i_s(k \cdot \Delta x, l \cdot \Delta y)$. More precisely, Δx and Δy must be chosen such that $\frac{1}{\Delta x} > 2 \cdot B_x$ and $\frac{1}{\Delta y} > 2 \cdot B_y$.

The quantities B_x and B_y are the bandwidths in the x- and y-direction, whereas $\frac{1}{\Delta x}$ and $\frac{1}{\Delta y}$ are called the sampling frequencies in the corresponding direction. Figure 1 illustrates the Fourier transforms of $i(x, y)$ and $i_s(k \cdot \Delta x, l \cdot \Delta y)$. Notice that $I_s(f_x, f_y)$, the Fourier transform of i_s, is obtained by repeating $I(f_x, f_y)$ at frequencies that are integer multiples of $\frac{1}{\Delta x}$ and $\frac{1}{\Delta y}$. The Sampling Theorem ensures that different repeated versions of $I(f_x, f_y)$ do not overlap each other so that an exact reconstruction is possible.

In practice, $i(x, y)$ may not be band-limited, i.e., B_x and B_y may tend to infinity. For such a signal, it is theoretically impossible to have an exact reconstruction from its sampled version. The question then becomes how we can still sample the signal without losing too much information. The technique usually adopted is to approximate $i(x, y)$ by a band-limited signal $i_1(x, y)$ and then to sample $i_1(x, y)$ yielding $i_1(k, l)$. This technique is called *anti-aliasing* since it prevents different repeated versions from being overlapped. The approximation of $i(x, y)$ by $i_1(x, y)$ can be achieved by a *low-pass filtering*, i.e., keeping the low-frequency components of $i(x, y)$ unchanged while eliminating its high-frequency components (large values of f_x and f_y). A simple version of low-pass filtering consists in assigning to $i_1(x, y)$ the mean value of $i(x, y)$ over a neighborhood centered at position (x, y).

As the above technique was derived from the frequency domain, it is legitimate to ask what it means in the spatial domain, and particularly how important it is with respect to document images. To get an insight of the problem, let us examine an example where the main concern is to preserve positional information of *edge points*. Positional information is very important since an error in position may yield an error in size, which in turn induces an error on an object's shape. In Fig. 2, $i(x)$ represents a line of an image with two abrupt changes, corresponding to edge points. Such a signal is present everywhere on the page you are reading; the edge points represent the transitions from black to white and vice versa. It is well known that this signal is not band-limited [4]. The function $i(k)$ is obtained by sampling $i(x)$ without anti-aliasing filtering, whereas $i_1(k)$ is the sampled version of $i_1(x)$, which is a low-pass filtered version of $i(x)$. Let us consider the following simple edge detector operating on the sampled signals $i(k)$ and $i_1(k)$. The samples are linearly interpolated and edge points are those where the interpolated signal takes the value of $\frac{1}{2}$. It is clear from Fig. 2 that $i_1(k)$ provides much more precise edge locations than $i(k)$. The edge detector we use here is not a conventional one[a], it merely serves the purpose of showing that $i_1(k)$ preserves more information than $i(k)$ does, illustrating the importance of the anti-aliasing technique in sampling document images.

In practice, the anti-aliasing filtering is achieved electronically by the digitiser (i.e., camera or scanner) and the only action the user can take is to check whether it has been well designed. This can be done by submitting a calibration image containing high frequency components to the digitiser. The resulting image should not contain any Moiré patterns for a well designed digitiser [6].

The anti-aliasing technique is especially useful for the *subsampling* (or downsampling) of a digital image, initially sampled at high resolution, to a lower resolution. The image is first blurred by a low-pass filtering and then subsampled. For simplicity let us assume that $\Delta\ x = \Delta\ y = \Delta$. If we use a Finite Impulse Response low-pass filter of length $L \cdot \Delta$, the Sampling Theorem says that the subsampling frequency $(1/\Delta_{sub})$ must be at least twice the cutoff frequency of the low-pass filter, which is of the order of $1/(L \cdot \Delta)$. That is, $(1/\Delta_{sub}) > 2/(L \cdot \Delta)$ or $L > 2 \cdot (\Delta_{sub}/\Delta)$, where Δ_{sub}/Δ is the subsampling factor. For instance, if we want to downsample the signal by a factor 3, the low-pass filter must compute each output sample value by averaging over at least 6 input sample values. In [7], a subsampling factor of 2 is achieved by averaging over five input samples via a Gaussian-like low-pass filter. A

[a]See [5] for an example of subpixel accuracy edge detection.

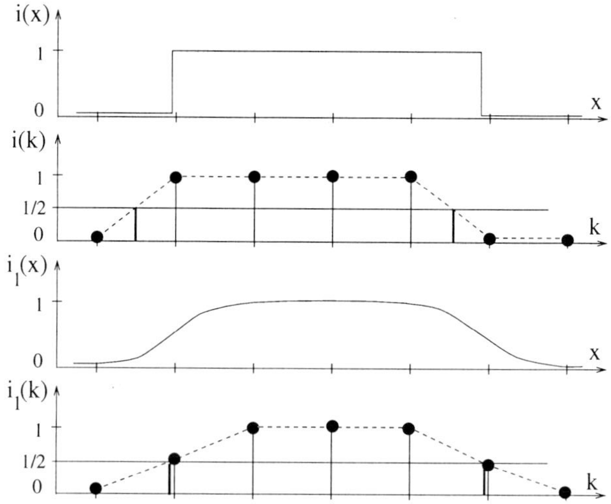

Fig. 2. Preservation of positional information.

precise treatment can be found in [4]. This kind of technique is currently used for character classification in one of the most accurate commercial Optical Character Recognition (OCR) products [8].

Now that we have reviewed the background of the sampling process, let us examine the problem of determining the spatial resolutions. Generally speaking, the higher the resolution is, the more the information is preserved. However, a too high resolution may also reveal the texture of the paper which is of little interest to document analysis. Moreover, high resolution also requires more sophisticated and expensive equipment, and produces a huge amount of data which may not be necessary for the purpose at hand. Eventually, the choice of spatial resolution is determined by two factors, namely, the contents of the document and the purpose of the subsequent operations. Indeed, a text page printed with a large font requires a lower resolution sampling to remain readable than the same text printed with a smaller font. If the operation that follows the sampling is binarization, we need a higher resolution than if the gray-levels are preserved [9, 10]. Simon proposes to choose Δ such that after binarization the thinnest line structure in the document image contains at least 3 pixels in width so as to avoid artefacts in subsequent operations [11]. In the literature, different spatial resolutions have been proposed for various applications. For instance, Dengel *et al.* use a very low resolution (75 dots per inch (dpi)) in their study of block segmentation of business letters [12], while high-accuracy machine printed OCR systems utilise a spatial resolution from 300 to 400 dpi for "normal" texts [13].

The variety of spatial resolutions used by different authors suggests that a complex application involving many subtasks may work best by using one appropriate resolution for the each subtask. This requires an initial high-resolution sampling; the lower resolutions may then be obtained using the downsampling technique described above (see [14] for more detail).

In short, the sampling process is a necessary step in converting a document into electronic form. It can cause a loss of information. However, this loss can be controlled by anti-aliasing, the parameters of which are governed by the Sampling Theorem. The same technique is also applicable when we want to downsample a digital image, initially sampled at a higher resolution. Sampling theory does not dictate the resolution at which a document must be sampled. This is eventually application-dependent.

2.2. *Quantization*

Samples produced by the sampling process take on continuous real values and must therefore be quantized into a finite number of gray-levels so that they can be processed by a digital computer. If the number of gray-levels, L, is equal to two, quantization is also called *binarization*. Binarization is a popular operation in document image analysis for two reasons. First, most document images look like having only two tones, namely, black and white, so the idea of binarization seems natural. Second, binarization provides results which drastically simplify the subsequent operations, such as segmentation and recognition. Therefore, we will first consider the general quantization briefly and then concentrate on binarization.

Physically, quantization is accomplished by an electronic device called "analog-to-digital converter", which is incorporated in modern digitisers. Figure 3 shows a typical quantizer's characteristic function. This particular quantizer converts each real sample value into one of eight discrete levels, which are coded by 3 bits using the natural binary code. The values v_j and r_j are called *decision* and *reconstruction levels*, respectively. In some digitisers, the user has the possibility of choosing the number of gray-levels (equal to the number of reconstruction levels), which is usually an integer power of two. In other systems, this number is fixed to a rather large value, typically 256. The quantizer of Fig. 3 is *uniform* in that the decision levels are regularly spaced; general quantizers as those studied below do not necessarily follow this strict rule.

It happens that the user decides to perform the quantization in two steps. In the first step, a large number, say 256, of gray-levels is used so that the quantizer's characteristic function is almost linear up to saturation. In the second step, a smaller number, say 2 to 8, of gray-levels is used upon the result of the first step. The main interest of this two-step approach is that the first step provides us with a numerical representation that allows sophisticated algorithms, e.g. histogramming, to determine an optimal characteristic function for the second step.

The most commonly used optimality criterion is the mean square error (MSE) between the sample value $\mathbf{v}$ and its reconstruction level $r(\mathbf{v})$.

$$E\{\mathbf{e}^2\} = E\{[\mathbf{v} - r(\mathbf{v})]^2\} = \int_0^{\infty} [v - r(v)]^2 p(v) dv \tag{2.1}$$

where $p(v)$ is the probability density function of the random variable $\mathbf{v}$, which can

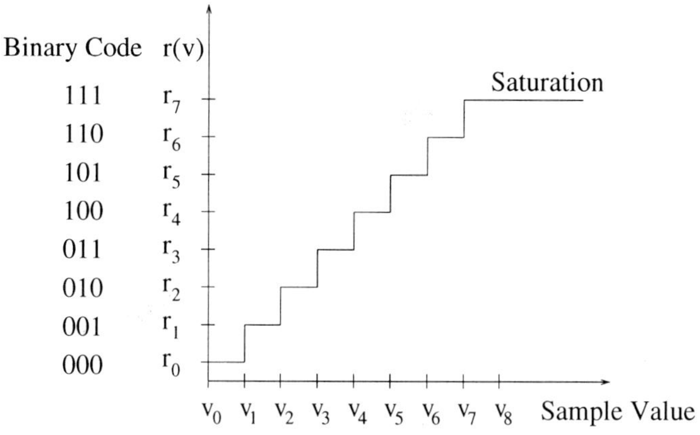

Fig. 3. Quantizer's characteristic function.

be either known *a priori* for a given class of documents or approximated by the histogram of the sample values in the two-step approach. For a given number L of gray-levels, the MSE can be expressed by:

$$E\{\mathbf{e}^2\} = \sum_{j=0}^{L-1} \int_{v_j}^{v_{j+1}} (v - r_j)^2 p(v) dv \tag{2.2}$$

since $r(v) = r_j$ is constant within the interval $[v_j, v_{j+1}]$.

For a given $p(v)$ and a fixed number of reconstruction levels L, the decision levels $v_j, j = 1,.., L-1$ and reconstruction levels $r_j, j = 0,.., L-1$ that minimize the MSE obey the following relations [15]:

$$v_j = \frac{r_{j-1} + r_j}{2}; \qquad j = 1,.., L-1 \tag{2.3}$$

$$r_j = \frac{\int_{v_j}^{v_{j+1}} v p(v) dv}{\int_{v_j}^{v_{j+1}} p(v) dv}; \quad j = 0,.., L-1 \tag{2.4}$$

However, no closed-form solutions exist unless some approximations are accepted. Otherwise, numerical methods must be used, which lead to the tabulation of v_j and r_j for many standard parametric distributions, such as uniform, Gaussian, Laplacian and Rayleigh [1, 2].

The particular but important case of $L = 2$, where we wish to binarize the image, is now examined. The MSE becomes

$$E\{\mathbf{e}^2\} = \int_{v_0}^{v_1} (v - r_0)^2 p(v) dv + \int_{v_1}^{v_2} (v - r_1)^2 p(v) dv \tag{2.5}$$

Assuming that $p(v)$ can be estimated from the histogram, and v_0 and v_2 correspond to v_{min} and v_{max} respectively, there remain three parameters to be computed, namely, r_0, r_1 and v_1. The parameter v_1 is called *binarization threshold.* Moreover,

$r_0(v_1)$ and $r_1(v_1)$ that minimize the MSE, for a given value of v_1, are simply the mean values over the corresponding interval (Eq. (2.4)):

$$r_0(v_1) = \frac{\int_{v_0}^{v_1} vp(v)dv}{\int_{v_0}^{v_1} p(v)dv} \tag{2.6}$$

$$r_1(v_1) = \frac{\int_{v_1}^{v_2} vp(v)dv}{\int_{v_1}^{v_2} p(v)dv} \tag{2.7}$$

Therefore it is sufficient to vary v_1 from v_0 to v_2, compute the MSE by replacing r_0 and r_1 by $r_0(v_1)$ and $r_1(v_1)$, respectively, and choose the v_1^* that minimizes the MSE.

Otsu proposed a strictly equivalent but computationally simpler criterion based on discriminant analysis [16]. In this formulation, the MSE is equal to the within-class variance $\sigma_W^2(v_1)$. If $\sigma_W^2(v_1)$ is added to the between-class variance $\sigma_B^2(v_1)$, the total variance σ_T^2 (independent of v_1) is obtained. Therefore, instead of minimizing the MSE, Otsu's algorithm maximizes the between-class variance:

$$v_1^* = arg\ max\{p_0(v_1)[\mu_0(v_1) - \mu_T]^2 + p_1(v_1)[\mu_1(v_1) - \mu_T]^2\} \tag{2.8}$$

where

$$p_0(v_1) = \omega(v_1) \tag{2.9}$$

$$p_1(v_1) = 1 - \omega(v_1) \tag{2.10}$$

$$\mu_0(v_1) = \mu(v_1)/\omega(v_1) \tag{2.11}$$

$$\mu_1(v_1) = \frac{\mu_T - \mu(v_1)}{1 - \omega(v_1)} \tag{2.12}$$

$$\mu_T = \mu(v_2 = v_{max}) \tag{2.13}$$

and

$$\omega(v_1) = \int_{v_0}^{v_1} p(v)dv \tag{2.14}$$

$$\mu(v_1) = \int_{v_0}^{v_1} vp(v)dv \tag{2.15}$$

Equation (2.8) can be simplified to

$$v_1^* = arg\ max\{\frac{[\mu_T \cdot \omega(v_1) - \mu(v_1)]^2}{\omega(v_1)[1 - \omega(v_1)]}\} \tag{2.16}$$

Once v_1^* has been determined, the reconstruction levels r_0 and r_1 can be computed using Eqs. (2.6) and (2.7), or equivalently:

$$r_0^* = \mu(v_1^*)/\omega(v_1^*) \tag{2.17}$$

$$r_1^* = \frac{\mu_T - \mu(v_1^*)}{1 - \omega(v_1^*)} \tag{2.18}$$

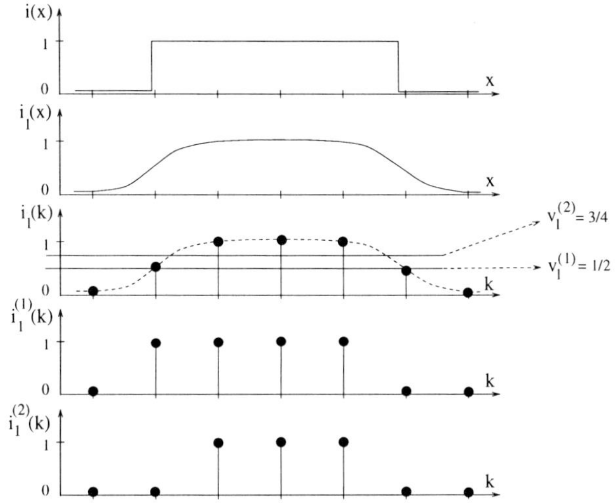

Fig. 4. Positional distortion due to binarization.

There exist other quantization criteria, e.g. entropic, that are also based on the histogram of gray-levels. The histogram can be computed from the whole document image or from a restricted local neighborhood around the pixel of interest. Histograms are first-order spatial statistics, and extensions to second-order statistics using co-occurrence matrices have also been investigated for binarization. More detailed discussions can be found in [17, 18].

In general, the study of quantization effects is difficult because of the nonlinearity of the quantizer's characteristic function. If L is large, say more than 16, its effects can practically be neglected for most document images. However, for $L = 2$, its effects becomes noticeable and can cause inadmissible distortions in some applications. Since binarization is very common in the field of document analysis, we will now examine its effects on a typical case in some detail.

Let us take the same example as that of Sec. 2.1. We assume that the sampling of $i(x)$ is achieved via $i_1(x)$, which is a low-pass filtered version of $i(x)$, see Fig. 4; $i_1(k)$ is the sampled signal but not yet binarized (or sampled and quantized with a very large value of L). Let us assume that the binarization threshold v_1 is computed by Otsu's algorithm and takes the value $v_1^{(1)} = 1/2$. The binarized version of $i_1(k)$ is denoted by $i_1^{(1)}(k)$ and is identical to $i(k)$, which was obtained by sampling directly $i(x)$ without anti-aliasing filtering. Recall that $i(k)$ was shown to be poor in terms of preserving positional information. The function $i_1^{(2)}(k)$ is another binarized version obtained by assuming that Otsu's algorithm yields another threshold value $v_1^{(2)} = 3/4$. It is easy to see that in the worst case the binarization step can yield a positional error of $\pm\Delta/2$. Thus, the total amount of error on the width can be as high as Δ. If $i(x)$ represents the cross section of a thin line structure with the true width of 2Δ, the worst case corresponds to a relative error of 50%!

In brief, quantization effects can be neglected for most document images if we use more than 4 bits per sample ($L = 16$ gray-levels). For documents that contain natural images, like identity cards with photographs, L should be larger. Binarization, which is a particular case of quantization with $L = 2$, is useful in many respects. It reduces the amount of data to be stored (see the next section), and simplifies subsequent operations, such as segmentation and recognition. However, it may introduce an important loss of information for small line structures, such as those found in small size machine printed characters.

2.3. *Image Coding*

A document image sampled and quantized may contain a huge amount of data, which may cause problems in storage, transmission and processing. This section addresses image coding as a means of alleviating these problems. We limit our discussions to *lossless* coding schemes, in contrast with *lossy* or *perceptual* schemes. Lossless coding means that the original image can be perfectly reconstructed from the encoded version. Moreover, only *binary* images will be treated due to their predominant role in the field of document image analysis.

We will present two key issues in image coding. The first is nearby-neighbor *modeling*. For instance, adjacent pixels of an image tend to have the same value, i.e., black or white. This property is sometimes called "spatial redundancy". One particularly simple yet efficient way to take advantage of spatial redundancy is to consider *runs* of contiguous black or white pixels instead of the pixels themselves. This results in the *run-length* coding scheme. The second issue is *entropic coding*. It exploits the fact that some *spatial configurations* may appear more often than others. The idea consists in using short *code words* for frequent configurations and longer code words for rare configurations, resulting in less data to be stored on the average. In the following we will use a simple example to illustrate the run-length coding (Sec. 2.3.1) and one entropic coding scheme known as Huffman code (Sec. 2.3.2).

2.3.1. *Run-length coding*

Run-length coding consists in representing the image, line by line, by the positions and lengths of contiguous black runs. A run is thus specified by a pair of numbers. The first of these numbers indicates the position of the first black pixel with respect to the left border of the image, while the second represents the number of contiguous black pixels following the first one. Each line is terminated by a special number, say 0, representing the EOL (end of line). A blank line is thus represented by a single EOL.

In Fig. 5, the first line is blank and thus represented by an EOL. The first run of the second line starts at the second column, has seven black pixels, and is therefore represented by the pair (2,7). Similarly, the second run of the second line is represented by (13,2). Since the second line has only two runs, an EOL follows

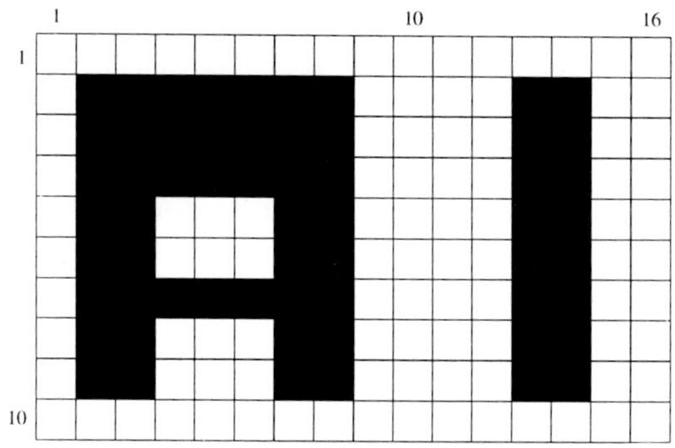

Fig. 5. A binary image example.

the pair (13,2). The process continues until we get to the last line. Finally, we get the following result (EOL is encoded to 0):

```
0
(2,7)  (13,2)  0
(2,7)  (13,2)  0
(2,7)  (13,2)  0
(2,2)   (7,2)  (13,2)  0
(2,2)   (7,2)  (13,2)  0
(2,7)  (13,2)  0
(2,2)   (7,2)  (13,2)  0
(2,2)   (7,2)  (13,2)  0
0
```

Thus the image of Fig. 5 is represented by a sequence of 50 numbers. If we use four bits to represent each number so that the whole range [0,13] is covered, the resulting code will have 4 x 50 = 200 bits, which is longer than the original canonical representation (10 x 16 = 160 bits)! However, we can observe that the range [0,13] is not fully used and in fact only the set of four numbers {0,2,7,13} is necessary. Therefore we need only two bits to encode this set by assigning, for instance,

```
0  -> 00
2  -> 01
7  -> 10
13 -> 11
```

The assignment must of course be appropriately stored or transmitted. Now the image can be represented by only 2 x 50 = 100 bits. This kind of assignment constitutes in fact a primitive form of entropic coding (see the next section).

There exist many different versions of run-length coding. For instance, instead of encoding the position of a run by its absolute distance to the left border, we can also encode it by the relative distance to the previous run of the same line. Due to their simplicity and lossless nature, run-length codes can also be directly used in some image processing operations such as noise filtering. For instance, very short runs that have no neighbors in the previous line nor in the next line can be eliminated without referring to the canonical representation. See [19] for more details.

Run-length coding is one way to exploit the spatial redundancy of images. It does not take into account the redundancy between adjacent lines. This between-line redundancy can be exploited by using *predictive coding* schemes, i.e., using the data of the previous line to predict that of the current line and encoding only the differences. For instance, the third line of Fig. 5 is identical to the second line and thus needs not be encoded; only an additional code word, reserved for this purpose, need to be stored or transmitted. A full exploitation of spatial redundancy is to encode only the contour points of the image. However, the decoder must then receive the whole image before it is able to decode the first line and therefore needs a larger buffer. A more detailed discussion of all these aspects can be found in [6].

2.3.2. *Huffman code*

Huffman code exploits the unequal distribution of symbol probabilities in a message. That is, frequent symbols are represented by short code words whereas longer code words are used for rare symbols. The result is a shorter average word length. The key point is therefore the design of code words that have variable lengths, based on the symbol probabilities. We will first present the encoding and then the decoding, which is not trivial as in the case of run-length coding.

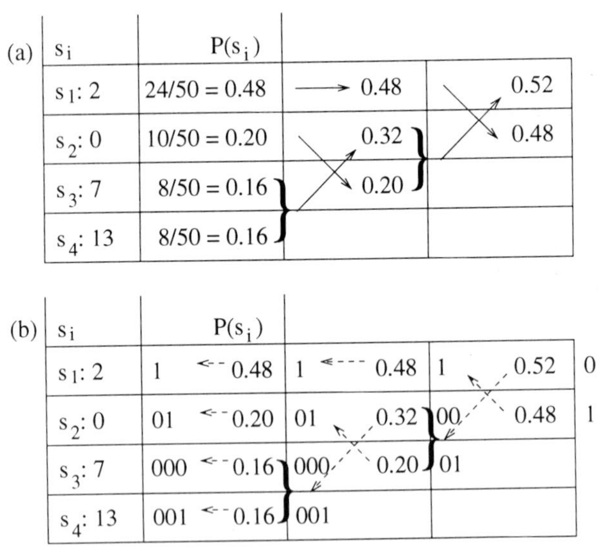

Fig. 6. An example of Huffman coding. (a) Reduction process. (b) Splitting process.

The construction of code words comprises two processes, namely, reduction and splitting [20]. Let us consider as an example the set of numbers {0,2,7,13} resulting from run-length coding (see Fig. 6). It can be seen that they are not equally frequent. Let us reorder them in decreasing order of probability (relative frequency), and assign the symbols $\{s_i; i = 1, .., 4\}$ to them; see the first and second columns of Fig. 6(a). In the reduction process, we combine the two least frequent symbols into a new symbol whose probability is the sum of the two probabilities. This reduces the number of symbols by one. The new set of symbols is again reordered in decreasing order of probabilities. This completes the first iteration of the reduction process. We continue the process until there remain only two symbols, see Fig. 6(a). In the splitting process, symbols that were artificially created are now recursively split. During the splitting process, the code words are gradually constructed by appending the bits 0 and 1 to the splitting results, see Fig. 6(b). Eventually, we get a set of code words having variable lengths $\{s_1 : 1, s_2 : 01, s_3 : 000, s_4 : 001\}$. The average length per symbol is:

$$\begin{aligned} L_{average} &= \sum_i l_i \cdot P(s_i) \\ &= 1 \cdot 0.48 + 2 \cdot 0.20 + 3 \cdot 0.16 + 3 \cdot 0.16 \\ &= 1.84 \; bits/symbol \end{aligned} \tag{2.19}$$

Recall that the run-length encoding of the image of Fig. 5 produced 50 symbols. Thus the Huffman encoding of the run-length codes results in a total of $1.84 \cdot 50 = 92$ bits. This is to be compared with the 160 bits required for the canonical representation. Note that the 100 bits obtained at the end of the last section was already a kind of entropic encoding in that the unused numbers between 0 and 13 had no code words.

The decoding is performed via the *decoding tree*, which is also built during the splitting process. Each split is mapped to a test in the tree to determine whether the next bit is 0 or 1, see Figs. 6(b) and 7. Each leaf of the tree represents one symbol. The decoding process starts at the root of the tree, i.e., node a. The paths are determined by the received bit stream. Whenever a leaf is reached, a symbol is emitted and we go back to the root and start the decoding of a new symbol.

As an example, let us encode the first two lines of Fig. 5 into a bit stream and then decode it. The run-length encoding produces the following sequence of numbers: 0,2,7,13,2,0. In terms of symbols, we get (see the first column of Fig. 6(a)) $s_2 s_1 s_3 s_4 s_1 s_2$. By using the Huffman code of Fig. 6(b), the sequence is converted into the bit stream: 011000001101. The decoding process makes use of the decoding tree and starts at the root. As the first bit is 0, we go to node b. The second bit is 1. Therefore we go to symbol s_2, emit it, and go back to the root. Because the third bit is 1, we go to symbol s_1, emit it, and go back to the root. This process is continued until all bits have been read. Eventually we get the sequence of symbols $s_2 s_1 s_3 s_4 s_1 s_2$, which corresponds to the sequence of numbers 0,2,7,13,2,0. The conversion of this run-length code back to the original image is trivial.

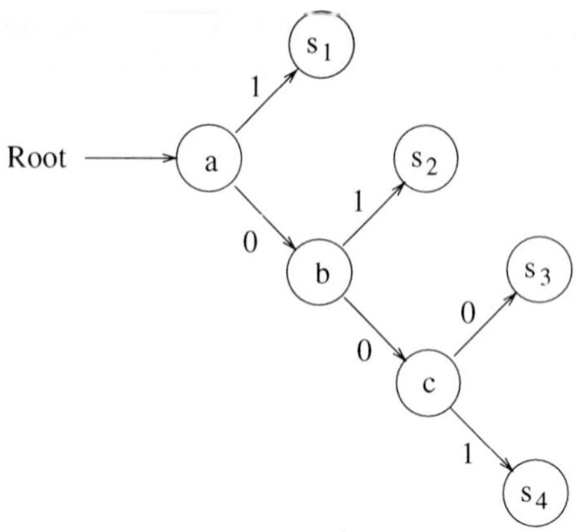

Fig. 7. Decoding tree.

Up to the middle of the 1970's, Huffman's code was considered the best since it approaches the theoretical limit given by the entropy of the symbol source. In practice, the Huffman code is optimal only if the symbol probabilities are positive integer powers of (1/2). This limitation was recently overcome by a new family of codes, known as *arithmetic codes* [21].

Image coding has a long history and presently many practical coding schemes are standardized by the International Consultative Committee for Telegraphy and Telephony (CCITT) and the International Standards Organisation (ISO). Under the auspices of these two organisations, the JBIG (Joint Bi-level Image Expert Group) elaborated a *recommendation* or *norm* for binary images and the JPEG (Joint Photographic Experts Group) for gray-level images [22]. As an example, the method proposed by JBIG utilises progressive spatial resolution buildup instead of run-length codes, followed by an adaptive arithmetic coder instead of the static Huffman coder used in our example. At the resolution of 200 dpi, JBIG yields a compression ratio between 5 and 62 on various business documents. Generally, the compression ratio increases with the spatial resolution.

3. Image Transformation

Image transformation is an image-to-image operation, i.e., the input image is transformed into an output image. Thus, the transformation is characterized by the input-output relationship, the nature of which depends on the goal of the operation. Geometrical transformations (Sec. 3.1) can serve the purpose of correcting the distortions due to image acquisition and the normalization of eccentric handwriting. Filtering (Sec. 3.2) is a fundamental operation in image processing and can help improve figure-background separation (Sec. 3.3). Object boundary detection (Sec. 3.4) and thinning (Sec. 3.5) are two commonly used operations in document image analysis. They constitute the interface from pixel to higher level representation.

Our exposition will be based on pixel representation. Some image transformations, such as noise suppression and thinning, can also be applied directly on the run-length representation described in Sec. 2.3.1 [19].

3.1. *Geometrical Transformations*

A document image acquired through a digitizer may exhibit various kinds of geometrical distortions, which can be corrected by appropriate geometrical transformations. For instance, a document may not be well aligned on a scanner, thus yielding a *skewed* (rotated) digital image, see Fig. 8. The acquisition by camera may exhibit, in addition to rotation, an optical distortion due to the imperfect lens system, see Fig. 9. Geometrical transformations can also be used to normalize handwritten scripts produced in an eccentric way, see Fig. 10. In technical drawings, characters and numerals may be written in different orientations and therefore their recognition requires a prior rotation, see Fig. 11.

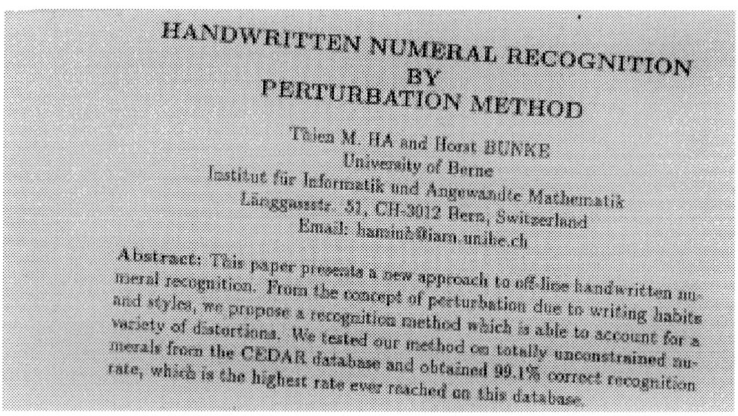
HANDWRITTEN NUMERAL RECOGNITION
BY
PERTURBATION METHOD

Thien M. HA and Horst BUNKE
University of Berne
Institut für Informatik und Angewandte Mathematik
Länggassstr. 51, CH-3012 Bern, Switzerland
Email: haminh@iam.unibe.ch

Abstract: This paper presents a new approach to off-line handwritten numeral recognition. From the concept of perturbation due to writing habits and styles, we propose a recognition method which is able to account for a variety of distortions. We tested our method on totally unconstrained numerals from the CEDAR database and obtained 99.1% correct recognition rate, which is the highest rate ever reached on this database.

Fig. 8. Part of a skewed document.

In order to achieve a proper correction, many steps are necessary. First, we must determine the type of distortion and make a mathematical model. In the case of Figs. 8 and 11, the distortion type is clearly a rotation and a precise rotation model exists from standard mathematics. Unfortunately, the type of distortion is less simple for Fig. 9, where the optical distortion is highly nonlinear, let alone the case of Fig. 10 in which the distortions are due to human writing habits and styles. Second, we must determine the parameter values of the transformation, e.g. the rotation angle. This is highly application-dependent. In Fig. 8, the rotation angle can be determined globally for the whole document whereas in Fig. 11, it must be locally determined for each group of characters. Last, the correction itself can be achieved via a resampling of the original digital image. Section 3.1.1 addresses the general methodology, Sec. 3.1.2 the parameter estimation and Sec. 3.1.3 the resampling problem.

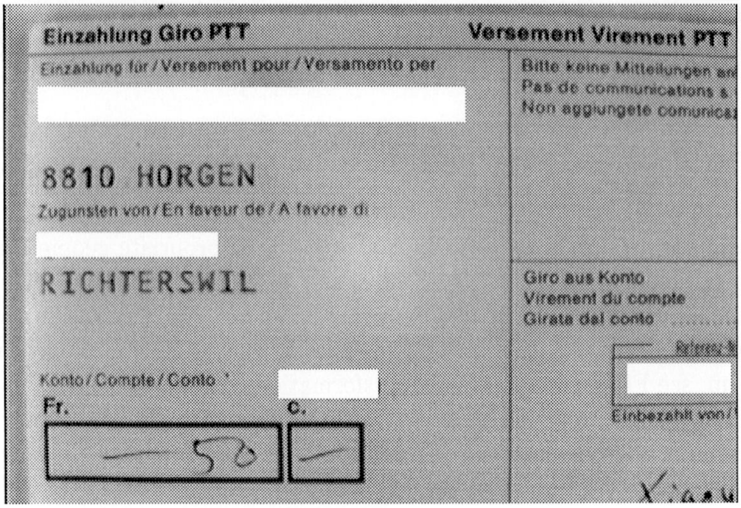

Fig. 9. An optically distorted document.

Fig. 10. Examples of handwritten numerals.

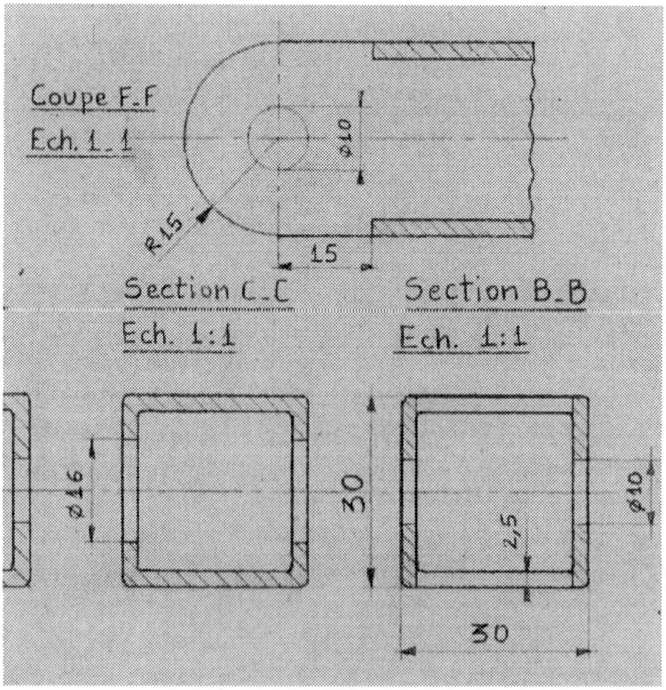

Fig. 11. A technical drawing.

3.1.1. *Basic methodology*

Geometrical transformations can be formulated in a fairly general way. Let $i(x, y)$ be the original image and $i'(x', y')$ its distorted version. The two images are related by the equations

$$x' = f_1(x, y) \tag{3.20}$$
$$y' = f_2(x, y) \tag{3.21}$$

and the original image can be expressed in terms of the distorted image as follows.

$$\begin{aligned} i(x, y) &= i'(x', y') \\ &= i'[f_1(x, y), f_2(x, y)] \end{aligned} \tag{3.22}$$

The distortion type is characterized by the nature of $f_1(.,.)$ and $f_2(.,.)$. *Affine* (linear) transformations cover translation, scaling, rotation, and slant [23]. These can be represented in matrix notation:

$$\begin{pmatrix} x' \\ y' \end{pmatrix} = \begin{pmatrix} a_{11} & a_{12} \\ a_{21} & a_{22} \end{pmatrix} \begin{pmatrix} x \\ y \end{pmatrix} + \begin{pmatrix} b_1 \\ b_2 \end{pmatrix} \tag{3.23}$$

Table 1 summarizes the properties of affine transformations.

Table 1. Affine transformations.

Type	Properties		Meaning
Translation	$a_{ij} = 0$	; i, j = 1, 2.	
Scaling	$a_{12} = a_{21} = 0$		
Rotation	$a_{11} = \cos\alpha$ $a_{21} = \sin\alpha$	$a_{12} = -\sin\alpha$ $a_{22} = \cos\alpha$	α : rotation angle
Slant	$a_{11} = 1$ $a_{21} = 0$	$a_{12} = \mathrm{tg}\,\beta$ $a_{22} = 1$	β : slant angle

There exist many families of nonlinear transformations, such as projective and polynomial. These transformations are useful in modeling the distortion produced by a camera system [24]. In the field of document analysis, the optical axis of the camera system is generally arranged to be perpendicular to the document plane. Therefore, projective transformations are not necessary. In contrast, polynomial transformations remain very useful to model the distortions due to the lens system, see Fig. 9. Two common types of optical distortion are *barrel* and *pincushion*, see Fig. 12. Both are *radial* distortions and can be approximated by [25]:

$$r' = C_m \cdot r + C_d \cdot r^3 \tag{3.24}$$

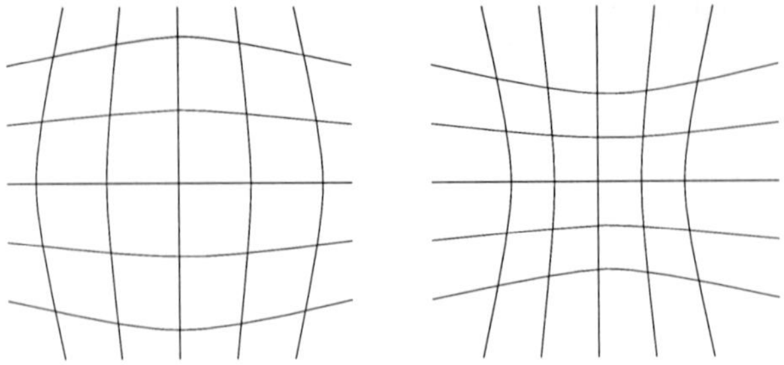

Fig. 12. Optical distortions due to lens imperfection.

where

$$r' = \sqrt{x'^2 + y'^2} \;, r = \sqrt{x^2 + y^2}$$

C_m is the magnification and C_d the distortion coefficient. The latter coefficient is negative for barrel and positive of pincushion distortion. The second degree coefficient is zero due to the symmetry of the lens system. Figure 9 is clearly due to a barrel distortion. Equation (3.24) can be rewritten in polynomial form [25]:

$$\begin{pmatrix} x' \\ y' \end{pmatrix} = C_m \begin{pmatrix} x \\ y \end{pmatrix} + C_d \begin{pmatrix} (x^2 + y^2)x \\ (x^2 + y^2)y \end{pmatrix} \tag{3.25}$$

In general, polynomial transformations can be used to approximate a variety of geometrical distortions. For instance, they have been used to model handwriting styles and habits [26].

3.1.2. *Parameter estimation*

Due to the variety of possible applications of geometrical transformations, only two cases will be discussed in this section. The first consists in estimating the skew angle of a text page and the second in approximating the radial distortion coefficient of a lens system.

The skew angle estimation algorithm considered in the following is based on hypothesis-testing. That is, we project the image along a number of axes, compute the orientation-dependent histograms and look for the direction that maximizes an alignment criterion [27]. For a graphical illustration, see Fig. 13. First, the histogram under a projection angle α is computed.

$$H(y_l; \alpha) = \sum_{k=-\infty}^{\infty} i(x_k, y_l) \tag{3.26}$$

$$= \sum_{k=-\infty}^{\infty} i'(cos\alpha \cdot x_k - sin\alpha \cdot y_l, sin\alpha \cdot x_k + cos\alpha \cdot y_l) \tag{3.27}$$

with the conventions $i'(x'_k, y'_l) = 0$ if (x'_k, y'_l) is outside the document, and $i'(x'_k, y'_l) = i'[round(x'_k), round(y'_l)]$ otherwise. Then, we define the alignment criterion $A(\alpha)$

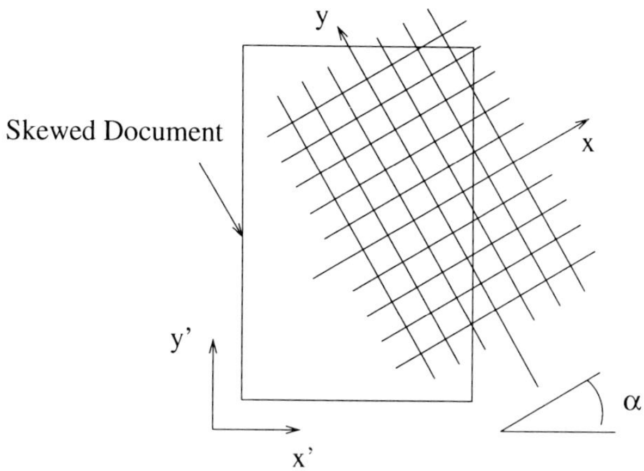

Fig. 13. Skew angle estimation.

based on the histogram variation between consecutive lines along the direction specified by α as follows:

$$A(\alpha) = \sum_{l=-\infty}^{\infty} [H(y_l;\alpha) - H(y_{l+1};\alpha)]^2 \tag{3.28}$$

Finally, the estimated skew angle is given by

$$\alpha^* = arg\ max\ A(\alpha) \tag{3.29}$$

This method is of practical interest because it can operate directly on the gray-level image; no prior binarization nor segmentation is required. More details can be found in [27]. See [28, 29, 30] for other skew correction methods. For an example of slant estimation applied to handwriting recognition, see [31].

The estimation of a lens distortion coefficients according to Eq. (3.25) is now examined. The method consists of many steps. First, a calibration image $i(x, y)$, e.g. a square grid, is submitted to the acquisition system, and a distorted version $i'(x', y')$ is acquired. Then, a number N_{CP} of *control points* are chosen, e.g. cross points of the grid; N_{CP} must be greater than the number of unknown coefficients in Eq. (3.25). Next, the correspondences $(x_j, y_j) \leftrightarrow (x'_j, y'_j), j = 1, .., N_{CP}$ are set up. Finally, the unknown coefficients are determined by a *least squares* method. Equation (3.25) can be rewritten in matrix notation as follows:

$$Z' = Z \cdot C \tag{3.30}$$

where

$$Z' = \begin{pmatrix} x' \\ y' \end{pmatrix}; Z = \begin{pmatrix} x \mid (x^2 + y^2)x \\ y \mid (x^2 + y^2)y \end{pmatrix}; C = \begin{pmatrix} C_m \\ C_d \end{pmatrix} \tag{3.31}$$

For each pair of points $\{(x_j, y_j) \leftrightarrow (x'_j, y'_j)\}$, we get one equation system $Z'_j = Z_j \cdot C + R_j$, where R_j is the measurement error vector. The least squares solution minimizing $\sum_{j=1}^{N_{CP}} R_j^T R_j$ is given by:

$$\hat{C} = \left(\sum_{j=1}^{N_{CP}} Z_j^t Z_j \right)^{-1} \cdot \left(\sum_{j=1}^{N_{CP}} Z_j^t Z_j' \right) \tag{3.32}$$

where the superscript t denotes the matrix transposition [32, 24].

This estimation method is global in that the distortion over the whole image field is modeled by the same coefficients. More precise, local distortion models have also been proposed (see, for instance [33]). Very high precision calibration procedures are discussed in [34].

3.1.3. *Digital image resampling*

The last step in geometrical correction is the reconstruction of the ideal undistorted image from the distorted version, via Eqs. (3.20), (3.21) and (3.22). These equations need be adapted to the digital images $i(k{\cdot}\Delta x, l{\cdot}\Delta y)$ and $i'(k'{\cdot}\Delta x', l'{\cdot}\Delta y')$. For a given point $(x = k \cdot \Delta x, y = l \cdot \Delta y)$, the corresponding point (x', y') obtained by Eqs. (3.20) and (3.21) generally does not lie on the grid defined by $\{(k' \cdot \Delta x', l' \cdot \Delta y')\}; k', l' \in N$. Therefore, some approximations are necessary.

The simplest approximation is provided by the *nearest neighbor* method. That is,

$$i'_{NN}(x', y') = i'[\Delta x' \cdot round\left(\frac{x'}{\Delta x'}\right), \Delta y' \cdot round\left(\frac{y'}{\Delta y'}\right)] \tag{3.33}$$

Figure 14 shows the corrected image of Fig. 9. The zigzag effects due to the nearest neighbor approximation are clearly seen on this figure. It is clear that these effects would have catastrophic consequences for subsequent operations, such as OCR.

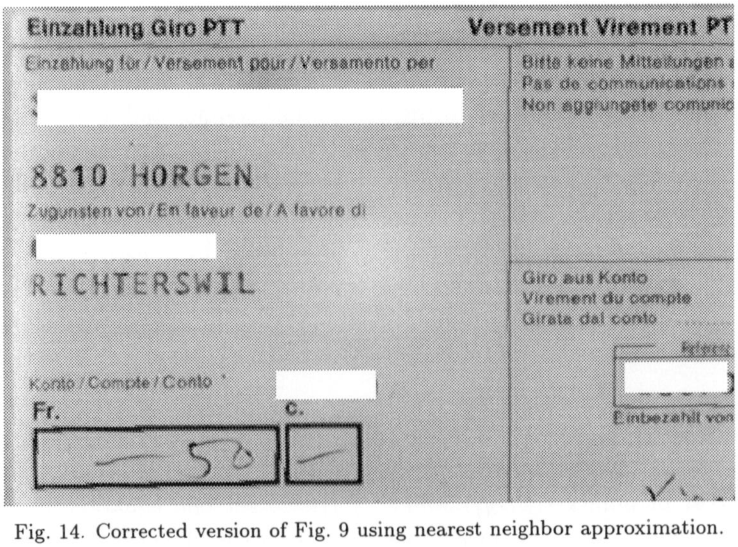

Fig. 14. Corrected version of Fig. 9 using nearest neighbor approximation.

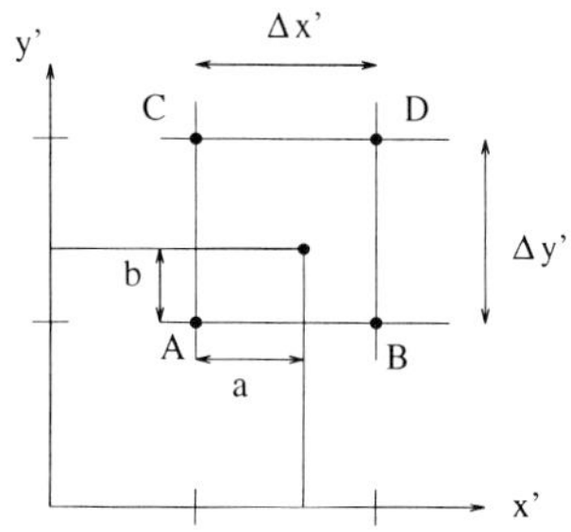

Fig. 15. Bilinear interpolation method.

A better method is *bilinear interpolation.* That is, we first locate the rectangle $ABCD$ surrounding (x', y') (see Fig. 15), and then compute $i'(x', y')$ by:

$$i'_{BL}(x', y') = i'_{AB} + \left(\frac{b}{\Delta y'}\right)(i'_{CD} - i'_{AB}) \tag{3.34}$$

where

$$i'_{AB} = i'(A) + \left(\frac{a}{\Delta x'}\right)[i'(B) - i'(A)] \tag{3.35}$$

$$i'_{CD} = i'(C) + \left(\frac{a}{\Delta x'}\right)[i'(D) - i'(C)] \tag{3.36}$$

Figure 16 shows the corrected version of Fig. 9 using bilinear interpolation. We note that the zigzag effects have disappeared.

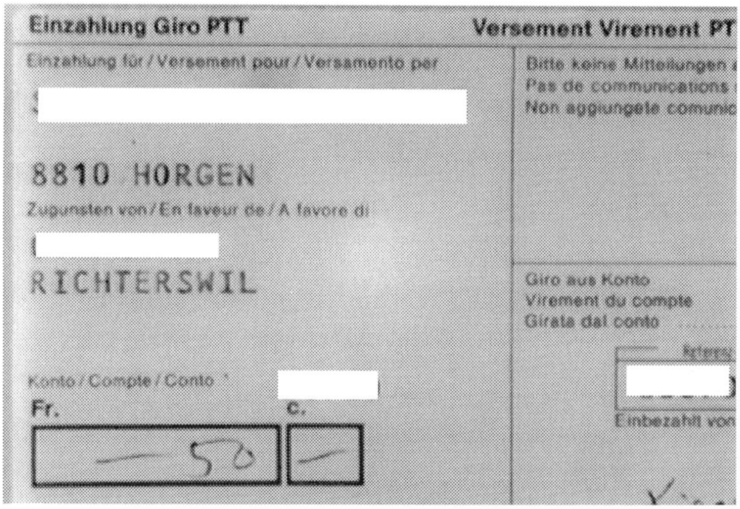

Fig. 16. Corrected version of Fig. 9 using bilinear interpolation.

Bilinear interpolation is generally sufficient for most applications in the field of document analysis. However, if the corrected image has a lower sampling rate than that of the distorted image, the resampling should operate on a low-pass filtered version of the distorted image to avoid aliasing.

3.2. *Filtering*

This section introduces both linear and nonlinear digital filtering. The purpose is to provide the basic tools for subsequent sections on the applications of image filtering to document images. Filtering consists in transforming an input image $i(k,l)$ into an output image $i_o(k,l)$, which is also called the filtered version of $i(k,l)$. The ouput value $i_o(k,l)$ is usually a function of input values in a local neighborhood around position (k,l). The linearity, respectively nonlinearity, of this function defines linear, respectively nonlinear, filters. Section 3.2.1 presents linear filters. Then two classes of nonlinear filters will be discussed in Secs. 3.2.2 and 3.2.3, namely, rank order filters and morphological filters, respectively.

3.2.1. *Linear filtering*

We consider only the class of Finite Impulse Response (FIR) filters. They are characterized by the *convolution* equation:

$$i_o(k,l) = i(k,l) * f(k,l) = \sum_{k'=-\infty}^{\infty} \sum_{l'=-\infty}^{\infty} i(k-k', l-l') \cdot f(k', l') \tag{3.37}$$

where $f(k,l)$ is the filter *impulse response*, also called *mask* or *template*. Both $i(k,l)$ and $f(k,l)$ are of finite extent, i.e., they take on the value zero outside their support region. Therefore, the summation in Eq. (3.37) is limited to a finite subset of indices. For instance, for a 3 x 3 centered mask, the range of summation indices is $k', l' = -1, 0, 1$. The following impulse responses are commonly used in image analysis [2]:

$$\frac{1}{16}\begin{bmatrix} 1 & 2 & 1 \\ 2 & 4 & 2 \\ 1 & 2 & 1 \end{bmatrix} \quad \textit{(smoothing or lowpass filter)} \tag{3.38}$$

$$\frac{1}{4}\begin{bmatrix} 1 & 0 & -1 \\ 2 & 0 & -2 \\ 1 & 0 & -1 \end{bmatrix} \quad \textit{(vertical edge sensitive filter)} \tag{3.39}$$

$$\frac{1}{8}\begin{bmatrix} -2 & 1 & -2 \\ 1 & 4 & 1 \\ -2 & 1 & -2 \end{bmatrix} \quad \textit{(discrete Laplacian edge detector)} \tag{3.40}$$

$$\frac{1}{8}\begin{bmatrix} -1 & 2 & -1 \\ -2 & 4 & -2 \\ -1 & 2 & -1 \end{bmatrix} \quad \textit{(vertical line sensitive filter)} \tag{3.41}$$

$$\frac{1}{8}\begin{bmatrix} 1 & -2 & 1 \\ -2 & 5 & -2 \\ 1 & -2 & 1 \end{bmatrix} \quad \textit{(enhancement filter)} \tag{3.42}$$

Neighborhoods larger than 3 x 3 are also possible [1, 2].

Properties of linear filtering are well known and described in many comprehensive text books, see [1, 2, 24]. Generally speaking, linear filters are usually used in conjunction with other nonlinear operations. For instance, the Laplacian edge detector must be followed by a zero-crossing detection procedure to locate edge points. For line detection, many filters each sensitive to a specific direction, must be applied and the results are combined by a *max*-operator to determine the local dominant orientation. Moreover, it may be necessary to use multiple spatial resolution to detect lines having different widths [35]. These examples illustrate the role of linear filters in processing document images: they are components of complex algorithms but rarely used alone. Besides, many analysis tasks can be achieved by alternative, simpler methods. For instance, line detection can be performed by binarization and thinning (see Sec. 3.5), at least for good quality document images.

3.2.2. *Rank-order filtering*

Rank order filters are nonlinear and can be understood by considering a 3 x 3 neighborhood around the pixel of interest $i(k,l)$. Let $S_{k,l}$ be the set of gray-levels in the neighborhood of (k,l):

$$S_{k,l} = \{i(k',l') \mid |k'-k| \leq 1, |l'-l| \leq 1\} \tag{3.43}$$

and $R_{k,l}$ its ordered sequence:

$$R_{k,l} = \{r_1 \leq r_2 \leq \ldots \leq r_9 \mid r_j \in S_{k,l}\} \tag{3.44}$$

The following rank order operations are known from the literature:

$$i_o(k,l) = r_1(erosion) \tag{3.45}$$

$$i_o(k,l) = r_9(dilation) \tag{3.46}$$

$$i_o(k,l) = r_9 - r_1(contour\ detection) \tag{3.47}$$

$$i_o(k,l) = r_5(median) \tag{3.48}$$

The median filter is the most popular rank-order filter, since it is able to eliminate impulse noise without blurring the input image. In general, rank order filters can be applied to both gray-level and binary images (see Sec. 3.3.1 for an example). We will see in the next section that there exist similarities between rank-order filters and morphological filters.

3.2.3. *Morphological filtering*

Morphological filtering is based on set-theoretic concepts and extracts object features by choosing an appropriate *structuring element* and a suitable morphological operator [36, 37]. It considers images as sets and manipulates them using logical operations such as union and intersection. The most intuitive morphological operator is *hit-or-miss.* That is, we scan the input image and compare at each position

```
. . . . . . .          |          . . . . . . .
. 1 1 1 1 1 .        1 1 1        . 1 1 1 1 1 .
. . 1 . 1 1 .     -- 1 . 1 --     . . 1 1 1 1 .
. 1 1 1 1 1 .        1 1 1        . 1 1 1 1 1 .
. . . . . . .          |          . . . . . . .

Input Image     Structuring Element     Output Image
```

Fig. 17. Hit-or-miss operation.

the neighboring pixels with the structuring element. If there is a perfect match, a predefined value is assigned to the output image at that position; otherwise the output can be set to the same value as that of the input or the complement of the predefined value. For instance, the 3 x 3 structuring element in Fig. 17, where "1" indicates that the pixel belongs to the structuring element, can be used to fill in small holes of a binary image. Hit-or-miss operators are suitable for simple tasks but not appropriate for complex ones. In the following, we present the four basic operations that are the most important building blocks of morphological filtering.

Let A be a set representing a binary image, i.e., $A = \{(x_j, y_j) \mid i(x_j, y_j) = 1\}$, and B a structuring element. $(A)_b$ is defined as the translation (shift) of A by vector $b = (x_b, y_b)$, i.e., $(A)_b = \{(x_j, y_j) + (x_b, y_b) \mid (x_j, y_j) \in A\}$. The four basic morphological operations of A by B are defined as follows:

$$\textit{Dilation} \quad A \oplus B = \bigcup_{b \in B} (A)_b \tag{3.49}$$

$$\textit{Erosion} \quad A \ominus B = \bigcap_{b \in B} (A)_{-b} \tag{3.50}$$

$$\textit{Closing} \quad A \bullet B = (A \oplus B) \ominus B \tag{3.51}$$

$$\textit{Opening} \quad A \circ B = (A \ominus B) \oplus B \tag{3.52}$$

Figure 18 illustrates the dilation operation and Fig. 19 the four basic morphological operations.

Dilation is an expansion and erosion is a shrinking of the original image. Closing by a disk structuring element — the 3 x 3 structuring element in Fig. 19 is a rough approximation — smoothes the contour and fills in small holes. Opening by a disk structuring element also smoothes the contour and eliminates impulse noise.

We have briefly presented morphological operations on binary images. Extensions to gray-level morphological operations can be found in [38]. Recent advances in this field are accounted for in [39]. Other families of nonlinear filters, e.g. polynomial, are discussed in [40].

3.3. *Figure-Background Separation*

Most document images result from the process of printing or writing on a uniform background, such as white paper. The separation of figure — text or drawings — from background is one of the most fundamental operations in image processing.

```
. 1 1 1 .
. 1 1 1 .                     |
. 1 1 . .                -- 1 1 1 --
. . . . .                     |

Original Image           Structuring Element

1 1 1 . .                . 1 1 1 .                . . 1 1 1
1 1 1 . .                . 1 1 1 .                . . 1 1 1
1 1 . . .                . 1 1 . .                . . 1 1 .
. . . . .                . . . . .                . . . . .

Image translated         Image translated         Image translated
by (-1,0)                by (0,0)                 by (1,0)

1 1 1 1 1
1 1 1 1 1
1 1 1 1 .
. . . . .

Dilated Image = Union of translated versions
```

Fig. 18. Illustration of dilation by a symmetric structuring element.

It is a prerequisite to subsequent operations, such as segmentation and labeling. Section 3.3.1 addresses the separation process based on gray-level thresholding, assuming a uniform background, in both noise-free and noisy environments. In Sec. 3.3.2, we suppress the uniform background assumption and consider the separation of figure from textured background.

3.3.1. *Gray-level thresholding*

If the document is of "good" quality and the background uniform, the separation can be performed by directly using the binarization method described in Sec. 2.2. Examples are the cases of Figs. 10 and 16, which result in Figs. 20 and 21. (In Fig. 21, some characters do touch each other because the spatial resolution is insufficient. But the same document scanned at a higher resolution gives an almost perfect separation [9, 10].) Unfortunately, when we apply this method to the image of Fig. 11, the result is not satisfactory (see Fig. 22). Lines of low contrast are missed. By manually adjusting the threshold[b], it is possible to recover the missing lines. However, the price is that we now introduce noisy pixels from the background (see Fig. 23). The adjustment method used here consists of choosing a threshold such that a given percentage, say 10%, of pixels have gray-levels lower than this threshold (p-tile method) [17]. In the following, we present two approaches to alleviate this problem. For simplicity, we will only illustrate global thresholding methods; it should be clear from Fig. 11 that local thresholding methods are more appropriate for this image [17].

The first approach consists in filtering the image *before* binarization. Figure 24

[b]The conversion of engineering drawings to Computer Aided Design (CAD) form is not yet a completely automatic process [41].

```
. . . . . . . . . .
. 1 1 1 1 1 1 1 . .
. 1 . 1 1 . . . . .
. 1 1 1 1 1 . . . .
. 1 1 1 1 1 . . . .
. 1 1 1 1 1 . . . .
. 1 1 1 1 1 . . . .
. 1 . 1 1 . . . . .
. . . . . . . . 1 .
. . . . . . . . . .

Original Image
```

```
      |
    1 1 1
 -- 1 1 1 --
    1 1 1
      |

Structuring Element
```

```
1 1 1 1 1 1 1 1 1 .
1 1 1 1 1 1 1 1 1 .
1 1 1 1 1 1 1 1 1 .
1 1 1 1 1 1 1 . . .
1 1 1 1 1 1 1 . . .
1 1 1 1 1 1 1 . . .
1 1 1 1 1 1 1 . . .
1 1 1 1 1 1 1 1 1 1
1 1 1 1 1 1 . 1 1 1
. . . . . . . 1 1 1

Dilated Image
```

```
. . . . . . . . . .
. . . . . . . . . .
. . . . 1 . . . . .
. . . . 1 . . . . .
. . 1 1 1 . . . . .
. . 1 1 1 . . . . .
. . . . . . . . . .
. . . . . . . . . .
. . . . . . . . . .
. . . . . . . . . .

Eroded Image
```

```
. . . . . . . . . .
. 1 1 1 1 1 1 1 . .
. 1 1 1 1 1 . . . .
. 1 1 1 1 1 . . . .
. 1 1 1 1 1 . . . .
. 1 1 1 1 1 . . . .
. 1 1 1 1 1 . . . .
. 1 1 1 1 . . . . .
. . . . . . . . 1 .
. . . . . . . . . .

Closed Image
```

```
. . . . . . . . . .
. . . 1 1 1 . . . .
. . . 1 1 1 . . . .
. 1 1 1 1 1 . . . .
. 1 1 1 1 1 . . . .
. 1 1 1 1 1 . . . .
. 1 1 1 1 1 . . . .
. . . . . . . . . .
. . . . . . . . . .
. . . . . . . . . .

Opened Image
```

Fig. 19. Illustration of four basic morphological operations.

is the result obtained by using the linear smoothing filter given by Eq. (3.38) before binarization. It can be seen that noise is attenuated, but line structures are thickened making some of them touch each other. Nonlinear filters, such as median, can also be used. Figure 25 is the result obtained by a median filter, followed by binarization. As can be easily seen, Figs. 24 and 25 differ only in small details.

The second approach attempts to eliminate noise *after* binarization. A median filter applied to a binary image becomes a *majority* filter, which is able to fill small holes, close short gaps, and eliminate impulse noise. Figure 26 shows the result of this filter applied to the image of Fig. 23. It can be seen that Fig. 26 contains more noisy pixels than Figs. 24 and 25, although the visual differences are small. More sophisticated methods are discussed in [17, 42, 43].

Fig. 20. Binarized version of Fig. 10.

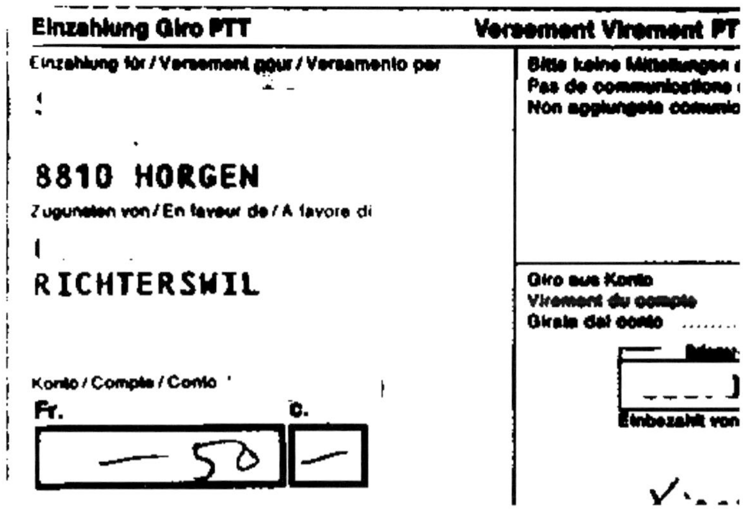

Fig. 21. Binarized version of Fig. 9.

3.3.2. *Textured background*

Many modern text editors provide the possibility to produce textured background. For instance, a *highlighted* text is designed to draw the attention of human readers to some text regions, see Fig. 27(a). Unfortunately, it creates many problems for OCR. In the following, we describe a simple and efficient technique based on morphological filtering to tackle the problem of separating figure from textured background [44].

First, the image is binarized, see Fig. 27(b), using the method of Sec. 2.2 for instance. Second, an erosion operation (Eq. (3.50)) is applied using a 3 x 3 structuring element, see Fig. 27(c). Last, a dilation (Eq. (3.49)) by the same structuring element is performed on the eroded image, see Fig. 27(d). That is, the binary image has been opened (Eq. (3.52)). It can be seen from Fig. 27(d) that if the size of textural elements is smaller than those of the structuring element and of text, the technique works well. Otherwise, more sophisticated processing is required [44].

3.4. *Boundary Detection and Representation*

After separation from background, objects are localized, measured and eventually classified. An object is completely specified by its boundary, which can be *traced* by the algorithm of Fig. 28 [2].

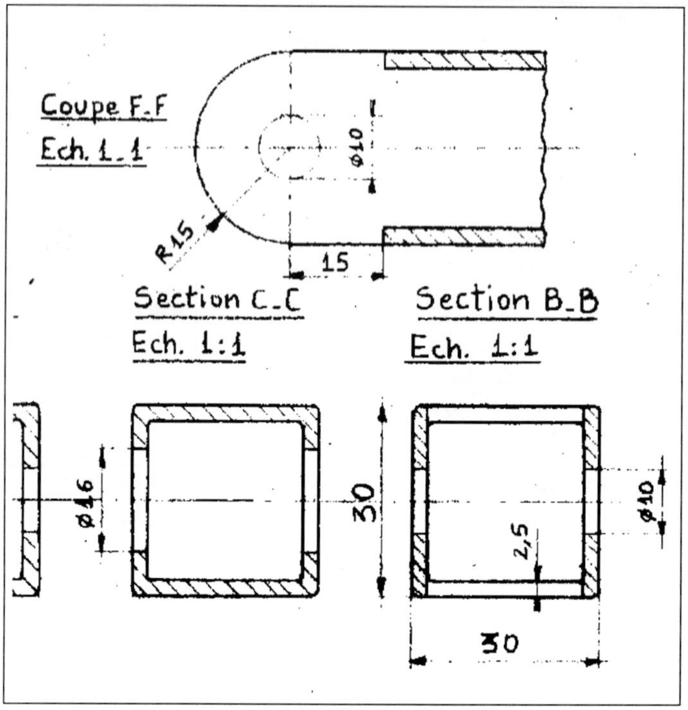

Fig. 22. Binarized version of Fig. 11 by Otsu's algorithm.

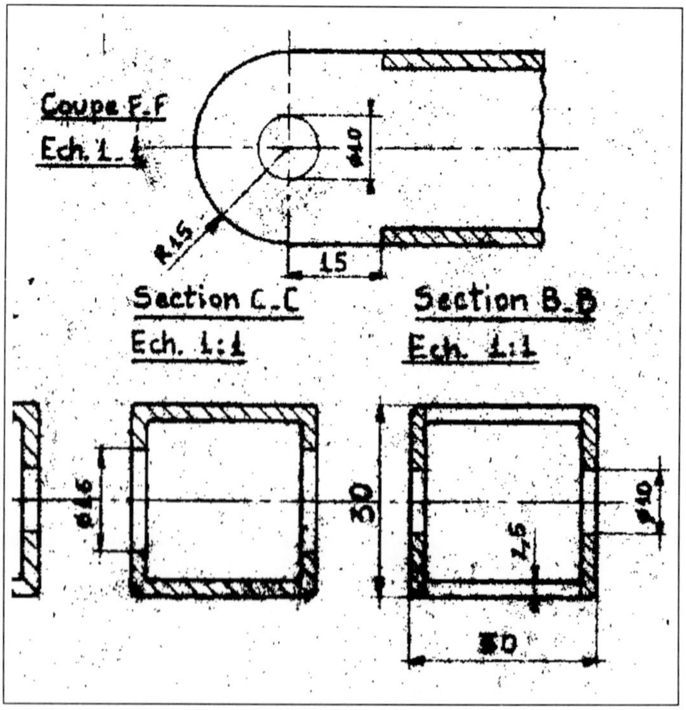

Fig. 23. Binarized version of Fig. 11 by the p-tile method (p=10%).

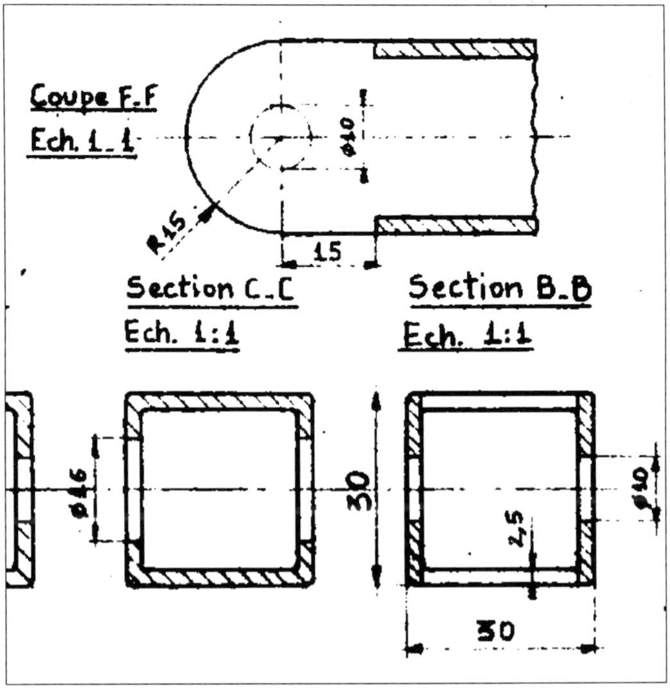

Fig. 24. Result of binarization of a smoothed version of Fig. 11 by the p-tile method (p=10%).

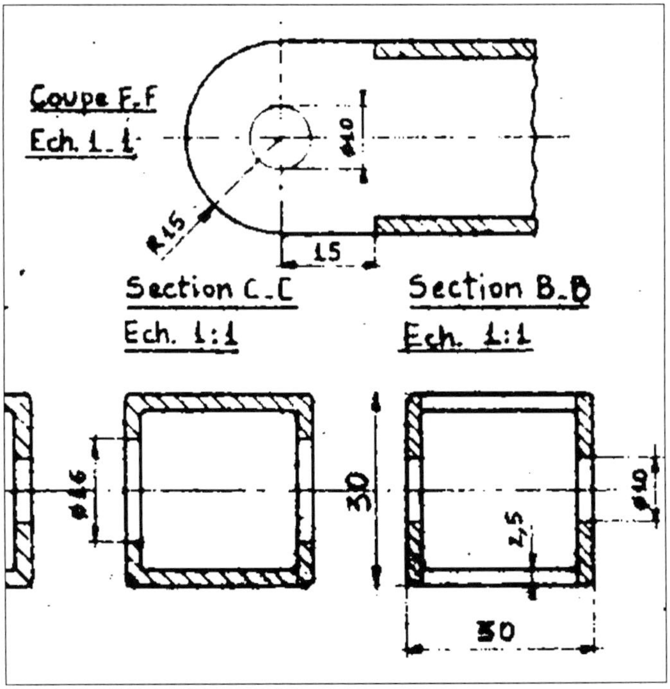

Fig. 25. Result of binarization of a median filtered version of Fig. 11 by the p-tile method (p=10%).

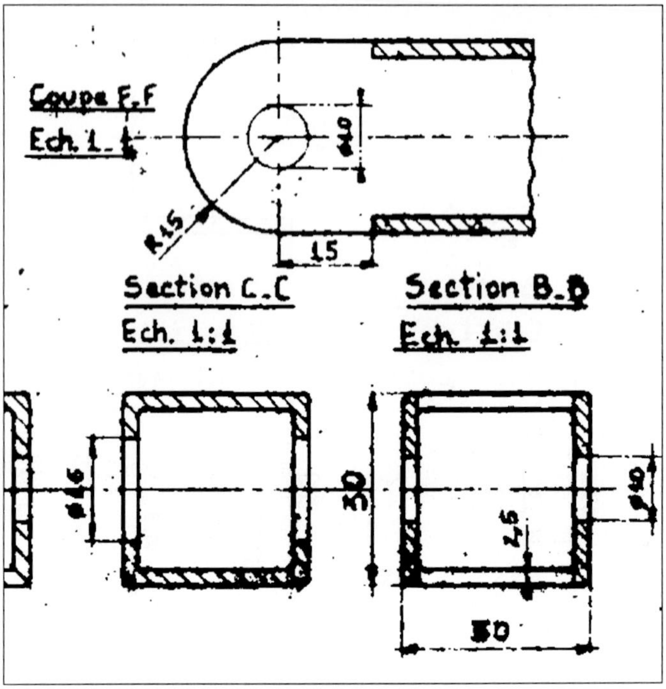

Fig. 26. Result of a majority filter on Fig. 23.

Figure 29 illustrates the tracing of a simply connected object. We notice that the same pixel, e.g. pixel 1, is encountered several times. However, multiple occurrences of the same pixel can be easily eliminated. In general, the image may contain many objects, including holes, which in turn may contain smaller objects. For such a case, the tracing must be recursively repeated to detect all outer and inner boundaries of all objects. Figure 30 shows the tracing of an image containing multiple and nested objects. Note that tracing all outer and inner boundaries constitutes in fact a segmentation of the input image into regions (see Sec. 4 for more details).

A boundary is specified by the x- and y-coordinates of its pixels, called boundary pixels. A more compact representation is the Freeman *chain code*. Only the x- and y-coordinates of the first boundary pixel are stored; the following boundary pixels are coded using their direction relative to the previous boundary pixel. Figure 31 shows the eight chain codes, each of which can be represented by three bits. As an example, the boundary pixels of Fig. 29 are encoded as follows:

0 7 7 1 7 5 4 4 4 3 3 1

or

000 111 111 001 111 101 100 100 100 011 011 001

Boundary pixels obtained by tracing always represent a closed curve. In other words, the sequence of x- and y-cooordinates is periodic. This means that the boundary can be represented by a Fourier series, the coefficients of which charac-

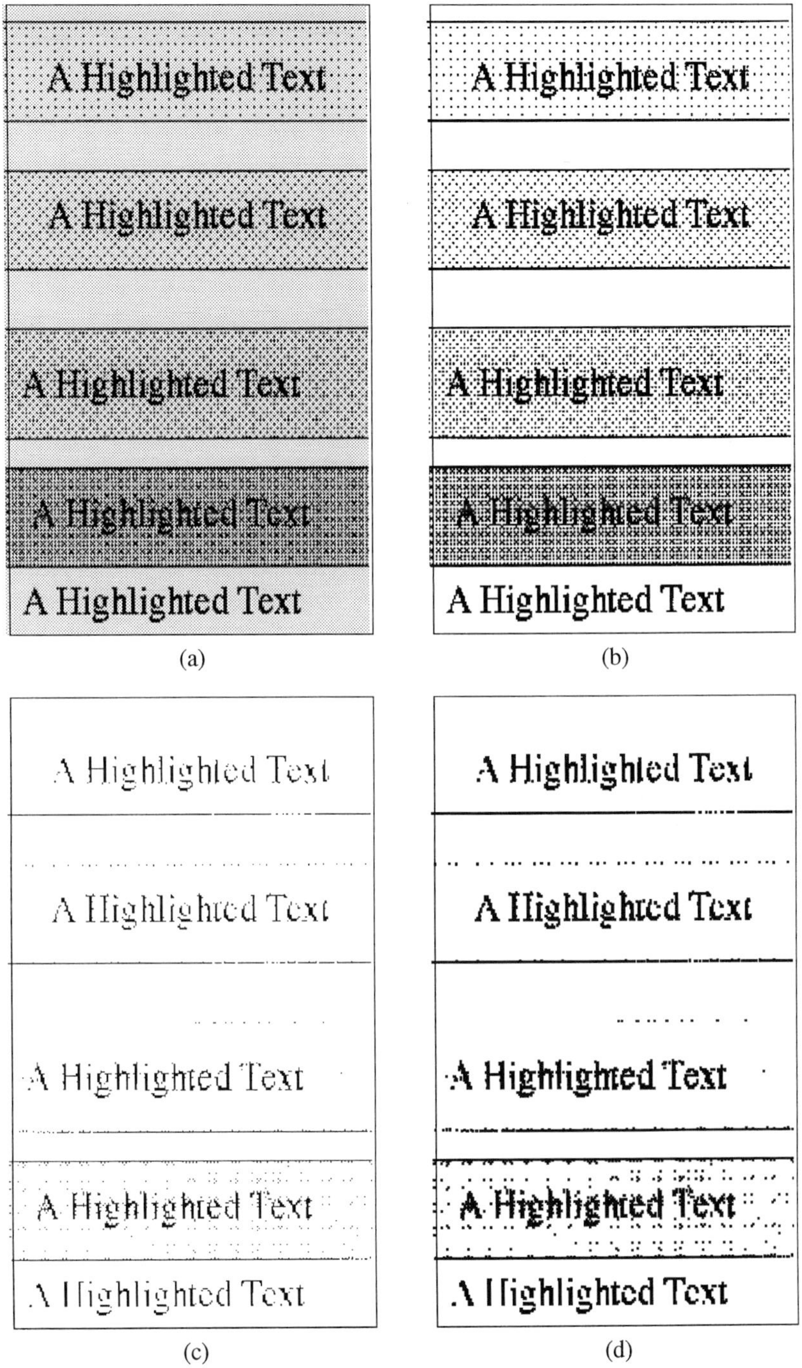

Fig. 27. Separation of text from textured background by morphological filtering. (a) A text with textured background. (b) Binarized version of (a). (c) Eroded version of (b). (d) Opened version of (b).

```
1. Scan the image until a black pixel is encountered. Call it Pixel 1.
2. REPEAT
           IF current pixel is black
           THEN backtrack
           ELSE turn right
   UNTIL Pixel 1 is met
```

Fig. 28. Boundary Tracing Algorithm.

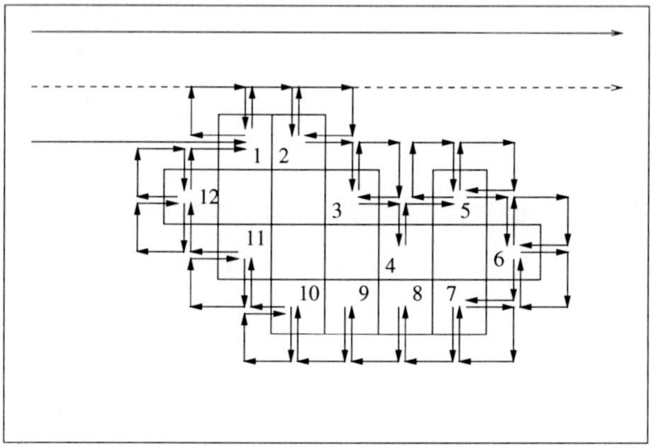

Fig. 29. Tracing a simple boundary.

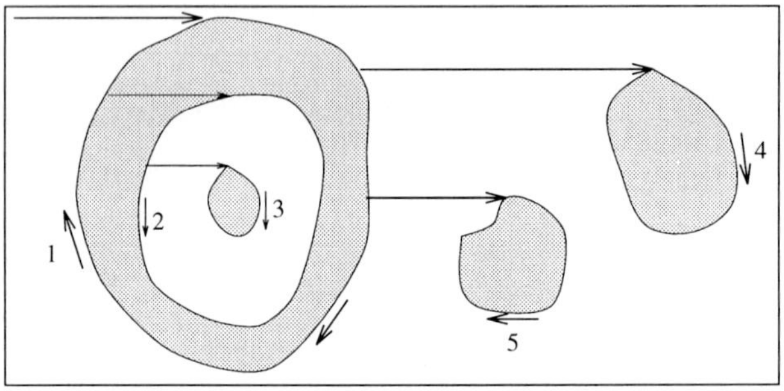

Fig. 30. Tracing boundaries of multiple and nested objects.

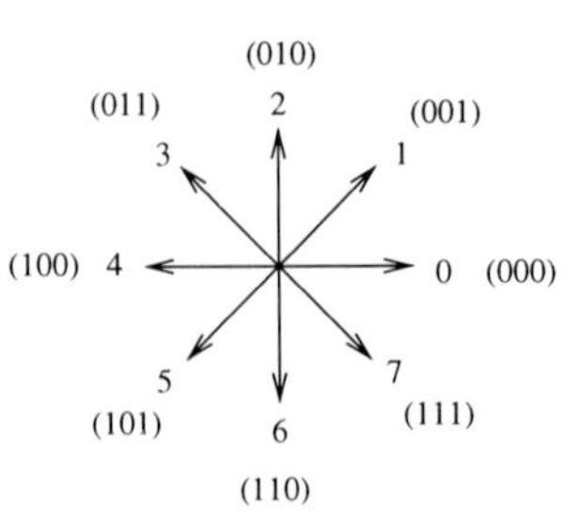

Fig. 31. Freeman chain codes.

terize the closed curve. Furthermore, these coefficients can be combined into *Fourier Descriptors* exhibiting interesting position- and rotation-invariant properties [45, 1].

3.5. *Thinning and Structural Representation*

Most objects in document images have line structures. In certain applications where the line thickness is of secondary importance, a convenient alternative to representing an object by its boundary is line representation. It is of practical interest because it can simplify many subsequent operations in structural analysis. Line representation can be obtained by a *thinning* process.

The basic idea of thinning is to repeatedly delete object boundary pixels so as to reduce the line width to one pixel. This must be done without locally disconnecting the object (splitting the object into two parts) nor deleting line end points. In other words, thinning can be seen as conditional deletion of boundary pixels. A simple parallel algorithm, insensitive to contour noise, is as follows [46, 47].

P_3	P_2	P_9
P_4	P_1	P_8
P_5	P_6	P_7

Fig. 32. Neighborhood convention for thinning.

Using the neighborhood convention of Fig. 32, let $NT(P_1)$ be the number of zero (white) to nonzero (black) transitions in the ordered sequence $< P_2, P_3, .., P_9, P_2 >$, and $NZ(P_1)$ be the number of nonzero neighbors of P_1. Pixel P_1 is deleted (set to zero) if

$$2 \leq NZ(P_1) \leq 6 \tag{3.53}$$

and

$$NT(P_1) = 1 \tag{3.54}$$

and

$$P_2 \cdot P_4 \cdot P_8 = 0 \quad or \quad NT(P_2) \neq 1 \tag{3.55}$$

and

$$P_2 \cdot P_4 \cdot P_6 = 0 \quad or \quad NT(P_4) \neq 1. \tag{3.56}$$

The process is repeated until there are no more changes in the image.

Figure 33 shows the thinned version of Fig. 20. A large number of thinning algorithms, differing from one to another by the deletion conditions, have been proposed [48]. In a recent study, a comparison of 20 different algorithms was conducted, taking into account criteria such as reconstructibility, computation speed, similarity to the reference skeleton, quality of the skeleton, connectivity after thinning, and the

Fig. 33. Thinned version of Fig. 20.

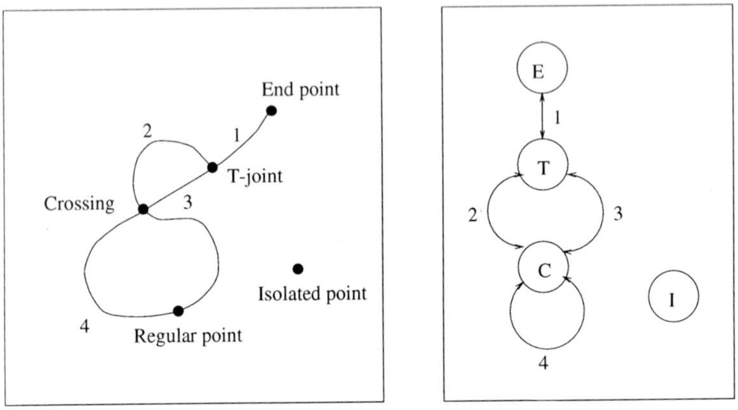

(a) (b)

Fig. 34. (a) Illustration of regular and singular points. (b) Structural representation of (a).

degree of parallelism [49]. Unfortunately, the main conclusion of this study is that there is no panacea for thinning.

In general, the thin lines produced by a thinning algorithm contain two types of black pixels, namely, *regular* and *singular* points. A regular point has 2 neighboring black pixels. A singular point may have 0, 1, 3, or 4 neighboring black pixels. These numbers are also called the order of the black pixel. Thus, a pixel of order 0 is an isolated point, of order 1 an end point, of order 2 a line point, of order 3 a T-joint, and a pixel of order 4 is a crossing point, see Fig. 34(a).

A convenient way to describe the structure of a thinned image is a *graph* representation [50]. *Nodes* are associated with singular points and *arcs* with thin lines between nodes, if any, see Fig. 34. More precisely, a node is specified by its type (isolated, end point, T-joint, or crossing) and position (x- and y-coordinates). An arc between two nodes can be represented by the chain code described in Sec. 3.4 or by some approximations, such as polygones or B-splines [19, 1, 51].

A simple algorithm to find a *polygonal approximation* of a curve is as follows. Approximate the curve by the line segment joining its end points A and B. If the distance from the farthest curve point C to segment AB is greater than a predefined tolerance threshold, join A to C and C to B. Repeat the procedure for new segments AC and BC until the predefined tolerance is satisfied [52]. Another algorithm for the approximation of curves by line segments has been described in [53].

4. Image Segmentation

Segmentation is the process of dividing an image into regions, each susceptible to containing a single object or a group of objects of the same type. For instance, an object can be a character on a text page or a line segment in an engineering drawing; a group of objects can represent a word or two line segments that touch each other. In document image analysis, four commonly used segmentation algorithms are *connected component labeling*, *X-Y-tree decomposition*, *run-length smearing*, and *Hough transform*. They will be described in more detail in the following sections.

4.1. *Connected Component Labeling*

This standard technique assigns to each connected component of the binary image a distinct label. The labels are usually natural numbers starting from one to the total number to connected components in the input image [54]. The algorithm scans the image from left-to-right and top-to-bottom. On the first line containing black pixels, a unique label is assigned to each contiguous run of black pixels. For each black pixel of the next and succeding lines, the neighboring pixels on the previous line and the pixel to the left are examined (see Fig. 35(a)). If any of these neighboring pixels has been labeled, the same label is assigned to the current black pixel; otherwise the next unused label is used. This procedure continues to the bottom line of the image.

Upon completion of this process, a connected component may contain pixels having different labels because when we examined the neighborhood of a black pixel, e.g. pixel "?" in Fig. 35(c), the pixel to its left and those on the previous line might have been labeled differently. (In our example, we decide to use the label of the left neighbor.) Such a situation must be detected and remembered. After the scanning process, the labeling is completed by unifying conflicting labels and reassigning unused labels. For an illustration of the complete procedure, see Fig. 35.

4.2. *X-Y-Tree Decomposition*

The X-Y-tree decomposition, also called the Iterative Projection Profile Cuttings method, is a popular segmentation algorithm in the context of document image analysis [55, 56, 57]. The basic idea behind the algorithm is the exploitation of the fact that most document images have a vertical and/or horizontal structure. Indeed, a normal text page is usually composed of different horizontal text lines (see Fig. 36). This observation immediately leads to the idea of horizontally projecting the black pixels on a vertical axis (see Fig. 36). The resulting profile clearly indicates the line structure of the page. A simple analysis of the profile, like comparison of the projection profile with a threshold, segments the page into blank bands and text bands. A blank band corresponds to a set of contiguous rows having less than n black pixels, and a text band to a set of contiguous rows having a least n pixels if the comparison's threshold is set to n. Subsequently, each text band can be vertically projected on a horizontal axis yielding the positions of the characters within this

```
. . . . .
. P P P .
. L ? . .
. . . . .
```

(a) Neighborhood of '?': P = previous line; L = left neighbor.

```
. . . . . . . . . . . . . . .
. . . * * . . . * * * . . . .
. . . * * . . . * * * . . . .
. . * * * * . * * * * . . . .
. . . * * * * * * * * . . . .
. . . . . * * * * . . . . . .
. . . . . . * * * . . * . . .
. . . . . . * * * . . * * . .
. . * * . . * * * . . * * . .
. . * * . . . . . . . . . . .
. . . . . . . . . . . . . . .
```

(b) Original binary image.

```
. . . . . . . . . . . . . . .
. . . 1 1 . . . 2 2 2 . . . .
. . . 1 1 . . . 2 2 2 . . . .
. . 1 1 1 1 . 2 2 2 2 . . . .
. . . 1 1 1 ? * * * * . . . .
. . . . . * * * * . . . . . .
. . . . . . * * * . . * . . .
. . . . . . * * * . . * * . .
. . * * . . * * * . . * * . .
. . * * . . . . . . . . . . .
. . . . . . . . . . . . . . .
```

(c) Labeling in progress.

```
. . . . . . . . . . . . . . .
. . . 1 1 . . . 2 2 2 . . . .
. . . 1 1 . . . 2 2 2 . . . .
. . 1 1 1 1 . 2 2 2 2 . . . .
. . . 1 1 1 1 1 1 1 1 . . . .
. . . . . 1 1 1 1 . . . . . .
. . . . . . 1 1 1 . . 3 . . .
. . . . . . 1 1 1 . . 3 3 . .
. . 4 4 . . 1 1 1 . . 3 3 . .
. . 4 4 . . . . . . . . . . .
. . . . . . . . . . . . . . .
```

(d) Scanning completion.

```
. . . . . . . . . . . . . . .
. . . 1 1 . . . 1 1 1 . . . .
. . . 1 1 . . . 1 1 1 . . . .
. . 1 1 1 1 . 1 1 1 1 . . . .
. . . 1 1 1 1 1 1 1 1 . . . .
. . . . . 1 1 1 1 . . . . . .
. . . . . . 1 1 1 . . 2 . . .
. . . . . . 1 1 1 . . 2 2 . .
. . 3 3 . . 1 1 1 . . 2 2 . .
. . 3 3 . . . . . . . . . . .
. . . . . . . . . . . . . . .
```

(e) After label unification and reassignment.

Fig. 35. Connected component labeling.

band. Figure 36 shows the vertical projection of the first band. Thus, in the case of the text page of Fig. 36, the following procedure can be used for extracting the characters:

1. Compute the horizontal projection profile for the entire page.
2. Analyse the projection profile to extract the lines.
3. For each line, compute the vertical projection profile.
4. Analyse each projection profile obtained in step 3 to extract the characters.

More generally, however, it may be necessary to perform more than two projections to reach the characters. Figure 37 shows a spatial configuration (characters are symbolically represented by boxes) in which some characters are reached only after the third projection (horizontal). It can be readily seen that more complex documents may require even more projections to extract the individual characters. Therefore, in order to be sure that all indivisible zones are reached, the general algorithm should alternatively project the document vertically and horizontally until two consecutive projections yield the same result.

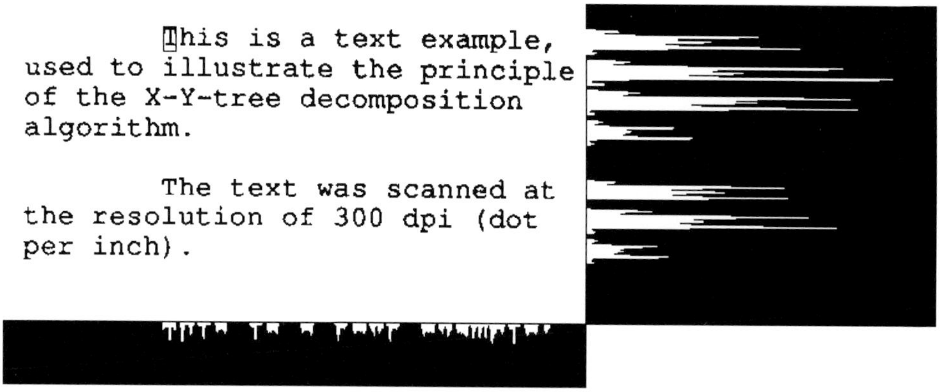

Fig. 36. Horizontal projection of the page and vertical projection of the first line.

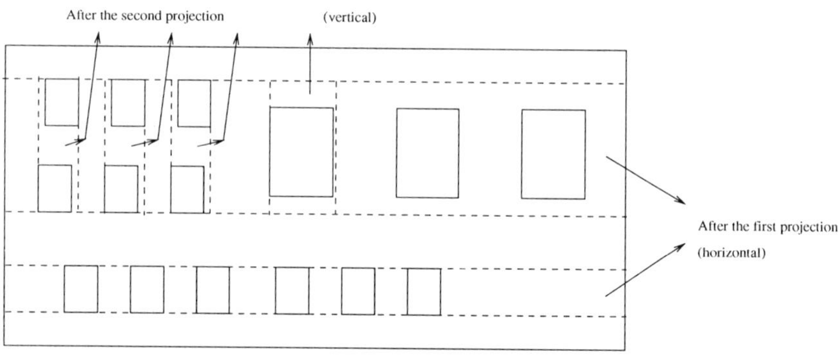

Fig. 37. Example of a document requiring three projections.

The natural data structure to store the result of the algorithm is a tree whose nodes represent rectangular zones. Each node may have many children each representing a sub-zone of the parent's zone. The terminal nodes (leaf-nodes) are those which are not further decomposable. They represent the indivisible zones and are called *atomic entities*. Figure 38 shows an example of text where characters are represented by boxes, and its tree representation.

A problem with the X-Y-tree decomposition algorithm arises in the presence of some particular graphics forms. Figure 39 depicts two typical forms of graphics, namely, close-form and open-form, that prevent the algorithm from succesfully decomposing the image. In the presence of such graphics forms in the document, it is necessary to localize them and apply the connected component analysis [57].

In the above description, the X-Y-tree decomposition is purely considered as a low-level algorithm in the sense that no high-level knowledge of the document is required to perform the segmentation. Some authors make use of high-level knowledge to guide the decomposition. The method described in [55] embeds the picture grammar and [58] introduces publication-specific knowledge in the segmentation process. This combination of low- and high-level information results in an elegant

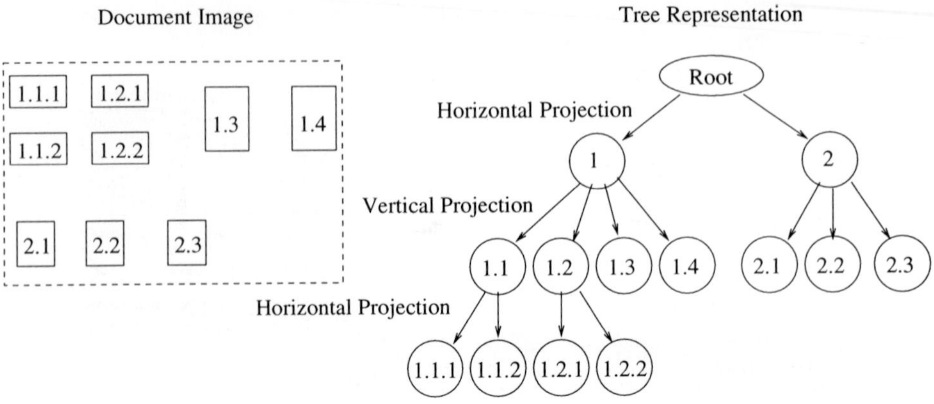

Fig. 38. A document and its tree representation.

Fig. 39. Typical forms that cannot be decomposed by the X-Y-tree. (a) close form. (b) open form.

solution but requires that the document layout has a well-defined structure. More importantly, an error in the data due to noise may irremediably lead to a segmentation failure [56]. In the approach where low- and high-level processing are separated, the problem of noisy data is postponed to the later stage of labeling. The advantage is that the amount of data to deal with is then much smaller than in the original representation.

4.3. *Run-Length Smearing*

Like X-Y-tree decomposition, the run-length smearing (RLS) algorithm requires a prior skew correction. Let us consider a line consisting of 0's (white pixels) and 1's (black pixels). The RLS first detects all white runs (contiguous 0's) of the line and then converts those runs, whose length is shorter than a predefined threshold T, to black runs. Black runs remain unchanged. For instance, with $T = 3$, the RLS converts the line

0 0 0 1 1 0 1 0 1 1 1 0 0 1 0 0 0 1 1 1 0 0

into

0 0 0 1 1 1 1 1 1 1 1 1 1 1 0 0 0 1 1 1 1 1

To obtain a segmentation, the RLS algorithm is first applied line-by-line and then column-by-column, yielding two distinct bitmaps, which are eventually combined by the logical AND operation. Figures. 40(a), 40(b), 40(c) and 40(d) show the original

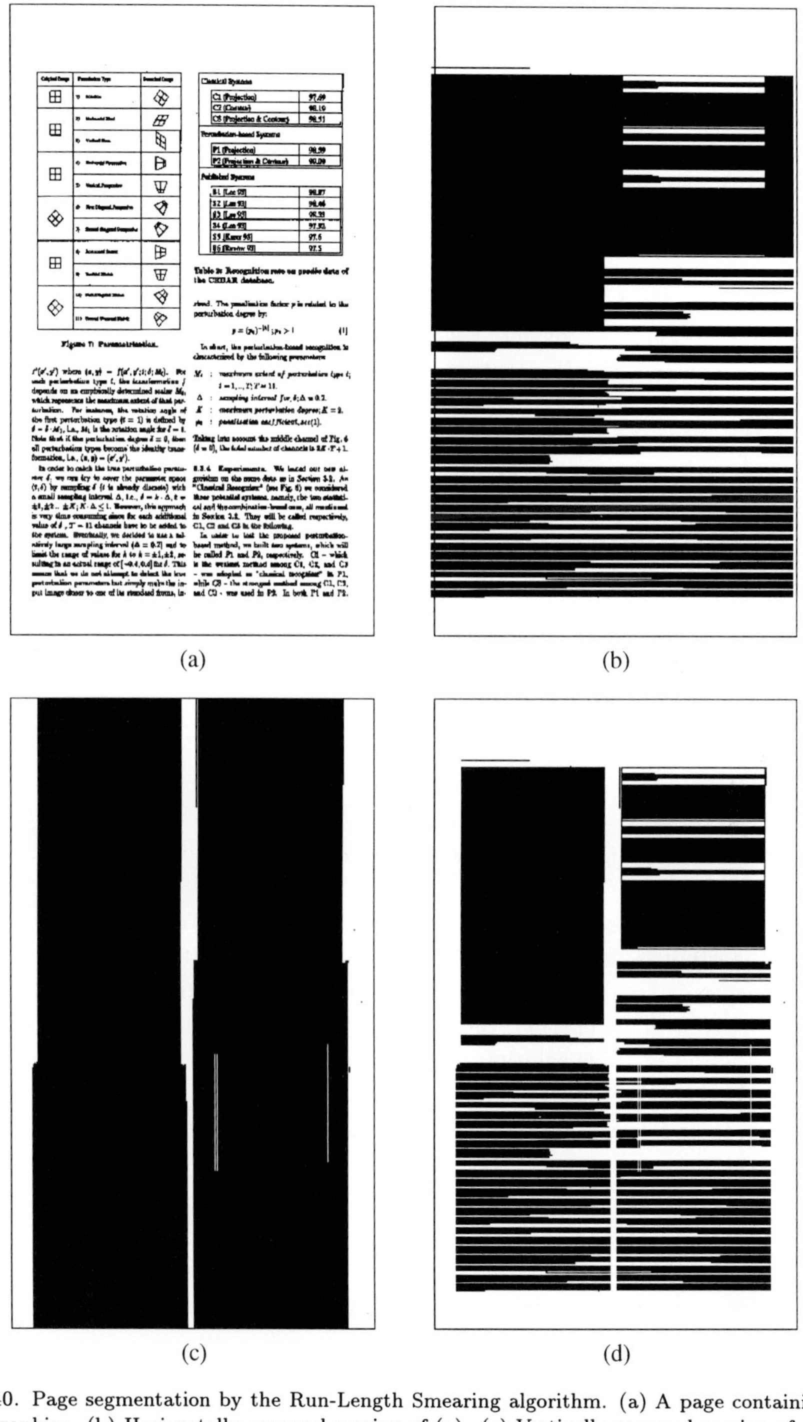

Fig. 40. Page segmentation by the Run-Length Smearing algorithm. (a) A page containing text and graphics. (b) Horizontally smeared version of (a). (c) Vertically smeared version of (a). (d) Segmentation result of (a) by the Run-Length Smearing algorithm.

binary image, its horizontal ($T_h = 100$) and vertical ($T_v = 100$) smeared versions, and the final result, respectively. The thresholds are application-dependent, but are rather uncritical [59].

The result of the RLS algorithm can be used as input to the connnected component labeling, yielding a set of regions. Those having very large sizes in both x- and y-coordinates usually correspond to graphics objects whereas text lines mostly produce horizontally elongated regions.

4.4. *Hough Transform*

Unlike the previously considered segmentation methods, where distinct objects must be disconnected, the Hough transform works succesfully even when different objects are connected to each other. the Hough transform is best explained through the example of line extraction.

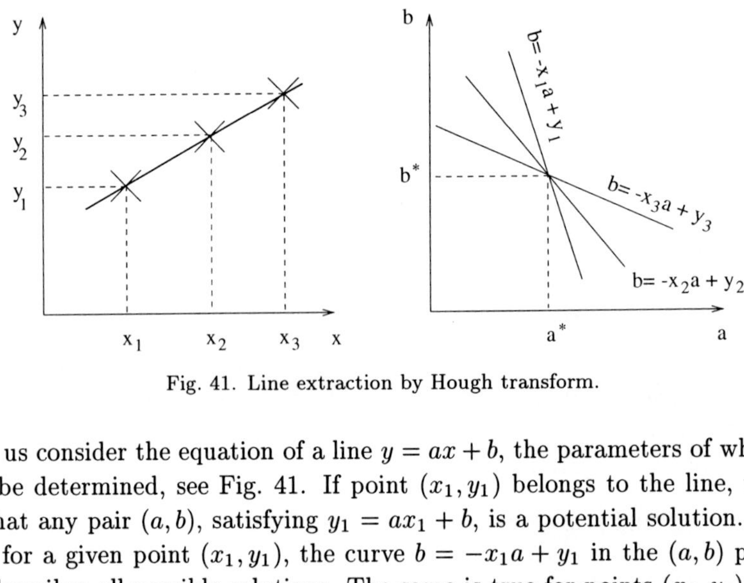

Fig. 41. Line extraction by Hough transform.

Let us consider the equation of a line $y = ax + b$, the parameters of which (a, b) are to be determined, see Fig. 41. If point (x_1, y_1) belongs to the line, then it is clear that any pair (a, b), satisfying $y_1 = ax_1 + b$, is a potential solution. In other words, for a given point (x_1, y_1), the curve $b = -x_1 a + y_1$ in the (a, b) parameter space describes all possible solutions. The same is true for points (x_2, y_2), (x_3, y_3), and so on. If the n points (x_1, y_1), (x_2, y_2), .., (x_n, y_n) lie on the same line in the image space, their corresponding curves must intersect each other at the same point (a^*, b^*) in the parameter space. Based on this observation, the Hough algorithm works as follows:

1. Quantize the (a, b) parameter space into cells.

2. Form an accumulator $Acc(a, b)$ and initialize its components to zero.

3. For each black pixel (x, y) in the image space, all cells of $Acc(a, b)$ that satisfy $b = -xa + y$, are incremented by one.

4. Detect local maxima in the (a, b) parameter space. Each maximum corresponds to a set of collinear points in the image space.

In practice, the value of a is infinite for vertical lines, which creates problems in Step 1. A more appropriate line parametrization is $r = x \cdot cos\theta + y \cdot sin\theta$, where r is the distance from the origin of the (x, y) space to the line and θ is the angle of the normal to the line. This produces a sinusoidal curve in the (θ, r) parameter space for each point (x, y). The Hough algorithm remains unchanged. Many practical implementation aspects of the Hough transform, including parameter quantization and local maxima detection, are described in [60].

The Hough transform can be applied to any parametric curve, e.g. circle, ellipse, and even to non-parametric curves [50, 61, 62]. It is robust against noise and is especially useful in the analysis of engineering drawings, where most curves are line segments or circle arcs. Furthermore, text strings in engineering drawings can also be extracted via Hough transform by considering each string as a line whose elements are characters [60].

5. An Introduction to Feature Extraction

One of the goals of document image analysis is to classify characters and symbols into classes. Feature extraction is one step that may help in easing the classification task. Feature extraction is traditionally a subject of study in pattern recognition and can be divided into two approaches, namely, statistical and structural [63, 64, 50, 65, 66]. The aim of this section is to present some simple and effective feature extraction methods that are widely used in document image analysis.

5.1. *Geometrical Features*

Simple geometrical features can be very effective in many applications. For instance, the sizes in the x- and y-directions of a connected component may be sufficient to distinguish characters from graphical parts, see Fig. 39. Below is a non-exhaustive list of geometrical features.

1. Sizes in x- and y-direction, and their (aspect) ratio
2. Perimeter
3. Area
4. Maximum and minimum distances from the boundary to the center of mass
5. Number of holes
6. Euler number = (Number of connected components) – (Number of holes)
7. Compactness = $(Perimeter)^2/(4\pi \cdot Area)$
8. Signatures (horizontal and vertical projections of black pixels)

More details about these features and others can be found in [50]. In document image analysis, these features are commonly used for the preclassification of objects into broad categories, such as character and graphics. Afterwards, components in the character category are sent to an OCR module whereas those in the graphics category are analyzed by a graphics recognition module [60]. Apart from the purpose of preclassification, some of the above features, e.g. aspect ratio and signatures, can significantly contribute to the final classification of characters as well [26].

5.2. *Moments*

Objects can also be described by their moments, which are defined by

$$M_{p,q} = \int\int_{\mathcal{D}} i(x,y) \cdot x^p y^q \, dxdy \tag{5.57}$$

where p and q are the non-negative integer degrees of the moment, $i(x,y)$ is the gray-level image of the object, and $\mathcal{D}$ is the spatial extent of the object. Note that if the image is binary, and takes on the value one for black and zero for white pixels, then $M_{0,0}$ is simply the area of the object whereas $M_{1,0}$ and $M_{0,1}$ are the x- and y-coordinates of the center of mass. It can be shown that the infinite set of moments $\{M_{p,q}; p,q = 0,1,2,...\}$ uniquely determines $i(x,y)$ [1]. However, for large values of p and q, the moments become very sensitive to noise and in practice, only small values of p and q are used.

The moments defined by Eq. (5.57) are position dependent and therefore cannot be directly used for shape comparison (except for the area). Another definition, yielding *central* moments, is usually adopted.

$$M^c_{p,q} = \int\int_{\mathcal{D}} i(x,y) \cdot (x-\bar{x})^p (y-\bar{y})^q \, dxdy \tag{5.58}$$

where

$$\bar{x} = \frac{M_{1,0}}{M_{0,0}} \quad and \quad \bar{y} = \frac{M_{0,1}}{M_{0,0}} \tag{5.59}$$

The set of second degree ($p+q=2$) central moments, namely, $M^c_{2,0}$, $M^c_{0,2}$ and $M^c_{1,1}$, defines the inertial ellipse of the object and yields, through the eigen vectors, the main orientation θ of the object.

$$\theta = \frac{1}{2} arctan \left[\frac{2 \cdot M^c_{1,1}}{M^c_{2,0} - M^c_{0,2}} \right] \tag{5.60}$$

In particular, it can be shown that the following combinations of central moments are independent of the orientation of the object.

$$\phi_1 = M^c_{2,0} + M^c_{0,2} \tag{5.61}$$

$$\phi_2 = (M^c_{2,0} - M^c_{0,2})^2 + 4 \cdot (M^c_{1,1})^2 \tag{5.62}$$

Similar invariants exist for higher degree moments [67, 1, 66]. Due to its rotation-invariant property, combinations of moments (Zernike moments) are very useful for OCR [68].

5.3. Orthogonal Transforms

Linear transformations can be used to extract features from an image. That is, instead of using the input binary- or gray-levels as raw features, we use their linear combinations. Formally, let f be the vector whose components represent the gray-levels of the input image, and g, a vector each component of which is a linear combination of components of f. The linear transformation can be expressed as a matrix multiplication.

$$g = \mathcal{T} \cdot f \tag{5.63}$$

where $\mathcal{T}$ is a matrix of constant components. Many standard transformations, such as Fourier, Cosinus, Hadamard, can be used for $\mathcal{T}$ [1, 2]. These transformations have fast computation algorithms; but since they are generally defined, they may not be suited to a given application.

The Karhunen-Loeve transformation (KLT) is, in contrast, obtained from the experimental data of a particular application [65]. The KLT is defined as follows.

$$g = \mathcal{T}^t \cdot (f - \bar{f}) \tag{5.64}$$

where the superscript "t" denotes matrix transposition, and

$$\bar{f} = \frac{1}{N} \sum_{i=1}^{N} f_i \tag{5.65}$$

is the sample mean vector of all available f_i. Each column of $\mathcal{T}$ is an eigen vector of the sample covariance matrix

$$C = \frac{1}{N} \sum_{i=1}^{N} (f_i - \bar{f})^t (f_i - \bar{f}) \tag{5.66}$$

Each eigen vector corresponds to an eigen value. It is common to choose only those $M < N$ eigen vectors that correspond to the largest eigen values. In such a case, a reduction of the number of features is achieved. The KLT is widely used for feature extraction or reduction in character classification problems [69].

5.4. Relational Descriptions

Simple relational descriptions of objects were already shown in Figs. 34 and 38. More generally, relational descriptions can be used to represent spatial relationships between distinct objects or groups of objects. For instance, we can define a character as a basic object. A text line is then a group of characters that are horizontally aligned. Consecutive lines form a paragraph, and so on.

Relational description is a first and important step in document understanding because a document contains objects arranged in a "certain" way and not arbitrarily. This arrangement, also called *layout structure*, can naturally be represented by a relational description. Therefore, a relational description of the image provides a

patches, *Computer Vision, Graphics, and Image Processing* **47** (1989) 385–394.

[34] H.A. Beyer, Accurate calibration of CCD-cameras, *IEEE Computer Society Conf. on Computer Vision and Pattern Recognition,* Champaign, Illinois, June 1992, 96–101.

[35] J.A.F. Leite and E.R. Hancock, Statistically combining and refining multi-channel information, in *Progress in Image Analysis and Processing III,* ed. S. Impedovo (World Scientific, 1993) 193–201.

[36] J. Serra, *Image Analysis and Mathematical Morphology* (Academic, 1982).

[37] J. Serra, Introduction to mathematical morphology, *Computer Vision, Graphics, and Image Processing* **35** (1986) 283–305.

[38] R.M. Haralick, S.R. Sternberg, and X. Zhuang, Image analysis using mathematical morphology, *IEEE Trans. on Pattern Analysis and Machine Intelligence* **9** (1987) 532–550.

[39] P. Salembier, ed., Special Issue on Mathematical Morphology and its Applications to Signal Processing, *Signal Processing* **38** (1994).

[40] M. Gabbouj, ed., Special Issue on Nonlinear Digital Signal Processing, *Signal Processing* **38** (1994).

[41] A.J. Filipski and R. Flandrena, Automated conversion of engineering drawings to CAD form, Special Issue on Optical Character Recognition, *Proc. of the IEEE,* eds. T. Pavlidis and S. Mori, (1992) 1195–1209.

[42] J.C. Cheng and H.S. Don, Segmentation of bilevel images using mathematical morphology, *Int. J. of Pattern Recognition and Artificial Intelligence* **6** (1992) 595–628.

[43] M. Kamel and A. Zhao, Extraction of binary character/graphics images from grayscale document images, *CVGIP: Graphical Models and Image Processing* **55** (1993) 203–217.

[44] S. Liang and M. Ahmadi, A morphological approach to text string extraction from regular periodic overlapping text/background images, *CVGIP: Graphical Models and Image Processing* **56** (1994) 402–413.

[45] G.H. Granlund, Fourier preprocessing for hand print character recognition, *IEEE Trans. on Computer* **21** (1972) 195–201.

[46] D. Rutovitz, Pattern recognition, *J. Royal Statistical Society,* series A **129** (1966) 504–530.

[47] R. Stefanelli and A Rosenfeld, Some parallel thinning algorithms for digital pictures, *J. of ACM* **18** (1971) 255–264.

[48] L. Lam, S.W. Lee, and C.Y. Suen, Thinning methodologies – A comprehensive survey, *IEEE Trans. on Pattern Analysis and Machine Intelligence* **14** (1992) 869–885.

[49] S.W. Lee, L. Lam, and C.Y. Suen, A systematic evaluation of skeletonization algorithms, *Int. J. of Pattern Recognition and Artificial Intelligence* **7** (1993) 1203–1225.

[50] D.H Ballard and C.M. Brown, *Computer Vision* (Prentice-Hall, 1982).

[51] L. O'Gorman and R. Kasturi, *Document Image Analysis* (IEEE Computer Society Press, 1995).

[52] U.E. Ramer, An iterative procedure for the polygonal approximation of plane curves, *Computer Graphics and Image Processing* **1** (1972) 244–256.

[53] K. Wall and P.-E. Danielsson, A fast sequential method for polygonal approximation of digitized curves, *Computer Vision, Graphics, and Image Processing* **28** (1984) 220–227.

[54] A. Rosenfeld and A.C. Kak, *Digital Picture Processing* (Academic Press, 1976).

[55] G. Nagy and S. Seth, Hierarchical representation of optically scanned documents, *Proc. of the 7th Int. Conf. on Pattern Recognition,* Montreal, July 1984, 347–349.

[56] S.N. Srihari and G.W. Zack, Document image analysis, *Proc. of the 8th Int. Conf. on Pattern Recognition,* Paris, France, Oct. 1986, 434–436.

[57] R.G. Casey and K.Y. Wong, Document-analysis, systems and techniques, in *Image Analysis and Applications,* eds. R. Kasturi and M. Trivedi, (Marcel Dekker, 1990) 1–35.

[58] M. Viswanathan, Analysis of scanned documents – A syntactic approach, *Pre-Proc. of the Workshop on Syntactic and Structural Pattern Recognition,* Murray Hill, New Jersey, U.S.A., June 1990, 450–459.

[59] F.M. Wahl, K.Y.Wong, and R.G. Casey, Block segmentation and text extraction in mixed text/image documents, *Computer Graphics and Image Processing* **20** (1982) 375–390.

[60] L.A. Fletcher and R. Kasturi, A robust algorithm for text string separation from mixed text/graphics images, *IEEE Trans. on Pattern Analysis and Machine Intelligence* **10** (1988) 910–918.

[61] P.D. Picton, Hough transform references, *Int. J. of Pattern Recognition and Artificial Intelligence* **1** (1987) 413–425.

[62] J. Illingworth and J. Kittler, A survey of the Hough transform, *Computer Vision, Graphics, and Image Processing* **44** (1988) 87–116.

[63] R.O. Duda and P.E. Hart, *Pattern Classification and Scene Analysis* (John Wiley & Sons, 1973).

[64] J.R. Ullmann, *Pattern Recognition Techniques* (Butterworths, 1973).

[65] K. Fukunaga, *Introduction to Statistical Pattern Recognition,* second edition (Academic Press, 1990).

[66] R.J. Schalkoff, *Pattern Recognition: Statistical, Structural and Neural Approaches* (John Wiley & Sons, 1992).

[67] M.R. Teague, Image analysis via the general theory of moments, *J. of Optical Society of America* **70** (1980) 920–930.

[68] M. Sabourin, A. Mitiche, D. Thomas, and G. Nagy, Classifier combination for hand-printed digit recognition, *Proc. of the Second Int. Conf. on Document Analysis and Recognition,* Tsukuba, Japan, Oct. 1993, 163–166.

[69] J.L. Blue, G.T. Candela, P.J. Grother, R. Chellappa, and C.L. Wilson, Evaluation of pattern classifiers for fingerprint and OCR applications, *Pattern Recognition* **27** (1994) 485–501.

[70] H. Bunke, Syntactic and structural pattern recognition, in *Handbook of Pattern Recognition and Computer Vision,* eds. C.H. Chen, L.F. Pau, and P.S.P. Wang (World Scientific, 1993) 163–209.

Handbook of Character Recognition and Document Image Analysis, pp. 49–78
Eds. H. Bunke and P. S. P. Wang

CHAPTER 2

PATTERN CLASSIFICATION TECHNIQUES BASED ON FUNCTION APPROXIMATION

ULRICH KRESSEL and JÜRGEN SCHÜRMANN
Daimler-Benz AG, Research and Technology
P.O. Box 2360, 89013 Ulm, Germany

If pattern generation is viewed as a stochastic process, which usually is appropriate for optical character recognition, speech recognition, and similar tasks, pattern classification can be based on the idea of function approximation. Starting from decision theory, it is shown in this chapter that minimum error decisions are obtained by the classifier if it approximates the a posteriori probabilities ruling the pattern source (Bayesian approach). Since for practical problems the properties of the stochastic pattern source are only implicitly given by the learning set, this set should truly represent the data population underlying the recognition task. Furthermore, the approximating functions should be taken from a family of functions which have — at least in principle — the power of approximating any desirable function (universal approximator). This contribution discusses in some detail the important paradigms for classification based on function approximation: polynomial classifier (including the linear approach) and multilayer perceptron, which are both global approximation schemes; radial basis functions, which are local approximators similar to nearest neighbor techniques; and finally classifiers which are based on certain distribution assumptions. All these paradigms, including different variants, are introduced based on a common notation, and special characteristics of the different approaches are pointed out using the classification of handwritten digits as a practical example.

Keywords: Pattern classification; Function approximation; Decision theory; Bayes classifier; A posteriori probability; Least mean squares approximation; Universal approximator; Polynomial classifier; Multilayer perceptron; Radial basis function; Nearest neighbor classifier; Gaussian classifier; Clustering techniques; Principal component analysis; Optical character recognition.

1. Introduction

Before getting started with any detailed classification techniques, we think it is necessary and helpful

- to identify the classification task in the processing chain of a typical pattern recognition problem (Sec. 1.1),
- to differentiate between various approaches to classification (statistical versus rule based; supervised versus unsupervised; see Sec. 1.2),
- and to state why statistical pattern classification strongly depends on learning from examples (Sec. 1.3).

Based on these introductory remarks, an overview (Sec. 1.4) of the present chapter should help to guide the reader into the details of pattern classification techniques.

1.1. *Segmentation, Feature Extraction, and Knowledge Based Processing*

In most applications, the *classification* of patterns is only part of the overall recognition system, embedded in a sophisticated network of operations such as sensor data preprocessing, segmentation, feature extraction, and knowledge based processing [1]. The task of pattern classification is giving names to observations, thus establishing the bridge between the subsymbolic and the symbolic worlds of information processing.

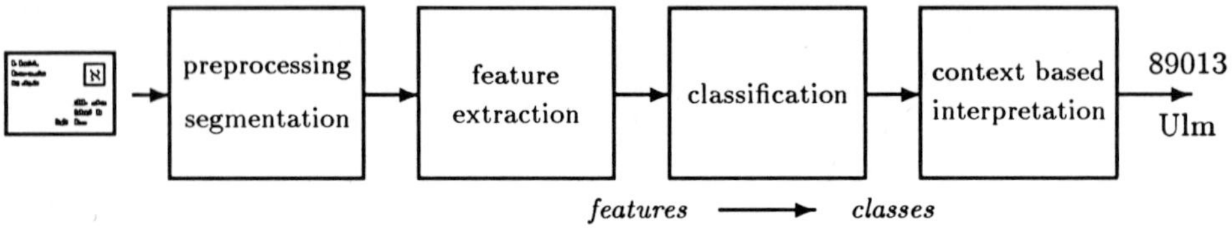

Fig. 1. Typical processing chain for pattern recognition shown by the example of zip code and address recognition for mail sorting.

These relations are best exemplified by choosing the example of recognizing the addresses on postal mail pieces. The first steps are scanning and preprocessing of the address image, resulting in a two-value raster image with dark text and white background pixels. Segmentation is based on a geometrical analysis in order to identify the address block as an arrangement of lines (i.e. to find the regions of interest), and then to divide these lines into single classifiable objects, characters and digits. The last task is usually carried out by a connected component analysis [2], making use of the hypothesis that connected components one-to-one represent isolated characters. This is a simplification which is true for most cases, but which causes many problems if it is violated; for example, think of the difficulties to segment handwritten text (script) into single characters. Feature extraction is highly important for every classification problem. A good choice of features is often half the solution. In feature extraction the domain dependent knowledge of the human system designer comes into play, since the task is to describe the relevant properties of the objects to be recognized as precisely as possible with a fixed number, as small as possible, of feature variables. There exists a close connection between feature extraction and classification, since the efforts going into feature extraction and classification can be exchanged to a certain degree. For (Latin) character recognition, however, a relatively simple set of features is appropriate: size-normalized, gray-valued images with 16×16 pixels. After classification the possibly ambiguous estimations of the pattern classifier are processed further making use of contextual constraints, such as N-gram statistics, dictionaries, and application dependent knowledge.

1.2. *Function Approximation, Rule-Based Approaches, and Clustering Techniques*

Since the "renaissance" of neural networks, at the latest, *function approximation* is the predominant approach to most classification problems including OCR (for explanation and more details see Sec. 2). Besides the function approximation, which is based on statistical concepts, there exist other approaches such as decision trees, rule based methods, syntactic grammars and fuzzy techniques [3, 4, 5, 6] for certain classification tasks. These methods, however, are not covered in this chapter. One important difference, which can sometimes turn out to be an advantage, is that for all these other concepts the classification knowledge is more explicitly expressed than in the statistical approach and it can therefore be interpreted by the designer and manually modified. Yet, these techniques usually require much more input during the design process than just examples for learning, and mostly do not yield classification results as good as the statistical methods (at least in OCR).

Furthermore, we want to differentiate between supervised and unsupervised learning in (statistical based) classification, depending on whether the examples have class labels or not. The unsupervised case, which is also called clustering, seems not very important for OCR problems at first glance, since we usually know which classes are to be discriminated, and thus can easily label the examples. In Sec. 4, we will see, however, how clustering techniques are interwoven with the function approximation based approach to classification.

1.3. *Stochastic Processes and Learning from Examples*

The concept of a stochastic source is ideally suited for modelling any real-world pattern classification task. The pattern itself is represented by a stochastic variable. The *pattern source* is able to generate a whole variety of different patterns, each with a certain probability of being observed. All we need to know is the probability law ruling the stochastic pattern source.

In practical applications, however, these stochastic laws are never explicitly known. All we have available, instead, is a collection of samples drawn from the pattern source which can be used to recover approximately the unknown probability laws. This approach is called *learning from examples* and is the common core of all pattern classification techniques. What we are actually doing is recovering unknown functions from pointwise observations in feature space given by the individual samples of the learning set. Recovering unknown functions from pointwise observations corresponds directly to interpolation between points of support and to extrapolation into their neighborhood. Many of the techniques in pattern classification are simple, intuitive and straightforward if they are applied in low-dimensional feature spaces — such as in one or two dimensions — but become more complicated and less intuitive in those high-dimensional spaces which we usually have to deal with in realistic pattern classification tasks (e.g., in the following example of digit recognition there are 16×16 dimensions with $256^{(16\times16)} \approx 10^{617}$ possible patterns).

For the comparison of different classification approaches throughout this chapter, we use the recognition of handwritten digits as an example [7]. Given are 10000 size-normalized, gray-valued images (1000 per class) for learning and another 10000 images for testing (generalization). For examples see Fig. 2 on the next page.

Fig. 2. Examples of handwritten digits; the spatial resolution is 16×16 with 8 bit quantization.

1.4. *Chapter Overview*

In Sec. 2 we will explain the relation that exists between statistical decision theory and the concept of function approximation. It will be shown that both, risk minimization (minimum error optimization) and least mean square optimization of a suitably chosen target vector, lead to the same solution, namely, discriminant functions approximating the a posteriori probability function of the pattern source. Classifiers, therefore, should be universal approximators (Sec. 2.3), adapted by a learning set. We will further differentiate between global approximation schemes (such as polynomial classifier and multilayer perceptron) and local approximation methods (such as radial basis function and nearest neighbor classifier). In Secs. 3 and 4 all these paradigms are defined using a common notation, the equations for the parameter optimization are derived, special characteristics are pointed out, and modifications to the basic methods are discussed. In contrast to the universal approximation approach, it is shown in Sec. 5 how the discriminant functions are calculated if the class conditional distributions are known to be Gaussian. We conclude in Sec. 6 with a comparison of the discussed techniques for handwritten digit recognition, an outlook to even more complex classification systems based on these methods, and some remarks about the relations between "classical" statistical pattern recognition and neural networks.

2. Decision Theory and Function Approximation

From the mathematical point of view, there are actually three different theories involved in the classification task: decision theory (showing that the minimum error decision functions are the a posteriori probabilities; see Sec. 2.1), approximation theory (proving that there exist different function classes being universal approximators; see Sec. 2.3), and optimization theory also called learning theory (telling us how to adapt the free parameters of the chosen function classes; see Secs. 3 and 4). As a result, classification is based on function approximation. Figure 3 illustrates this relation by the sketch of a two-dimensional (Euclidean) feature space with samples from three different classes. The classification corresponds to the projection of the samples into target vectors (decision space), here with three components each representing the membership to one class.

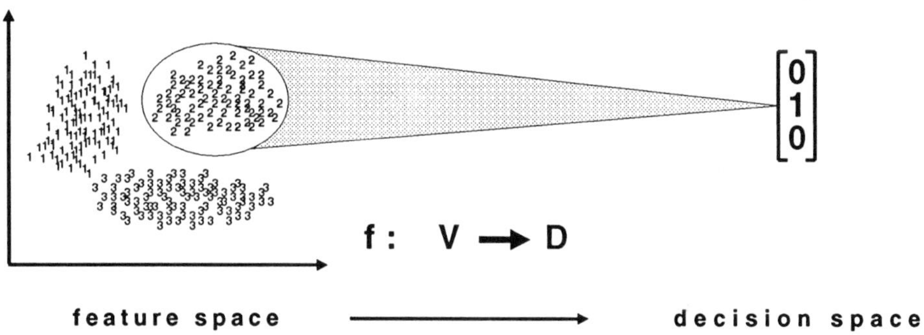

Fig. 3. Classification based on function approximation.

2.1. *Risk Minimization and Bayes Classifier*

Given the joint distribution $P(\omega_k, \mathbf{v})$ of the feature vector $\mathbf{v}$ and the possible classes ω_k of the pattern source, the total risk R can be defined as the mathematical expectation[a] of the costs (mean costs):

$$R = E\{\text{costs}\} = \sum_{\mathbf{v}} \sum_{k} \mathbf{C}(\omega_k, d(\mathbf{v})) \cdot P(\omega_k, \mathbf{v}) . \tag{2.1}$$

The cost matrix $\mathbf{C}$ gives the costs for the (not necessarily correct) decision $d(\mathbf{v})$, with ω_k being the true class.

Despite the fact that the costs of a wrong decision may vary with the specific confusion, often only the number of errors is of interest, which leads to the following simplified cost matrix:

$$\mathbf{C}(\omega_k, d(\mathbf{v})) = \begin{cases} 0 & \text{right decision ,} \\ 1 & \text{wrong decision .} \end{cases} \tag{2.2}$$

[a]In Eq. (2.1) the symbol $\sum_{\mathbf{v}}$ is used for both summing or integration, depending on whether the variable $\mathbf{v}$ is discrete or continuous.

Equation (2.1) is rewritten using the Bayesian formula $P(x,y) = P(x|y) \cdot P(y)$, at first. Considering the cost assumption in Eq. (2.2) leads to (with $d(\mathbf{v}) \doteq \omega_d$):

$$\begin{aligned} R &= \sum_{\mathbf{v}} P(\mathbf{v}) \left[\sum_k \mathbf{C}(\omega_k, d(\mathbf{v})) \cdot P(\omega_k|\mathbf{v}) \right] \\ &= \sum_{\mathbf{v}} P(\mathbf{v}) \left[\left(\sum_k P(\omega_k|\mathbf{v}) \right) - P(\omega_d|\mathbf{v}) \right] \\ &= \sum_{\mathbf{v}} P(\mathbf{v}) \left[1 - P(\omega_d|\mathbf{v}) \right] . \end{aligned} \tag{2.3}$$

From Eq. (2.3) it becomes clear that in order to minimize the risk R the pattern classifier must choose the class ω_d which has the highest a posteriori probability:

$$R \doteq \min_{d(\mathbf{v})} \quad \rightsquigarrow \quad P(\omega_d|\mathbf{v}) \doteq \max_k . \tag{2.4}$$

This solution, which is also intuitively appealing, is known in the literature as Bayes classifier.[b] Summarizing, we can state: *The Bayes classifier minimizes the number of errors by using the a posteriori probabilities of the pattern source as discriminant functions.*

In the case of nonuniform costs for confusion (e.g. in technical diagnosis systems) the discriminant functions must be derived directly from Eq. (2.1), taking the different cost assumptions explicitly into account. The a posteriori probabilities, however, remain the essential constituents of the discriminant functions. A further special case (e.g. in medical screening tests) is the assumption that the confusion costs for the different true classes ω_k are inversely proportional to their a priori probabilities: $\mathbf{C}(\omega_k, d(\mathbf{v})) \sim 1/P(\omega_k)$. This leads to the so-called maximum likelihood classifier, which compares not the a posteriori probabilities $P(\omega_k|\mathbf{v})$ but the class conditional probabilities $P(\mathbf{v}|\omega_k)$ for the best decision and aims at achieving the same error rate for each class, instead of minimizing the total sum of errors.

2.2. *Least Mean Squares Approximation*

In classification problems usually the K classes (and therefore also the decision vectors) are coded as K-dimensional unit vectors $\mathbf{y}$ (1-out-of-K coding), instead of just a scalar class index. This, on the one hand, avoids any kind of special neighborhood relations between the classes, and on the other hand, opens a further approach for approximating the discriminant vectors. This time, the optimization criterion is the minimization of the mean squared distance between the class vectors $\mathbf{y}$ and the discriminant function $\mathbf{d}(\mathbf{v})$:

$$\mathrm{E}\{\, |\mathbf{y} - \mathbf{d}(\mathbf{v})|^2 \,\} \doteq \min_{\mathbf{d}(\mathbf{v})} . \tag{2.5}$$

[b] It must be noted that so far no assumptions about the stochastic process were needed, especially not the hypothesis of class conditional Gaussian distributions (what is sometimes, however misleading, also referred to as Bayes classifier in the literature; we will deal with this assumption in Sec. 5).

The final decision $d(\mathbf{v})$ is the class associated with that unit vector closest to $\mathbf{d}(\mathbf{v})$ in decision space, which corresponds to just taking the maximum component of $\mathbf{d}(\mathbf{v})$ as classication result. Note that in Eq. (2.5) the discriminant function $\mathbf{d}(\mathbf{v})$ is any arbitrary function, and therefore the whole problem belongs to the domain of calculus of variation.

The general solution of Eq. (2.5) is found immediately if one remembers that the mean vector of a distribution is exactly that point whose Euclidean distance to all members of the ensemble is minimum in the mean. Considering that here the measured feature vector $\mathbf{v}$ is known as a condition, the solution of Eq. (2.5) is:

$$\mathbf{d}_{opt}(\mathbf{v}) = \mathrm{E}\left\{\mathbf{y}\,|\,\mathbf{v}\right\} \tag{2.6}$$

$$= \sum_{k=1}^{K} \mathbf{y}_k \cdot P(\mathbf{y}\,|\,\mathbf{v}) = \begin{pmatrix} P(\omega_1\,|\,\mathbf{v}) \\ \vdots \\ P(\omega_K\,|\,\mathbf{v}) \end{pmatrix}. \tag{2.7}$$

Equation (2.6) is known as regression function and Eq. (2.7) follows from the 1-out-of-K coding of the classes. And again (compare Bayes classifier), we are led to *the a posteriori probabilities as discriminant functions using the least mean squares approximation.* Or more exactly stated: least mean squares approximation (LMS) of an arbitrary function $\mathbf{d}(\mathbf{v})$ with $\mathbf{y}$ as target vector (representing the true class membership ω_k) results in a function that turns out to be the least mean squares approximation to $\mathbf{P}(\mathbf{v})$ (being the vector function of the K a posteriori probabilities).

This interesting fact is the justification for the use of least mean squares techniques in pattern classification. If we replace the arbitrarily flexible function $\mathbf{d}(\mathbf{v})$ with any specific parameterized function $\mathbf{d}(\mathbf{v}, \mathbf{W})$, we are also generating approximations to $\mathbf{P}(\mathbf{v})$, but maybe with less precision. For this approximation, we return to the optimization criterion in Eq. (2.5), replace the mathematical expectation $\mathrm{E}\{\cdots\}$ by the sample mean $\sum_L \cdots$ for the given learning set L, and search for the optimum parameters $\mathbf{W}$ of the now parameterized function $\mathbf{d}(\mathbf{v}, \mathbf{W})$.

2.3. *Universal Approximators and Learning Set*

From approximation theory [8, 9] we know that there exist certain function classes being so-called *universal approximators.* They can approximate any given function arbitrary well, provided the number of free parameters is sufficiently high. Thus, we can reduce the task of arbitrary function approximation (calculus of variation) to the "easier" problem of parameter optimization (including the task of choosing an appropriate number of parameters) without losing any quality, at least theoretically. Therefore, the effort to select the best-suited family of functions seems to be somewhat superfluous from the mathematical point of view if all of them are universal approximators. But since the other properties of the different families of functions are quite different, it may well be worthwhile to look for the most appropriate function structure.

After having selected a specific function structure from the class of universal approximators, the next important question is how to find the optimum number of parameters given only a finite learning set. This question is one of the basic issues and the main concern in classification based on function approximation, since a compromise has always to be found between the number of parameters and the number of samples in the learning set. And the mathematical treatment manifests the necessary compromise, calling it the ***bias/variance dilemma*** [10]. In order to at least illuminate this dilemma, we look at the generalization error $E_{gen}(M, L)$. Here, M denotes the number of parameters and L the number of samples. The generalization error $E_{gen}(M, L)$ is defined as the mean distance between the found approximation and the true a posteriori probabilities. This error can be split into two components, at least theoretically: the approximation error E_{app}, caused by the finite approximator, and the estimation error E_{est}, which is caused by the finite learning set.

The approximation error E_{app} depends on the function which has to be approximated and the chosen (universal) approximator, as might be expected. Nevertheless, it can be shown [11] that under quite general assumptions the approximation error vanishes proportionally to $M^{-s/d}$, where s is a smoothness index and d is the dimension of the approximation function (or the pattern source). The phenomenon that the approximation error decreases much more slowly (exponential) with an increasing dimension d, is known as the curse of dimensionality. The only way to counteract this is to impose a certain smoothness constraint on the approximated function, which is also called blessing of smoothness. The estimation error E_{est} decreases with the size L of the given data set, but increases with the complexity of the chosen approximation class, which can be measured by the Vapnik-Chervonenkis (abbr. VC) dimension [12]. The VC dimension again is closely related to the number M of free parameters, e.g. they are identical for the polynomial classifier.

Summarizing, we can state that there are two counteracting forces: the bias or the approximation error caused by the finite approximation length, and the variance or the estimation error caused by the finite learning set. The important task for the designer of a practical system now is to find the best compromise. The mathematical treatment of this compensation shows, however, that there is almost always a lack of data (i.e. ill-posed problem) at least in high-dimensional spaces [13].

Before we look at different approximation schemes (see Secs. 3 and 4), we like to stress, as a final remark, that the *collection of the learning set* deserves at least the same effort and care as the selection and optimization of the approximation scheme. The learning set should not only be as large as possible (even if its collection might be very time consuming) but also should represent the given task as truly as possible. For example, in the first NIST (US National Institute of Standards and Technology) test on OCR [14, 15] the learning set was collected from census forms, while the independent test set was written by uncooperative test persons. This caused a significant difference between the statistical characteristics of the learning and the test set and, therefore, led to quite unsatisfactory results.

3. Global Approximation

For a detailed discussion of (universal) approximators we selected the four most prominent architectures (at least for OCR): polynomial classifier (PC), multilayer perceptron (MLP), radial basis function (RBF), and nearest neighbor classifier (NNC). Furthermore, we divided these methods into global and local approximation schemes (certainly not the only ordering criterion that could be used). In the context of the present chapter, *global approximation* is defined as having not vanishing, global interactions within the approximation structure. These global interactions, of course, have influence on the generalization ability as well as on the additional learning of new examples.

First, we introduce the polynomial classifier (Sec. 3.1), whose attractive advantage is the analytical solution for parameter adaption. Second, we try to summarize all important facts about the multilayer perceptron (Sec. 3.2), which nowadays is the "work horse" of neural networks for many classification tasks including OCR.

3.1. *Polynomial Classifier*

The *polynomial classifier* [16] (also known as Π-Σ units [17] or functional link net [18] in more recent literature) consists actually of two layers. In the first layer we find the polynomial (multiplicative) combinations of the original features. These combinations are also referred to as enhanced features and are described mathematically with the so-called polynomial structure list $\mathbf{x}(\mathbf{v})$. Each component $x_s(\mathbf{v})$ of this structure list is a selected product $\prod_s$ of various features v_m: $x_s(\mathbf{v}) = \prod_s v_m^i$. The second layer is a linear combination (coefficient matrix $\mathbf{W}$) of the basic functions described above. Mathematically, the polynomial classifier (PC, see also Fig. 4) is defined by:

$$\mathbf{d}_{PC}(\mathbf{v}) \quad = \quad \mathbf{W}^{\mathrm{T}} \cdot \mathbf{x}(\mathbf{v})\,. \tag{3.1}$$

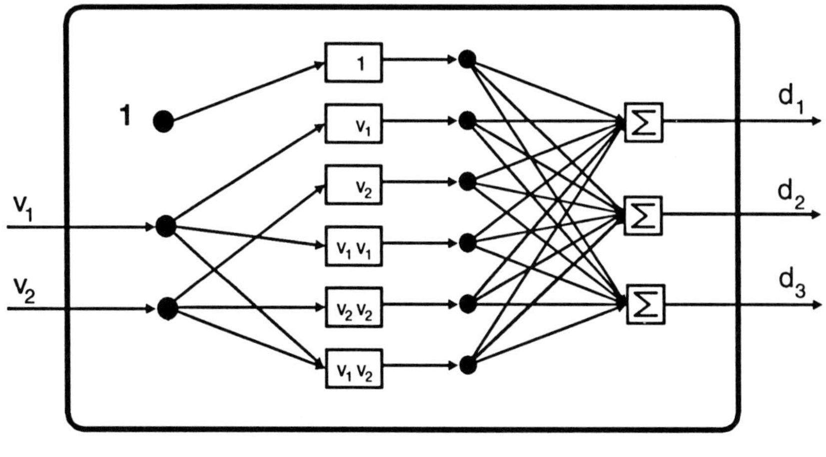

Fig. 4. Complete quadratic polynomial classifier for two features and three classes.

The universal approximator characteristic is guaranteed by a sufficiently long polynomial structure list, and it is proved by the approximation theorem of Weierstraß [8], similarly to the well known one-dimensional series expansion.

While the polynomial structure list $\mathbf{x}(\mathbf{v})$ must be chosen by the designer (e.g. complete quadratic as in Fig. 4), the adaption of the weight matrix $\mathbf{W}$ is done automatically from the learning set. One important advantage of the PC is that there exists an analytical solution for the weight matrix $\mathbf{W}$:

$$\mathrm{E}\{\mathbf{x}\mathbf{x}^{\mathrm{T}}\} \cdot \mathbf{W} = \mathrm{E}\{\mathbf{x}\mathbf{y}^{\mathrm{T}}\} . \tag{3.2}$$

This key equation is usually solved by a Gaussian elimination procedure resulting in the weight matrix $\mathbf{W}$. Based on the chosen pivot strategy and without incurring any extra cost, we get a complete rank order among the set of given polynomial terms with respect to their discriminative power, and also the elimination of those terms that are linearly dependent on others. This procedure ensures that Eq. (3.2) can always be successfully solved, even if $\mathrm{E}\{\mathbf{x}\mathbf{x}^{\mathrm{T}}\}$ does not have the full rank [16].

The key equation (3.2) corresponds to the so-called least mean squares pseudo inversion of an overspecified system of linear equations (more examples than coefficients per class). For example, in MatLab [19], the solution can be simply obtained by `W = (X * X') \ (X * Y')` if the training samples are column-wise arranged in `X` and `Y`. If the system of linear equations contains linear dependencies and a unique solution is enforced by the elimination of redundant variables, this technique is called generalized inversion.

The moment matrices $\mathrm{E}\{\mathbf{x}\mathbf{x}^{\mathrm{T}}\}$ resp. $\mathrm{E}\{\mathbf{x}\mathbf{y}^{\mathrm{T}}\}$ of Eq. (3.2) are defined to be mathematical expectations. These are usually not known exactly, but can easily be estimated from the learning set by replacing the expectation operator with the corresponding sample mean $\sum_L \mathbf{x}^{(\lambda)}\mathbf{x}^{(\lambda)^{\mathrm{T}}}$. For large classification problems with large numbers of classes to be discriminated — typical of OCR applications — it is highly recommended to construct a database of moment matrices and to calculate the moment matrices only once on a per class basis since the computation of moment matrices is the most time consuming operation of the whole PC adaptation procedure. Additionally, moment matrices usually require less storage space than the corresponding feature vectors. For the parameter adaption of a specific classifier, the moment matrices can then be merged by a linear combination with arbitrary weights, just as sample sets can be merged with the same weights.

At first glance, a severe disadvantage of the PC concept seems to be that the polynomial structure list (length S) grows rapidly with the polynomial degree G and the number of features M: $S = \binom{M+G}{G}$. There are, however, several ways to avoid this problem. One solution is to work with incomplete polynomials, usually based on the rank ordering of the PC adaptation procedure. Here, the original feature set is first ranked by calculating the linear classifier, and then specific higher order polynomial terms are designed being combinations of the most promising features only [20]. There are also a number of heuristics applicable, which support the use of other specific feature combinations (e.g. in OCR combinations of features, i.e. pixels, which are spatially related).

Another, and often the better, way is to reduce the feature set dimensionality M. The most powerful tool for this purpose is the (linear) principal component analysis (PCA[c]), which compresses all the relevant information into the first components of the transformed feature vector. Even if the PCA produces a reduced feature set under the optimization criterion of minimum reconstruction error — in the sense of reconstructing the original feature set from the PCA transformed features — it obviously conserves what is necessary for discriminating between the different classes of the classification problem. Particularly in character and digit recognition, the PCA has proven an extremely useful tool (see also the comparison in Table 1 in Sec. 6).

The polynomial approach also embraces the simple *linear classifier*, i.e. polynomial classifier with degree one or multilayer perceptron without hidden layers. It is always recommended to try this type of classifier first in order to get an impression of the complexity of the given classification task. A most valuable by-product of the linear classifier adapted by pseudo inversion is the rank order among the given features and the elimination of linearly dependent features.

The PC concept has the property that its basis functions, from which the estimations of the a posteriori probabilities are gained by a linear combination, run quickly to plus or minus infinity outside the learning interval, i.e. in that regions of the feature space where learning examples have not been presented. This allows one to detect garbage patterns — those patterns which may occur in practical applications but do not belong to the learned classes — by checking the outcome of the estimation procedure. To the input of garbage patterns the PC normally responds with estimations $\mathbf{d}(\mathbf{v})$ clearly outside the zero-to-one interval. This is a property which is especially valuable for OCR-systems, since it can never be excluded that character images not belonging to the set of learned classes may be presented to the recognition system, such as alpha characters to a digit recognizer. Pattern recognition systems are always trained under the closed world assumption and this assumption is violated in such a case. For further discussion of this issue see also Sec. 6.

3.2. *Multilayer Perceptron*

There are many versions of the *multilayer perceptron* (MLP) and its error backpropagation learning rule found in the literature. Therefore, we start here with the basic definitions of the "classic" MLP [17] and then motivate some of the modifications.

The smallest building block of the MLP is the neuron, which calculates a linear combination of its input values and modifies the result with a nonlinear (usually sigmoidal) activation function: $out_j = \sigma\left(bias_j + \sum_i w_{ij} \cdot in_i\right)$. All neurons that

[c]PCA is also called Karhunen-Loève transformation, subspace method, or singular value decomposition; see various other chapters in this handbook.

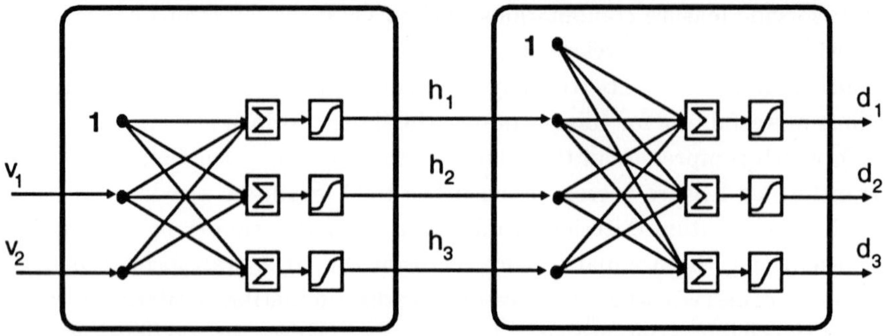

Fig. 5. Multilayer perceptron with one hidden layer and three hidden neurons.

depend on the same inputs are combined into a single layer, the next larger building block of the MLP. The MLP itself, as the name already suggests, consists of several layers (see Fig. 5). Mathematically, this structure resembles a recursion of vectorial functions, each corresponding to a layer. To simplify the notation, it is useful to treat the bias term as an additional weight to be multiplied by the constant term one. Representing this extension by a tilde, we arrive at the following compact formula for a multilayer perceptron with H layers:

$$\mathbf{d}_{MLP}(\mathbf{v}) = \sigma\left(\mathbf{W}^{\mathrm{T}}_{<H>} \cdots \tilde{\sigma}\left(\mathbf{W}^{\mathrm{T}}_{<1>} \cdot \tilde{\mathbf{v}}\right)\right). \tag{3.3}$$

Similarly to the polynomial classifier, the designer has the choice of the number of layers and the number of neurons in the so-called hidden layers. Since the variables of the input layer are the given features and the variables of the output layer represent the classes, both are fixed for a given classification problem.

The weights themselves, however, are adapted based on the learning set following again the least mean squares approach: $\mathrm{E}\left\{|\mathbf{y} - \mathbf{d}_{MLP}(\mathbf{v})|^2\right\} \doteq \min$. Since there is no known analytical solution for the determination of the weight matrices $\mathbf{W}_{<h>}$ in Eq. (3.3), a gradient descent is usually applied. The gradient can be calculated using the chain rule for concatenated functions:

$$\frac{\partial\, \mathrm{F}^{(\lambda)}}{\partial\, w_{ij}^{<h>}} = \frac{\partial\, \mathrm{F}^{(\lambda)}}{\partial\, o_j^{<h>}} \cdot \frac{\partial\, o_j^{<h>}}{\partial\, a_j^{<h>}} \cdot \frac{\partial\, a_j^{<h>}}{\partial\, w_{ij}^{<h>}}. \tag{3.4}$$

Here, $\mathrm{F}^{(\lambda)} = |\mathbf{y}^{(\lambda)} - \mathbf{d}(\mathbf{v}^{(\lambda)})|^2$ is the error contribution of the example λ from the learning set, a_j is the activation value of the neuron j, and o_j is the output $o_j = \sigma(a_j)$ of the neuron. Because of $a_j^{<h>} = \sum_i w_{ij}^{<h>} \cdot o_i^{<h-1>}$ the last term can be rewritten as:

$$\frac{\partial\, a_j^{<h>}}{\partial\, w_{ij}^{<h>}} = o_i^{<h-1>}. \tag{3.5}$$

For the sigmoid function $\sigma(x) = 1/(1+e^{-x})$, which is also called logistic function, the second term[d] of Eq. (3.4) reduces to:

$$\frac{\partial\, o_j^{<h>}}{\partial\, a_j^{<h>}} = o_j^{<h>} \cdot \left(1 - o_j^{<h>}\right) . \tag{3.6}$$

The remaining first term in Eq. (3.4) can be easily computed for the output layer, whereas for any hidden layer the chain rule must be applied recursively:

$$\frac{\partial\, \mathrm{F}^{(\lambda)}}{\partial\, o_j^{<h>}} = \begin{cases} \dfrac{\partial\, \mathrm{F}^{(\lambda)}}{\partial\, d_j} = -2 \cdot \left(y_j - d_j\right) & \text{output layer,} \\ \displaystyle\sum_m \frac{\partial \mathrm{F}^{(\lambda)}}{\partial\, o_m^{<h+1>}} \cdot \frac{\partial\, o_m^{<h+1>}}{\partial\, a_m^{<h+1>}} \cdot \frac{\partial\, a_m^{<h+1>}}{\partial\, o_j^{<h>}} & \text{hidden layers.} \end{cases} \tag{3.7}$$

Although Eq. (3.7) looks quite tedious, it turns out that it is possible to calculate all error gradients rather efficiently in a single backward run after all neuron outputs have been evaluated in a forward run. This is where the name error backpropagation comes from. Because of the reuse of intermediate results, this procedure is similar to the butterfly algorithm for the fast Fourier transform.

The weight update Δw_{ij} is given by:

$$\Delta\, w_{ij} = \alpha \cdot \left(- \frac{1}{L} \sum_{\lambda=1}^{L} \frac{\partial\, \mathrm{F}^{(\lambda)}}{\partial\, w_{ij}} \right) . \tag{3.8}$$

The proportionality factor α is called the learning rate, since it controls the adaption speed. It is sometimes time dependent, and also weight dependent: $\alpha_{ij}(t)$.

The summation index λ in Eq. (3.8) includes, according to the optimization criterion in Eq. (2.5), the complete learning set. This update rule is also referred to as batch learning (epoch update). For very large and, therefore, somehow redundant learning sets (as they occur, for example, in OCR), however, it is advantageous to apply the weight update after each individual sample (corresponding to $L = 1$). This so-called singular update (pattern update) turns the gradient procedure into a stochastic approximation, which also might help to avoid local minima. For parallel computers sometimes block-wise updates are even better (with a block size $1 \ll B \ll L$), where about the same number of samples per class should be contained in each block.

The three equations (3.3), (3.4), and (3.8) define the MLP together with its adaption rule, which is sometimes called "vanilla" backprop. It can be shown [8] that a MLP with a single hidden layer (as in Fig. 5) is a universal approximator, provided the number of hidden neurons is not limited. Note, that the counting of layers is not unique in the literature, depending on whether the input layer is counted or not. The solution adapted here is to count explicitly the number of hidden layers.

[d]Note for the symmetric activation function $\tanh(x) = (e^x - e^{-x}) / (e^x + e^{-x})$, which is sometimes used instead of $\sigma(x)$, we get the following result for the differentiation: $1 - (o_j^{<h>})^2$.

All modifications of the MLP concern either the structure or/and the learning rule. In the structure of the MLP sometimes direct connections between the input and the output (short cuts) are included, to model linear dependencies between input and output more directly and therefore more easily. If there are strong local dependencies between the input features, it might be sufficient to use receptive fields instead of fully connected layers. This method is possible for all layers, leading to a pyramid-like structure of the MLP and compacting the desired information more and more from one layer to the next. One advantage of the receptive fields is that the number of weights, i.e. the number of free parameters, is significantly reduced (hopefully without too much effect on the approximation ability). Additionally, the respective weights of the "parallel" receptive fields might be forced during the learning process to be identical (weight sharing, restricted weights) in order to further reduce the number of free parameters. Summarizing, almost all modifications to the structure of the MLP try to get the most flexible approximation with the lowest possible number of parameters. Therefore, they try to alleviate the bias/variance dilemma (see Sec. 2.3) by reducing the number of free parameters [21, 22].

Modifications to the learning rule are almost countless, attacking one or more of the known problems for error backpropagation: convergence speed, local minima, and overfitting (which is actually more a problem of the structure). There are many methods (quickprop, conjugate gradient, rprop [23, 24, 25]) which try to improve the convergence speed of batch learning based on second order statistics and/or adaptive learning rates. Our practical experience in OCR, however, shows that singular (on-line) learning is usually much faster than batch learning and, therefore, these improvements seem to be of more academic interest. The computing time required for the adaption of an MLP is still very high compared to the polynomial classifier. Local minima do not seem to be too critical for practical tasks in OCR compared to more artificial demonstration problems such as the XOR problem. Again, singular update resolves some of the problems encountered for batch update. Finally, the risk of overfitting can be reduced by using learning rules with weight decay and/or weight elimination (see also optimum brain damage/surgery [26, 27, 28]). The idea is to add an additional term to the optimization criterion for punishing large weights, and/or to check the influence which the single weights have on the result. Another idea is to use stopping criteria [29] for interrupting the training procedure whenever the approximation performance, as measured with an independent validation set, begins to deteriorate. While these methods seem to be important in cases where the MLP is used for time series prediction (where the number of samples is usually small), it is more efficient for OCR — at least in our experience — to invest additional time in collecting samples, or to spend computational effort in reducing the number of input features by the principal component analysis and in designing the MLP structure.

In summary, we can state for the MLP that a large number of practical applications from quite different fields have proven its effectiveness and robustness regarding structure selection and parameter optimization, but for large problems (such as OCR) these favorable properties must be paid for by very long adaption times (sometimes in the order of cpu days). The clear structure of the MLP, its resemblance to neuro-biological models, and the intuitive behaviour of its learning rule somehow inspired many researchers to invent numerous variations and (mostly heuristic) solutions to almost any problem encountered with the MLP, often however ignoring, at least partly, what is known about the analytical background and the mathematical consequences (such as the bias/variance dilemma). More detailed information on modifications of the MLP (including motivations) can be found in textbooks [30, 31, 32] or in recent conference proceedings [33, 34, 35, 36].

4. Local Approximation

Local approximation schemes approximate the desired function in the neighborhood of the given points of support (e.g. samples of the learning set). Similar to the sampling theorem, which holds for equidistant samples, the quality of the approximation is tightly related to the number and positions of the support points. Moreover, the curse of dimensionality requires an exponentially growing number of samples if the dimension of the feature space is increased. Therefore, we should be very careful with extrapolations from the two-dimensional case, which is usually used for demonstration [37], to higher dimensions, which are typical for OCR. The advantage of local approximation, however, is that there is no global interference during the learning process, and that the information about the classification task can be more easily located and interpreted at the points of support (prototypes).

In Sec. 4.1 we first want to discuss the radial basis functions, which belong to the class of universal approximators. These functions, including many modifications, have recently attracted a large share of interest in the neural network literature. Second, we want to show in Sec. 4.2 how the well known nearest neighbor classifier fits into the function approximation approach, even if it is not a universal approximator. In Sec. 4.3, we also discuss some clustering techniques, which are necessary to find good reference vectors for the radial basis function approach as well as for the nearest neighbor methods.

4.1. *Radial Basis Function*

The *radial basis function* (RBF) network [38] consists of two layers similar to the PC (see Sec. 3.1). This time, the first layer calculates the squared Euclidean (or some other) distances z between the input vector $\mathbf{v}$ and the given set of reference vectors $\mathbf{r}_i$ (also called prototypes). Each distance is then weighted by an exponential decreasing function (e.g. Gaussian function): $\varrho_i(z) = e^{-z/c_i}$. The second layer provides a linear combination of these (radial) basis functions (see Fig. 6). Mathematically, the k-th discriminant function is defined as follows:

$$d_{RBF,k}(\mathbf{v}) \;=\; \sum_i w_{ik} \cdot \varrho_i\left(\left\|\mathbf{v} - \mathbf{r}_i\right\|^2\right) , \tag{4.1}$$

where $\mathbf{r}_i$ are the reference vectors and ϱ_i are the local interpolation functions.

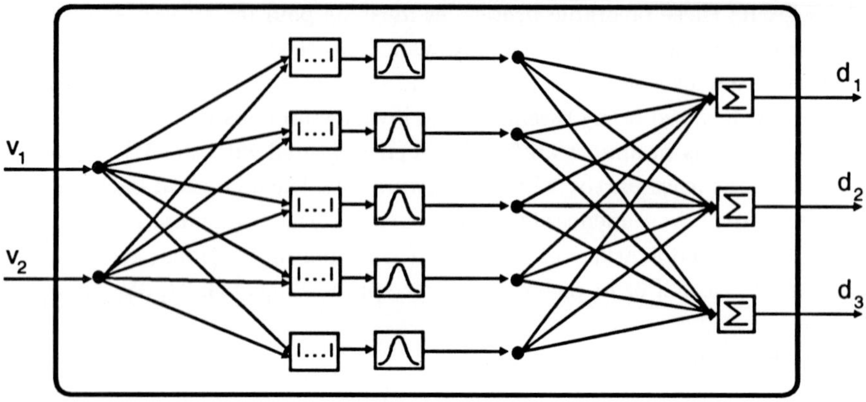

Fig. 6. Radial basis function network with five reference vectors.

Contrary to the two approaches discussed so far, PC and MLP, we have here a set of parameters (reference vectors, local interpolation functions with smoothing factors and possibly non-Euclidean distance functions, and a linear combination) with quite different properties, which must be appropriately selected and adapted. Often the adjustment of these parameters is carried out independently, whereas only the simultaneous optimization of all of them will give optimum results. The overall optimization criterion again is the least mean squares approach: $\mathrm{E}\left\{|\mathbf{y}-\mathbf{d}_{RBF}(\mathbf{v})|^2\right\} \doteq \min$. In order to show the influence of the different types of parameters and the possibilities for their adaption more clearly, we will discuss them separately in the following.

The number of reference vectors is, similar to the number of hidden neurons for the MLP, a design decision. Again, a sufficiently large number of reference vectors assures the universal approximator characteristic [9]. However, by far not all of the samples of the learning set should be used as reference vectors, since this tends to overload the computer resources (storage and computing time) for large learning sets, which on the contrary are required to avoid overfitting. The basic idea, from the statistical viewpoint, is to select such representatives that each stands for a partial stochastic process, while many of these processes are combined to model the given distribution as a mixture of simpler densities. Practically, those prototypes can be found by unsupervised clustering techniques (such as vector quantization and self organizing feature maps), by class-wise clustering, or by supervised methods (e.g. learning vector quantization) [39, 40, 41, 42]. The set of reference vectors obtained thus forms the basis for the RBF network, but these prototypes can be modified during the overall adaption of the RBF by a gradient search.

The appropriate selection of the interpolation functions ϱ_i, which map the distance functions $\left\|\mathbf{v}-\mathbf{r}_i\right\|^2$ into the hidden variables of the RBF network, provides the largest potential for optimization. In the simplest case, an exponentially decreasing interpolation function with a smoothing factor common to all interpolation functions is chosen: $\varrho_i(z) = e^{-z/C}$. This structure already leads to a universal approximator. More appropriate, however, is to have individual smoothing factors c_i for each reference vector $\mathbf{r}_i$. These constants c_i are initialized according to certain rules (e.g. c_i equal to the distance of the reference vector to its closest neighbor) and can be optimized by a gradient descent.

Furthermore, the Euclidean distance function can be replaced by a special distance matrix. Often the inverse covariance matrix, which is also known as the Mahalanobis distance, is used for this purpose; see also Sec. 5. There is either one matrix common to all reference vectors, or an individual distance matrix for each prototype. However, this approach introduces another $\binom{M+1}{2}$ additional parameters (possibly per reference vector) to be appropriately adapted. This might be prohibitively expensive if the input dimension M is high. Therefore, sometimes only the diagonal of the distance matrix is used, leading to ellipsoids parallel to the axes for equal distance values. It is easily seen that the number of free parameters for the interpolation function grows rapidly from 1 (common smoothing factor), over R (individual factor for each reference vector), and $R \times M$ (axis parallel ellipsoids for each prototype), to $R \times \binom{M+1}{2}$ (individual distance matrix for each reference vector). This must be compared with the $R \times M$ parameters needed for representing the reference vectors and the $R \times K$ coefficients of the linear combination. It is necessary to find a sensible compromise between these possibilities, and of course the number of all free parameters must be considered for a fair comparison if we are going to compare the different structures.

In order to get an impression of the distribution of reference vectors in a high-dimensional space (here $M = 256$; digit recognition), in Fig. 7 the maximum, mean, and minimum distance between the reference vectors (found by divisive vector quantization, see also Fig. 8 in Sec. 4.3) is plotted for a growing number of reference vectors. We see that the maximum distance between any two reference vectors grows slowly with the number of reference vectors, whereas the minimum distance is almost constant for more than 20 reference vectors. The smoothing factor c (identical for all reference vectors) of the Gaussian interpolation function $\varrho_i(z) = e^{-z/c}$ was optimized by a gradient search. From the relation between the smoothing factor c and the distances z of the reference vectors we can approximately estimate the influence of the interpolation function. For the case of 100 reference vectors we calculate the influence as $e^{-0.3} = 0.7$, $e^{-1.7} = 0.2$, and $e^{-3.2} = 0.04$ for the minimum, mean, and maximum distance, respectively.

Finally, the weight matrix $\mathbf{W}$ for the linear combination can be calculated by a pseudo inversion (in the same way as for the PC), though often in the literature a gradient descent is used.

The radial basis function approach has recently received much interest in the literature [43, 44, 45, 46]. It establishes in a certain sense the bridge from the simple nearest neighbor approach (see the next section) over special neighborhood based algorithms (such as RCE [47]) to the comprehensive theory of support point approximation. Caused by the interfering dependencies of the parameters (and their selection), the optimum adaption of the RBF requires more input and more

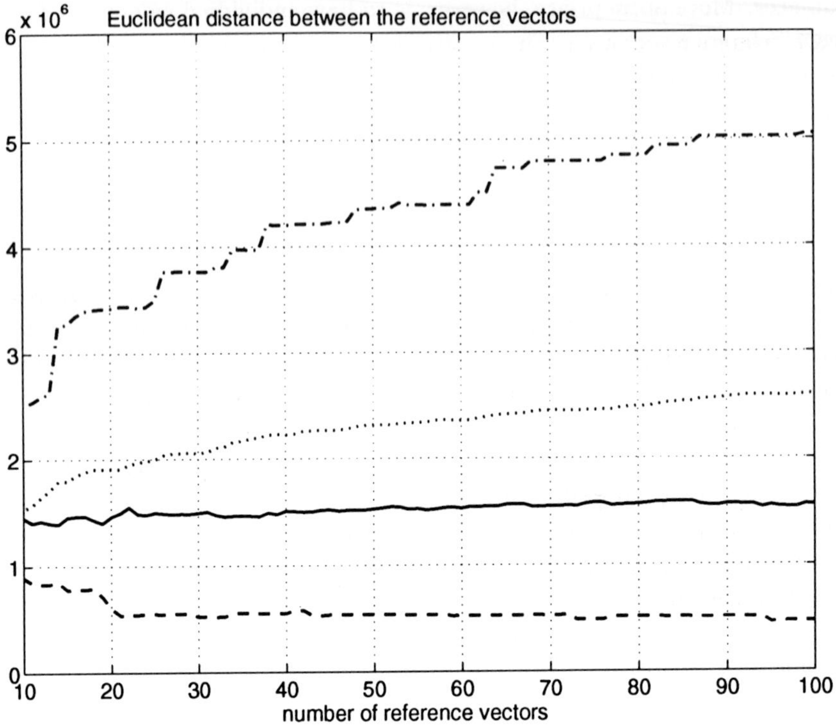

Fig. 7. Optimum value of the smoothing factor c (—) for the RBF network in comparison with the maximum (– · –), mean (···), and minimum distance (– –) between the reference vectors depending on the number of prototypes for the example of handwritten digit recognition.

experience from the designer than the MLP and PC, for example. This large variety of choices and the dependent adaption of the parameters also complicates the fair comparison of RBF classifiers to other approaches. One attractive advantage of the RBF is that, due to the locality of the set of basis functions, this technique is almost immune against garbage patterns — those patterns which do not belong to the learned classes — since all estimations vanish to zero outside those regions in feature space that are occupied by samples of the learning set.

Before we finish this section, we want to outline what in our opinion seems to be the best way of making use of the RBF idea for classification. From low-dimensional examples it can easily be seen that the local interpolation functions resemble Gaussian shaped functions (bell curves), whereas the a posteriori probabilities $P(\omega_k|\mathbf{v})$ are more like flat plateaus (on the level one) inside their respective class regions and exhibit more or less sharp ridges at the class borders. Therefore, a large number of support points is needed to model this behaviour adequately. The situation changes remarkably if we try instead to directly model the class conditional probabilities $P(\mathbf{v}|\omega_k)$ with the given exponentially decreasing local interpolation functions. This idea to model the distribution function of each class by the (linear) combination of simpler distributions is known from the literature [48] as mixture-of-Gaussians approach, but does not fit at first glance in the function approximation scheme based on least means squares approximation (as it was introduced in Sec. 2.2). The Bayes theorem, however, establishes the relation between the a posteriori probabilities $P(\omega_k|\mathbf{v})$ and the class conditional probabilities $P(\mathbf{v}|\omega_k)$:

$$P(\omega_k\,|\,\mathbf{v}) \;=\; \frac{P(\mathbf{v}\,|\,\omega_k)\,\cdot\,P(\omega_k)}{P(\mathbf{v})} \;=\; \frac{P(\mathbf{v}\,|\,\omega_k)\,\cdot\,P(\omega_k)}{\sum_i P(\mathbf{v}\,|\,\omega_i)\,\cdot\,P(\omega_i)}\,. \tag{4.2}$$

Therefore, we can initialize the reference vectors, the interpolation functions and the linear combination of the RBF with the results (center, covariance matrix and weight) obtained from a class-wise cluster analysis, thus modelling the class conditional probabilities $P(\mathbf{v}|\omega_k)$. Next, we can interpret the denominator in Eq. (4.2) as a normalization-to-one operator and implement it accordingly. This normalization operation (and consideration of the class probabilities $P(\omega_k)$) changes the mixture-of-Gaussians approach into an approximation of the a posteriori probabilities $P(\omega_k|\mathbf{v})$. In this structure all parameters can now be (simultaneously) optimized further obeying the least mean squares approximation, e.g. by a modified gradient descent search; similar ideas are proposed in [44].

4.2. *Nearest Neighbor Classifier*

The *nearest neighbor classifier* (NNC) certainly belongs to the most basic methods known for classification [49] and has a long tradition not only in the field of pattern recognition. We will concentrate here on the relations which exist between the nearest neighbor approach and the concept of function approximation based classification.

The very basic definition of the nearest neighbor classifier ($k = 1$) is that a set of reference samples $\mathbf{r}_i$ with known class membership is given and that as decision for an unknown sample $\mathbf{v}$ the class label of its closest neighbor is taken:

$$\mathbf{d}_{NNC}(\mathbf{v}) \;=\; \text{class}\{\,\mathbf{r}_i\,\} \qquad \Longleftrightarrow \qquad \min_{\mathbf{r}_i}\,\left\|\mathbf{v}-\mathbf{r}_i\right\|^2\,. \tag{4.3}$$

This is the special case ($k = 1$) of a more general approach, which inspects the whole neighborhood around the given unknown sample $\mathbf{v}$ consisting of its k nearest neighbors. This k nearest neighbors classifier derives its decision from some kind of majority vote among these k neighbors. This coincides to a certain degree with a local approximation of the a posteriori probabilities $P(\omega_k|\mathbf{v})$ in the neighborhood of the unknown sample vector $\mathbf{v}$, since the relative frequencies of the different classes within the set of the k nearest neighbors are estimations of the a posteriori probabilities at this location. This concept is also closely related to approximating probability density functions by counting occurrences of events in specific intervals, i.e. generating histograms, which is a well known technique for estimating probabilities. One fundamental problem with the conventional histogram approach, however, is that the arrangement of the histogram bins (number, position, and individual size if they are not equidistant over a certain region) must be determined from the beginning. If the intervals are too wide, spatial resolution is lost and if they are too narrow, no reliable statistical estimation of probabilities is possible. This problem is perfectly avoided by the nearest neighbor approach, since the bin size is adaptive, determined just by the given number k of events (neighbors).

From this point of view the relations to the concept of function approximation based on radial basis functions also become visible. The RBF technique carries the class information (probability), which is known for a support point (center of the RBF), into its neighborhood by a smoothly interpolating function. The nearest neighbor approach uses the same support points to derive local estimations of the a posteriori probabilities. In the special case of $k = 1$, however, these estimations are constant in a certain neighborhood of the support point. This neighborhood is known in computational geometry as Voronoi cell. Even if these views on the NNC (as a special histogram technique for a probability estimation) might be surprising, they are, in our opinion, most useful to clarify the relations to the approximation techniques discussed before.

Next, we are interested in the approximation (classification) qualities of the NNC. It was shown [50] that the NNC has an error rate at most twice as large as the optimum (unrestricted) Bayesian classifier in the limit for infinite sample sets:

$$P_{error,\ NNC} \leq 2 \cdot P_{error,\ Bayes} \tag{4.4}$$

In practice, this fact is often used to get estimates on the optimum achievable error rate for a given classification problem, though the available learning set is of course always finite, and not the error rate but the large computing resources needed by the NNC are usually the limiting factor for its application [7].

Naturally, there are many ideas and methods to overcome the limitations and problems encountered with the NNC. We can order them into three categories, namely distance function, decision function and — most important — placement of prototypes, but we will discuss them here only briefly: Since the distance function is usually Euclidean, the scaling of the axes, spanning the feature space, should be appropriately chosen. Normalization respective to each axis (no bias and equal variance) of the data is usually the first step. Moreover, the usage of the Mahalanobis distance (see Sec. 5) may help if the features are (highly) correlated. Yet, the impact of the choice of the distance function decreases if the number of prototypes is increased. Another important point of influence is the type of decision function chosen, which embraces the number k of nearest neighbors involved in the decision and the mode of the decision (majority vote, weighted decision, list of alternatives, and possibly rejections). The number and placement of the reference vectors, however, certainly play the decisive role. The goal is to achieve the lowest generalization error with as few prototypes as possible. Since there are so many suggestions found in the literature to approach this goal, we will discuss in the next section only the underlying principle of most of these ideas, namely cluster analysis.

4.3. *Clustering Techniques*

The main goal of any clustering procedure is to find, for a given (large) set of samples, a small number of prototypes (cluster centers), representing the whole set as well as possible, according to a certain optimization criterion. The assignment of the individual samples to the corresponding prototypes at the same time defines a partition of the sample set into subsets, which are called clusters.

Having OCR in mind, we are interested only in *clustering techniques* which can cope with large sample sets (size L) and produce a small number I of prototypes: $L \gg I > 1$. Therefore, neither a complete search (enumeration) of all partitions $P(I, L)$ is possible nor are procedures applicable which need the distance matrix between all samples (such as single link or complete link). Furthermore in order to keep the discussion easier, we restrict the considerations here to the case of the Euclidean distance.

The optimization criterion (called variance criterion [51]) can now be defined as follows, where $\mathbf{r}_i$ are the prototypes and $\mathbf{C}_i$ are the respective clusters:

$$\sum_{i=1}^{I} \sum_{\lambda \in C_i} \left\| \mathbf{v}_\lambda - \mathbf{r}_i \right\|^2 \doteq \min_{P(I,L)} . \tag{4.5}$$

Expressed in words, Eq. (4.5) states that the partition $P(I, L)$ is optimum for which the sum of the Euclidean distances of the samples $\mathbf{v}_\lambda$ to the respective prototypes (mean of the cluster C_i) is minimum. In source coding (vector quantization [39]) this optimization criterion corresponds to the minimum quantization error. Furthermore, it can be shown that the partition determined by Eq. (4.5) is a minimum distance partition (i.e., each sample belongs to the closest prototype), and that all partitions are divided by linear hyperplanes. A disadvantage of the variance criterion, however, is that the optimum partition is not invariant against scaling, since it relies on the Euclidean distance.

The problem of clustering can usually be split in two subtasks: first, finding the optimum partition for a given number I of prototypes, and second, finding the "right" number of prototypes. For the first task, the common approach is an iterated partitioning and calculating of means — known as k-means[e] algorithm. In the partitioning phase each sample is assigned to the closest prototype, whereas in the calculation-of-means phase a new mean is calculated for each cluster. The procedure is started with some (e.g. randomly chosen) prototypes. The variance criterion is then monotonically decreased by alternatively iterating the two steps (partition/mean), until convergence is reached or a stopping criterion is activated. The minimum found by this procedure, however, is not necessarily the global optimum but may be a local one depending on the initial choice of prototypes. Nevertheless, for any practical application this solution is usually sufficiently good. Besides this method, there also exist exchange procedures [52], optimizing the partition by exchanging single members between the clusters. Even if this exchange idea seems to

[e] We use here I instead of K for the number of prototypes, since K denotes in this chapter the number of classes.

be more time consuming, it has almost the same behaviour as the partition/mean procedure (and is sometimes also referred to as k-means algorithm).

Finally, there exist also iterative procedures (similar to the singular update for backpropagation, and therefore also called online k-means algorithm; all those iterative algorithms can be viewed as being stochastic approximations), which might be applied, for example, if the learning set is so large that it does not fit into the main memory of the computer. In this iterative approach, only the closest reference vector $\mathbf{r}_i$ is partially moved (according to the learning factor α) after each presentation of a sample $\mathbf{v}_\lambda$ into the direction of this specific sample: $\mathbf{r}_i(t+1) = \mathbf{r}_i(t) + \alpha \cdot (\mathbf{v}_\lambda - \mathbf{r}_i(t))$. This iterative method (again with various modifications) is often found in neurally inspired methods (such as learning vector quantization and self organizing feature map [41, 42]), but converges only for $\alpha(t \to \infty) \to 0$ and has similar problems with local minima as the (batch) k-means algorithm.

The second subtask for clustering techniques is the choice of the "right" number of prototypes. Mathematically, there is no intuitive criterion for this choice since the variance criterion of Eq. (4.5) goes to zero with an increasing number of prototypes. The "right" solution, therefore, must be found by the designer having other criteria (such as error rate or number of classes) at hand. Basically, there are two methods

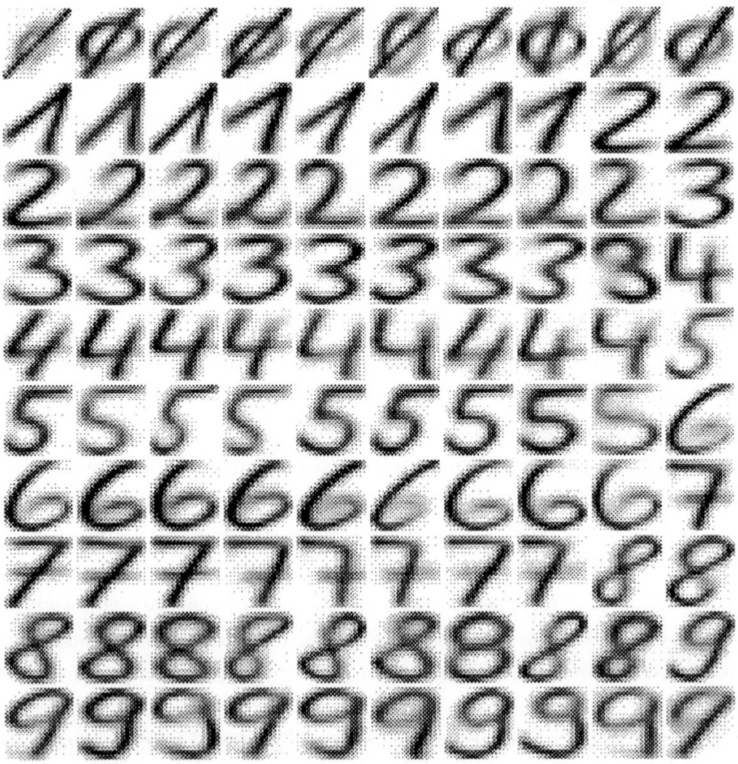

Fig. 8. Prototypes (100) generated by unsupervised cluster analysis (divisive vector quantization) for the handwritten digits — afterwards sorted by classes.

to find a good number of prototypes: divisive or agglomerative. For the divisive approach, the whole learning set is viewed as one cluster, which is successively split into more and more clusters (see for an example the divisive vector quantization of handwritten digits in Fig. 8). For the agglomerative approach, on the other hand, more and more samples are grouped into clusters, either by adding samples to existing clusters or by creating new clusters. One advantage of the divisive procedure is that the intermediate results are partitions of the complete sample set and therefore better guide the designer to the desired solution. Depending on the application and the final goal (e.g. cluster analysis or classification) there are numerous variations and heuristics to find the "optimum" clusters [52, 53].

5. Distribution-Based Approximation

Only in a few practical cases, for example in some applications in communications, the stochastic properties of the pattern source are known analytically and the discriminant functions can be calculated accordingly. More often, it is only hypothesized that the pattern source has certain stochastic properties — frequently even without the possibility to prove this characteristic by an appropriate test. Nevertheless, it is usually sufficient to just implement the discriminant functions and to test whether the respective classifier produces satisfying results. By far the most often applied hypothesis is that the pattern source produces Gaussian distributed samples. This case is discussed more thoroughly in this section.

For the Gaussian case, we hypothesize that the class conditional probabilities $P(\mathbf{v}|\omega_k)$ are M-dimensional normally distributed:

$$P(\mathbf{v}|\omega_k) = \mathcal{N}_{\mathbf{v}}(\boldsymbol{\mu}_k, \mathbf{K}_k) = \frac{1}{\sqrt{(2\pi)^N \det \mathbf{K}_k}}\, e^{-\frac{1}{2}(\mathbf{v}-\boldsymbol{\mu}_k)^{\mathrm{T}} \mathbf{K}_k^{-1} (\mathbf{v}-\boldsymbol{\mu}_k)},$$

$$\text{where } \boldsymbol{\mu}_k = \mathrm{E}\{\mathbf{v}\,|\,\omega_k\} \text{ and } \mathbf{K}_k = \mathrm{E}\{(\mathbf{v}-\boldsymbol{\mu}_k)(\mathbf{v}-\boldsymbol{\mu}_k)^{\mathrm{T}}\,|\,\omega_k\}\,. \tag{5.1}$$

Obeying the Bayesian approach (minimizing the number of errors, see Sec. 2.1), we have to use the a posteriori probabilities $P(\omega_k|\mathbf{v}) = P(\omega_k) \cdot P(\mathbf{v}|\omega_k) \,/\, P(\mathbf{v})$ as discriminant functions. Since only the maximum of the discriminant functions is searched as class decision, we can modify the a posteriori functions with any monotonic transformation. Using the logarithm to remove the exponential dependencies and neglecting constant terms, we finally arrive at the following discriminant function for the class k of the *Gaussian assumption Bayes classifier* (GBC):

$$d_{GBC,k}(\mathbf{v}) = 2\ln P(\omega_k) - \ln\left[\det \mathbf{K}_k\right] - (\mathbf{v}-\boldsymbol{\mu}_k)^{\mathrm{T}} \mathbf{K}_k^{-1} (\mathbf{v}-\boldsymbol{\mu}_k)\,. \tag{5.2}$$

Analyzing Eq. (5.2) further, we see that the Gaussian assumption Bayes classifier has quadratic discriminant functions with respect to the features $\mathbf{v}$.

In practical applications the expectation operator $\mathrm{E}\{\cdots\}$ is replaced by the sample mean $\sum\cdots$ of the learning set. For high-dimensional feature spaces and a large number of classes, however, the number of parameters and also the required computing resources for the GBC become prohibitively large ($K \cdot M$ for the means $\boldsymbol{\mu}_k$ and $K \cdot \binom{M+1}{2}$ for the covariance matrices $\mathbf{K}_k$). That large number of parameters also leads, at least for smaller learning sets, to the known problems of overfitting.

One common hypothesis is therefore that the covariance matrices are identical for all classes: $\mathbf{K}_k = \mathbf{K}$. This assumption is fulfilled, for example, if the feature vectors $\mathbf{v}$ are produced by adding Gaussian noise to predefined signals $\boldsymbol{\mu}_k$ (e.g., in communications applications). The discriminant functions of Eq. (5.2) reduce then, after some calculations, to:

$$d_{GBC(\mathbf{K}_k=\mathbf{K}),k}(\mathbf{v}) \quad = \quad \ln P(\omega_k) \; - \; \tfrac{1}{2}\,\boldsymbol{\mu}_k^{\mathrm{T}}\,\mathbf{K}^{-1}\,\boldsymbol{\mu}_k \; + \; \boldsymbol{\mu}_k^{\mathrm{T}}\,\mathbf{K}^{-1}\,\mathbf{v}\,. \tag{5.3}$$

For this special case of identical covariance matrices $\mathbf{K}$, the discriminant functions become linear with respect to the feature vector $\mathbf{v}$, consisting of a class specific constant: $\ln P(\omega_k) - \frac{1}{2}\,\boldsymbol{\mu}_k^{\mathrm{T}}\,\mathbf{K}^{-1}\,\boldsymbol{\mu}_k$, and the coefficient vector: $\boldsymbol{\mu}_k^{\mathrm{T}}\,\mathbf{K}^{-1}$. Comparing Eq. (5.3) with the Mahalanobis distance [54], we see that both lead to the same result, if the a priori probabilities are equal: $P(\omega_k) = 1/K$, since this constant term can then be omitted.

The simplest case of a Gaussian assumption is given if the covariance matrix is hypothesized to be the identity matrix: $\mathbf{K}_k = \sigma^2\mathbf{I}$. This corresponds to the situation, where the single features are uncorrelated and have uniform variance σ^2. In this case the discriminant function is further simplified to:

$$d_{GBC(\mathbf{K}_k=\sigma^2\mathbf{I}),k}(\mathbf{v}) \quad = \quad \sigma^2 \ln P(\omega_k) \; - \; \tfrac{1}{2}\left|\boldsymbol{\mu}_k\right|^2 \; + \; \boldsymbol{\mu}_k^{\mathrm{T}}\,\mathbf{v}\,. \tag{5.4}$$

The coefficient vector of this linear discriminant function is now identical to the class mean $\boldsymbol{\mu}_k$, leading to the name matched filter. Further if the a priori probabilities are equal, Eq. (5.4) is equivalent to the Euclidean distance and the respective classifier is then equivalent to a nearest neighbor classifier with one prototype (the mean) for each class. This relation manifests once more, that quite different argumentation chains might lead to the same classifiers (simplified NNC or further restricted GBC).

6. Conclusion

Having presented so many different approaches (PC, MLP, RBF, NNC, and GBC) for pattern classification, the first question from the reader will probably be: *Which one is the best?* However, it was also shown that three of them (PC, MLP, and RBF) are universal approximators and will therefore yield similar error rates at least in the limit for an infinite learning set and a sufficiently large number of parameters. Nevertheless, we give in Table 1 a comparison of the different approaches for the task of handwritten digit recognition (10 classes, 10000 samples each for the learning and the test set; see also Fig. 2). The most important criterion for the comparison of classifiers is certainly the error rate on an independent test set. For a fair and sensible comparison of the error rates, the number of free parameters of each classifier must also be taken into consideration, since this is the switch to turn for getting the desired results.

Table 1. Comparison of the discussed classifier structures.

Classifier	Specification	Parameters	Error rate Learning set	Test set
PC	Linear	2313	8.6%	10.0%
PC	Incomplete quadratic	9225	0.6%	2.2%
PC	PCA & complete quadratic	10280 & 7749	0.8%	1.6%
MLP	256–40–10	10690	0.3%	2.5%
MLP	PCA & 40–200–10	10280 & 10210	0.3%	1.8%
MLP	256–160–10	42730	0.2%	2.1%
RBF	40 prototypes	10640	7.3%	7.9%
RBF	80 prototypes	20560	5.4%	5.8%
RBF	160 prototypes	42560	3.6%	3.9%
NNC	40 prototypes	10240	9.5%	9.9%
NNC	Complete learning set	2560000	0.0%	2.2%
GBC	Matched filter	2560	12.7%	13.4%

Looking closer at Table 1, we first find a linear classifier (in the section PC), which serves as a point of reference (10.0% error rate on the test set). Since the sum-equal-one condition holds, we must calculate only $K-1$ discriminant functions: $9 \cdot (256+1) = 2313$ parameters. Aiming at about 10000 free parameters, which is a reasonable number for handwritten digit recognition, we first adapted an incomplete quadratic PC, yielding already very promising results (2.2% error rate). As mentioned before, a principal component analysis (PCA) [55] is often used for OCR, to reduce the feature dimensionality. Here we reduced the original gray-valued image with 256 features (pixels) to 40 dimensions (needing $256 \cdot 40 + 40 = 10280$ parameters for the transformation), and then adapted a complete quadratic PC ($9 \cdot \binom{42}{2} = 7749$ parameters), leading to the best result (1.6% errors) of the whole comparison. For the fully connected MLP we always used only one hidden layer and trained it by error backpropagation (singular update) until the lowest error rate on the test set was reached [56]. For the MLP we used various initializations and never encountered problems with local minima. Again the version with the principal component analysis was best (1.8% error rate). We notice further the very good adaption to the learning set (only 0.3% errors). For the RBF, we used one optimum smoothing factor (found by gradient descent) for all reference vectors (obtained by a divisive vector quantization of all patterns, see also Figs. 7 and 8) and adapted the coefficients of the linear combination by a pseudo inversion. We chose this relatively simple set-up, since it already has the potential for universal approximations. The error rates indeed improve with an increasing number of reference vectors, but are still unsatisfactory compared to the PC or MLP. However, there are more promising results in the literature [57], reported for similar problems. We hope to reach those improvements with the mixture-of-Gaussians and normalization-to-one approach, after having applied it very successfully to low-dimensional demonstration problems. Furthermore, we listed two results for the NNC. The first result for 40 reference vectors (identical to the RBF support points) shows the difference between the NNC ($k = 1$) and the RBF (9.9% versus 7.9% error rate). The unsupervised clustering procedure to find the reference vectors for the RBF, however, is certainly not the best selection for an NNC (see, for example, LVQ). The second result, using the whole learning set and majority decision among $k = 3$ neighbors (here the optimum value out of the set $k = [1, 3, 5, 7, 11]$), can serve as a practical estimate for the Bayesian error rate (half of the result of 2.2% errors), even if the number of parameters is prohibitively large for a practical application. The GBC is included in the comparison only in its simplest version (identity matrix as covariance matrix leads to a matched filter approach) since otherwise the covariance matrices are too large.

Summarizing the comparison, we do not want to favour any specific approach (especially not for all possible classification problems), but like to stress that there are other important factors for the design decision, such as adaption time, robustness, manual tuning required during adaption, availability, and rejection behaviour. For the task of character recognition at least, we have experienced very satisfying characteristics of the PC in all these categories. This also has been proven in the first NIST competition [14]. Furthermore, we have to note that the numbers in Table 1 are not intended to start a competition, but rather should serve as a guideline. All mentioned approaches exhibit a certain trade off between performance and computational effort, and various fine tuning possibilities exist for all of them, which certainly pay off many times in practical applications such as automatic address reading.

Moreover, the classifiers discussed, PC, MLP, and RBF, are often only the basic building blocks in more *complex classification systems* for demanding applications. The single classifiers can then be viewed as experts, which are often applied only to dedicated subtasks. There are several different constellations known to structure the interaction between classifiers: hierarchical structures, nets for pairwise classification, gated responsibility, or majority voting by classification [58, 59, 60].[f] But only the adequate combination and interaction of the basic classifiers (which may be chosen from different paradigms, or even have not to be statistical classifiers) lead to the desired performance improvements. Furthermore, we want to mention here that for the combination of the classifier votes, and also for further decisions in the whole recognition process, the output of the single classifiers must be comparable. This property is not necessarily given since all discussed classifiers only approximate the a posteriori probabilities or even work with nonlinear monotonic mappings of them (such as the GBC). Therefore, the probability axioms, which say that: $0 \leq d_k \leq 1$, $\sum_k d_k = 1$, and d_k corresponds to the frequency of right decisions, are not necessarily fulfilled. For the MLP only the zero-to-one interval is

[f]For a more comprehensive treatment of this subject see also the chapter by J. Franke in this book.

guaranteed, whereas for the PC the sum-equal-one condition is met. One solution is to satisfy the first two conditions by a normalization operation, which however does not pay attention to the third condition. In our experience, it is more appropriate to stabilize the third condition by an additional, specially adapted transformation of the discriminant values (Bayesian stabilization), which automatically takes care of the other two conditions. Included in this procedure is also the rejection of those patterns, which violate the closed-world-assumption of the classifier (i.e. they come from regions of the feature space, which are not occupied by learning examples). The a posteriori probabilities are undefined in this case, since $P(\mathbf{v}) = 0$.

Summarizing our contribution, we can state that the stochastic based function approximation approach for classification has received a great deal of interest by designers of practical applications. Especially driven by the buzz word "neural networks", many methods are now available directly as software tools and are widely applied. For many classification problems (in OCR and elsewhere) the possibility to apply easily even complex nonlinear approximation schemes, which are adapted only by learning from examples, yields successful solutions and reduces expensive development time. However, some of the hype, which have been postulated after the first success of neural networks and relied upon the assumption that artificial neural networks are technical replicates of the human ones, are now replaced by mathematical groundwork. The goal of this chapter (see also [61]) is to give a *unifying framework* for all classification techniques based on the concept of function approximation, including the "classical" statistical methods as well as the neural network approaches.

References

[1] J. Schürmann *et al.*, Document analysis: From pixels to contents, *Proc. of the IEEE* **80**,7 (1992) 1101–1119.

[2] E. Mandler and M. Oberländer, One-pass encoding of connected components in multi-valued images, *Proc. 10th IAPR Int. Conf. on Pattern Recognition*, Atlantic City, 1990, 64–69.

[3] K. Fu, *Syntactic Methods in Pattern Recognition* (Academic Press, New York, 1974).

[4] L. Breiman *et al.*, *Classification and Regression Trees* (Chapman and Hall, New York, 1993).

[5] J. Quinlan, *C4.5: Programs for Machine Learning* (Morgan Kaufmann, San Mateo, 1993).

[6] B. Kosko, *Neural Networks and Fuzzy Systems* (Prentice Hall, Englewood Cliffs, 1992).

[7] U. Kreßel, The impact of the learning-set size in handwritten-digit recognition, in *Artificial Neural Networks (Proc. Int. Conf. on Artificial Neural Networks)*, eds. T. Kohonen *et al.* (North-Holland, Amsterdam, 1991) 1685–1689.

[8] H. White *et al.*, *Artificial Neural Networks: Approximation and Learning Theory* (Blackwell, Cambridge, 1992).

[9] J. Park and I. Sandberg, Universal approximation using radial-basis-function networks, *Neural Computation* **3**,2 (1991) 246–257.

[10] S. Geman, E. Bienenstock and R. Doursat, Neural networks and the bias/variance dilemma, *Neural Computation* **4**,1 (1992) 1–58.

[11] G. Lorentz, *Approximation of Functions* (Chelsea Publishing Company, New York, 1986).

[12] V. Vapnik, *Estimation of Dependences Based on Empirical Data* (Springer-Verlag, New York, 1982).

[13] P. Niyogi and F. Girosi, On the relationship between generalization error, hypothesis complexity, and sample complexity for radial basis functions, A.I. Memo 1467, Massachusetts Institute of Technology, 1994.

[14] R. Wilkinson *et al.*, *The First Census Optical Character Recognition Systems Conference*, National Institute of Standards and Technology, Gaithersburg, 1992.

[15] L. Bottou *et al.*, Comparison of classifier methods: a case study in handwritten digit recognition, *Proc. 12th IAPR Int. Conf. on Pattern Recognition*, Jerusalem, 1994, 77–82.

[16] J. Schürmann, *Polynomklassifikatoren für die Zeichenerkennung: Ansatz, Adaption, Anwendung* (Oldenbourg Verlag, München, 1977).

[17] D. Rumelhart, J. McClelland, and the PDP Research Group, *Parallel Distributed Processing* (MIT Press, Cambridge, 1986).

[18] Y.-H. Pao, *Adaptive Pattern Recognition and Neural Networks* (Addison-Wesley, New York, 1989).

[19] *MatLab Reference Guide*, The MathWorks Inc., Natick, 1993.

[20] S. Farlow, ed., *Self-Organizing Methods in Modeling: GMDH Type Algorithms* (Marcel Dekker, New York, 1984).

[21] Y. Le Cun *et al.*, Backpropagation applied to handwritten zip code recognition, *Neural Computation* **1**,1 (1989) 541–551.

[22] K. Fukushima and S. Miyake, Neocognitron: A new algorithm for pattern recognition tolerant of deformations and shifts in position, *Pattern Recognition* **15**,6 (1982) 455–469.

[23] S. Fahlman, Faster-learning variations on back-propagation: an empirical study, *Proc. of the 1988 Connectionist Models Summer School*, eds. D. Touretzky, G. Hinton, and T. Sejnowski (Morgan Kaufmann, San Mateo, 1989) 38–51.

[24] W. Schiffmann, M. Joost, and R. Werner, Optimization of the backpropagation algorithm for training multilayer perceptrons, Technical Report, University of Koblenz, Institute of Physics, 1992.

[25] M. Riedmiller and H. Braun, A direct adaptive method for faster backpropagation learning: The RPROP algorithm, *Proc. of the IEEE Int. Conf. on Neural Networks*, ed. H. Ruspini (San Francisco, 1993) 586–591.

[26] A. Weigend and N. Gershenfeld, eds., *Times Series Prediction: Forecasting the Future and Understanding the Past* (Addison-Wesley, Reading, 1993).

[27] Y. Le Cun, J. Denker, and S. Solla, Optimal brain damage, *Advances in Neural Information Processing Systems 2 (NIPS-89)*, ed. D. Touretzky (Morgan Kaufmann, San Mateo, 1990) 598–605.

[28] B. Hassibi and D. Stork, Second order derivatives for network pruning: optimal brain surgeon, *Advances in Neural Information Processing Systems 5 (NIPS-92)*, eds. S. Hanson, J. Cowan and C. Giles (Morgan Kaufmann, San Mateo, 1993) 164–171.

[29] F. Hergert *et al.*, Domain independent testing and performance comparisons for neural networks, *Artificial Neural Networks 2 (Proc. Int. Conf. on Artificial Neural Networks)*, eds. I. Aleksander and J. Taylor (North-Holland, Amsterdam, 1992) 1071–1075.

[30] R. Hecht-Nielsen, *Neurocomputing* (Addison-Wesley, Reading, 1989).

[31] J. Hertz, A. Krogh, and R. Palmer, *Introduction to the Theory of Neural Computation* (Addison-Wesley, Redwood City, 1991).

[32] S. Haykin, *Neural Networks: A Comprehensive Foundation* (Macmillan, New York, 1994).

[33] S. Gielen and B. Kappen, eds., *ICANN'93, Proc. of the Int. Conf. on Artificial Neural Networks* (Springer-Verlag, London, 1993).

[34] M. Marinaro and P. Morasso, eds., *ICANN'94, Proc. of the Int. Conf. on Artificial Neural Networks* (Springer-Verlag, London, 1994).

[35] J. Cowan, G. Tesauro, and J. Alspector, eds., *NIPS'93, Advances in Neural Information Processing Systems 6* (Morgan Kaufmann, San Francisco, 1994).

[36] G. Tesauro, D. Touretzky and T. Leen, eds., *NIPS'94, Advances in Neural Information Processing Systems 7* (Morgan Kaufmann, San Francisco, 1995).

[37] J. Blue *et al.*, Evaluation of pattern classifiers for fingerprint and OCR applications, *Pattern Recognition* **27**,4 (1994) 485–501.

[38] T. Poggio and F. Girosi, Networks for approximation and learning, *Proc. of the IEEE* **78**,9 (1990) 1481–1497.

[39] Y. Linde, A. Buzo, and R. Gray, An algorithm for vector quantizer design, *IEEE Trans. on Commun.* **28**,1 (1980) 84–95.

[40] T. Kohonen, *Self-Organization and Associative Memory* (Springer-Verlag, Berlin, 1989).

[41] SOM Programming Team, SOM-PAK: The Self-Organizing Map Program Package, Version 1.2, Copyright: T. Kohonen, J. Kangas, and J. Laaksonen, Helsinki University of Technology, Espoo, 1992.

[42] LVQ Programming Team: LVQ-PAK: The Learning Vector Quantization Program Package, Version 2.1, Copyright: T. Kohonen *et al.*, Helsinki University of Technology, Espoo, 1992.

[43] F. Girosi, M. Jones, and T. Poggio, Regularization theory and neural networks architectures, *Neural Computation* **7**,2 (1995) 219–269.

[44] J. Moody and C. Darken, Fast learning in networks of locally-tuned processing units, *Neural Computation* **1**,2 (1989) 281–294.

[45] J. Platt, A resource-allocating network for function interpolation, *Neural Computation* **3**,2 (1991) 213–225.

[46] D. Specht, Probabilistic neural networks and the polynomial adaline as complementary techniques for classification, *IEEE Trans. on Neural Networks* **1**,1 (1990) 111–121.

[47] D. Reilly, L. Cooper, and C. Elbaum, A neural model for category learning, *Biological Cybernetics* **45** (1982) 35–41.

[48] R. Redner and H. Walker, Mixture densities, maximum likelihood and the EM algorithm, *SIAM Review* **26**,2 (1984) 195–239.

[49] B. Dasarathy, ed., *Nearest Neighbor (NN) Norms: NN Pattern Classification Techniques* (IEEE Comp. Soc. Press, Los Alamitos, 1990)

[50] T. Cover and P. Hart, Nearest neighbour pattern classification, *IEEE Trans. on Inf. Theory* **13** (1967) 21–27.

[51] H. Bock, *Automatische Klassifikation* (Vandenhoeck & Ruprecht, Göttingen, 1974)

[52] H. Späth, *Cluster-Formation und -Analyse: Theorie, FORTRAN-Programme, Beispiele* (Oldenbourg, München, 1983)

[53] B. Fritzke, Wachsende Zellstrukturen — ein selbstorganisierendes neuronales Netzwerkmodell, Ph.D. Thesis, F.-A.-Universität, Erlangen-Nürnberg, 1992.

[54] R. Duda and P. Hart, *Pattern Classification and Scene Analysis* (Wiley, New York, 1973)

[55] K. Fukunaga, *Introduction to Statistical Pattern Recognition* (Academic Press, San Diego, 1990)

[56] R. Henkel and U. Kreßel, Konfigurieren und Trainieren von Multilayer-Perzeptrons am Beispiel der Ziffernerkennung, *Mustererkennung 1992*, eds. S. Fuchs and R. Hoffmann (Springer-Verlag, Berlin, 1992) 353–360.

[57] S. Grudszus, Adaptive Lernverfahren für 'Generalized Radial Basis Functions' Neuronale Netzwerke — Vergleich und Weiterentwicklung, Master Thesis, Allgemeine und Theoretische Elektrotechnik, F.-A.-Universität Erlangen-Nürnberg, 1993.

[58] J. Franke, Zur Entwicklung hierarchischer Klassifikatoren aus der entscheidungstheoretischen Konzeption, *Mustererkennung 1983*, (VDE-Verlag, Berlin, 1983) 261–265.

[59] M. Jordan and R. Jacobs, Hierarchical mixtures of experts and the EM algorithm, *Neural Computation* **6**,2 (1994) 181–214.

[60] T. Hrycej, *Modular Learning in Neural Networks: A Modularized Approach to Neural Network Classification* (Wiley, New York, 1992)

[61] J. Schürmann, *Pattern Classification: A Unified View of Statistical and Neural Approaches* (Wiley, New York, 1996).

Handbook of Character Recognition and Document Image Analysis, pp. 79–101
Eds. H. Bunke and P. S. P. Wang

CHAPTER 3

COMBINATION OF MULTIPLE CLASSIFIER DECISIONS FOR OPTICAL CHARACTER RECOGNITION

LOUISA LAM
Hong Kong Institute of Education
Northcote Campus, 21 Sassoon Road, Hong Kong

YEA-SHUAN HUANG
Computer & Communication Research Laboratories
Industrial Technology Research Institute, Hsinchu, Taiwan 31015

and

CHING Y. SUEN
Centre for Pattern Recognition and Machine Intelligence
Concordia University, Montreal, Quebec H3G 1M8, Canada

In order to improve recognition results, decisions of several classifiers can be combined. The combination can be accomplished in different ways depending on the types of information produced by the individual classifiers. This chapter considers combination methods that can be applied when the information is provided at both the abstract and measurement levels.

For abstract-level classifiers, the combination methods discussed in this chapter consist of majority vote, weighted majority vote with weights derived from a genetic search algorithm, Bayesian formulation, and Behavior-Knowledge Space method. To combine the decisions of measurement-level classifiers, a multi-layer perceptron is used.

Theoretical considerations and experimental results on handwritten characters are presented, and the results show that combining multiple classifiers is an effective means of producing highly reliable decisions for both categories of classifiers.

Keywords: Combination of classifiers; OCR; Majority vote; Genetic algorithm; Neural networks.

1. Introduction

In the recognition of handwritten characters and words, there has been a recent movement towards combining the decisions of several classifiers in order to arrive at improved recognition results. This is due to a number of reasons, among which are the demands imposed by real-life applications and the availability of a wide variety of algorithms. Practical applications demand highly reliable classifications, which are extremely difficult for a single algorithm to achieve. Since many algorithms are available for these tasks, it is logical to consider the use of several classifiers to achieve higher reliability.

The combination can be implemented using different strategies. In [1] and [2], the combined decision is obtained by a majority vote of the individual classifiers, and variations of this scheme are implemented in [3] and [4]. When the individual classifiers output ranked lists of decisions, these rankings can be used to derive combined decisions by the highest rank, Borda count, and logistic regression methods ([5], [6]). From these ranked lists, a nonparametric procedure can be used to combine the classification results and a measure of confidence assigned to that decision [7]. Further developments in obtaining a combined decision include statistical approaches [8], formulations based on Bayesian and Dempster-Shafer theories of evidence ([9], [8], [4]), and neural networks ([10], [11], [12]). Fuzzy theory has also been used to integrate the information obtained from multiple classifiers [13]. Other authors ([14], [15]) use a polynomial classifier to combine the results of multiple classifiers, using the output of the individual classifiers as features. In all these cases, it was found that using a combination of classifiers can result in remarkable improvements in the recognition performance, and this is true regardless of whether the classifiers are independent or make use of orthogonal features.

In general, the methods of combining multiple classifier decisions depend on the types of information produced by the individual classifiers. This article considers combination methods that can be applied when this information is provided at both the abstract and measurement levels. In the former case, each classifier e outputs a unique label or class for each input pattern, while in the latter instance, e produces a measurement value for each label.

For abstract-level classifiers, the combination methods discussed in this article consist of:

(i) majority vote,
(ii) weighted majority vote with weights derived from a genetic search algorithm,
(iii) Bayesian formulation, and
(iv) the Behavior-Knowledge Space method.

Among all the combination methods for abstract-level classifiers, majority vote is the simplest to implement, and its simplicity has permitted theoretical analysis ([16], [17]). In Sec. 2, we consider extensions of this method to cases in which classifiers are assigned unequal weights based on their performance. These weights are obtained through the optimization of an objective function for the combined decision by using a Bayesian formulation and a genetic algorithm. In Sec. 3, we compare the experimental results obtained from these three methods on seven classifiers trained on a large set of handwritten numerals.

The behavior-Knowledge Space (BKS) method, which can be considered to be a refinement of the Bayesian method, is described in Sec. 4, where a comparison of their results is also presented.

In Sec. 5, we describe a method of combining the decisions of measurement-level classifiers by a neural network. Using this process, the output values of each classifier are transformed into a form of "likeness" measurement, and the resulting measurements are aggregated by a multi-layer perceptron to produce the final classifications. We conclude with some observations and remarks in Sec. 6.

2. Voting Methods

In combining decisions of pattern classifiers, the method used depends on the nature of the output produced by each recognition algorithm. If this output consists of a single assigned class, then many of the combination methods would not be applicable, and one of the most suitable means of arriving at a combined decision would be some form of voting. In addition to a simple majority vote (in which all votes have equal weight), the votes can be weighted so that each classifier carries the same weight for all pattern classes, or the weights can be determined according to the performance of each classifier on each class.

We will study the two weighting procedures and compare their results. In the former case, the assignment of weights will be based on optimizing the value of an objective function through a genetic algorithm. For the second system of weighting, a Bayesian combination rule [4] will be used, and the value of a parameter will be determined so that the same function is optimized. In this way, we can realistically compare the results of the procedures on the same sets of data.

2.1. *Optimization of Objective Function F*

In pattern recognition, the recognition (correct) and substitution (error) rates are often used to measure the performance of a classifier. Ideally, one would like to maximize the recognition rate and minimize the substitution rate, but this is very difficult to achieve in practice. When there is a third option of rejecting the input sample in case of uncertainty, it is a common experience that the procedures used to reduce the error rate would also lead to higher rejection (and lower recognition) rates. On the other hand, it is impractical to eliminate the reject option (and force the classifier to decide on the identity of every sample), since the decisions made in uncertain cases would cause a disproportionate increase in the error rate.

This being the case, it would be natural for a measurement of classifier performance to contain some trade-off factor between the recognition/rejection and error rates. For example, at the first and second IPTP competitions in Japan [12], the cost of an error was set at 10 times that of a rejection, so the precision index was established as $Rejection + 10 * Error$.

For our present experiment, the objective function to be maximized was defined as $F = Recognition - \beta * Error$, where β has values 10, 15, 20, 25, and 30. Obviously, the value of β varies with the accuracy or reliability desired for a particular application. The function F is related to the precision index defined above; for example, when $\beta = 10$, maximizing F is equivalent to minimizing $Rejection + 11 * Error$. These high levels of reliability are set for our experiment because they can be sustained by the classifiers considered.

2.2. *Genetic Algorithm*

First proposed by Holland [18], genetic algorithms have been found to be robust and practical optimization methods. A genetic algorithm begins with an initial set (also called a population) of randomly generated potential solutions to an optimization problem. The value of an objective function (fitness value) of each solution is evaluated, and the "best" solutions are selected for survival. Then the genetic algorithm manipulates these selected solutions in its search for better solutions. Each solution is encoded into a binary string (chromosome), so that new encoded solutions can be generated through the exchange of information among surviving solutions (crossovers) as well as sporadic alterations in the bit string encodings of the solutions (mutations).

This method is applied to our problem to obtain an optimal set of weights, one for the vote of each classifier across all pattern classes. Optimality is defined according to the value of F described above, which is also used as the fitness value. To ensure that fitter strings have proportionally higher chances of surviving in the subsequent generation, the selection mechanism is implemented by a roulette scheme [19] after the fitness values have been linearly scaled so that the maximum scaled fitness value is 1.5 times the average fitness value of the population. The scaling mechanism was implemented because the variance in string fitness values becomes small after several generations, with the result that all strings would have approximately the same number of offsprings without scaling, thereby neutralizing the propagation of fitter strings. Since a linear scaling may cause low fitness values to be mapped onto negative scaled values, the chromosomes having fitness values below a certain threshold f_t are eliminated. In this case $f_t = f_{ave} - 2.5\sigma$ where f_{ave} is the average fitness of the population, and σ is the standard deviation of fitness values in the population. So if f is the fitness value of a string, its scaled value f' is given by

$$f' = \frac{f - f_t}{f_{max} - f_t} \times (1.5 \times f_{ave} - f_t) + f_t, \tag{1}$$

where f_{max} is the maximum fitness of the population before scaling.

The population size used is 50, and each gene occupies 10 bits, so the weight of each classifier ranges between 0 and 1.023. The probabilities of crossover and mutation are 0.9 and 0.05, respectively. Control parameters in these ranges have been proposed by several researchers to guarantee good performance on carefully chosen testbeds of objective functions [20]. This is a robust process, in the sense that the same optimal fitnesses would result upon replications of the process starting from random weights. The distributions of weights that produce the optimal solutions may vary with each repetition, but the procedure would result in the same maximum values of the objective function F, as well as the same recognition and error rates. Figure 1 shows an example of the behaviors of the maximum and average fitness values through the generations.

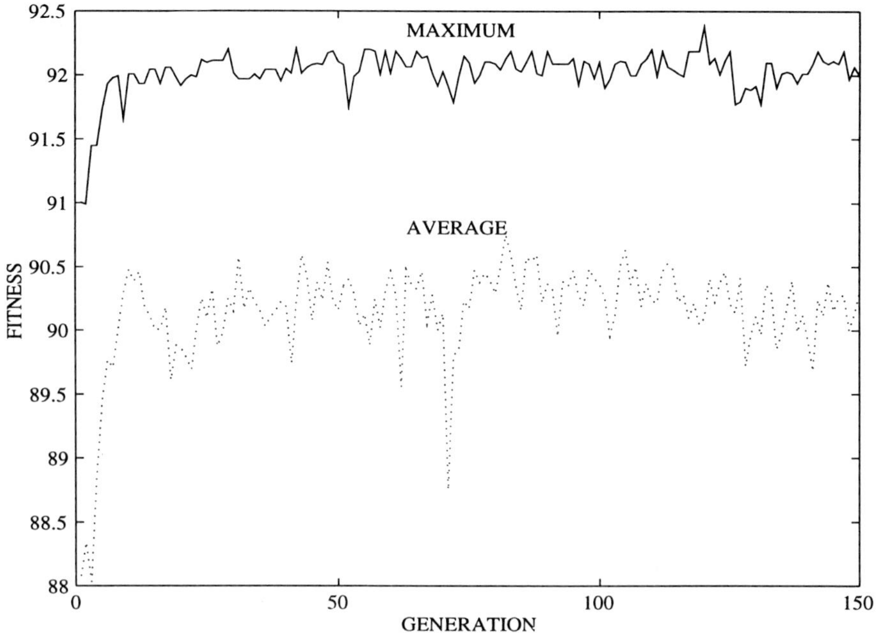

Fig. 1. Maximum and average fitness values of genetic algorithm.

From Fig. 1, it can be observed that the fitness values do not vary greatly through the generations. This is due to the relatively large number (seven) of classifiers being combined, and the high individual performances. In other words, whenever a weighted majority of the votes are in agreement, then the group decision would be quite reliable. For this reason, varying the weights does not create dramatic improvements. On the other hand, any gain in performance is useful for practical applications.

2.3. *Bayesian Combination Rule*

The genetic algorithm implemented assigns a weight to the vote of each classifier (also called an *expert*), and this weight would be applied to all patterns regardless of the decision made by the expert. Another method of determining the weights is through the Bayesian decision rule, which takes into consideration the performance of each expert on the training samples of each class. In particular, the confusion matrix C of each classifier on a training set of data would be used as indications of its performance. For a problem with M possible classes plus the reject option, C is an $M \times (M + 1)$ matrix in which the entry C_{ij} denotes the number of patterns with actual class i that is assigned class j by the classifier when $j \leq M$, and when $j = M + 1$, it represents the number of patterns that are rejected.

From the matrix C, we can obtain the total number of samples belonging to class i as the row sum $\sum_{j=1}^{M+1} C_{ij}$, while the column sum $\sum_{i=1}^{M} C_{ij}$ represents the total number of samples that are assigned class j by this expert. When there are K experts, there would be K confusion matrices $C^{(k)}$, $1 \leq k \leq K$. Consequently, the conditional probability that a pattern x actually belongs to class i, given that expert k assigns it to class j, can be estimated as

$$P(x \in C_i \mid e_k(x) = j) = \frac{C_{ij}^{(k)}}{\sum_{i=1}^{M} C_{ij}^{(k)}}. \tag{2}$$

For any pattern x such that the classification results by the K experts are $e_k(x) = j_k$ for $1 \leq k \leq K$, we can define a belief value that x belongs to class i as

$$bel(i) = P(x \in C_i \mid e_1(x) = j_1, ..., e_K(x) = j_K). \tag{3}$$

By applying the Bayes' formula and assuming independence of the expert decisions [4], $bel(i)$ can be approximated by

$$bel(i) \doteq \frac{\prod_{k=1}^{K} P(x \in C_i \mid e_k(x) = j_k)}{\sum_{i=1}^{M} \prod_{k=1}^{K} P(x \in C_i \mid e_k(x) = j_k)} \tag{4}$$

for $1 \leq i \leq M$.

For any input pattern x, we can assign x to class j if $bel(j) \geq bel(i)$ for all $i \neq j$ and $bel(j) > \alpha$ for a threshold α. Otherwise x is rejected, and it is also rejected if $e_k(x) = M + 1$ for all k (i.e., if x is rejected by all classifiers). The results obtained from this method depend on the value of α chosen. As α increases, so does the degree of certainty expected of the decision; therefore the error rate would decrease, but the recognition rate would be also lower. Figure 2 gives an example of the behavior of the recognition and error rates when α varies from 0.1 to 0.99999999.

Given that the results depend on the choice of α, and because we would like to compare optimal results obtained from different methods, the value of α was chosen to maximize the value of the same objective function F. In Fig. 2, the optimal value of α is 0.999952 when $\beta = 10$, and this value of α represents the threshold above which the ratio of the changes in the recognition and error rates would exceed β. In this figure, each dotted line contains the points yielding the same value of F, and it can be seen that the maximum value of F obtained is approximately 96. More detailed experimental results are presented in Table 5 of Sec. 3.3.

3. Experimental Data and Procedure

The combination methods described in this section can be applied to any category of patterns or classifiers, therefore they can be tested on any type of data. Our

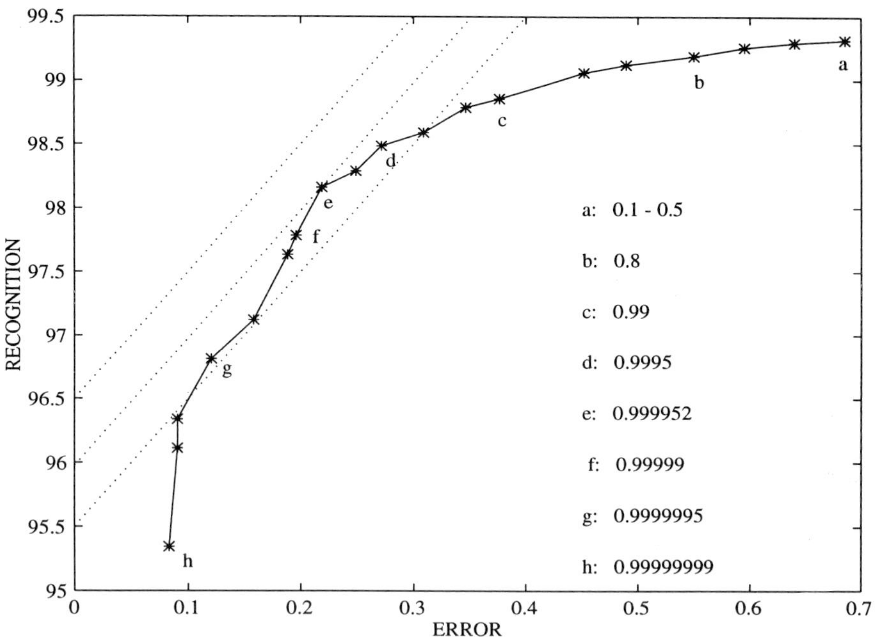

Fig. 2. Recognition results for different values of α (shown as labelled).

experiment was performed on seven classifiers used to process a large collection of handwritten numerals. This database consists of 46451 numerals collected by the Industrial Technology Research Institute (ITRI) of Taiwan. Of these, 24427 were used to train the individual classifiers, and the other 22024 samples form the test set for the recognizers. Since the combination methods should first be trained on representative classification results, only the results of the test set are used in our study. This set was subdivided into 50 subsets at ITRI, with 5 subsets per class. The first 3 subsets of each class were assigned to set A and the rest to set B, with the result that 13272 samples belong to set A and 8752 to B.

The seven classifiers used to recognize this data cover a wide variety of approaches. The features used include the pixels of the pattern, contour and algebraic features, together with structural information derived from a skeleton. The classifications are based on tree classifiers, dynamic programming, relaxation and exhaustive matching, as well as a neural network. More details about these experts are contained in ([21], [2], [22]). These classifiers have not been trained to the same extent on the training set of 24427 samples, and their individual performances on sets A and B are shown in Table 1.

Given that both sets A and B had been test sets for the recognizers and the division into two sets was arbitrary, the differences in results between the two sets indicate the presence of more distorted samples in set A. However, no effort was made to manipulate the data in order to obtain a more even partition.

Table 1. Performance of individual classifiers on handwritten numerals.

	Set A			Set B		
Expert	Recognition	Error	Obj. F.	Recognition	Error	Obj. F.
e1	82.791	3.187	50.919	84.004	3.005	53.953
e2	91.132	1.982	71.316	92.207	1.874	73.469
e3	93.264	1.575	77.517	93.864	1.165	82.210
e4	87.176	1.831	68.867	88.425	1.714	71.287
e5	94.929	0.799	86.942	95.007	0.857	86.437
e6	95.999	0.821	87.786	96.023	0.697	89.054
e7	93.716	5.327	40.446	95.212	4.273	52.479

The classification results of set A were used to train the combination methods in the following ways:

(i) To derive optimal weights (using the genetic algorithm) that should be assigned to the different classifiers for the weighted majority vote. As stated in Sec. 2.2, the search process is continued through 150 generations, with 50 chromosomes per generation. Each chromosome is decoded into a set of seven weights which are then normalized so that their sum equals one. Each weight is assigned to the vote of the corresponding classifier. The decisions of these seven classifiers on each sample of set A are combined using weighted majority vote, from which the recognition and error rates of the combined decision on set A can be obtained. The fitness value of the chromosome is the value of F derived from these rates. The optimal weights are the ones that produce the maximum value of F over the 150 generations.

(ii) For the Bayesian method, the confusion matrix obtained from set A is used to estimate the probabilities $P(x \in C_i \mid e_k(x) = j_k)$. Using these probabilities, the recognition results of set A are processed by the Bayesian combination rule for different values of the parameter α, and the α that produces the maximum value of F is approximated. The optimal α is located through a successive refinement of the scope of the search, which is a reasonable procedure because of the discrete nature of the problem.

When training is complete, the results are tested on set B in two ways:

(a) The optimal weights obtained from (i) are assigned to the respective classifiers, and the weighted majority rule is applied to obtain recognition results for the combination.

(b) For the Bayesian method, the probabilities and optimal value of α described in (ii) are used to combine the decisions of the seven classifiers.

Experimental results are given below.

3.1. *Results of Genetic Algorithm*

This algorithm was trained by combining the decisions of the classifiers on set A in order to obtain a set of weights that would maximize the function $F = Recognition - \beta \times Error$, where $\beta = 10$, 15, 20, 25 and 30. For each value of β, the search begins with 50 sets of random weights and this is propagated through 150 generations. This process is performed 9 times for each value of β, with different starting weights. Consequently, 45 cycles of search had been completed (on a total of 337,500 sets of weights). For each β, the recognition and error rates that give $F_{max}(\beta)$ among all the 9 cycles are determined. For the five values of α, this procedure produces the four sets of results shown in Table 2. Among these results, the second set appears far more frequently as optimal with the smaller β's; this is logical given that its error rate is the highest, which means that the fitness of this result would decrease as β increases.

Table 2. Optimal results from genetic algorithm.

Result	Recognition	Error	Reject
1	96.9033	0.1507	2.9460
2	96.9711	0.1582	2.8707
3	96.8279	0.1507	3.0214
4	96.6772	0.1432	3.1796

Of the results shown in Table 2, the first set of values gives the maximum F for all values of β, and therefore they are considered optimal for our experiment. It should be noted that the set of weights producing a value of F may not be unique, due to the fact that F does not vary continuously with the weights, but is piecewise constant. In other words, F changes only when the weights have shifted sufficiently to change the majority vote for some patterns. In addition, different sets of weights may create different combined decisions in a small number of cases, but result in equal recognition and error rates. For example, the optimal results are actually obtained from the two sets of weights shown in Table 3.

Table 3. Weights producing optimal results.

Expert	Weight 1	Weight 2
e1	0.0359	0.0642
e2	0.1882	0.1986
e3	0.0898	0.0892
e4	0.0918	0.1028
e5	0.2219	0.2122
e6	0.2532	0.2310
e7	0.1191	0.1021

It is instructive to note the following points:

(i) e1 is usually assigned the lowest weight by this search algorithm. Given that its performance is low, the result supports the validity of the process.
(ii) Results 4 of Table 2 are obtained only when $\beta = 30$, or the cost of an error is very high (31 times that of a rejection). In this case, the optimal result produces a lower error rate, and e7 was actually assigned the lowest weight, which is consistent with its high error rate.
(iii) e3 is usually assigned a low weight (directly above that of e1), which cannot be explained by its performance alone. Since the reason can be that this classifier does not make much contribution to the combined result (in the sense that the vote of this expert may not change the combined decision of the other experts in most cases), an attempt was made to verify this hypothesis.

The decisions of the six experts without e3 were combined using the entire procedure described above in this subsection, as well as those of six experts with e3 replacing e2. In other words, the search process is propagated through the entire 45 cycles for each set of six experts, and the optimal results are determined as described above. These results, shown in Table 4, indicate that the combined performance is better with e2 than e3. Given that e2 definitely has weaker individual performance than e3 (as shown in Table 1), it can be inferred that e2 is more effective as a complement to the other classifiers, and this point has been identified by the genetic algorithm.

Table 4. Optimal results from combining six experts.

Combination	Recognition	Error	Reject
e1, e2, e4-e7	96.5642	0.1507	3.2851
e1, e3, e4-e7	96.4210	0.1658	3.4132

The optimal weights shown in Table 3 are then applied to set B, and the results are summarized later in Table 6.

3.2. *Results of Bayesian Combination*

Using the procedure described in Sec. 2.3, the classifier decisions are combined using the Bayesian method. It should be noted that the results are sensitive to the value of α chosen. Since higher values of α require higher levels of confidence, the error rate would be a decreasing function of α, as is the recognition rate. The function F, however, would increase to a maximum, then decrease. The decrease begins at the point where the trade-off between the recognition and error rates involve a factor larger than β. Table 5 shows some results obtained from set A for different values of α, when $\beta = 10$.

For each value of β, the α that yields the maximum F is determined. Logically, higher values of β impose higher costs on errors, from which it follows that the optimal solutions would be more reliable. In order to achieve this, α also needs to

Table 5. Results for different values of α.

Value of α	Recognition	Error	$F(\beta = 10)$
0.5	99.3144	0.6856	92.4584
0.8	99.1863	0.5500	93.6864
0.95	99.0582	0.4520	94.5378
0.995	98.7870	0.3466	95.3213
0.9995	98.4856	0.2712	95.7734
0.99995	98.1617	0.2260	95.9015
0.999995	97.6343	0.1884	95.7508
0.9999995	96.8131	0.1205	95.6076

have a larger magnitude. This is in fact the case, except that for $\beta = 20$, 25 and 30, results have stabilized and the same $\alpha = 0.9999999$ and optimal recognition results are obtained. When $\beta = 10$, $\alpha = 0.999952$. These α's were then used as thresholds for set B, and the combined recognition results are given in Table 6. In this table, the results of a simple majority vote are also included.

Table 6. Results of combinations by voting methods.

	Set A				Set B			
Method	Recogn.	Error	Rej.	Obj. F	Recogn.	Error	Rej.	Obj. F
Majority vote	96.233	0.196	3.571	94.274	96.778	0.160	3.062	95.178
Bayesian 1 (α=0.999952)	98.162	0.218	1.620	95.977	97.784	0.571	1.645	92.071
Bayesian 2 (α=0.9999999)	96.338	0.090	3.572	95.434	96.550	0.366	3.084	92.655
Genetic algorithm	96.903	0.151	2.946	95.396	97.075	0.228	2.697	94.790

An examination of the patterns misclassified by the combinations shows the following:

(i) As expected, all the errors of Bayesian 2 are contained in those of Bayesian 1.

(ii) Of the substitutions made by majority vote, all except one are also misclassified in the same way by weighted majority vote derived from the genetic algorithm.

(iii) Of the 14 misclassifications made by majority vote, 10 are common to all the classifiers.

The 14 misclassified samples from set B are shown in Fig. 3, where the class of each sample is indicated, and the recognition result is given in parentheses.

3.3. *Analyses of Voting Results*

From Table 6, it can be seen that the results of a majority vote follow the general trend of the individual classifiers in that the results of set B are better than those

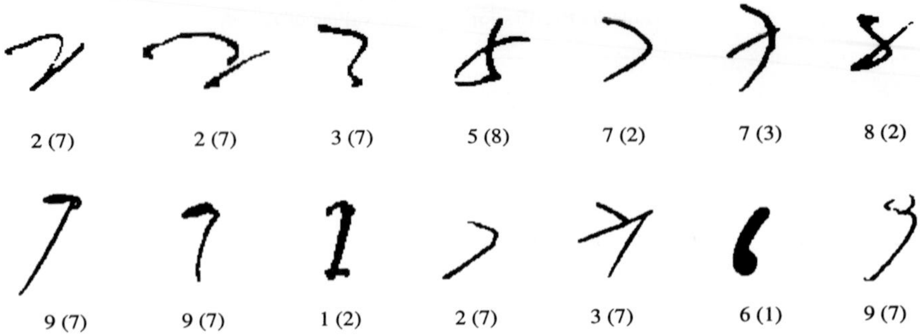

Fig. 3. Patterns misclassified by majority vote.

of A in every aspect, which is reasonable since no training of the combination is involved. This is not the case for the other two combination methods when the error rates are considered. These methods are trained to produce optimal results on set A, while their performances on set B are lower.

If we consider the results of Table 6, it is clear that for any $\beta \geq 4$, or for any reasonable trade-offs between the recognition and error rates, simple majority vote produces the best results on set B. This is a reflection of the size and representative capability of the training data. From the results on set A, it is evident that using more parameters in the combination process can produce more optimal results, provided the parameters are accurate reflections of the data set. For this reason, the optimal values of $F(\beta = 10)$ are 94.274, 95.396 and 95.977 for majority vote, genetic algorithm, and Bayesian 1 respectively, and this ordering agrees with the extent to which the combination methods are modeled after set A. Consequently, for datasets with characteristics close to those of A, the same pattern of behavior would be expected. However, when these characteristics are not the same, adapting the parameters to set A can result in "overfitting", and this effect is observed in the results of set B, even though the 2 sets of data were collected together under the same conditions. For this second set, the optimal values of F are in reverse order to those of A, showing that it is more difficult to generalize as the number of parameters increases.

When we examine the results of the combination methods, it is evident that a training set of 13272 samples is insufficient to establish accurate values of $bel(i)$ for $0 \leq i \leq 9$, for the Bayesian method. These values are calculated from the confusion matrices of the classifiers, each of which has 100 entries (not counting the rejections) for the classification of numerals. The off-diagonal entries, representing erroneous identifications, usually have very low magnitudes. Paradoxically, the better the classifier, the smaller would be these entries. Consequently, $bel(i)$ would be partly based on rather scant evidence. This problem is especially serious when one recognizer makes a particular misclassification only in the test set. In this case, the corresponding factor in the numerator of $bel(i)$ (obtained from the training set) would be zero, and hence the combined decision cannot be correct regardless of the decisions of the other classifiers.

The problems arising from size and uniformity of the datasets have less pronounced effects on the genetic algorithm, because only a few parameters need to be determined from the training data. In our problem, only the seven optimal weights (one for each classifier) have to be defined. The less specific nature of the training process has also resulted in a closer proximity between the results of sets A and B.

Moreover, the genetic algorithm is capable of identifying dependencies among classifiers, by assigning lower weights to those that are less effective in influencing the group decision in the optimal direction. This is a very useful feature, since this kind of knowledge is not available when classifiers are designed and implemented. In general, it is a non-trivial problem to decide whether classifiers are independent, since there can exist many possible overlaps between feature sets and classification methods. Even after a recognizer has been trained and tested, and its individual performance is known, it is still difficult to know how significant a role it would assume as a member of a group. The genetic algorithm can help to obtain such information, which can lead to simpler and more efficient multiple classifier systems.

In terms of computing time involved, it is certain that genetic algorithms take much longer to train, in order to search for optimal weights. However, once these weights are obtained and applied to the system, then the weighted majority vote is much faster than the Bayesian combination at the recognition stage.

4. The Behavior-Knowledge Space (BKS) Method

A thorough analysis of the reason why most abstract-level combination methods (such as majority vote [2], Bayesian [4] and Dempster-Shafer [23]) require an assumption that classifiers behave independently of each other reveals that either they treat each classifier equally, or they derive probabilities from the confusion matrix of a single classifier. Naturally, both situations require the assumption that the decisions of classifiers are independent. To avoid this assumption, the information should be derived from a knowledge space which can *concurrently* record the decisions of all classifiers on each learned sample. Since this knowledge space records the behavior of all classifiers, we call it "Behavior-Knowledge Space". Simply speaking, the BKS method derives its final decisions from a behavior-knowledge space, which we will describe next.

4.1. *Behavior-Knowledge Space*

A Behavior-Knowledge Space (BKS) is a K-dimensional array of cells, where each dimension corresponds to the decision of one classifier. When the problem involves M pattern classes, each classifier has at most $M+1$ possible decision values chosen from the set $\Lambda = \{1, \cdots, M+1\}$. The intersection of the decisions of individual classifiers occupies one cell of the BKS, and each cell accumulates the number of incoming samples for each class. The cell which is the intersection of the classifiers' decisions of the current input is called the *focal* cell.

Table 7 gives an example of a two-dimensional BKS. In this table, cell (i, j) is the focal cell when $e(1) = i$ and $e(2) = j$, where $e(i)$ is the decision of classifier i. Each cell contains three kinds of data: (1) the total number of incoming samples, (2) the best representative class, and (3) the total number of incoming samples for each class. Learned samples distributed to one cell/class are called incoming samples of that cell/class.

Table 7. Two-dimensional Behavior-Knowledge Space.

BKS =

e(1) \ e(2)	1	⋯	j	⋯	11
1	(1,1)	⋯	(1,j)	⋯	(1,11)
⋮	⋮	⋮	⋮	⋮	⋮
i	⋮	⋮	(i,j)	⋮	⋮
⋮	⋮	⋮	⋮	⋮	⋮
11	(11,1)	⋯	(11,j)	⋯	(11,11)

Symbols used in defining a BKS are:

BKS = a K-dimensional behavior-knowledge space,

BKS$(e(1), \cdots, e(K))$ = a cell of BKS, where classifier 1 gives its decision as $e(1)$, $\cdots$, and classifier K gives its decision as $e(K)$,

$n_{e(1)\cdots e(K)}(m)$ = the total number of incoming samples belonging to class m in BKS$(e(1), \cdots, e(K))$,

$$T_{e(1)\cdots e(K)} = \text{the total number of incoming samples in BKS}(e(1), \cdots, e(K)),$$

$$= \sum_{m=1}^{M} n_{e(1)\cdots e(K)}(m), \tag{5}$$

$$R_{e(1)\cdots e(K)} = \text{the best representative class of BKS}(e(1), \cdots, e(K)),$$

$$= \{j \mid n_{e(1)\cdots e(K)}(j) = max_{1 \le m \le M} n_{e(1)\cdots e(K)}(m).\} \tag{6}$$

The following is an example of a two-classifier BKS, which illustrates a situation where the classifiers give different decisions. Suppose that for sample x, the decision of the first classifier is 4 and that of the second classifier is 9, i.e. $e(1) = 4$ and $e(2) = 9$. Obviously, the relevant focal cell is BKS$(4, 9)$. Suppose that for this focal cell, there are incoming samples only from classes 4 and 9, and they are given by

$$\begin{cases} n_{49}(4) = 15, \\ n_{49}(9) = 5, \\ n_{49}(m) = 0, \text{ when } m \notin \{4, 9\}. \end{cases}$$

Then, $T_{49} = 20$ and $R_{49} = 4$.

Interestingly, the semantic meaning of the BKS is clear. For example, in the above, it means that when classifier 1 recognizes x to be 4 and classifier 2 to be 9, there is a 75% probability that the input x belongs to class 4, and also a 25%

probability to class 9. Obviously, for any focal cell, if rejection is not allowable, then the class with the highest probability is the best and the safest to choose as the final decision.

4.2. *Two Stage Operations and Decision Rule*

The BKS method operates in two stages: knowledge modeling and decision making. The knowledge-modeling stage uses the learning set of samples with both genuine and recognized class labels to construct a BKS; then the values of $T_{e(1)\cdots e(K)}$ and $R_{e(1)\cdots e(K)}$ of each cell BKS$(e(1),\cdots,e(K))$ are computed by Eqs. (5) and (6). The decision-making stage, according to the constructed BKS and the decisions offered from the individual classifiers, enters the focal cell and makes the final decision by the following rule

$$E(x, e(1), \cdots, e(K)) = \begin{cases} R_{e(1)\cdots e(K)} & \text{when } T_{e(1)\cdots e(K)} > 0 \text{ and} \\ & \frac{n_{e(1)\cdots e(K)} R_{e(1)\cdots e(K)}}{T_{e(1)\cdots e(K)}} \geq \lambda, \\ M+1 & \text{otherwise}, \end{cases} \tag{7}$$

where λ is a threshold ($0 \leq \lambda \leq 1$) which controls the reliability of the final decision. The knowledge-modeling stage needs to be executed only once for the learning set of patterns, but the decision-making stage will be performed on each test pattern. Obviously, this decision is based on the Bayesian decision rule because it assigns a pattern to the class having the highest probability that the pattern belongs to the class.

Many good properties can be derived from this method. However, most of them exist only in the context that the classifiers' behavior stored in the behavior-knowledge space is unbiased. Detailed discussions of these advantageous properties can be found in [24].

4.3. *Results of BKS Method*

In this experiment, we compare the results of combining abstract-level classifiers by the Bayesian and BKS methods. For both of these methods, large amounts of data are required for the knowledge modeling stage. This requirement has already been shown for the Bayesian method in Sec. 3.3, and is due mainly to the need for determining the $M \times (M+1)$ entries of the confusion matrix for each classifier, where M is the number of pattern classes. For the BKS method, this need is even more acute, given that it involves $(M+1)^K$ focal cells each containing $M+2$ items of information, K being the number of classifiers to be combined. In order to obtain a large data set for knowledge modeling, a leave-one-out estimation [25] is adopted, so that an unbiased comparison can be performed.

With a leave-one-out estimation, all samples but one are learned, and then the unlearned one is tested; the same operation is repeated until every sample in the data set has been left out and tested. Experimentally, the leave-one-out estimation has been found to be approximately unbiased for any data distribution. However, usually it suffers from extremely excessive computation, as it computes the distribution for each repeated run. Fortunately, both the Bayesian and BKS methods store their information in matrices, thus only a small portion of these matrices should be updated in each run. Therefore, both methods can be efficiently implemented for this estimation.

For this experiment, three classifiers (called e_1, e_2 and e_3) were chosen as experts. These three classifiers are different from those described in Sec. 3, and they are selected for this experiment because they output measurement values as their decisions. For this reason, they are also suitable for experimentation in the combination method to be described in the next section. These classifiers had been trained on 5074 samples of the training set, and their individual performances on the combined sets A and B (22024 samples) are shown in Table 8. In this table, the "reliability" is also given, and it is defined as

$$Reliability = \frac{Recognition}{Recognition + Error}. \tag{8}$$

Table 8. Performance of individual classifiers on 22024 test samples.

	Recognition	Error	Reliability
e_1	89.77	10.23	0.8977
e_2	90.50	9.50	0.9050
e_3	91.67	8.33	0.9167

When the three classifiers are combined by majority vote, the recognition, error, and rejection rates obtained are 92.76%, 4.30%, and 2.94% respectively. For the Bayesian and BKS methods, threshold parameters are involved in making their decisions, and Table 9 and Fig. 4 show the tabular and graphical representations of the corresponding recognition performances for different threshold values. It is easy to observe that the BKS method performs better than the Bayesian method.

Table 9. Results of combining three classifiers with different thresholds.

(a) The Bayesian method

α	Recogn.	Error	Reject	Rel.
0.000	94.74	5.26	0.00	0.9474
0.200	94.26	4.99	0.75	0.9497
0.500	93.62	4.22	2.17	0.9569
0.700	93.07	3.26	3.67	0.9662
0.900	91.85	2.52	5.63	0.9732
0.950	89.74	1.97	8.29	0.9785
0.995	84.38	1.10	14.52	0.9871

(b) The BKS method

λ	Recogn.	Error	Reject	Rel.
0.00	95.01	4.27	0.72	0.9570
0.10	93.67	3.43	2.90	0.9646
0.30	91.41	2.01	6.58	0.9785
0.45	89.08	1.61	9.30	0.9822
0.60	87.60	1.14	11.26	0.9872
0.80	84.03	0.91	15.06	0.9893
0.95	81.30	0.85	17.84	0.9896

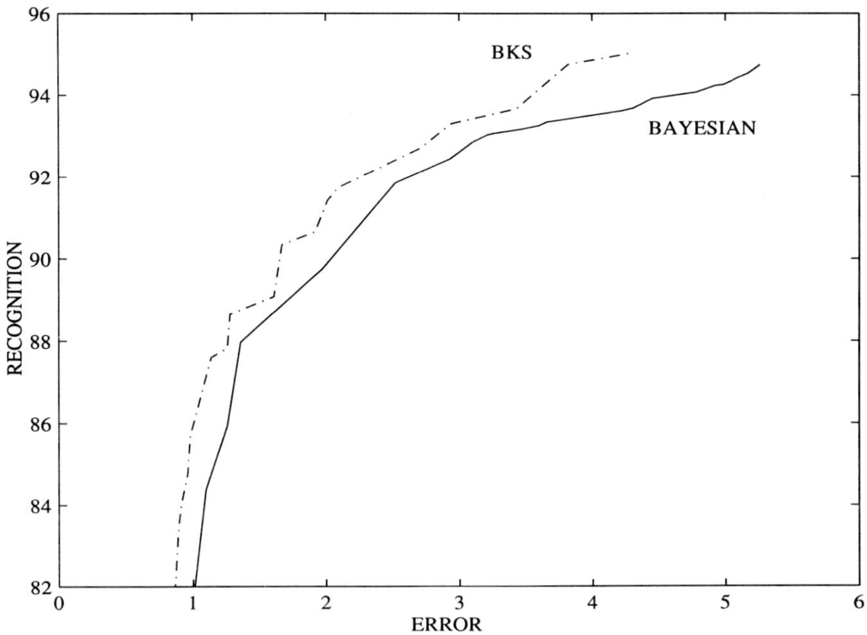

Fig. 4. Performances of two combination methods by a leave-one-out estimation.

5. Neural-Network Method for Measurement-Level CME

In a multi-expert recognition system, individual classifiers function not only as *character classifiers* but also as *feature extractors* [26]. When they function as feature extractors, their outputs become the features for later classification. From this point of view, measurement-level CME (Combination of Multiple Experts) becomes a generic pattern recognition problem, and the combination function E indeed turns out to be a pattern classification function. Intrinsically, neural networks are suitable to serve as combination functions, because they contain the following four valuable characteristics:

(i) they behave as collective systems;
(ii) they can infer subtle, unknown relationships from data;
(iii) they can generalize, meaning that they can respond correctly to patterns that are only similar to the original training data; and
(iv) they are nonlinear, that is, they can solve some complex problems more accurately than linear techniques do.

As a matter of fact, all the four characteristics specify exactly the desired functions of CME.

5.1. *Data Transformation*

We note that it is not valid to input the measurement values of classifiers to a neural net directly, because there are different meanings in the measurement values supplied by various classifiers. For example, some classifiers may output distance, while others output similarity or confidence. For a distance, the smaller the measurement value is, the closer a certain class corresponds to the input pattern x; but for similarity or confidence, they have the opposite meaning, i.e. the larger the measurement value is, the closer a class corresponds to x. Besides, even if the outputs of two classifiers have the same meaning, they may be in quite different scales, e.g., one may range from 0.0 to 1.0, while the other from 100 to 10000. Because of the possibility of different meanings and scales, a transformation or normalization becomes essential before measurement data can be combined effectively. Obviously, the object of data transformation is to convert the output of individual classifiers into a new form having the same meaning and scale. We call this new form *likeness* measurement, which means that the possibility of a class having a pattern is proportional to the likeness value which ranges from 0.0 to 1.0. Figure 5 shows a block diagram of this new consideration of measurement-level CME.

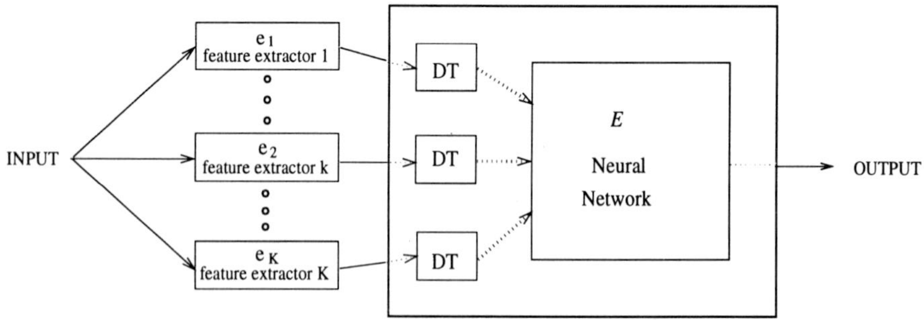

DT denotes data transformation

Fig. 5. A new look for measurement-level CME.

5.2. *Multi-layer Perceptron with Generalized Delta Rule*

Since the ground-truth class of each learning sample is known already, the learning process is supervised. Among different models of neural nets, the multi-layer perceptron with the *Generalized Delta Rule* (GDR) is chosen to serve as the combination function E because it has been used successfully in various pattern recognition applications with good recognition results. Figure 6 shows a three-layer perceptron with one input layer, one hidden layer and one output layer, where the transformed measurement values are fed to the input layer and $O_1, \ldots,$ and O_M are the output values of the output layer. Except for the input-layer nodes, the net input to each node is the sum of the weighted outputs of the nodes in the previous layer and the output of this node is the value produced by a nonlinear activation function f. The number of input layer nodes equals that of the total transformed measurement values, and the number of output layer nodes equals the total number of M classes.

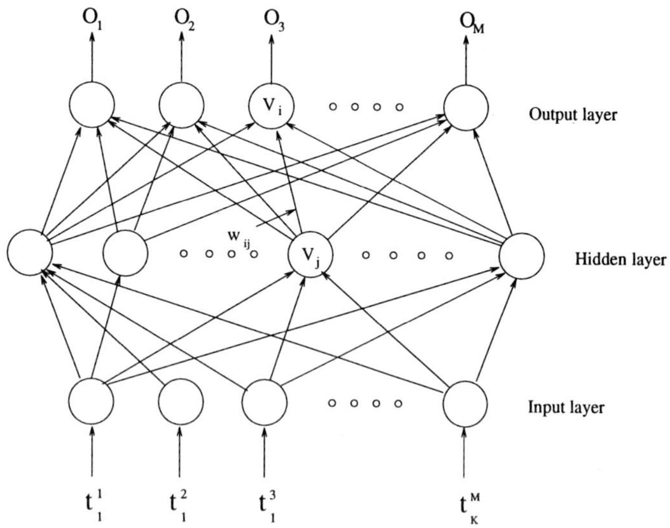

Fig. 6. Diagram of a Three-Layer Perceptron.

GDR is an iterative gradient algorithm designed to minimize the mean square error between the actual and desired outputs of a multi-layer perceptron. Through a set of learning samples, it can find the best weights w_{ij} automatically, enabling this network to exhibit optimal classification ability.

5.3. *Decision Rule*

After weight training, for the recognition of a testing sample x, the values of the output layer will generally fall into one of the three situations described below:

(i) x is classified: only one output node is active, which represents the recognized class of x;
(ii) The net is confused about the class of x: more than one output node are active;
(iii) The net fails to make a decision as to the class of x: none of the output nodes is activated.

Considering these three situations, the final decision rule is defined as

$$E(x) = \begin{cases} j & \text{if } O_j = Max_1 \text{ and } (Max_1 - Max_2) \geq \alpha, \\ M+1 & \text{otherwise,} \end{cases} \tag{9}$$

where α is a threshold, $Max_1 = max_{i\in\Lambda}O_i$, $Max_2 = max_{i\in\Lambda-\{j\}}O_i$. When α is large, only samples of situation (i) will be recognized, and samples of situations (ii) and (iii) will be rejected.

5.4. *Experimental Results for Measurement-Level CME*

A three-layer perceptron with 30 input, 20 hidden, and 10 output nodes was used in this experiment. The results of experts e_1, e_2, and e_3 (described in Sec. 4) are also used for this experiment. The results of set A described in Sec. 3 (consisting of 13,272 samples) was used to train the network, and the 8752 samples of set B are used to test the recognition performance of the trained net.

In Sec. 5.1, we noted that a data transformation is necessary before different measurement values can be applied as input to a multi-layer perceptron, and this point has been verified here. For this purpose, measurement values with and without data transformations were input to train and test the perceptron on Sets A and B, respectively. Without data transformations, the perceptron performs quite poorly and produces only 20.06% recognition on test set B with no rejections. Obviously, the perceptron could not converge with the original measurement values produced by the three classifiers. Using the data transformation functions proposed in [27], the results are shown in Table 10. By comparison with the results in Table 9, it can be seen that the perceptron performs much better than either the Bayesian or the BKS methods. For example, the perceptron achieves 97.58% recognition without rejections, while the corresponding rates for the Bayesian and BKS approaches are 94.74% and 95.01%, respectively. These results support the notion that measurement-level decisions are more informative than abstract-level ones, and that with a proper data transformation, measurement-level CME's can also perform better than abstract-level ones.

Table 10. Recognition results of measurement-level CME based on a multi-layer perceptron.

	Set A				Set B			
α	Recognition	Error	Reject	Rel.	Recognition	Error	Reject	Rel.
0.0	98.63	1.37	0.00	0.9863	97.58	2.42	0.00	0.9758
0.1	98.43	1.02	0.56	0.9898	97.19	2.06	0.75	0.9792
0.2	98.17	0.90	0.93	0.9909	96.65	1.55	1.79	0.9842
0.3	97.81	0.71	1.48	0.9928	95.68	1.20	3.12	0.9876
0.4	97.45	0.62	1.94	0.9937	95.08	0.99	3.93	0.9897
0.5	96.75	0.53	2.73	0.9946	94.14	0.79	5.07	0.9917

6. Concluding Remarks

In this chapter, we have described methods for combining the decisions of classifiers, at both the measurement and abstract levels. When these decisions are in the form of measurements (such as distances, similarities or confidence values), they can be regarded as pattern features. With this viewpoint, combining the decisions of the classifiers becomes a generic pattern recognition problem. As a result, many developed classification techniques can be used to produce the combined decision, and a three-layer perceptron was selected in this work. However, measurement-level decisions should be transformed properly so that the perceptron can be well trained. Experiments have shown that very high levels of performance can be achieved by this method, and this performance is much better than those obtained by the Bayesian and BKS approaches.

Abstract-level decisions contain less information, but they can be provided by any classifier, since measurement-level decisions (as well as decisions in the form of ranked lists) can be reduced to abstract-level ones, but not the other way around. For this category of decisions, we have discussed four combination methods which make different demands on the system. Among these methods, majority vote is the simplest to implement, and its requirements (of time and memory) are negligible. It has the further advantage that theoretical analyses can be made of this method, so that certain facets of its behavior can be deduced. For example, we can confidently predict that an even number $2n$ of classifiers would produce more reliable combined recognition results than can be obtained by adding another classifier, or by eliminating one of the classifiers. This conclusion is valid whether the classifiers are independent or not. For this and other properties of this method, the reader is referred to [16] and [17].

As we refine this voting process through increasingly specific knowledge acquisition and modeling, we consider (in that order) weighted majority vote, Bayesian formulation, and the BKS method. The demands on memory also increase in that order, until we arrive at the exponential requirements of the BKS method. More importantly, these methods also impose heavy demands on the quality and size of the training data. In Sec. 3.3, it has been noted that highly specific modeling requires large volumes of representative data; otherwise overfitting may occur, and the generalization capability would become diminished.

Increasing the size of the training set is a simpler concern, provided that time and effort are available for data acquisition. It is much more difficult to obtain a training set that is truly representative, in quality and complexity, of data in general. The problem is compounded by the fact that the degree of recognizability of patterns actually depends on the features and the classifier used. This being the case, it remains a challenging problem as to how one can partition a database into training and test sets that would be considered equal in difficulty for classifiers in general. It would be even more challenging to obtain or select a training set that could give optimal performance for unknown test data.

In conclusion, our experimental results demonstrate that combining the decisions of multiple classifiers does increase the reliability of the decision. The choice of the combination method would depend upon the size and representational capability of the training data, and the ability of the system to meet with the requirements of each method.

Acknowledgements

The authors wish to thank Raymond Legault, Ke Liu, and Christine Nadal of CENPARMI, Lawrence Pang of ICSA, together with Kai-Hsiang Chou and Lo-Ting Tu of ITRI, for the use of their recognition results.

This research was supported by the Natural Sciences and Engineering Research Council of Canada, the National Networks of Centres of Excellence program of Canada, and the FCAR program of the Ministry of Education of the province of

Québec. The second author is a member of the Chinese Character Recognition Project (project no. : 35N7100) supported by MOEA, Taiwan, R.O.C.

References

[1] C.Y. Suen, C. Nadal, T.A. Mai, R. Legault, and L. Lam, Recognition of totally unconstrained handwritten numerals based on the concept of multiple experts, *Proc. Int. Workshop on Frontiers in Handwriting Recognition*, Montréal, Canada, April 1990, 131–143.

[2] C.Y. Suen, C. Nadal, R. Legault, T. A. Mai, and L. Lam, Computer recognition of unconstrained handwritten numerals, *Proc. IEEE* **80** (1992) 1162–1180.

[3] P.D. Gader, D. Hepp, B. Forester, and T. Peurach, Pipelined systems for recognition of handwritten digits in USPS ZIP codes, *Proc. U.S. Postal Service Advanced Technology Conference*, 1990, 539–548.

[4] L. Xu, A. Krzyzak, and C.Y. Suen, Methods of combining multiple classifiers and their application to handwritten numeral recognition, *IEEE Trans. on Systems, Man and Cybernetics* **22** (1992) 418–435.

[5] T.K. Ho, A theory of multiple classifier systems and its application to visual word recognition, Doctoral Dissertation, State University of New York at Buffalo, 1992.

[6] T.K. Ho, J.J. Hull, and S.N. Srihari, Decision combination in multiple classifier systems, *IEEE Trans. on Pattern Analysis and Machine Intelligence* **16** (1994) 66–75.

[7] J.D. Tubbs and W.O. Alltop, Measures of confidence associated with combining classification results, *IEEE Trans. Systems, Man, and Cybernetics* **21** (1991) 690-692.

[8] J. Franke and E. Mandler, A comparison of two approaches for combining the votes of cooperating classifiers, *Proc. 11th Int. Conf. on Pattern Recognition*, The Hague, Netherlands, Sept. 1992, vol. 2, 611–614.

[9] E. Mandler and J. Schuermann, Combining the classification results of independent classifiers based on the Dempster/Shafer theory of evidence, in *Pattern Recognition and Artificial Intelligence*, eds. E.S. Geselma and L.N. Kanal (North Holland, Amsterdam, 1988) 381–393.

[10] D.S. Lee and S.N. Srihari, Handprinted digit recognition: A comparison of algorithms, *Pre-Proc. 3rd Int. Workshop on Frontiers in Handwriting Recognition*, Buffalo, USA, May 1993, 153–162.

[11] D.-S. Lee, A theory of classifier combination: the neural network approach, Doctoral Dissertation, State University of New York at Buffalo, 1995.

[12] T. Noumi, T. Matsui, I. Yamashita, T. Wakahara, and T. Tsutsumida, Results of the Second IPTP Character Recognition Competition and studies on multi-expert handwritten numeral recognition, *Proc. 4th Int. Workshop on Frontiers in Handwriting Recognition*, Taipei, Taiwan, Dec. 1994, 338–346.

[13] F. Yamaoka, Y. Lu, A. Shaout, and M. Shridhar, Fuzzy integration of classification results in a handwritten digit recognition system, *Proc. 4th Int. Workshop on Frontiers in Handwriting Recognition*, Taipei, Taiwan, Dec. 1994, 255–264.

[14] J. Franke, Statistical combination of multiple classifier adapted on image parts, *Proc. 1st European Meeting on Postal Technology (JET POSTE)*, Nantes, France, 1993, 566-572.

[15] J. Franke and M. Oberländer, Writing style detection by statistical combination of classifiers in form reader applications, *Proc. 2nd Int. Conf. on Document Analysis and Recognition*, Tsukuba, Japan, Oct. 1993, 581–584.

[16] L. Lam and C.Y. Suen, A theoretical analysis of the application of majority voting to pattern recognition, *Proc. 12th Int. Conf. on Pattern Recognition*, Jerusalem, Israel, Oct. 1994, 418–420.

[17] L. Lam and C.Y. Suen, Increasing experts for majority vote in OCR: theoretical considerations and strategies, *Proc. 4th Int. Workshop on Frontiers in Handwriting Recognition*, Taipei, Taiwan, Dec. 1994, 245–254.

[18] J.H. Holland, *Adaptation in Natural and Artificial Systems* (Univ. of Michigan Press, Ann Arbor, 1975).

[19] D. Goldberg, *Genetic Algorithm in Search, Optimization and Machine Learning* (Addison-Wesley, Reading, Mass., 1989).

[20] M. Srinivas and L. M. Patnaik, Genetic algorithms: A survey, *Computer* (1994) 17–26.

[21] K. Liu, Y.-S. Huang, and C. Y. Suen, Image classification by classifier combining technique, *Proc. SPIE Conf. on Neural and Stochastic Methods in Image and Signal Processing III*, 1994, 210–217.

[22] K.-H. Chou, L.-T. Tu, and I.-S. Shyu, Performance analysis of a multiple classifiers system for recognition of totally unconstrained handwritten numerals, *Proc. 4th Int. Workshop on Frontiers in Handwriting Recognition*, Taipei, Taiwan, Dec. 1994, 480–487.

[23] M.D. McLeish, P. Yao, and T. Stirtzinger, A study on the use of belief functions for medical expert systems, *J. of Applied Statistics* **18** (1991) 155–174.

[24] Y.S. Huang, Combination of multiple classifiers for the recognition of totally unconstrained handwritten numerals, Doctoral Dissertation, Concordia University, 1994.

[25] P.A. Devijver and J. Kittler, *Pattern Recognition – A Statistical Approach* (Prentice-Hall, London, 1982).

[26] F.F. Soulie, E. Vinnet, and B. Lamy, Multi-modular neural network architectures: applications in optical character and human face recognition, *Int. J. Pattern Recogn. Art. Intell.* **5** (1993) 721–755.

[27] Y.S. Huang, K. Liu, and C.Y. Suen, The combination of multiple classifiers by a neural network approach, *Int. J. Pattern Recogn. Art. Intell.* **9** (1995) 579–597.

CHARACTER RECOGNITION

Handbook of Character Recognition and Document Image Analysis, pp. 103–121
Eds. H. Bunke and P. S. P. Wang

CHAPTER 4

ISOLATED HANDPRINTED DIGIT RECOGNITION

JÜRGEN FRANKE*
Daimler-Benz AG, Research and Technology
P.O. Box 2360, 89013 Ulm, Germany

This chapter describes some experiments with the polynomial classifier approach. First we make some general remarks concerning this approach and explain some useful preprocessing algorithms. In the main part of this chapter, three different approaches are introduced and evaluated with some experiments on three different standardized data sets of handprinted digits. First, simple linear classifiers are constructed and tested in order to have a point of reference for the different data sets. With the knowledge gained from these linear classifiers, more sophisticated polynomial structures are constructed to enhance the performance. Secondly, the effects of feature extraction are demonstrated with the Karhunen-Loève transformation. In addition, some results in iterative learning are given. Third, for a fixed classifier approach different structures of the classifier system are explained and the performance effects are demonstrated on the same data sets. The efforts for these different systems in the training and testing stage are explained. At the end some ideas on how to reduce the computational effort for these systems are given.

Keywords: Feature extraction; Preprocessing; Polynomial classifier; Classifier structures; Classifier iteration; Hierarchical classifier; Net classifier.

1. Introduction

This contribution is not intended to give an exhaustive overview of the different preprocessing, feature extraction and classification techniques, but to show the effects of tuning one special classification approach and to enhance the performance by structuring the classifier system. The chosen classification approach is the polynomial classifier [1, 2, 3]; see also Chap. 2 of this book. The resulting effects can also be gained with other classifier approaches and are not inherent to the polynomial classifier, but to the selection of features and the design of the whole classifier system.

First we make some general remarks concerning the polynomial classifier approach and the preprocessing algorithms. The experiments are subdivided into three different parts. The first part explores different polynomial classifiers directly adapted to the used data sets, without any specific effort in structuring these classifiers. These experiments are conducted simply to establish a point of reference for the different data sets and the more sophisticated classifiers. Furthermore, they show how much performance can be gained by using quadratic instead of linear classifiers and how such incomplete quadratic polynomial structures can be constructed

*Phone: +49-731-505-2355, Fax: 0731-5054113, e-mail: franke@dbag.ulm.DaimlerBenz.COM.

in a recursive process.

Next, the effects of improving the feature extraction or, in this contribution, feature compression are shown. Here the principal axis (Karhunen-Loève) transformation is used and the effect of this linear sub-space transformation is shown in some experiments. With these classifiers the effects of classifier iteration are also shown.

Then different structured classifiers are described. Each of these classifiers is developed from the same statistical data and the difference between them resides only in the structure of the classifier system itself:

- single-stage classifier (one classifier for the whole problem)
- two-stage hierarchical classifier (four special classifiers on specific subsets of the whole problem and one classifier to activate the relevant classifiers which follow)
- complete hierarchical classifier (a tree of classifiers adapted to different subsets of the digits) [4, 5]
- network of two by two classifiers for each pair of classes

The different structures are described and their effects on the performance are demonstrated. In addition, the simplicity of arranging new classifier systems for the polynomial classifier approach is shown — which is one advantage of polynomial classifiers in contrast with other approaches, for example, multilayer perceptrons.

At the end of this chapter some ideas on how to reduce the evaluational effort of hierarchical and network classifiers are given.

All classifier approaches are tested with three different data sets of handprinted digits:

- CENPARMI data set [6, 7, 8, 9]: The training and test set contain 4000 and 2000 digits, respectively. The classes are equally distributed.
- BS data set [7]: This data set consists of 2213 digits and is available from USPS and is distributed by SUNY, Buffalo.
- ITRI data set: The training and test set contain 4000 and 3000 digits, respectively. The classes are equally distributed. This data set was made available to us by the Industrial Technology Research Institute, ITRI, Taiwan.

Since the BS data set could only be used as a test set, not every experiment was conducted with it.

2. The Polynomial Classifier

In this section the basic ideas of polynomial classifiers are explained and the preprocessing steps of our approach are explained. For a more detailed explanation of this approach and the similarities and differences to other approaches, see Chap. 2 by Kreßel and Schürmann in this book.

2.1. General Approach

The task of classifier adaptation is equivalent to finding the best function $\mathbf{d}$ according to a suitable optimization criterion. Common mean-square-optimization leads to

$$E\left\{|\mathbf{y}(\mathbf{x}) - \mathbf{d}(\mathbf{x})|^2\right\} \stackrel{!}{=} \min_{\mathbf{d}} , \qquad (2.1)$$

where $E\{\cdots\}$ stands for the expectation operator and is called residual variance, $\mathbf{d}(\mathbf{x})$ is the result of the classification, and $\mathbf{y}(\mathbf{x})$ is the desired output for the feature vector $\mathbf{x}$.

The approximation of the vector function $\mathbf{d}$ can be made, for example, by a functional classifier or a multilayer perceptron. The essential idea in the functional classifier approach is to connect the input variables in a non-linear way $f_j(\mathbf{x})$ — thus enhancing the features — but to use only a linear function for the output layer (matrix $\mathbf{A}$):

$$d_i(\mathbf{x}) = \sum_j a_{ji} \cdot f_j(\mathbf{x}) = \mathbf{a}_i^T \cdot \boldsymbol{f}(\mathbf{x}) \qquad \text{or} \qquad \mathbf{d}(\mathbf{x}) = \mathbf{A}^T \boldsymbol{f}(\mathbf{x}) .$$

Therefore, once the functional connections have been generated, the functional classifier becomes merely a linear classifier of the enhanced features $f_j(\mathbf{x})$, and therefore, the mathematics of solving linear problems can be applied. The only two steps required to determine the coefficients of $\mathbf{A}$ are, first, to compute the cross-correlation matrix $E\{\boldsymbol{f}(\mathbf{x})\,\mathbf{y}^T\}$ and the moment matrix of the enhanced features $E\{\boldsymbol{f}(\mathbf{x})\ \boldsymbol{f}(\mathbf{x})^T\}$ from the learning set, and second, to simply solve a set of linear equations:

$$E\left\{\boldsymbol{f}(\mathbf{x})\,\boldsymbol{f}(\mathbf{x})^T\right\} \cdot \mathbf{A} = E\left\{\boldsymbol{f}(\mathbf{x})\,\mathbf{y}^T\right\} . \qquad (2.2)$$

The solution to this equation is found only by matrix inversion which is a well understood mathematical technique. Therefore, no iterative learning is necessary. Additionally, it is assured that the coefficients are optimal for the actual task or learning set.

In this chapter, the function set used for the functional classifier approach is the set of polynomials. For example, $f_j(\mathbf{x}) = x_k$ is a linear feature; $f_j(\mathbf{x}) = x_k \cdot x_l$ is a feature of second order, specially a quadratic feature if $f_j(\mathbf{x}) = x_k^2$.

2.2. Preprocessing on the Character Objects

Before classification all characters are first processed with some algorithms to normalize the binary images. The algorithms mentioned below are described in greater detail in [10].

- **slant normalization**, to reduce the variability of the writing style
- **stroke width normalization**, to reduce the variability of the writing instrument
- **size normalization**, — 16*16 8-bit — to reduce the variability of the characters' size

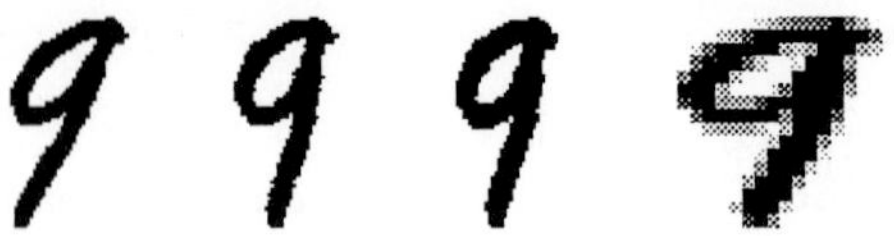

Fig. 1. The images show the effects of the different preprocessing algorithms. The images from left to right are: original, slant normalized, stroke width normalized, size normalized image.

The slant and stroke width normalizations are used only to reduce the variability of the characters. These algorithms are very helpful in improving the recognition performance of the adapted classifiers but they are not necessary for the polynomial classifier approach. Whereas size normalization is essential for this classification approach, because each input feature vector to this classifier must be of fixed length.

To improve a classifier it is also ingenious to divide the classes of the recognition task into sub-classes, where the distributions of the classes themselves are not unimodal. Then the designer of the system has to decide how many and which sub-classes should be defined. To solve this problem, suitable clustering techniques should be used. After determining the proper sub-classes, the data set has to be labelled according to this new sub-class definitions. This labeling of the image with their shape labels can be done automatically, for example, with quickly adaptable nearest neighbor classifiers trained with some images per shape class.

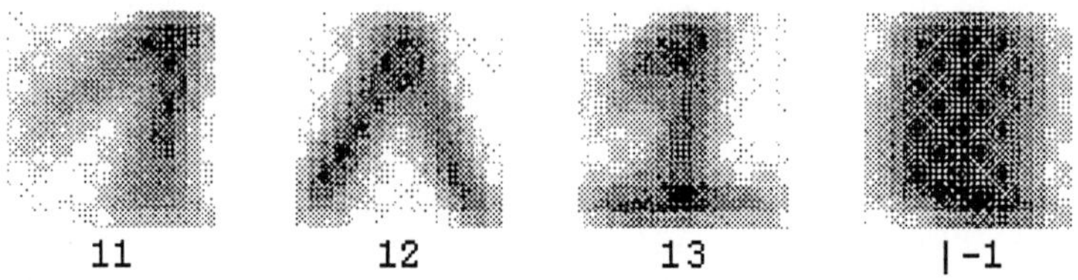

Fig. 2. Different shape classes of the digit 1. The images are the mean images calculated out of some hundred images per shape class. The image with the label |-1 is the size normalized image of the character 1 written as a single stroke.

For the character recognition process the classes are the different shape-classes of the characters (for example, for the class 1 in handwriting: shape-classes "11", "12", "13" and "|-1").

2.3. *Equivalence of Data Sets and Moment Matrices*

In the case of the functional classifier, since the main information for calculating the coefficients of the functional equation is condensed in the moment matrices $\mathrm{E}\{\boldsymbol{f}(\mathbf{x})\boldsymbol{f}(\mathbf{x})^T\}$, see Fig. 3 and $\mathrm{E}\{\boldsymbol{f}(\mathbf{x})\mathbf{y}^T\}$, the adaptation (inversion of the matrix) can be separated from the calculation of the statistical moments. Since there exists an equivalence between mixing sets of samples and mixing the corresponding moment matrices by linear combination with weights given by a mixing rule, all mixing processes can be done at the level of moment matrices. Computing moment matrices is the most time consuming part of the adaptation procedure. Therefore, parallel to the database of raster images of the learning samples, a corresponding database of moment matrices is established from which the matrices for Eq.(2.2) can be obtained with little effort.

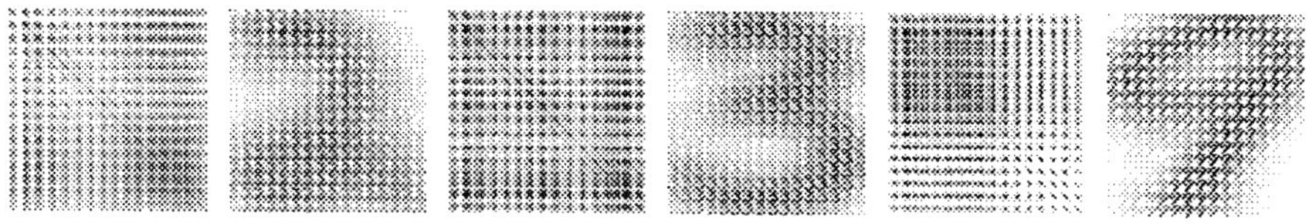

Fig. 3. The images show the class specific matrices $E\{f(\mathbf{x})f(\mathbf{x})^T\}$ of the digits 2, 3 and 9. The matrices are calculated to adapt the classifier PCL256_SW13_CP, see Sec. 3, using the CENPARMI learning set. First each matrix is shown in its natural representation and then in a reordered form, showing that the statistical information stored in the matrices has some kind of self similarity.

Therefore, this time consuming task can be done independently of the current problem, if the statistical moments are calculated for each class separately. Those matrices can be used for multiple tasks (e.g. numeral classifier, combined classifier for numerals and letters) and for different structural approaches (e.g. single-stage or hierarchical classifiers). Consequently functional classifiers for special purposes can be calculated very quickly and cost effectively.

This advantage of mixing the moment matrices for different system requirements is used in Sec. 4.

3. Unstructured Polynomial Classifiers

In this section only the *simplest* kinds of polynomial classifiers are described. In this context *simplest* means that the structure of the classifier is a single-stage classifier. Therefore, the classification task is done by a polynomial classifier with as many equations as there are distinctive classes — for the case of digits see Fig. 4.

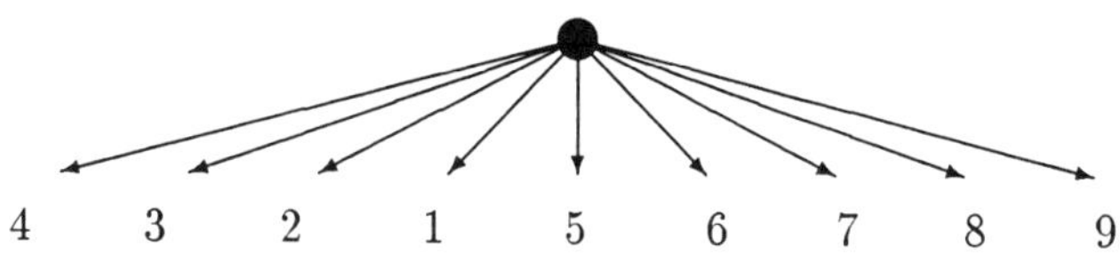

Fig. 4. Single-stage classifier for the digits 1,...,9.

In all the experiments described in this section the classifiers were adapted to (and only to) the CENPARMI or the ITRI training set respectively. In Table 1, the results of four different classifier approaches are given first. These classifiers are described in [6]. They are not polynomial classifiers but four different rule based systems with sophisticated and totally different preprocessing procedures generating skeletons and contours, and some with relaxation techniques.

The first adapted classifier is called PCL256_CP. It is a **P**olynomial classifier which is **C**omplete and **L**inear in **256** features — the 16*16 size normalized gray value pixels after the slant and size normalization — adapted to the **CenParmi** data set. This classifier is the basic reference for the recognition task. Because this classifier is a pure linear classifier the performance is not very good, see Table 1. To improve this classifier a **S**troke **W**idth normalization is added to the preprocessing. The aim of the procedure is to standardize the digits to a stroke width of about **13**% relative to the width respectively the height of the character. Therefore, the parameters for the dilation or erosion to obtain this stroke width are different in the horizontal and vertical directions. As a result, after size normalization the stroke width of the character is the same in each direction. To these digits again a **C**omplete **L**inear classifier in **256** features (PCL256_SW13_CP) is adapted. This additional preprocessing has the effect of reducing the forced error rate by about 0.6%, see Table 1.

In all the tables in this chapter the following abbreviations are used:

- **Sub**: Substitution, error rate percent;
- **Rej**: Rejection, percentage of digits which were rejected;
- **Rec**: Recognition = $(100. - Rej - Sub)$, percentage of correctly recognized digits;
- **Rel**: Reliability = $\frac{100.*Rec}{Rec+Sub}$, rate of correctly recognized digits among the accepted;
- **Forced**: Substitution rate on the test set without rejection;
- **Training**: Forced substitution rate on the training set.

Table 1. Performance of the one stage classifiers on the CENPARMI data set.

Classifier	Rec	Sub	Rej	Rel	Forced	Training
Expert #1	86.05	2.25	11.70	97.45		
Expert #2	93.10	2.95	3.95	96.98		
Expert #3	92.95	2.15	4.90	97.74		
Expert #4	93.90	1.60	4.50	98.32		
PCL256_CP	85.60	3.90	10.70	95.64	9.25	6.90
PCL256_SW13_CP	85.45	3.15	11.40	96.45	8.60	6.20
PIQ255_SW13_CP	93.10	3.10	3.80	96.78	5.10	2.95
PIQ1072_SW13_CP	94.45	1.20	4.35	98.75	3.05	0.13

3.1. *Sophisticated Polynomial Structures*

To improve the polynomial classifiers significantly, higher order terms for the polynomial approach are required. But there is a problem when the number of features is large. If the functional approach is polynomial, the number of terms will grow with the power of the polynomial. For example, for $n = 200$ input variables the number of terms of the complete quadratic polynomial is $\binom{n+2}{2} = 20301$, which is prohibitive for practical purposes.

Therefore, there are two possible solutions for the designer of a classifier:

- choosing only the relevant combinations of the features or
- applying some feature extraction methods

The polynomial classifier approach offers a tool for determining a rank order of the features showing how important each term is for the classification task [1] . This rank order can be computed by solving Eq. (2.2). Therefore, the classifier designer gets useful information about the importance of the features (combinations) and it is possible to calculate a linear classifier first and using its rank order list for the construction of incomplete quadratic (or higher degree) polynomials with very good performance. In the experiments the following classifiers are structured by this method due to the rank order list of PCL256_SW13_CP:

- **PIQ255_SW13_CP** is a **P**olynomial classifier with an **I**ncomplete **Q**uadratic structure of length **255**. 128 pixel features are selected based on the rank order of the linear classifier and 127 products of second degree are built due to the rank of each feature. The combinations are only allowed between different pixels and therefore, no product of a pixel with itself are built.
- **PIQ1072_SW13_CP** is constructed like PIQ255_SW13_CP but with different parameters. Its length is **1072** — 175 linear terms and 897 quadratic terms. The polynomial structures are shown in Fig. 5.

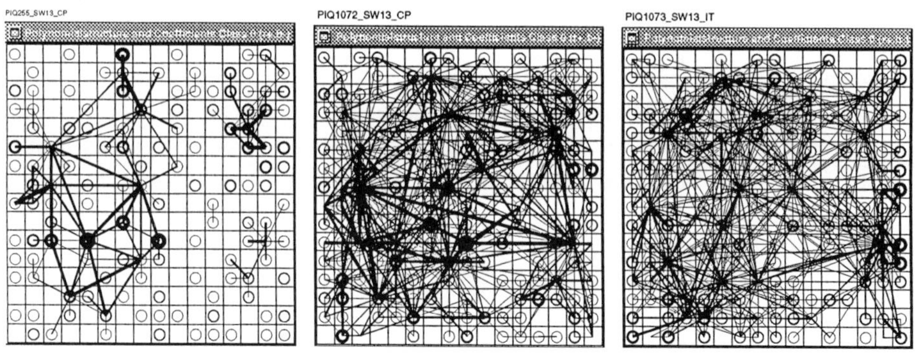

Fig. 5. Plot of the polynomial structure of classifier PIQ255_SW13_CP, PIQ1072_SW13_CP and PIQ1073_SW13_IT. In a 16 × 16 raster image, circles indicate the used linear terms and lines indicate the quadratic terms (products) of the corresponding picture elements of the polynomial. The line thickness corresponds to the magnitude of the coefficients of the adapted classifier for the equation for the digit 0.

PIQ255_SW13_CP is constructed to show the effects of using quadratic terms in the polynomial without enlarging the length of the polynomial. Compared with the linear classifier the error rate drops from 6.20% to 2.95% and from 8.60% to 5.10% on the learning and test set respectively. This improvement is gained without adapting more parameters to the classification. The effect is only due to the quadratic combinations. The classifier PIQ255_SW13_CP is now better than Expert #1 and with a threshold to produce a reject rate of 3.80%, its results are comparable to

Expert #2, see Table 1.

The classifier PIQ1072_SW13_CP improves the results on the learning set dramatically. The error rate on the test set is reduced by more than 2% but the relation between the error rates on the learning and test sets is no longer acceptable — the ratio between both error rates is too high. The classifier starts to be overadapted, see [11]. Nevertheless, the results of the classifier are better than Experts #3 and #4. We expect that this classifier could be improved on a larger learning set, like in [11], where the learning set consists of 1000 characters per class instead of 400 per class in this data set.

Table 2. Performance of the one stage classifiers on the ITRI data set.

Classifier	Rec	Sub	Rej	Rel	Forced	Training
PCL256_SW13_IT	84.83	4.57	10.60	94.90	9.94	5.23
PCL256_SW13_I_IT	86.00	3.70	10.30	95.90	8.03	6.24
PIQ1073_SW13_IT	93.41	1.45	5.14	98.50	3.40	0.27

In Table 2 similar results are given on the the ITRI data set. Again the classifier PCL256_SW13_IT is a linear classifier. In this case it is trained on **ITRI** training set and tested on the test set. The difference in the forced recognition error rate is much higher than for the same structured classifier on the CENPARMI data set. This indicates that the training and test sets are not very similar in their statistical appearance. Therefore, we performed an experiment vice versa. PCL256_SW13_I_IT (I=Inverse) is a linear classifier trained on the **test** set and tested on the **training** set. This classifier has a much better relation in the performance between its test and training sets. This experiment confirms the above assumption on an unequal distributed data set.

Again an incomplete quadratic classifier (PIQ1073_SW13_IT) is constructed which has a much better performance but is already in an overfitting state.

3.2. *Polynomial Classifiers Applied to Extracted Features*

The second possible solution to the problem of increasing polynomial length is to compress the feature vector. For this purpose we use the principal axis transformation [12]. By using this transformation we conducted some experiments reducing the number of features (measurements) from 256 down to 10 and up to 40 in steps of 10. The principal axis transformation was adapted to the training set of 4000 characters.[a] The principal axis transformation is determined by the eigenvectors corresponding to the highest eigenvalues for the moment matrix $\mathrm{E}\{\boldsymbol{f}(\mathbf{x})\boldsymbol{f}(\mathbf{x})^T\}$. The underlying optimization criterion for the principal axis transformation is to minimize the reconstruction error on the learning set. Each of these eigenvectors itself can be interpreted as a feature vector in the feature space. For the used 16*16

[a]The principal axis transformation was calculated out of the statistical data of the classifier PCL256_SW13_CP.

size normalized features the eigenvectors can be plotted as size normalized images, see Fig. 6.

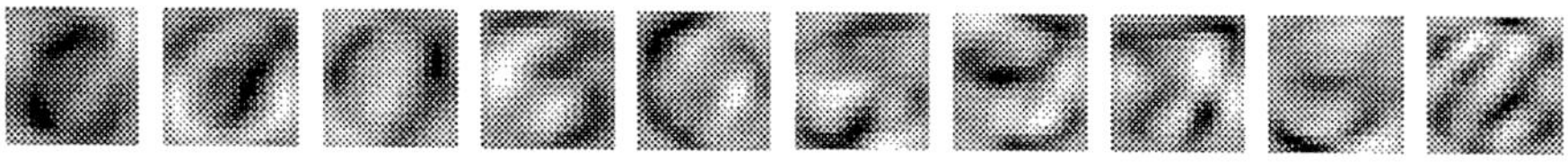

Fig. 6. The first 10 eigenvectors printed as size-normalized images.

It can be seen that the first eigenvectors try to extract global features like circles and strokes — in the middle or at the border of the image — and that the higher eigenvectors try to extract features with higher frequencies in the image resulting in a more detailed reconstruction ability, see Fig. 7.

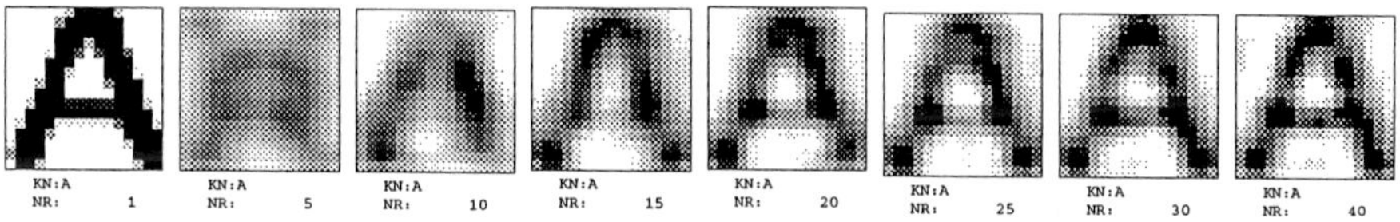

Fig. 7. Original image and different reconstructed ones, using 5, 10, 15, 20, 25, 30 and 40 principal axis transformed features.

The names of the classifiers in Table 3 indicate that the used polynomial classifier is Complete **Q**uadratic in, for example, **10 P**rincipal **A**xis **T**ransformed features calculated out of the **C**EN**P**ARMI training set. The number in the second half of the name, for example, 65, 230, is the length of the polynomial. All these classifiers include the stroke width normalization in the preprocessing.

Table 3. One stage classifiers with principal axis transformation on the CENPARMI data set.

Classifier	Rec	Sub	Rej	Rel	Forced	Training
PIQ255_SW13_CP	93.10	3.10	3.80	96.78	5.10	2.95
CQ10PAT_65_CP	92.25	4.10	3.65	95.70	5.80	6.00
CQ20PAT_230_CP	95.10	1.05	3.85	98.90	2.70	1.55
CQ30PAT_495_CP	95.45	0.90	3.65	99.10	2.65	0.65
CQ40PAT_860_CP	95.55	0.65	3.80	99.30	2.40	0.13
PIQ1072_SW13_CP	94.45	1.20	4.35	98.75	3.05	0.13

Compared with PIQ255_SW13_CP as a reference, CQ20PC_230_CP has a much better performance, although the length of the polynomial is 10% shorter. Also CQ30PAT_495_CP is comparable to PIQ1072_SW13_CP, even though the length of the first polynomial is less than half that of the second. This comparison shows that good feature selection methods have a high impact on the recognition performance of a classifier system, and therefore, the use of principal axis transformed features is preferable to creating sophisticated polynomial structures in the original feature space.

It can be observed that, when the length of the polynomial increases by a certain factor, the error rate decreases by about the inverse of that factor on the learning set, see Table 4. This is no more valid for the test set, because the learning set is too small and overadaptation takes place.

Table 4. Effects of the length of the polynomial on the error rate.

Ratio of the length	Inverse ratio of the error rate	
	On learning	On testing
230/ 65 = 3.54	6.00/1.55 = 3.87	5.80/2.70 = 2.15
495/230 = 2.15	1.55/0.65 = 2.39	2.70/2.65 = 1.02
860/495 = 1.74	0.65/0.13 = 5.00	2.65/2.40 = 1.10

In the limit, when the polynomial length reaches the number of samples of the learning set, the error rate will be zero, because it is possible to determine the coefficients in the polynomial in such a way that the classifier can estimate the class membership exactly — if each feature vector is unique. This would lead to an arbitrarily large inverse ratio on the learning set, which is normally impossible for the test set.

But for large learning sets, as a rule of thumb, we can estimate the effects on the error rate by the inverse ratio of the polynomial lengths for a given classifier approach.

Similar results can be observed for the ITRI data set, see Table 5.

Table 5. One stage classifiers with principal axis transformation on the ITRI data set.

Classifier	Rec	Sub	Rej	Rel	Forced	Training
PCL256_SW13_IT	84.83	4.57	10.60	94.90	9.94	5.23
CQ10PAT_65_IT	92.79	3.34	3.87	96.50	5.51	3.67
CQ20PAT_230_IT	94.73	1.33	3.94	98.60	3.07	1.38
CQ30PAT_495_IT	95.36	0.97	3.67	99.00	2.60	0.60
CQ40PAT_860_IT	95.56	0.87	3.57	99.10	2.44	0.25
PIQ1073_SW13_IT	93.41	1.45	5.14	98.50	3.40	0.27

3.3. *Improving Classifiers by Iteration*

In the above experiments the classifiers are only adapted to the learning set without any special effort to improve the performance by retraining the classifiers with problem characters. Problem characters are those that are misclassified by the original classifier or generate bad confidence values. By extracting such patterns from the learning set by the adapted classifier itself it is possible to retrain the classifier with the following procedure.

The extracted patterns are used to calculate the moment matrices

$E\{f(x)f(x)^T\}$ and $E\{f(x)y^T\}$. These matrices — M_P —, which store the statistical information about the problem set, can be mixed linearly with the matrices calculated out of the original learning set — M_O. The resulting iterated matrices — M_I — can be used to calculate the iterated classifier. Because this mixing procedure

$$M_I = (1 - \alpha)M_O + \alpha M_P$$

is a pure linear procedure, it is easy to calculate several classifiers with different weights α. In Table 6 the effects of several iterations are shown.

Table 6. Iterated classifiers on the CENPARMI data set.

Classifier	Forced	Training
CQ10PAT_65_CP	5.80	6.00
CQ10PAT_65_CP_IT20	5.45	5.35
CQ10PAT_65_CP_IT30	5.20	5.18
CQ10PAT_65_CP_IT40	5.15	4.98
CQ20PAT_230_CP	2.70	1.55
CQ20PAT_230_CP_IT10	2.40	1.03
CQ20PAT_230_CP_IT20	2.45	0.78

The iteration is done with different weights indicated, for example, by IT20: The problem matrices are mixed with the original matrices with the weight of 20%, i.e., $\alpha = 0.2$. The weights in the mixing procedure are relevant for the performance of the resulting classifiers. As we can see the optimal weight is much smaller for the classifier with the better original performance corresponding with the length of the polynomial. For the classifier CQ10PAT_65_CP the weight can be chosen to be 40%, whereas for the classifier CQ20PAT_230_CP the weight should be chosen as about 10%. Nevertheless the iteration technique also depends very much on the size of the training data set, i.e., the size of the problem data set, which is much smaller for the classifier CQ20PAT_230_CP.

In this section we have seen that there are many tools to improve classifiers, and the designer of a system has to decide which tools are appropriate for his problem:

- selection of quadratic features by the rank order of a preadapted classifier
- feature compression, e.g. principal axis transformation
- iteration techniques (reinforced training with problem characters)

4. Different Structures for Classifier Systems

In this section the effects of structuring the same polynomial classifier are shown. For these classifiers, identical statistical data and the same polynomial structure is used. As mentioned in Sec. 2.3, the statistical information can be stored in the moment matrices $E\{f(x)f(x)^T\}$ and $E\{f(x)y^T\}$ and can be mixed based on the requirements of the problem. This advantage of the polynomial classifier approach is used in this section.

4.1. *Higher Order Classifiers by Classifier Structures*

The functional classifier approach as sketched in Sec. 2 leads to a single-stage procedure. The feature vector **x** is mapped into the discriminant vector **d** in one step, see Fig. 4.

There are, however, a number of different structural approaches to distinctively improve the performance of such systems. By partitioning the set of classes into subsets, a hierarchy of classifiers can be constructed, which operate very much like a decision tree; see Fig. 8.

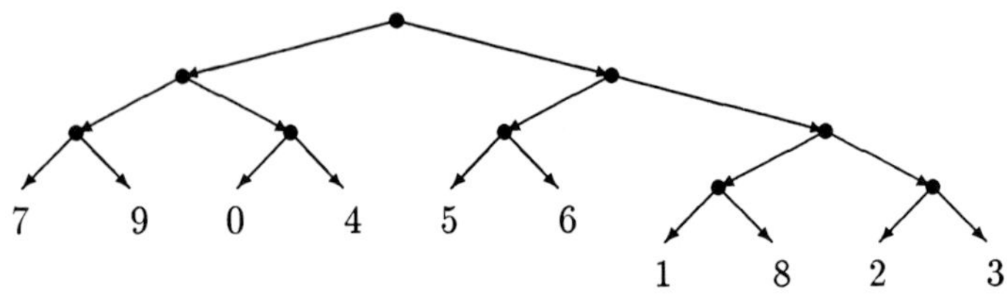

Fig. 8. Binary tree classifier.

The main difference with decision trees in the traditional sense is that conditional probabilities rather than hard decisions are forwarded along the tree branches. These probabilities are repeatedly multiplied from one level to the next and estimations of the a posteriori probabilities are obtained at the leaf nodes [5]. If the classification at the nodes is done with the true a posteriori probabilities instead of their approximations, the order of hierarchy will have no influence on the performance at all. Unfortunately, this is impossible in practice. Therefore, there is a potential for improving recognition performance by partitioning the overall problem into subproblems and designing suitable hierarchical structures. This is due to the fact that the majority of node classifiers within the classifier tree operate on small subsets of the set of classes and improve recognition accuracy by specialization. This kind of classifier architecture has one additional advantage: It allows the computational effort to be directed to those branches of the classifier tree which most probably will generate the relevant results and thus, saving redundant operations.

A close look at the operations of the hierarchical classifier shows further that multiplying the estimations along the paths is equivalent to multiplying the individual discriminant functions. This results in overall functional (polynomial) discriminant functions of a much higher degree than those from which the classifier tree was composed [4], see Fig. 9.

However, since the coefficients are optimized separately at each node, the multiplied functionals do not necessarily have the overall-optimal coefficients. This emphasizes the importance of properly structuring the classifier hierarchy by clustering procedures. In the experiments described in Sec. 4.2 we used the complete link clustering procedure performed on the mean feature vectors for the classes to be distinguished.

The effort of calculating such hierarchical classifiers is not much higher than that

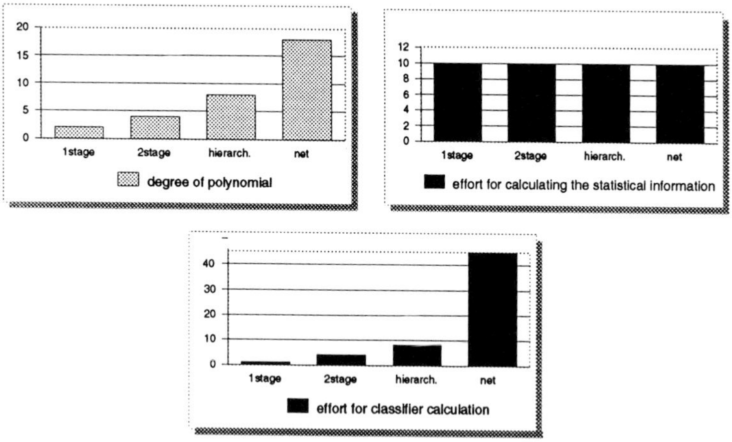

Fig. 9. Growth of the theoretical degree of the overall polynomial function for different structures. The number of classes is 10 and the degree of the polynomials in each node is 2 in this example. In the second and third diagram the effort for calculating the relevant statistical information for different structured classifiers and the number of matrix inversions to calculate the needed classifiers are given.

of calculating a single-stage classifier, because the relevant statistical information is condensed in the moment matrices which only have to be mixed along the path of the tree from the bottom up to the top [1], see Fig. 9.

The number of matrix inversions merely grows up, see Fig. 9, but this is not the most time consuming task during the classifier generation. Therefore, it is possible to generate very high dimensional polynomials with small effort.

An approach between the fully hierarchical binary tree and the single-stage classifier is the two-stage classifier, widely used in our address-reading machines, see Fig. 10. It uses the advantages of building up a hierarchy of specialized classifiers (experts or specialists for a subtask of the overall problem). On the other hand, the structure is not as complicated as the fully hierarchical tree. Also for this kind of classifier, the use of good clustering algorithms is essential for the performance.

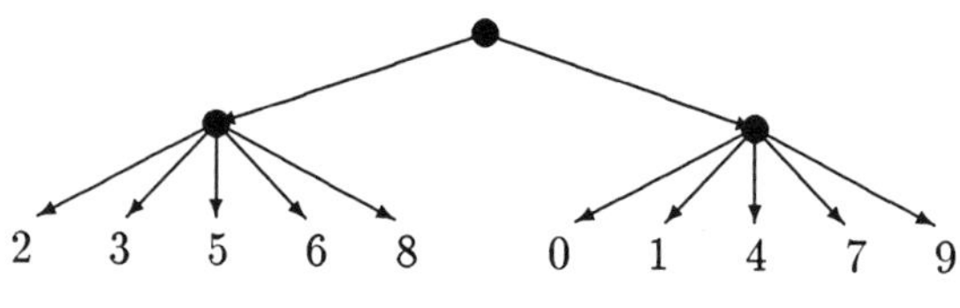

Fig. 10. Two-stage classifier.

Another special classifier design is the classifier net, constructed from as many individual classifiers as there are pairs of classes [13], see Fig. 11. Each of the individual classifiers is trained on the class specific subsets of the overall learning set in order to separate the two corresponding classes. The estimations of these individual classifiers are mathematically combined in order to yield again estimations for the a posteriori probabilities. Since the number of two by two classifiers grows quadratically $\binom{k}{2}$, redundant pairs must be detected and removed before the whole net is calculated.

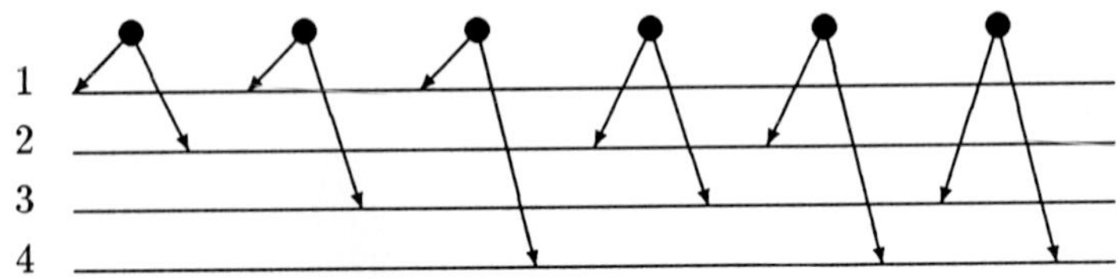

Fig. 11. Net of two by two classifiers for the digits 1, 2, 3 and 4. All of the classifiers' estimations have to be combined in an appropriate way.

For both approaches — hierarchy and net — the possibility of generating the necessary moment matrices at each node by mixing them out of the class specific matrices is essential to efficiently calculate different classifier structures.

4.2. *Experiments with Different Structured Classifiers*

In the experiments described in this section the four different structural approaches (single-stage, two-stage, fully hierarchical, net) were realized using a very large data set (over 1,000,000 digits) taken from a real application (address reading machines from AEG-ElectroCom for the United States Postal Services, USPS). Therefore, these classifier systems were not adapted to any of the above data sets at all. Based on this identical training set, five types of classifiers were developed and compared. They all were created by mixing the class specific moment matrices due to the structure of the system. Only in the 6th experiment was the preprocessing changed for the test set, but not on the learning set: We merely added the stroke width standardization SW13 which we also used in the experiments described in the previous section.

The 5 (6) different classifiers (experiments) are:

- **SHAPE_1st** is a single-stage classifier adapted to 22 predefined shape-classes. Therefore, the 10 meaning-classes of the digits were analyzed and for some classes up to 4 sub-classes are defined, see Sec. 2.2.
- **SHAPE_2st** is a two-stage classifier. The 22 shape-classes are divided into 4 sub groups and the classifier at the top has to decide which sub-classifier(s) has (have) to be activated for the current pattern.
- **SHAPE_H**[b] is a fully hierarchical classifier for the 22 shape-classes as shown in Fig. 8.
- **MEAN_N** is a net of 45 two by two classifiers, which are adapted to each pair of the 10 meaning-classes, mixing the moment matrices of different shapes first.

[b]The result for SHAPE_H on the training set is not available.

- **SHAPE_N** is a net of 231 two by two classifiers. Each classifier is adapted to each combination of shape-classes. Even the classifiers to distinguish between two shape-classes of the same numeral are calculated.
- **SHAPE_N_SW13** is identical to SHAPE_N. Only the stroke width standardization is added to the preprocessing, but the classifier is not adapted to this.

Although these classifiers have no information about the learning set of the CENPARMI data set, the obtained results are comparable with Expert #4 or even better, see Table 7. But this is due to the use of our very large database for the learning procedure. Therefore, the classifiers have a very good generalization ability.

Table 7. Different structured classifiers tested with the CENPARMI data set.

Classifier	Rec	Sub	Rej	Rel	Forced	Training
			CENPARMI data set			
SHAPE_1st	93.65	2.15	4.20	97.76	4.45	4.88
SHAPE_2st	93.75	1.75	4.50	98.17	4.00	4.00
SHAPE_H	93.95	1.95	4.10	97.97	3.80	
MEAN_N	94.75	1.40	3.85	98.54	3.05	2.58
SHAPE_N	95.05	1.30	3.65	98.65	2.70	2.70
SHAPE_N dif. threshold	94.40	1.20	4.40	98.75	2.70	2.70
SHAPE_N_SW13	94.45	1.10	4.45	98.85	2.45	2.58
			BS data set			
SHAPE_1st	91.87	1.08	7.05	98.84	3.62	
SHAPE_2st	93.45	1.04	5.51	98.89	3.25	
SHAPE_H	94.67	1.08	4.25	98.87	2.76	
MEAN_N	95.26	1.08	3.66	98.87	2.17	
SHAPE_N	97.02	0.90	2.08	99.08	1.67	
			ITRI data set			
MEAN_N_SW13	85.58	4.44	9.98	95.07	9.51	7.57
SHAPE_N_SW13	85.80	4.10	10.10	95.44	9.38	7.03

It is very interesting to observe that the error rates for some classifiers on the training and test sets are identical: SHAPE_2st 4.00 % and SHAPE_N 2.70%. This confirms that both, training and test sets, were drawn from the same statistical process and have the same properties, and that the classifiers are not adapted to the training set.

When we compare only the forced recognition rates of these classifiers — different only in their structure — we see a large improvement from a 4.45% error rate down to 2.70%. Therefore, applying classifiers of advanced structural design to a given problem improves the performance without any additional effort on feature selection or enlarging the data set.

The results on the BS data set of this experiments are also shown in Table 7. Again, the performance improves with the improved complexity of the classifier structure. The results are graphically shown in Fig. 12.

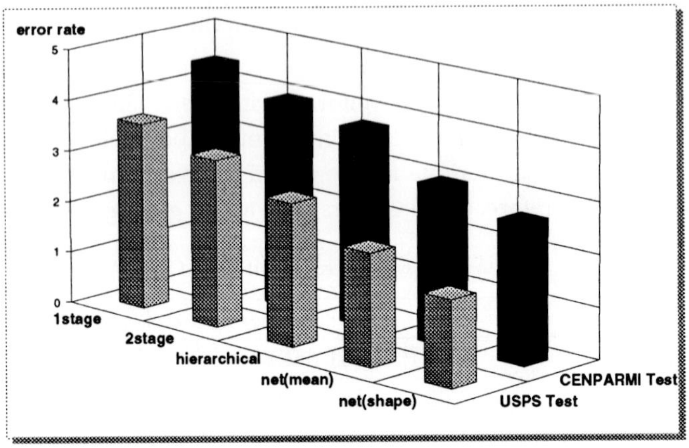

Fig. 12. Performance on the CENPARMI and the USPS test set

For the ITRI data set we only did experiments with the net classifiers, see Table 7. Because the CENPARMI and BS data sets are from US postal applications and the proprietary classifiers were adapted for the USPS market, the results are very similar and good. The ITRI data set was from a totally different cultural background, and therefore, the results are not very good. In this case an iteration of the existing classifiers to the different writing styles would be necessary.

5. Reduction Strategies

Using nets of two by two classifiers has the problem of increasing computational effort. The number of classifiers grows quadratic with the number of classes. In the following, some strategies to reduce the effort in the classification phase are described and tested by applying the classifier MEAN_N on the whole CENPARMI data set (6000 characters).

To understand the different strategies some remarks are necessary to show how the different classifiers are combined in a classifier network. At the beginning each class gets the initial net estimation of 1 (100% probability of being possible). Then one node in the network is evaluated and its estimations are multiplied with the net estimation. In each step, therefore, the current net estimation is gained by multiplying the node estimation of the current node with the preceding net estimation. Therefore, if some node generates an estimation of 0.0 for one class then the corresponding net estimation is set to 0.0 as well and this class has furthermore no possibility getting a higher net estimation at all.

The following strategies to reduce the number of evaluated nodes are investigated:

- **all:** All nodes are evaluated.
- **all≠0:** If the current net estimation for a special class is under a certain limit (0.001 in these experiments), then all corresponding classifiers (those who generate estimations for this class) are no longer evaluated at all.
- **max.net.est.:** After an arbitrary first classification, that node is evaluated next, which corresponds to that class with the current maximum net estimation. The procedure stops, when all nodes are evaluated which correspond to the current maximum net estimation. During this process the current max.net estimation class changes, but at the end all nodes for the winning class are evaluated.
- **cur.win.:** After an arbitrary first classification, that node is evaluated next, which generates estimations for the current winning class (winning in the preceding classification node), i.e., the class with the current highest net estimation is not investigated further, but the current winning class. This procedure runs only once through the list of pairs.

The results of the experiments are shown in Table 8. They demonstrate that it is possible to reduce the effort of net classifiers to the level of a one stage classifier (see Fig. 13) without losing any performance quality. The differences in the error rates are not significant.

Table 8. Classification results on the whole CENPARMI data set by reducing the effort in evaluating the net classifier MEAN_N; # eqs = mean number of evaluated nodes (= equations); err = forced error rate.

MEAN_N	All	All≠0	Max.net.est.	Cur.win.
# eqs	45.0	26.3	15.9	9.0
err	2.77	2.70	2.68	2.72

Also for hierarchical classifiers not the whole classifier tree is evaluated, but only those paths which still have the possibility to generate the maximum output. Therefore, the effort to compute an estimation for a given pattern can be reduced compared with a single stage classifier. Again it is essential that the number of

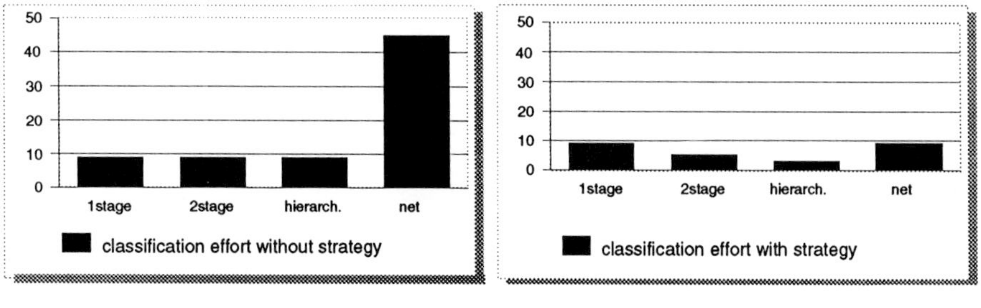

Fig. 13. Effort for applying the different structured classifiers with and without a processing strategy. The number of equations to be computed for a flat classifier is equal to the number of classes minus 1; note that the last equation is linearly dependent on the others.

equations to be evaluated in each node is equal to the number of classes minus 1. The last equation is linearly dependent on the others, see Fig. 13.

6. Conclusion

In this chapter different possibilities of constructing functional classifiers are considered. It is shown that quadratic combinations have much effect on the recognition quality. Besides, some experiments on feature compression showed that "good" features have an even higher effect on the performance. Nevertheless, a good design of a system includes some iteration techniques to force the adaptation procedure to learn problem characters. In the fourth section it is shown that arranging the same statistical information — stored in class specific moment matrices — and the same functional approach in different structured classifier systems has a significant impact on the performance of such a system. At the end some ideas are given to reduce the computational effort without losing performance.

Because of the modularity of polynomial classifier systems — the statistical information is stored separately from the pure adaptation — it is possible to adapt quickly to various specific problems. This advantage in flexibility for the functional classifiers is shown in the above experiments described in Sec. 4.

Nevertheless, the designer of a character recognition system has to decide between many system parameters to optimize his or her system.

References

[1] J. Franke, On the functional classifier, *Proc. of First Int. Conf. on Document Analysis and Recognition*, St. Malo, 1991, 481–489.

[2] J. Schürmann, *Polynomklassifikatoren für die Zeichenerkennung* (Oldenbourg Verlag München, 1977).

[3] J. Schürmann, A multi-font word recognition system for postal address readings, *IEEE Trans. on Comp.* **C-27** (1978) 721–732.

[4] J. Franke, Zur Entwicklung hierarchischer Klassifikatoren aus der entscheidungstheoretischen Konzeption, *5th Symp. of DAGM*, German Association on Pattern Recognition, Karlsruhe, 1983, 261–265.

[5] J. Schürmann and W. Doster, A decision theoretic approach to hierarchical classifier design, *Pattern Recognition* **17**,3 (1984) 359–369.

[6] C.Y. Suen, C. Nadal, R. Legault, T.A. Mai and L. Lam, Computer recognition of unconstrained handwritten numerals, *Special Issue of Proc. of IEEE* (1992).

[7] P.D. Gader, D. Hepp, B. Forester, T. Peurach and B.T. Mitchell, Pipelined systems for recognition of handwritten digits in USPS zip codes, *Proc. U.S. Postal Services Advanced Technology Conf.*, Washington, 1990, 539–548.

[8] C.Y. Suen, C. Nadal, R. Legault, T.A. Mai and L. Lam, Recognition of handwritten numerals based on the concept of multiple experts, *Proc. Int. Workshop on Frontiers in Handwriting Recognition*, 1990, 131–144.

[9] J. Franke, L. Lam, R. Legault, C. Nadal and C.Y. Suen, Experiments with the CENPARMI data base combining different classification approaches, *Proc. of 3rd Int. Workshop on Frontiers in Handwriting Recognition*, Buffalo, 1993, 305–311.

[10] N. Bartneck, T. Bayer, J. Franke, E. Mandler, M. Oberländer and J. Schürmann, Document analysis — From pixel to contents, *IEEE-Proc.: Special Issue on OCR and Document Analysis*, 1992, 1101–1119.

[11] U. Kreßel, The impact of the learning-set size in handwritten-digit recognition, *Artificial Neural Networks, Proc. Int. Conf. on Artificial Neural Networks*, eds. T.Kohonen *et al.*, Amsterdam, 1991, 1685–1689.

[12] H. Niemann, *Pattern Analysis and Understanding* (Springer Verlag, New York, 1989).

[13] L. Bernhardt, Zur Klassifizierung vieler Musterklassen mit wenigen Merkmalen, *5th Symp. of DAGM*, Karlsruhe, 1983, 255–260.

Handbook of Character Recognition and Document Image Analysis, pp. 123–156
Eds. H. Bunke and P. S. P. Wang

CHAPTER 5

SEGMENTATION-BASED CURSIVE HANDWRITING RECOGNITION

M. SHRIDHAR
University of Michigan-Dearborn
Dearborn, Michigan 48128, USA

and

F. KIMURA
Mie University, Tsu City, Japan

In this chapter, word recognition algorithms based on segmentation of a word image into primitive components (characters and sub-characters) and concurrent concatenation and recognition of these primitive components will be described. Two approaches to word recognition are discussed: i) context-free recognition where the recognition system yields an optimum letter string and a lexicon is used as a post-processor to select the best match and ii) lexicon directed recognition, where a presegmented word image is matched against all the words of the lexicon to obtain the best match. While word recognition may be based on context-free or lexicon directed techniques, numeral string recognition such as ZIP Code recognition or courtesy amount recognition in a bank check etc. is always based on context-free techniques. The recognition of words in a document follows a hierarchical scheme as described below: i) remove tilt (skew) of the document, ii) extract lines of words from document, iii) remove slant from each line, iv), extract words from each line, v) presegmentation of each word into primitive components (characters and sub-characters), vi) concatenation of components followed by character recognition to recognize each word. In this chapter, lexicon directed word recognition will be discussed in detail. The extension to context-free recognition will be illustrated. Results based on extensive testing with handwritten addresses obtained from the United States Postal Service are presented.

Key Words: Handwritten Word Recognition, Numeral String Recognition, Character Recognition, Likelihood Estimation, Context-free Recognition, Lexicon-directed Recognition

1. Introduction

The process by which people recognize handwritten characters, words and documents has been the subject of intense interest and investigation by researchers from very diverse fields. A good understanding of the mechanism of human recognition of handwritten documents will have a significant impact on the development of machines capable of recognition and interpretation of handwritten documents. However, the human recognition process is quite complex and it incorporates information extracted at different levels: characters, whole words, key words, and contextual processing. The efficiency of human recognition of handwriting can be attributed to the effective integration of multiple cues and exploitation of redundancies contained in most documents. However, if the goal of this study is to develop machines that are capable of automatic transcription of handwritten documents, then one must recognize the immense difficulty of adopting the human recognition process. In this chapter the primary focus will be on the development of practical approaches to handwriting recognition. The word "document" is used in a very general sense. Thus, a document will include characters, words, phrases, sentences and whole paragraphs. There are two main approaches to handwriting recognition: (a) techniques based on holistic approaches whereby an entire word or a character string is recognized as a unit and (b) techniques based on extraction and recognition of characters (also referred to as segmentation-recognition approach) contained in a word or a string. Due to the focus on practical approaches, this chapter will present an in-depth overview of recognition techniques based on segmentation-recognition.

The chapter will be organized as follows:

(1) A taxonomy of the handwriting recognition processes

(2) Image normalization processes

(3) Image presegmentation

(4) Context-free recognition of primitives and concatenation of primitives

(5) Lexicon driven recognition based on word matching

(6) Case study

2. Taxonomy

An important goal of this chapter is to provide a taxonomy of handwriting recognition. Handwriting recognition, in its most general form is the transcription of a handwritten document into machine written text. The process starts at the document level and goes through the following steps: (i) extraction of lines from the document (ii) extraction of words from the line (iii) holistic or character based recognition of words, including punctuation marks such as periods, commas, colons and semicolons, apostrophes etc. (iv) a post-processing step for integration of contextual and a priori knowledge to improve and enhance the recognition process. Figure 1 displays samples of handwritten documents with the characteristics described above.

Techniques for the extraction of lines and extraction of words would be properly classified as pre-processing steps. Additionally, in handwritten documents, it is not uncommon to have words, phrases or even complete sentences underlined or framed. In such cases, it would become necessary to incorporate underline and frame line removal in the pre-processing step. In general, a handwritten document can have one or more of the following characteristics:

(i) *Slant of the writing:* It is very common to find a distinct slant in the writing habits of most humans. Slant is measured as an angle with respect to a vertical frame of reference. Slant angles are typically in the range of -45° to $+45^{\circ}$. Often, the slant is not uniform across the document causing further complications in the analysis of the document.

(ii) *Skew or tilt:* A skew occurs when the lines of words are at an angle to the horizontal frame. This often occurs either during imaging of a misaligned document or more typically, when the writer introduces a skew owing to his/her inability to write on a reference line that is often not physically present.

(iii) *Underlines:* It is very common to encounter underlines in many handwritten documents. The underlines which are often undulating and not straight are intended to emphasize some key features of the document. Underlines and frame lines are also present when a person fills in a form with information in designated lines or boxes.

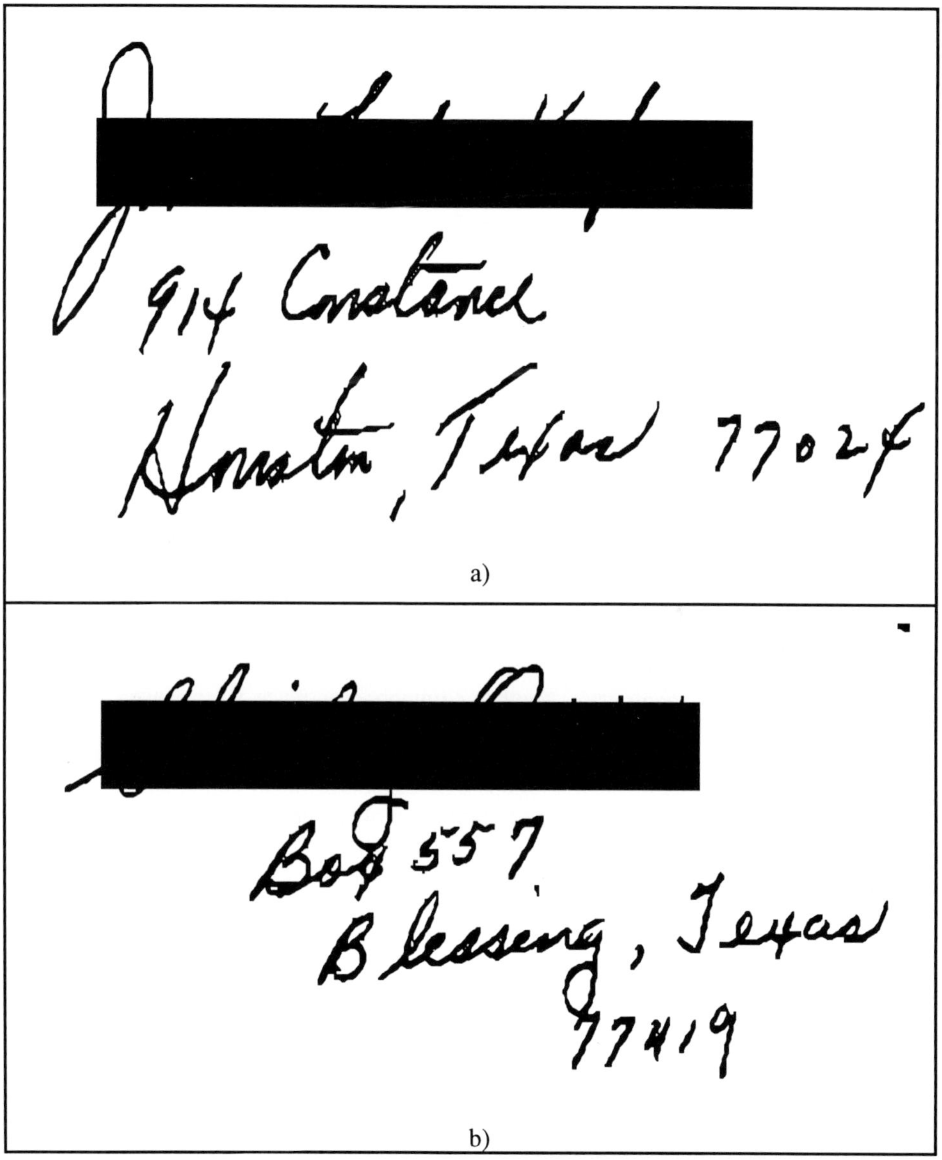

Fig. 1 Images of handwritten address blocks (names blocked out for reasons of privacy) a) Overlapping line fields b) Connected line fields

(iv) *Overlapped lines:* This is a serious problem in handwriting recognition. Due to limited spacing and the generally poor writing habits of people, words from one line intrude into adjacent lines, often intersecting with words in the lines above and below. Such interferences are typically caused by ascender (l, t, b, h, etc.) and descender characters (f, g, q, y, etc.) in the words.

v) *Discrete handwriting:* Most people write in a discrete format. Thus a word in a document may consist of more than one connected component. Also, a typical handwritten word may have discrete components (single isolated characters) as well as cursive components consisting of several characters. Often these components of a single word are spatially separated. The main problem with discrete writing is the confusion with regard to the location of precise word boundaries in a line of handwritten text.

(vi) *Imprecise punctuation:* In handwritten documents, it is often not easy to recognize punctuation marks as they are not precisely rendered during writing. Thus a comma may be mistaken for a character owing to both its size and location.

(vii) *Broken characters:* This is a very common occurrence in handwritten documents. Thus the horizontal bar of "T" may be physically disconnected from the vertical limb. This also occurs for characters such as "A","B", "D", "H", "R", etc.

(viii) *Similarly shaped words:* In handwritten words, it would often be very difficult to distinguish the word "clean" from "dean", when written in a cursive mode. Contextual interpretation would be needed to resolve this type of confusion.

(ix) *Ligatures:* Again due to the diverse writing styles of people, long and sometimes unusual ligatures connecting adjacent characters often add confusion to the identity of the word. This is especially problematic with words in which one encounters "w", "u", "v", etc. Other examples include unwanted connections such as in "tt", "ff".

(x) *Overlapping characters:* Another common occurrence is the overlapping of character fields within a word. Thus a "t" or a "T" with a long horizontal limb often overlaps wit characters in adjacent positions, making it quite difficult to extract characters. Such overlaps also occur with such combinations as "gh", "gy", etc. especially when the bottom loop of "g" or "y" is quite wide.

It is in the context of the above observations that one must approach the goal of developing machines capable of reading handwritten documents. It is important to note that even without most of the problems cited above, handwriting recognition would still be a formidable challenge.

2.1. Recognition Strategies

Word recognition algorithms may be classified into the following categories:

(i) Holistic approach

(ii) Analytic approach

Character extraction approach

External character segmentation

Internal character segmentation

(iii) Feature sequence matching

Hidden Markov Model

The holistic approach generally utilizes shape features extracted from the word image and attempts to recognize the entire word from these features. The analytic approach segments the word image into primitive components (typically characters). Character segmentation prior to recognition is called external character segmentation, while concurrent segmentation and recognition is called internal character segmentation. Feature sequence matching extracts features sequentially and derives word identity from this sequence. The Hidden Markov Model has been used widely for recognition based on feature sequences. It must be conceded that recognition based on Hidden Markov Models is often classified as a holistic approach.

For more details on the application of Hidden Markov Models to handwriting recognition, the reader is referred to the chapter by A. Kundu in this book.

2.2. Holistic approach

In the holistic approach, a word is recognized as a unit and techniques are derived for recognition of the entire word without attempting to analyze the letter content of the word. A set of features (strokes, holes, arcs, ascenders, descenders etc.) characterizing the entire word is derived and used in recognition. It is generally accepted that holistic methods are feasible only when the number of words to be recognized is small (typically less than 100). One of the earliest papers in holistic recognition is the classic work of Frishkopf and Harmon of Bell Laboratories [1]. In this approach a word is represented in terms of its horizontal and vertical extremes. In this context, an extreme is defined as a point at which one finds a horizontal (vertical) maximum or minimum. Thus a word is represented as an ordered list of extremes. Recognition is based on the best match between test features and features derived from dictionary words. Although the test inputs were obtained in an on-line mode, recognition was essentially off-line. Another early paper in holistic recognition was from Earnest [2], who used off-line recognition strategies to on-line test data. Earnest used features extracted from the middle zone of the words, ascenders and descenders.

Recent investigations indicate that the features used in these earlier works continue to be the principal features. However, the methods of comparison have changed significantly. Using distance measurements, a quantitative measure of fit between test words and dictionary words is obtained, thus allowing for multiple choice recognition. Some recent works in this area [3,4] have used Dynamic Programming techniques to optimally align features (ascenders, descenders, directional strokes, loops etc.) with features of lexicon words. Using different features derived from the contours of the word images, Leroux [5] proposed a similar scheme. Gorsky [6] proposed the derivation of prototypes called "holographs" that are derived for each lexicon word. These holographs are derived from lines extracted from thinned binary images of word samples provided by different writers. A correlation measure is used to obtain distances between the test word and the lexicon words. Hidden Markov Models where a word is represented as a matrix of transition probabilities of feature occurrences have been proposed for handwritten word recognition [7-9].

As noted earlier, holistic approaches can be used in two principal environments: 1) when the words to be recognized are prespecified and when the number of words is small 2) when the main objective is the reduction of lexicon size by eliminating obvious mismatches, thereby facilitating a more accurate, but computationally more intensive technique to be used for final word recognition

2.3. Character Extraction Approach

In this approach which is also described as a character based approach, algorithms are derived to extract and recognize the individual characters of the word. There are three principal issues that need to be considered:

(1) In cursive writing, it is very difficult if not, impossible to extract the characters of the word. If one considers words containing the characters "w", "m", "d" and letter pairs "rn", "nr", "un" "iv" etc., it is evident that many segmentations leading to identifiable characters are possible. Also in cursive writing, it is often difficult to distinguish the letter "o" from "a" especially, when ligatures are involved.

(2) Erroneous recognition of characters extracted from the word image can lead to incorrect word recognition. It is more typical to encounter letter strings that do not constitute a legitimate word. In such cases, it would be necessary to incorporate a post-processing stage to select the closest words from a lexicon, using expression matching techniques.

(3) The availability of a lexicon of words that also contains the true word is crucial to developing efficient techniques for word recognition. Fortunately in certain applications such as check processing and address recognition, a suitable lexicon can be generated. However, in the case of numeral strings, a lexicon is not usually available.

It is also necessary to recognize that segmentation is not a local process; rather it is dependent on both the previous extracted character (and its identity) and the likely character that follows the current character. Contextually, it is clear that if the current character is "q" (this character could easily have been recognized as "g"), then the previous character is most likely a vowel and the following character is with a probability of 1, the character "u". However, if the previous character were "g", then the next character could be any legitimate letter. Di-gram and/or tri-gram analysis would be needed to eliminate ambiguous letter strings.

This chapter will describe in detail word recognition strategies based on segmentation-recognition.

3. Word Recognition Based on Segmentation-Recognition

In this section, word recognition algorithms based on segmentation of a word image into primitive components (characters and sub-characters) and concurrent concatenation and recognition of these primitive components will be described. There are two approaches to word recognition:

(i) Context-free recognition where the recognition system yields an optimum letter string and a lexicon is used as a post-processor to select the best match

(ii) Lexicon directed recognition, where a presegmented word image is matched against all the words of the lexicon to obtain the best match.

While word recognition may be based on context-free or lexicon directed techniques, numeral string recognition such as ZIP Code recognition or courtesy amount recognition in a bank check etc. is always based on context-free techniques. The recognition of words in a document follows a hierarchical scheme as described below:

(i) remove tilt (skew) of the document
(ii) extract lines of words from document
(iii) remove slant from each line
(iv) extract words from each line
(v) presegmentation of each word into primitive components (characters and sub-characters)
(vi) concatenation of components followed by character recognition to recognize each word.

In this chapter, lexicon directed word recognition will be discussed in detail. The extension to context-free recognition will be illustrated.

3.1. Preprocessing (Line/word segmentation)

In this section, segmentation of lines in a document, and segmentation of words in a line as well as related techniques such as tilt and slant correction are discussed. *Line segmentation* is defined as the process of extracting the individual lines of words from a document. *Word segmentation* is defined as the process of extracting words from a given line. *Character segmentation* is defined as the process of extracting the individual characters that constitute the word unit.

Horizontal projection of a document image is most commonly employed to extract the lines from the document. If the lines are well separated, and are not tilted, the horizontal projection will have well separated peaks and valleys. These valleys are easily detected and used to determine the location of boundaries between lines. This simple strategy fails if the lines are tilted with respect to the horizontal axis of the image plane. The peaks and valleys would not be distinctive for tilted document images, and the text lines cannot be separated by horizontal boundary lines. There are two approaches to handle the problem of tilt. In the first approach the tilt is estimated and the document image is corrected for tilt prior to line segmentation. In the second approach, tilt correction is applied to each line after line segmentation has been performed.

Vertical projection of a line image is employed in the extraction of words in a line, analogous to the process of line segmentation. Again this simple strategy fails if the words have a slant with respect to the vertical axis of the image plane. To resolve the problem, it is necessary to perform slant estimation and correction. Slant correction is effective and widely used as a preprocessing operation for character segmentation in a word. Slant estimation and correction are also useful in detection and recognition of printed italic letters.

Some documents are handwritten along pre-printed or hand-drawn underlines, which interfere with the segmentation and recognition of words and characters. In such cases, it is necessary to detect and eliminate the underlines. A simple underline elimination algorithm is also described in this section.

3.2. Tilt correction

3.2.1. Tilt correction before line segmentation

(1) Crossing point method [10]

Tilt correction is generally performed in two steps. In the first step, the tilt of the document is estimated and in the second step shear transformation is applied to remove the tilt. Tilt or skew is estimated by finding the direction in which the projection of the document has maximum separability with regard to the peaks and valleys in the projection profile. In the crossing point method, only crossing points are counted to obtain the projection. The crossing point is a pixel with value "1" adjacent to its left pixel with value "0". The use of the crossing points rather than the entire foreground pixels is advantageous both in improving the separability of the projection and in saving computation time.

The variance of the number of crossing points is used as a simple measure of separability. To find the direction which maximizes the separability measure, multiple projections in different directions differing by one or two degrees are calculated within the range of the expected tilt. Once the tilt of the image is estimated, tilt correction is implemented as a shear transformation.

It is worth observing that this straightforward enumerative search for maximum separability is more efficient than expected, if it is implemented properly: All the multiple projections are calculated in a single raster scan. Only the crossing points are projected in multiple directions. The mapping is performed incrementally method without multiplications. If the document image is quite large and has sufficient resolution, the raster scan can be performed for every two pixels on alternate scan lines. This interleaving is equivalent to processing a down sampled image, and the process is more efficient, unless the down sampled image itself is needed in succeeding processes.

During the process of tilt estimation, several other characteristics of the document image, e.g. the number of lines, the interval, and the average height of characters, can be estimated, which are of great use in subsequent operations. Figure 2 shows an example of tilt correction. In this example, tilt is estimated as the direction which maximizes the variance of crossing counts was selected over a range varying from -8° to $+8^{o}$.

2) Hough transformation

Hough transformation is one of the most common techniques for detection of line segments in an image. It maps the original image to θ-ρ parameter plane, and a line in the original image forms a cluster in the parameter plane. Once the location of the clusters are determined, the tilt of each line and the average tilt is easily estimated. The Hough transformation is very attractive and useful because it is available to detect not only solid lines but also broken lines and even text lines consisting of characters and words. This generality however sacrifices the processing efficiency and the implementation has to be performed carefully, utilizing domain specific knowledge:

To reduce the processing time only border points or crossing points are required to be mapped to the parameter plane instead of the entire foreground pixels. Further reduction is achieved by restricting the range of θ in the parameter plane. It should be noted that if the transformation is performed only for specified θ, e.g. every 2° from -8° to $+8^{o}$, the Hough transformation approach is equivalent to the crossing point method described above. A section of the parameter space at a specified θ is simply the pixel projection in θ direction.

3.2.2. Tilt correction after line segmentation

Techniques for tilt estimation in an entire document such as an address block generally yields an average value; individual lines in the document may still exhibit residual tilt that needs to be removed for more accurate word recognition. A different approach is needed to estimate tilt in a single line of words. It is also observed that the method described in the previous section may yield incorrect tilt estimates, when the number of lines in the document image is less than two. A common approach to estimate tilt in a single-line image is the use of a least-squared error line fit to the bottom profile of the line image (ignoring any descenders in profile derivation). The slope of the resulting line yields a good estimate of the tilt in a single line image.

3.3. Underline elimination

(1) Projection method

Many techniques have been proposed for the detection and elimination of underlines. These include methods based on Hough transform and morphological operations with suitable structuring elements. An approach based on a line fit to the bottom profile of a single line image has also been used to eliminate underlines characterized by low curvature segments.

In a handwritten document consisting of several lines of data, vertical extents of underlines are simply estimated in the horizontal projection of the tilt corrected document image. Within the vertical extents, short vertical runs which are isolated in the extent are removed.

A novel algorithm using a morphological approach has been proposed by Liang et al. [29] for removing interference strokes, including underlines from word images. This method is capable of removing hand and machine drawn underlines, even when these underlines cut across the characters of the word image.

3.4. Line segmentation

3.4.1. Line segmentation before tilt correction

(1) Zone method [11]

The algorithm divides an input image into vertical zones. For each zone, a horizontal projection is computed. A vertical extent of the projection with non-zero values form a block. Blocks which are horizontally adjacent and vertically overlapping are connected to form text lines. Heuristics are used to split and/or join blocks. Connected components that are located entirely in blocks from a single line are assigned to that line. Connected components that span more than one text line are split into two or more components and placed in separate text lines.

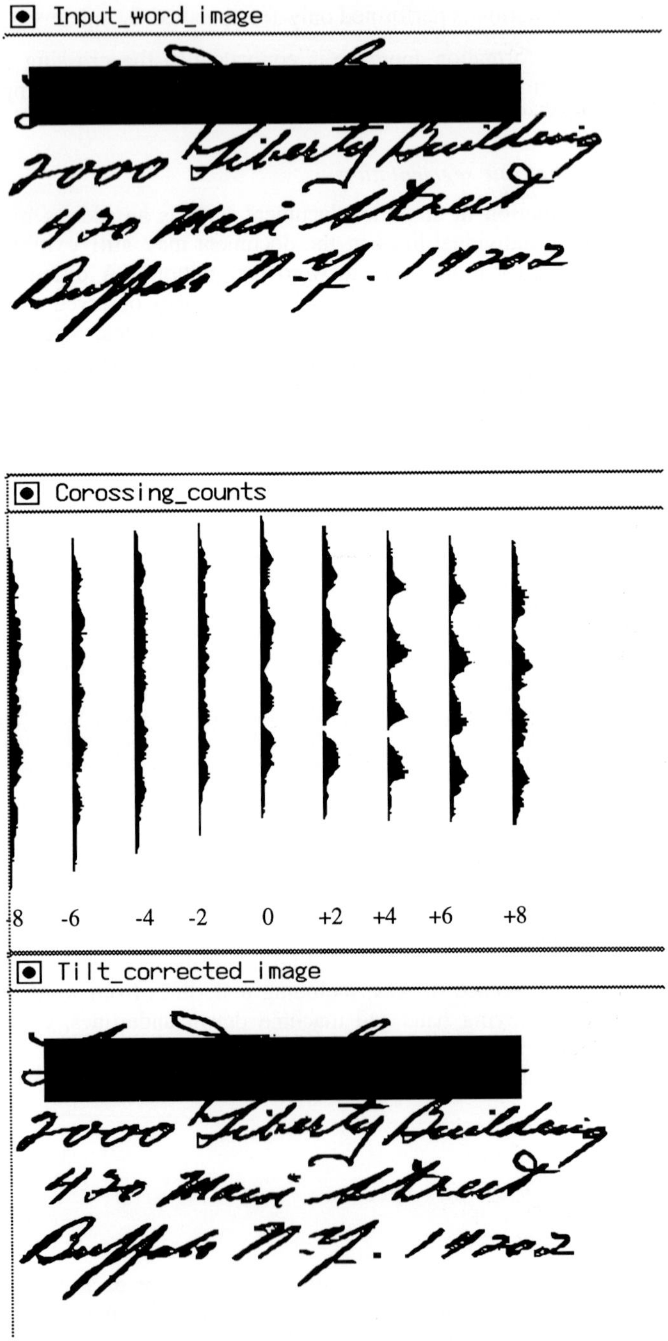

Fig. 2 Example of tilt correction.

(2) Morphological method [12]

In the morphological method, core regions of the image are generated using a morphological operation. These core regions generally fill in the body of each word, but eliminate ascender and descender strokes. For the core images, a technique similar to the zone method is applied. Because of the preceding morphological operation, less heuristics are required to split or merge blocks (to form text lines) than for the zone method. Zones in the morphological method are assumed to have overlapped areas, while in the zone method, zones are mutually disjoint.

3.4.2. Line segmentation after tilt correction

(1) Projection method

As described earlier, the horizontal projection of a document image is calculated, and the valley points are detected and used to determine the location of boundaries between lines. Some valley points may be merged or removed based on heuristics such as the expected interval of lines. If two valley points are closer than a given threshold, they are merged. The advantage of the projection method is its robustness in dealing with documents containing connected lines due to extenders. The disadvantage is the underlying assumption that line boundaries are horizontal.

(2) Component clustering method [13]

Line segmentation can be considered as a problem of clustering for the connected components. Each connected component is mapped into a lower dimensional space, e.g. into a two-dimensional space (plane) in terms of their vertical extents (y_{min}, y_{max}) (Figure 3(c)).

The clusters are detected using typical clustering algorithms such as the K-means clustering [14]. The K-means clustering requires initially K clusters or K centers of clusters. These initial values can be obtained in the process of tilt correction. The projection method may be employed to obtain these initial values, if necessary. To suppress the undesirable influence of small noise components or large components including multiple lines connected by extenders, the extended version of the K-means clustering called the weighted K-means clustering, is known to be useful. In the weighted K-means clustering, the center of a cluster is the weighted mean of the samples. The weight of each connected component is defined so that the closer the height of the component is to the estimated character height, the larger its weight is.

If the number of clusters is not uniquely estimated, but with range, e.g. 6 ± 1, the K-means clustering is applied for all possible values of K. Among the results, those which have poor separability are discarded. Those which do not satisfy spatial constraints required by valid document lines are also discarded. The remaining clusters yield the number and the position of the lines and the components are assigned to these clusters (lines). Multiple line components occupying a vertical extent of more than two lines are split at the line boundary (Figure 3(b)). The advantage of the component clustering method is its ability to construct complex line boundaries.

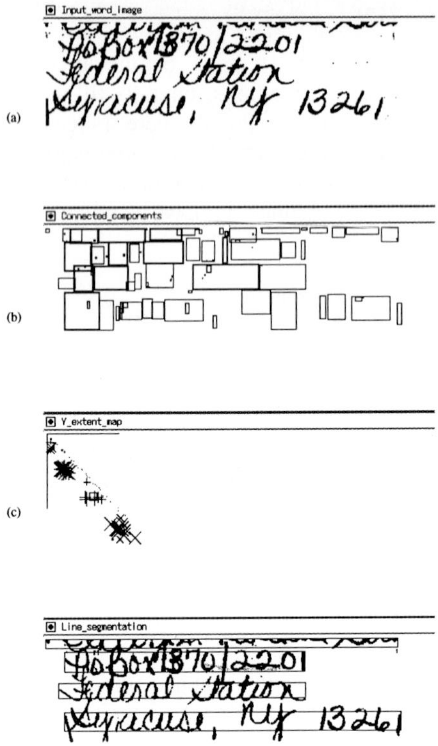

Fig. 3 Example of line segmentation (Connected components with thick bounding boxes are subdivided multiline components).

3.5. *Slant correction*

There are two principal approaches for estimating the slant of a word. These include the projection method and the chain code method. A brief description is provided below.

(1) Projection method

The average slant of characters in a word or in a line is estimated by the analysis of slanted vertical projections (histograms) at various angles [28]. The average slant is found by looking for the greatest positive derivative in all of the slanted projections.

(2) Chain code method [24]

In contrast with the tilt estimation, the average slant of characters in a word or in a line is easily estimated using the chain code of the border pixels. The average slant is given by

$$\theta = \tan^{-1}\left(\frac{n_1 + n_2 + n_3}{n_1 - n_3}\right) \tag{1}$$

where n_i is the number of chain elements at an angle of i times 45° (/ or | or \). Shear transformation is then applied to correct the slant.

It is interesting to see why this simple expression gives a good estimate of the average slant. To estimate the slant of characters, only vertical and near vertical edges are useful and horizontal edges are of no use. If horizontal chain elements are removed, the borders of the characters are separated into chain segments having its average slant between 45° to 135°. Since each average slant of these chain segments is calculated by (1), the overall average is also calculated by (1).

(a)

(b)

(c)

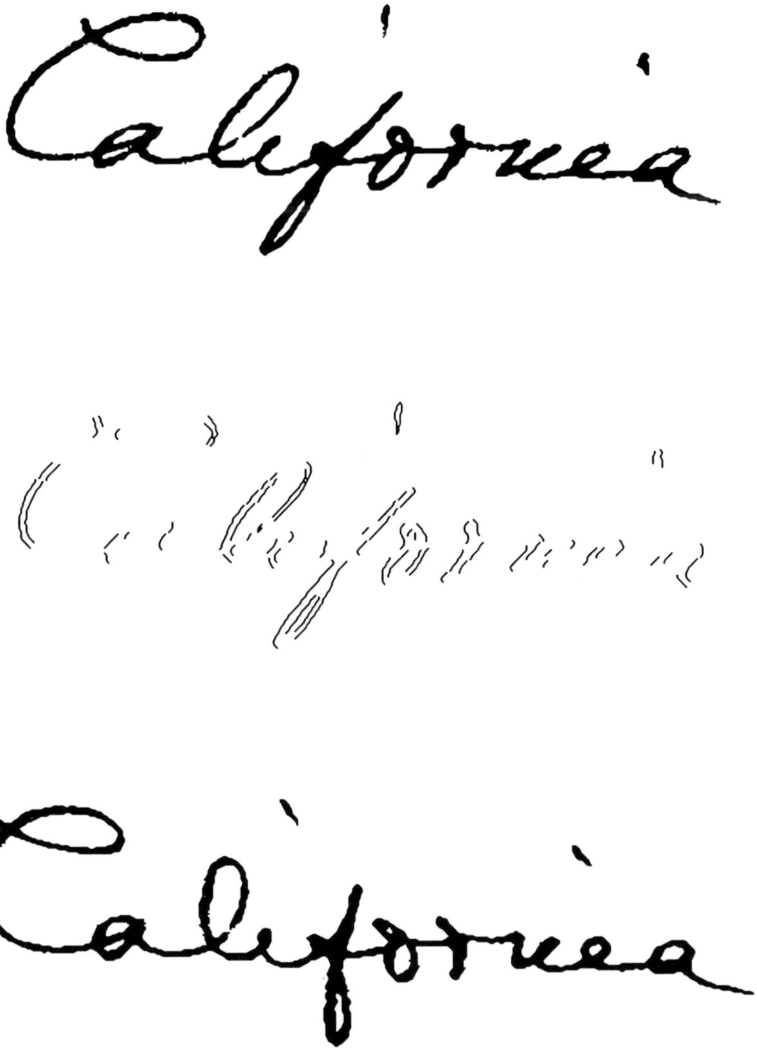

Fig. 4 Input word image (a), non-horizontal border pixels (b), and slant corrected word image (c)

Figure 4 shows an example of slant correction of a word. Figure 4(b) shows the non-horizontal chain elements, and Figure 5 illustrates the calculation of the average slant of non horizontal chain elements using (1).

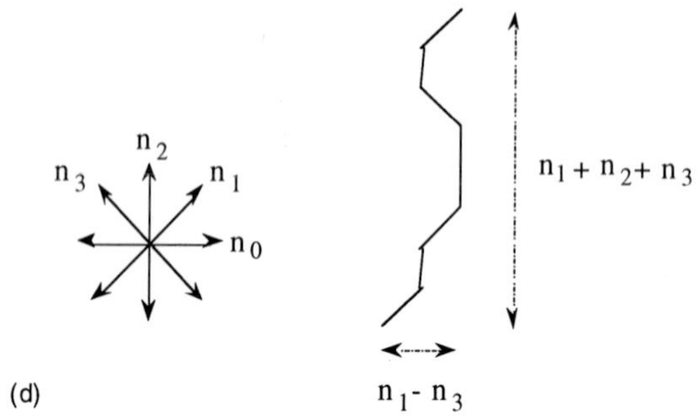

Fig. 5 Input word image (a), non-horizontal border pixels (b), slant corrected word image (c), and a chain segment (d).

3.6. Word segmentation

(1) Gap analysis [13]

Words are assumed to be separated by a space, a comma, or a period. The space detection algorithm detects the spaces by classifying each gap between the character segments as "between field gap" or "within field gap" respectively. If the gap is wider than a threshold, the gap is classified as "between field gap", otherwise as "within field gap". The threshold, based on the distribution of the gap width for text lines is found by applying a standard technique such as Otsu's method[7].

Exact segmentation of a handwritten line field is very difficult unless it is integrated with the word recognition process. It is interesting to observe that human beings usually employ this approach very efficiently for word recognition In a typical integrated word segmentation process, the word segmentation is assumed to yield over segmented word images, where some words can be split and divided into sub-words. These (pre) segments are merged again into a whole word through multiple application of the word recognition algorithm to a set of successive segments. To obtain an over-segmented word image, the line is subject to further segmentation, if under segmentation is anticipated. If the number of words in a line is too few or the estimated length of a word is too long, the word is divided at the maximum within-field gap. This procedure is repeated until no further subdivision is necessary.

(2) Convex hull [17]

Different metrics can be employed to measure the spatial gaps with varying degrees of accuracy [17]. Convex-hull metric requires computation of convex-hulls for each of the components in a line. The distance between the convex hulls of the components along the line joining their centers of gravity is used as a distance measure.

4. Recognition Algorithm

Before segmenting a word into its character components, slant estimation and correction are applied to the word image. Segmentation points (character boundaries) are then detected for splitting the word at these segmentation points. As is the case with word segmentation, most character segmentation techniques are designed to generate over-segmented character images. The character segments are merged into a whole character in the succeeding process of character recognition or word recognition.

Detection of segmentation points are based on the shape analysis of the word image. Contour analysis, profile analysis, and run-length analysis are most commonly used for this purpose.

4.1 Character segmentation

(1) Contour analysis [15, 18]

The contour analysis method is suitable to obtain over segmented character images. Possible segmentation points are detected through local extrema analysis of the upper contour of the word image. Among the local minima, those that are not deep enough from the adjacent local maxima are sequentially removed. In order to obtain characters separated by vertical lines, segmentation points determined in the previous step are often shifted horizontally to the right or the left as follows:

If the minimal point is not open vertically upward, the point is shifted to the right or to the left of this point depending on where the number of runs and the total length of runs is minimal. Figure 6 (a) illustrates this process.

(2) Profile analysis [19]

Instead of the upper contour of a word, the upper profile is analyzed. Here the upper profile is defined to be the set of the topmost foreground pixels in each column. Post processing is required to find some segmentation points overshadowed by character segments such as the long "t" bar.

(3) Run length analysis [20]

Ligatures between "or", "on", ..., occasionally do not have a valley point in the upper contour. To detect and split these ligatures, a single-run stretch is detected and split at the middle point. The run is vertical streaks of one or more black pixels, and the single-run is the unique run on a single vertical line. The single-run stretch is a horizontal stretch of single-runs shorter than a threshold determined depending on the average stroke width. Figure 7 illustrates this analysis. Among these single-run stretches, those which have well defined peaks and valley points in the upper contour are removed. The remaining single-run stretches are split at the middle point.

(4) Disjoint box segmentation [15, 18]

Characters and character segments are more easily handled if their bounding boxes are mutually disjoint. If over segmentation is permitted, horizontal overlapping of character segments is resolved by a simple algorithm. A word image is split vertically at each pre-segmentation point and is separated into horizontally non-overlapping zones. A connected component analysis is applied to the split image to detect the boxes enclosing each connected component. These boxes are usually disjoint and do not include parts of other connected components (Figure 6(b)). The disjointedness of the bounding boxes is necessary for high speed feature extraction (described in section 4.4).

Fig. 6 (a) Example of pre-segmentation points, (b) disjoint box segmentation, (c) optimum character segmentation

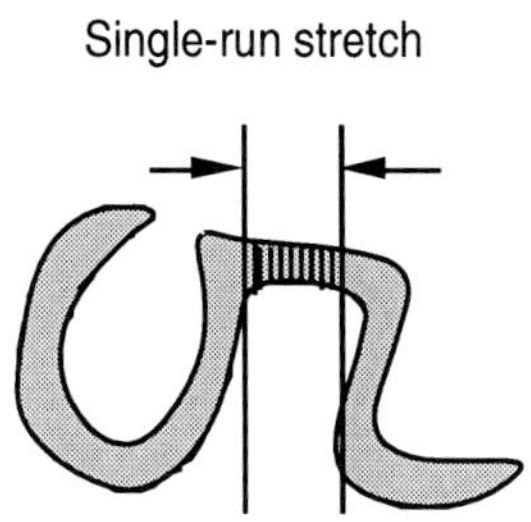

Fig. 7 Single-run stretch between "o" and "r"

4.2 Word matching algorithm

4.2.1. Lexicon directed algorithm [15, 18]

The number of boxes (or segments) obtained by the disjoint box segmentation is generally greater than the number of characters in the word image. In order to merge these segments into characters so that the final character segmentation is optimal, dynamic programming (DP) is applied, using the total likelihood of characters as the objective function. The likelihood of each character is given by a discriminant function described in 4.3.2. To apply the DP technique, the boxes are sorted left to right according to the location of their centroids. If two or more boxes have the same x coordinates at the centroids, they are sorted top to bottom. Numbers at the right upper right-hand corner of boxes in Figure 6(b) show the order of the sorted boxes. It is worth observing that the disjoint box segmentation and the box sorting process reduce the segmentation problem to a simple Markov process, in most cases. For example, in figure 6 (b), boxes 1 to 4 correspond to the letter "C" of California, box 5 to "a", box 6 to "l" ... and so on. This assignments of boxes to letters can be represented as

```
i --> 1  2  3  4  5  6  7  8  9 10
Ai          C  a  l  i  f  o  r  n  i  a
j(i) -->    4  5  6  8  9 10 11 13 15 16
```

where i denotes the letter number, $j(i)$ denotes the number of the last box corresponding to the i-th letter. Note that the number of the first box corresponding to the i-th letter is $j(i-1)+1$. Given $[j(i), i=1,2,..,n]$ the total likelihood of the character is represented by

$$L = \sum_{i=1}^{n} l(A_i, j(i-1)+1, j(i)) \tag{2}$$

where $l(A_i, j(i-1)+1, j(i))$ is the likelihood for the i-th letter.

In the lexicon directed algorithm, an ASCII lexicon of possible words is provided and the optimal character segmentation is found for each lexicon word. All lexicon words are then ranked regarding the optimal likelihood per character (L^*/n) to select the best candidate word. The optimal assignment (the optimal segmentation) which maximizes the total likelihood is found by using the dynamic programming technique described below.

The optimal assignment $j(n)^*$ for n-th letter is the one such that

$$L^* = L(n, j(n)^*) = \max_{j(n)} L(n, j(n)) \tag{3}$$

where, $L(k, j(k))$, the maximum likelihood of partial solutions given $j(k)$ for the k-th letter, is defined and calculated recursively by

$$L(k, j(k)) = \max_{j(1),j(2),\ldots,j(k-1)} \{\sum_{i=1}^{k} l(A_i, j(i-1)+1, j(i))\}$$

$$= \max_{j(1),j(2),\ldots,j(k-1)} [l(A_k, j(k-1)+1, j(k)) + \sum_{i=1}^{k-1} l(A_i, j(i-1)+1, j(i))\}]$$

$$= \max_{j(k-1)} [l(A_k, j(k-1)+1, j(k)) + \max_{j(1),j(2),\ldots,j(k-2)} \{\sum_{i=1}^{k-1} l(A_i, j(i-1)+1, j(i))\}] \tag{4}$$

$$= \max_{j(k-1)} [l(A_k, j(k-1)+1, j(k)) + L(k-1, j(k-1))] \tag{5}$$

$$L(0, j(0)) = 0, j(0) = 1, 2, \ldots, m$$

Starting from (5), all $L(k,j(k))$'s are calculated for k=1,2,...,n using (4) to find $j(n)^*$ using (3). The rest of $j(k)^*$'s ($k = n$-1,n-2,...,1) are found by back-tracking a pointer array representing the optimal $j(k\text{-}1)^*$'s which maximizes $L(k,j(k))$ in (4).

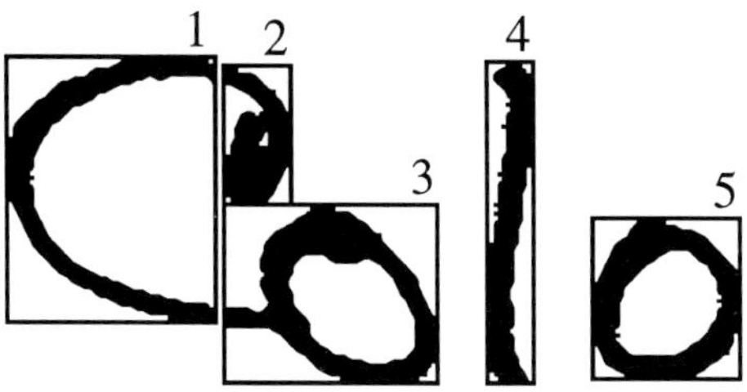

Figure 8 Segmentation - Recognition with DP

Table 1 Table of likelihood values

5	-	-	-	**6.71**	
4	-	-	**4.87**	4.57	
3	-	**3.00**	3.25	-	L(k,j(k))
2	**1.65**	3.11	-	-	
1	1.90	-	-	-	
0					
k->	**1**	**2**	**3**	**4**	
Letter	**C**	**o**	**l**	**o**	

Table 2 Search for optimum segmentation

	5	-	-	-	**4**	
	4	-	-	**3**	3	
j(k)	3	-	**2**	2	-	j(k-1)* for given k and j(k)
	2	**0**	1	-	-	
	1	0	-	-	-	
	0					
		1	**2**	**3**	**4**	Letter k->
		2	**3**	**4**	**5**	j(k) for k-th letter
		C	**o**	**l**	**o**	Letter

The example illustrated in figure 8, table 1 and table 2 shows the values of L and $j(k\text{-}1)^*$ for given k and $j(k)$. In this example, $L(n,j(n))^* = L(4,5) = 6.71$ and $j(n)^*=5$. Succeeding $j(k)$'s are 4, 3, 2, 0 respectively. The box "0" corresponds to a virtual box standing for the last box of the letter preceding the first letter "C". Figure 8 illustrates the word matching procedure based on segmentation-recognition with DP.

4.2.2. Lexicon free algorithm

A lexicon free word recognition algorithm is easily obtained from the lexicon directed algorithm by simple modification. In the lexicon directed word matching, character likelihood is calculated for a single letter in a specified position of a lexicon word. While in the lexicon free word matching, the total likelihood for an input word is given by

$$L = \sum_{i=1}^{n} l(A_i^*, j(i-1)+1, j(i)) = \sum_{i=1}^{n} \underset{A_i}{\mathrm{Max}}\{l(A_i, j(i-1)+1, j(i))\} \tag{6}$$

instead of (2). The character likelihood for all letters are calculated and the maximum value and the associated letter A_i^* are determined. The word matching process is applied only once for an input word, and the recognition result is given by $A_1^*A_2^*...A_n^*$. When the word length n is unknown, an upper bound is estimated and used as the value of n. It is very convenient that the likelihood array L(k,j(k)) and the pointer array for a word of length n include s all the entries for shorter words.

The lexicon free algorithm is required when no lexicon is available or for numeral string recognition. It is also efficient and suitable if the input word is written neatly and segmentation and recognition of characters are relatively accurate.

(4) K-best path algorithm [21]

The lexicon free algorithm can be extended so that it generates not only the optimal word interpretation but also the K best interpretations for a specified K. Among the K interpretations, invalid interpretations are removed through lexicon search and the best valid interpretation is selected.

(5) Expression matching [30,31]

Expression matching is the process of matching a character string against words in a lexicon and computing a measure (called "edit distance") indicative of the degree of match between the given string and the words in the lexicon. Spell checking operations use expression matching to determine alternative words to correct misspelled words. This process uses three operations:

(i) Deletion, where certain letters in the string are dropped to obtain the edit distance between the letter string and a lexicon word of lesser length. A penalty is added for every letter dropped.

(ii) Insertion, where letters are added to the string to obtain the edit distance between the string and a longer word in the lexicon.

(iii) Substitution, where one letter in the string is replaced by another letter to obtain the edit distance. It is noted that the process of substitution is equivalent to a combination of deletion and insertion.

Edit distance is the minimum cost of using the three operations of deletion, insertion and substitution to match a given letter string to a word in the given lexicon. Dynamic programming is used in deriving this minimum cost. Final word selection is based on the smallest edit distance between the letter string and the words in the lexicon. The principal advantage of this process is the ease with which the string generated by the recognition algorithm can be matched against lexicon words, even when the length of the string is different from the length of the lexicon word.

In handwritten word recognition, expression matching is used as a postprocessing operation to determine the best match between the lexicon words and the optimum letter string derived in a context-free recognition mode.

4.3. *Character recognition*

Most character recognition techniques have been proposed and studied primarily for isolated numerals and characters. Among these techniques, those having the following characteristics are desirable for handwritten word recognition.

(1) High speed recognition

(2) Capability of likelihood calculation

(3) Automated learning

(4) Robustness for wide variations in writing style

It is known that these properties are possessed by statistical and neural classifiers rather than syntactic and structural classifiers. In the following subsections, typical techniques for statistical classification and feature extraction suitable for handwritten word recognition are described. More details about character recognition may be found in the chapters by J. Franke, U. Kressel and J. Schürmann in this book.

4.3.1. Feature extraction

In statistical and neural classification, a feature vector representing the shape of the character is composed in the process of feature extraction. Generally, features of characters such as medial axis, contour, cavities and loops are first extracted and then the feature vector is calculated from these features. Among these features, the contour (edge) of the character image is desirable for high speed feature extraction because of the relative computational efficiency of the contour extraction. The shape of a contour is easily characterized as a feature vector by using the local chain code histogram. The shape information of a character contour is also characterized using the gradient of the character image. These two basic procedures are described in the following. A special technique for high speed chain code feature extraction is described in section 4.4.1.

(1) Chain code feature [13]

Local chain code histograms of the character contour are used as a feature vector (Figure 9). The rectangular frame enclosing a character is first divided into 7 * 7 blocks. In each block, a chain code histogram of the character contour is calculated. The feature vector is composed of these local histograms. Since contour orientation is quantized to one of four possible values (0° or "-", 45° or "/", 90° or "|" or 135° or "\"), a histogram in each block has four components. After the histogram calculation, the 7 * 7 blocks are down sampled with Gaussian filter into 4 * 4 blocks. Thus the feature vector has 64 elements when all the 16 blocks are included.

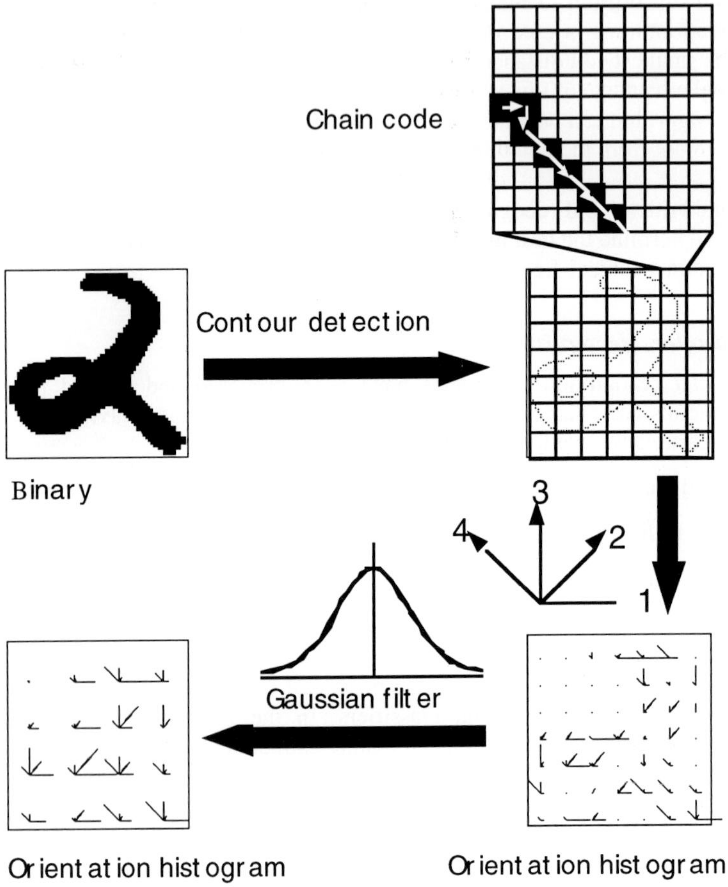

Fig. 9 Feature extraction from contour chain code.

Height normalization is simply performed by multiplying each vector component by the ratio of the standard height to the actual height of the letter.

The variable transformation is performed by taking the square root of each component. The use of power transformation to obtain Gaussian-like distribution from a distribution of causal (positive valued) features is described in [22].

(2) Gradient feature

The desired accuracy of the character classifier employed in internal segmentation will normally be constrained by the requirement on computation time. However, once the internal character segmentation is completed, a high accuracy character classifier can be applied to each segmented character to obtain a more accurate character likelihood with relatively small additional processing time. As long as the character segmentation is correct and the accuracy of the post classifier is higher, the accuracy of word recognition will be improved.

A high accuracy character classifier is implemented using a high dimensional feature vector. A 400-D feature vector is obtained in a manner similar to the 64-D feature vector. The number of blocks is initially 9 * 9 and subsequently down sampled to 5 * 5. The quantization level for orientation is 16 directions instead of 4 used in the 64-D case. To obtain 16 directions, a Gaussian filter and a Roberts gradient filter are applied to the character image to obtain a gradient image. The arc tangent of the gradient is quantized into 16 directions and the strength of the gradient is accumulated for every direction level in each of the 9 * 9 blocks. Full details of high dimensional feature extraction may be found in [23]. Gradient-based feature is also described and studied in [24], where directions of the gradient are quantized into 12 bits with an OR operation to obtain a binary-valued vector.

4.3.2. Likelihood calculation

(1) Subspace method

In the subspace method feature vectors of a class are assumed to be distributed along a hyperplane (subspace), and the hyperplanes are different from class to class. To determine the class of unknown feature vector, the distances to all hyperplanes are calculated to find the nearest hyperplane. The square of the distance is given by

$$g(X) = |X - M|^2 - \sum_{i=1}^{k} \{\Phi_i^T (X - M)\}^2 = \sum_{k+1}^{n} \{\Phi_i^T (X - M)\}^2 \tag{7}$$

where X denotes the input feature vector, M denotes the sample mean vector for each character class, and Φ_i denotes the eigenvectors of the sample covariance matrix.

This discriminant function is the squared error (truncation error) of the KL expansion of a feature vector X in terms of the first k dominant eigenvectors of the covariance matrix (Figure 10). The value is the square of the distance from X to the (minimum mean square error) fitting hyper plane to the distribution. Figure 11 shows the decision boundary for two two-dimensional distributions classified by Eq. (7). In this example the number of eigenvectors used is one for each class i.e. $k = 1$. Although this decision boundary consists of a pair of lines, it is not a linear boundary, but a degenerate quadratic boundary. For three or higher dimensional distributions, the degeneration rarely occurs.

The CLAFIC (Class-Featuring Information Compression) [25] has been known as effective similarity measures for pattern recognition. The CLAFIC is a special form of Eq. (7) for norm independent feature vectors, the class of which is independent from the norm. For example, a vector X which represents a gray scale character image is a norm independent feature vector. In this case, the norm of X represents the contrast of the character image and does not affect the class to which X belongs (Figure 12). Another example of a norm independent feature vector is the power spectrum used in speech recognition [26]. Since the norm independent feature vector X and $-X$ belong to the same class, the distribution is symmetric with respect to the origin and the mean vector is zero (vector).

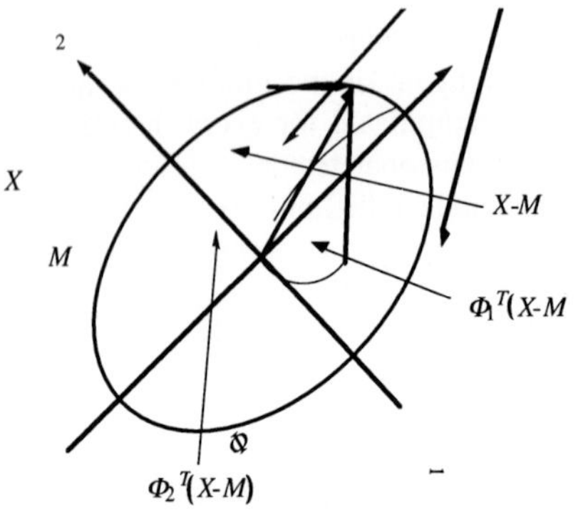

$$X = \Phi_1{}^T(X\text{-}M)\,\Phi_1 + \Phi_2{}^T(X\text{-}M)\,\Phi_2 + M$$
$$\tilde{X} = \Phi_1{}^T(X\text{-}M)\,\Phi_1 + M$$
$$|\,\Delta X\,|^2 = \left\{\Phi_2{}^T(X\text{-}M)\right\}^2$$

Fig. 10 Squared error of K-L expansion

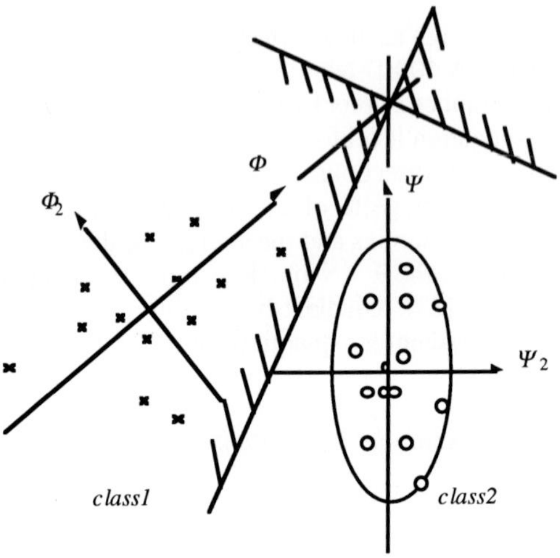

$$g_1(X) = \left\{\Phi_2{}^T(X\text{-}M_1)\right\}^2$$
$$g_2(X) = \left\{\Psi_2{}^T(X\text{-}M_2)\right\}^2$$

Fig. 11 Decision boundary of K-L expansion based classification

Thus, Eq. (7) is simplified to

$$g(X) = |X|^2 - \sum_{i=1}^{k}(\Phi_i^T X)^2 \tag{8}$$

If the norm of X is normalized to 1, it is further simplified to

$$g(X) = 1 - \sum_{i=1}^{k}(\Phi_i^T X)^2 \tag{9}$$

The second term is the CLAFIC. Full details may be found in [27].

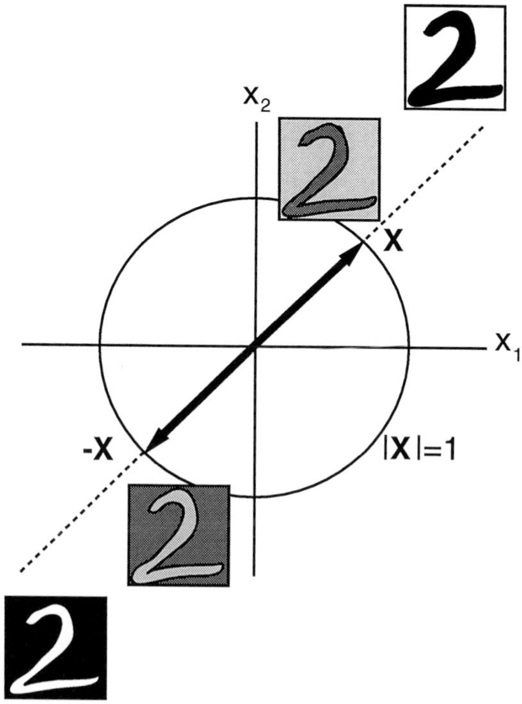

Fig. 12 Example of norm independent feature vector

(2) Neural network

Neural networks provide an alternative approach to character recognition. They are very useful and often provide more accurate recognition, especially when the probability distribution of the features are not known á priori. Most features that are used in character classifiers perform equally well with neural networks. The principal drawback of neural networks is the unavailability of guidelines for selecting the structure of the neural network. This includes the number of nodes in the hidden layers and the number of hidden layers. Convergence of the learning algorithm is not always assured. Furthermore, there is always the danger of overtraining a neural network, such that its performance in actual recognition applications is not assured.

4.4 High speed processing

4.4.1 High speed feature extraction [15.18]

One critical point in segmentation recognition techniques, using dynamic programming is the speed of feature extraction, because it is applied to a number of alternative segmentations. The use of the cumulative orientation histogram enables one to realize high speed chain code feature extraction. Border following and orientation labeling are performed only once to an input word image, and the orientation feature vector of a rectangular region including one or more boxes is extracted by a relatively small number of arithmetic operations. Each border pixel has one of four labels representing horizontal, vertical, and two diagonal orientations. These pixels constitute four separate border pixel images each of which is processed independently as follows:

The cumulative orientation histogram is defined by

$$C(i,j) = C(i-1,j) + C(i-1,j-1) + c(i,j)$$
$$C(i,0) = C(0,j) = 0 \tag{10}$$

where $c(i,j)$ denotes a border pixel image.

Using this cumulative histogram, the number of border pixels within a specified rectangular region is calculated by

$$n(i_{min}, j_{min}, i_{max}, j_{max}) = C(i_{max}, j_{max}) - C(i_{min}-1, j_{max}) - C(i_{max}, j_{min}-1) + C(i_{min}-1, j_{min}-1) \tag{11}$$

where (i_{min}, j_{min}) and (i_{max}, j_{max}) are the coordinates of the upper left-hand corner and the lower right-hand corner of the rectangular region.

For example, the number of border pixels in the rectangular region defined by the upper left-hand coordinates (4,2) and lower right-hand coordinates (8,4) in the following illustration is calculated as $n(4,2,8,4) = 9 - 0 - 2 + 0 = 7$. Note that the calculation involves only one addition and two subtractions for any rectangular region with arbitrary size. When a letter consists of a single box (of a connected component), e.g. "a", "l", "f", "o", "r", "a" as in figure 6, the extraction of the feature vector is quite simple.

j\i	0	1	2	3	4	5	6	7	8	9
0	0	0	0	0	0	0	0	0	0	0
1	0	0	0	0	0	0	0	0	0	0
2	0	0	1	1	1	1	0	0	0	0
3	0	0	0	0	0	0	0	0	0	0
4	0	0	0	0	1	1	1	1	1	0
5	0	0	0	0	0	0	0	0	0	0

c(i,j)

j\i	0	1	2	3	4	5	6	7	8	9
0	0	0	0	0	0	0	0	0	0	0
1	0	0	0	0	0	0	0	0	0	0
2	0	0	1	2	3	4	4	4	4	4
3	0	0	1	2	3	4	4	4	4	4
4	0	0	1	2	4	6	7	8	9	9
5	0	0	1	2	4	6	7	8	9	9

C(i,j)

The calculation of the border pixel number using (11) is simply repeated for each 4 by 4 sub rectangular region for the four orientations. When a letter consists of multiple boxes, e.g. "C", "I", "n", "i", (Figure 6) each box is divided into the intersections with sub-rectangles. The border pixel number is calculated for these intersections and summed up for each sub rectangle.

4.4.2. Demand driven character recognition [18]

Since the segmentation recognition process is repeated for all lexicon words having the word length within the estimated word length interval, the processing time increases proportionally to the lexicon size. This is a main drawback of the lexicon directed word recognition approach. However, this problem can be alleviated if repetitions of the same feature extraction and character recognition for a large lexicon, are avoided.

Daemons for feature extraction and character likelihood calculation are helpful to avoid unnecessary calculations. The character likelihood daemon stores the computed likelihood values in a table, when the likelihood is first required and calculated. It just returns the value when the same character likelihood is required for different words. The feature extraction daemon works in a similar manner.

4.4.3 Word length estimation [18]

Estimation of input word length is helpful in reducing both the size of the lexicons as well as the processing time for word recognition. It also reduces the possibility of confusion between words with different word lengths. The upper bound and lower bound of the word length are estimated and the lexicon words not having the word length within this range are skipped and not used in the segmentation recognition process.

Therefore, under estimation (or over estimation) of the upper bound (lower bound) excludes the possibility of correct word recognition, while tighter bounds are preferable in achieving higher speed and accuracy. The upper bound and lower bound are estimated depending on the result of the presegmentation as follows. Input word image is split vertically at each presegmentation point to have horizontally non-overlapping zones.

Zones whose width are smaller than a threshold are regarded as a part of their preceding zones. The number of zones is used as the upper bound B_u of the word length. The lower bound B_l is determined by

$$B_l = \text{trunc}((B_u + 1)/2) \tag{12}$$

where trunc(x) denotes the integer part of x.

5. Handwritten Address Interpretation System (Case study)

In this section, the authors present two case studies illustrating the application of recognition algorithms discussed earlier. The first case study deals with the performance evaluation of lexicon directed segmentation-recognition algorithm, using real world handwritten word images. The second case study describes the performance of an integrated handwritten address recognition system that requires detection and recognition of ZIP code field, City/State field, street number field and street name/PO Box field and finally the correct nine-digit ZIP code. The interesting feature of this study is the lack of any á priori information about the nature of the address. Addresses may contain a PO Box and/or street number/name fields. The integrated system is required to determine the type of field present and determine the nine-digit ZIP code.

The handwritten address interpretation system consists of subsystems for preprocessing, ZIP code line recognition, street line recognition, P. O. Box line recognition, and delivery point code determination (Figure 13). The preprocessing subsystem applies tilt correction, line segmentation, slant correction, and word pre-segmentation. The ZIP code line recognition subsystem generates several ranked ZIP code candidates.

The street line recognition subsystem generates several ranked pairs of street numbers and street names for given 5-digit ZIP code. If the top candidate pair is accepted with sufficient confidence, it is sent to the DPC determination subsystem together with the 5-digit ZIP code.

The P. O. Box line recognition subsystem generates several ranked P. O. Box number for given 5-digit ZIP code. If the top candidate is accepted with sufficient confidence, it is sent to the DPC determination subsystem with the 5-digit ZIP code.

If the top candidate is a 5-digit unique ZIP code with sufficient confidence, it is encoded directly in the DPC determination subsystem. If the top candidate is a 9-digit ZIP code on mail piece with sufficient confidence, it is also directly encoded to DPC in the DPC determination subsystem.

The DPC determination subsystem encodes given information from each subsystem to a DPC. If no valid DPC is obtained, and the 5-digit ZIP code has sufficiently high confidence, it is accepted. Otherwise it is rejected. Each subsystem is described in subsequent sections.

5.1. ZIP Code Line Recognition Subsystem

The ZIP code is first assumed to be at the last field of the last line. If the likelihood of the detected ZIP code is less than a threshold, up to two preceding lines are assumed successively to be the ZIP code line until a ZIP code with sufficient likelihood is detected. In actual pre-segmented images, ZIP code fields are often split and divided into several pieces, which have to be merged again into a field. This problem is resolved through multiple use of the word recognition algorithm to a set of successive pre-segments. The word recognition algorithm employs a lexicon free word matching described in section 4.2.

5.2. Street Line Recognition Subsystem

The street line recognition system consists of three parts:

(i) the first part deals with the detection and recognition of the street number field,

(ii) the second part deals with the generation of a lexicon of street names by accessing the ZIP + 4 database with the ZIP code and street number as indices,

(iii) and the third part deals with the recognition of street name through the use of word matching algorithm described earlier.

5.2.1. Street Number Location and Recognition

The street number is assumed to be the first field of the street line. If ZIP code line includes only the ZIP code, the second preceding line is first assumed to be the street line, otherwise the immediate preceding line is assumed to be the street line. The rest of the street number location works in the same way as in the ZIP code location. The word recognition algorithm employs a lexicon free word matching.

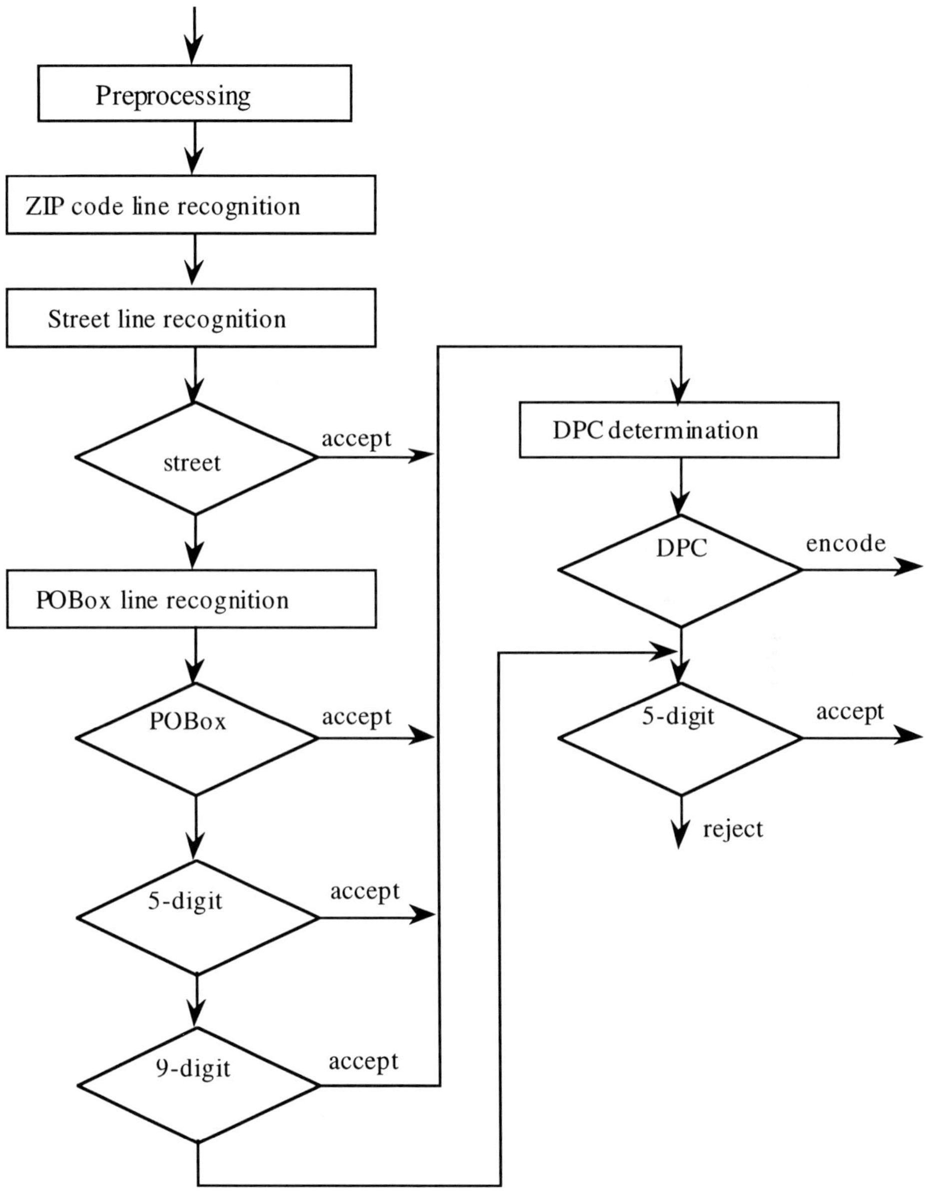

Fig. 13 Block diagram for handwritten address interpretation.

5.2.2. Street Name Recognition

A street line recognition system is composed of the ZIP/street number recognition system and the lexicon directed word recognition algorithm. The lexicon is generated through the ZIP+4 directory search for a given pair of ZIP code and street number. The street name recognition is performed in the long word lexicon scheme, i.e. the pre-directional, street name, and the suffix are concatenated in a word, and is dealt as a single word. The word images in a street line except the street number image are supplied as a single word image to the word recognition algorithm. The word recognition algorithm employs a lexicon directed word matching described in section 4.2.

6. Performance Evaluation - Case Study

6.1. Performance Evaluation of Word Recognition

A total of 2998 word images extracted from the "bha" database (handwritten address block data collected at Buffalo, New York) were used for this test. The test images included city, street, state, personal, and business names, including abbreviated forms. The style of writers ranged from strictly printed to strictly cursive, and from upright to very slanted. The distribution of writing styles reflected the corresponding distribution for the mail stream.

Each word image in the test database was associated with three separate ASCII lexicons of size 10, 100, and 1000, respectively. The lexicons were generated through a random selection process from a statistically significant sample containing city, street, state, and personal and business name words. Each of the three lexicons included the ASCII representation of the associated word image, spelt exactly as it appeared in the image. Thus if the word Illinois was wrongly spelt as Illionis, then the lexicon contained the word Illionis. Table 1 shows the cumulative correct recognition rate. The character classifier was designed (trained) using the character samples extracted from state name and city name word images in the "Bd" database. The number of characters used for classifier design was 22606 (435 per character in average) and the correct character recognition rate was about 74.2% for the design samples. The top correct recognition rate was 98.01%, 95.46%, and 91.49% for lexicons of size 10, 100, and 1000 respectively. Figure 14 shows examples of correctly recognized words. The speed of word recognition was 2.0, 2.5, and 3.5 sec/word for each lexicon on a SUN SPARC Station 2.

Table 1 Cumulative correct recognition rate of word recognition

Rank of correct word	Lexicon size 10	Lexicon size 100	Lexicon size 1000
1	98.01%	95.46%	91.49%
2	98.80%	96.70%	91.78%
5	99.60%	97.86%	94.89%

6.2. Performance Evaluation of an Integrated Address Interpretation System

The integrated address interpretation system was designed to determine the 9-digit ZIP code by locating and recognizing the ZIP code , street number and/or the PO Box fields.

The USPS 5-digit City/State/ZIP directory consisting of 100,000 records was used to generate a lexicon of city names for city name recognition. The USPS ZIP+4 address directory consisting of 26 million records was used to generate a lexicon of street names for street name recognition. The performance of the integrated system was evaluated using "bha" test samples. All the samples from bha_6000 to bha_7603 were used for this test. Table 2 summarizes the performance at different operating points specified in column 1 of the table. The error rate for one set of operating points was 1.12% with 50.19% encode rate. With a different set of operating points an error rate of 0.87% with 43.12% encode rate was obtained. In other words the system could be tuned to achieve a specified error rate.

Table 2. Error vs. Encode rate

t_{t1}	t_{t2}	Encode rate	Error	Correct
20.0	5.0	50.19 (803)	1.12 (9)	98.88 (794)
40.0	7.0	43.12 (690)	0.87 (6)	99.13 (684)

7. Conclusion

Word recognition algorithm using the segmentation-recognition approach is shown to be robust, accurate and commercially feasible. Context-free recognition is shown to be feasible for numeral string recognition, while lexicon directed approach is recommended for word recognition.

The performance of the integrated system developed for the US Postal Service exceeded the performance specifications set by USPS for processing handwritten addresses. The integrated recognition system incorporates several novel features such as tunability for adjusting error-rejection rates, lexicon truncation to achieve low error rate and high processing speed. However, it should be noted that this study was preliminary in nature. More exhaustive testing is necessary to tune the system's performance. The effects of image resolution on the performance must be investigated.

In conclusion, it can be stated that handwriting recognition is a feasible technology and can be used with advantage in many commercial applications such as address recognition, forms processing, check processing etc.

Acknowledgment

The authors would like to thank Dr. John Tan and Dr. Binh Phan of USPS for their support of this work and Mr. Gilles Houle of TRW, Inc. for his many critical suggestions.

Fig. 14 Examples of correctly recognized words

References

[1] L. S. Frishkoff and L. D. Harmon, Machine reading of cursive script, C Cherry (Ed), Information Processing, Butterworth, London, 1961, 300-315.

[2] L. D. Earnest, Machine recognition of cursive writing, C Cherry (Ed), *Information Processing*, Butterworth, London, 1961, 462-466.

[3] B. Plessis, A. Siscu, E. Menu, and J. W. V. Moreau, Isolated handwritten word recognition for contextual address reading, *Proc. USPS 5-th Advanced Technology Conf.*, Nov. 1992, 579-580.

[4] T. Paquet and Y. Lecourtier, Handwriting recognition: Application on bank cheques, *Proc. of 1st Intl. Conf. Document Anal. and Recog.*, St. Malo, France, Sept. 1991, 749-750.

[5] M. Leroux, J. C. Salome, and J. Badard, Recognition of cursive script words in a small lexicon, *Proc. 1st Intl. Conf. Document Anal. and Recog.*, St. Malo, France, Sept. 1991, 774-775.

[6] N. Gorsky, Off-line recognition of bad quality handwritten words using prototypes, *Fundamentals in Handwriting Recognition*, S. Impedovo (Ed), NATO ASI Series F: Computer and Systems Sciences, Vol. 124, Springer-Verlag, (1994), 199-217.

[7] L. R. Rabiner and B. H. Juang, An introduction to Hidden Markov Models, *IEEE ASSP Magazine*, Jan. 1986, 4-16.

[8] A. Kundu, Yang He, and P. Barl, Recognition of handwritten word: first and second order Hidden Markov Model based approach, *Pattern Recognition*, Vol. 22, No. 3, (1989), 283-297.

[9] Mou-Yen Chen, Amlan Kundu and Sargur N. Srihari, Variable duration Hidden Markov Model and morphological segmentation for handwritten word recognition, *IEEE Trans. Image Processing*, Vol. 4, No. 12, Dec. 1995, 1675-1688.

[10] Y. Ishitani, Document skew detection based on local region complexity, *Proc. of 2nd ICDAR*, (1993),.49-52.

[11] V. Govindaraj, A. Shekhawat, and S. N. Srihari, Interpretation of handwritten addresses in US mail stream, *Proc. of 3rd IWFHR*, (1993), 197-206.

[12] M. J. Ganzberger and R .M. Rovner, D. J. Hepp, A. M. Gillies, C. D. Lake, and C. R. Wood, A system for handwritten address interpretation, *Proc. of 5th Advanced Technology Conference*, (1991), 337-351.

[13] F. Kimura, Y. Miyake, and M. Shridhar, Zip code recognition using lexicon free word recognition algorithm, *Proc. of 3rd ICDAR* , (1995), 906-910.

[14] A. Devijver and J. Kittler, Pattern Recognition, , Prentice/Hall International, London (1982), 409-410.

[15] F. Kimura, M. Shridhar, and Z. Chen, Improvements of a lexicon directed algorithm for recognition of unconstrained handwritten words, *Proc. of 2nd ICDAR*, (1993), 18-22.

[16] N. Otsu, A threshold selection method from gray-level histograms, *IEEE Trans. SMC*, Vol. SMC-9, No. 1, (1979), 62-66.

[17] S. N. Srihari, V. Govindaraju, and R. K. Srihari, Handwritten text recognition, *Proc. of 4th IWFHR*, (1994), 265-274.

[18] F. Kimura, S. Tsuruoka, M. ShrIdhar, and Z. Chen Context directed handwritten word recognition for postal service applications, *Proc. of 5th Advanced Technology Conference*, (1992), 199-213.

[19] E .Lecolinet and J. Crettez, A grapheme-based segmentation technique for cursive script recognition, *Proc. of 1st ICDAR*, (1991), 740-748.

[20] R .M. Bozinovic and S. N. Srihari, Off-line cursive script word recognition, *IEEE Trans. PAMI*, Vol. 11, No. 1, Jan. 1989, 68-83.

[21] C. R. Nohl, C. J. Burges, and J. I. Ben, Character-based handwritten address word recognition with lexicon, *Proc. of 5th Advanced Technology Conference*, (1992), 167-180.

[22] K. Fukunaga, *Introduction to Statistical Pattern Recognition*, Second Edition, Academic Press, New York (1990), 76-78.

[23] T. Wakabayashi, S. Tsuruoka, F. Kimura and Y. Miyake, Accuracy improvement through increased feature size in handwritten numeral recognition, *Systems and Computers in Japan* , Vol. 26, No. 8 Scripta Technica, Inc., (1995), 35-44.

[24] D. Lee and S. N. Srihari, Handprinted digit recognition: A comparison of algorithms, *Proc. of 3rd IWFHR*, (1993), 153-162 .

[25] S. Watanabe and N. Pakvasa, Subspace method of pattern recognition, *Proc. 1st Int. Joint. Conf. on Pattern Recognition*, Oct. - Nov. 1973, 25-32.

[26] E. Oja, Subspace Methods of Pattern Recognition, *Research Studies Press*, England, 1983

[27] F. Kimura, Y. Miyake, and M.Shridhar, Relationship among quadratic discriminant functions for pattern recognition, *Proc. of 4th IWFHR* , (1994) 418-422.

[28] D. Guillevic and C. Y. Suen, Cursive script recognition: a sentence level recognition scheme, *Proc. of 4th IWFHR*, (1994), 216-223.

[29] S. Liang, M. Ahmadi and M. Shridhar, Segmentation of interference strokes using morphological approach, *Proceedings of 3-rd Int'l Conf. Document Anal. and Recog.*, Montreal, Canada, Aug. 1995, 1042-1046.

[30] F. Kimura, M. Shridhar and N. Narasimhamurthi, Lexicon directed segmentation - recognition procedure for unconstrained handwritten words, *Proceedings of third International Conf. on Frontiers in Handwriting Recog.*, Buffalo, May 1993, 122-131.

[31] H. Bunke, A fast algorithm for finding the nearest neighbor of a word in a dictionary, *Report of Institut fur Informatik und Angewandte Mathematik*, Universitat Bern, Switzerland, Nov. 1993.

Handbook of Character Recognition and Document Image Analysis, pp. 157–182
Eds. H. Bunke and P. S. P. Wang

CHAPTER 6

HANDWRITTEN WORD RECOGNITION USING HIDDEN MARKOV MODEL

AMLAN KUNDU
US WEST Advanced Technologies, Boulder, CO 80303, USA

This paper discusses a number of handwritten word recognition (HWR) systems where the segmentation uncertainty, the character transition information, and the shape ambiguity are well characterized by a stochastic model called the Hidden Markov Model (HMM). Some representative schemes and their typical experimental performance are described. It is concluded that despite the difficulties regarding feature extraction and limited available databases, the HMM could be a dominant technique in the field of HWR.

Keywords: Off-line handwriting recognition, hidden Markov models (HMM), continuous density HMM, non-ergodic HMM, multi-level HMM, modified Viterbi algorithm, segmentation, morphology.

1 Introduction

In on-line handwriting recognition, the machine recognizes the writing while the user writes. Off-line handwriting recognition, by contrast, is performed after the writing is completed. The off-line problem is more difficult as the temporal information, such as the number of strokes, the order of the strokes, etc., is generally not available. In this section, the relevance of HMM to the HWR problem is outlined.

1.1 Model for Handwritten Words

Real world processes generally produce observable outputs which can be characterized as signals. Whatever their form, signals are of interest only because of the information they contain. The signal model is usually a compact parametric model with few parameters that often works very well in practice. Such models offer better insights into the signals which facilitate the processing and classifications of these signals. We provide the following rationale for the hidden Markov model as our preferred signal model in relation to HWR problem.

- Markov models: In any written language, the information is primarily sequential in nature. The Markov models can successfully code the sequential information. For example, a word can be regarded as a Markov chain of individual characters; a sentence is then another Markov chain of words, etc.

- "Hidden" states: Hidden Markov Model (HMM) is a doubly stochastic process with an underlying Markov process that is not directly observable (it is hidden), but can only be observed through another set of stochastic processes that produce the sequence of observed symbols. It can handle more general cases as it allows one more degree of freedom than the Markov model. In handwriting, the writer translates the language to the written text with added ambiguities, e.g., writing style variation, ambiguity caused by the connected letters etc.. These ambiguities can be accounted for by the symbol probability, i.e., the relationship between the observed symbols and the "hidden" states.

1.2 Types of Handwritten Words

Handwritten words, based on the form of written communication, can be divided into two categories: cursive scripts and hand-printed characters. In practice, a combination of these two forms is seen frequently. Based on the nature of writing and the difficulty of segmentation processes, five stages [1] have been defined to typify different handwriting styles – Figure 1 gives examples of these stages (except stage 1, defined as boxed discrete characters, which can be regarded as a special case of stage 2). Hand-printed character recognition (stages 1, 2 and 3) is simpler

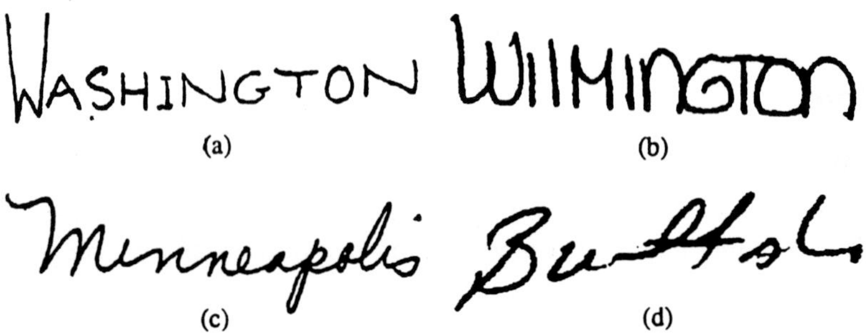

Figure 1: Different types of handwritten words: (a) Stage 2: spaced discrete characters (WASHINGTON); (b) Stage 3: run-on discretely written characters (WIlMINGTON); (c) Stage 4: pure cursive script (Minneapolis); (d) Stage 5: mixed cursive, discrete, and run-on discrete (BuffALo).

due to the absence or near absence of segmentation problems and fewer variations at the character level [2]. The strategies for cursive script recognition (stage 4) can be roughly classified into three categories. In the first category, the word is segmented into several characters and character recognition techniques are applied to each segment, e.g., [3]. This method depends heavily on the accuracy of the segmentation points found. However, such an accurate segmentation technique is not yet available, and may need to combine the interaction of word segmentation and character recognition. In the second category, the whole words are recognized without doing any kind of formal segmentation, e.g., [4, 5]. The global shape is analyzed to hypothesize the words. This is not a practical approach for the case of a large lexicon. The third category is a compromise solution between the above two schemes. It does a loose segmentation to find a number of potential segmentation points in the pre-segmentation procedure [6]. The final segmentation and the word length are determined later in the recognition stage by the help of lexica. The HMM based HWR systems usually belong to the third category.

2 Hidden Markov Model

The theory of hidden Markov models was introduced in the late 1960's by Baum *et al.* [7, 8]. Shortly afterward, the HMM's were applied to automatic speech recognition at CMU [9] and IBM [10]. Nowadays, the HMM's have become the predominant approach in speech processing and recognition [11, 12]. Recently, the HMM has been applied to many other real world problems, for example, shape recognition, texture segmentation, eye movement, target tracking [13], sonar signal identification etc. More recently, the HMM's have also been used for handwriting recognition with success as described in the later sections.

2.1 The Theory of Hidden Markov Models

Markov Dependency: The joint probability $P(Q)$ of a sequence of states $Q = \{q_1, \ldots, q_T\}$ can be defined as (assuming that the occurrences of the states depend only on their immediate preceding states, i.e., first-order Markov chain)

$$P(Q) = P(q_1)P(q_2 \mid q_1)P(q_3 \mid q_2) \cdots P(q_T \mid q_{T-1}) \ . \tag{1}$$

Similarly, n-th order Markov chains can be defined if we assume that q_t depends only on n immediate preceding states.

Hidden Markov Model: An HMM is a doubly stochastic process with an underlying Markov process that is not observable (the states are hidden), but can only be observed through another set of stochastic processes which are produced by the Markov process (the observations are probabilistic functions of the states). Let's assume that a sequence of observations $O = (o_1, \ldots, o_T)$ is produced by a Markov state sequence $Q = (q_1, \ldots, q_T)$ where each observation o_t is from the set of M observation symbols $V = \{v_k; 1 \leq k \leq M\}$ and each state q_t is from the set of N states $S = \{s_i; 1 \leq i \leq N\}$. Thus, an HMM can be characterized by

$$\begin{aligned}
\Pi &= \{\pi_i\}, \text{ where } \pi_i = P(q_1 = s_i) \text{ is the initial state probability;} \\
A &= \{a_{ij}\}, \text{ where } a_{ij} = P(q_{t+1} = s_j \mid q_t = s_i) \text{ is the state transition probability;} \\
\Gamma &= \{\gamma_j\}, \text{ where } \gamma_j = P(q_T = s_j) \text{ is the last state probability;} \\
B &= \{b_j(k)\}, \text{ where } b_j(k) = P(o_t = v_k \mid q_t = j) \text{ is the symbol probability;}
\end{aligned} \tag{2}$$

and they satisfy the probability constraints

$$\sum_{i=1}^{N} \pi_i = 1; \ \sum_{j=1}^{N} a_{ij} = 1 \ \ \forall\, i; \ \sum_{j=1}^{N} \gamma_j = 1; \ \sum_{k=1}^{M} b_j(k) = 1 \ \ \forall\, j. \tag{3}$$

The last state probability Γ is included in this definition. In a Markov state sequence, the last state probability models the different probable final states as the initial state probability does for the initial states. Or, one could interpret this probability as a transition probability to an imaginary "final" or "absorbing" state. This kind of definition makes HMM more robust for many general real-world applications [13] although it does not appear in most HMM related literature. We will denote the HMM by a compact notation $\lambda = \{\Pi, A, \Gamma, B\}$.

2.2 An Example

To understand the concept of HMM, consider the following simplified example. We are asked to estimate the daily weather condition, dry or rainy (2 states). Assume that today's weather only depends on the weather of yesterday, but not on other days in the past or the future (first-order Markov chain). Our equipment cannot tell exactly whether it is dry or rainy, but the humidity can be measured accurately (output observation). So, we have the following definitions:

States (Weather): S = { s_1=dry, s_2=rainy }
Symbols (Humidity): V = { v_1=0%, v_2=25%, v_3=50%, v_4=75%, v_5=100% }

Note that the humidity, which is a continuous variable, is quantized to five levels using some pattern matching rule. Thus, the probability of changing weather (or staying with the same weather) represents the state transition probability, while the probability of the humidity for each weather condition is interpreted as the symbol probability. Suppose that we have measured the daily humidity for a week, i.e., the observation sequence, which is given by

$$O = 0\%, 25\%, 25\%, 50\%, 25\%, 75\%, 50\%$$

Now, if we have the state transition, initial and the symbol probabilities at our disposal, the HMM can help us estimate the weather condition for this week by means of Viterbi algorithm (Appendix A.2.). Observe that we started with quantitative observations such as humidity, and ended up with such qualitative decisions as rainy day or sunny day. In HWR, the input is the raw word image which provides nothing but quantitative information. From this information, we wish to find the constituent letters, i.e., qualitative information, of the word image. In HMM, we identify the symbols with the quantitative observations and the states with the qualitative information we wish to arrive at. This interplay of symbols and states, quantitative versus qualitative information makes the HMM such a grand model for many recognition tasks.

2.3 The Three Problems for HMM's

In the study of HMM's, there are three basic problems.

1. Given an observation sequence $O = o_1, \ldots, o_T$ and a model $\lambda = \{\Pi, A, B, \Gamma\}$, how do we find $P(O \mid \lambda)$? This is the scoring problem.

2. Given the training data $O = o_1, \ldots, o_T$, how do we find $\lambda = \{\Pi, A, B, \Gamma\}$ such that $P(O \mid \lambda)$ is maximum? This is the training problem.

3. Given $\lambda = \{\Pi, A, B, \Gamma\}$ and an observation sequence $O = o_1, \ldots, o_T$, how do we find the optimal state sequence $Q = q_1, \ldots, q_T$? This is the recognition problem.

In the sequel, we will discuss the solutions to these problems.

- **Scoring Problem**: A forward variable $\alpha_t(j)$ is defined as the probability of having the state q_j at time t generating the observation sequence $O_1^t = o_1, \ldots, o_t$. It can be computed iteratively as (assuming $o_t = v_k$)

$$\alpha_t(j) = \begin{cases} \pi_j b_j(k) & \text{if } t = 1 \\ \sum_{i=1}^{N} \alpha_{t-1}(i) a_{ij} b_j(k) & \text{otherwise} \end{cases} \tag{4}$$

Then it can be shown that

$$P(O \mid \lambda) = \sum_{i=1}^{N} \alpha_T(i) \gamma_i \tag{5}$$

This algorithm is called the *forward algorithm* [11].

- **Training Problem**: In the conventional training of HMM, the *Maximum Likelihood* (ML) is usually used as the optimization criterion. That is, given the training sequence $O = o_1, \ldots, o_T$, we want to adjust the model parameters $\lambda = (\Pi, A, \Gamma, B)$ such that $P(O \mid \lambda)$ is maximized. An iterative procedure, which is known as the Baum-Welch algorithm [7], is usually used for this purpose (Appendix A.1). This kind of algorithm, however, can only reach a local optimum. So the performance depends heavily on the initial guess of the parameters. In general, the HMM's can be trained by the Baum-Welch algorithm with satisfactory performance.

- **Recognition Problem**: In many situations, our objective is to find the optimal state sequence of an HMM. That is, given the model $\lambda = \{\Pi, A, \Gamma, B\}$ and an observation sequence $O = o_1, \ldots, o_T$, we want to find the optimal state sequence

$$Q^* = \arg\max_Q P(Q \mid O, \lambda) = \arg\max_Q P(Q, O \mid \lambda) \tag{6}$$

A *dynamic programming procedure*, called the Viterbi algorithm [14] (see Appendix A.2), can solve the recognition problem efficiently.

2.4 Implementation Issues

Continuous versus Discrete Model: Up to this point, we have considered only the case of discrete symbol HMM's, where the observations are chosen from a finite alphabet. There might be serious degradations associated with discretization, i.e., quantization, when the signals are continuous in nature [12]. Hence, it would be advantageous to consider the continuous HMM's where the observations are continuous signals (or vectors). For such a model, the symbol probability $b_j(k)$ is often replaced by the probability density function (pdf) of the form

$$b_j(x) = \sum_{m=1}^{M_j} c_{jm} \cdot \mathcal{N}[x, \mu_{jm}, U_{jm}], \ 1 \leq j \leq N \tag{7}$$

where $\mathcal{N}(.,.)$ represents a Gaussian distribution with mean vector μ_{jm} and covariance matrix U_{jm} for the m-th mixture component in state j, x is the vector being modeled, M_j is the number of Gaussian component $\mathcal{N}(.,.)$ in state j, and c_{jm} is the mixture coefficient for the m-th Gaussian component in state j. The mixture gains satisfy the stochastic constraint $\sum_{m=1}^{M_j} c_{jm} = 1, \ 1 \leq j \leq N$ with $c_{jm} \geq 0, \ 1 \leq j \leq N, \ i \leq m \leq M_j$, and the *pdf* is properly normalized, i.e., $\int_{-\infty}^{\infty} b_j(x)dx = 1, \quad 1 \leq j \leq N$. This *pdf* can be used to approximate, arbitrarily closely, any finite, continuous density function of practical importance [12].

Tied Transitions: For simplification, we usually assume that the output observation o_t is associated with the current state q_t only. An alternative is to assume that the observation emitted at time t depends only on the transition taking place at that time (from q_{t-1} to q_t).

$$P(o_t \mid o_1, \ldots, o_{t-1}, q_1, \ldots, q_t) = P(o_t \mid q_{t-1}, q_t) \tag{8}$$

The latter assumption, however, increases the size of the symbol probability matrix from NxM to N^2xM, and much more training samples are required to estimate the model parameters. To overcome this problem, one might consider "tying" the transitions, i.e., allowing different transitions to share the same symbol probability. An example of using the tied transitions is the MLHMM* character model (Sec. 5).

State Duration: In conventional HMM's, the inherent duration probability, $P(d \mid q_i)$, associated with state q_i with self transition a_{ii}, is of the form [12]:

$$P(d \mid q_i) = a_{ii}^{d-1}(1 - a_{ii}) = \text{prob. of } d \text{ observations while in state } q_i \tag{9}$$

This exponential distribution of state durations is inappropriate if the states of the HMM are to represent linguistically meaningful components [15, 12]. A Variable Duration Hidden Markov Model (VDHMM) is given by $\lambda = \{\Pi, A, \Gamma, B, D\}$;

$$D = \{p_i(d)\}, \text{ where } p_i(d) = P(duration = d \mid s_i) \text{ is the state duration probability} \tag{10}$$

and $\{\Pi, A, \Gamma, B\}$ are as defined in Eqs. (2). For more details related to VDHMM, please refer to [16, 17].

Scaling Problem: From Eq. (4), it can be seen that the forward parameter $\alpha_t(i)$ consists of a product of many terms, where each term is less than 1. Thus, $\alpha_t(i)$

*Please also see Appendix A.3.

starts to head exponentially to zero as t increases, i.e., we are faced with an underflow problem. In recognition of handwritten words, the observation sequence is not very long. Hence, by transforming multiplication into logarithmic addition, the scaling problem could be solved. Also, the Baum-Welch algorithm could be implemented with a built-in scaling property.

2.5 Implementation Strategies

In practical pattern recognition problems, there are two ways of building the HMM's [18]. One way is to build one or more HMM's for each class of pattern. Given an observed pattern, we calculate the path probability against each model, and classify it to the class whose model leads to the maximum path probability. This kind of model has been highly successful for speech recognition, especially for the recognition of isolated words [12]. We call this classical approach the *model discriminant* HMM (MD-HMM) since the patterns are classified by different models. Another approach is to build one *path discriminant* HMM (PD-HMM) for all the classes and use different paths to distinguish one class from the others [15, 13]. The advantages and disadvantages of each strategy are discussed below.

1. **Speed/memory and dictionary size:**
   If we model each word as an HMM, the MD-HMM is a reasonable approach for vocabularies of up to a few hundred words, and it becomes prohibitive in terms of memory and computation time when the size of the dictionary is 1000 words or bigger [15]. The PD-HMM, on the other hand, is likely to be independent of the dictionary size as one HMM is built for the whole dictionary.

2. **Accuracy of modeling dictionary constraint:**
   Since many modeling constraints can be easily implemented in the MD-HMM approach, the MD-HMM often performs better than the PD-HMM [13].

3. **Other considerations:**
   In the HWR problem, if one model is created for each word, we need a host of training samples to train that model (this requirement is not easily met at present). Also, the system with this kind of training requirement has "poor portability" - portability means the ability to adapt as the dictionaries change. For example, an HWR system that is trained to recognize the street names with the ZIP code 14228 cannot be used to recognize the street names with other ZIP codes. In the case of PD-HMM, however, the only need for a changing dictionary is to recompute the transition probability.

Another alternative is to combine the two approaches — PD-HMM and MD-HMM — and this composite approach has also been successfully used [13, 11].

2.6 Optimization Criteria

For MD-HMM, various optimization criteria, such as the traditional maximum likelihood (ML), maximum mutual information (MMI) [12], minimum discrimination information (MDI) and others [13] have been used to train the HMM's. In PD-HMM, on the other hand, the states are usually transparent during training and they are semantically meaningful. The reestimation, as expected, may produce a better model according to certain mathematical optimization criteria. However, it does not preserve the correspondence of the states to individual members of the alphabet (assuming that is the semantic meaning of the state); and yields less accurate recognition results [13, 15].

3 HMM's for HWR: An Overview

In this section, we review works on HWR using the HMM's. We start with the Single Contextual Hidden Markov Model (SCHMM) that was introduced in [19] to recognize the hand-printed words, i.e., handwritten words that are naturally segmented. When the letters of the words are naturally segmented, and if these letters are identified as states [19], there are a finite number of predetermined states, for example, 26 for English. The general handwritten words, however, are usually not naturally segmented into letters, and a word segmentation algorithm has to be used for such task. At present, there exists no good segmentation algorithm which is likely to separate all the letters perfectly, and without any spurious segmentation points. In [20], a more general framework that is applicable to cursive, non-cursive, naturally segmented or any other type of handwritten words is proposed. In this approach, a morphology based segmentation algorithm is first used to divide the word image into a sequence of segments, which could signify a whole letter, a partial letter or joint letters. The sequence of segments is then recognized by an HMM type stochastic network which can take care of the problem of touching and broken characters. Since touching characters are not guaranteed or required to be split by the segmentation algorithm, the number of states, which depends on the training set, may go up to 6,241 [20] for handwritten English words. Consequently, the state assignment for a large training set is rather complicated. Also, this individual segment based recognition system might never know how well a character is formed by combining several consecutive segments. Nevertheless, the scheme described in [20] has clearly shown that the application of HMM to a large vocabulary HWR problem is indeed much more complex than one described in [21].

To overcome the problems of the SCHMM system, a new system using a Continuous Density Variable Duration Hidden Markov Model (CDVDHMM) [22] is proposed. With the help of an enhanced segmentation algorithm, which is required to split all the touching characters (of course, this leads to more spurious segmentation points), the CDVDHMM defines the 26 letters in the alphabet as 26 different states, and this number is fixed and much smaller than the system described in [20]. Consequently, the recognition speed is also improved. The implementation and experiments for the CDVDHMM system are discussed in [13, 20] and are outside the scope of the present chapter. However, the main ideas are briefly sketched in Sec. 5. An alternative to the CDVDHMM in HWR is the NEHMM (Non-Ergodic HMM) based system proposed [23]. The NEHMM system follows the MD-HMM strategy but the model parameters can be derived from the statistics of the CDVDHMM.

This strategy appears to perform better than the CDVDHMM strategy at a slower speed. A combination using both CDVDHMM and NEHMM can be considered as a trade-off between performance and speed.

The problem with the VDHMM system is reliable computation of duration probabilities given the limited amount of databases available at the present time. An interesting idea has been presented in [18] to avoid the computation of duration probabilities. Because of over-segmentation, this scheme considers many different subsets of the segmentation points. Each subset leads to one distinct observation sequence. Thus, each handwritten word may be considered as a number of different realizations of the HMM, i.e., a number of different observation sequences. The recognition task is to find the best segmentation, i.e., find the subset that contains the correct segmentation points, and find the associated optimal state sequence, i.e., the letter sequence of the word. This philosophy is similar to that of VDHMM. The added complexity of computing the duration probability in each state is avoided in this approach by making a simplifying but realistic assumption which is stated as follows: Assuming that a character can be broken into at most 4 segments, there are four discrete duration probabilities for each state. Instead of assigning pre-computed duration probabilities to each state, only one duration will be picked (during recognition) by matching one, two, three and four consecutive segments to the symbols in the feature space, and finding the best match and its corresponding number of segments. In this way, the computation of duration probability in each state is avoided without sacrificing the advantage of VDHMM. However, the structure of the Viterbi algorithm used during recognition is substantially altered. The overall performance of this scheme, as expected, is quite similar to the VDHMM based word recognition system [18].

In the previous approaches (SCHMM, CDVDHMM, and NEHMM), the models are actually semi-hidden Markov models, i.e., the states of HMM's are transparent during training. Because the reestimation algorithms such as the Baum-Welch algorithm does not preserve the correspondence of the states to their semantic meanings, it is not suitable for the semi-hidden Markov models. In [13, 24], a novel approach is described using a Multi-Level Hidden Markov Model (MLHMM) which is a doubly embedded network of HMM's, i.e., character is modeled by an HMM while a word is a higher-level HMM. The HMM belongs to the MD-HMM strategy at the character level. Since the states are not assigned any semantic meaning at the character level, the re-estimation algorithm is applicable. For the word model, on the other hand, both the MD-HMM and PD-HMM strategies are applicable. Another major difference between this new system and the previous approaches is the *output-independence assumption* of the HMM (Sec. 2). The details of this approach are described in [13] and are briefly outlined in Sec. 5.

Although we will elaborate the SCHMM, VDHMM and MLHMM systems in the rest of this chapter, there are many other works that are equally important and have greatly influenced the evolution of handwriting recognition using HMM. We will review the related contributions briefly. The discussions here are, by no means, exhaustive. Today, there is a growing interest in applying HMM's to the problem of document analysis and recognition; and a large body of literature is being published in reputed journals and conference proceedings. Several promising research achievements have been presented at recent conferences and workshops, for example, ICPR [25, 26], ICASSP [27, 28, 22, 29, 30, 31], IWFHR'93 [32, 33, 34, 35], etc. Historically, the hidden Markov model has been used in text recognition in as early as 1980 by Cave and Neuwirth [36], who have analyzed machine-printed text using HMM. For on-line recognition of handwriting, the HMM was first used in [21] where the approach followed the basic HMM scheme - each word is modeled by an HMM, like the one used in the recognition of isolated digits of speech. The success of these attempts, however, was limited to constrained experiments: single writer, small and fixed vocabulary, small test samples, etc. The application of HMM's to the more general problem of handwriting recognition involving large dictionaries, off-line data, unconstrained style, etc., was introduced in [19, 20, 13]. The basic problems of handwriting recognition are common to all languages, but the special features, constraints, etc. for each language need to be considered as well. For example, the large set of Chinese characters and their complicated combination of strokes make the recognition task difficult. In [37], the projection profiles are first obtained from each Chinese character image. The HMM's are then used to model the sequence of the histogram of the projection profiles. To counter the serious loss of stroke information after the projection of the character image [35], the Regional Projection Contour Transform (RPCT) is proposed to transform the character image into the contour of four feature maps. As the pattern transformed by RPCT has only one outer contour and does not contain any internal contour, the HMM's used for 2-D planar shape recognition such as the one proposed in [38] are directly applicable. Another interesting application of HMM to off-line HWR problem was recently introduced in [39].

Besides handwriting recognition, HMM's have also been used to analyze document images. Vlontzos and Kung have proposed a multilevel structure of HMM's for the recognition of machine (or hand-) printed text [40]. Kuo and Agazzi have successfully spotted key words in a poorly printed document image using the pseudo 2-D HMM [29], where the word image is modeled by a hierarchical structure composed of vertical and horizontal HMM's. A complete scheme for encoding the printed document using HMM's, from the pixel level to characters, words, and whole documents, is recently proposed by Kopec and Chou [41].†

The above review clearly indicates that the field of document understanding and, in particular, handwriting recognition using HMM are undergoing an exciting phase of research and development. Indeed, the HMM has the potential to become one of the most dominant techniques in this field.

†For more on other techniques related to handwriting and cursive script recognition, the readers are referred to the chapter by M. Sridhar and F. Kimura in this book.

4 Preprocessing, Segmentation and Feature Extraction

The HMM is a model for 1-D signals while handwriting is a 2-D signal. Fortunately, the progression of handwriting in all known languages proceeds either from left to right, or from right to left. Utilizing this constraint, it is possible to represent a handwritten word as a sequence of small segment images, i.e, observations, in the right order. This sequence representation is straightforward for speech because of its strong temporal constraints. For handwriting, especially for off-line recognition, it is not trivial; and the need for a good segmentation algorithm cannot be overstressed. Another crucial aspect in making HMM successful is good representation of the frames (segments) in terms of features. For handwriting, unfortunately, no satisfactory representation exists for the observation frames; and this remains an open research direction.

Finally, one has to select a good unit of the signal to be modeled by the HMM. Like in speech, the words are also the natural units for handwriting. Moreover, since the models are created by the given lexicon, "portability" is another serious consideration. The "portability" — adaptation to changing dictionary — is essential in most handwriting recognition applications. This section describes the typical but essential algorithms that are used to process the word image before it is an input to an HWR system.

4.1 Preprocessing

The most important preliminary processings are slant normalization and noise reduction (Fig. 2).

- Slant Normalization:
$$X_1 = SNORM(X_0, \theta) \tag{11}$$
where X_0 is the original word image, and θ is the slant estimated by the program mentioned in [6]. As per the suggestion given in [6], this slant normalization needs to precede other preprocessing.

- Noise Reduction: A typical noise reduction scheme could be
$$X_2 = (X_1 \bullet B_1) \circ B_2 \tag{12}$$
where B_1, B_2 are 3×3 and 2×2 disks, respectively. The morphological closing ($\bullet$) is used here to fill narrow channels and thin lakes, then an opening ($\circ$) is used to smooth the contours and suppress small islands (noise). ‡

4.2 Segmentation

For unconstrained handwritten words, due to enormous variations in the writing styles, there is no algorithm which could provide only the correct segmentation points. Over-segmentation is an essential feature of most segmentation algorithms.

‡More details on morphological image processing can be found in the chapter by T. M. Ha and H. Bunke in this book.

The segmentation algorithms also have to parse the segments in the right order. The task of the HWR system is to pick the right segmentation points during recognition. One successful algorithm is briefly described in the following [13]. The criteria for this segmentation algorithm are: (1) Each complete character can be segmented into at most 4 parts; (2) all touching characters must be split; (3) no "null" state is allowed. The algorithm first partitions the object pixels into two classes, singularity and regularity. Basically, the backbones of the character bodies can be represented by the singular strokes while the regular strokes are the connections of these bones. The singular strokes are located by an opening operation. The structured element used here is a one-pixel-wide vertical line whose length is decided by the stroke width [13]. Then an iterative dilation with a 3-pixel-long vertical line structured element is used to group these singular strokes. The regular strokes are formed by subtracting the image of singular strokes (Fig. 2(d)) from the word image (Fig. 2(c)). In Fig. 2(e), each regular stroke is represented by a one-pixel-wide vertical line. These are the candidates for segmentation points (SP). These steps are followed by verification of the segmentation points. The preliminary segments are found by subtracting the SP's (Fig. 2(e)) from the word image (Fig. 2 (c)). The verification procedure is done by finding the right-most segment (RM) and left-most segment (LM) and then tracing the sequence of qualified segments. The tracing procedure can be implemented by the Viterbi algorithm with properly assigned transition costs, or the geodesic morphology based algorithm [13]. Finally, these segments are fine tuned using some concavity information (Figs. 2(f-h)).

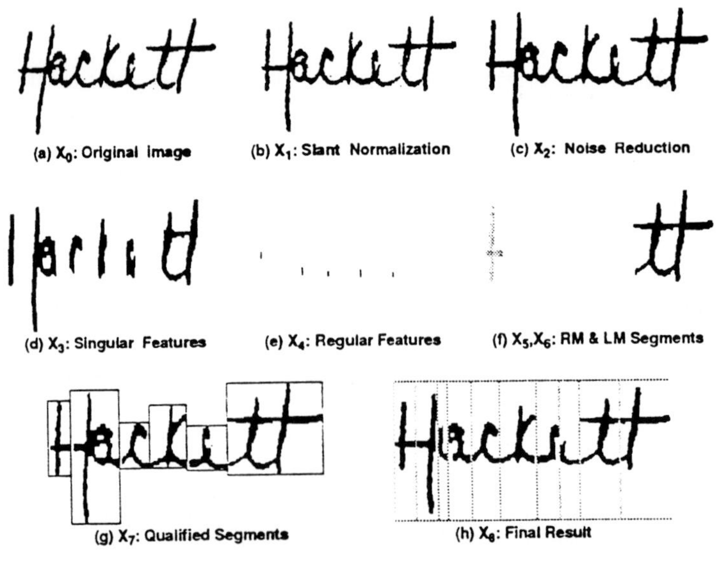

Figure 2: An example of pre-processing and segmentation.

4.3 Feature Extraction

To represent each segment image by a more compact form, i.e., a feature vector, a mathematical model of such images with a finite number of parameters is desired. Unfortunately, no satisfactory mathematical model exists yet for such signals. As a result, there is no universally accepted set of feature vectors in document image understanding. Almost all the researchers rely on heuristic features [13]. It is clearly understood, however, what specific information the features should capture. The features that capture topological and geometrical shape information, both globally and locally, are desired most. The features that capture the spatial distribution of black pixels (assuming that white pixels represent the background) are also very important. Some features related to positional information of the segment vis-a-vis others are also desired. A good mixture of these features is expected to work well. Using experimental results as the selection criterion, the following thirty-five features are used for the definition of symbols in [13, 20]. These features are: 3 moment features, 2 zero-crossing features (one along the horizontal and the other along the vertical direction), 11 regional and global topological features such as T-joint, X-joint, end-point and no-of-segments, and 19 spatial distribution of pixel. While the moment features provide information about the global shape, the topological features capture many regional and local attributes of the shape of character. The features related to the spatial distribution of pixels capture information about writing style variations. Needless to say, this feature selection technique is empirical and largely subjective.

5 HMM Based HWR Systems

This section briefly describes some of the details for the successful implementation of HMM based HWR systems.

5.1 A HWR System for Naturally Segmented Words

This problem is, by far, the simplest one in HWR because the segmentation points are naturally given. Such a system would only need preprocessing and feature extraction but no segmentation. The sequence of feature vectors are used to train the HMM in the training mode and to recognize an unknown word in the recognition mode. The output is either the correctly recognized word or a small set of words that includes the correct word.

5.1.1 States and symbols for handwritten words

In English, there are 26 letters in the alphabet; these 26 letters correspond to 26 states, i.e., N=26. The states could then be identified as q_1= a or A, q_2= b or B and so on. In addition, 10 numbers (0,1,..,9) could also be incorporated into the model. The next problem is the identification of the number of possible symbols (M) and their representation v_1, v_2,..,v_M. As the writing style varies among individuals, there are many possible representation of the same letter, say r. This might suggest a Markov model with an increased number of states. However, different letters may look alike. For example l written by one person may look like e written by another person. That is to say, a set of features (a symbol) that represents the letter l of one person may represent the letter e of another person. If l is counted as 12th state and e is counted as 5th state, the same symbol can appear in both states (state 12 and 5). This is the rationale behind the use of the hidden Markov model rather than the Markov model with an increased number of states. As there are many variations of l itself depending on the writer's identity, writing equipment variation etc., to represent l under all conditions of variations, a relatively large number of symbols are required. Using the vector quantization algorithm, the optimum set of M symbols, i.e., the codebook, is generated such that the average distortion in replacing each of the training set vectors s_i , i=1,..,I, by the closest entry in the codebook , $\hat{s_m}$, is minimum. The distance measure used in [19] is the unweighted Euclidean distance measure; and M, the optimum number of symbols, is generally determined experimentally.

5.1.2 Calculation of model parameters

State Transition Probability: In Cryptography, the English language is often modeled as a Markov source with 26 states. The 1-state transition probabilities and initial probabilities are then determined by statistical studies [19]. Since the states in our proposed HMM are the letters of English language, the a_{ij} probabilities are readily determined.

Initial State Probability: The meaning of the initial probability π_j is the probability that a word could start with a letter depicting the state "j". Using a commonly used (Merriam-Webster) dictionary and ignoring the word frequency distribution, the initial probability for letter 'z' is π_{26} =0.002, and so on.

Symbol Probability:

$$b_j(k) = \frac{\gamma_j(k)}{\gamma_j} \tag{13}$$

In this equation, γ_j = total number of times the letter depicting the state j appear in the sample; $\gamma_j(k)$ = total number of times the symbol k is observed while the state is "j". The symbol probability could approach the actual probability only when the sample size is sufficiently large.

5.1.3 Recognition of unknown words

For a given test word , each unknown letter in the word is first transformed into its representative symbol by matching it with all the symbols. Assuming that the test word is T time units long, this procedure yields an observation sequence of length T given by $O_1, O_2, .., O_T$. For example, if T=4, the observation sequence could be symbol # 50, 23, 11, 72 from the set of 90 symbols determined experimentally. The recognition problem is to find the best sequence of states, i.e., letters, given these observed symbols. If $P(X \mid O)$ is the probability of choosing the state sequence X given the observed symbol sequence O, the optimal state sequence corresponds to the maximum value of this probability. The solution can be obtained by the 'Viterbi algorithm' (Appendix A.2).

5.1.4 Experimental results

During the learning phase, 2500 letters written by different writers of different sex and age with very diverse ethnic backgrounds are used. Using 2500 letters, a VQ codebook of 90 symbols is generated; and the symbol probability distribution is calculated. Other model probabilities are calculated using the procedure as outlined. In one recognition experiment, 20 words written by different writers are used to test the recognition system. Up to 91% recognition accuracy is reported. A slightly better result is obtained with a second order model. For the implementation of the second order model and other details, the readers are referred to [19].

5.2 A General HMM Based HWR System

Since the segmentation points are not given, we need to find them before feature extraction. Also, as before, we are interested in modeling all the handwritten words in the dictionary by one PD-HMM where the states are identified with the letters. Naturally, the HMM structure has to account for the segmentation ambiguity, i.e., spurious segmentation points. The variable duration HMM (VDHMM) is ideally suited for this purpose. Here, each state is allowed to have variable durations with corresponding probabilities. During recognition, the HMM would try to fuse a number of consecutive segments together to best match a character (in the symbol space), and the whole sequence to a string of characters.

5.2.1 CDVDHMM statistics

In this HWR system, the discrete state duration probability $P(d|q_i)$ is estimated from the training samples with $d = 1, 2, 3, 4$ and $i = 1, 2, \ldots, 26$. The reason is that the segmentation algorithm segments each handwritten character into at most four segments. The CDVDHMM statistics are computed from two sources: training images and a given dictionary.

State probabilities: Since the 26 letters of the alphabet are defined as the states of our CDVDHMM, it is straightforward to compute the initial π_i, transition a_{ij} and last-state γ_j probabilities as:

$$\pi_i = \frac{\text{number of words beginning with } \mathcal{L}(q_i)}{\text{total number of words in the dictionary}} \tag{14}$$

$$a_{ij} = \frac{\text{number of trans. from } \mathcal{L}(q_i) \text{ to } \mathcal{L}(q_j)}{\text{number of transitions from } \mathcal{L}(q_i)} \tag{15}$$

$$\gamma_j = \frac{\text{number of words ending with } \mathcal{L}(q_j)}{\text{total number of words in the dictionary}} \tag{16}$$

where the $\mathcal{L}$ function transforms a state to its representative member of the alphabet. To calculate the state duration probability, we need to perform the segmentation procedure over all the training images. Inspecting the segmentation results, the state duration probability $P(d|q_i)$ is estimated as

$$P(d|q_i) = \frac{\text{number of times that } \mathcal{L}(q_i) \text{ is split into } d \text{ parts}}{\text{number of times that } \mathcal{L}(q_i) \text{ appears}} \tag{17}$$

Mixture Gaussian distribution for symbols: With so many different writing styles for handwritten words, a single Gaussian distribution is definitely not good enough to model each character symbol. Hence, the mixture Gaussian distribution is used in this system (see Sec. 2 for details). Each Gaussian distribution in the feature space is expected to represent one among many different writing styles of the characters. As suggested in [12], because of the limited amount of training data available, a small constant ρ is added to the diagonal elements of the covariance matrix to prevent it from becoming singular. It is relevant to mention that the observation O in VDHMM is composed of one or several segments. The symbol probability is defined as

$$b_j(o_1 o_2 \cdots o_d) = b_j(O_1^d)^d \tag{18}$$

where O_1^d is the image built by merging segment images $o_1, o_2, \ldots, o_d$ together. The power of d in Eq. (18) is used to balance the symbol probability for a variable number of segments.

5.2.2 *Modified Viterbi algorithm (MVA)*

The objective of the recognition phase is to find the optimal state sequence I^* given a sequence of observations O and model parameter λ, i.e.,

$$I^* = \arg\max_I [\Pr(I \mid O, \lambda)] \tag{19}$$

where $$\max_I \Pr(I \mid O, \lambda) = \max_I \frac{\Pr(O, I \mid \lambda)}{\Pr(O)} = \max_{1 \le i \le N} \frac{\delta_T(i)\gamma(i)}{\Pr(O)}$$

$$\delta_t(j) = \max_{1 \le i \le N} \max_{1 \le d \le D} \delta_{t-d}(i) a_{ij} P(d|q_j) b_j(O_{t-d+1}^t)^d \tag{20}$$

Equations (19–20) suggest the Viterbi algorithm for finding the best path. In [20], two MVA's, which provide an ordered list of the best L state sequences, are described. The first one is a parallel version which simply extends the Viterbi net to a three-dimensional array where the third dimension represents the choice. On the other hand, the serial version, which searches the $(l+1)$th globally best path based on the previous l best paths, can be more efficiently programmed on a conventional machine. These two MVA's, are adapted to VDHMM's by incorporating the duration probability [13]. A full mathematical description of the algorithm is outside the scope of this paper, but an outline is given in Appendix A.2.

5.2.3 *Post-processing*

The output of the Viterbi algorithm for a PD-HMM system is not guaranteed to be a legal word from the given dictionary. A post-processing step is necessary and its objective is to find

$$W_j^* = \arg\max_{1 \le j \le J} \{\sum_{l=1}^{L} \Pr(W_j \mid I^{l\text{-th}})\} \tag{21}$$

assuming a J-word dictionary ($W_1, W_2, \ldots, W_J$) and L character strings ($I^{1st}, \ldots, I^{L\text{-th}}$) from the modified Viterbi algorithm. Usually, a word-length filter with $\pm 30\%$ of the estimated word length is used as its range to trim the dictionary size. Then the string edit distance [43] of the output words is computed against all the words in the dictionary. That is, $\Pr(W_j \mid I^{l-\text{th}})$ in Eq. (21) is calculated as

$$\Pr(W_j \mid I^{l-th}) = w^{l-\text{th}} \cdot min\_edit\_distance(W_j, I^{l-\text{th}}) \tag{22}$$

The normalized path probability associated with state sequence $I^{l-\text{th}}$ is used as the weight factor $w^{l-\text{th}}$. A dynamic programming approach described in [44] gives an efficient way to find the minimum edit distance (*min_edit_distance*) between W_j and $I^{l-\text{th}}$. If the state sequence $I^{l-\text{th}}$ exactly matches W_j in the given dictionary, that is, *min_edit_distance* = 0, this word is said to be *directly recognized* as W_j. Otherwise, hypotheses based on the weighted edit distances to all the dictionary words are generated. To calculate the edit distance, the error probabilities of insertion, deletion and substitution for a certain letter (or a pair of letters for conditional error probabilities) are to be estimated in advance. For the lack of complete knowledge at this point, a simple empirical guess of these probabilities [13] is used.

5.2.4 Experiments

To evaluate the performance of the CDVDHMM based HWR system, the USPS databases are used in the experiments. In the following experiment, among 1583 available images, 1489 images are used for training, and 94 images are used for testing. In the recognition phase, the dictionary with the associated word frequency is used to extract the initial state, last state and state transition probabilities. In Table 1, a 271-word dictionary is used as the lexicon. Given the limited databases

Table 1: Cityname recognition using the 271-word dictionary.

ρ	0.05	0.08	0.10	0.12	0.15
Dir. Recog.	36.2%	40.4%	40.4%	43.6%	45.7%
1 Hypoth.	63.8%	69.1%	70.2%	70.2%	67.0%
2 Hypoth.	70.2%	77.7%	76.6%	77.7%	75.5%
5 Hypoth.	78.7%	83.0%	83.0%	83.0%	83.0%
20 Hypoth.	87.2%	89.4%	88.3%	87.2%	88.3%

available at present and the sensitivity of HMM training to databases, the performance of CDVDHMM based system is commendable. Figure 3 depicts a typical output of the CDVDHMM based HWR systems. Observe that, for a well-written word, the word is recognized even when the dictionary size is 1000. For a badly written word with a similar dictionary, the 18th hypothesis could recover the correct word.

Model Optimization: The re-estimation algorithm, such as the Baum-Welch algorithm (based on maximum likelihood criterion), may produce a better model according to a certain mathematical optimization criterion. However, it does not preserve the correspondence of the states to individual members of the alphabet; and yields less accurate recognition results. In the next subsection, we describe an approach for HWR where the states at the character level are not assigned any semantic meaning, and the re-estimation algorithm is applicable.

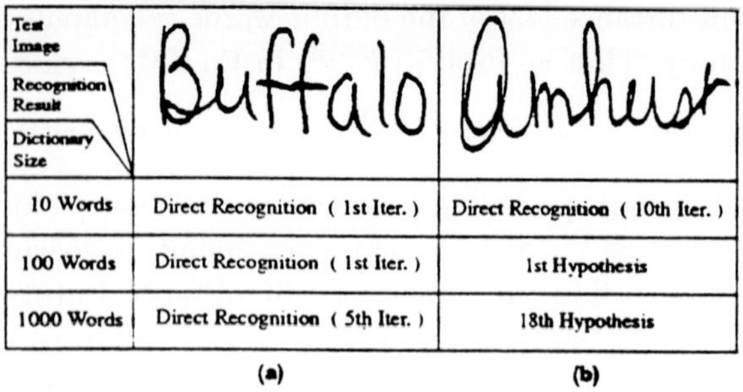

Figure 3: Typical recognition output of CDVDHMM based system for (a) a well-written word, (b) a not so well-written word.

5.3 MLHMM Based HWR System

The MLHMM is a doubly embedded network of HMM's where each character is modeled by an HMM while a word is modeled by a higher-level HMM. The major difference between the new system and the previous approaches is the *output-independence assumption* of the HMM [12]. In this new model, we associate the observation with the transition. As the states are not assigned any semantic meaning, the re-estimation algorithm is applicable. Also, the concept of "tied transition" is utilized in this HMM [12] to address the problem of limited training data.

5.3.1 HMM topology and training

Using the segmentation algorithm described in [22, 13], most of the characters are segmented into three segments or less (ligature is the 4th segment). Based on these observations, the character model as shown in Fig. 4(a) is found to be quite suitable. The five observation segment labels in character models are: W – the whole characters; L, M, R – the left, the middle, and the right part of the characters, respectively; N – null segments, i.e., ligatures.

For the word model, in the MD-HMM approach, the character models are cascaded while retaining only one N observation between characters (Fig. 4(b)). The training procedures for 26 character models are described below.

Initialization: By examining the segmentation results of each character, the segments are manually characterized as W, L, M, R, or N. The transition probability a_{ij} is initialized by calculating the occurrences of pairs of states. That is,

$$a_{ij} = \frac{\text{number of transitions from } s_i \text{ to } s_j}{\text{number of transitions from } s_i} \quad 0 \le i, j \le 3 \tag{23}$$

All the models start with the initial state s_o. To estimate the symbol probability $b_{ij}(k)$, the k-means clustering algorithm is applied to group the training samples (feature vectors of segment images [13, 24]) into several clusters. A global codebook is used for all the character models. The symbol probability for each character model can be initialized by counting the occurrences of the symbols in each cluster. More precisely,

$$b_{ij}(k) = \frac{\text{number of times } symbol(i,j) \text{ appears in cluster } k}{\text{number of times } symbol(i,j) \text{ appears}} \tag{24}$$

where $symbol(i,j)$ maps the observation emitted at transition $i \rightarrow j$ to one of $\{ W, L, M, R, N \}$. Here, $b_{i,j}(k)$ is referred to as the probability of observing symbol k with state transition $i \rightarrow j$. With these initially estimated parameters, the Baum-Welch algorithm is applied to increase the likelihood of the models using their representative training sequences.

5.3.2 Model normalization/co-occurrence smoothing

In the recognition phase of MD-HMM, we can simply classify the given observation sequence to the class whose model has the largest best-path likelihood, i.e.,

$$w^* = \arg \max_{1 \leq w \leq W} P_w \theta_w \;; \quad P_w = P(I_w^*, O \mid \lambda_w) \tag{25}$$

P_w is the best-path likelihood of model λ_w given the observation sequence O, and this likelihood can be found by the Viterbi algorithm [12]. The problem, basically, is to find a weight factor θ_w associated with each model λ_w such that the overall recognition result (based on the training set) is optimal. We choose the range for the weight factors to be [−M,M], and sequentially optimize for each HMM starting from the model that has the worst recognition performance. This algorithm does not guarantee the optimal set for the weight factors, but it ensures that the overall recognition rate of the training set would increase [13].

Compared with the number of free parameters, the number of training data is usually insufficient. In practice, smoothing of the parameters after re-estimation is essential if enough training data are not available [11]. Co-occurrence smoothing is based on the formula of computing $P_{co}(u \mid v)$, the co-occurrence probability of symbol u appearing given symbol v. By definition,

$$P_{co}(u \mid v) = \frac{\sum_{w=1}^{W} \sum_{i=1}^{N} \sum_{j=1}^{N} a_{ij}^w b_{ij}^w(u) b_{ij}^w(v) P(w)}{\sum_{k=1}^{M} \sum_{w=1}^{W} \sum_{i=1}^{N} \sum_{j=1}^{N} a_{ij}^w b_{ij}^w(k) b_{ij}^w(v) P(w)} \tag{26}$$

Here, W is the total number of HMM's; N is the number of states for each HMM; M is the total number of clusters; $a_{ij}^w, b_{ij}^w(k)$, and $P(w)$ are the transition probability, symbol probability, and the *a priori* probability for model λ_w. $P_{co}(u \mid v)$ can be loosely interpreted as "*when symbol v is observed, how often symbol u is observed in similar contexts*". These probabilities are obtained from the parameters after re-estimation, except $P(w)$ which is usually assumed to be uniformly distributed. By using the co-occurrence probability $P_{co}(u \mid v)$, smoothed parameters can be obtained as ($\lambda = 0.8$)

$$\bar{b}_{ij}^w(u) = \sum_{k=1}^{M} P_{co}(u \mid k) b_{ij}^w(k) \tag{27}$$

$$b_{ij}^w(k) = \lambda b_{ij}^w(k) + (1 - \lambda) \bar{b}_{ij}^w(k), \quad 0 < \lambda < 1. \tag{28}$$

5.3.3 Recognition

Given the word image, a sequence of segment images is obtained using the sub-character segmentation algorithm proposed in Sec. 4 [22, 13]. Each segment image is then assigned its symbol number by matching against the codebook of symbols generated during the training phase. Finally, this sequence of symbols and the associated dictionary for this word are applied to the word model. Both HMM strategies (MD-HMM and PD-HMM) are used in the word models.

MD-HMM Approach: Here, we need to build one HMM for every word in the dictionary. The word model is constructed by cascading the character models, as shown in Fig. 4. While the state transition (A) and symbol (B) probabilities are obtained from each of the individual character models, the initial state (Π) and the last state (Γ) probabilities for word model can be simply assigned as

$$\pi_i = \begin{cases} 1.0 & \text{if } s_i = F_i \\ 0.0 & \text{otherwise} \end{cases} \tag{29}$$

$$\gamma_i = \begin{cases} 1.0 & \text{if } s_i = L_i \\ 0.0 & \text{otherwise} \end{cases} \tag{30}$$

Here, F_i is the first state of the first character in the word; and L_i is the last state of the last character in the word. With the given sequence of observation symbols, the best-path probability for each word model can be computed using the Viterbi algorithm. This probability is then normalized (all in *log* form) by $\mathcal{L}(w) = P_w + \theta(w)$ where $\theta(w)$ is the weight factor for word w obtained by averaging the weight factors from every character model in this word. The word model which leads to the highest normalized likelihood will be chosen.

PD-HMM Approach: The word recognition algorithm for our PD-HMM approach is based on the level-building algorithm [12] and post-processing. For the details, interested readers are referred to [13, 24].

5.3.4 Experiments

The training character images are extracted from 3,103 training word images (USPS Database-IV) while the testing character images are extracted from another 1,034 word images. In total, 20,5129 character images are used for training while another 6,609 character images are used for testing. After re-estimation, smoothing and normalization, up to 80% character recognition accuracy is achieved.

The proposed systems are further evaluated using 3,000 test word images. Despite better modeling, the performance of this new HWR system is only comparable to that of the previous systems largely because the features used are more suitable for complete characters than individual segments (strokes) of the character. Moreover, it is found that the PD-HMM version of MLHMM has obtained nearly 95% recognition rate at the top 30 choices even with a 1000-word dictionary. Thus, the PD-HMM version of MLHMM could be a good frontier word filter, i.e., the

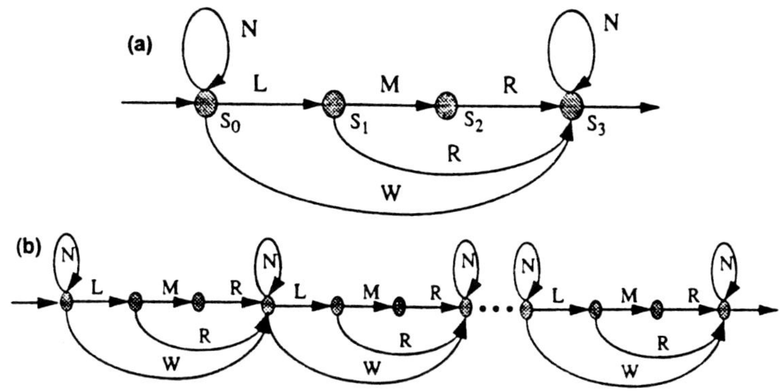

Figure 4: (a) A character is modeled as an HMM with 4 states, 5 observation symbols and 7 transitions. (b) For the MD-HMM approach, the word models are built by cascading the character models while eliminating one 'N' symbols between characters.

PDHMM based system is first used to pick the most likely words and to shrink the dictionary. The MDHMM (like the one described in [23]) based system is then used to reorder the recognition results for better accuracy. This technique overcomes the slow speed of the MDHMM based system.

6 Conclusions

In this chapter, a number of HWR systems using different implementations of HMM's are described. Considering the difficulty in recognizing unconstrained handwriting and the limited data available, the performance of the HMM based system is remarkable. In the coming years, researchers in this field have to deal with the following problems: (1) creation of sufficient and representative databases, (2) preferably more-or-less universally accepted feature vectors, (3) language specific post-processing operations to eliminate wrong choices by incorporating context and semantic information and (4) refinement of the HMM's to address trainability, portability, complexity and performance issues.

References

[1] C. C. Tappert, C. Y. Suen, and T. Wakahara, The state-of-the-art in on-line handwriting recognition, *IEEE Trans. Pattern Anal., Machine Intell.* 12 (1990) 787–808.

[2] V. K. Govindan and A. P. Shivaprasad, Character recognition - a review, *Pattern Recognition J.* 23 (1990) 671–683.

[3] R. G. Casey and G. Nagy, Recursive segmentation and classification of composite characters, *Proc. Int. Conf. on Pattern Recognition*, 1982, 1023–1025.

[4] R. Farag, Word level recognition of cursive script, *IEEE Trans. Computer* 28 (1979) 172–175.

[5] J. J. Hull and S. N. Srihari, "A computational approach to word shape recognition: Hypothesis generation and testing," *Proc. IEEE Conference on Computer Vision and Pattern Recognition (CVPR)*, 1986, 156–161.

[6] R. M. Bozinovic and S. N. Srihari, Off-line cursive word recognition, *IEEE Trans. Pattern Anal., Machine Intell.* 11 (1989), 68–83.

[7] L. E. Baum, An inequality and associated maximization technique in statistical estimation for probabilistic functions of Markov processes, *Inequalities*, 3 (1972) 1–8.

[8] L. E. Baum and T. Petrie, Statistical inference for probabilistic functions of finite state Markov chains, *Ann. Math. Stat.* 37 (1966) 1554–1563.

[9] J. K. Baker, The dragon system - an overview, *IEEE Trans. Acoust., Speech, Signal Processing* 23 (1975) 24–29.

[10] F. Jelinek, Continuous speech recognition by statistical methods, *IEEE Proc.* 64 (1976) 532–556.

[11] K.-F. Lee, H.-W. Hon, and R. Reddy, An overview of the SPHINX speech recognition system, *IEEE Trans. Acoust., Speech, Signal Processing* 38 (1990) 35–45.

[12] L. R. Rabiner, A tutorial on hidden Markov model and selected applications in speech recognition, *IEEE Proc.* 77(1989) 257–286.

[13] M.-Y. Chen, Handwritten word recognition using hidden Markov model, Ph.D. Thesis, State University of New York at Buffalo, 1993.

[14] G. D. Forney, Jr., "The Viterbi algorithm," *IEEE Proc.*, 61 (1973) 268–278.

[15] A. Ljolje and S. E. Levison, "Development of an acoustic-phonetic hidden Markov model for continuous speech recognition," *IEEE Trans. Signal Processing*, 39 (1991) 29–39.

[16] J. D. Ferguson, Variable duration models for speech, in *Proc. of the Sym. on the Application of Hidden Markov Models to Text and Speech*, ed. J. D. Ferguson, Princeton, 1980, 143–179.

[17] S. E. Levinson, "Continuously variable duration hidden Markov models for automatic speech recognition," *Computer Speech and Language*, 1 (1986) 29–45.

[18] Y. He, *Planar shape and handwritten word recognition using hidden Markov models*. Ph.D. Thesis, State University of New York at Buffalo, 1991.

[19] A. Kundu, Y. He, and P. Bahl, "Recognition of handwritten word: First and second order hidden Markov model based approach," *Pattern Recognition J.*, 22 (1989) 283–297.

[20] M.-Y. Chen, A. Kundu, and J. Zhou, "Off-line handwritten word recognition using a hidden Markov model type stochastic network," *IEEE Trans. Pattern Anal., Machine Intell.* 16 (1994) 481–496.

[21] R. Nag, K. H. Wong, and F. Fallside, "Script recognition using hidden Markov model," in *Proc. IEEE Int. Conf. on Acoustics, Speech, Signal Processing*, Tokyo, Japan, 1986. 2071–2074.

[22] M.-Y. Chen, A. Kundu, and S. N. Srihari, Unconstrained handwritten word recognition using continuous density variable duration hidden Markov models, *Proc. IEEE Int. Conf. on Acoustics, Speech, Signal Processing*, Minneapolis, MN, May 1993, V.105–108 (also see IEEE Trans. on Image Processing, 4 (1995) 1675–1688).

[23] M.-Y. Chen and A. Kundu, "A complement to variable duration hidden Markov model," in *Proc. IEEE Int. Conf. on Image Processing*, Austin, Texas, Nov. 1994, 174–178.

[24] M.-Y. Chen and A. Kundu, "Multi-level hidden Markov model handwritten word recognition," in *Proc. IEEE Int. Conf. on Acoustics, Speech, Signal Processing*, Detroit, Michigan, May 1995, 2623–2626.

[25] C. B. Bose and S.-S. Kuo, "Connected and degraded text recognition using hidden Markov model," in *Proc. Int. Conf. on Pattern Recognition*, 1992, 116–119.

[26] J. J. Hull, "A hidden Markov model for language syntax in text recognition," in *Proc. Int. Conf. on Pattern Recognition*, 1992, (B)124–127.

[27] E. Levin and R. Pieraccini, "Dynamic planar wraping for optical character recognition," in *Proc. IEEE Int. Conf. on Acoustics, Speech, Signal Processing*, San Francisco, CA, March 1992, 149–152.

[28] F. R. Chen, L. D. Wilcox, and D. S. Bloomberg, "Word spotting in scanned images using hidden Markov models," in *Proc. IEEE Int. Conf. on Acoustics, Speech, Signal Processing*, Minneapolis, Minnesota, April 1993, V.1–4.

[29] S.-S. Kuo and O. E. Agazzi, "Machine vision for keyword spotting using pseudo 2-d hidden Markov models," in *Proc. IEEE Int. Conf. on Acoustics, Speech, Signal Processing*, Minneapolis, Minnesota, April 1993, V.81–84.

[30] K. S. Nathan, J. R. Bellegarda, D. Nathamoo, and E. J. Bellegarda, "On-line handwriting recognition using continuous parameter hidden Markov models," in *Proc. IEEE Int. Conf. on Acoustics, Speech, Signal Processing*, Minneapolis, Minnesota, April 1993, V.121–124.

[31] T. Starner, J. Makhoul, and G. Chou, "On-line cursive handwriting recognition using speech recognition methods," in *Proc. IEEE Int. Conf. on Acoustics, Speech, Signal Processing*, Adelaide, Australia, April 1994, V.125–128.

[32] E. J. Bellegarda, J. R. Bellegarda, D. Nahamoo, and K. S. Nathan, "A probabilistic framework for on-line handwriting recognition," in *Proc. Int. Workshop on Frontiers in Handwriting Recognition*, Buffalo, New York, May 1993, 225–234.

[33] S. Bercu and G. Lorette, "On-line handwritten word recognition: An approach based on hidden Markov models," in *Proc. Int. Workshop on Frontiers in Handwriting Recognition*, Buffalo, New York, May 1993, 385–390.

[34] T. Caesar, J. Gloger, A. Kaltenmeier, and E. Mandler, "Recognition of handwritten word images by statistical methods," in *Proc. Int. Workshop on Frontiers in Handwriting Recognition*, Buffalo, New York, May 1993, 409–416.

[35] H.-S. Park and S.-W. Lee, "Off-line recognition of large-set handwritten hangul with hidden Markov models," in *Proc. Int. Workshop on Frontiers in Handwriting Recognition*, Buffalo, New York, May 1993, 51–61.

[36] R. L. Cave and L. P. Neuwirth, "Hidden Markov models for English," in *Proc. of the Sym. on the Application of Hidden Markov Models to Text and Speech*, ed. J. D. Ferguson, Princeton, 1980, 16–56.

[37] B.-S. Jeng, F.-H. Liu, S.-W. Sun, and T.-M. Wu, "Hidden-Markov-model based optical-character recognition – a novel approach," in *Proc. IEEE Int. Sym. on Information Theory*, San Diego, CA, Jan. 1990, 162.

[38] Y. He and A. Kundu, "2-D shape classification using hidden Markov model," *IEEE Trans. Pattern Anal., Machine Intell.* 13 (1991) 1172–1184.

[39] H. Bunke, M. Roth, and E. G. Schukat-Talamazzini, "Off-line cursive handwritten word recognition using hidden Markov models," *Pattern Recognition J.* 28 (1995) 1399–1413.

[40] J. A. Vlontzos and S. Y. Kung, "Hidden Markov models for character recognition," *IEEE Trans. Image Processing* 1 (1992) 539–543.

[41] G. E. Kopec and P. A. Chou, "Document image decoding using Markov source models," *IEEE Trans. Pattern Anal., Machine Intell.* 16 (1994) 602–617.

[42] H. S. Baird, "Feature identification for hybrid structural/statistical pattern classification," *Computer Vision, Graphics, and Image Processing* 42 (1988) 318–333.

[43] T. Okuda, E. Tanaka, and T. Kasai, "A method for the correction of garbled words based on the Levenshtein matrix," *IEEE Trans. Inform. Theory* 25 (1976) 172–178.

[44] R. A. Wagner and M. J. Fischer, "The string-to-string correction problem," *J. ACM* 21 (1974) 168–173.

Appendix A.1. Baum-Welch Re-estimation Algorithm

Like $\alpha_t(j)$, we define $\beta_t(j)$ as $P(O_{t+1}^T|q_t = s_j, \lambda)$. Thus,

$$\beta_t(j) = \sum_{j=1}^{N} \beta_{t+1}(j) a_{ij} b_j(o_{t+1}) \tag{31}$$

The initial conditions are

$$\alpha_1(i) = \pi_i b_i(o_1) \; ; \; \beta_T(i) = \gamma_i \; ; \; \forall i \tag{32}$$

We then define

$$\delta_t(i) = \frac{\alpha_t(i)\beta_t(i)}{\Pr(O|\lambda)} \; ; \quad \xi_t(i,j) = \frac{\alpha_t(i) a_{ij} b_j(o_{t+1}) \beta_{t+1}(j)}{\Pr(O|\lambda)} \, . \tag{33}$$

Using these definitions, the HMM parameters are computed as

$$\hat{\pi}_i = \delta_1(i) \ ; \ \hat{a}_{ij} = \frac{\sum_{t=1}^{T-1} \xi_t(i,j)}{\sum_{t=1}^{T-1} \delta_t(i)}. \tag{34}$$

$$\hat{\gamma}_i = \delta_T(i) \ ; \ \hat{b}_j(k) = \frac{\sum_{t=1, o_t=v_k}^{T-1} \delta_t(j)}{\sum_{t=1}^{T-1} \delta_t(j)}. \tag{35}$$

Using the current values of $a_{ij}, b_j(k)$ and π_i, δ and ξ etc., are evaluated which are then used to recompute the A, B, Π parameters iteratively.

Appendix A.2. Modified Viterbi Algorithm

A formal technique for finding the best single state sequence is the Viterbi algorithm [14, 12]. The formal statement of the algorithm follows:

- Initialization: for $1 \leq i \leq N$

$$\delta_1(i) = \pi_i b_i(o_1) \ ; \ \Psi_1(i) = 0 \tag{36}$$

- Recursion: for $2 \leq t \leq T, 1 \leq j \leq N$

$$\delta_t(j) = \max_{1 \leq i \leq N} [\delta_{t-1}(i) a_{ij}] b_j(o_t) \ ; \ \Psi_t(j) = \arg \max_{1 \leq i \leq N} [\delta_{t-1}(i) a_{ij}] \tag{37}$$

- Termination:

$$P^* = \max_{1 \leq i \leq N} [\delta_T(i) \gamma_i] \ ; \ i_T^* = \arg \max_{1 \leq i \leq N} [\delta_T(i) \gamma_i]. \tag{38}$$

- Path backtracking: for $t = T-1, T-2, \ldots, 1$

$$i_t^* = \Psi_{t+1}(i_{t+1}^*)$$

The Viterbi algorithm is an efficient search for the globally best path. In HWR, a considerable performance improvement can be obtained if the knowledge of the globally second best path, the third best path, etc., can be utilized during post-processing. Two modified Viterbi algorithms, which provide an ordered list of the best L state sequences, are developed in [13]. The first one is a parallel version which simply extends the Ψ and δ to another dimension where the extra dimension represents the choice L.

1. Initialization: Extend the two-dimensional array used in standard Viterbi algorithm to the third dimension and initialize.

2. Recursion: Extend the trellis structure to the third dimension while performing the Viterbi recursion. In the 3-D structure, the nodes in the first plane is same as those of the 2-D trellis structure generated by the standard Viterbi algorithm which represent the best scores (highest probability) to reach that node and the corresponding transition from the previous stage; the nodes in the second plane represent the second best scores to reach that node and the corresponding transition from the previous stage; and so on to other higher planes.

3. Termination: All terminal nodes in all the planes are sorted in descending order according to the node probability.
4. Backtracking: Perform the backtracking according to the order of the sorted list; terminate the procedure after the desired number of state sequences are found.

This straightforward modification of the Viterbi algorithm is not very efficient. Given an N-state system, we need to search $N \times L$ previous nodes at every time slot and every current state; and select only L out of the $N \times L$ possible paths while all the others are discarded as redundant. If N is much larger than L, this is particularly unsuitable. To overcome this problem, another version which is called the serial version of modified Viterbi algorithm is also developed [13].

Appendix A.3. Glossary of HMM Related Terms

- (CD)VDHMM – (Continuous density) Variable duration HMM.
- SCHMM – Single contextual HMM.
- MLHMM – Multi-level HMM.
- NEHMM – Non-ergodic HMM.
- MDHMM – Model discriminant HMM.
- PDHMM – Path discriminant HMM.

Handbook of Character Recognition and Document Image Analysis, pp. 183–225
Eds. H. Bunke and P. S. P. Wang

CHAPTER 7

OVERVIEW AND SYNTHESIS OF ON-LINE CURSIVE HANDWRITING RECOGNITION TECHNIQUES

ISABELLE GUYON*
AT&T Bell Laboratories
955 Creston Road, Berkeley, CA 94708, USA

MARKUS SCHENKEL†
Swiss Federal Institute of Technology (ETH)
CH-8092 Zürich, Switzerland

and

JOHN DENKER‡
AT&T Bell Laboratories
Holmdel, NJ 07733, USA

In this chapter, we analyze several on-line cursive handwriting recognition systems. We find that virtually all such systems involve (a) a preprocessor, (b) a trainable classifier, and (c) a language modeling post-processor. Such architectures are described within the framework of Weighted Finite State Transductions, previously used in speech recognition by Pereira *et al.* We describe in some detail a recognition system built in our laboratory. It is a writer independent system which can handle a variety of writing styles including cursive script and handprint. The input to the system encodes the pen trajectory as a time-ordered sequence of feature vectors. A Time Delay Neural Network is used to estimate a posteriori probabilities for characters in a word. A Hidden Markov Model segments the word in a way which optimizes the global word score, taking a lexicon into account. The last part of the chapter is devoted to bibliographical notes.

Keywords: Cursive, handwriting, recognition, on-line, character recognition, hidden Markov models, finite state automata, finite state transductions, neural networks, WFSA, WFST, HMM, graph, Viterbi, TDNN, grammar, UNIPEN, segmentation, product graph, INSEG, OUTSEG, script, handprint.

1. Introduction

"On-line" handwriting recognition refers to the case where we know the pen coordinates as a function of time and not only a static image of words written on paper. This task has become increasingly important over the past few years. Until the beginning of the nineties, on-line handwriting recognition research was mainly academic. The situation has changed with the rapid growth of the pen computing

*isabelle@research.att.com
†schenkel@isi.ee.ethz.ch
‡jsd@research.att.com

industry, and in particular with the appearance on the market of the first personal communicators.

The mediocre quality of handwriting recognition has been a major obstacle to the success of pen computers. Users report that it is "too inaccurate, too slow, and too demanding of user attention". The entire pen computing industry will turn its back on handwriting and revert to *popup keyboards* unless major advances in handwriting recognition are made.

The most challenging version of this problem is writer-independent mixed-style handwriting recognition for large vocabularies:

- *Writer independent* means that the system can be used directly by any writer, without requiring an adaptation or training session. Potential users when they purchase a system expect high "walk up" recognition accuracy and it is therefore very important to guarantee a high degree of writer independence. The adaptation of the machine to a particular writer style can then be used to increase the accuracy of recognition.
- *Mixed style* means that the system can handle both handprint and cursive or even a mixture of both within the same word. This covers all natural writing styles and does not place on the writer any constraints (such as writing in boxes, combs, etc.).
- *Large vocabulary* means a vocabulary containing all the common words (25,000 words at least). Large vocabularies are required to be able to use handwriting for text entry applications.

In this chapter, we describe and analyze several techniques for on-line cursive and mixed style recognition for the Latin alphabet which permit reaching high levels of writer independence and using large vocabularies.

2. Common Features of Most Systems

In this section, we describe a fairly general framework for the problem of on-line cursive and mixed style handwriting recognition; Section 3 will consider the pros and cons of various designs for some of the building blocks. Traditionally, on-line recognizers consist of a preprocessor, a classifier (which estimates probabilities for the different categories of characters or other subword units) and a language-modeling postprocessor (often a Hidden Markov Model). The system usually has adjustable parameters whose values are determined during a training session. Postprocessors incorporating language models are of special importance; without them the error rate would be unacceptable (typically around 30% on a per character basis).

2.1. *System Architecture*

The overall architecture of a *generic* cursive recognition system can be found in Fig. 1.

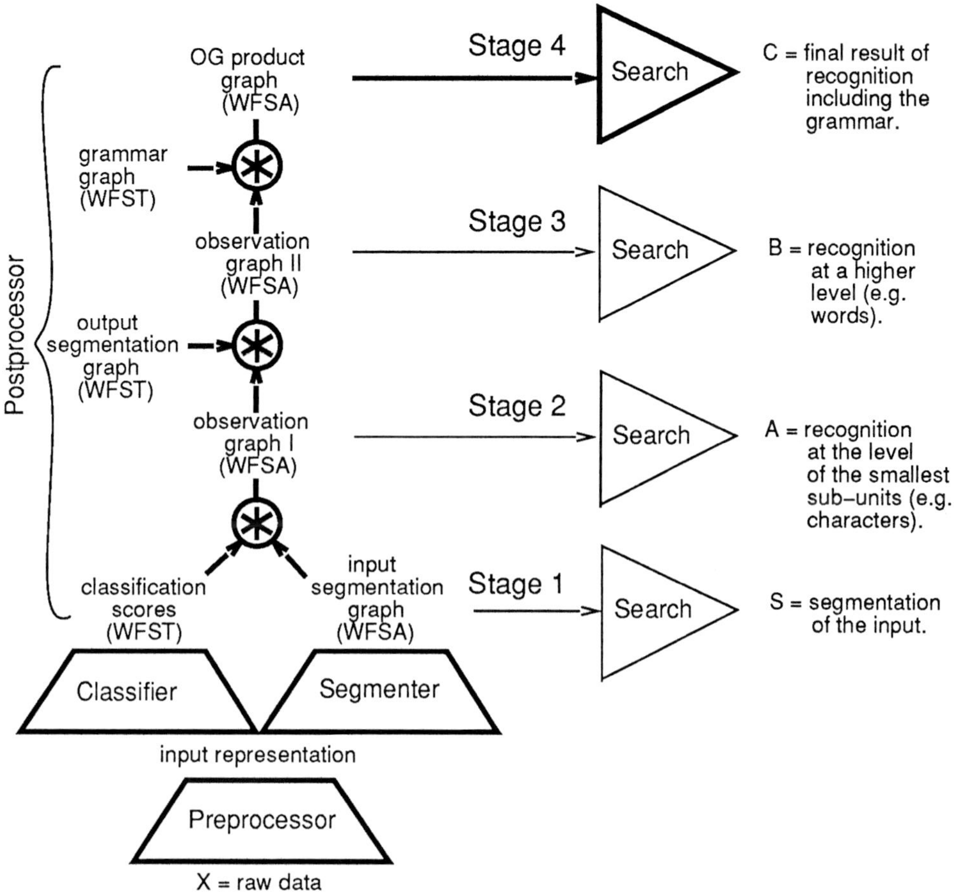

Fig. 1. Overview of a generic cursive recognition system. Most cursive recognition systems use graph algorithms to determine simultaneously the best recognition and segmentation. The overall process can be decomposed into intermediate steps. The lower level graphs are usually not searched. If they would be searched, the results would be an intermediate recognition result, less refined than the last level. For instance, the input segmentation graph would provide a low level segmentation (S) which does not take the classification scores or grammar into account. There are two main approaches to segmentation: *in*put *seg*mentation (*inseg*) and *out*put *seg*mentation (*outseg*). An *inseg* system (Fig. 7) uses an elaborate input segmentation graph build from segmentation points heuristically determined but has no output segmentation graph. An *outseg* system (Fig. 8) has a trivial input segmentation graph and an elaborate output segmentation graph. Graphs are composed with the transduction product method (see text). WFSA stands for Weighted Finite State Automaton and WFST for Weighted Finite State Transducer (see text).

The **input** to on-line cursive systems is generally a sequence of regularly time spaced pen coordinates: $X = [[x_1, y_1, z_1][x_2, y_2, z_2]...[x_t, y_t, z_t]...]$ where t is the time: $t = 1, ..., T$. A typical sampling rate is 100 points per second. The z coordinate tells whether the pen is "up" or "down" (i.e., touching the writing surface or not); more detailed pressure information is usually too unreliable to be useful.

The **preprocessor** accepts the raw input data X and improves its signal quality, enforces invariances, and detects relevant features. The output of the preprocessor is the "recognizer input representation". Many representations have been proposed, putting more or less emphasis on the temporal information (see Secs. 3.2 and 3.4). No consensus on representation has emerged so far.

The **segmenter** splits the recognizer input representation into "segments" that may or may not overlap. For a character-based recognizer, for instance, each segment corresponds to a "tentative character".[a] A simple example of segmentation is shown in Fig. 2.[b] There are usually many more segments than characters in the input. This is why the method is sometimes called "over-segmentation". Whether a segment coincides with a handwritten character is determined later with the help of the scores of the classifier.

For each segment, the **classifier** produces scores for each class associated with the characters of a chosen alphabet.[c] Scores typically represent either probabilities or negative-log-probabilities. A variety of classifiers have been proposed in the literature [16].[d] No one technique has been outstandingly successful, although many of the best classifiers are based on neural network techniques.

As explained in more detail in the next section, the **postprocessor** uses a graph algorithm to determine the most probable segmentation and recognition, taking into account a language model. Most language models used for cursive handwriting recognition are implementable by Weighted Finite State Automata or WFSA. In Fig. 3, we show two examples of language models implementable by WFSAs that are commonly used: A lexicon trie (or trie) and simple Markov model or n-gram. Lexicon trees are used to store lists of words with their frequency. Character n-grams are used to predict the next character given a window $(n-1)$ of past characters. Word n-grams are used to predict the next word given a window of $(n-1)$ past words. Grammars implementable by WFSAs are called "regular". They offer speed and accuracy advantages over other more complex grammars [35]. In the long run, as computing power and algorithms improve, many hopes are pinned on improving recognition by modeling the syntax and semantics of the language with more complex grammars.

---

[a] For some recognizers, segments may correspond to sub-characters, sub-words or words. For simplicity, we discuss the "character" case; the same line of reasoning applies to other cases.
[b] More examples of cursive word segmentation are shown in Fig. 13.
[c] In some approaches that classify sub-characters shapes (e.g. strokes) instead of characters, the classification stage may be trivial (e.g. a simple clustering). The classification at the character level takes place at a later stage, but the basic architecture is similar.
[d] See also the chapter by U. Kressel and J. Schürmann in this book.

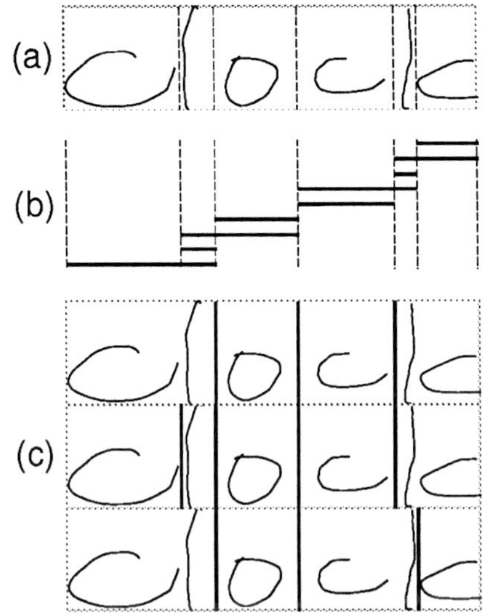

Fig. 2. Handwriting segmentation: (a) Tentative cuts are proposed by the segmenter to split the input representation into segments (dashed vertical lines).(b) Examples of segments delineated by two tentative cuts (bold horizontal lines). (c) Examples of segmentations (bold vertical lines).

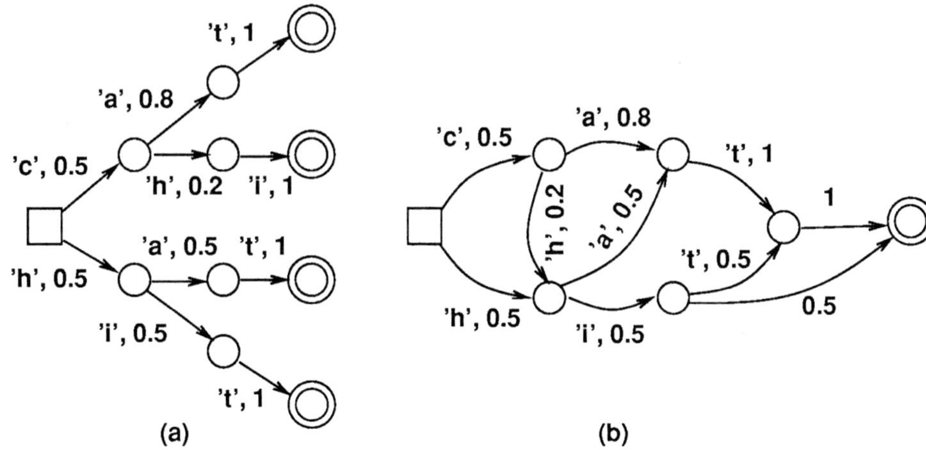

Fig. 3. Regular grammars: Regular grammars can be implemented by Weighted Finite State Automata. Nodes (circles) represent states. Arcs (arrows) represent transitions. The weights on the arcs represent conditional probabilities P(next-char|state). The overall score for a word is given by the product of the weights of the arcs traversed. (a) Lexicon tree for the words 'cat', 'chi', 'hat' and 'hit'. Each node (state) can be reached from a single path starting at the root node (square). Therefore, each state "remembers" all the past characters since the beginning of the word. In this example, the score for 'cat' is given by P('cat') = P('c') P('a'|'c') P('t'|'ca') $= 0.5 \times 0.8 \times 1 = 0.4$. (b) Markov model of order $(n-1)$ or n-gram. Each state remembers at most $(n-1)$ past characters. In this example, the score for 'cat' is given by P('cat') = P('c') P('a'|'c') P('t'|'c') $= 0.5 \times 0.8 \times 1 = 0.4$.

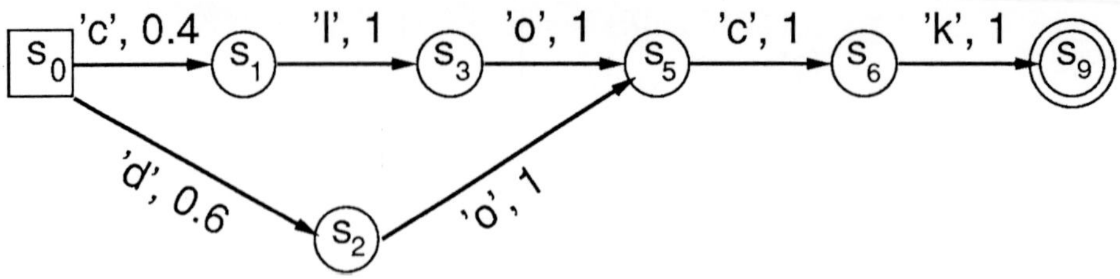

Fig. 4. Graph-based segmentation and recognition: The graph of a WFSA used for the recognition of the word of Fig. 2, incorporating classification scores, segmentation information, and a language model. The best path through the graph gives the final segmentation and recognition.

2.2. *Graph-based Segmentation and Recognition*

Most cursive recognition systems formulate segmentation and recognition as a "best path through graph" problem. It is useful to unite the various approaches using the formalism of Weighted Finite State Automata (WFSAs) and Weighted Finite State Transducers (WFSTs) proposed in Ref. [55].

2.2.1. *Graph search techniques*

The fundamental idea of graph-based segmentation and recognition is to build a Weighted Finite State Automaton (WFSA) which incorporates classification scores, segmentation information, and a language model (Fig. 4). A WFSA is conveniently represented by a graph. It has finitely many states (the nodes of the graph), and a finite alphabet. Transitions between states (the arcs of the graph) are associated with the emission (or acceptance) of a character[e] and with a weight.[f]

If the weights are normalized like probabilities, the weight for a sequence of transitions is the product of the weights of the individual transitions; if, alternatively, the weights are normalized like costs (negative-log-probabilities), the weight for a sequence of transitions is the sum of the individual weights. Even though for computational reasons the negative-log-probability representation is often preferred, our discussion will use plain probabilities. A typical weight is

$$P(s', c|s) \tag{2.1}$$

which is the probability[g] of the transition along the arc $s \to s'$, with emission (or acceptance) of a character c.[h]

[e] We use the word character whether or not the recognizer is character-based; replace character by sub-character, sub-word or word if other atoms of handwriting are used.

[f] This representation of WFSAs with weights on the arcs only is completely general and equivalents can be found with representations with weights on nodes or on both nodes and arcs. For a proof, see Ref. [10]. For people familiar with HMMs, the WFSA framework corresponds to "discrete" HMMs. The equivalent of WFST for "continuous" HMMs has been proposed in Ref. [5].

[g] Note that any set of scores satisfying the axioms of probability theory can rightfully be called a probability.

[h] For notation simplicity, we omit a number of conditioning variables on our conditional probabil-

The overall score for a path in the graph is given by the product of the weights of the arcs traversed:

$$P(s_1, c_1, s_2, c_2, \ldots, s_n, c_n) = \prod_{k=1}^{n} P(s_k, c_k | s_{k-1}) \quad , \tag{2.2}$$

where by convention $P(s_1, c_1|s_0) = P(s_1, c_1)$. Equation (2.2) embodies the assumption that the model is first-order Markovian.[i]

The probability of a given output string is given by summing over all possible ways of producing that string, i.e., by summing over all paths in the graph that produce the given string:

$$P(c_1, c_2, \ldots, c_n) = \sum_{\{s_1, s_2, \ldots, s_n\}} P(s_1, c_1, s_2, c_2, \ldots, s_n, c_n) \quad . \tag{2.3}$$

The most probable string is the recognition result. The forward algorithm [56] is an efficient dynamic programming technique to compute the above sum (2.3). An approximation often made is to pick the best single path in the graph as the recognition result. This amounts to replacing the sum in Eq. (2.3) by its largest term; the Viterbi algorithm [56] makes this approximation. Approximate search procedures to find the most probable string are often preferable for computational reasons. These include beam-search procedures and the $A*$ algorithm [21, 11]. Various "fastmatch" techniques can be used to narrow down the search space [21, 61, 41].

2.2.2. *Transduction cascade decomposition*

The Weighted Finite State Automaton (WFSA) used for recognition can be decomposed as a cascade of Weighted Finite State Transductions (WFST) corresponding to several intermediate stages of recognition.

Formally, a WFST is a mapping from pairs of strings over two alphabets to weights. WFSTs are represented by automata which accept a string of characters and emit another one. The arcs of the automata are labeled by pairs of characters $\lambda : \mu$, where λ is the character accepted and μ the character emitted.

Given two graphs, one can form their product, as defined below. The product of two WFSTs is another WFST. The product of a WFSA with a WFST is a WFSA. Let us denote by $s \xrightarrow{\lambda} s'$ the transition between state s and state s' with emission (or acceptance) of a character λ and $P(s', \lambda|s)$ the associated weight. Consider the two graphs:

---

ities. When we write, for instance, $P(s', c|s)$, we really mean $P(s', c|s, X, \mathcal{R}, \mathcal{S}, \mathcal{L}, \mathcal{D}, \mathcal{A}, \mathcal{U})$, where X is the input, $\mathcal{R}$ is the family of classifiers, $\mathcal{S}$ is the family of segmenters, $\mathcal{L}$ is the family of language models, $\mathcal{D}$ is the training data, $\mathcal{A}$ is the training algorithm, and $\mathcal{U}$ is the rest of the universe.

[i] This means that s_k depends on $s_{k-t} (1 < t \leq k)$ only through s_{k-1}. Markov models of higher order can be brought back to first order Markov models, by increasing the number of states (nodes) of the WFSA.

$$\text{WSFA1: } s \xrightarrow{\lambda} s', \quad P(s', \lambda|s) \tag{2.4}$$

$$\text{WSFT: } t \xrightarrow{\lambda:\mu} t', \quad P(t', \lambda : \mu|t) \tag{2.5}$$

The product of WSFA1 and WSFT has states and transitions defined as:

$$\text{WSFA2: } st \xrightarrow{\mu} s't', \quad P(s't', \mu|st) = \textstyle\sum_\lambda P(s', \lambda|s)P(t', \lambda : \mu|t) \tag{2.6}$$

where the labels are composed with the rule $(\lambda, \lambda : \mu) \to \mu$. Throughout the paper, we use the notation:

$$\text{WFSA2} = \text{WFSA1} \circledast \text{WFST} \tag{2.7}$$

to denote the transduction product. An example is shown in Fig. 5.

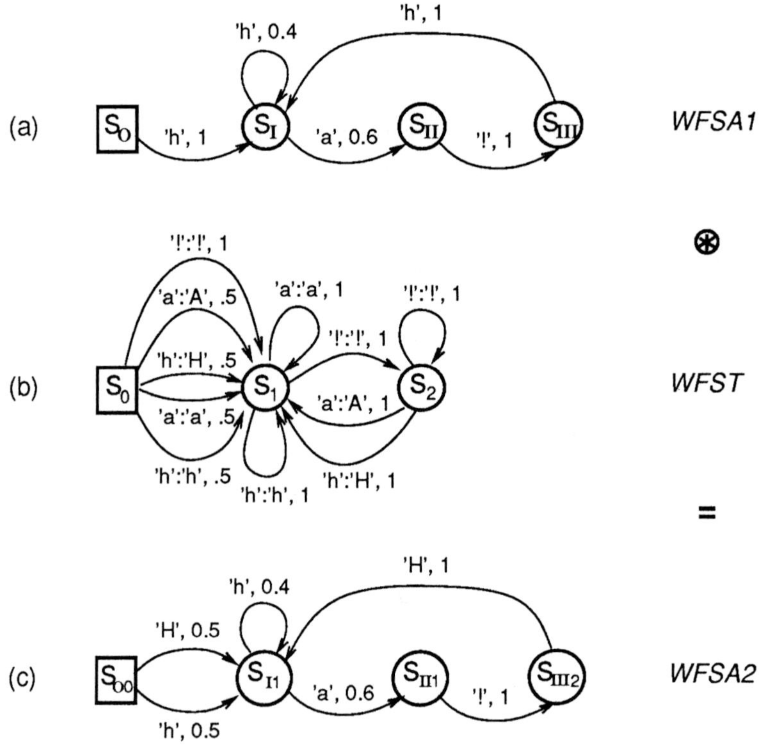

Fig. 5. Transduction product: (a) WFSA1. (b) WFST. (c) WFSA2 = WFSA1 ⊛ WFST.

In the general case, the product of two graphs has states composed of pairs of states of the original graphs. The product graph transitions emit and/or accept labels belonging to the Cartesian product of the labels emitted and/or accepted by the states of the original graphs. Similarly, the corresponding weights are a function of the pair of the corresponding original weights (e.g. sum or product). Examples of products of graphs are shown in Figs. 17, 18, 19 and 20. For a more general description of operations on WFSAs and WFSTs, see Ref. [55].

2.2.3. *Search in a product graph*

Interestingly, in order to search for the best path in a product graph it is not necessary to explicitly build it. Those nodes in the product that are visited can be constructed "on-the-fly" during the search procedure (see Fig. 6).

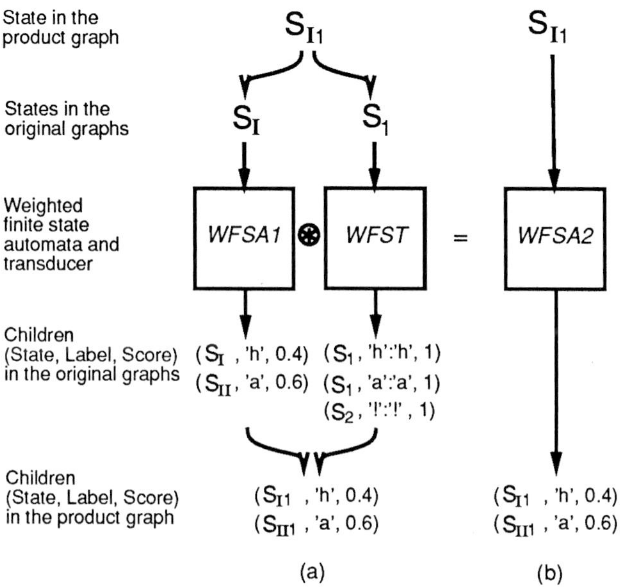

Fig. 6. Search in a product graph: This example is related to that of Fig. 5. Searching for the best path in a product graph does not require actually building the graph. Calculations can be done "on the fly". A "child state" is defined to be one that can be reached in one transition from a given state. The automata receive the following query: what are the children states of a given state, the label emitted during the transition and the corresponding probabilities? (a) WFSA1 and WFST are queried separately. The answers are combined to form the product states. Building WFSA2 explicitly is avoided, (b) WFSA2 is queried directly.

2.2.4. *Decomposition of a generic WFSA used for cursive recognition*

In our block diagram of Fig. 1, each stage corresponds to a transduction. There are two complementary strategies of segmentation that we refer to as *inseg* and *outseg*. Readers familiar with Hidden Markov Models (HMMs) will recognize their approach as an *outseg* strategy, whereas reader familiar with Neural Networks and "over-segmentation" techniques will recognize their approach as an *inseg* strategy. Both strategies are complementaries and can indeed be combined, although only a few attempts have been made. In the description that follows, we precise which stages are present or absent in the *inseg* and *outseg* systems:

1. The first stage yielding the "input segmentation graph", considers only the low level segmentation information (such as pen-lifts, points of high curvature,

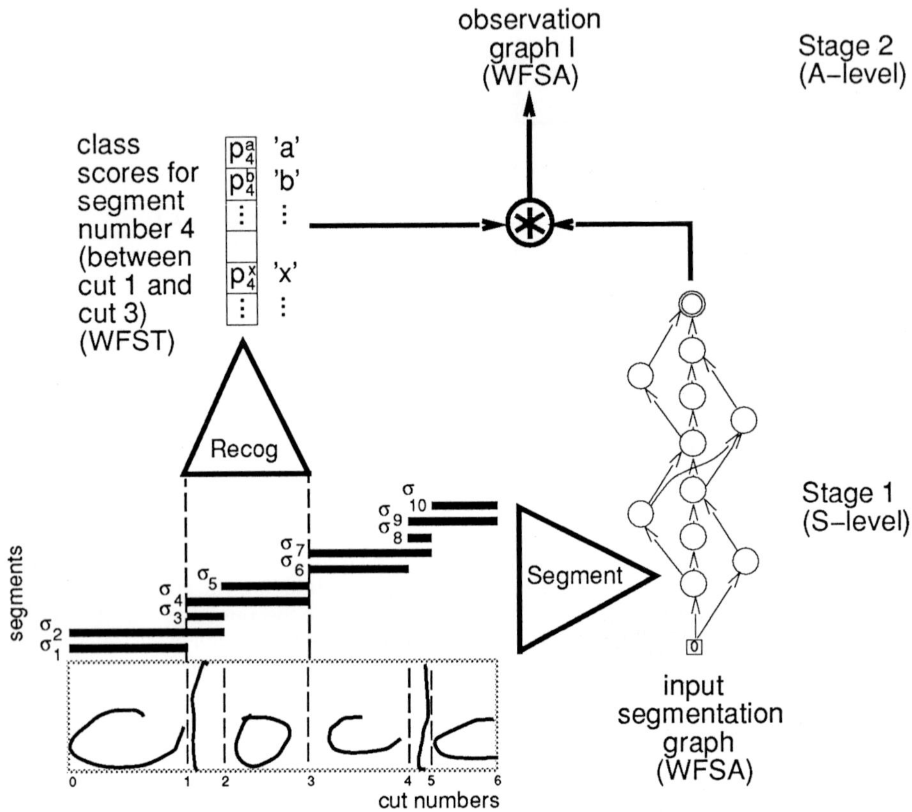

Fig. 7. Central stages of *inseg* type systems: The tentative cuts in the input representation are based on the dynamics of the pen (e.g. pen-lifts, pen velocity) and/or geometrical clues (e.g. spaces, corners). The input segmentation graph is non trivial, which gives the name *inseg* (for input segmentation) to the method. To simplify the diagram, we show examples of handprinted characters but the principles apply as well to cursive handwriting. The details of the various stages are shown in Figs. 19 and 20.

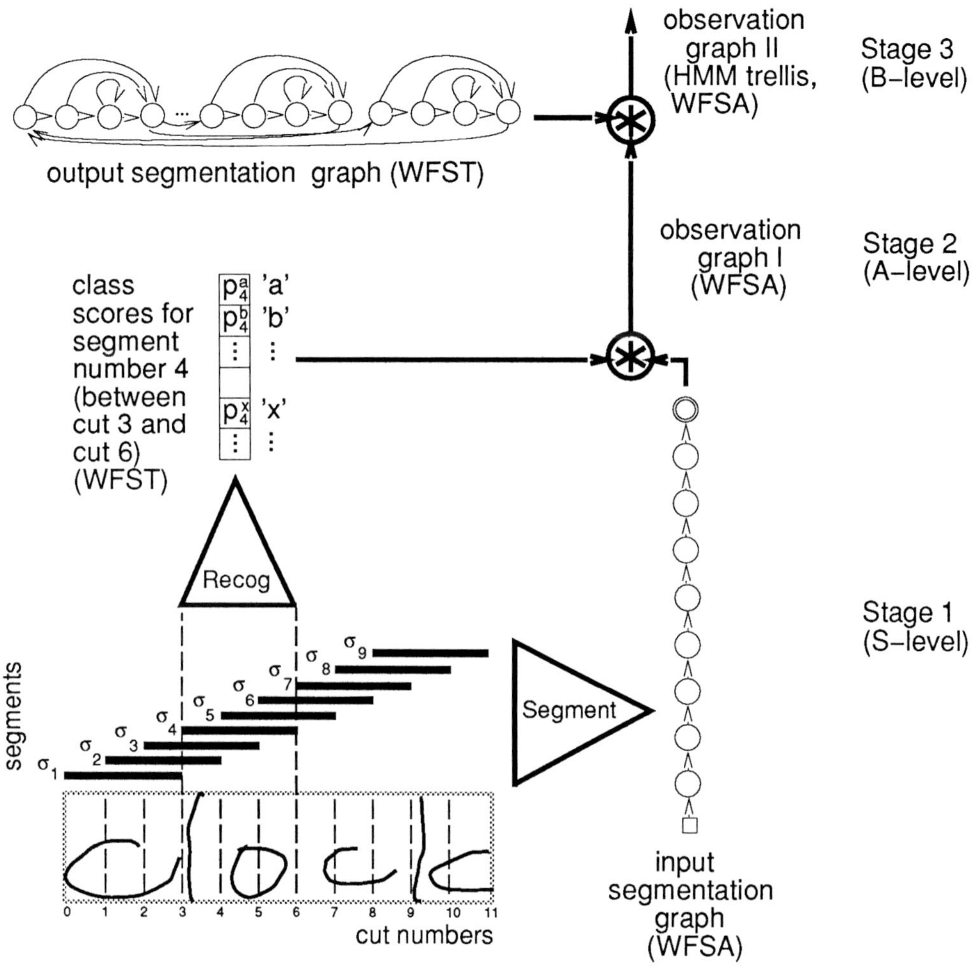

Fig. 8. Central stages of *outseg* type systems: The input segmentation graph is trivial. The segmentation is postponed until after the recognition, which gives the name *outseg* (for output segmentation) to the method. To simplify the diagram, we show examples of handprinted characters but the principles apply as well to cursive handwriting. The details of the various stages are shown in Figs. 17 and 18.

etc.). This is where the segmentation really takes place for *inseg* systems (see Fig. 7). By contrast, the first stage is trivial in the case of *outseg* systems (see Fig. 7 and Section 3.9). For people familiar with classical Hidden Markov Model approaches (HMM), HMMs are particular cases of *outseg* systems: the input segmentation graph consists of a chain of states representing the succession of time frames.

At the first stage level, the input alphabet is composed of triplets of numbers (x_t, y_t, z_t) and the output alphabet of segment labels $\sigma_1, \sigma_2, ...,$. The first stage provides weights $P(S|X)$ for all possible segmentations $S = \sigma_1\sigma_2...,$ of the input X. These weights embody the quality or plausibility of segmentation, but no shape recognition.

2. The second stage combines the results of classification (shape recognition) to those of segmentation, yielding to the "observation-graph I". The input alphabet is $\sigma_1, \sigma_2, ...,$ and the output alphabet may be alphanumeric characters or other types of sub-character, sub-word or word labels (depending on the granularity of segmentation) that we denote by $\alpha_1, \alpha_2, ...$. Spaces or transitions may be included in that alphabet. The second stage provides weights $P(A|S)$ for all possible interpretations $A = \alpha_1\alpha_2...,$ corresponding to the segmentations $S = \sigma_1\sigma_2...,$. These weights embody the scores of the classifier or shape recognizer.

3. The third stage adds a high-level model of segmentation (such as a model of character durations or a model of combination of strokes into characters), yielding the "observation-graph II". This third stage is absent from *inseg* systems (see Fig. 7 and Section 3.9) because they rely only on the first stage for segmentation. It is where the segmentation really takes place for *outseg* systems and classical Hidden Markov Model approaches (see Fig. 8 and Section 3.9). The input alphabet $\alpha_1, \alpha_2, ...,$ and the output alphabet $\beta_1, \beta_2, ...,$ are usually composed of either alphanumeric characters or word labels. The third stage provides weights $P(B|A)$ for all possible interpretations $B = \beta_1\beta_2...,$ corresponding to the lower level strings $A = \alpha_1\alpha_2...,$. These weights embody the knowledge of the grouping of sub-units into units: if the α's correspond to stroke sub-units, the β's may correspond to character or word units; if the α's are characters, the β's may be words. The third stage may include several sub-stages, e.g. strokes-to-characters followed by characters-to-words.

4. The last stage adds a grammar (a linguistic model or a simple lexicon), yielding the observation-grammar product graph ("OG product-graph"). The input alphabet $\beta_1, \beta_2, ...,$ and the output alphabet $c_1, c_2, ...,$ are both either alphanumeric characters or word labels. The last stage provides weights $P(C|B)$ for all possible interpretations $C = c_1c_2...,$ corresponding to the lower level strings $B = \beta_1\beta_2...,$. These weights embody the knowledge about the language.

Table 1. Factors of a transduction cascade for a generic recognition system.

Factor	Type of machine	Graph name	Characters accepted/emitted
$P(S\|X)$	WFSA	input segmentation	σ
$P(A\|S)$	WFST	classification	$\sigma : \alpha$
$P(A\|X)$	WFSA	observation I	α
$P(B\|A)$	WFST	output segmentation	$\alpha : \beta$
$P(B\|X)$	WFSA	observation II	β
$P(C\|B)$	WFST	grammar	$\beta : c$
$P(C\|X)$	WFSA	og product	c

The final WFSA in Fig. 1 is the observation-grammar product graph ("OG product graph"). It calculates $P(C|X)$ which can be factored as:

$$P(C|X) = \sum_{\{B,A,S\}} P(C|B)P(B|A)P(A|S)P(S|X) \tag{2.8}$$

where the factors represent the successive intermediate steps of recognition, as summarized in Table 1. Each step is conditionally independent of the previous step but may be differently conditioned on various external sources of information, such as the families of classifiers, segmenters and language models, the training data, and the training algorithms.

2.3. *Training*

Training consists of adjusting the parameters of the system to get the best recognition performance by using the examples of a training set.

Often some parameters of the system are adjusted manually, while others are trained automatically. Usually, the classifier contains most of the trainable parameters. As computers become more and more powerful, there is a tendency to rely more and more on the training data and to adopt a statistical classifier with many (hundreds of thousands) of adjustable parameters. Neural networks and decision trees are popular methods. Researchers originating from the speech community often prefer Gaussian mixtures and Radial Basis Functions (RBF).

The most popular training method is gradient descent which involves making incremental changes to the weights according to the negative gradient of the training error:

$$\Delta w = -\eta \frac{\partial E}{\partial w} \quad , \tag{2.9}$$

where η is a proportionality constant, the "learning rate".

Maximum likelihood techniques can be included in this formalism, by choosing E to be $-\log L$, where L is the likelihood function:

$$L = \prod_k P(X^k|C^k,\mathcal{U}) \quad , \tag{2.10}$$

where (X^k, C^k) are the training examples (data, class) and $\mathcal{U}$ represents other conditioning variables, including the family of classifiers, the family of segmenters, the family of language models, and the training data, the training algorithms.

Provided that all the stages are differentiable, it is possible to optimize globally the recognition system by back-propagating gradients, similarly to the back-propagation algorithm used to train neural networks [57] and the forward-backward procedure (or Baum-Welsh algorithm) used to train Hidden Markov Models [56]. This means that the weights of the segmentation graphs and of the classifier, plus other preprocessing parameters eventually, are all simultaneously optimized to minimize the training error.[j] Examples of global training applied to handwriting recognition are found in [13, 4].

When the sum over all paths in Eq. (2.3) is replaced by a maximum, a two-step algorithm, similar to an Expectation-Maximization or EM procedure [12], is used. In the Expectation step, the weights of the system are frozen and best segmentation for each training example (X^k, C^k) is determined. In the Maximization step, the weights are modified using gradient descent (Eq. (2.9)) to maximize the likelihood (minimize the training error), using the segmentation thus defined. Steps E and M are iterated until convergence.

With either method mentioned above, no prior segmentation of the data is needed. For speed of convergence however, it is often recommended to pre-train the classifier on a set of isolated characters.

Techniques that improve recognition independently of the classifier family include: capacity control via cross-validation or Structural Risk Minimization [69], data cleaning [48], voting among several classifiers [72][k], using an objective function which emphasizes learning of atypical patterns (works only on clean data) [25], and using a distortion model to enlarge the training set artificially [15].

3. Which Solution to What Problem?

In this section we discuss particular difficulties of on-line cursive handwriting and describe alternative designs of the main building blocks of the recognition systems.

3.1. *Intrinsic Ambiguity of Handwriting*

By definition, cursive handwriting contains connected characters which render recognition and segmentation problems inseparable. However, some recognition difficulties can be attributed to either shape recognition problems or segmentation problems.

Shape recognition is concerned with assigning class probabilities to the input given by the geometrical shape of the written character — including its form, relative position, and relative size.

Ideally, differences between several drawings of the same character would be less significant than differences between drawings of different characters. Unfortunately,

[j] This "global" optimization of the system does not imply that a global minimum of the training error is reached. Actually, gradient descent methods are prone to yielding local minima. This problem is alleviated with "stochastic gradient" techniques for which parameters are updated at each pattern presentation with the local gradient, as opposed to being updated at each "epoch" with the global gradient calculated with all the training patterns.

[k] See also the chapter by L. Lam *et al.* in this book.

there are a number of intrinsic shape ambiguities in the Latin alphabet: 0 (zero) versus uppercase O (as in Oscar), 1 (one) versus uppercase I (as in India) versus lowercase l (as in Lima), 2 (two) versus Z (as in Zulu), ; (semicolon) versus j (as in Juliet) which can be resolved only with the help of syntactic or semantic context. Less severe are size (e.g. C–c, O–o, S–s, U–u, V–v, W–w, X–x and Z–z) and position ambiguities (e.g. P–p, Y–y and 9–g) which can sometimes be resolved by comparing relative sizes and positions of characters. Aside from these intrinsic ambiguities, some writers write similarly pairs of characters such as a–u, u–n, M–m, m–n, F–f, r–v, 2–L, 3–z, 4–Y, 6–G, 6–b, i–I, j–J, l–L, g–S, k–K etc. Shape ambiguities are illustrated in Fig. 9.

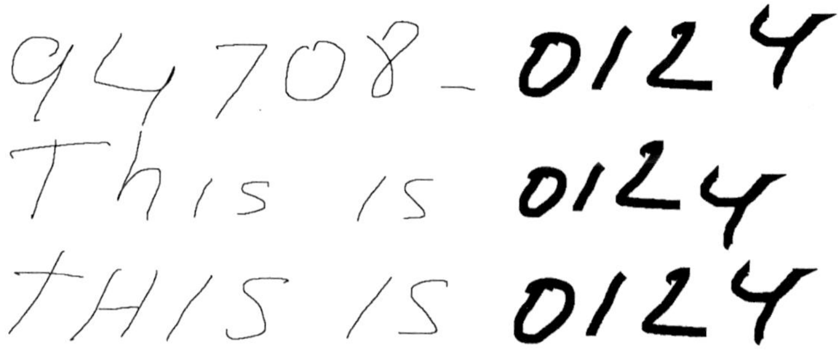

Fig. 9. Ambiguity of character shapes: The same four handwritten characters have different interpretations depending on their relative sizes and positions and on the syntactic and semantic context [54].

Character segmentation is the task of identifying character boundaries. In the case of consistently spaced characters, segmentation is fairly easy and can be done independently from the recognition process by setting a threshold on the space between strokes. For cursive and mixed-styles writing, segmentation points are harder to determine and can be very ambiguous. Segmentation cannot be done independently of recognition (Fig. 10).

One way to circumvent shape and segmentation ambiguities is to design a new alphabet with unambiguous shapes, each of which consists of a single stroke or "unistroke" [23]. Although this approach seems to defeat the purpose of using the pen as a more natural computer user interface than the keyboard, the company Palm Computing has introduced recently to the market a product called Graffiti which uses a unistroke alphabet that has been reasonably well accepted.

3.2. *On-line versus Off-line*

Strictly speaking, *on-line* recognition means that recognition takes place as the user is writing. In the handwriting recognition jargon, *on-line* also refers to a special type of input data containing information about the pen trajectory as given, for instance, by a digitizing tablet (Fig. 11 (a)). A typical application of on-line handwriting recognition is pen computer interfaces. Pen computers impose con-

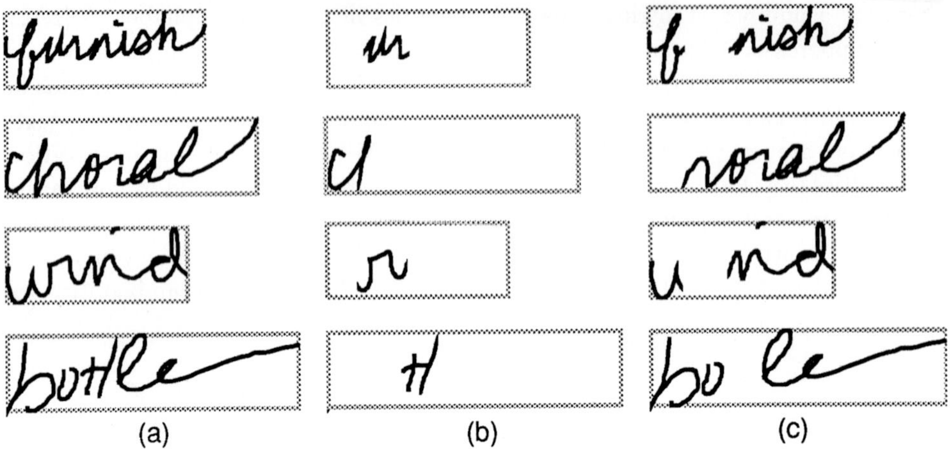

Fig. 10. Ambiguity of segmentation and benefit of using context: (a) Examples of cursive writing. (b) Segmentation ambiguities that cannot be resolved without context. (c) Contexts that narrow down considerably the number of candidate characters that could fit in the gap.

straints on the speed of the processor (usually a low power PC) and the memory availability (usually less than 500 Kbytes). A speed of at least 10 characters per seconds is required for the user to feel comfortable.

In contrast, *off-line* recognition takes place sometimes long after handwriting. The raw input data is most often a digitized image or "pixel map", obtained, for instance, with a scanner (Fig. 11 (b)). This is why "off-line" character recognition and Optical Character Recognition (OCR) are usually synonyms. Typical OCR applications include post office address readers and bank check readers. The recognizers must often process hundreds of characters per second but the customers can usually afford fast processors and large amounts of memory.

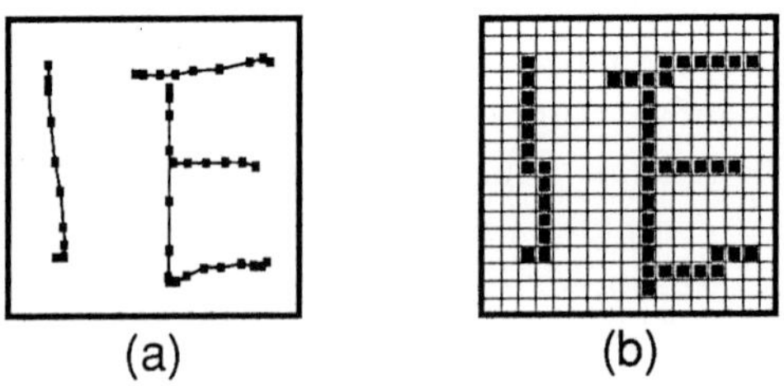

Fig. 11. On-line and Off-line data. (a) On-line data consist of a sequence of time ordered coordinate points; the writing trace is known. (b) Off-line data consist of image pixel data. They are space ordered and the writing trace is unknown.

There is a natural temptation to convert pen trajectory data to pixel images and process them with an "off-line" recognizer. But, the on-line handwriting recognition problem has a number of distinguishing features which must be exploited to get the best results:

- **preprocessing** operations such as normalization, smoothing, deslanting and deskewing and dehooking and **feature extraction** operations such as the detection of line orientations, corners, loops and cusps are easier and faster with the pen trajectory data than on pixel images.

- **discrimination** between optically ambiguous characters (for example 'j' and ';') may be facilitated with the pen trajectory information.

- **segmentation** operations are facilitated by using the dynamic information and the pen-lifts.

- **immediate feedback** can be given by the writer whose corrections can be used to further train the recognizer.

For those reasons, on-line recognition is usually found easier than off-line recognition. In an experiment [45], on-line handwritten uppercase characters were converted to off-line data. The error rate increased from originally 4% to 10% resulting in a significantly superior recognition performance for on-line data.

In unfavorable cases, however, temporal information can actually complicate the recognition. For example the letter 'E' can be written in one to four strokes and various stroke orders and still produce the same static image. Late-written 'i' dots and 't' crosses can cause a problem because they can be far apart in time from the rest of the corresponding character. A reasonable compromise is to perform the preprocessing steps on the pen trajectory data but then generate a time independent representation [4].

Conversely, on-line techniques can be used for off-line recognition. This requires reconstructing temporal information from static off-line data (e.g. [14]). However, this approach is still more a curiosity than a competitive way of handling off-line recognition.

3.3. *Cursive versus Handprint*

There is a debatable distinction between "cursive handwriting" and "natural unconstrained handwriting". Indeed, one is necessarily a subset of the other, but is arguable whether it is desirable to build recognizers for pure cursive styles, knowing that many people write in mixed styles which incorporate both cursive and handprinted styles (see Fig. 12). For that reason, we have concentrated our attention on the "mixed style" problem rather that pure cursive.

The problem of segmentation is harder for cursive and mixed style handwritings than for handprint. Since the writer is not constrained to lift the pen between characters, there is a quasi continuum of possible segmentation points, as opposed to a limited number of pen lifts. Two strategies are commonly used (Fig. 13):

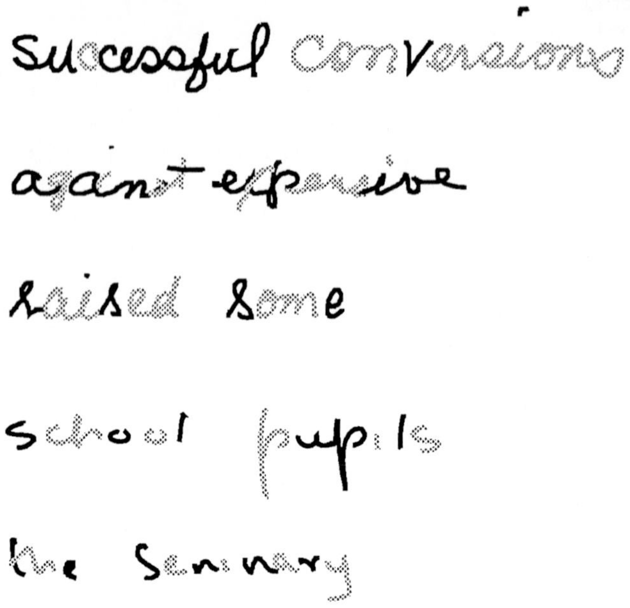

Fig. 12. Mixed styles: Many writers do not have a pure cursive or handprinted style. The figure shows handwriting samples with various amounts of pen lifts which are indicated by alternating black and grey ink.

- **Making heuristic provisional segmentation points**, usually based on pen lifts, extrema of the speed of the pen or on points of high curvature. Hypothetical characters are then obtained by grouping 1 to n elementary segments defined by these provisional segmentation points, where n does not exceed 4 or 5.
- **Making regularly spaced provisional segmentation points** along the pen trajectory. Hypothetical characters are then obtained by grouping a fixed number of elementary segments defined by these provisional segmentation points. This is equivalent to sweeping a window of fixed size over the input.

3.4. *Local versus Global Representations*

After all low level preprocessing operations have been performed (i.e. normalization, smoothing, deslanting and deskewing, dehooking etc.) the system must generate a data representation which is suitable for the classifier. Representations can be either very faithful to the original data, in which case most of the work is left to the classifier or, on the contrary, composed of sophisticated high level features.

3.4.1. *Local representations*

The representations that are most faithful to the data preserve the finely sampled pen trajectory information. An advantage of such representations is that they do

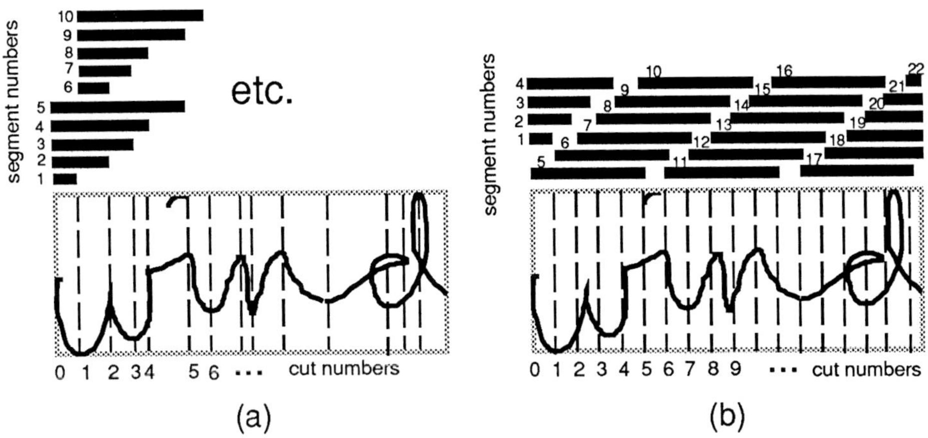

Fig. 13. Cursive handwriting segmentation: The input representations are split into segments to be presented to the classifier. Each segment (indicated by a bold horizontal bar) is delineated by two tentative cuts (indicated by vertical dashed lines). Input representations are generally not images and tentative cuts are detected using the pen trajectory information. These two-dimensional representations are only shown to help the reader visually. (a) Tentative cuts generated at the zeros of the vertical velocity. (b) Tentative cuts regularly spaced.

not make any early and potentially "information destructive" decisions. This is particularly important if the segmentation decisions are postponed until the last stages of the system, as in the *outseg* architecture (see Section 3.9).

Preprocessing for pen trajectory representations commonly includes resampling to increase, reduce, or standardize the number of points. Resampling might preserve the initial time spacing or produce points regularly spaced in arc length. Because of inconsistencies in writing speeds, the latter solution is preferable. One difficulty is to determine the appropriate sampling step. For instance, the sampling step can be a fixed fraction of the character core height (height of a character such as 'a', 'e', 'o', or'u' which have no descenders nor ascenders) if that one is accurately determined.

Many features of the pen-trajectory have been suggested as being particularly relevant for recognition [34, 46, 61]; standard choices include local direction and local curvature. Pen trajectory representations suffer from the difficulty of detecting and reordering delayed strokes and dealing with multiple variants of the same character which differ in pen trajectory but not shape. A solution to that problem is to do all the preprocessing steps using the pen trajectory information but then arrange the features in a two-dimensional map [4].

3.4.2. *Global representations*

Another approach which preserves the pen trajectory ordering, but not the fine sampling, is to encode sub-characters with global features.

Several authors base their method on a pre-segmentation of the data into "ballistic" strokes, that are pen traces delimited by zeros of the speed of the pen (zeros of the vertical, tangential or angular velocity) and pen lifts. Each ballistic stroke is then encoded with a small number of parameters (typically between 3 and 10 parameters). For instance, strokes can be modeled as arcs of circles with only 3 parameters: the tangent at the first point, the radius and the total arc length. In Ref. [68], using signal processing arguments, the author concludes that 4 parameters per stroke are sufficient to obtain a faithful encoding. The original data, usually sampled at 100 points per seconds, is very redundant. Substantial compression rates of 5 to 10 have been achieved by using such handwriting models [33, 1].

Compressed representations have the advantage of alleviating overfitting problems from which the classifier may suffer: classifiers with a small number of adjustable parameters require less training data. However this argument is weak, especially since large databases are becoming publicly available [26][l] . The danger of making early decisions about segmentation is that character segmentation points are not necessarily a subset of the ballistic stroke segmentation points (Fig. 14).

Fig. 14. Danger of using a segmentation method independent of recognition: Ballistic strokes are often used as atoms of handwriting. Unfortunately, some ballistic strokes (e.g. the bold pen traces on the figure) may belong to several characters.

3.5. *Word-based, Character-based or Stroke-based Recognition*

Most of our discussion so far was centered on the idea of recognizing characters. Characters seem to be a natural atom of handwriting because we are taught in school how to handwrite characters. However, other sub-divisions of handwriting may be better for recognition purpose, depending on the nature of the task:

- **Word recognition**: for small vocabularies (no more than a few hundred words), it is justified to recognize entire words. Problems of normalization such as core height detection are avoided. Global features on the word shape can then be used. Segmentation is simple and based on word spacing. Simple

[l] See also the chapter by I. Guyon, R.M. Haralick, J.J. Hull, and I. Phillips in this book.

template matching can eventually be used for writer dependent word recognition.

- **Stroke-based recognition**: many characters share common sub-characters. The idea is to create an alphabet of sub-characters and base the recognition on those sub-characters.
- **Glyph recognition**: atoms of handwriting can also be subwords consisting of more than one character. This notion is particularly relevant to cursive recognition where many frequently written groups of characters are severely distorted and cannot be segmented into individual character shapes (Fig. 15).

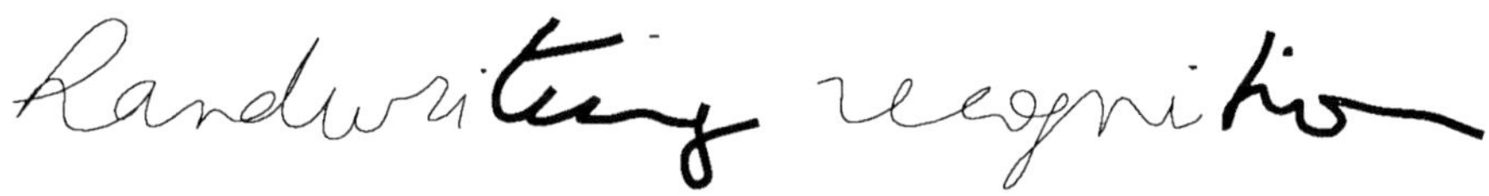

Fig. 15. Subword units as handwriting atoms: The bold pen traces show word endings whose characters are difficult to recognize individually but that are not ambiguous as a whole. This suggests that some subword units (or glyphs) could be used as handwriting atoms.

3.6. *Invariant Representations versus Invariant Metrics*

Most systems rely on the preprocessing to enforce invariances with respect to translation, scale, skew, and stretching. Another strategy is not to normalize the data but use a classifier which is insensitive to such geometrical distortions [64, 71]. Assume, for instance, that we are interested in building a word recognizer based on template matching. The "tangent distance" [64] method allows us to find the closest template, up to given geometrical transformations, including translation, scaling and hyperbolic transformations (Fig. 16).

3.7. *Direct Estimation of a posteriori Probabilities versus Analysis by Synthesis*

Another debatable question is whether the recognizer should be a model of handwriting generation.

Handwriting generation can be described in the classical "noisy channel" framework which has been successfully applied to speech recognition for a number of years. The writer's brain produces strings of characters C with a probability $P(C)$. These strings are then "distorted" by the motor control system which generates handwriting characterized by a probability density function $p(X|C)$. $p(X|C)$ is the probability density function over the geometrical shapes X that can occur given the intended character sequence C.

Analysis by synthesis is a recognition method which consists of approximating $p(X|C)$ usually with a Hidden Markov Models and $P(C)$ by a statistical language

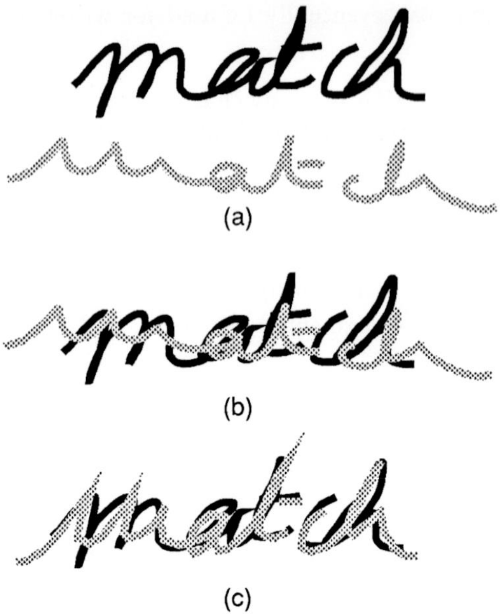

Fig. 16. Invariant metric: (a) Original words. (b) Words superimposed. (c) Words matched with the "tangent distance".

model. Using Bayes rule we obtain the desired posterior probabilities:

$$P(C|X) = \frac{p(X|C)P(C)}{p(X)} \quad . \tag{3.11}$$

The trouble is that this framework does not apply well to a large number of recognition systems which do not make any attempt at modeling handwriting, i.e. do not compute an approximation of $p(X|C)$.

Fortunately, the transduction cascade framework presented in Section 2.2 is consistent with the direct estimation of $P(C|X)$. Furthermore, the same framework applies as well to the analysis by synthesis method and $p(X|C)$ can be decomposed into several intermediate synthesis steps.

3.8. *Weight Arithmetic*

When using graph algorithms, there are a number of subtle issues with weight normalizations and weight arithmetic. We list some of them here in an arbitrary order and leave some questions open:

- **Normalization of the classifier scores**: Not all classifiers provide estimates of probabilities or at least well behaved probability-like scores that add up to one. Graph algorithms do not require using probabilities. It is often better **not** to normalize the scores provided by the classifier since there may be good reasons why they do not add up to one (e.g. there is an implicit extra class for "meaningless" patterns which accounts for the lost probability mass). It is only important to determine whether the scores are multiplicative or additive [13].

- **Correction of "wrong" priors**: Sometimes the training data has very biased frequencies of words, characters, or groups of characters. As a consequence, some classes may have systematically too low or too high scores. Normalizing the score of each class by its average value over positive training examples of that class is one possible way to alleviate that problem. A better solution is to design the training set in a way that avoids this problem.
- **Normalization of the weights of WFSTs**: WFSTs implement conditional probabilities. Consequently, one should not expect that the weights of the outgoing arcs of a given node add up to one.
- **Normalization of the weights of the OG product graph**: In the transduction cascade framework, either the observation graph is a WFSA and the grammar graph a WFST (in the case of the direct estimation of a posteriori probabilities) or vice versa (in the case of the analysis by synthesis). In practice, both graphs are often represented as WFSA graphs which emit (or accept) characters of the same alphabet. The weight normalizations of the product graph are usually not performed.
- **Normalization with respect to the amount of ink consumed**: A classical problem is that of segmentation graphs which favor short strings of characters. This arises because string probabilities are obtained by multiplying the probabilities of its constituents. Therefore, the fewest factors, the most probable! This problem is automatically corrected if the weights of the segmentation graph are globally optimized during training with a discriminant criterion. If the weights are set "by hand", the amount of ink consumed should be accounted for.
- **Multiplicative or additive weights**: Whether the scores represent probabilities or log-probabilities is not the only criterion to decide on their arithmetic. For instance, the observation graph and the grammar graph can be thought of as two "experts" which estimate independently $P(C)$ under different conditions. The overall recognition system could combine the expertise of these two experts by weighing their opinions $P_o(C)$ and $P_g(C)$ according to how much they can be trusted: $\gamma P_o(C) + (1-\gamma) P_g(C)$. This contrasts with the approach presented in Section 2.2 which suggests to combine the scores as: $P_o(C)P_g(C)$.

3.9. *Input versus Output Segmentation*

The question is: should we take advantage of the geometrical and dynamical properties of the pen trace to pre-segment handwriting before any recognition takes place (*inseg* strategy, Fig. 7) or should we make as few assumptions about segmentation as possible before recognition and perform recognition and segmentation simultaneously (*outseg* strategy, Fig. 8). Few studies have been made to compare the two approaches using the same classifier and the same data [60, 70]. In principle, *outseg* should be better for cursive since it is almost impossible to decide where the cuts are before recognizing the characters. However, we do not know of any systematic study on cursive recognition which has demonstrated that this is actually true.

The input segmentation graph of the *outseg* method is trivial (Fig. 17). All the knowledge of the designer is put into the architecture of the output segmentation graph (Fig. 18).

The input segmentation graph of the *inseg* method is nontrivial (Fig. 19). It connects segments using geometric rules. The simplest rules are to connect segments delineated by cuts (i, j) to segments delineated by cuts (j, k), thus forbidding reusing the same ink in more than one character.

Computationally, *inseg* is more economical because only a few segments are considered for classification. This results not only in fewer calls to the classifier, but also in a smaller observation graph and observation-grammar product graph (Fig. 20). Therefore, searching for the final segmentation and recognition solution is faster for *inseg* than for *outseg*, unless some clever search pruning techniques are implemented.

For best recognition accuracy, a combination of *inseg* and *outseg* approaches may be the solution [20, 27], but combined approaches are still in their infancy.

This figure is related to Fig. 7. The notation (σ_7 : 'a') means that the classification automaton accepts segment σ_7 and emits character 'a'. The weights of the observation graph are products of weights of its two parent graphs. For instance: $P(s_7, 'a'|s_5) = P(s_7, \sigma_7|s_5)P_7^a$.

3.10. *Writer Adaptation*

User expectations with respect to handwriting recognition accuracy are around 1% character error rate. Error rates larger than 2% character error are intolerable: they require too much user attention to make corrections and slow down text entry considerably [28]. Even for limited size vocabulary, such low error rates are very difficult to achieve (see Section 2.5). They are probably out of reach of writer independent recognizers, for the time being.

Although "walk-up" recognition accuracies may be quite poor, users do adapt to a particular recognizer and often get the illusion that the recognizer improves over time, even if in reality it is the user who improves. Commercial products have been reluctant to introduce any other form of adaptation because of the danger of degrading the recognition performance rather than improving it. It is indeed very difficult to design a writer adaptation training algorithm which is robust against misuse such as intentional or non-intentional introduction of meaningless patterns.

Technically, writer adaptation is not difficult to achieve under laboratory conditions [49, 17]. The basic idea is to retrain superficially the system by modifying

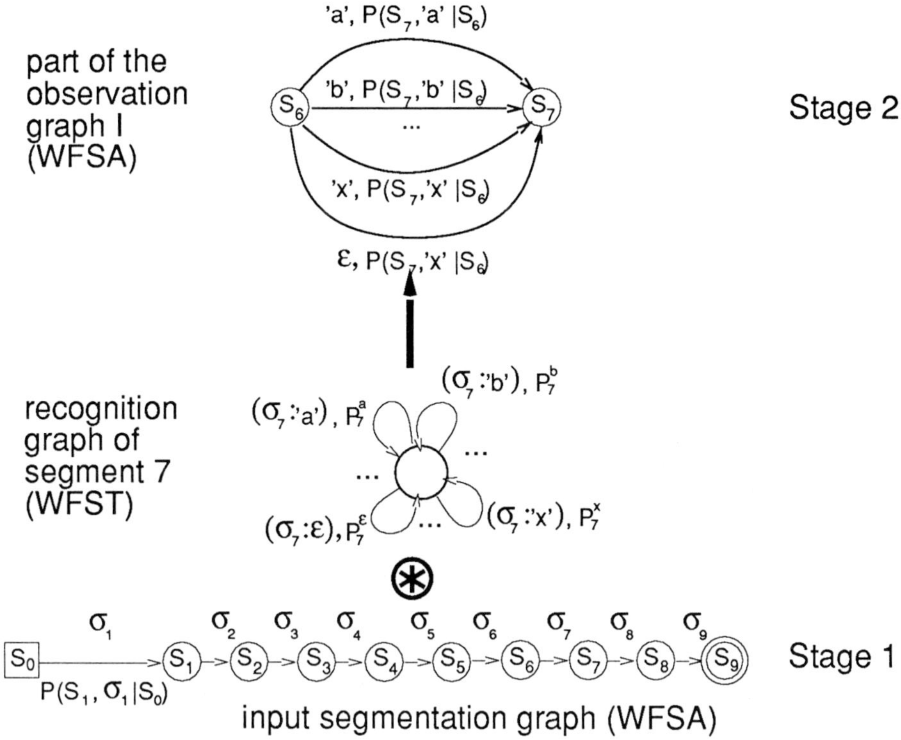

Fig. 17. Calculation of an *outseg* observation graph I: This figure is related to Fig. 8. The notation ε represents an empty character (no character observed). ε characters are emitted by segments that are not well aligned with and actual handwritten character (e.g. ligatures).

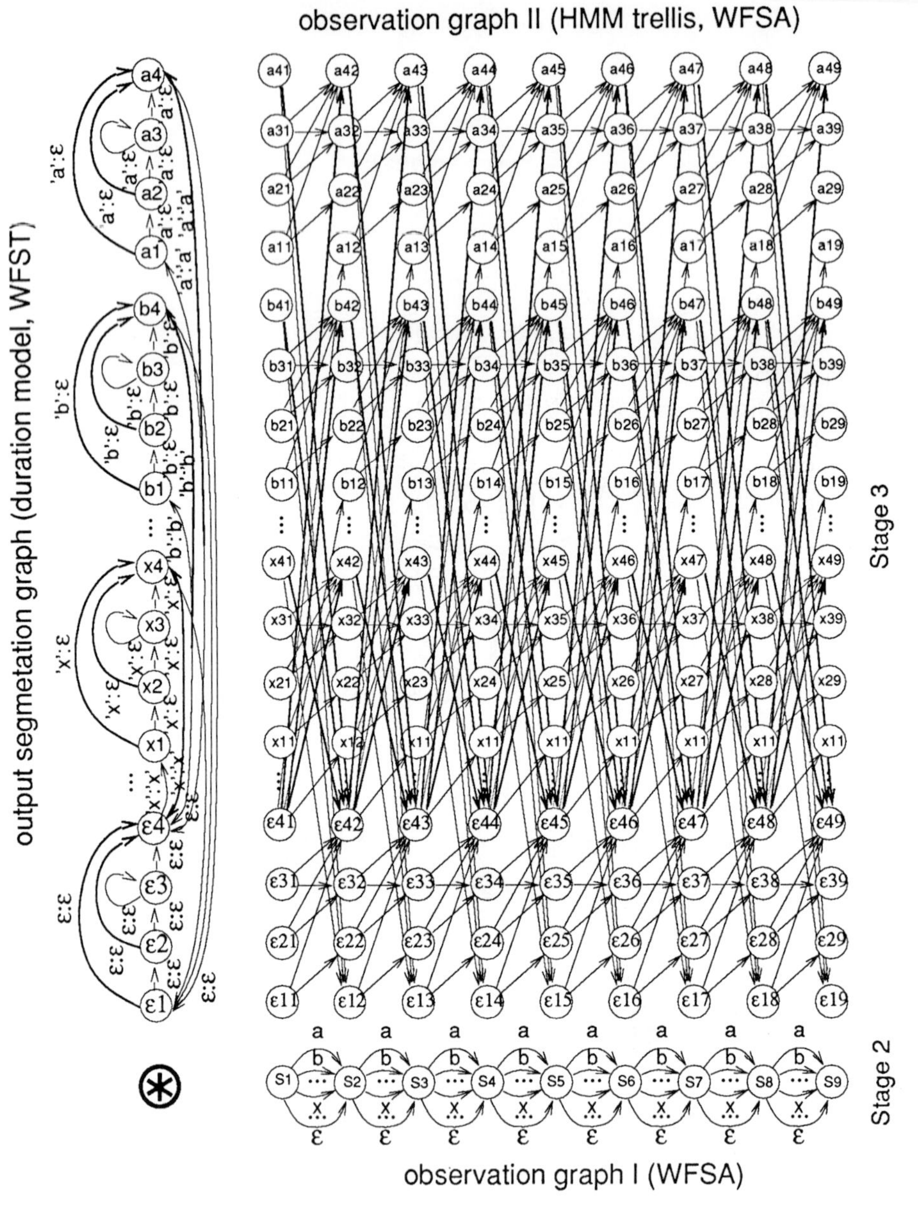

Fig. 18. Calculation of an *outseg* observation graph II (HMM trellis): This figure is related to Fig. 8. In this particular example, the output segmentation graph is a duration model such as the ones used in Section 2.4. It forces alternating between a character model and a model of transitions between characters or ligatures.

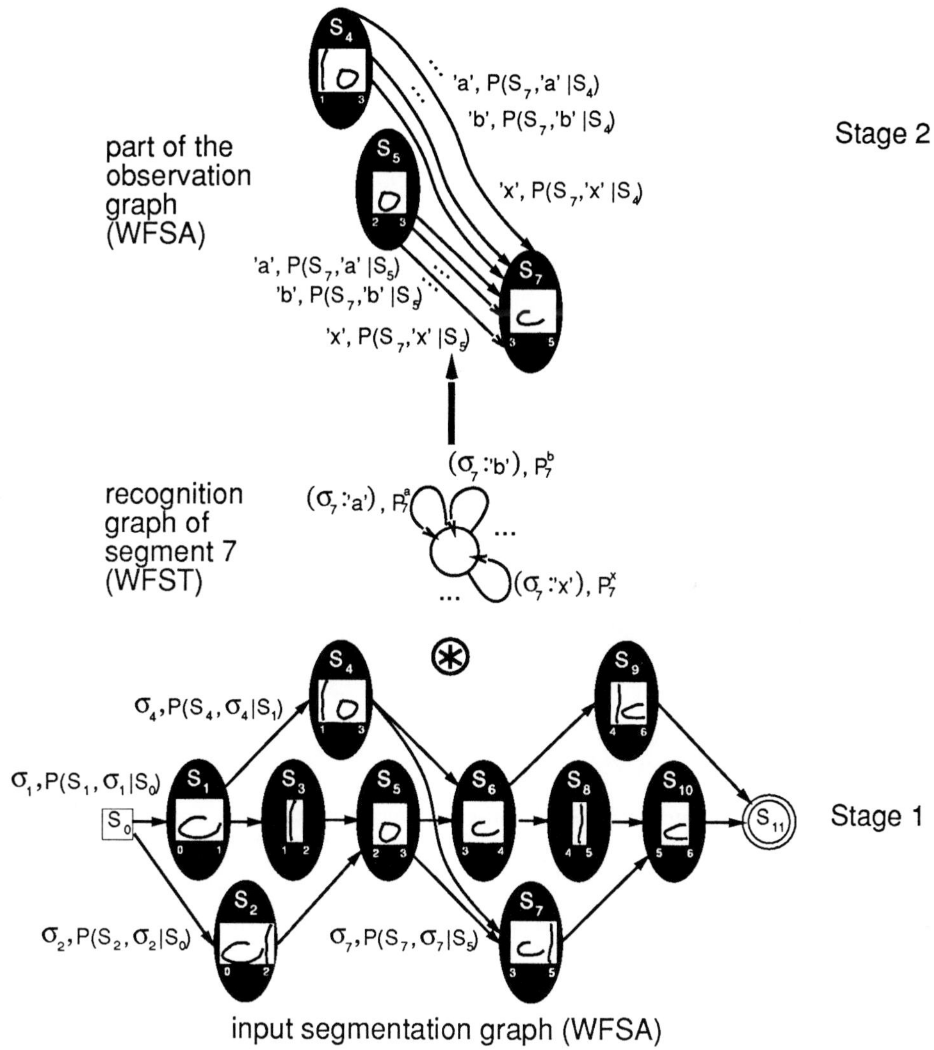

Fig. 19. Calculation of an *inseg* observation graph:

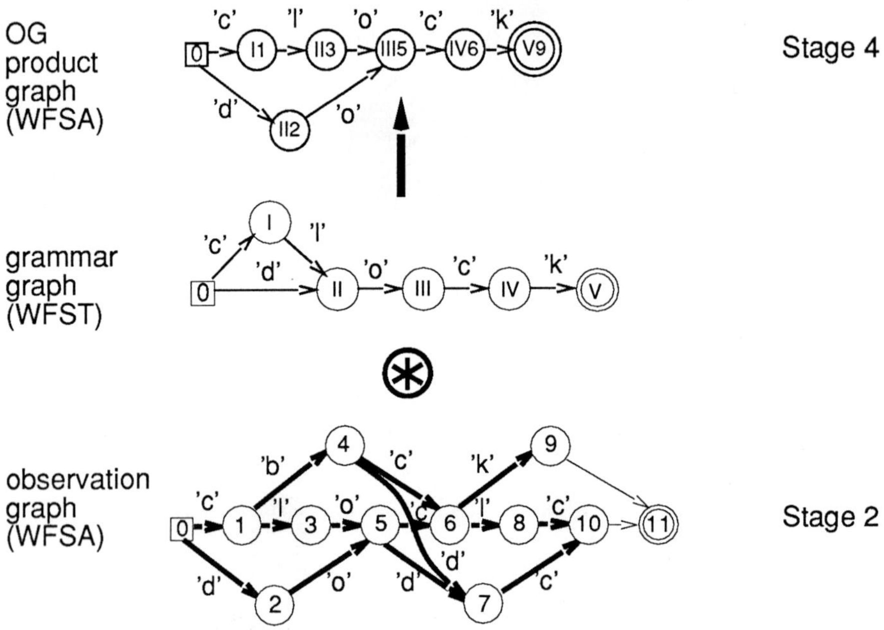

Fig. 20. Calculation of an *inseg* observation-grammar product graph: This figure is related to Fig. 7. In the observation graph, a bold arrow represents multiple connections, one per class. Only the character corresponding to the best score is represented. The grammar graph is a transducer which accepts and emits the same characters. The notation '*c*' : '*c*' is therefore simplified to '*c*'. The weights are not represented. Note that stage 3 (observation graph II) is absent from *inseg* systems.

only the weights of the upper levels.

In real world applications, writer adaptation is harder to deploy. The adaptation engine must be designed jointly with an appropriate user interface. Ideally, adaptation should be transparent to the user. The system would continuously learn from its mistakes, the corrections of the user and the redundancy of the language. This remains to be seen.

4. Example of an Entire System Design

In this section, we describe an entire system design as an illustration of the main concepts presented in the previous sections. The overall system is a word recognizer, consisting of a trainable character classifier and a language model. This gives us the flexibility of changing the language model, e.g. changing the vocabulary, without retraining the character classifier.

The system was designed to provide acceptable "walk-up" recognition accuracy — without forcing the writer to go through a training session and without imposing any restriction in writing style. The preprocessing and the classifier are designed to avoid making potentially wrong segmentation decisions at an early stage in the recognition process.

4.1. *Description of the System*

The main building blocks of our system, shown in Fig. 21, are a Time Delay Neural Network (TDNN) and Hidden Markov Model (HMM). The architecture is of the *outseg* type (see Section 3.9 for details). In Fig. 22, we show a schematic graph interpretation of the system in terms of transduction cascade.

The **data collection device** (Fig. 23) provides pen trajectory information as a sequence of (x, y) coordinates at regular time intervals (10–15 ms). The pen trajectory of one word is collected. For each word, the **preprocessing** begins by estimating the height of the *middle area* (or *core height*) with two linear regressions through the local minima and maxima of the y-coordinate, excluding the outliers. This results also in an estimate of the orientation. After the word has been normalized, oriented and smoothed, we perform a resampling of the data to points regularly spaced in arc length. The spacing between points is a fixed fraction of the core height. We then compute local features of the pen trajectory at each point: slope, curvature and speed [24, 60]. Disjointed components such as diacritical marks, the dot on i, and the cross on t are removed and treated as additional features. The result of this preprocessing is a finely sampled sequence of time-ordered feature vectors at regular intervals in space. This representation is very faithful to the original data and is suitable both for cursive and printed handwriting.

After the preprocessing is performed, a **Time Delay Neural Network** is used to estimate posterior probabilities of occurrence for each character given a part of the input data whose length corresponds approximately to one character. This operation is repeated with segments of the input which overlap considerably and

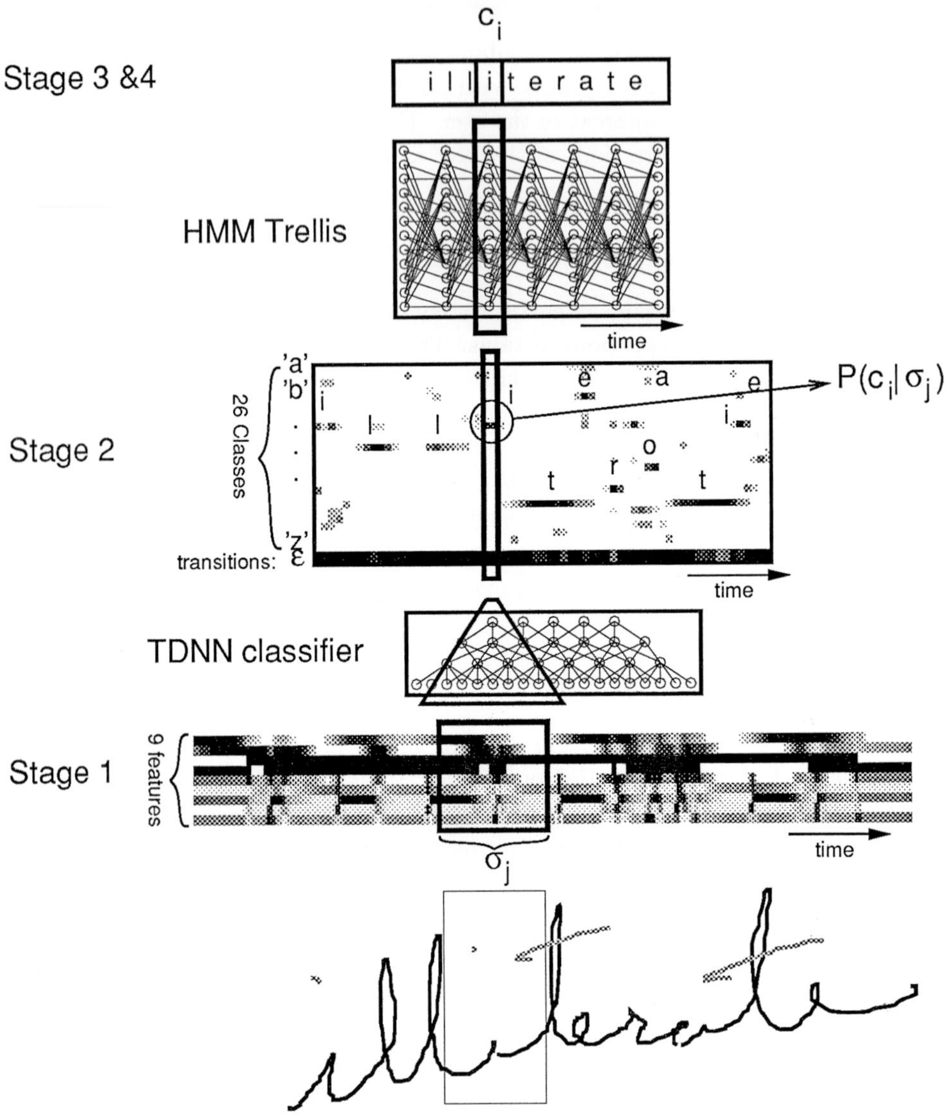

Fig. 21. A TDNN-HMM hybrid system. After the preprocessing, a time ordered sequence of feature vectors is presented to the Time Delay Neural Network (TDNN). The darkness indicates the amplitude of the coefficients and the color (black or white) their sign. The TDNN estimates a posteriori probabilities for characters in a word. A Hidden Markov Model (HMM) segments the word in a way which optimizes the global word score. If the HMM does not use any language model or grammar (no stage 4), the word "lliteroti" is the best interpretation provided by the system. If the HMM is forced to chose words from an English lexicon (stage 3 and 4 combined), the correct word "illiterate" is found.

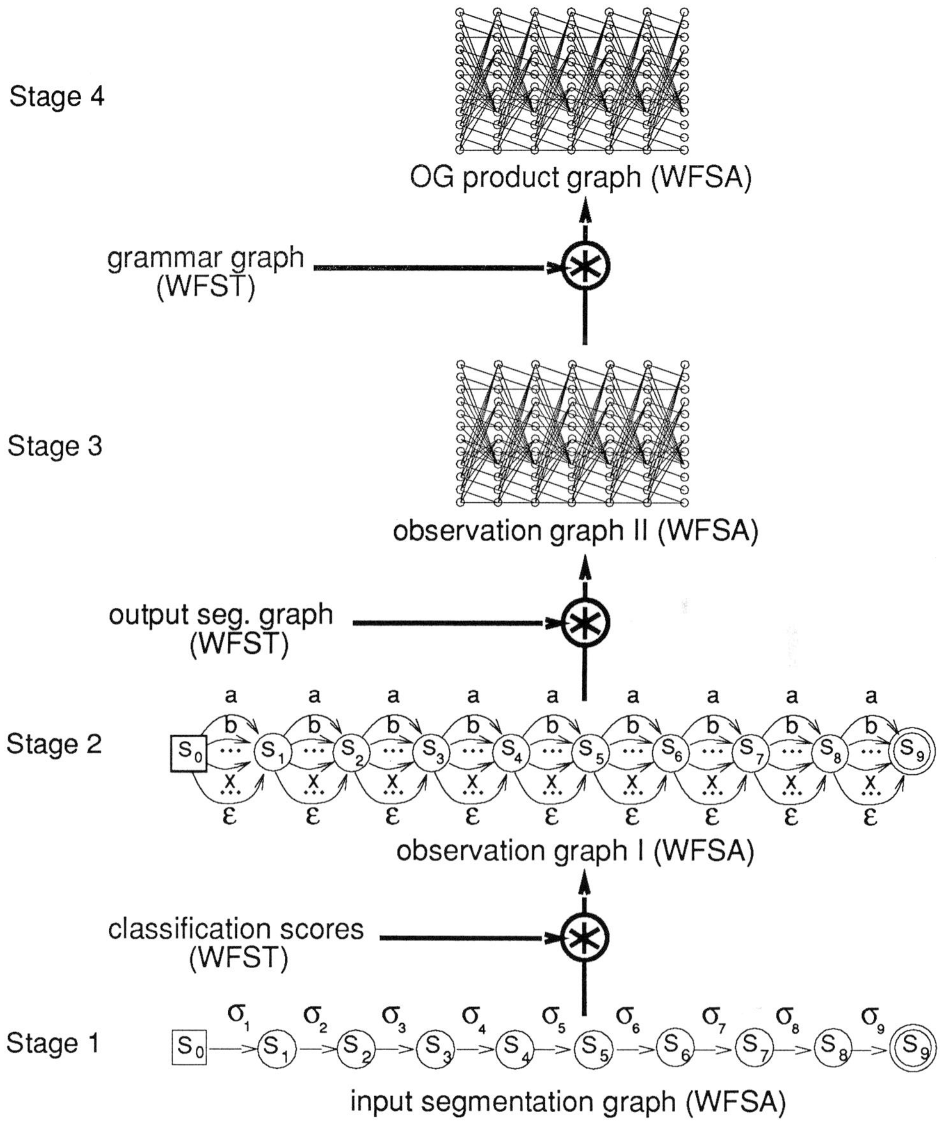

Fig. 22. Graph interpretation of Fig. 21. Each stage is obtained from the preceding one by a transduction product.

are shifted in time, thus creating probability estimates $P(c_i|\sigma_j)$ for each character c_i given the input segment σ_j at time step j. An additional category, the ε category, is introduced to model transitions between characters. The observation probability for ε is $P(\varepsilon|\sigma_j) = 1 - \sum_i P(c_i|\sigma_j)$ which represents the probability that there is no character in the segment σ_j.

The TDNN is a multi-layer feed-forward network with shared weights and with connections constrained to be "local" in time. Its successive layers perform successively higher-level feature extraction. There is one output unit for each character-class, and its activation level represents the score for that class. TDNNs, which were previously applied to speech recognition [39], are well suited for sequential signal processing. This allows us to use a representation which preserves the sequential nature of the data, in contrast to approaches based on pixel-map representation.

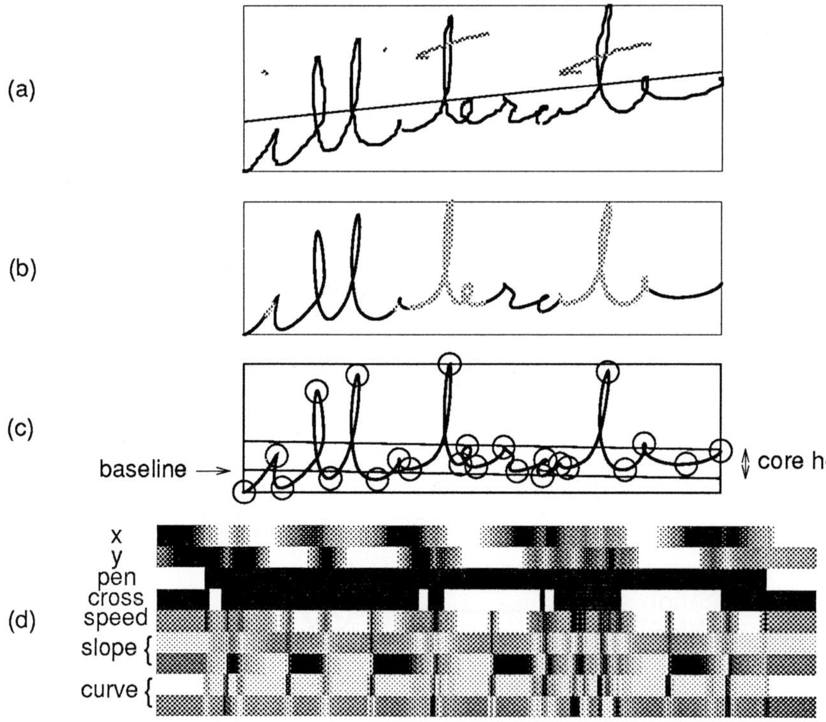

Fig. 23. A preprocessing suitable for TDNNs. (a) Word orientation obtained by linear regression. (b) Smoothing and removal of i dots and t bars. Grey areas indicate the projection of the shade of i dots and t bars. (c) Detection of the baseline and the core height by fitting two lines with the local extrema of the y-coordinate, excluding the outliers. (d) The final representation encoding local features of the pen trajectory regularly spaced in arc length.

It may seem computationally expensive to perform so many classifications, most of which correspond to segments of data not aligned with a character. Thanks to the convolutional structure of the Time Delay Neural Network (TDNN) used here, most of the calculations between two consecutive overlapping segments are identical and are thus done only once.

We train the network with a "position invariant learning" method [36]. During training, we add a layer which transforms the outputs of the TDNN into a single score vector indicating presence or absence of a given character anywhere in the input field. The presence indicator of character c_i is computed as:

$$P(c_i) = 1 - \prod_j (1 - P(c_i|\sigma_j)) \tag{4.12}$$

which approximates the probability that character c_i appears *at least once* in the input field. This allows us to train the network with one target vector containing only information about presence or absence of a character anywhere in the input field.

Although this target vector contains no character order information anymore, position guessing is enforced by the architecture of the TDNN and the redundancy of the data. It self-organizes the activities of the units in a competitive fashion to infer the positional information and approximates the posterior probabilities $P(c_i|\sigma_j)$ (see [36] for details). Therefore no manual segmentation is needed, word labels can be used directly.

The outputs of the TDNN are processed by a **Hidden Markov Model** which uses a duration model as shown in Fig. 24. The probability of remaining exactly d time steps in the same character model after entering it is given by $\rho(d) = \prod_{i=1}^{d-1} p_i(1-p_d)$. With a simple recursion formula, one can determine p_i from values of $\rho(d)$. We truncate the distribution by limiting the model to a finite number of states and add a self loop to the last state. This is equivalent to replacing the tail of the distribution by an exponential. For $\rho(d)$ we chose a Poisson distribution. The parameter of the distribution (both mean and standard deviation) is determined by estimating character durations from the training data statistically.

The observation probabilities for the HMM at each time step are given by the Neural Network and the transition probabilities between states are the parameters of the duration model. In terms of WFST, the duration model is the "output segmentation graph". The details of the graph products are found in Figs. 17 and 18.

The Viterbi algorithm [56] is used to find the best path in the HMM trellis and determine the best recognition and segmentation.

A natural way of introducing a **lexicon** in this process is to restrict the solution space of the HMM to words included in a given lexicon. Unfortunately this means building the product graph between the observation graph and the grammar graph, or OG product graph (see example in Fig. 20). If the grammar graph is a large lexicon tree, the OG product graph gets prohibitively large and the Viterbi algorithm gets prohibitively slow. One way of keeping calculations tractable is to narrow the lexicon search to those words which are closest in edit-distance (or "Levenhstein

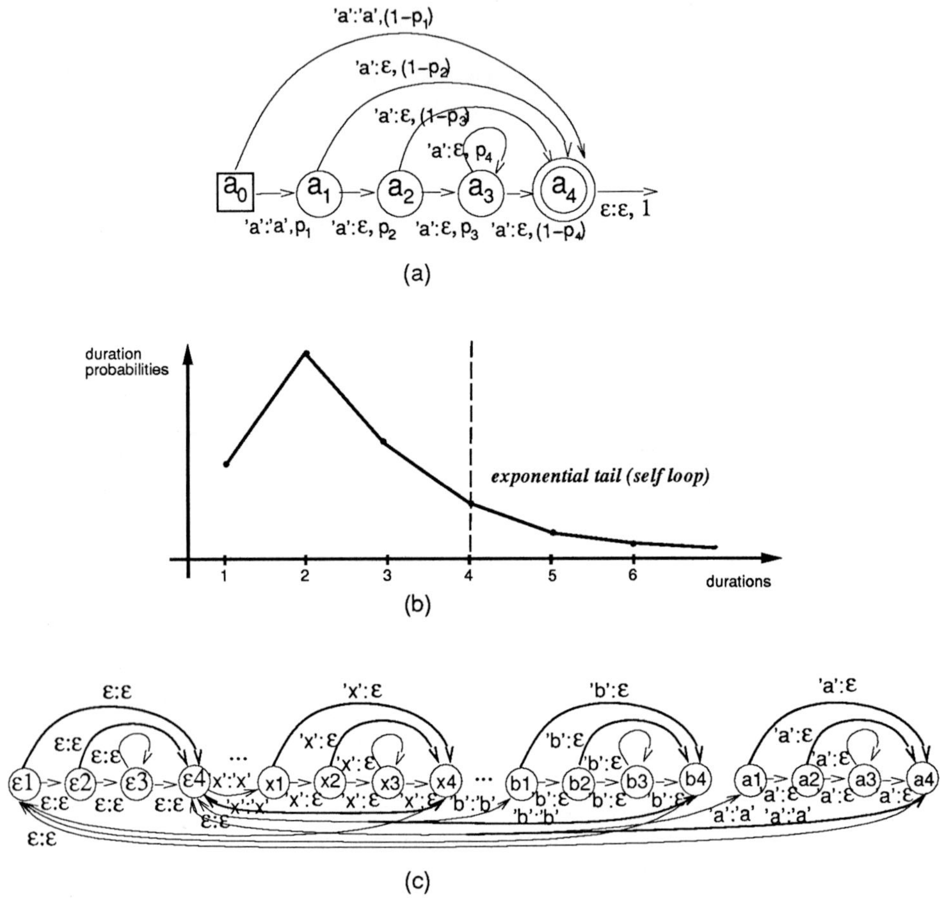

Fig. 24. Duration model. (a) Duration model for character 'a'. The ε character is the empty character. Character 'a' is accepted and emitted at the first transition; then, characters 'a' are "eaten", that is accepted with the emission of an ε character. (b) The corresponding durations. (c) The entire duration model used as "output segmentation graph".

distance" [42]) to the most probable word obtained at stage 3, without the lexicon. With this method, the word list is typically reduced to 1 to 5 words. Since the calculation time of the edit-distance is a negligible fraction of the total recognition time, the recognition speed is increased substantially. Even more importantly, the speedup is at the expense of very little performance degradation.

In our example of Fig. 21, the words in the 25,000 word lexicon of the $UNIX^{(TM)}$ *spell* dictionary that are closest to "lliteroti" are: "alliterate", "illiterate" and "literate". They are all at an edit distance of three. For instance, "illiterate" is obtained from "lliteroti" by inserting an 'i', substituting 'o' for 'a' and 'i' for 'e'. Edit distances can be computed with the Viterbi algorithm [56]. We show in Fig. 25 the reduced grammar graph thus obtained.

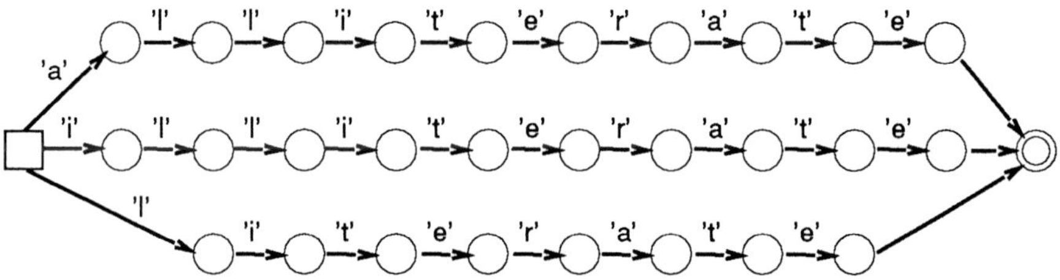

Fig. 25. A reduced grammar graph obtained with the fastmatch technique. This figure represents the grammar graph used for the example of Fig. 21. Out of a 25,000 word lexicon, only 3 words are closest in edit distance to the best guess without lexicon "lliteroti". This allows considerable pruning of the grammar graph, and consequently of the OG product graph.

4.2. *Recognition Performance*

Several networks were trained for uppercase, cursive, and mixed-case recognition tasks. The networks used have three to four convolutional layers and a total of 8,000 to 10,000 weights. Only the best results for the various tasks are reported here. Explanations on the parameters chosen are found in Ref. [61].

Training was done on up to 30,000 uppercase, 27,000 lowercase and 27,000 cursive words. Training took up to 60 back-propagation cycles through the training set which corresponds to two weeks on a SUN Sparc10. The system was tested on 607 words from a disjoint set of writers. The $UNIX^{(TM)}$ *spell* dictionary containing 25,000 English words was used to generate lexicons.

The best performance using a separate network for each writing style category is summarized in Table 2. Character and word error rates are reported with and without lexical constraints. Table 3 gives the results obtained for a single mixed style network.

Note that there is no severe loss in cursive recognition performance when using a single network for all writing styles. This shows that the approach presented is able to handle mixed style writing.

On average, only 4 words are checked when the word list is restricted to words

Table 2. Performance on the test set. Best performance for the uppercase and cursive task. For each task a separate network was trained.

	No Lex.		25,000 Word Lex.	
Task	% Char.	% Word	% Char.	% Word
Upper	7.5	37.1	1.1	3.1
Cursive	26.7	75.7	11.1	20.8

Table 3. Performance on the test set. Best performance for the uppercase, lowercase and cursive task using a single network.

	No Lex.		25,000 Word Lex.	
Task	% Char.	% Word	% Char.	% Word
Upper	13.0	55.3	2.0	4.9
Lower	15.2	56.1	3.3	6.8
Cursive	32.7	82.8	13.8	24.7

that are closest in edit-distance; if the second closest words are also included, about 57 words are checked. Compared to the full search where all 25,000 words in the lexicon would have to be checked, a speedup factor of 400 to 6000 is gained by using the proposed fast-matcher. The recognition of an average seven character word takes about 2 seconds on a Sparc10/40.

As expected, the size of the lexicon has a big influence on the performance, as is shown in Table 4. Changing the lexicon from 1000 words to 25,000 words changes the error rate by about a factor of three.

Table 4. Performance for various lexicon sizes with a network trained on cursive data only.

Lexicon	% Char.	% Word
Unrestricted	32.7	82.8
25000 Words	11.1	20.8
1000 Words	5.7	7.3

5. Bibliographical Notes

The history of cursive on-line handwriting recognition goes back to the early 1960s. Many attempts to build recognition systems have been made; good reviews on the different approaches are given in [43, 6, 67, 53, 40].

In precursor papers around 1962 and 1972, Harmon [31, 32] uses a set of five geometrical features to classify characters with a decision tree. The segmentation is based on a rough localization of "landmarks" and an estimate of character width. The system is not trained and needs carefully written words. He reports a 40% character error rate on the specimen sentences of five persons. Mermelstein and Eden [50] propose a model for the generation of handwriting based on the concept of ballistic strokes, segmented at points of zero Y-velocity. They derive one of the first model-based recognition methods, but do not yet combine recognition and segmentation. For a vocabulary of 59 similar words they report results on 100 words written by four subjects. Their word error rate varies from 9%, when training and testing on the same writer to 31%, when testing on novel writers.

In the early 70s, a big change was introduced by realizing the importance of the segmentation being part of the recognition process. In his off-line recognition system, Sayre [59] considers each "dip" in the writing as a possible provisional segmentation point. Segments encoded with geometrical features are categorized. All possible combinations which may form an acceptable character are listed in a lexicon. The postprocessor selects the most likely word given the recognition results and letter trigram statistics. An on-line system based on similar ideas is presented by Ehrich and Koehler [18]. They report 29% word rejection rate with an error rate close to zero, using a 300 word lexicon of seven-letter words.

As larger data sets become available, more emphasis is put on statistical techniques and Neural Networks. More recently Schomaker *et al.* [63] and Morasso *et al.* [51] use stroke-based or character-based syntactic recognition methods. Their new contribution is that their feature representation is subject to training. They use an unsupervised learning technique, Kohonen's self-organizing maps [37]. They both report only writer dependent results ranging between 17 and 40% word error rate, using training data sets of 400–600 words and test sets of 100–200 words from a lexicon of 4000 words.

Another important step is the introduction of Hidden Markov Models (HMM). Traditionally used in speech recognition [56], HMMs are applied to character recognition since 1980 [52, 38, 47, 9, 8]. With HMMs, recognition-based segmentation is more rationalized: segmentation and recognition are optimized simultaneously during training, therefore it is no longer necessary to segment the data manually. Recent work of Fujisaki *et al.* [22] report writer dependent results by training on 1600 distinctly written characters and testing on 161 connected handwritten words. They obtain 6.8% word error rate with a 2,171 word lexicon. Ha *et al.* report writer independent results [30] with 4,783 training characters manually segmented from 6 writers and 728 test words from 3 writers. They obtain 26.4% word error rate with a 25,000 word lexicon.

Classical HMMs use Gaussian mixtures to estimate emission probabilities. As demonstrated in speech recognition applications, Neural Networks advantageously replace Gaussian mixtures [3, 7] because of their inherent discriminant power. Several groups [60, 70, 46, 4, 61] have demonstrated the applicability of the combination Neural Network/HMM to on-line handwriting recognition of run-on hand-printed and cursive words. The results reported in this paper are typical of the performance of such systems: ranging from 3% word error rate for uppercase printed words to 20% word error rate for cursive words, with a 25,000 word lexicon.

In an effort to reduce the dimension of input space and thus limit the number of parameters to be estimated during training, most authors perform preprocessing which makes early and irreversible decisions about segmentation and recognition [50, 18, 6, 62, 66, 58, 68, 1, 65]. The typical method consists of pre-segmenting the data at local extrema of the velocity and to encode each stroke with the coefficients of a geometric or a dynamic model of the trajectory, such as the one proposed in [33]. When large data sets are available, it is possible to rely more on the training data and to avoid such early decisions which may limit a priori the performance of the system. Finely sampled local and redundant feature representations [24, 2, 20, 46, 4, 61, 44] overcome this problem.

A completely different approach is taken by whole-word recognition systems which do not attempt to segment at all and take words as basic units such as [19]. As whole-word systems need a class label for each *word* to be recognized, they are limited to small lexicon sizes due to the necessary training data and computational load.

Over time, the systems have moved from being writer dependent to writer independent by increasing the number of writers in the training data. Writer independent systems are praised because they do not impose a training session to the user prior utilization. However, to boost recognition accuracy, several authors have proposed to fine-tune writer independent systems to the particular style of a given writer with a small number of examples (writer adaptation) [49, 17].

Research on language models is evolving slowly. Most systems use simple lexicons, n-gram models, or a combination of those [29].[m]

The recognition rates reported in the literature are generally optimistic and difficult to interpret or compare. The tasks reported vary strongly in difficulty according to writing styles, quality of writing, data sets, lexicon sizes etc. The task chosen by the authors usually demonstrates the strength of their system. Objective benchmarks are needed to compare systems on various, well-defined tasks [26].[n]

6. Conclusion and Perspectives

Only a few years ago, cursive handwriting recognition seemed out of reach. Today the dream has become a reality. Yet, recognizers currently available are still disappointing to users. There is a wide margin for improvement which should

[m] For more details on this topic, see the chapter by A. Dengel *et al.* in this book.
[n] See also the chapter by I. Guyon, R.M. Haralick, J.J. Hull, and I. Phillips in this book.

challenge researchers and developers.

To be able to read cursive writing, humans make extensive use of contextual information. The keys to progress in that field will be to understand how to design and use elaborate language models and how to introduce adaptation to particular writer styles. The success of incorporating such contextual information in speech recognition systems is an encouragement for handwriting recognition researchers to pursue in that direction.

The classical distinction between *statistical pattern recognition* and *syntactic and structural pattern recognition* has become practically irrelevant for cursive recognition since most systems borrow techniques from both fields: they have a lot of adjustable parameters that are optimized during a training session and use *graph algorithms* to integrate all the stages of the recognition process.

The Weighted Finite State Transduction framework provides a unification to various graph-based segmentation and recognition frameworks that have been proposed in the literature. In the future, it may become a powerful tool since it allows merging various design concepts such as "input" segmentation and "output" segmentation. Having a single consistent framework facilitates the system *global optimization* during training. The modular decomposition into *recognition stages* permits us to manipulate only small graphs. One can avoid building the large *product graph* of the overall system since training and search can be performed by building "on-the-fly" only the nodes that are needed.

Finally, there is often a large discrepancy between the error rate obtained in laboratory experiments and those obtained on the field. Recognizers should be tested, as far as possible, in realistic conditions of utilizations or at least on realistic test data. With projects such as UNIPEN [26], it will be possible to exchange a wide variety of data and organize public competitions.[o]

Acknowledgements

We gratefully acknowledge discussions with our colleagues at AT&T, Y. Bengio, C. Burges, Y. Le Cun, C. Nohl, and F. Pereira. We are particularly thankful to Y. Bengio, C. Burges, J. Hu and C. Nohl for their comments on this paper.

[o]See also the chapter by I. Guyon, R.M. Haralick, J.J. Hull, and I. Phillips in this book.

References

[1] A. Alimi and R. Plamondon, Performance analysis of handwriting stroke generation models, *Third Int. Workshop on Frontiers in Handwriting Recognition* (Buffalo, May 1993) 272–283.

[2] E. J. Bellagarda, J. R. Bellagarda, D. Nahamoo, and K. S. Nathan, A probabilistic framework for on-line handwriting recognition, *Third Int. Workshop on Frontiers in Handwriting Recognition* (Buffalo, May 1993) 225–234.

[3] Y. Bengio, R. deMori, G. Flammia, and R. Kompe, Global optimization of a neural network-hidden Markov model hybrid, *IEEE Trans. on Neural Networks* **3**, 2 (1992) 252–259.

[4] Y. Bengio, Y. LeCun, C. Nohl, and C. Burges, LeRec: A NN/HMM hybrid for on-line handwriting recognition, *Neural Computation* **7**, 6 (1995) 1289–1303.

[5] Y. Bengio and P. Frasconi, An input/output HMM architecture, *Advances in Neural Information Processing Systems* **7**, eds. G. Tesauro, D.S. Touretzky, and T.K. Leen (MIT Press, Cambridge, MA, 1995) 427–434.

[6] M. Berthod, On-line analysis of cursive writing, *Computer Analysis and Perception* **1**, eds. C, Y, Suen and R, De Mori, (CRC Press, 1982) 55–81.

[7] H. Bourlard and N. Morgan, *Connectionist Speech Recognition* (Kluwer Academic, Boston, MA, 1993).

[8] C. J. C. Burges, O. Matan, Y. Le Cun, D. Denker, L. D. Jackel, C. E. Stenard, C. R. Nohl, and J. I. Ben, Shortest path segmentation: A method for training neural networks to recognize character strings, *IJCNN'92* **3** (Baltimore, 1992).

[9] J. Camillerapp, G. Lorette, G. Menier, H. Ouladj, and J-C. Pettier, On-line and off-line methods for cursive script recognition, *From Pixels to Features III*, eds. S. Impedovo and J-C. Simon (North-Holland, 1992) 273–287.

[10] E. Charniak, *Statistical Language Learning* (MIT Press, Cambridge, MA, 1993).

[11] T. H. Cormen, C. E. Leiserson, and R. L. Rivest, *Introduction to Algorithms* (MIT Press, Cambridge MA, 1992).

[12] A. P. Dempster, N. M. Laird, and D. B. Rubin, Maximum likelihood from incomplete data via the EM algorithm, *J. Royal statistical Society Series B* **39** (1977) 1–38.

[13] J. Denker and C. J. C. Burges, Image segmentation and recognition, *The Mathematics of Generalization* (Addison Wesley, 1994).

[14] D. S. Doermann and A. Rosenfeld, Recovery of temporal information from static images of handwriting, *Proc. of Conf. on Computer Vision and Pattern Recognition CVPR* (Champaign, IL, 1992) 162–168.

[15] H. Drucker, R. Shapire, and P. Simard, Boosting performance in neural networks, *Advances in Pattern Recognition Systems using Neural Network Technologies*, eds. I. Guyon and P.S.P. Wang, (World Scientific, Singapore, 1993) 61–75.

[16] R.O. Duda and P.E. Hart, *Pattern Classification And Scene Analysis* (Wiley and Son, 1973).

[17] L. Duneau and B. Dorizzi, On-line cursive script recognition: A system that adapts to an unknown user, *Proc. of 12th Int. Conf. on Pattern Recognition* (Jerusalem, Oct. 1994) 24–28.

[18] R. W. Ehrich and K. J. Koehler, Experiments in the contextual recognition of cursive script, *IEEE Trans. on Computers* **24**, 2 (1975) 182–194.

[19] R. F. H. Farag, Word-level recognition of cursive script, *IEEE Trans. on Computers* **28**, 2 (1979) 172–175.

[20] N. S. Flann and S. Shekhar, Recognizing on-line cursive handwriting using a mixture of cooperating pyramid-style neural networks, *World Congress on Neural Networks* (1993).

[21] *Advances in speech signal processing*, eds. S. Furui and M. M. Sondhi, (Marcel Dekker, New York, 1992).

[22] T. Fujisaki, K. Nathan, W. Cho, and H. Beigi, On-line unconstrained handwriting recognition by a probabilistic method, *Third Int. Workshop on Frontiers in Handwriting Recognition* (Buffalo, May 1993) 235–241.

[23] D. Goldberg and C. Richardson, Touch-typing with a stylus, *Proc. of INTERCHI'93* (Amsterdam, 1993) 80–87.

[24] I. Guyon, P. Albrecht, Y. Le Cun, J. Denker, and W. Hubbard, Design of a neural network character recognizer for a touch terminal, *Pattern Recognition* **24**, 2 (1991) 105–119.

[25] I. Guyon, N. Matić, and V. Vapnik, Discovering informative patterns and data cleaning, *AAAI Workshop on Knowledge Discovery in Databases, KDD'94* (Seattle, WA, July 1994) 145–156. Also in *Advances in Knowledge Discovery and Data Mining*, eds. U. Fayyad, *et al.* (MIT Press, 1995) 181–203.

[26] I. Guyon, L. Schomaker, R. Plamondon, M. Liberman, and S. Janet, UNIPEN project of on-line data exchange and recognizer benchmarks, *12th Int. Conf. on Pattern Recognition* (Jerusalem, Israel, 1994).

[27] I. Guyon, J. Bromley, N. Matić, M. Schenkel, and H. Weissman, Penacée: A neural network system for recognizing on-line handwriting, *Models of Neural Networks* **3**, eds. E. Domany, *et al.* (Springer Verlag, New York, 1994).

[28] I. Guyon and C. Warwick, Handwriting as computer interface, *Joint EC-US Survey of the State-of-the-Art in Human Language Technology*, NSF, 1996 (to appear).

[29] I. Guyon and F. Pereira, Design of a linguistic postprocessor using variable memory length Markov models, *Proc. ICDAR'95* (Montreal, Canada, Aug. 1995) 454–457.

[30] J. Y. Ha, S. C. Oh, J. H. Kim, and Y. B. Kwon, Unconstrained handwritten word recognition with interconnected hidden Markov models, *Third Int. Workshop on Frontiers in Handwriting Recognition* (Buffalo, May 1993) 455–460.

[31] L. D. Harmon, Automatic reading of cursive script, *Optical Character Recognition*, eds. G. L. Fisher, D. K. Pollock, B. Raddack, and M. E. Stevens (Spartan, Washington, DC, 1962) 151–151.

[32] L. D. Harmon, Automatic recognition of print and script, *Proc. of the IEEE* **60**, 10 (1972) 1165–1176.

[33] J.M. Hollerbach, An oscillation theory of handwriting, *Biological cybernetics* **39** (1981) 139–156.

[34] J. Hu, M. K. Brown, and W. Turin, Handwriting recognition with hidden Markov models and grammatical constraints, *Fourth International Workshop on Frontiers in Handwriting Recognition* (Taipei, Dec. 1994).

[35] F. Jelinek. Up from trigrams! *Proc. of the Eurospeech Conf.* (1991) 1037–1040.

[36] J. Keeler, D. E. Rumelhart, and W-K. Leow, Integrated segmentation and recognition of hand-printed numerals, *Advances in Neural Information Processing Systems* **3**, eds. R. P. Lippmann, Moody J. E., and Touretzky D. S. (Morgan Kaufmann, San Mateo, CA, 1991) 557–563.

[37] T. Kohonen, *Self-Organization and Associative Memory*, 2nd edition, (Springer-Verlag, New York, 1984).

[38] A. Kundu and P. Bahl, Recognition of handwritten script: A hidden Markov model based approach, *Int. Conf. on Acoustics, Speech & Signal Processing* (1988) 928–931.

[39] K. J. Lang and G. E. Hinton, A time delay neural network architecture for speech recognition, Technical Report CMU-cs-88-152, Carnegie-Mellon University, Pittsburgh PA, 1988.

[40] E. Lecolinet and O. Baret, Cursive word recognition: Methods and strategies, *Funda-*

mentals in Handwriting Recognition **124**, ed. S. Impedovo (Springer Verlag, NATO-ASI series F, 1994) 235–263.

[41] A. Leroy, Lexicon reduction based on global features for on-line handwriting, preprint, 1995.

[42] V. I. Levenshtein, Binary codes capable of correcting deletions, insertions and reversals, *Cybernetics and Control Theory* **10**, 8 (1965) 845–848.

[43] N. Lindgren, Machine recognition of human language, part III – Cursive script recognition, *IEEE Spectrum* (May 1965) 104–116.

[44] R.F. Lyon and L.S. Yaeger, On-line hand-printing recognition with neural networks, *Fifth Int. Conf. on Microelectronics for Neural Systems and Fuzzy Systems* (Lausanne, Switzerland, February 1996).

[45] E. Mandler, R. Oed, and W. Doster, Experiments in on-line script recognition, *Proc. 4th Scandinavian Conf. on Image Analysis* **1** (Champaign, IL, 1985) 75–86.

[46] S. Manke and U. Bodenhausen, A connectionist recognizer for on-line cursive handwriting recognition, *Proc. ICASSP-94* (Adelaide, Australia, 1994).

[47] O. Matan, C. J. C. Burges, Y. Le Cun, and J. Denker, Multi-digit recognition using a space dispacement neural network, *Advances in Neural Information Processing Systems* **4**, ed. J. E. Moody (Morgan Kaufmann, San Mateo, CA, 1992) 488–495.

[48] N. Matić, I. Guyon, L. Bottou, J. Denker, and V. Vapnik, Computer aided cleaning of large databases for character recognition, *Proc. of 11th Int. Conf. on Pattern Recognition* (Amsterdam, The Netherlands, Aug. 1992) 330–333.

[49] N. Matić, I. Guyon, J. Denker, and V. Vapnik, Writer adaptation for on-line handwritten character recognition, *Second Int. Conf. on Pattern Recognition and Document Analysis* (Tsukuba, Japan, October 1993) 187–191.

[50] P. Mermelstein and M. Eden, Experiments on computer recognition of connected handwritten words, *Information and Control* **7** (1964) 255.

[51] P. Morasso, L. Barberis, S. Pagliano, and D. Vergano, Recognition experiments of cursive dynamic handwriting with self-organizing networks, *Pattern Recognition* **26**, 3 (1993) 451–460.

[52] R. Nag, K. H. Wong, and F. Fallside, Script recognition using hidden Markov models, *Proc. ICASSP* (Tokyo, May 1986) 2071–2074.

[53] F. Nouboud and R. Plamondon, On-line recognition of handprinted characters: survey and beta tests, *Pattern Recognition* **23**, 9 (1990) 1031–1044.

[54] E. P. Pednault, A hidden Markov model for resolving segmentation and interpretation ambiguities in unconstrained handwriting recognition, Technical Memorandum BL0113520-930115-01TM, AT&T Bell Labs., Holmdel NJ, 1993.

[55] F. Pereira, M. Riley, and R. Sproat, Weighted rational transductions and their application to human language processing, *Proc. ARPA Natural Language Processing Workshop* (1994).

[56] L. R. Rabiner, A tutorial on hidden Markov models and selected applications in speech recognition, *Proc. of the IEEE* **77**, 2 (Feb. 1989) 257–285.

[57] D. E. Rumelhart, G. E. Hinton and R. J. Williams, Learning internal representations by error propagation, *Parallel Distributed Processing I: Explorations in the Microstructure of Cognition*, eds. D. E. Rumelhart, *et al.* (Bradford Books, Cambridge, MA, 1986) 318–362.

[58] D. Rumelhart and et al., Integrated segmentation and recognition of cursive handwriting, *Third NEC Symposium Computational Learning and Cognition* (Princeton, NJ, 1992).

[59] K. M. Sayre, Machine recognition of handwritten words: A project report, *Pattern Recognition* **5** (1973) 213–228.

[60] M. Schenkel, H. Weissman, I. Guyon, C. Nohl, and D. Henderson, Recognition-based

segmentation of on-line hand-printed words, *Advances in Neural Information Processing Systems* **5**, eds. S. J. Hanson et al. (Morgan Kaufmann, San Mateo, CA, 1993) 723–730.

[61] M. Schenkel, I. Guyon, and D. Henderson, On-line cursive script recognition using time delay neural networks and hidden Markov models, Machine Vision and Applications **8**, ed. R. Plamondon (Springer Verlag, 1995) 215–223.

[62] L. Schomaker and H.L. Teuling, A handwriting recognition system based on properties of the human motor control system *Int. Workshop on Frontiers in Handwriting Recognition* (Montreal, 1990) 195–211.

[63] L. Schomaker, Using stroke- or character-based self-organizing maps in the recognition of on-line, connected cursive script, *Pattern Recognition* **26**, 3 (1993) 443–450.

[64] P. Simard, Y. Le Cun, and J. Denker, Efficient pattern recognition using a new transformation distance, *Advances in Neural Information Processing Systems* **5**, eds. S. Hanson, *et al.* (Morgan Kaufmann, San Mateo, CA, 1993) 50–58.

[65] Y. Singer and N. Tishby, Dynamical encoding of cursive handwriting, *IEEE Conf. on Computer Vision and Pattern Recognition* (1993).

[66] K. A. Sizov, Recognition of symbols and words written by hand, *Proc. of the Int. Conf. on Document Analysis and Recognition* **2** (Saint-Malo, France, 1991).

[67] C. C. Tappert, C. Y. Suen, and T. Wakahara, The state of the art in on-line handwriting recognition, *IEEE Trans. on Pattern Analysis and Machine Intelligence* **12**, 8 (1990) 787–808.

[68] H.-L. Teuling, Invariant handwriting features useful in cursive-script recognition, Technical Report Nijmegen Institute for Cognition and Information, Nijmegen, The Netherlands, 1993.

[69] V. Vapnik, *Estimation of Dependences Based on Empirical Data* (Springer Verlag, New York, 1982).

[70] H. Weissman, M. Schenkel, I. Guyon, C. Nohl, and D. Henderson, Recognition-based segmentation of on-line handprinted words: Input vs. output segmentation, *Pattern Recognition* **27** 3 (1994) 405–420.

[71] G. Wilfong, F. Sinden, and L. Ruedisueli, On-line recognition of handwritten symbols, Technical Memorandum 11228-940801-06TM, AT&T Bell Labs., Murray Hill, NJ, 1994.

[72] L. Xu, K. Adam, and C. Y. Suen, Methods of combining multiple classifiers and their applications to handwriting recognition, *IEEE Trans. Syst. Man and Cybernetics* **22**, 3 (1992) 418–435.

Handbook of Character Recognition and Document Image Analysis, pp. 227–258
Eds. H. Bunke and P. S. P. Wang

CHAPTER 8

TECHNIQUES FOR IMPROVING OCR RESULTS

ANDREAS DENGEL, RAINER HOCH, FRANK HÖNES, THORSTEN JÄGER,
MICHAEL MALBURG, ACHIM WEIGEL
German Research Center for Artificial Intelligence — DFKI GmbH
P.O. Box 2080, 67608 Kaiserslautern, Germany

In this chapter, we give an overview of the state-of-the-art techniques for improving recognition results of OCR systems. OCR results may contain segmentation as well as classification errors due to low image quality. Such errors can often be corrected by contextual post-processing. We will present the most important techniques for the post-processing of OCR results: voting techniques, lexical post-processing as well as techniques that consider the word or document context. Voting techniques combine the recognition results from multiple OCR devices, typically without utilizing any contextual knowledge. Other post-processing techniques are able to correct remaining OCR errors by employing various sources of contextual knowledge. Lexical post-processing, for example, makes use of knowledge about valid words of natural language. More sophisticated techniques integrating knowledge about the word context or even the entire document context can also be applied to further improve the quality of OCR results. The most useful is the incorporation of knowledge about valid word sequences. In general, post-processing of recognition results considerably improves the OCR accuracy if various kinds of contextual knowledge beyond the level of individual characters are utilized.

Keywords: OCR; Contextual post-processing; Voting; Dictionary look-up; Dictionary organization; Markov models; Similarity measure; Classifier combination; Word transitions; Probabilistic grammars; Document context.

1. Introduction

Generally, text may be seen as a collection of atomic parts which are the symbols of an alphabet. The meaning of the text, however, is determined by semantic units such as words, word combinations, phrases, sentences or even the entire context of a document. Traditional OCR systems suffer from the problem that they come to local decisions for the classification of characters. Thus, OCR systems are forced to use contextual information at different levels for improving the accuracy of recognition results where the ultimate goal would be a 100% recognition accuracy of any text.

Most OCR systems incorporate some form of post-processing to correct recognition errors. Voting, for example, deals with the combination of competitive recognition results of different recognizers to become better than the best of the individual recognizers under consideration. Other approaches attempt to utilize knowledge sources to improve recognition results of machine-printed, hand-printed or even cursive script which are usually employed by humans for reading. In this chapter, we want to address, as completely as possible, the most important issues for improving text recognition results, including both the theoretical and practical aspects.

Figure 1 illustrates the different techniques that will be discussed in the remainder of this chapter. For better understanding, all intermediate results are also depicted. The underlying triangles indicate the successive reduction of ambiguities and gain in information quality during character processing as well as word processing. However, the complexity of knowledge increases from level to level.

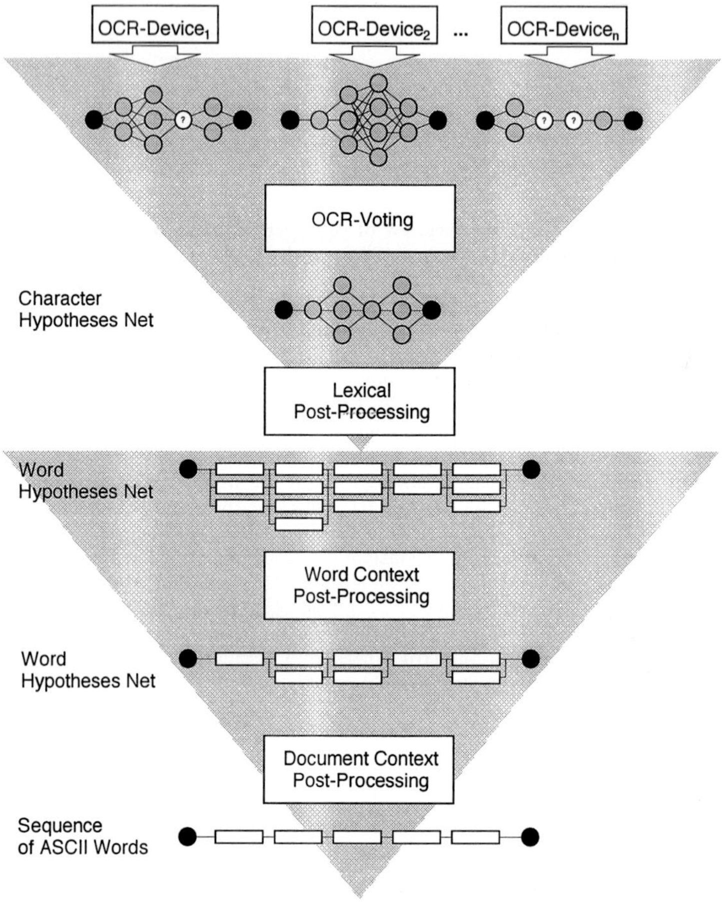

Fig. 1. Different techniques for improving OCR results.

In this chapter, we will discuss following techniques in more detail:

- Voting. Voting techniques combine the recognition results from multiple OCR devices given by separate character hypotheses nets. We will address the integrity of voting components, alternatives for representing results, approaches for combining results ranging from simple techniques to complex probabilistic methods, and restrictions imposed on the basic OCR devices. The output of voting is a character hypotheses net, in the optimal case without any ambiguities.

- Lexical post-processing. This section describes bottom-up methods (e.g., binary or probabilistic n-grams, Markov models) and top-down methods including dictionary look-up techniques (hashing, tries), but also error correction models for character sequences (e.g., Levenshtein distance or probabilistic models). The output of this step are word hypotheses nets for subsequent contextual post-processing.
- Word context. Two approaches for contextual post-processing on the word level are presented. Statistical approaches considering transition probabilities of word categories are discussed in the first section. Then, we explain syntactic parsing techniques which exploit the linguistic structure of a text. Both approaches as well as combined approaches will be presented. Using the word context, irrelevant word alternatives of the hypotheses net can be eliminated.
- Document context. As a last source of information for the verification of word hypotheses, the relations between different document parts are considered. In general, two alternative approaches are known. The first approach is mainly based on the geometric structure of the entire document, whereas the second one relies on the text coherence structure of the whole document.

2. Combining Results from Multiple OCR Devices (Voting)

OCR devices today offer an efficient and reliable classification of characters. Most devices attain high recognition accuracy, independent of the predominant font type or size. However, even the best devices barely reach 100% recognition accuracy when faced with poorly-printed or poorly-copied dense text. Recognition and error rates of current OCR devices are documented in [1]. Here, the average recognition accuracy for the tested devices was 97.5%. For most applications, such as database queries or extraction of document semantics, they require a higher recognition accuracy to produce the best possible results,[a] recognition accuracy must therefore be further improved.

Formerly, research was mainly directed by developing more sophisticated techniques and algorithms for pure classification. Also, features relevant for the classification process were intensively studied. In recent years a new technique for improving OCR results, called voting, was proposed. Voting is no "classical" recognition approach, instead, recognition results from different classifiers are combined, normally without further evaluation of the character's image. This approach is mainly motivated by the observation that different OCRs make different errors. The objective of classifier combination is to focus on the individual classifiers' strengths and avoid their weaknesses and thus produce more reliable recognition results than any of the individual classifiers. To do so, the most challenging task is the development of an optimal combination strategy based on an adequate result representation formalism.

[a]Typically, the required recognition accuracy is 99.8% or higher.

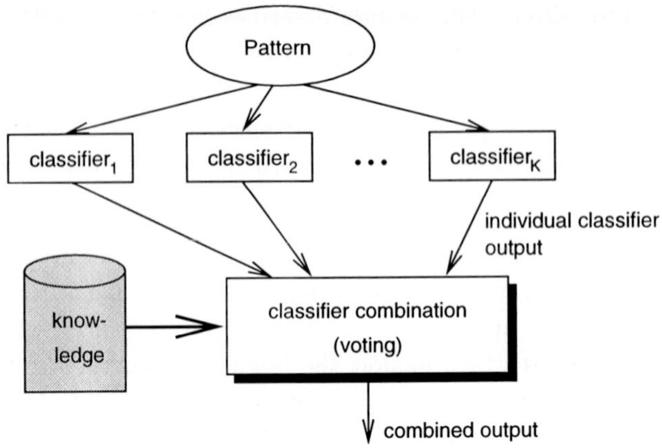

Fig. 2. The global architecture of a classifier combination system is shown. The knowledge base is optional.

In the last few years, the principles of combination, of result representation, and the use of probabilistic reasoning have been increasingly refined. However, these advances do not affect the global architecture of a system for classifier combination as shown in Fig. 2. Generally, the input of voting is the output directly produced by the individual classifiers which is combined and sometimes supported by conflict resolution knowledge. In the following subsection, we discuss some issues which are fundamental for classifier combination problems but not limited to OCR or text recognition in general.

The subject of multiple classifier combination is treated in detail in the chapter by L. Lam *et al.* in this book. As it is an important technique for improving OCR results, we also address it in the present chapter.

2.1. *Basics*

There are four main aspects which the developer of a voting machine has to consider:

1. degree of integrity
2. representation of classifier results
3. combination of classifier results
4. methodological and representational restrictions

Before describing these aspects in greater detail, we will introduce a more formal notation suitable for most types of recognition problems (cf. [2]):

Let P be a pattern space consisting of M mutually exclusive sets $P = \cup_i C_i$ with each $C_i, \forall i \in \Lambda = \{1, 2, \ldots, M\}$ representing a set of specified patterns called a class. For a sample x from P, the task of a classifier e is to assign x one index $j \in \Lambda \cup \{M+1\}$ as a label. If $j = M+1$, x is rejected by classifier e, which means that e has no idea about x' class, otherwise, x is assumed to belong to class C_j. Thus, a classifier can be regarded as a function box receiving an input sample x and producing a label j, denoted by $e(x) = j$. In practice, classifiers may enrich their output by additional confidence values or less trusted hypotheses, within the representational alternatives.

2.1.1. *Degree of integrity*

The degree of integrity serves as a measure of how strongly the individual classifiers are directed by the voting algorithm. A high degree of integrity denotes that the voting machine directs all activations of the classifiers. It "knows" about strengths and weaknesses and runs only those classifiers which promise reliable results depending on the current situation. If the degree of integrity is low, the voting component has no effect on the basic classifiers and considers their output only.

2.1.2. *Representation of classifier results*

The second aspect concerns the representation of classifier results. Here, three alternatives are possible (cf. [3]):

1. **abstract level**: Each classifier e provides a single result/label j without any further information, i.e., $e(x) = j, \quad j \in \Lambda \cup \{M+1\}$.

2. **rank level**: Each classifier provides an ordered list of ranked results, where the first element is the most likely one, i.e., $e(x) = L$, where $L \subseteq \Lambda$ is an **ordered** list. A reject is denoted by $L = \emptyset$.

3. **measurement level**: Each classifier produces alternatives along with a real value indicating the recognition confidence. The values need not be taken from the interval $[0, 1]$ indicating a probability, but can also be distance measures to given reference patterns. In the latter case the values come from the interval $[0, \infty]$ and lower values indicate lower distance and thus higher confidence. Formally: $e(x) = B_e$, where $B_e = \{b(i_1); \ldots; b(i_n)\}, \{i_1; \ldots; i_n\} \subseteq \Lambda, n \leq M$ and $b(i_s)$ denotes the assessment of classifier e that x has label i_s. Rejection is denoted by $B_e = \emptyset$.

The different representation levels are exemplarily shown in Fig. 3. Obviously, the amount of data increases from alternative (1) to alternative (3). If the voting component has no influence on the result representation, it has to transform all the results in a consistent way. This produces no difficulties when decreasing the level of detail, since results from the rank or measurement level can be directly transformed into the abstract level by omitting all but the best result. Transformations from less detailed levels into more detailed ones are more difficult to perform, especially those resulting in the measurement level.

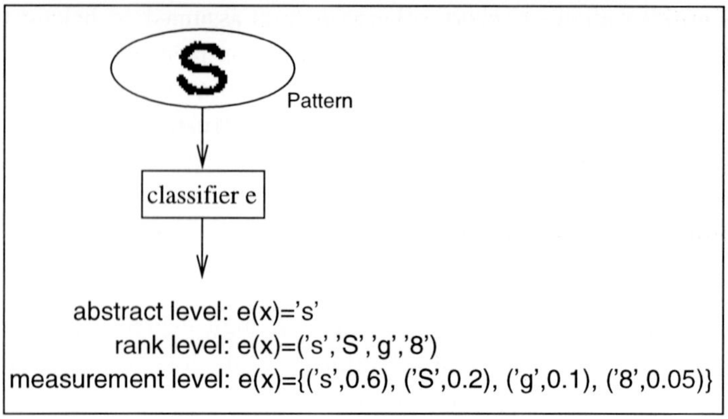

Fig. 3. Three different levels for representing individual classifier results. For sake of clarity, symbolic class names, like e.g., "s", are preferred over class indices.

2.1.3. *Combination of classifier results*

The representation selected strongly influences the underlying technique for combining the individual results but also the production of the final output. It's impossible to combine the results of individual classifiers represented differently without additional effort. Thus, we assume that all results are represented on the same level. The output of combination may be represented on a different level, but in most cases it will be the same.

Since a complete survey of combination strategies is far beyond the scope of this chapter, we will sketch only some of the more popular approaches and will provide references in the literature for the interested reader.

The most common method for combining results on the **abstract level** is the "majority vote". Here, the combined decision is obtained by assigning x the class label which is most often presented by the individual classifiers. In order to avoid less reliable results, several, more conservative, variations of the majority vote exist. For example, a combined result may be produced only when it receives more than Θ votes from the individual classifiers. The principle of the majority voting is shown in Fig. 4(a). Class indices $j \in \lambda$ are the output of the individual classifiers e_k and the combined classifier E.

For combination on the **rank level**, two principal methods are known: class set reduction and class set reordering. The goal of the reduction method is to extract a subset of classes which hopefully captures the correct class. In class set reordering, the objective is to derive a combined ranking of the given classes, such that the true class is ranked as close to the top as possible. Thus, for class set reduction the success criterion is two-fold: the size of the result set and the probability of inclusion of the true class in the result set. The only criterion for class set reordering is the rank position of the true class in the combined ranking.

Two approaches are proposed for class set reduction: the intersection approach and the union approach. Common to both approaches is the fact that they consider a neighborhood (classes ranked from the top down to a certain specified rank position) for each classifier. The result set for the intersection approach is the intersection of all these neighborhoods whereas the result set for the union approach is the union of all neighborhoods. According to the different approaches, the thresholds determining the neighborhood size for each classifier are estimated differently.

Two common methods for class set reordering are the Highest Rank method and the Borda Count method. The Highest Rank method simply assigns each class the best rank position of this class in any of the individual rankings. The main disadvantage of this method is that for a large number of classifiers many classes are ranked equally in the combined ranking. The Borda Count method assigns a Borda Count $B(C_i)$ to every class C_i, which is defined as

$$B(C_i) = \sum_{k=1}^{K} B_k(C_i) \tag{2.1}$$

where K is the number of classifiers, and $B_k(C_i), 1 \leq k \leq K, 1 \leq i \leq M$ denotes the number of classes in P which are ranked below C_i by classifier e_k in its individual ranking L_k. Subsequently, all classes are ranked according to their Borda Count values in decreasing order. An example is given in Fig. 4(b), where the individual outputs of the classifiers e_k are rankings of length four. All classes, represented by their indices $j \in \lambda$, are reranked according to their Borda Count values. Instead of reranking the classes, the output could also be the first choice of the new combined ranking. Thus, we can see that the output need not be represented on the same level as the input.

The **measurement level** provides class decisions along with confidence values for these classes. The confidence values can be similarity as well as distance measures. They need not be taken from the interval $[0, 1]$ expressing some kind of probability. One problem is the transformation of different confidence measures utilized by different classifiers. Another problem is the combination of the class confidence values to yield a combined confidence value for each class. For convenience, we assume that all classifiers utilize the same confidence values. Since the confidence values can be treated as vague information, all kinds of probabilistic reasoning can be employed for their combination. The two most popular methods are those based on Bayes' Theorem and the Dempster/Shafer Formalism (see [4, 5]). The first one utilizes the notion of probability and conditional probability, whereas the second one is based on the notion of belief functions. Both methods recalculate the confidence values for each class by applying Bayes' formula or by applying Dempster's rule of combination. A simple example is shown in Fig. 4(c). Here, we assume that all individual classifiers e_k are the so-called Bayes classifiers, approximating the postprobabilities $P(C_i|x)$, where x is an observed feature vector (for information on Bayes classifiers see [6]). One straightforward way to combine the outputs of the four classifiers is to approximate the postprobabilities by calculating the average postprobability, i.e.,

$$P(C_i|x) = \frac{1}{K}\sum_{k=1}^{K} P_k(C_i|x) \tag{2.2}$$

If necessary, the combined output can then be transformed to obtain a single class decision or a ranking as shown in Fig. 4(c). On the abstract level, the class with the highest postprobability is the combined output, here C_{21}; on the rank level, all classes are ordered according to their postprobabilities in decreasing order.

A thorough overview for combination strategies on the rank level is provided by [7]. Various combination methods for abstract or measurement level are discussed in [3].

2.1.4. *Methodological and representational restrictions*

Another important aspect of combination is the restriction concerning the methodology used by the individual classifiers to produce results as well as the representation of these results. Neglecting the technical aspects, a more "tolerant" voting component would combine results of classifiers using different classification approaches and utilizing different representation levels, while a more restrictive voting component requires, for example, a homogeneous representation or even a specific classification algorithm.

2.2. *State-of-the-Art*

Having discussed the basics of classifier combination in the previous subsection, we will now present voting techniques currently proposed by different researchers. The various techniques or complete systems will be classified according to the four main aspects, degree of integrity, representation of results, combination of results, and restrictions.

1. Ho, respectively Ho *et al.* (see [7, 8, 9]), have developed a classifier combination system for the task of word recognition.

    - integrity: Classifiers may be selected dynamically at classification time.
    - result representation: The individual classifiers output a ranking. Ho assumes that all classes are ranked with respect to the given sample x.
    - combination: Different combination methods are developed and their recognition rates are compared. All methods belong to the class set reordering approach, namely: Highest rank, Borda count, and Logistic regression (a special variant of the Borda Count with weights).
    - restrictions: The output of a classifier must be a ranking of all the respective classes.

Let E denote the "combined classifier"

Let K=4 be the number of classifiers

(a) $e_1(x)=17$ $e_2(x)=4$ $e_3(x)=17$ $e_4(x)=21$

class	no. of votes
4	1
17	2
21	1

→ abstract level: E(x)=17

(b)

$e_1(x)$	$e_2(x)$	$e_3(x)$	$e_4(x)$
17	4	17	21
4	21	21	17
21	17	9	9
9	9	4	4

$B(4)=B_1(4)+B_2(4)+B_3(4)+B_4(4)=2+3+0+0=5$

$B(9)=0+0+1+1=2$

$B(17)=3+1+3+2=9$

$B(21)=1+2+2+3=8$

→ abstract level: E(x)=17; rank level: E(x) = 17, 21, 4, 9

(c)

i	4	9	17	21
$P_1(C_i\|x)$	0.3	0.1	0.4	0.2
$P_2(C_i\|x)$	0.3	0.05	0.1	0.2
$P_3(C_i\|x)$	0.08	0.1	0.3	0.5
$P_4(C_i\|x)$	0.02	0.04	0.1	0.8

$P(C_4|x)=(0.3+0.3+0.08+0.02):4=0.175$

$P(C_9|x)=0.0725$

$P(C_{17}|x)=0.275$

$P(C_{21}|x)=0.375$

→ abstract level: E(x)=21; rank level: E(x) = 21, 17, 4, 9

Fig. 4. Combining results of four individual classifiers. For simplicity, classes are denoted by their class index only. For each of the three levels of representation, i.e., abstract (a), rank (b), and measurement (c), an adequate combination method is presented. Where appropriate, alternative result representations are given.

2. Mandler and Schürmann [10] combine the output of character classifiers:
    - integrity: The voting component combines only the output of the individual classifiers.
    - result representation: The results are represented on the measurement level.
    - combination: Dempster/Shafer theory of evidence is used. For comparison of different approaches see also [11].
    - restrictions: The voting component imposes tight restrictions on the individual classifiers. All classifiers have to be nearest neighbor classifiers, ensuring that their confidence values are a distance measure.

3. Huang and Suen [2] discuss the influence of the chosen representation on the overall performance:
    - integrity: The voting component combines only the output of the individual classifiers.
    - result representation: The results are represented on the measurement level.
    - combination: The combination is performed with the Linear Confidence Aggregation Method (LCA). In the first step, the different measures are transformed into a single measure. These measures are added for each class, yielding a confidence value for this class. The class with the highest confidence value greater than a given threshold will be output.
    - restrictions: There are no other restrictions, even the measures may be chosen freely. Although for their experiment only two polynomial classifiers were integrated, Huang and Suen showed that their LCA method performs better than majority voting or Bayes based voting.

4. Rice *et al.* [12] presented the results of a test of two voting systems in their 1994 report. The system developed at ISRI is characterized as follows:
    - integrity: The voting component has no effect on the individual classifiers.
    - result representation: ISRI deviates from the given scheme. Here, only strings, or more precisely substrings, are used for voting. The voting component has only to process those regions, where output differs.
    - combination: Majority voting is used.
    - restrictions: No restrictions are imposed on the classifiers.

5. Lee and Srihari [13] proposed neural networks to combine decisions of multiple experts.
    - integrity: The voting component has no effect on the individual classifiers.

- result representation: Classifiers may represent their results on any of the three levels. Before actually combining the results, a set of transformation functions is applied to map each classifier output to a unit square in the N-dimensional real space.
- combination: The proposed basic "decision combination neural network" is a multilayer perceptron with no hidden layer. The network is trained by backpropagation.
- restrictions: No restrictions are imposed on the classifiers.

6. Bartell *et al.* [14] proposed a classifier combination procedure to combine expert/classifier opinions in the domain of information retrieval.

    - integrity: The voting component has no effect on the individual classifiers.
    - result representation: The classifiers provide results on the measurement level. These results are documents together with a numerical estimate of the relevance of documents to a query.
    - combination: The combination is based on a linear combination model. Model parameters are optimized globally, thus, the classifiers need not be independent.
    - restrictions: The voting component imposes no further restrictions on the classifiers.

The above mentioned approaches are chosen as examples, to sketch some of the current work done in the classifier combination field. Other approaches can be found in [15, 16, 17, 18, 19]. For a short discussion see also [7], pp. 9–15. The topic of classifier combination is strongly related to the field of decision fusion (see [20]). For further details see the chapter by L. Lam *et al.* of this book.

2.3. *Summary*

Classifier combination, or voting, is one way to improve OCR results. This is achieved by combining the output from all independent classifiers working in parallel. We have presented the four main aspects of fundamental relevance for any voting algorithm, namely integrity, result representation, result combination, and classifier restrictions. Various combination strategies have been presented and we have the seen the great influence of the chosen representation on the actual combination.

Classifier combination offers a promising approach to attain more reliable recognition results, as is confirmed by many researchers. For example, Rice *et al.* [12] determined progress in error reduction at about 40%. They also observed that the error reduction rate drops when the individual classifiers perform badly. In such a case, the conflict arising from many different votes cannot be resolved by the voting.

Although voting is no panacea, it offers encouraging performance improvements, especially in, but not limited to, the area of OCR.

3. Lexical Post-Processing

In general, lexical post-processing has the task of verifying OCR results using lexical knowledge and of generating a ranked list of possible word candidates when the input is noisy. This ranking process includes some distance or similarity measure between the misspelled input word and the respective word candidates.

The lexical knowledge about the legal words of a language can be represented in different ways, either by probabilities of letter transitions (Markov models), by n-grams [21], or by an exact representation of all legal words in a dictionary.

According to the distinct knowledge sources involved, lexical post-processing of OCR results is generally based on three fundamental approaches [22]: bottom-up methods, top-down methods and hybrid methods. In the following, we will briefly discuss these approaches.

3.1. *Bottom-up Methods (Markov Models)*

Bottom-up methods for lexical postprocessing can be characterized by the information they use for their task. Instead of using full knowledge about the allowed words of the vocabulary they approximate this knowledge by enumerating the possible character transitions, sometimes integrating their probabilities. In this section we want to give a short overview of such techniques. The motivation for bottom-up techniques is the redundancy of a language. According to [23] only 70% of all 2-grams (letter pairs) occur in English textbooks. The percentage of 3-grams (triples of letters) occurring is even much smaller [21]. It is exactly this phenomenon which is used by bottom-up techniques for error recognition, ambiguity reduction, and error correction.

Hanson and Riseman proposed a method for contextual postprocessing based on *positional binary n-grams* [24]. In the following we will explain their method using positional binary 2-grams. For this method it is assumed, that each input word is contained in a dictionary L. A *positional binary 2-gram array* is defined by:

$$D_{i,j}(a,b) = \begin{cases} 1 & \text{if letter a occurs at position } i \text{ and letter b at position } j \\ & \text{of some word in L} \\ 0 & \text{otherwise} \end{cases}$$

The information stored in these matrices can be used to locate classification errors and to correct some of them. Assume a sample word consists of the following sequence of characters $x_1, x_2 \ldots x_m$, where each x_i is a member of the alphabet A. We have identified an error, if there is some $D_{i,j}(x_i, x_j) = 0$. In this case we have identified a character pair which does not occur at the corresponding positions in any word in the dictionary.

With this method we can never be sure of finding an error even if $x_1x_2 \ldots x_m$ is not a word in the dictionary. This happens if $x_1x_2 \ldots x_m$ is not an allowed word but contains only positional binary 2-grams which occur in some dictionary words.

Positional binary 2-grams can also be used for error correction. Let us consider an input sample $x_1x_2x_3x_4x_5$. Assuming only one single error at position 3, the following 2-grams might identify it: $D_{1,3}(x_1, x_3)$, $D_{2,3}(x_2, x_3)$, $D_{3,4}(x_3, x_4)$, $D_{3,5}(x_3, x_5)$. If at least two of them are equal to 0, position 3 is located as an error. A correction of this error is possible if there is exactly one character x, for which $D_{1,3}(x_1, x)$, $D_{2,3}(x_2, x)$, $D_{3,4}(x, x_4)$, and $D_{3,5}(x, x_5)$ is equal to 1.

The definition of positional binary 2-grams can be straightforwardly expanded to n-grams with $n \geq 2$. Larger n's do have much more expressive power, because they approximate the dictionary more exactly, but they need a huge amount of storage.

Binary n-grams only indicate if a letter combination occurs or not. More specific information is used if the probability of letter combinations is also used. A well known approach integrating such information is described in the following [25].

Again let the observed word be $X = x_1x_2 \ldots x_m$. The probability that a letter sequence $Z = z_1z_2 \ldots z_m$ has caused X can be expressed by the use of Bayes decision theory

$$P(Z|X) = \frac{P(X|Z)P(Z)}{P(X)} \tag{3.3}$$

where $P(X|Z)$ is the probability of observing X under the condition that Z is the right input word. $P(Z)$ is the *a priori* probability of Z and $P(X)$ is the probability of the string X.

The objective now is to find that letter sequence Z, which maximizes Eq. (3.3), or, because X is independent of Z, which maximizes

$$P(X|Z)P(Z) \tag{3.4}$$

The *a priori* probability $P(Z)$ is approximated by assuming that words are generated by an *nth-order Markov source*, i.e.

$$P(Z) = P(z_m|z_{m-1} \ldots z_{m-n}) \ldots P(z_2|z_1\varepsilon)P(z_1|\varepsilon) \tag{3.5}$$

where $P(z_i|z_{i-1} \ldots z_{i-n})$ is the probability of observing z_i under the condition that the previously observed characters are $z_{i-1} \ldots z_{i-n}$. The $P(z_i|z_{i-1} \ldots z_{i-n})$ are the approximate representation of the vocabulary by transition probabilities of characters. ε denotes a delimiter between words. Then $P(z_2|z_1\varepsilon)$ describes the probability that z_2 is observed under the condition that z_1 is observed previously as the first character of the word.

Storing all possible values for $P(X|Z)$ is impractical because it requires a large amount of storage. Therefore the term is approximated under the assumption of conditional independence among the $x_1, x_2, \ldots, x_m$ by

$$P(X|Z) = \prod_{j=1}^{m} P(x_j|z_j) \tag{3.6}$$

To compute Eq. (3.6) a further knowledge source, namely, the *confusion probabilities* $P(x_j|z_j)$ are necessary. $P(x_j|z_j)$ is the probability of observing character x_j when the right character is z_j.

Finding the word Z which maximizes $P(X|Z)P(Z)$ can be efficiently computed with the *Viterbi algorithm* ([26, 27]). The Viterbi algorithm uses a dynamic programming approach and is described in the following for 2th-order Markov sources.

Viterbi Algorithm for 2th-order Markov Sources

Input: $X = x_1x_2 \ldots x_m$
for all characters a from the alphabet A do
    $C_1(a) = P(x_1|a)P(a|\varepsilon)$
endfor
for $i = 2$ to m do
    for all characters a from the alphabet A do
        $bestPred = c \in A$ where
            $C_{i-1}(c)P(x_i|a)P(a|c) = \max\{C_{i-1}(b)P(x_i|a)P(a|b)|b \in A\}$
        $C_i(a) = C_{i-1}(bestPred)P(x_i|a)P(a|bestPred)$
        $Path_i(a) = Concat(Path_{i-1}(bestPred), a)$
    endfor
endfor
return $Path_m(a)$ where $C_m(a) = \max\{C_m(b)|b \in A\}$
end

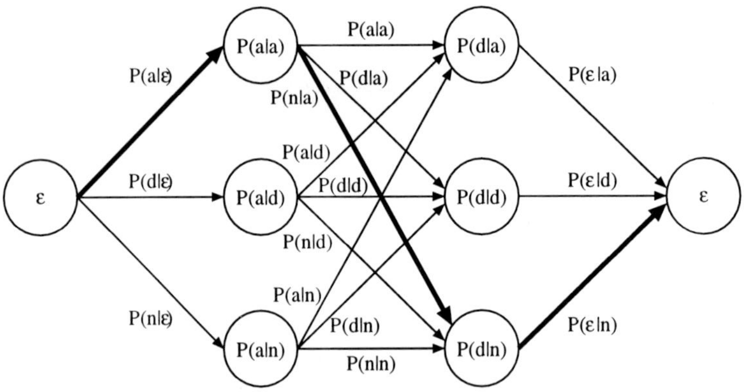

Fig. 5. Alternative graph for the Viterbi algorithm.

In order to demonstrate, how the Viterbi algorithm works, we give a short example. Assuming a three letter alphabet $A = \{a, d, n\}$ and an observed string $X = ad$. Figure 5 shows the trellis, which is traversed by the Viterbi algorithm. The node values represent the confusion probabilities and the values at the edges represents the transition probabilities of the Markov source. Any path from the start node to the end node represents a sequence of letters, e.g., the dark path in Fig. 5 represents the letter sequence *an*. The aim is to find that letter sequence which maximizes the product of the probabilities of its corresponding path. This is done by processing the vertical node layers step by step from left to right.

Modifications of the Viterbi algorithm have been proposed for the sake of computational efficiency such as using logarithms of probabilities and a summation of them rather than a product [28], or by defining a variable number of alternatives by which the observed characters can be substituted ([29, 30]).

To handle typical OCR anomalies, such as fragmented and overlapping characters, a more complete model, that can accommodate multiple character substitutions, insertions, and deletions is required [31].

In [30] the binary n-gram technique is compared with an approach which uses transition probabilities. It was shown that the binary n-gram technique is good for error detection. However, the statistical approach is better for error correction.

3.2. *Top-down Methods (Dictionary Look-up)*

3.2.1. *Dictionary organization*

Top-down methods achieve the validation of OCR results by matching them against a dictionary containing all forms of the words which are to be considered legal words of the language. In its simplest form, the corresponding OCR words are only considered valid when they exist in the dictionary (exact matching). Otherwise, if the input word is not in the dictionary, correction techniques applying a similarity measure are needed for the improvement of recognition results (best matching).

The organization of the dictionary has a large effect on the performance of dictionary look-up. Access time is a crucial point when the dictionary size is large, for example, containing more than a hundred thousand words. Thus, a multitude of dictionary organizations and corresponding access techniques have been developed. A survey of adequate data structures for the representation of dictionaries is given in [32, 33, 22]. Harris [34] also compares different dictionary structures (binary search, indexing, tries, hashing) and furthermore indicates a possible syntax for lexical entries. However, sophisticated techniques which are adapted to the special needs of character recognition dealing with noisy input are hardly found in the literature.

Two dictionary organizations are very popular for the validation of OCR results: tries (i.e., letter or character node trees) and hash tables. They will be discussed in more detail below. Other dictionary organizations such as sequential lists or more sophisticated tree structures are discussed elsewhere [33].

Tries: A trie [b] is an important data structure for the contextual post-processing of OCR results [35, 36]. A trie is a particular tree structure that contains only one character at each node [37]. Each node may have ordered links to descendant nodes whose number maximally equals the number of the letters of the underlying

[b] The technical term *trie* has its origin in the word "re*trie*val" [37].

alphabet. In particular, the root of a trie has up to n descendants where n corresponds to the first letters of a fixed vocabulary. Word access in a trie is strictly performed character by character beginning at the root of the trie (cf. Fig. 6). Nodes that represent the last character of a word are always marked by a special flag stopping the search process.

Tries are attractive because of their simple and compact storage allocation. Common prefixes are stored exactly once. In addition, they allow the integration of sophisticated error correction algorithms for dealing with noisy input, although problems arise when the beginning of a word is erroneous. For large dictionaries the trie is not the best choice because considerable time would be spent traversing links from one character to the next. Another disadvantage is the waste of storage due to storing additional pointers and housekeeping information. Trie optimizations have been proposed, for example, in [38, 39, 40, 41], etc.

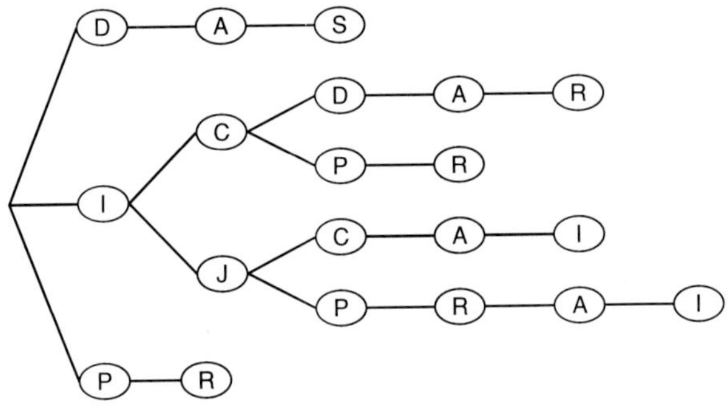

Fig. 6. Trie representation for the dictionary words {DAS, ICDAR, ICPR, IJCAI, IJPRAI, PR}.

Hash tables: A hash table is a random access device for looking up by a key, or code value, computed from the characters within the word. The idea in hashing is to select certain aspects of a search key, for example, distinct characters of an input word, and to use this partial information for the search process. Thus, the partial information represents the input for a special hash function that computes the corresponding address or index of the word entry in the hash table (cf. Fig. 7). If distinct keys are mapped by the hash function onto the same index of the hash table, methods for collision resolution, e.g., the chaining of entries or open addressing, are needed. For more details about hashing see [32].

The advantage of hashing is the extremely fast access to dictionary entries by computing the hash code. A large number of string comparisons as would be needed in searching a sequential list or tree structure is not needed. If a good or even a perfect hash function with a huge hash table is applied, dictionary entries can be accessed by only one comparison. However, the problem with hash tables is that they are rather static data structures. In addition, OCR errors in the word key have to be modeled in advance which is not always possible. Here, adequate error models are essential to tolerate different kinds of recognition errors [42, 43].

While general hash table methods have been extensively developed over the last two decades and are well explored, there is a pressing need for sophisticated hashing which is tailored to improve character recognition. Here, only a few papers can be found in the literature [44, 45, 43]. While Doster and Sch"urmann [46, 44] also propose an n-fold hashing technique via the left and right half of a word, the approach of Hoch and Kieninger [47, 48, 49] is much more flexible allowing the integration of different hash functions as well as the definition of contextual views. Kohonen and Reuhkala [45, 42] present a scheme for redundant hashing based on trigrams as significant word features. This procedure was primarily developed for dealing with distorted phonemic strings in speech recognition.

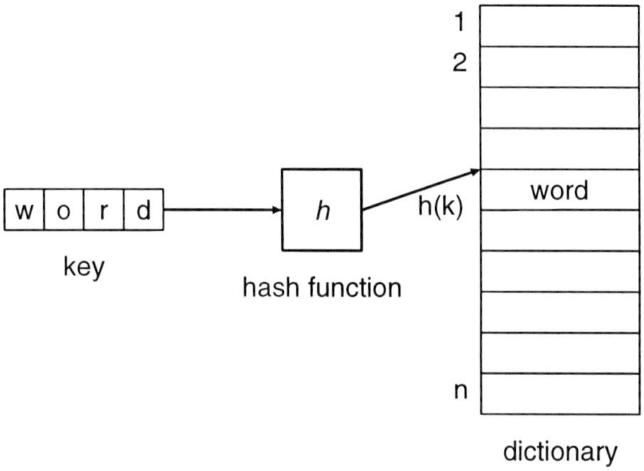

Fig. 7. Principle of hash addressing.

3.2.2. *Dictionary partitioning*

Dictionary look-up is a rather straightforward task. But dictionary access time is a serious problem when the dictionary size exceeds a few thousand or even a hundred thousand words. This problem can be tackled in three ways: by efficient dictionary access techniques (e.g., hash tables, see above); by reducing input words to their stems and searching the respective stems in the dictionary (morphological analysis); and by dictionary partitioning schemes [50].

Because technological advances have made the storage demands of large dictionaries less critical, the current tendency is to store all inflectional forms of dictionary entries separately for the validation of OCR results. Although morphological analysis is needed for text analysis tasks it becomes a problem when input words are noisy. On the other hand, the coverage of a dictionary containing all inflectional forms can be essential [51].

The partitioning of large dictionaries is a well-known strategy to improve dictionary access time and to reduce the number of potential word candidates. In [52] dictionary partitions are based on three major criteria: the word length, the word envelope (ascenders, descenders) and a combination of well recognized characters. In [51] and [53] concepts are shown which partition the dictionary physically by frequency where the most frequently used words of the language are stored in main memory. Hoch and Kieninger [47] propose hierarchical dictionary partitions with respect to distinct criteria such as the structure of a document or textual information by the definition of so-called lexical views. Their approach is based on redundant hash addressing [48].

Beuvron and Trigano [54] also describe hierarchical partitions of the dictionary that are based on phonetic codes allowing errors of exactly one character. A hierarchical structuring of the dictionary for address recognition can be found in [55].

3.2.3. *Dictionary-based error correction*

So far, we have discussed efficient dictionary look-up techniques for the validation of OCR results, i.e., the detection of recognition errors. However, the correction of OCR errors is a much harder problem. For the correction of OCR errors, adequate word candidates of the dictionary have to be generated and ranked either by a lexical similarity measure or a probabilistic estimate of the likelihood of the corrections.

A well-known lexical similarity measure is the string edit distance or the Damerau–Levenshtein metric [56]. String editing methods measure the similarity between two strings by the minimum number of basic edit operations between a misspelled string and a dictionary entry. Wagner and Fisher [57] define the minimum edit distance as the minimum number of substitutions, insertions and deletions to transform one string into another. Each of these operations can be weighted by a constant cost. Wagner and Fisher apply dynamic programming techniques to solve this problem efficiently [57, 41]. Often, the edit operations are extended to model generalized substitutions, context dependent costs [41], spelling errors or OCR errors [58].

Alternatively, similarity key techniques define a string similarity measure based on features which are extracted from the input words. Using hashing techniques a number of words similar to the input word can be retrieved [42]. Kohonen gives an excellent overview on general similarity measures such as the Hamming distance or the cosine measure [59].

Another approach to approximate string matching and the respective similarity measures is through a probabilistic model for the generation of the noisy input word from the correct word of a dictionary [60, 61]. Kashyap and Oommen [62] propose a method that models patterns for multiple insertion, deletion and substitution errors with joint event probabilities. The likelihood function for a word is a recurrence relation which is similar to dynamic programming methods for the computation of weighted edit distances.

3.2.4. *Hybrid methods*

Hybrid methods aim to combine the advantages of top-down and bottom-up techniques [63, 36]. Markov models in the form of the Viterbi algorithm or its variants are used to generate likely word alternatives to the input. At each step, the corresponding substrings are then validated against the dictionary to check whether they constitute valid word prefixes or not. Hence, incorrect word alternatives are avoided. If the input word, however, is not found in the dictionary, the Markov models can be applied to obtain the most likely word candidates.

3.3. *Summary*

Both bottom-up and top-down techniques have been frequently used for lexical post-processing. Bottom-up techniques are very fast whereas dictionary-based approaches can become computationally inefficient, especially if a large vocabulary is used. On the other hand, methods using information about the valid words in a dictionary reveal much better capabilities for error detection and error correction but they have problems with unknown input words.

4. Word Context Post-Processing

Post-processing techniques based upon word context start with complete recognized words at least. Techniques starting from word shape tokens, or character shape tokens respectively, have also been proposed but will not be treated in this chapter. For techniques starting with recognized words, there are two possible approaches. Firstly, the recognized word itself as a string is used, or secondly, additional information available for the word is used. Such additional information can be a word's frequency, its part-of-speech and so on.

The goal of this kind of post-processing is the verification or falsification of given word hypotheses according to their context. More generally speaking, an *a posteriori* weighting of hypotheses with respect to context should be calculated. An extension to this goal, as is sometimes proposed, is the prediction of syntactic or other categories for not yet recognized words.

From the linguistic point of view, a technique for contextual post-processing can incorporate a multitude of different knowledge sources: frequencies of single words and word combinations; compounds and idioms; linguistic structures like phrases, sentences, etc. and so on. Overviews of possible knowledge sources for post-processing can be found in [64, 65, 66]. The structure of this section will not follow linguistic motivation but instead group techniques according to the kind of knowledge exploited for post-processing:

1. purely statistical approaches, especially Markov models;
2. purely symbolic (or syntactic) approaches relying on grammars as the central means of description;
3. possible combinations of the preceding two types of techniques;

4. techniques based on combinations of words occuring very frequently.

We give a brief description of the evaluation methods for the described techniques in the fifth subsection. The section concludes with a short summary.

4.1. *Statistical Approaches*

The use of statistical approaches, especially Markov models, for word syntax is quite similar to their use for character combinations described above in Section 3.1. In principle, instead of characters we now take words. Thus, we now have n-grams of word sequences, e.g., 2-grams for pairs of words; *a priori* probabilities of single words; and confusion probabilities for n-grams of word sequences.

As the number of words is much higher than the number of characters, a straightforward transformation of the model from character to word level is not possible. At character level, there will be 26×26 transitions in a 2-gram. At word level, there will be at least several thousand words to be represented which leads to more than a million transitions.

Therefore, certain word descriptions, e.g., syntactic category or part-of-speech, are used instead of individual words. A full syntactic description would not only comprise the syntactic category but also word inflection attributes like number, case, gender, and probably more syntactic information. Therefore, a "full" syntactic categorization is not possible, since the number of classes would be too high. Usually, a coarse classification is used which takes the most important syntactic attributes into account.

Which classification is chosen for the n-gram model may have a significant influence on the quality of the resulting system. Hull has shown in an experiment [67], that the error rate of a system may be improved by a factor of 4 by only changing the class assignment for words (that is not exactly what he shows, but is closely related). Kucera and Francis [68] propose a word classification for American English which is widely used. It was proposed together with the so-called Brown Corpus. Its word classes are a kind of compromise between syntactic categories and inflectional attributes. For example, each form of the verb "to do" has its own class tag, whereas all forms of adjectives are attached to one class tag.

Thus, we can describe language syntax with sequences of such class tags. For a "full" description of language it would be necessary to model variable-length n-grams since sentences can have any number of words. The statistical approaches known do not support variable-length n-grams but only transition orders 2 or 3.

Comparisons of the different transition orders are given in [67] and [69]. Hull states that the transition order of the Markov model has no significant influence on the quality of the system, whereas Keenan *et al.* agree with this for certain "parameter" adjustments only. The differences between the two experiments are the system architectures and the evaluation methods. The main architectural difference is, that Keenan *et al.* [69] do not really use Markov models. Roughly simplified it works as follows: They use the same information as the Viterbi algorithm does, i.e., frequencies for (part-of-speech) word transition, and, additionally part-of-speech occurrence probabilities. Calculations of the best scoring result is done by a kind of windowing technique which locally sums up the best scoring transitions. Further on, they can take into account a so-called grammatical frequency factor (GFF) for frequencies of word-tag pairs (like Markov models do), and a lexical probability factor (LPF) as a confidence calculated from the OCR results. Their main result is that the incorporation of both factors GFF and LPF causes 3-grams to be significantly better than 2-grams. This may result from the incorporation of two measures (LPF and part-of-speech occurrence) which are not used in Markov models.

4.2. *Syntactic Parsing Techniques*

A totally different method of post-processing is given by syntactic parsing techniques. According to the Chomsky-hierarchy of grammars, we distinguish between approaches using regular or context-free grammars. In addition, each grammar type may be enhanced by a special type of constraint (called feature equations) which can be defined for the non-terminal symbols of the grammar. Using a syntactic grammar for post-processing of OCR involves the same lexical information on word category or part-of-speech as mentioned above for Markov models. The effect of using these techniques is the same as for binary n-grams (cf. Sec. 3.1). Neighborhood sizes are reduced by dropping those alternatives not occuring in a valid parse. We will shortly describe three systems with different descriptive powers of the grammar formalism used: one regular, and two context-free, whereby one context-free uses feature equations to express additional constraints.

Regular parsing has, for example, been proposed in [70]. A regular grammar differs from a binary, i.e., non-probabilistic Markov model in the variable length of its constituents described, and in that they allow certain kinds of recursion. For example, two simple rules of a noun phrase grammar can be noted like this:

$$NP => D + N.$$
$$NP => D + A + N.$$

where NP=noun phrases, N=(common) noun, D=determiner, A=adjective. Given the input sentence:

The young girl is leaving home.

a parsing system may proceed as follows. First, a syntactic lexicon provides the needed information for these words: "the" is D (determiner), "young" is A (adjective), "girl" is N (common noun), and so on. Secondly, the parser looks for rules matching the categories of the input words. The first rule of our example grammar ($NP => D + N$) does fit for the first word "the" since "the" has category D. But it does not fit for the second word "young" which has category A while category N is required. In taking the second rule we will have more luck: the sequence of categories for the first three input words "the young girl" is $D + A + N$ and thus identical with the rhs of the rule.

The strategies for parsing regular expression are well known and widely used, for example, as part of the UNIX command language. More interestingly, Crowner and Hull [70] acquire the grammar rules automatically. This is quite important since most grammars are hand-crafted which needs a lot of manual effort.

Context-free grammars, a more complex description of syntax, are used in [71]. They incorporate not only the syntactic, but also the semantic component (more precisely: a case-frame parser) of a machine-translation system into their text recognition approach. Their parsing technique is known as left-corner parsing, a strategy, where alternating bottom-up and top-down parsing steps start from the left, proceeding in reading direction to the right end of the utterance. The advantage of context-free grammars in comparison to regular ones is their higher descriptive power which involves a higher, but still tractable, complexity in analysis.

The last purely syntactic approach to be mentioned is that in [72]. A context-free grammar is annotated with so-called feature equations defining constraints on the grammar rule's constituents. This is also a technique adapted from natural language processing and currently the state-of-the-art for syntactic processing of natural language. A feature equation serving as a constraint on the applicability of the rule can, for example, define the agreement of the grammatical person between subject and predicate of a sentence. That means, sentences such as "You go" and "She goes" are accepted in contrast to ungrammatical sentences such as "She go" and "You goes". In the address recognition and interpretation system described in [72] this type of constraint is used to define the semantic interdependencies within printed addresses, for example, the agreement between the genders of title and person.

4.3. *Combined Syntactic and Statistical Techniques*

The third group of techniques, combined syntactic and statistical ones, comprise the descriptive power and coverage of structural approaches together with the frequency based efficiency of the statistical ones. Here, we have to distinguish between two principally different kinds of approaches.

The first method is defined as a cascade of the two techniques: "standard" post-processing of OCR from Markov model techniques preceding syntactic processing. Its central gain is the reduction of complexity in parsing, since the relatively slow parsing component only gets such hypotheses as input having a high probability. An encouraging evaluation of this technique — in comparison to a purely statistical post-processing — is given in [73]. They propose the integration of Abney's so-called chunk-parsing [74] as a syntactic method. Chunks are a certain type of syntactic structures consisting of relatively fixed sequences of words. In particular, chunks are phrases headed by function words like prepositions, articles, auxiliaries, etc. For example, the phrases "the young girl" and "is leaving home" are chunks. Chunk parsing is a method for finding such chunks.

Another combination of syntactic and structural approaches is the integration of statistical based search strategies into the parsing process itself. In such approaches, each grammar rule is assigned a confidence score indicating its priority in comparison to other rules. One such technique for probabilistic context-free grammars is proposed in [75] where the parsing is a derivate of the Cocke–Kasami–Younger algorithm. Extensions to this approach also score the constituents of the rule's right-hand-sides which can be seen as the corresponding bottom-up confidences.

We now adapt the example from Section 4.2 on regular noun phrase rules:

$$NP(25\%) => D(45\%) + N(38\%).$$
$$NP(35\%) => D(40\%) + A(47\%) + N(15\%).$$

The scores at the lhs say that 25%, or 35% respectively, of all occurring NPs will be constituted like the rule's rhs tells us. The scores of the rhs have a bottom-up meaning. For example, the N of the first rule has score 38% which means: if you find an N in parsing, it will, in 38% of all cases, be integrated in the NP-construction described here. In using these scores while parsing, usually the rule with the higher score is preferred over others. For our example sentence "The young girl is leaving home." the failing application of the first rule from Sec. 4.2 could thus be avoided.

4.4. *Techniques based on Word Collocations*

Another type of post-processing technique makes use of word collocations (sometimes also called *word co-occurrences*). In general, a word collocation is a set, mostly a pair, of words which occur together within a certain distance with a high statistical significance. One can further distinguish between different types of words, e.g., by only handling pairs of verb and noun, such as the idiom "take a ride" instead of the simple verb "ride". In [76], word collocations are treated for content words, i.e., nouns, verbs, and adjectives, up to a distance of four words, which has been discovered to be linguistically adequate. A method for acquisition of collocations is also given in [76] by using both general and domain specific corpora. The relevance of the training corpus, i.e., the source of the collocations, is discussed there, too.

The usefulness of this approach will be illustrated by the following example (from [76]). In the two phrases

John loves to drink strong tea.
Mary appreciates driving powerful cars.

the words "strong" and "powerful" cannot be exchanged although their meaning is approximately the same. The switched use of "strong cars" and "powerful tea" would rarely occur in a real-world text. Possible misreadings, e.g., "string" instead of "strong" can thus be rejected by using word collocation frequencies. Other examples like

The tea was very strong.
Rachel is very strong and she likes to drink tea.

support the above mentioned assertion that a distance of four words is common upper limit for word collocations.

4.5. *Evaluation Methods*

After having described the different techniques in this field, possible evaluation methods for these techniques are sketched. The possibilities for evaluating the methods are different from those in Sec. 2 and 3. In Sec. 2 on voting techniques an integration of the recognition results of different recognizers is performed in order to make these results comparable. In Sec. 3 on lexical post-processing word hypotheses are generated on the basis of the character hypotheses. In this section, a post-processing of already generated hypotheses is performed which should lead to somehow "better" graded hypotheses with respect to their context. What "better" can be is the topic of this subsection.

In general, all evaluation methods compare the situation before and after post-processing. For this comparison, an assessment function is used which assesses a given set of word hypotheses, e.g., with regard to the correct hypothesis. We now present three slightly different assessment functions relying on, firstly, the number of hypotheses, secondly, their ranking, and, thirdly, the number of errors.

In the simplest function, the number of hypotheses at a certain word position (also called *neighborhood size*) is counted before and after post-processing, i.e., there is no weighting of hypotheses. The average neighborhood size (cf. [67, 77]) is defined as the average number of word hypotheses

$$ANS = \frac{1}{N} \sum_{i=1}^{N} n(i)$$

where N is the number of word positions in the given text and $n(i)$ is the neighborhood size at position i.

An assessment of a post-processing method can thus be given by the percentual difference between ANS before and after pre-processing. In the same way, an error rate can be given by counting those word positions, where the correct word hypotheses have been deleted.

A second method for evaluation requires a ranking of the hypotheses which may be derived from a given probability. Thus, an evaluation of methods can be done by calculating the average rank of, firstly, the correct, and, secondly, the wrong hypotheses for a text (cf. [69]). Similarly to the method mentioned above, a comparison of the ranks before and after post-processing is done.

A third method for evaluation is useful if the incorporated post-processing technique proposes new word hypotheses. In this case, it is appropriate to count the number of OCR errors corrected and the superfluous hypotheses added by the method.

4.6. *Summary*

For word context post-processing of OCR results, four main types of techniques can be distinguished: statistical, syntactic, combined statistical and syntactic, and those based on word collocations. The first type, statistical approaches, are often used and quite well established whereas the second type, syntactic approaches, are less customary in document analysis. Unfortunately, comparisons of syntactic parsing approaches to Markov models or other techniques are seldom given, and, thus, hard to judge their worth. Therefore we only will conclude this section by a general assessment.

Syntactic methods have in general low coverage, but high computational complexity. Additionally, the development of grammars normally is quite expensive since an automated acquisition is not easy. But, for small, strongly structured domains, like addresses, structural approaches are feasible. On the other hand, statistical methods can be implemented quickly and their weights (confusion probability, etc.) can be acquired quite easily. But, they have the big disadvantage that correct choices are often missed since other hypotheses are locally more probable. Still, they are more robust than syntactic methods, and often claimed to be better suited to the recognition task.

These two basic techniques can be combined in different ways: by cascading respective components or by integrating statistical techniques within a syntactic parser. Such combinations use the descriptive power of syntactic approaches together with the efficiency of the statistical ones. Techniques using word collocations are slightly different from the others. They can also be seen as a combination of statistical and syntactic approaches since word collocations are frequently occuring structures of several words.

For the evaluation of such contextual post-processing techniques several metrics are in use. They all have in common the comparison of some graded state before and after the post-processing phase.

5. Document Context

In this section, the use of the whole document as the context for post-processing of OCR results is used. Unfortunately, no systematic experiments in this field are known, but only specialized applications for mostly narrow domains. Therefore, we prefer to give an overview of possible methods instead of listing those few techniques proposed in the literature.

Document context can be seen from two different points of view: firstly, document context is given by the logical structure of the document and its interdependencies with the geometrical layout structure; secondly, the context is given by the document text and, for this case, is the same as the linguistic context. In a broader sense of context, one could also mention the document language itself, since this information can be seen as a (necessary) context enabling the understanding of the document. Corresponding techniques aimed at language determination will not be treated here.

5.1. *Geometric Context*

The central goal of using geometric context for post-processing of OCR results is to establish an expectation towards the contents of a given logical object [78, 79]. Thus, an adequate dictionary or even specialized post-processing for certain objects can be used to improve the given OCR results. Geometric context can help only in the case of well-structured documents. For simply-structured documents like the plain text blocks of a book page there is no structural context which could be exploited. For structured documents, the basic post-processing strategy is as follows. A layout object is assigned a predefined logical label (cf. the Chapter by D. Dori *et al.* of this book) which itself is associated with a certain linguistic or other structure. In the simplest case, this structure may be represented as a set of valid words, i.e., a special dictionary. The structure can also be a syntactic description of the typical contents of the logical object.

For example, the recipient address on a business letter can be located and post-processed by such techniques. Valid words for an address are typically names of persons, streets, cities, etc., besides a few keywords like "Mister", "P.O. Box" and "c/o". By guided dictionary look-up, perhaps using logical or partitioned dictionaries, this kind of knowledge can easily be exploited [47, 48, 49]. Syntactic denotations of textual contents have already been described in Sec. 4 on n-gram models and grammars. In performing a syntactic analysis with Markov models or syntactic parsers, a verification or falsification, respectively, of word hypotheses can be done in a straightforward manner.

For the analysis of recipient addresses, this procedure is incorporated in the system described in [72]. Therein, the expectation yielded from logical labeling concerning the address block is the starting point for its syntactic analysis and, thus, contextual post-processing. More restricted, but very similar in principle is the approach in [80]. "Graphical" parsing with a context-free grammar is performed on the geometric structure of an envelope's address together with rules for word labeling. The result of this algorithm is the location of the zip-code in order to perform a specialized OCR on it. Forms, another strongly structured type of document, is the application area in [81]. Guided by the keywords heading the form fields, strong constraints are given for the field's content, e.g., sex or marital status of persons.

5.2. *Textual Context*

Textual context, our second interpretation of the term "context" is equally powerful, but harder to deal with. This subsection will give an impression of possibilities in using linguistic knowledge by going beyond the techniques of Sec. 4. Unfortunately, there are no well explored techniques for the improvement of OCR results known in this field. The central idea of the following is to exploit existing redundancy with a document's text.

From the purely linguistic point of view, a relatively well explored research area is the field of text coherence structure and discourse. Several rules for the formalization of text structure have been established which will be explained for the following artificial example in order to highlight the important items.

John has bought a new book. Mary saw this book and asked John to lend it to her. Therefore, John gave the book to Mary.

Let us take a closer look at this sentence by syntactic means. The object of the first sentence "the book" occurs both in the second and third sentence. The subject of the first sentence becomes the object in the second sentence but again the subject in the third sentence. The same holds for Mary with switched roles. Now, this is an extreme example, but it shows the themes of previous sentences being taken up again in following sentences. This is an important and quite established phenomenon of the natural language's text syntax.

Suppose a system consists of a syntactic analyzer, itself feeding a case-frame parser for analyzing sentence structures; e.g., the one proposed in [71] as described in Secs. 4.3 and 4.7. Results from case-frame parsing comprise information such as that mentioned here: subject, action, direct and indirect object, and so on. By incorporating the rules of text structure, subsequent case-frame representations of sentences can be checked against probable and invalid text syntax rules. For example, a sequence of identical subject fillers is highly improbable since it offends the human sense of euphony and writing style. On the other hand, a sequence with at least one entry passed on to the subsequent frame with an alternating role gives a high probability for the interpreted recognition results to be correct. In the example above, the subject in the first sentence becomes the object in the second sentence, and so on.

Showing the applicability of such an approach, we give one last example.

John bought a new book. Mary likes cooking french cuisine.

Standard human reaction in this case is like "what connection is there between the two sentences?". The rules of text coherence are violated. A following sentence such as:

Mary enjoyed the book on Provençale cooking John gave her.

"restores" text coherence in linking together several items which used to be isolated from each other.

5.3. *Summary*

The techniques for post-processing based on the whole document context are not yet very well explored. The only relatively wide-spread technique is that using restricted logical dictionaries. This means, a specialized dictionary contains only those words which are expected to occur in a given context. Unexplored but promising seems the exploitation of properties of linguistic structures. As we described, using text coherence structure restricts sentence sequences by certain rules. Observing these rules mainly guarantees that the text is comprehensible.

6. Conclusion

OCR results usually contain errors because of character classification and segmentation problems. For the correction of recognition errors, OCR systems apply contextual post-processing techniques. Such techniques utilize knowledge sources beyond the level of individual characters, for instance, character transition probabilities, the word context as well as knowledge about the structure or the contents of a document in which the word occurs. Some contextual post-processing techniques assume that the set of valid words is contained in a lexicon, others use probabilistic measures of character transitions or involve a broader context of words. Alternative approaches combine the results of different character classifiers.

In this chapter we have discussed the most important techniques for the post-processing of OCR results: voting techniques, lexical post-processing and techniques considering the word or document context for the correction of OCR errors.

The principal idea behind voting algorithms is to combine the results of multiple character classifiers using a decision function. Thus, voting techniques can take advantage of the strengths of the individual classifiers and avoid their weaknesses. Nevertheless, the quality of the distinct classifiers directly determines the final output quality of the voting algorithm.

The well-known application of lexical knowledge for contextual post-processing compares dictionary-based (top-down) and statistical approaches (bottom-up). The advantage of statistical over dictionary-based methods is computational time and memory utilization. On the other hand, lexical knowledge about entire words is more accurate when using a dictionary.

Finally, contextual post-processing of OCR results can also take into account knowledge about the context of words. Examples are the frequencies of word combinations, word collocations, syntactic or semantic knowledge as well as structural information and expectations about the contents of a document. These additional knowledge sources can be used for the reduction of potential word candidates. In general, human knowledge sources contain many challenging research issues for the improvement of OCR results which still remain to be investigated.

It is important to know that the above techniques are not restricted to the contextual post-processing of recognition results of machine-printed texts. They can also be applied in other domains such as spell checking, the recognition of cursive script, or, in part, for speech recognition.

References

[1] S.V. Rice, F.R. Jenkins, and T.A. Nartker, The fourth annual test of OCR accuracy, in *1995 Annual Report*, UNLV Information Science Research Institute.

[2] Y.S. Huang and C.Y. Suen, Combination of multiple classifiers with measurement values, *Proc. of the 2nd ICDAR*, Japan, Oct. 1993, 598–601.

[3] L. Xu, A. Krzyzak, and C.Y. Suen, Methods of combining multiple classifiers and their applications to handwriting recognition, *IEEE Trans. on Systems, Man, and Cybernetics* 22,3 (1992), 418–435.

[4] P. Krause and D. Clark, *Representing Uncertain Knowledge* (Kluwer Academic Publishers, Dordrecht, The Netherlands, 1993).
[5] G. Shafer, *A Mathematical Theory of Evidence* (Princeton University Press 1976).
[6] R.O. Duda and P.E. Hart, *Pattern Classification and Scene Analysis* (Addison Wesley, New York, 1973).
[7] T.K. Ho, A theory of multiple classifier systems and its application to visual word recognition, Technical Report 92-12, Dep. of Computer Science, State University of New York at Buffalo, Buffalo, New York, May 1992.
[8] T.K. Ho, J.J. Hull, and S.N. Srihari, On multiple classifier systems for pattern recognition, *Proc. of the 11th IAPR*, The Hague, The Netherlands, 1992, 84–87.
[9] T.K. Ho, J.J. Hull, and S.N. Srihari, Decision combination in multiple classifier systems, *IEEE Trans. on Pattern Analysis and Machine Intelligence (PAMI)* 16, 1 (1994), 66–75.
[10] E. Mandler and J. Schürmann, Combining the classification results of independent classifiers based on the Dempster/Shafer theory of evidence, in *Pattern Recognition and Artificial Intelligence*, ed. by E.S. Gelsema and L.N. Kanal (Elsevier Science Publishers B.V., North Holland, 1988) 381–393.
[11] J. Franke and E. Mandler, A comparison of two approaches for combining the votes of cooperating classifiers, *Proc. of the 11th IAPR*, The Hague, The Netherlands, 1992, 611–614.
[12] S.V. Rice, J. Kanai, and T.A. Nartker, The third annual test of OCR accuracy, in *1994 Annual Report*, UNLV Information Science Research Institute.
[13] D. Lee and S.N. Srihari, A theory of classifier combination: The neural network approach, *Proc. of the 3rd ICDAR*, Montreal, Canada, Aug. 1995, 42–45.
[14] B.T. Bartell, G.W. Cottrell, and R.K. Belew, Automatic combination of multiple ranked retrieval systems, *Proc. of the 17th ACM-SIGIR*, Dublin, Ireland, July 1994.
[15] C.H. Chen and J.L. DeCurtins, Word recognition in a segmentation-free approach to OCR, *Proc. of the 2nd ICDAR*, Japan, Oct. 1993.
[16] X. Ling and W.G. Rudd, Combining opinions from several experts, *Appl. Artif. Intell.*, 3 (1989), 439–452.
[17] D. Lopresti and J. Zhou, Using consensus sequence voting to correct OCR errors, *Proc. of the DAS 94*, Kaiserslautern, Germany, Oct. 1994, 191–202.
[18] B. Plessis, A. Sicsu, L. Heutte, E. Menu, E. Lecolinet, O. Debon, and J.-V. Moreau, A multi-classifier combination strategy for the recognition of handwritten cursive words, *Proc. of the 2nd ICDAR*, Japan, Oct. 1993, 642–645.
[19] M. Sabourin, A. Mitchie, D. Thomas, and G. Nagy, Classifier combination for handprinted digit recognition, *Proc. of the 2nd ICDAR*, Japan, Oct. 1993, 163–166.
[20] B.V. Dasarathy, *Decision Fusion*, IEEE Computer Society Press, Los Alamos, California, 1994.
[21] C.Y. Suen, N-gram statistics for natural language understanding and text processing, *IEEE Trans. on Pattern Analysis and Machine Intelligence (PAMI)* PAMI-1, 2, (1979), 164–172.
[22] D.G. Elliman and I.T. Lancaster, A review of segmentation and contextual analysis techniques for text recognition, *Pattern Recognition* 23, 3/4 (1990), 337–346.
[23] L. McMahon and L. Cherry, Statistical text processing, *The Bell System Tech. J.*, 57, 6 (1978), 2137–2154.
[24] E.M. Riseman and A.R. Hanson, A contextual postprocessing system for error correction using binary n-grams, *IEEE Trans. on Computers*, c-23, 5 (1974), 480–493.
[25] S.N.S. Srihari *et al.*, Integrating diverse knowledge sources in text recognition, *ACM Trans. on Office Information Systems*, 1 (1983), 68–87.
[26] G.D. Forney, The Viterbi algorithm, *Proc. of the IEEE*, 61, 3 (1973), 268–278.

[27] S.N. Srihari, The Viterbi algorithm, *The Encyclopedia of Artificial Intelligence*, ed. S.C. Shapiro (J. Wiley, 1987), 1160–1162.

[28] R. Shingal and G.T. Toussaint, Experiments in text recognition with the modified Viterbi algorithm, *IEEE Trans. on Pattern Analysis and Machine Intelligence*, PAMI-1, 2 (1979), 184–192.

[29] W. Doster and J. Schürmann, An application of the modified Viterbi algorithm used in text recognition, *Proc. on the Fifth Int. Conf. on Pattern Recognition*, 1980, 853–855.

[30] J.J. Hull and S.N. Srihari, Experiments in text recognition with binary n-gram and Viterbi algorithms, *IEEE Trans. on Pattern Analysis and Machine Intelligence*, PAMI-4, 5 (1982), 520–530.

[31] R.M.K. Sinha *et al.*, Hybrid contextual text recognition with string matching, *IEEE Trans. on Pattern Analysis and Machine Intel. (PAMI)*, 15, 9, ep. 93, 915–925.

[32] D.E. Knuth, *The Art of Computer Programming*, Vol. III: Sorting and Searching (Addison-Wesley Publishing Company, Reading, Massachusetts, 1973).

[33] T.N. Turba, Checking for spelling and typographical errors in computer-based text, in ed. S.N. Srihari, *Computer Text Recognition and Error Correction*, Tutorial, IEEE Computer Society Press, Silver Spring, MD, 1985, 294–303 (cf. Srihari [82]).

[34] M.D. Harris, *Introduction to Natural Language Processing* (Reston Publishing Company Inc., Reston, Virginia, 1985).

[35] J.J. Hull, A Computational Theory of Visual Word Recognition, Technical Report 88-07, Dissertation, Department of Computer Science, State University of New York at Buffalo, 1988.

[36] S.N. Srihari, J.J. Hull, and R. Choudhari, Integrating diverse knowledge sources in text recognition, *ACM Transactions on Office Information Systems*, 1, 1 (1983), 68–87.

[37] E. Fredkin, Trie memory, *Commun. of the ACM*, 3, 9 (1960), 490–500.

[38] K. Maly, Compressed tries, *Commun. of the ACM*, 19, 7 1976), 409–415.

[39] A.W. Appel and G.J. Jacobson, The world's fastest scrabble program, *Commun. of the ACM*, 31, 5 (1988), 572–585.

[40] C.J. Wells *et al.*, Fast dictionary look-up for contextual word recognition, *Pattern Recognition*, 23, 5 (1990), 501–508.

[41] H. Bunke, String matching for structural pattern recognition, in eds. H. Bunke and A. Sanfeliu *Structural Pattern Analysis* (World Scientific Publication Co., Singapore, 1990) 119–144.

[42] E. Reuhkala, Recognition of strings of discrete symbols with special application to isolated word recognition, *ACTA Polytechnica Scandinavica, Mathematics and Computer Science Series*, no. 38, Dissertation, Helsinki, 1983.

[43] H. Takahashi, N. Itoh, T. Amano, and A. Yamashita, A spelling correction method and its application to an OCR system, *Pattern Recognition*, 23, 3/4 (1990), 363–377.

[44] J. Schürmann, Multifont word recognition system, *IEEE Trans. on Computers* c-27, 8 (1978).

[45] T. Kohonen and E. Reuhkala, A very fast associative method for the recognition and correction of misspelt words, based on redundant hash addressing, *Proc. of the Fourth Int. Joint Conf. on Pattern Recognition*, Kyoto, Japan, Nov. 1978, 807–809.

[46] W. Doster, Contextual post-processing system for cooperation with a multiple-choice character recognition system, *IEEE Trans. on Computers* c-26, 11 (1977), 1090–1101.

[47] R. Hoch and T. Kieninger, On virtual partitioning of large dictionaries for contextual post-processing to improve character recognition, *Proc. of the Second Int. Conf. on Document Analysis and Recognition (ICDAR'93)*, Tsukuba Science City, Japan, Oct. 1993, 226–231.

[48] R. Hoch, H.-G. Hein, and T. Kieninger, Using a partitioned dictionary for contextual post-processing of OCR-results, *Proc. of the 12th Int. Conf. on Pattern Recognition*

(ICPR'94), Jerusalem, Israel, Oct. 1994, 274–278.
[49] R. Hoch and T. Kieninger, On virtual partitioning of large dictionaries for contextual post-processing to improve character recognition, *Int. J. of Pattern Recognition and Artificial Intelligence* 10, 1996 (to appear).
[50] K. Kukich, Techniques for automatically correcting words in text, *ACM Computing Surveys*, 24, 4 (1992), 377–439.
[51] J.L. Peterson, Computer programs for detecting and correcting spelling errors, *Commun. of the ACM*, 23, 12 (1980), 676–687.
[52] R.M.K. Sinha, On partitioning a dictionary for visual text recognition, *Pattern Recognition*, 23, 5 (1990), 497–500.
[53] R.M.K. Sinha, Some characteristics for dictionary organization with digital search, *IEEE Trans. on Systems, Man, and Cybernetics* SMC-17, 3 (1987), 520–527.
[54] F.B. Beuvron and P. Trigano, Lexical architecture based on a hierarchy of codes for high speed string correction, *Proc. of the SPIE — The International Society for Optical Engineering, Applications of AI X: Knowledge-Based Systems*, vol. 1707, Orlando, Florida, April 1992, 290–298.
[55] K. Seino, Y. Tanabe, and K. Sakai, A linguistic post-processing based on word occurrence probability, in eds. S. Impedovo and J.C. Simon *From Pixels to Features III: Frontiers in Handwriting Recognition* (Elsevier Science Publications B. V., 1992).
[56] F.J. Damerau, A technique for computer detection and correction of spelling errors, *Commun. of the ACM*, 7, 3 (1964), 171–176.
[57] R.A. Wagner and M.J. Fischer, The string-to-string correction problem, *J. of the Association for Computing Machinery*, 21, 1 (1974), 168–173.
[58] A. Weigel and F. Fein, Normalizing the weighted edit distance, *Proc. of the 12th Int. Conf. on Pattern Recognition (ICPR'94)*, Jerusalem, Israel, Oct. 1994, 399–402.
[59] T. Kohonen, Content-addressable memories, *Springer Series in Information Sciences 1* (Springer-Verlag Berlin Heidelberg, 1980).
[60] P.A.V. Hall and G.R. Dowling, Approximate string matching, *Computing Surveys*, 12, 4 (1980), 381–402.
[61] R.M.K. Sinha, B. Prasada, G. Houle, and M. Sabourin, Hybrid contextual text recognition with string matching, *IEEE Trans. on Pattern Analysis and Machine Intelligence (PAMI)*, 15, 9 (1993), 915–925.
[62] R.L. Kashyap and B.J. Oommen, Spelling correction using probabilistic methods, *Pattern Recognition Lett.*, 2, 3 (1984), 147–154.
[63] R. Shingal and G.T. Toussaint, A bottom-up and top-down approach to using context in text recognition, *Int. J. Man-Machine Studies*, 11 (1979), 201–212.
[64] G. Nagy, What does a machine need to know to read a document?, *Proc. of the 1st Annual Symp. on Document Analysis and Information Retrieval*, Las Vegas, Nevada, USA, March 1992, University of Nevada, 1–10.
[65] S.N. Srihari, From pixels to paragraphs: the use of models in text recognition, *Proc. of the Second Annual Symp. on Document Analysis and Information Retrieval*, Las Vegas, Nevada, USA, April 1993, University of Nevada, 47–64.
[66] S.N. Srihari, From Pixels to Paragraphs: the Use of Contextual Models in Text Recognition, *Proc. of the Second Int. Conf. on Document Analysis and Recognition*, Tsukuba Science City, Japan, Oct. 1993, IEEE Computer Society Press, Los Alamitos, California, USA, 416–423.
[67] J.J. Hull, A hidden markov model for language syntax in text recognition, *Proc. of the 11th IAPR Int. Conf. on Pattern Recognition*, The Hague, The Netherlands, Aug./Sept. 1992, IEEE Computer Society Press, Los Alamitos, California, USA, 124–127.
[68] H. Kucera and W.N. Francis, *Computational Analysis of Present-Day American English* (Brown University Press, Providence, Rhode Island, USA, 1967).

[69] F.G. Keenan, L.J. Evett, and R.J. Whitrow, A large vocabulary stochastic analyser for handwriting recognition, *Proc. of the First Int. Conf. on Document Analysis and Recognition*, Saint-Malo, France, Sept./Oct. 1991, AFCET-IRISA/INRIA-Ecole Nationale Supérieure des Télécommunications, Rennes, France, 794–802.

[70] C. Crowner and J.J. Hull, A hierarchical pattern matcher and its application to word shape recognition, *Proc. of the Int. Conf. on Document Analysis and Recognition*, Saint-Malo, France, Sept./Oct. 1991, AFCET-IRISA/INRIA-Ecole Nationale Supérieure des Télécommunications, Rennes, France, 323–331.

[71] K. Kise, T. Shiraishi, S. Takamatsu, and H. Kusaka, Improvement of text image recognition based on linguistic constraints, *Proc. of the IAPR Workshop on Machine Vision Applications*, Tokyo, Japan, Dec. 1992, 511–514.

[72] M.H. Malburg and A.R. Dengel, Address verification in structured documents for automatic mail delivery, *Proc. of the First European Conf. Dedicated to Postal Technologies*, Nantes, France, June 1993, Service de Recherche Technique de la Poste, Nantes, 447–454.

[73] R.K. Srihari and C.M. Baltus, Incorporating syntactic constraints in recognizing handwritten sentences, *Proc. of the 13th Int. Joint Conf. on Artificial Intelligence*, Chambéry, France, Aug./Sept. 1993, Morgan Kaufmann Publishers, Inc., San Mateo, California, USA, 1262–1267.

[74] S. Abney, Parsing by Chunks, in eds. R. Berwick, S. Abney, and C. Tenny *Principle-Based Parsing* (Kluwer Academic Publishers, 1990).

[75] T. Hong and J.J. Hull, Text recognition enhancement with a probabilistic lattice chart parser, *Proc. of the Second Int. Conf. on Document Analysis and Recognition*, Tsukuba Science City, Japan, Oct. 1993, IEEE Computer Society Press, Los Alamitos, California, USA, 222–225.

[76] T.G. Rose and L.J. Evett, Semantic analysis for large vocabulary cursive script recognition, *Proc. of the Second Int. Conf. on Document Analysis and Recognition*, Tsukuba Science City, Japan, Oct. 1993, IEEE Computer Society Press, Los Alamitos, California, USA, 236–239.

[77] J.J. Hull, Incorporation of a markov model of language syntax in a text recognition algorithm, *Proc. of the 1st Annual Symp. on Document Analysis and Information Retrieval*, Las Vegas, Nevada, USA, March 1992, University of Nevada, 174–185.

[78] A. Dengel, R. Bleisinger, R. Hoch, F. Fein, and F. Hönes, From paper to office document standard representation, *IEEE Computer*, 25, 7 (1992), 63–67.

[79] A. Dengel, R. Bleisinger, R. Hoch, F. Hönes, M. Malburg, and F. Fein, OfficeMAID — A system for automatic mail analysis, interpretation and delivery, in eds. L. Spitz and A. Dengel *Document Analysis Systems* (World Scientific Publishing Co. Inc., Singapore, 1995), 52–75.

[80] R.J.N. Kalberg, G.H. Quint, and H. Scholten, Automatic interpretation of dutch addresses, *Proc. of the 11th IAPR Int. Conf. on Pattern Recognition*, The Hague, The Netherlands, Aug./Sept. 1992, IEEE Computer Society Press, Los Alamitos, California, USA, 367–370.

[81] K.L. Anderson and W.A. Barrett, Context spezification for text recognition in forms, *Proc. of the High-Speed Inspection Architectures, Barcoding, and Character Recognition*, Boston, Massachusetts, USA, Nov. 1990, SPIE-Proc. Series, Vol. 1384, The Society of Photo-Optical Instrumentation Engineers, Bellingham, Washington, USA.

[82] S.N. Srihari (ed.), Computer Text Recognition and Error Correction, Tutorial, IEEE Computer Society Press, Silver Spring, MD, 1985, 353 pages.

Handbook of Character Recognition and Document Image Analysis, pp. 259–284
Eds. H. Bunke and P. S. P. Wang

CHAPTER 9

MULTILINGUAL DOCUMENT RECOGNITION

A. LAWRENCE SPITZ*
Daimler Benz Research and Technology Center
1510 Page Mill Road
Palo Alto, CA 94304 USA

Most document recognition work to date has been performed on English text. Because of the large overlap of the character sets found in English and major Western European languages such as French and German, some extensions of the basic English capability to those languages have taken place. The only known commercial products for Han-based languages are for Japanese. Systems that can handle a mixture of languages or even single languages without prior identification of the language are unknown. Languages and their scripts have attributes that make it possible to determine the language of a document automatically. Detection of the values of these attributes requires the recognition of particular features of the document image and, in the case of Latin-based languages, the character syntax of the underlying language. We show the process of roughly classifying documents as either Latin-based or Han-based, and within those classes identify the Latin-based language as being one of 23, and the Han-based as being either Chinese, Japanese or Korean.

Keywords: Multilingual; Script recognition; Machine printed OCR; Language classification; Han-based languages; Latin-based languages; Asian scripts.

1. Introduction

There are many different languages in common use worldwide and many different scripts in which these languages are typeset. In an era of increasing international trade, where markets are established and manufacturing is done in areas where different languages are used, the capability of recognizing multilingual documents is both novel and useful. With such capability, many potential applications can be supported including cross-lingual access to patent information, business and regulatory information, document sorting in support of character recognition, translation and keyword finding in document images. The documents in question may each contain only one language but it may be one in which the user is less than fluent, or a single document may contain more than a single language. Dealing with multilingual documents raises many challenges including script identification, language determination, text reading direction, and differing character sets.

*spitz@rtna.daimlerbenz.com

In this chapter we concentrate on languages set in Roman (Latin) type (which include both European languages and non-European languages such as Swahili or Vietnamese). We also consider the Asian languages, Chinese, Japanese and Korean, which we will refer to as Han-based. Occasional examples involving other languages are cited in footnotes. The capabilities described here have been implemented in our laboratory in a demonstration system called Palace [1].

In Sec. 2, we describe some of the effects of the language of the document on the form of the document. We describe the pre-processing performed on page images in Sec. 3, including determination of text orientation, segmentation of individual text lines, registration of those lines, measurement of important line parameters, segmentation of words and segmentation of character cells. In Sec. 4 we describe the process of determining which of two broad script classes is present in the document: Han-based or Roman. In Sec. 5 we describe the process of language identification within the two basic script classes. In Sec. 6 we describe the ramifications of the multilingual nature of documents on the process of Optical Character Recognition.

2. Language and Document Images

The language of a document has (at least) two measurable effects on the image of that document. First is the attribute of script, or character set; second is the writing conventions of the language in question. In some instances, detection of the script is sufficient to classify the document by language. For example, presence of Hangul unambiguously identifies the document as containing Korean, while the presence of Roman script does not restrict the language classification to a great extent.

2.1. *Script*

There is a complex relationship between a language and its script. Roman (or Latin) script is used for many languages though it is augmented in various ways for different language classes. The Han script is strongly identified with the Chinese language but is widely used in other languages such as Japanese or Korean.[a] Both Japanese and Korean use other scripts as well.

Japanese uses four scripts: Kanji, two phonetic scripts (syllabaries), and Roman. One syllabary, Hiragana, is used for native Japanese words and for inflecting Chinese and Japanese words. The other syllabary, Katakana, is used for foreign words and for emphasis. Collectively, Hiragana and Katakana are known as Kana.

Korean uses Hangul for Korean words and for endings on Chinese nouns. Serbo-Croatian is a single spoken language which, in its written form, is called Serbian when it is set in Cyrillic script and Croatian when Roman script is used.

[a]Chinese characters are known as Hanzi in Chinese, Kanji in Japanese, and Hanja in Korean. In this chapter they may be referred to as Kanji even when the particular use is not restricted in Japanese.

2.2. *Character Positioning*

As the use of word processing and laser printing subsumes the use of typewriters and type-chain computer line printers, mono-spaced machine print is becoming less and less common in Latin-based documents. Proportional spacing, which has been used for more than a hundred years in press-print, has been adopted even by low-cost laser printers. Proportional spacing often results in kerning where the bounding boxes of adjacent characters overlap, though their renderings do not touch. Kerning is especially prevalent in italic typefaces.

However mono-spacing is still the norm for Chinese and Japanese documents because the character sets are inherently composed of fixed-width characters.[b]

Latin-based documents may be set either ragged-right or right justified. Han-based documents are always fully justified.

2.3. *Text Orientation*

Latin-based documents incorporate horizontal text. The lines of text are read left-to-right and the lines are read in a top-to-bottom sequence.[c] Asian documents may be set either horizontally or vertically. If they are horizontal, the individual lines are read as Latin-based documents are. If vertical, the lines are read top-to-bottom and the sequence of lines is from right-to-left.[d]

2.4. *Word Segmentation*

In Latin-based documents, words are delimited by wide white spaces either within a line or at the ends of lines. In Latin-based documents, words that are split across text lines are hyphenated. In Chinese and Japanese documents, word segmentation is a semantic and syntactic process [2]. Punctuation does indicate the presence of a word boundary, but most word boundaries are unmarked. Line boundaries cannot be used for word segmentation because words are split across lines without hyphenation.

In Japanese, the presence of a word boundary is often signalled by the transition from a string of Hiragana characters to Kanji.

In Korean documents, words are set with spaces but may be broken across text lines without hyphenation or any other indication.

2.5. *Character Forms*

Han-based scripts are characterized by larger numbers of character forms than are found in Latin-based script. With all possible diacritical marks, the number of Latin-based character forms is still much smaller than the number of Han-based

[b] Japanese may be set with a combination of half-width character cells, containing numerals and symbols such as %, and full-width character cells.

[c] Arabic and Hebrew also have horizontal text lines but they are read right-to-left.

[d] Except in Mongolian, which is not Han-based, where the line sequence is left-to-right.

forms.[e] In contrast, even basic high-school level Japanese incorporates more than 2000 characters. Modern Japanese literature might include as many as 3500 different characters and modern Chinese as many as 8000, assuming simplified/traditional pairs are not counted more than once. Representing large numbers of distinct characters implies much higher complexity per character. Increased complexity implies the need for high spatial resolution to resolve subtle differences between character forms.

Hangul, the Korean phonetic script, combines 10 basic vowel elements and 14 consonantal elements into characters, each of which represents a single syllable. In theory, these elements can combine in various ways to result in about 11,000 distinct characters. In modern practice 2300 distinct characters are in common use [3]. Chinese characters (usually representing nouns) may be inserted into running text.

Most Roman characters consist of a single connected component. The exceptions include **i**, **j**, **ü** and accented characters such as **ê**. Han-based characters may comprise many connected components.[f]

Japanese text will often supplement Kanji characters with small Hiragana characters placed above (to the right in vertical text) to indicate the pronunciation of obscure Kanji or the preferred pronunciation if there are multiple possible pronunciations. These characters are called Furigana or *rubi*.

Likewise, Korean text may indicate the pronunciation of Chinese characters using parenthesized Hangul characters following the Hanja.

2.5.1. *Capitalization*

In Latin-based documents, the first words of sentences and proper nouns and their associated adjectives are capitalized. In German, all nouns are capitalized as well. There is no analogous rendering in Han-based scripts.

2.5.2. *Symbols*

In various languages, symbols (including punctuation) are rendered differently. For example, quotation marks differ in English “ ”, French ≪≫, and Japanese 「」. The comma not only takes on differing forms across languages, but may be represented differently in horizontal and vertical Han-based text.

2.5.3. *Connectivity*

In all of the languages so far investigated, concepts such as the character cell — a rectangular or parallelogramic region containing the isolated mark or marks

[e] Vietnamese lacks f, j, w and z, but adds seven characters to the (English) Roman character set. The addition of tonal information increases complexity so that there are 18 forms of the letter “a” [3].

[f] Arabic letters may have up to four forms depending on whether the letter is word initial, word final, word medial or isolated.

comprising a character — are valid. However many scripts (e.g., Arabic, Telugu, Devanagari) have high connectivity where multiple characters or entire words are connected into large structures. Some Roman fonts contain ligatures connecting two or more characters, e.g., **fi**, **fl**, **ffl**.

2.6. *Interpolations*

Small amounts of *foreign* language — isolated words or phrases — may appear in documents of a single basal language. In Latin-based documents, the foreign nature of these occurrences is often signalled by a typeface change, typically to italic. In Japanese texts foreign words are indicated by the character set used: either Roman (called Romaji, and usually set with proportional spacing) or Katakana, as opposed to the Kanji and Hiragana used for native Japanese.

3. Text Image Processing

In general for document recognition, some image processing is required. Discussion will be limited here to that image processing specifically required to support the processing of multilingual documents.[g] It is presumed that text/graphics segmentation and other image-quality related processing required for single language documents is part of the systematic infrastructure. Nevertheless, the data structure (shown in Fig. 2) is generalized to incorporate salient features for the languages and scripts in question.

Several techniques described here are not only crucial to the multilingual aspects of documents, but are also of general utility.

3.1. *Skew Compensation*

Most techniques for page element segmentation are extremely sensitive to even small skew angles. The techniques described here are likewise dependent on text line orientation being aligned to the image coordinate system. To compensate for possible skew, skew angle detection based on Baird's technique [5] is performed. This technique finds the predominant alignment of fiducial points attached to the bottom centers of the bounding boxes of each connected component by measuring the variance of their population in a series of rotationally aligned bins. Once the skew angle has been determined, the individual connected components are translated such that their origins are properly aligned to the image coordinate system (but any rotation of individual components persists, an example is shown in Fig. 1).

All analysis is based on lists of connected components [6], an abstraction of the image, rather than on the binary bitmap itself. This representation is loss-less; that is, the original bitmap can be reproduced without error.

[g]The techniques used here have been described elsewhere and have found particular application in [1]. Other techniques, applicable across a broad range of documents, independent of language, are described elsewhere in this volume and in [4].

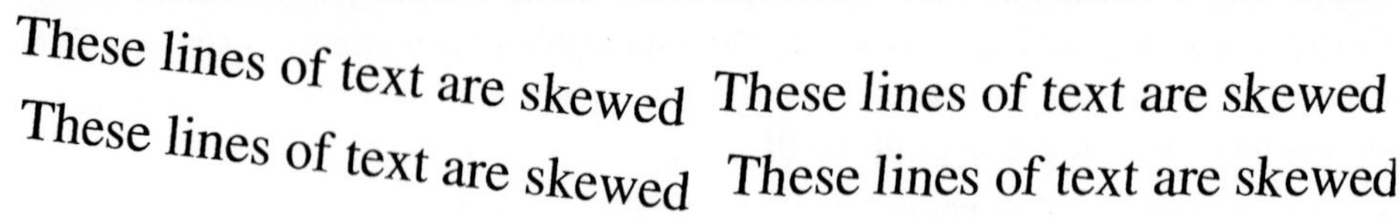

Fig. 1. Skewed and skew compensated text.

```
struct Line
{
          short int xorig, width;            /* xorig column relative *
          short int baseline;                /* column relative */
          short int xheight, ascender, descender;/*baseline relative */
          struct Line *next;
          struct Word *word;
};
struct Word
{
          short int xorig, width;            /* xorig line relative */
          struct Word *next;
          struct Ccell *ccell;
};
struct Ccell/* character cell */
{
          short int xorig, width;            /* xorig word relative */
          struct Ccell *next;
          struct Ccomp *ccomp;         /* connected component(s) pointer*
};
```

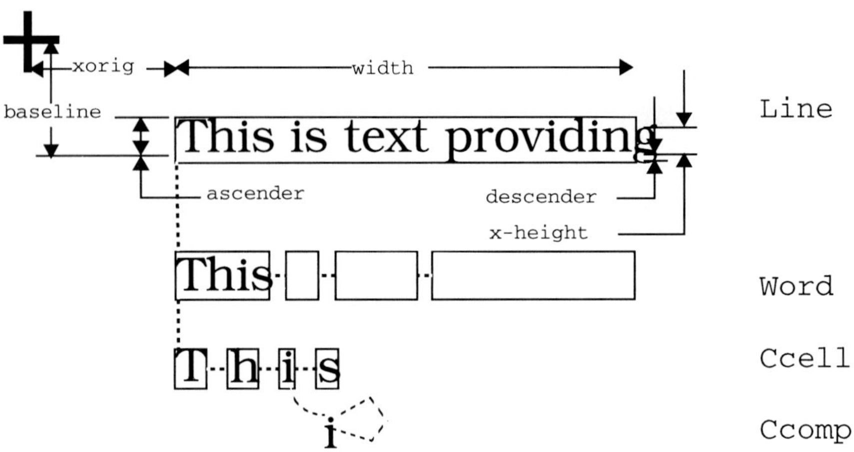

Fig. 2. Source code and spatial representations of line, word, and character cell data structure.

3.2. *Connected Component Generation*

The set of 8-connected components in the document image is calculated. The representation of each connected component includes the coordinates and dimensions of the bounding box and a list of the individual runs of black pixels that make up the component. The connected components are used as the basic image representation throughout the recognition process. The list of connected components provides an efficient data structure on which to perform image processing operations — not only those operations related to the positions of the components in different reference frames, but also the calculation of upward concavity and optical density required for script and Han-based language identification.

3.3. *Line Parameters*

Three spatial distributions are calculated: the number of connected components occupying each vertical position within the text line; the positions of the tops of connected components; and the positions of the bottoms of the connected components.

Starting from any vertical position within the peak value of the connected component distribution (which roughly corresponds to the vertical range from baseline to x-line) we search downward for a peak in the bottom distribution and label that peak position as the baseline position. This is in contrast to Kanai's method that relies on character prototypes for baseline prediction [7]. Note that in a text line that does not contain character descenders (or punctuation such as comma or semicolon that descend below the baseline), the baseline and the line bottom positions may be the same.

Starting from the baseline, we search upward for the peak in the connected component top distribution and label this position as the x-line or x-height position. Note that in the case of a text line containing a preponderance of capital letters, digits or lower case characters with ascenders, it may not be possible to accurately determine the x-height.

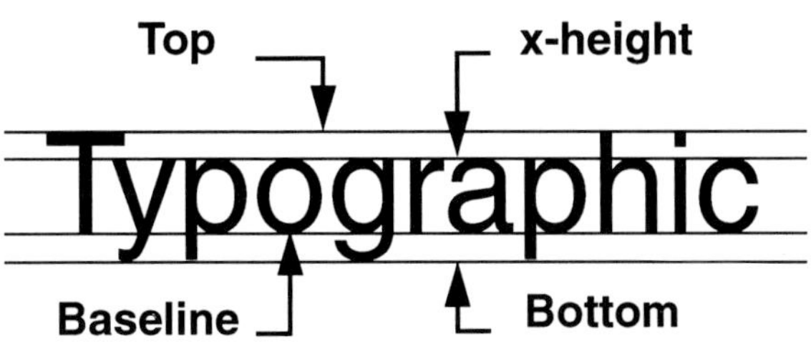

Fig. 3. A text image showing the text line parameter positions: Top, x-height, baseline and bottom.

Four horizontal lines define the boundaries of three significant zones on each text line (see Fig. 3). The area between the bottom and the baseline is the descender zone; the area between the baseline and the top of characters such as **x** is the x zone; and the area above the x-height level is the ascender zone.

3.4. *Text Orientation Detection*

Han-based language documents, in particular, may have either horizontal or vertical text layout. We use a modification of Ittner's technique [8,9] to find text line orientation. Ittner's algorithm is based on the observation that inter-character spacing is, in general, smaller than inter-line spacing. Ittner develops a minimal spanning tree connecting each connected component to its nearest neighbor, and calculates the predominant direction of the connecting branches.

3.5. *Text Line Processing*

Techniques for processing of images to enhance recognition processes are described in detail in [10] and [11]. Compensating for the dominant skew angle in an entire page image may not be sufficient adjustment to allow accurate text parameterization. Sometimes individual lines or small groups of lines have a skew angle relative to the orientation of the entire page. Additionally, artifacts often arise when processing photocopies of bound materials. In particular, text lines may curve toward the binding where the printed portion of the page does not contact the photocopier platen.

3.6. *Line Registration*

Registration is a process that aligns the baselines of characters on a text line. Since the actual desired baseline is difficult to characterize in the presence of descenders, we first calculate the modal bottom position for connected components. Components that do not lie at the bottom of their respective character cells (e.g., accents, **i** and **j** dots, and the upper components of **?!;:**) are classified as non-baseline components. For a connected component to qualify as a non-baseline component there must be another connected component below it. Quotation marks and apostrophes are, therefore, considered to be baseline components. While this may seem counter-intuitive since the bottoms of these connected components are so far above the baseline, this classification has not had any adverse effects on the performance of the algorithm. A left-to-right sequence of baseline component bottom positions constitutes a baseline profile.

The next step may be thought of as analogous to high-pass filtering of that baseline profile, allowing sharp discontinuities (due to adjacency of descenders and non-descenders) while eliminating the small variations in baseline position. For each baseline component, we measure and retain the relative vertical offset between the bottom of the connected component and its left baseline component neighbor. At the beginning of the line, we use the modal bottom position in place of the missing neighbor.

Now, we adjust the positions of the baseline connected components to the modal baseline which provides a perfect baseline alignment but temporarily masks the attribute of descending characters and apostrophes. Non-baseline components are moved distances equivalent to those baseline components with which they share character cells. Next we refer to the relative vertical offset information. In general, the relative offsets are distributed such that small values result from printing and scanning artifacts and font variations, but exceptional large values arise from the adjacency of descender characters to non-descenders. We take these large values and apply them to the vertical positions of the connected components associated with the descender characters, then reconstruct the textline registered to the baseline. We show an extreme example in Fig. 4. Note that individual connected components are still rotated relative to the coordinate system.

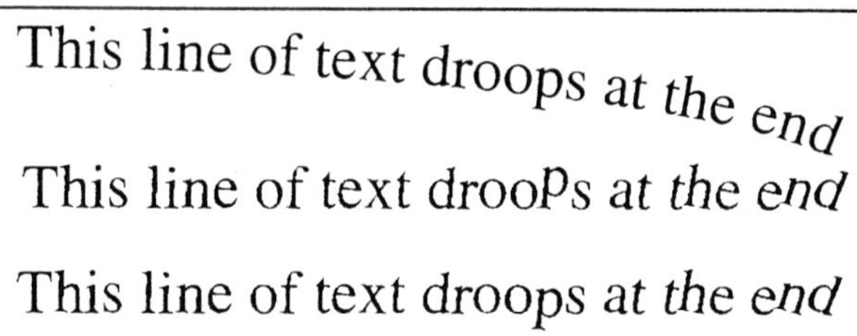

Fig. 4. Distorted, baseline-aligned and registered text line.

3.7. *Word Processing*

The beginnings and endings of lines are presumed to be word delimiters. Hyphenated words are reassembled by downstream processes if necessary. Within a line, spaces of significant width are found by looking first at the positions of components in lines. The inter-word space will be absent on lines containing a single word, or for example, in Japanese or Chinese text. Next, we examine the distribution of spaces detectable in the connected component list to find the word delimiting spaces.

For each text line, various attributes are determined. These include justification, leading and point size. In addition, values of x-height, ascender and descender length and stroke width provide information about font.

In a process directly analogous to line boundary determination, word boundary coordinates are determined by the inclusion of all connected components whose centers lie within the preliminary word dimension.

3.8. *Character Cell Processing*

Character cells are isolated within the spatial boundaries of each word. Vertical paths of white space divide the word rectangle into preliminary character cells that extend from the line top to the line bottom and are bounded on the left and right by inter-character spaces. These preliminary cells are later expanded to fully contain constituent components. Note that in a small number of instances where the character is horizontally disjoint, such as double quote or some Kanji 那 or Hangul characters 에 , more than one character cell per character will be generated. So far, this has not caused problems. In Japanese, for example, since text is set mono-spaced, it is trivial to join adjacent cells into a single cell where aberrant spacing is initially detected. In Korean, the inter-component space is much smaller than the inter-character space.

4. Script Classification

Scripts are divided into two broad classes: Roman and Han-based, including Kanji, Hiragana, Katakana and Hangul. This gross classification is accomplished on the basis of the vertical distribution of upward concavities.

It is trivial to examine sets of runs within the connected component to determine the presence and location of upward concavities. Where two runs of black pixels appear on a single scan line of the raster image, if there is a run on the line below that spans the distance between these two runs, an upward concavity is formed on the line (Fig. 5). The reference coordinate system used for defining the spatial distribution of concavities is the character cell baseline.

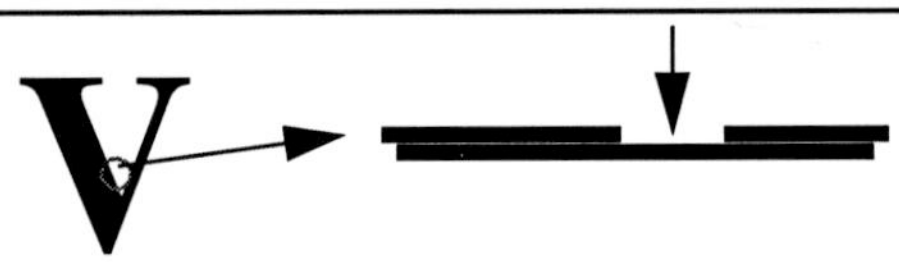

Fig. 5. Upward concavity defined by the spanning of a single run of the gap between a pair of runs on the scan line above.

Figure 6(a) shows the word **Laboratory** and the positions of the upward concavities. The **a**'s show two upward concavities near the baseline, while the **b** shows one concavity near the baseline and one well above the baseline. Figure 6(b) shows the positions of upward concavities in a single Kanji character.

For a Han-based script, the occurrence of upward concavities is significantly different from that for Roman script [12]. Because the more complex characters incorporate more instances of enclosed white space, there are many more concavity occurrences per character and the spatial distribution of these concavities is more random, whereas for Roman scripts the distribution is bi-modal.

Fig. 6. Locations of upward concavities in a word of Roman script text (a). Locations of upward concavities in a single, relatively complex, Kanji character (b).

Next we relate the positions of upward concavities to the position of a stable fiducial point. In instances where page layout information is not available or where there is a mixture of a small number of isolated characters, perhaps of varying size, we use the centroid of the associated connected component. Where text line information is available and where we do not already have the CCITT representation pre-calculated, we use the baseline position for the relevant text line as our fiducial level. It is sufficient to look at the vertical distribution of upward concavity position with respect to the fiducial level only.

Figure 7 shows vertical profiles of upward concavity population for typical Latin-based and Han-based documents. The distributions are normalized to the mean. In this figure, a single document sample is shown for each script.

Figure 8 shows display of the normalized distributions by language. The left column shows the Latin-based languages and the right column shows the Han-based

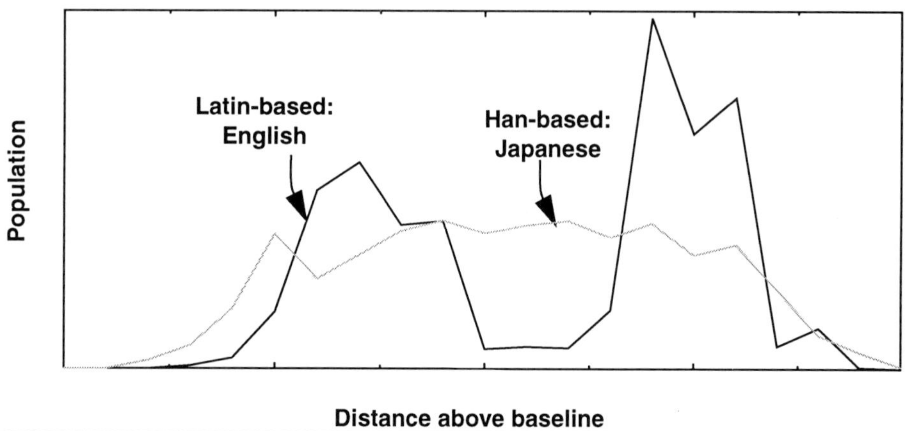

Fig. 7. Spatial distributions of upward concavity vertical distance with respect to baseline position for Latin-based and Han-based documents are characteristically different.

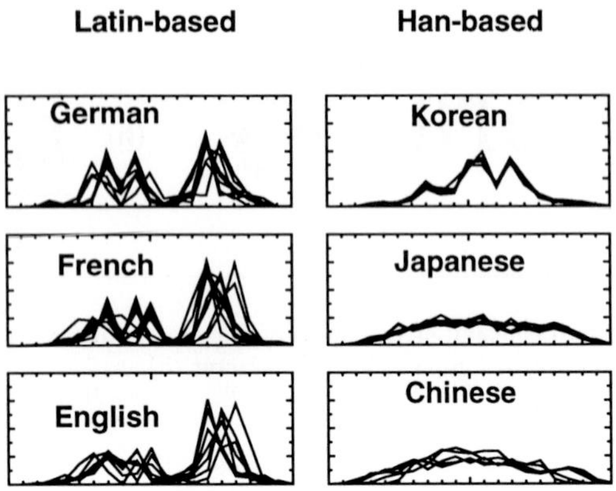

Fig. 8. Distributions of upward concavity vertical distribution with respect to baseline position from multiple documents in each of seven languages.

languages (Kanji, Kana and Hangul scripts). In this figure multiple document samples are shown for each script and language.

Discriminating between the two classes of distribution is relatively easy: we use a simple measure of variance. Latin-based script distributions show greater variance than for Han-based, reflecting the clustering of vertical position of upward concavities. The variance measure is insensitive to absolute offset between the distributions.

5. Language Classification

Han-based and Latin-based languages are classified using different techniques.

5.1. *Han-based Language Classification*

Examples of Japanese, Chinese and Korean text are shown in Fig. 9. Note that the Japanese characters are a mixture of relatively light characters (typically Kana)

原子核の質量は，原子の質量の大部分を占(し)めるので，

中國古時候，有一個人姓馬，名字叫書田。

새계에서 가장 정조관념이 굳다는 것이 중국의

Fig. 9. Text fragments from Japanese, Chinese and Korean documents.

and relatively dense characters (typically Kanji). The Chinese text is composed of predominantly relatively dense Kanji characters. The Korean text is made up of both locally dense and locally light characters.

5.1.1. *Optical density function*

The distribution of perceived optical density of the character cells is calculated. Within each character cell, the number of "on" pixels is counted by summing the lengths of the runs that comprise all of the connected components within the cell. This sum represents the mass of the character, and its value is represented across the entire horizontal span of the character cell. The inter-component and inter-character spaces, and the inter-word spaces found in Hangul, are ignored. The resulting function reflects the reading order distribution of optical density.

Re-examining the Japanese text sample, digitized at 300 pixels per inch, we derive the optical density function for a fragment of that image as shown in Fig. 10. The vertical axis represents the number of "on" pixels in the character cell, while the horizontal axis reflects the distance (not counting spaces) along the assemblage of text lines in units of pixels.

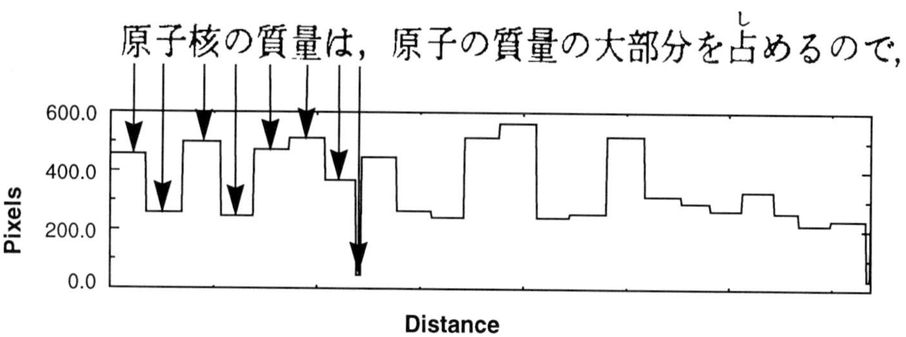

Fig. 10. The optical density function for part of a line of Japanese text.

5.1.2. *Optical density distribution*

Histograms showing the distribution of the density of Japanese, Chinese and Korean documents have characteristically different distributions. Note that the optical density function for Korean documents exhibits a distinct bi-modal nature, with the low density mode smaller than the high density mode. The distributions for the Japanese documents might also be characterized as bi-modal, but in this instance the relative heights of the modes are reversed: the low density mode is greater than the high density one. In the Chinese document there is only one significant mode.

5.1.3. *Classification*

We perform classification using Linear Discriminant Analysis (LDA).

In LDA, the **n**th Discriminant Variable is the linear combination of the original variables that maximizes the separation between groups, subject to the constraint that it must be orthogonal to all the previous variables. LDA examines a training set of data and uses it as a model to build a classification rule. Each document consists of a data vector that is based on the density histogram a and a variable that correctly identifies its language. LDA transforms the data vector to a new coordinate space where the variables have equal variance and are uncorrelated. The mean data vector for each language is then calculated. In general, new documents are identified by converting them to the new coordinate system and assigning them to the language group with the closest mean vector. LDA operates under the assumption that all language groups have roughly equal covariance matrices. The effectiveness of this method will depend on how badly this assumption is violated.

There remains the problem of choosing the right set of variables to represent each document. The obvious choice would be to apply LDA to the density distributions directly. However, this is not ideal, since the distribution contains a large number of highly correlated variables. It is difficult to accurately estimate a large covariance matrix with a training set of limited size. A common solution to this problem is to perform a principal components decomposition on the data matrix and replace a large number of variables by a smaller number of principal components. The first **k** principal components are linear combinations of the original variables that maximize the variance that can be explained in **k** dimensions.

A visual examination of the data suggests a second and probably more efficient approach. Clearly, the density distributions of the Asian languages are characterized by the relative areas in distinct regions of the distribution. A simple

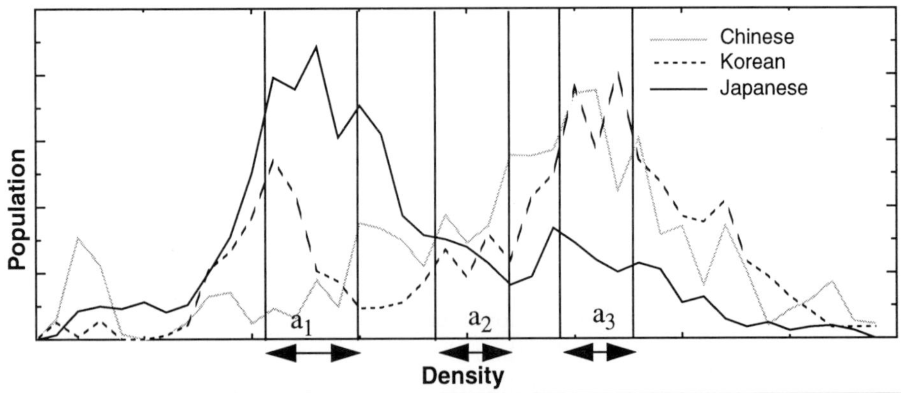

Fig. 11. Specific areas of the histograms showing the distribution of optical density are characteristic of different Han-based languages.

lower-dimensional summary of these distributions can be obtained by using the areas under the curve in three regions: $\mathbf{a}_1$, $\mathbf{a}_2$ and $\mathbf{a}_3$ roughly corresponding to the lower peak in the Korean and Japanese scripts, the null in the Korean script, and the peak in the Chinese and Japanese. We integrate the areas under the curves in these three regions (Fig. 11). This method provides an obvious three-dimensional summary of the data that is ideal for LDA. This method is likely to provide a good summary of the profile in a much lower dimension than principal components, since font differences shift the density distribution within a language group. Principal components is sensitive to this difference while maximum peak height is not.

We apply LDA to the multivariate area data to resolve into multiple language classes. Because we are attempting to select one of three classes, two LDA variables (V_1,V_2) are needed (Fig. 12).

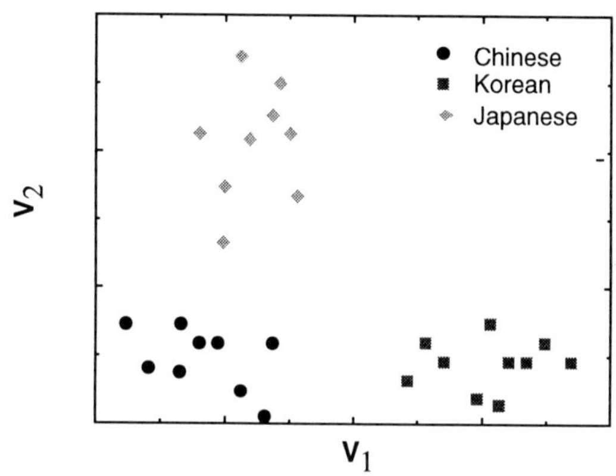

Fig. 12. Clusters are visible in the scatter plot of locations of the test documents in LDA space.

Classification is performed by selecting the closest centroid cluster in LDA space.

5.1.4. *Accuracy*

Within Asian script, separation of Korean from Chinese and Japanese using the technique described in this paper becomes problematic at small granularity, as the optical density distributions incorporate data from smaller and smaller samples.

At the present time we are able to reliably classify the script of a document as being either Han-based or Latin-based on the basis of as little as two lines of text, and we are able to classify the language of a document image containing Han-based script with perfect accuracy on text samples as small as six lines. Table 1 shows the increase in error rate as the number of lines decreases.

Table 1. The error rate of Han-based language classification relates to the number of lines of text processed.

Lines	Samples	Error rate (%)
10	72	0.0
8	95	0.0
6	140	0.0
4	217	0.5
2	447	0.2
1	910	15.8

5.2. *Latin-based Language Classification*

Nakayama & Spitz [13] developed a technique to accurately detect the language of a document (when that language was restricted to be one of English, French or German) based on the frequency of occurrence of particular word shapes. Sibun and Spitz [14] have extended this work to include more languages and to provide a mechanism for the automated selection of the optimal discriminating set of word shapes.

5.2.1. *Typography in this section*

The following examples use the following conventions: `mono-spaced` to represent input characters, **boldface** to represent the six character shape codes (**A**, **x**, **i**, **g**, **j**, **U**), and Sans-serif to represent typographic conventions.

5.2.2. *Character shape codes and word shape tokens*

Characterizations of the number of connected components in a character cell and, in some instances, their aspect ratios, contribute to their coding. Thus most

Table 2. Character shape codes representing individual alphabetic characters.

<table>
<tr><th>Character shape code</th><th>Character</th></tr>
<tr><td>A</td><td>A-Zbdfhkltß0-9#$&()/[]@{}|</td></tr>
<tr><td>x</td><td>acemnorsuvwxz</td></tr>
<tr><td>i</td><td>iáàâéèêîôûùñ</td></tr>
<tr><td>g</td><td>gpqyç</td></tr>
<tr><td>j</td><td>j</td></tr>
<tr><td>U</td><td>äëïöüÖÜ</td></tr>
</table>

characters, including all of the alphabetic characters, can be readily mapped from their positions relative to these baselines and x-heights to a small number of distinct codes (Table 2). Our method maps from the image to a shape-based representation. The basal representation is the *character shape code* (CSC) of which there are a small number. The decision tree defining the geometric relationships that result in the CSCs, including some punctuation, is shown in Fig. 13. These shape codes are aggregated into *word shape tokens* (WSTs) that are delimited by white space or punctuation. An example of the transformation from character codes to CSCs is shown in Fig. 14.

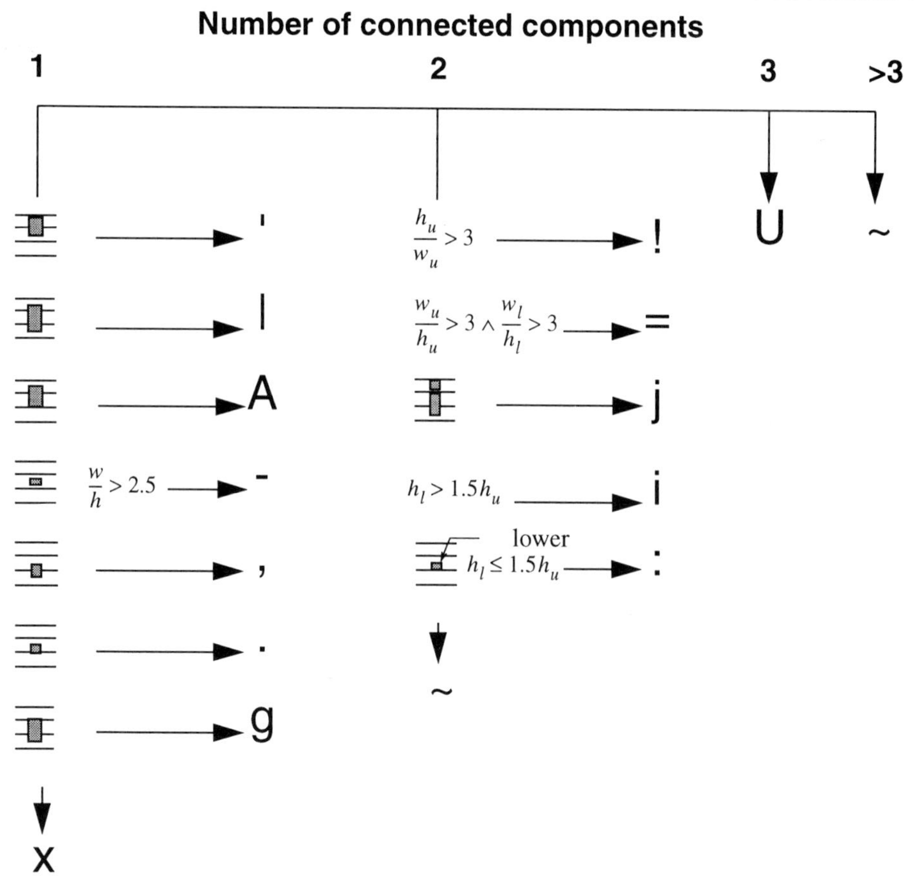

Fig. 13. Decision tree showing the process of defining the character shape code on the basis of the number of connected components, their position relative to the four fiducial lines: bottom, baseline, x-line and top, and morphology of the components. The character shape code $\sim$ represents the inability to classify the contents of a character cell.

Character codes

Confidence in the international monetary system was shaky enough before last week's action.

Character shape codes

AxxAAxxxx ix AAx ixAxxxxAixxxA xxxxAxxg xgxAxx xxx xAxAg xxxxgA Ax-Axxx AxxA xxxA'x xxAixn.

Fig. 14. Character code representation and character shape code representation.

Table 3. Most frequent word shape tokens in English, French and German: the top five for each language are shown; rankings of these are shown for the other languages when they fall in the top ten; shading indicates the characteristic token for each language.

	English		French		German	
WST	Rank	Word	Rank	Word	Rank	Word
AAx	1	the				
xA	2	of	7			
Ax	3	to	1	la le	6	
ix	4	is			10	
xxA	5	and			3	auf
xx	6		2	en	2	an
Axx	9		3	les	1	der das
xxx	8		4	aux	5	wer
gxx			5	pas		
Aix					4	die

5.2.3. *Typesetting effects*

Typesetters use different conventions. For example, in German text ü may be set as ue and ß may be set ss. Therefore, there may be several-to-one mappings of typeset information to CSCs, since ü maps to **U** and **ue** to **xx**. Likewise in French accents may or may not appear on upper case characters. Confusion due to typesetting convention is exacerbated in the character coded environment due to the unfortunate limitations of the ubiquitous ASCII character set and the lack of a single widely accepted international standard for representation of characters with diacritics (e.g., à, ç).

In some fonts for example, diacritics may actually touch the underlying character changing the shape representation of à from **i** to **A**.

Ambiguity results in the case of character strings such as `fi` or `ffl` where because the font and other typesetting parameters are unknown, it is, in general, not knowable whether this should be mapped as a single character or as multiple characters.

5.2.4. *Classification*

It is desirable to know the language of a document before trying to perform *optical character recognition* (OCR). We have found that we can classify the language of a document for a number of Latin-based languages. This technique relies on the high frequency of short words in Latin-based languages and on the diversity of their WST representations.

Our method of language determination is a form of statistical classification: our system learns how to discriminate a set of languages; then, for any input document, the system determines to which language the document belongs.

Our initial set of discriminable languages was English, French, and German. We tokenized these images following the procedure described above. We then ranked the frequency of WSTs across each corpus and noted the most frequent. By comparing these tokens for each of the languages, we were able to select one WST per language. Intuitively, each of these WSTs is characteristic of its language. The characteristic WST for English is **AAx**, for French is **Ax**, and for German is **Aix**.

There were two criteria for the initial selection of characteristic WSTs: relative frequency in one language and relative infrequency in the others. For English, the choice is easy: **AAx** constitutes 7% of the WSTs in the English corpus and is quite rare in the others. In the German corpus, **Aix** is selected even though it is not the most frequent WST because it is rare in English and French. While **Ax** is frequent in all corpora, it is overwhelmingly frequent in French, where it makes up 11% of the WSTs (vs. 4% for English and 2% for German). These differences in the distribution of the characteristic WSTs in the three corpora are sufficient to correctly identify each language almost every time. While it may seem fortuitous that in English **AAx** is virtually always a mapping of `the` or `The`, unique WSTs are more common in Latin-based languages than one might suppose (see Table 4).

We classified the language by applying LDA to the frequencies of occurrence of the characteristic WSTs.[h]

5.2.5. *Automated determination for many languages*

We have constructed a database of 755 one-page documents in 23 languages. They include virtually every European language written in the Roman alphabet and comprise 18 Indo-European languages and two Uralic languages as well as three

[h]In retrospect, the use of LDA in this context seems to have been a mistake, or at least, an example of the use of a sun-optimal tool, since there is nothing about the occurrence of components of language (letters, words, CSCs, etc.) that can be considered to be normally distributed. Pragmatically, however, it works reasonably well.

languages from disparate families: Turkish, Swahili, and Vietnamese. (See Table 4.)

To construct a set of discriminating features, we selected the five most frequent WSTs from each language. Because of overlap, this resulted in 24 WSTs. Some of these discriminating WSTs have a high frequency across languages; in fact, **xx** appears in the top five of 22 of the languages we examined. However, even when we consider 23 languages, there are eight WSTs appearing in the top five of one language that do not appear in the top five of any others. (This does not mean, of course, that these WSTs do not appear in other languages at all, but simply that they are relatively much less frequent.) The 24 WSTs comprise the set: **x**, **xx**, **xxx**, **xxxx**, **i**, **ix**, **xi**, **xix**, **A**, **AAx**, **Ax**, **AxA**, **AxAx**, **Axx**, **Axxx**, **xA**, **xxA**, **Ai**, **Aix**, **g**, **gx**, **xg**, **xxg**, **jx**.

Table 4. Language detection accuracy. The language type IE stands for Indo-European. The abbreviations shown are defined in ISO 639:1988 and are used as indices in Table 5.

Language	Acc (%)	Language type	ISO 639 abbr	Language	Acc (%)	Language type	ISO 639 abbr
Afrikaans	97	IE	af	Italian	95	IE	it
Croatian	100	IE	hr	Norwegian	95	IE	no
Czech/Slovak	44	IE	cs	Polish	100	IE	pl
Danish	96	IE	da	Portuguese	96	IE	pt
Dutch	100	IE	nl	Romanian	93	IE	ro
English	95	IE	en	Spanish	95	IE	es
Finnish	75	Uralic	fi	Swahili	97		sw
French	92	IE	fr	Swedish	98	IE	sv
Gaelic	86	IE	ga	Turkish	93		tr
German	97	IE	de	Vietnamese	100		vi
Hungarian	94	Uralic	hu	Welsh	97	IE	cy
Icelandic	96	IE	is				

5.2.6. *Accuracy*

As in the less comprehensive language sample, we again used LDA to build a statistical model of the language categorizations, and by cross validation we tested the accuracy of the model (see Table 5). Overall accuracy is better than 90%, while the accuracy for individual languages varies between 100% and 75%, with an outlier of 44% for Czech/Slovak. Examination of misclassifications proves somewhat instructive as can be seen in the confusion matrix in Table 5. For example, Dutch and Afrikaans are closely related languages, and the only error in either language is the categorization of one Afrikaans document as Dutch. Among the five Romance languages — French, Italian, Spanish, Portuguese, Romanian — nine of the ten classification errors are within that language family. For the Scandinavian language family — Danish, Norwegian, Swedish, and Icelandic — the pattern is less clear. Two Norwegian documents are classified as Icelandic, but the three other errors in that family are classifications outside of the family.

Table 5. Confusion matrix showing detection accuracy between languages. Numbers on the major diagonal indicate the number of correct classifications for each language. Numbers off the diagonal show classification errors.

	Detected Language																						Error Total
	en	ge	du	af	fr	it	sp	pt	ru	da	no	se	ic	ga	we	cr	cs	po	hu	fi	tu	sa	vi	
en	36	1			1																			2
ge		29												1										1
du			28																					0
af			1	29																				1
fr					23			1						1										2
it						35		1	1															2
sp							39	2																2
pt							1	25																1
ru						3			38															3
da					1					25														1
no											39		2											2
se												40								1				1
ic													23							1				1
ga						2			2				1	31										5
we															30							1		1
cr																33								0
cs									1							6	24	16	3	5				31
po																		28						0
hu																		2	29					2
fi																	6		2	24				8
tu							1													1	28			2
sa																					1	37		1
vi																							13	0
	0	1	1	0	2	5	2	4	4	0	0	0	3	2	0	6	6	18	3	9	1	1	0	69

6. Optical Character Recognition

At the present time, most commercial character recognizers are specifically designed for processing English text. Most are also able to handle a repertoire of other languages set in Roman font, but only on a document-by-document basis. There is also a small number of commercial products available for the processing of Japanese. Among commercial products, the ability to process documents that contain a mixture of Roman-font languages is rare;[i] none can process mixtures of Han-based and Latin-based languages.

Though some Japanese OCRs can process Romaji, the fact that these products are designed to handle mono-spaced text usually becomes obvious in the frequent errors made in proportionally spaced material.

Commercial recognizers, in general, require prior knowledge of the language of the document in order to specify the character set to be used and possibly to invoke a post-recognition spelling checker.[j]

Many researchers have developed character recognizers tuned to specific applications, but multilingual capability has not received much attention. Complexity as represented in the number of classes of symbols to be detected is the enemy of robustness. Morphological differences in character forms can be difficult or impossible to detect, particularly in the presence of printing and scanning artifacts.

A large number of classes is also the enemy of computational efficiency because the escalation of the number of classes results in a need to develop more features and more classifiers to resolve these features in spaces of high dimensionality. A classifier needs (in the theoretical limit) at least $F = C$-1 binary features to resolve C classes. In practice it may need many more. An OCR for English must process 90 or more symbols (alphabetics, numerals, punctuation) as well as font-dependent variants within a single symbolic class, e.g., aɑ, gg. As Baird and Fossey [15] point out, memory requirements are $O(CF)$ and computational time is $O(CF + log\, C)$. As the number of classes approaches the 20,000 needed for traditional Chinese, the computational burden becomes huge.

Perhaps the most flexible OCR yet to be developed, particularly in terms of its multilingual capabilities, is that described by Baird [16]. This system incorporates broad generality in the processing of differing character sets and language models and presents a very low barrier to the addition of new languages. Languages as diverse as English, Greek, Tibetan and Japanese[k] are processed using the same basic architecture.

This work has been extended by Ittner and Baird [18] to be able to process extremely small granularity language fragments such as those found in a cross-lingual dictionary. This is accomplished for a Russian-English dictionary by combining the symbol sets and detecting the locations of language changes by noting the sequence of recognized characters.

An introduction to the challenges in Arabic character recognition is given in [19] and in the chapter by A. Amin in this book.

---

[i]One commercial product can recognize Greek in addition to Roman-font languages.

[j]There is one product where the user can specify the mixture of languages to be found in the document, or specify "international" to make the character set all-inclusive. This product does no post-recognition context analysis.

[k]In a remarkable demonstration of the flexibility of this system, it has been used to recognize chess symbols while incorporating knowledge of the semantics of the game as a language model [17].

6.1. *Language Models Applied to OCR*

There are several types of language models which are used to improve the performance of OCR either by increasing the speed or accuracy of processing.

Character n-gram frequencies can be used to optimize the classification process by attempting to detect the most likely characters first. This process usually starts with unigrams (letter frequencies) and may be extended to bigrams or trigrams as individual characters are resolved. Applications of this form of language model is well-known in Latin-based languages [20,21] and has been successfully used to enhance the speed of a Japanese OCR [22]. In some instances the process is applied in reverse order, re-recognizing a previously processed character on the basis of the classification(s) of those that follow.

In the post-processing phase, conformance of the output of an OCR to a language model, usually specified by a highly inflected "spelling dictionary" is a well-developed technology for Latin-based languages. Word n-grams are also used, but most commonly in the case of $n = 1$, or word frequency. Most documents comprise a relatively few distinct words which are frequently repeated.

Because word segmentation in Chinese and Japanese is difficult, bi-gram language modeling has been successfully used to increase OCR accuracy in the absence of word boundary information [23,24].

Linguistic post-processing is discussed in greater detail in the chapter by A. Dengel *et al.* in this book.

6.2. *Character Codes*

The basic output of character recognition is character coded text streams. Since document recognition is a tool for transformation of data into a machine-processible domain, the choice of character code standards is influenced by the compatibility of components with which document recognition is integrated to form a useful system. ASCII is an adequate representation of the alphabetic characters found in English[l] but the extension of ASCII to European and other languages which basically use the Roman character set has introduced competing and incompatible coding standards.

Even the extension of ASCII to ISO Latin-1 (ISO-8859-1) fails to cover Polish, Czech, Hungarian and many other European languages along with Turkish and other important Latin-based scripts, leading to a plethora of ISO-8859-n Latin code pages.

Companies concerned with the problem of multilingual character sets have formed the Unicode consortium which has promulgated a character encoding standard designed to represent characters independent of language [25].

The situation with Chinese characters is similarly fragmented and exacerbated by the large number of characters and the minor differences which have crept in to their renderings, both over time and in the various languages in which they are

[l]Though ASCII does not have a representation of the dieresis in "coöperate", for example.

used. For example a single Chinese character is represented in different character code sets as shown in Table 6.

Table 6. Different character code standards for a single Chinese character (adapted from [26]).

<table>
<tr><th></th><th>Language</th><th>Standard</th><th>Code</th></tr>
<tr><td rowspan="5">生</td><td>Japanese</td><td>JIS X0208-1983</td><td>3224</td></tr>
<tr><td>Simplified Chinese</td><td>GB 2312-80</td><td>4190</td></tr>
<tr><td>Traditional Chinese</td><td>CNS 11643-86</td><td>A5CD</td></tr>
<tr><td>Korean</td><td>KS C5601-1987</td><td>6370</td></tr>
<tr><td>All</td><td>Unicode</td><td>751F</td></tr>
</table>

7. Conclusions

There is potential for increasing the number of languages classified by the techniques described here. The original work on Roman script language classification was limited to English, French and German. It has now been extended to more than 20 languages. The process of adding languages has been automated making it easy to add a new language once the document images constituting a training set have been scanned into a database.

Extension of the techniques applied to Latin-based script is likely to be productive, perhaps with minor modification, if also applied to Cyrillic, Greek or Hebrew scripts.

Extension of the Asian language capability is also feasible, again possibly with minor modification, to be able to classify languages such as Cambodian.

Development of techniques to recognize highly connected languages has not been initiated. Handling of Arabic, for example, depends not only on handling connectedness, but also on coping with the horizontal elasticity, called keshide, found in Arabic print.

Multilingual document recognition is a new technology. However there is increasing interest in the design of systems that can handle documents in more than a single language. Driven by the demands of increasing world trade, we can expect that multilingual capabilities will become more common in both the academic and the commercial worlds.

Acknowledgments

Several members of the research staff of the Fuji Xerox Palo Alto Laboratory contributed to this work. In particular, the author would like to thank Arlene Holloway, David Hull, Takehiro Nakayama, and Penelope Sibun for their contributions to the Latin-based language identification capability as well as Jean-Marie de La Beaujardière and Masaharu Ozaki for the design and implementation of the demonstration system which incorporates this technology.

Joseph Becker, Horst Bunke, Penelope Sibun and Karl Tombre have made many useful suggestions about the form and content of this chapter.

References

[1] A. L. Spitz and M. Ozaki, Palace: A multilingual document recognition system, in *Document Analysis Systems*, eds. A. L. Spitz and A. Dengel (World Scientific, Singapore, 1995) 16–37.

[2] Z. Wu and G. Tseng, ACTS: An automatic Chinese text segmentation system for full text retrieval, *J. Amer. Soc. Info. Sci.* **46** (1995) 83–96.

[3] A. Nakanishi, *Writing Systems of the World* (Tuttle, Rutland, Vermont, 1980).

[4] L. O'Gorman and R. Kasturi, *Document Image Analysis* (IEEE Computer Society Press, Los Alamitos, California, 1995).

[5] H. S. Baird, The skew angle of printed documents, *Proc. Conf. of the Soc. of Photographic Scientists and Engineers*, Rochester, 1987, 21–24.

[6] C. Ronse and P. A. Devijver, *Connected Components in Binary Images: The Detection Problem* (Research Studies Press, Letchworth, 1984).

[7] J. Kanai, Text line extraction using character prototypes, *Proc. IAPR Workshop on Syntactic and Structural Pattern Recognition*, Murray Hill, NJ, Jun. 1990, 182–191.

[8] D. J. Ittner, Automatic inference of textline orientation, *Proc. Symp. on Document Analysis and Information Retrieval*, Las Vegas, Apr. 1992, 123–133.

[9] H. S. Baird and D. J. Ittner, Language-free layout analysis, *Proc. 2nd Int. Conf. on Document Analysis and Recognition*, Tsukuba, Japan, Oct. 1993, 336–340.

[10] A. L. Spitz, Generalized line, word and character finding, in *Progress in Image Analysis and Processing III*, ed. S. Impedovo (World Scientific, Singapore, 1993) 377–383.

[11] A. L. Spitz, Text line characterization by connected component transformations, *Proc. SPIE*, San Jose, Feb. 1994, 97–105.

[12] A. L. Spitz, Script and language determination from document images, *Proc. Symp. on Document Analysis and Information Retrieval*, Las Vegas, Apr. 1994, 229–235.

[13] T. Nakayama and A. L. Spitz, European language determination from image, *Proc. Int. Conf. on Document Analysis and Recognition*, Tsukuba, Japan, Oct. 1993, 159–162.

[14] P. Sibun and A. L. Spitz, Language determination: Natural language processing from scanned document images, *Proc. Applied Natural Language Processing*, Stuttgart, Oct. 1994, 15–21.

[15] H. S. Baird and R. Fossey, A 100-font classifier, *Proc. First Int. Conf. on Document Analysis and Recognition*, St. Malo, France, Oct. 1991, 332–340.

[16] H. S. Baird, Anatomy of a versatile page reader, *Proc. of the IEEE, Special Issue on OCR* **80** (1992) 1059–1065.

[17] H. S. Baird and K. Thompson, Reading chess, *IEEE Trans. on Pattern Analysis and Machine Intelligence* **12** (1990) 552–559.

[18] D. J. Ittner and H. S. Baird, Programmable contextual analysis, in *Document Analysis Systems*, eds. A. L. Spitz and A. Dengel (World Scientific, Singapore, 1995) 76–92.

[19] H. Goraine, M. Usher and S Al-Emani, Off-line Arabic character recognition, *IEEE Computer* **25** (1992) 71–74.

[20] R. G. Casey and G. Nagy, An autonomous reading machine, *IEEE Trans. on Computers* **C-17** (1968) 492–503.

[21] R. M. K. Sinha and B. Prasada, Visual text recognition through contextual processing, *Pattern Recognition* **21** (1988) 463–479.

[22] K. Kigo, Improving the speed of Japanese OCR through linguistic processing, *Proc. 2nd Int. Conf. on Document Analysis and Recognition*, Tsukuba, Japan, Oct. 1993, 214–217.

[23] K. Mori and I. Masuda, Advances in recognition of Chinese characters, *Proc. 5th Int. Conf. on Pattern Recognition*, 1980, 692–702.

[24] T. Kawada, S. Amano and K. Sakai, Linguistic error correction of Japanese sentences, *Proc. COLING*, Tokyo, Sep.–Oct. 1980, 257–261.

[25] The Unicode Consortium, *The Unicode Standard: Worldwide Character Encoding* (Addison-Wesley, Menlo Park, 1991).

[26] N. Kano and A. Freytag, The international character set conundrum: ANSI, Unicode, and Microsoft Windows, *Microsoft Systems Journal*, Nov. 1994, 55–70.

ORIENTAL CHARACTER RECOGNITION

Handbook of Character Recognition and Document Image Analysis, pp. 285–304
Eds. H. Bunke and P. S. P. Wang

CHAPTER 10

AN OCR-ORIENTED OVERVIEW OF IDEOGRAPHIC WRITING SYSTEMS

J. KANAI and Y. LIU
Information Science Research Institute, University of Nevada, Las Vegas
Las Vegas, NV 89154, USA

and

G. NAGY
Electrical, Computer, and Systems Engineering Department, Rensselaer Polytechnic Institute
Troy, NY 12180, USA

Chinese ideographs evolved from pictures several thousand years ago and, unlike western alphabets, are still undergoing change. The large size and open-ended nature of the character set, the geometric complexity of some characters, the stroke-oriented structure, and the elusive relationship between symbol, sound, and shape, all have important implications not only for OCR and DIA, but also for other aspects of computer input/output and for text processing. After reviewing the more common character encoding methods, we examine Chinese typographic practices: orientation, reading order, punctuation, spacing, typefaces, and stylistic variants. We survey briefly alternative input and output methods and discuss their relevance to OCR and DIA. Ideographs (*Kanji*) also form an important part of the Japanese writing system which, however, includes several alphabetic (*Kana*) components. We proceed with a parallel treatment for Japanese symbols, layout, encoding, and typography, and point out similarities and differences with respect to both Chinese and Western writing systems.

Keywords: Chinese; Japanese; Encoding methods; Typography; Input methods; Output methods; Writing systems.

1. Introduction

Over one-half of the world's literate population uses an ideographic writing system. This part of the world is now entering the Information Age, bypassing intermediate stages like analog communications networks and large central computing centers. The pace of the technological revolution in Asia is fostered by rapid economic expansion and strong government support for high-tech ventures. Packet-switched computer networks are likely to become the most robust form of communication, equally resistant to natural and political disasters.

The Chinese writing system influenced many Asian countries and their cultures. *Hanzi* characters (Chinese ideographs) evolved from pictographs about 4,000 years ago. By definition, "an ideograph is a graphic symbol representing the idea of a thing without expressing its name" (The New Century Dictionary, 1938). However, some Chinese characters can be considered phonograms, which represent concepts that may be difficult to draw but sound similar, or determinatives, which facilitate certain distinctions. One study identified over 74,000 characters [1], and new characters can be invented. The evolution of this writing system is one of the most interesting chapters in the entire history of civilization.

Most of the languages related to Chinese now use a phonetic alphabet. As we shall see, reform movements have been historically an integral part of the evolution of ideographic writing systems, just as of alphabetic systems. (Examples of the latter are orthographic simplification in English, and the abandonment of Gothic characters in German.) The Japanese imported the Chinese writing system in the third and fourth centuries through Korean scribes. The Chinese *Hanzi* is called *Kanji* in Japan and *Hanja* in Korea. However, Chinese is a monosyllabic, uninflected, tonal language, which distinguishes from the polysyllabic, agglutinative, non-tonal Japanese and Korean languages. Not surprisingly, the writing system that developed in China did not prove entirely adequate elsewhere. Japan developed early in the eighth century two phonetic syllabaries (*Kana*) to help with inflections, but retained over 6,000 *Kanji*.

Korea devised a phonetic alphabet, called *hangul, hankul* or *onmun* (vernacular), in the fifteenth century, and retained over 1,800 Chinese characters for optional use. Hangul characters consist of a square cluster of two to five Hangul phonemic glyphs that represent a syllable. The glyphs correspond to the 24 basic letters of the modern Korean alphabet.

The Chinese writing system was also used during various periods in the regions now identified as Burma, Cambodia, Thailand, Vietnam, and Tibet, which have all abandoned it. Nevertheless, Chinese writing remained a matter of prestige and is still an integral part of the highly differentiated cultures of these countries. Being educated implies familiarity with early chronicles and literary works recorded in Chinese characters. Because ideographic (Chinese, Japanese) and cursive (Arabic, Bengali) writing systems do not readily lend themselves to key entry, DIA and OCR may assume greater importance in the East than in the West. Typing Chinese characters is about twelve times slower than typing an alphabetic language. Even typesetting is only half as fast — on a Linotype — as in another language [2]. Many research centers and universities have mounted large-scale efforts to develop workable systems for ideographic I/O.

From the point of view of DIA and OCR, and particularly with regard to reading order and the number and complexity of the shape of the symbols, Chinese and Japanese are very different from languages which use a phonetic alphabet, like English and Russian. (The intricate rules of character formation place Korean script in an intermediate position.) To give some idea of the number and complexity of the distinct components, Fig. 1 shows an instruction for filling out a U.S. custom form in English, Chinese, Japanese and Korean [3]. In Table 1, we offer some simple statistics.

Before arriving in the United States of America, each traveler or head of family is required to fill out a Customs Declaration Form.

Most of the questions can be answered with a "yes" or "no." The reverse side of this form must be signed and dated.

Please print legibly, using black or blue ink. Entries must be in ENGLISH and in ALL CAPITAL LETTERS.

The Customs Declaration form will be distributed during the flight.

(a)

在抵达美国前，每位旅客都需填写一张海
关申请表，但是每个家庭只需填写一张 。
报关表上多数问题只需用“ 不是 ”或“ 是 ”
即可回答清楚，请务必在背面签名并填上
申请日期 。
请你用蓝色或黑色墨水，以英文大写的正
楷仔细填写 。报关表会由空服员在飞机降
落前发给每一位乘客 。

(b)

米国の目的地へ到着する前に 、お客様には税関申告用紙への
ご記入をお願いしております 。用紙は着陸前にフライトアテ
ンダントがお配りいたします 。質問は 、ほとんどYES/NOで答
えられる簡単なものです 。裏面のサインと日付もお忘れのな
いようお願いいたします 。記入の際には 、黒又は青のインク
で 、全て英語の大文字で 、はっきりとご記入ください 。

(c)

미국도착전에, 각 여행자나, 가족의 대표자는 세관 신고서를 기재하여야 합니다. 대부분의
질문사항은, "예" 혹은 "아니오"로 대답하게 리어 있습니다. 이양식의 뒷면에 서명하시고
날짜를 기재 하십시오. 읽기 쉽게 검은색이나 파란색 펜을 사용하여 영문 대문자로 기재
하십시오. 세관신고서는 비행기내에서 나누 어 드립니다.

(d)

Fig. 1. An instruction for filling out a U.S. Customs Declaration Form in four languages: (a) English, (b) Chinese, (c) Japanese, and (d) Korean.

Table 1. Statistics obtained from text images in Fig. 1

	English	Chinese	Japanese	Korean
Total number of symbols (excluding space)	346	112	163	139
Number of unique symbols	44	93	92	78

The most commonly asked question about Chinese is: How many different characters are there? This number is no better defined than the number of distinct words in English or French. Nor is it important, with regard to the development of information processing technology, to be able to identify every grapheme. In the sequel we shall, however, discuss several well-defined sets of characters which have had digital codes assigned to them in a consistent manner. Although none of these sets have gained universal acceptance, collectively they represent almost all the characters in common use that are the target of efforts at automated transcription. Nevertheless, we must allow that over 74,000 different *Hanzi* have been enumerated [4]. Most are very seldom used (idle) characters: 33,357 are considered sufficient for almost every use [1]. Documents in China and Japan often incorporate roman characters and arabic numerals, while the converse is certainly not the case (yet). In fact, standard Chinese and Japanese character codes routinely accommodate ASCII symbols. Therefore Chinese and Japanese OCR generally include at least English OCR as a subcase.

Current DIA and OCR developments in China, Taiwan, Japan and Korea are proceeding on separate tracks, but cross-fertilization occurs at international conferences and through English-language journals. Research targeted at Chinese — and, to a lesser extent, Japanese — documents also thrives in Europe and America. Perhaps wide-scale adoption of Unicode [5] [6], which has provisions for encoding the characters of all written languages, will help to further unify this technical community and accelerate progress.

The remainder of this article consists of two parallel streams, one for Chinese and one for Japanese. We give a brief history of the evolution of each writing system and some relevant usage statistics, list current and proposed character encoding methods, describe typographic considerations that are specific to these languages, and discuss relevant aspects of ideographic computer input and output technologies. By way of conclusion we speculate on the nature of future progress in this lively and exciting domain of research.

2. The Evolution of Chinese Writing

Spoken Chinese may be divided into five major groups: *Mandarin,* and the *Wu, Fukienese, Cantonese,* and *Hakka* dialects. Each group has four to eight major subdivisions. *Mandarin* is a dialect spoken around Beijing, and has been designated as the "National Language" in both China and Taiwan. The difference between *Mandarin* and one of the south-eastern dialects, say the Cantonese spoken in Hong Kong, is about as great as between English and Dutch. The difference between two subtypes of *Mandarin*, say that of Mukden and that of Kunming, can be compared to that between the English of Chicago

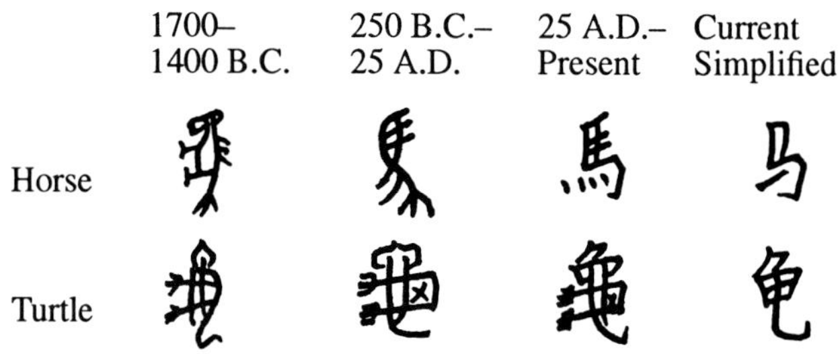

Fig. 2. Evolution of Chinese ideograph.

and the Cockney English of London [7]. There is a common language only in the sense of the written language.

The earliest known Chinese symbols are the Shell and Bone, which were engraved on ox bones and tortoise shells dated back to 1,400–1,200 BC and also found inscribed on both stone and bamboo. The form used until recently, known as "regular style", came into general use about 400 AD. In between, scholars recognize six or eight stylistic periods (e.g., Great Seal, Small Seal, Clerical, Grass, and True Style) — see Fig. 2. A wave of simplification was started by the Chinese Written Language Reform Committee in 1955, and resulted in the set of characters in current use in the People's Republic of China. The major reason for simplification was to accelerate both learning Chinese, and writing Chinese. It was guided by fourteen principles, and resulted in 515 simplified characters [2]. The simplified characters are now also used in Singapore and in Hong Kong.

Such simplification has been endemic throughout Chinese history: for instance, circles and complicated curves were gradually replaced by straight lines, parts of characters were eliminated, and common forms replaced archaic ones. Some of the simplified characters were already in use 2,000 years ago. At the same time, new inventions, concepts, and foreign terms necessitated the formation of new symbols through the embellishment or refinement of old ones. A rough equilibrium was maintained through periods alternating between reform and conservation. For instance, during the Ming dynasty (1368–1644), the classical examinations that were a prerequisite to success in the Chinese imperial civil service put a premium on complete mastery of the most complicated forms.

Until the modern period, the choice writing instrument was the brush. In Chinese poems, the shapes and meanings of the individual ideographs combine with their sequence and overall disposition for the desired poetic effect. Calligraphy is the art that blends painting and poetry

> A calligraphic piece must look lively and show both strength and fluidity. Saying that "the work penetrates the paper" is a compliment. To show strength, the hand must follow the writing brush, the wrist exerts substantial force, and the breath is synchronized with the rhythm of writing. To achieve fluency, the elbow (for large characters) or the wrist (for small characters) hangs loosely. Hooks are traversed quickly, while the brush halts for a pretty corner [8].

Table 2. Stroke-count distribution (total of 2,000 symbols).

Number of strokes	Number of characters
1-5	181
6-8	382
9-12	737
13-16	479
>16	221

Characters are always considered constructed of distinct strokes. Table 2 shows the stroke distribution of 2,000 of the most common traditional characters [2]. It is seen that 72% of the common characters have nine or more strokes. Less common characters will have an even higher stroke count.

The strokes in a given character are traced in a specified direction and sequence. The strokes can be classified into five groups: (1) dot, (2) horizontal, (3) vertical, (4) horizontal upward, and (5) downward drag as shown in Fig. 3. However, the position, length and direction of the strokes depends on their graphic context, and so one must recognize a character in order to classify its strokes according to this model. One attempt to classify the basic strokes into 33 complex stroke forms is set forth in [1]. The distinctive variation in stroke-width in some typefaces derives from the brush, which was rotated to vary its trace.

Fig. 3. Five basic strokes.

Certain stroke configurations are called radicals (or roots). Conventionally 214 radicals are recognized. The majority of the characters are so-called phonetic compounds that consist of a *phonetic* and a *signific* part: the radical is usually the latter. The radical may be horizontally or vertically adjacent to the rest of the character, surround it entirely, or interpenetrate it. The horizontal and vertical radical structures account for over 90% of all characters, but are less common in the most frequently used characters.

Radicals, like individual strokes, undergo substantial changes in appearance depending on the rest of the character. Some radicals (as well as many characters), have *variant* forms that bear no resemblance in shape to the main (*orthographic*) form but may occur more frequently — see Fig. 4. In certain characters, the assignment of the radical is purely arbitrary: several choices may be equally plausible on geometric grounds. Some compound characters, that look as though they contained a radical, are themselves radicals. The distribution of radicals, like every other distribution connected with natural language, is quite skewed: about 20 account for over half of the common characters.

一 壹 弌

Fig. 4. Orthographic and variant forms of the Chinese numeral 1.

The correspondence between characters and their meaning is many-to-many, and so is the mapping between shape and sound. Thus a given character may have several meanings, or several characters may have the same meaning. Different characters may sound alike (like homonyms). In fact, disregarding tones, there are only a few hundred possible different pronunciations of a syllable. On the other hand, polyphonetic characters have several different pronunciations determined by context.

An essential feature of spoken Chinese is the tonal quality. Tonal variations between relative pitch ranges encode meaning (the absolute pitch depends, of course, on the speaker). Recognized tones are (1) even (high and unwavering), (2) rising (questioning), (3) dipping (falling and rising), (4) falling (emphasis), and (5) light (untoned and unaccented). Each syllable is divided into an initial and a final segment. A single tonal quality, or movement of the pitch through time, applies to the entire syllable. Tones are not considered separate from the pronunciation, and words that differ only by tone are usually totally unrelated. Furthermore, the tones of syllables spoken in succession differ from the tones of the same syllables spoken in isolation.

Several systems, including diacritical marks, have been developed to represent tonal qualities. These markings depend, of course, on the dialect, but the standard system targets *Mandarin*. The Pin Yin phonetic symbols based on the Wade-Giles system are used on the Mainland, and the Zhu Yin (also spelled Ju Yin) system in Taiwan [1]. Reading these phonetic representations is currently not a major concern for OCR, but typing phonetic codes is a common method of data entry.

As already mentioned, a major difference between Chinese and western writing systems is the open-ended dynamic nature of the former. Examples of recently introduced simplified characters, and of recently created characters, are shown in Fig. 5.

(a) (b) (c) (d)

Fig. 5. Orthographic and new variants: (a) traditional, (b) simplified, (c) variant 1, and (d) variant 2.

Another important aspect is the need to accommodate foreign (primarily roman) scripts. Arabic numerals are also routinely used in scientific and commercial documents, but for small numbers, the Chinese characters are preferred.

Traditionally Chinese was written top to bottom, right to left, but within the last century horizontal lines read from left to right have been gaining ground.

Since the computer codification of Chinese characters is relatively new, character usage statistics are hard to come by. About 5,000 characters account for over 99% of the usage, and 2,000 suffice for 97% [1].

3. The Evolution of Japanese Writing

It is believed that the Japanese imported the Chinese writing system in the third and fourth centuries AD. Most *Kanji* (Chinese characters) have two kinds of pronunciation: the *on* reading is based on Chinese pronunciations and the *kun* reading is based on reading the character in native Japanese words.

Since Chinese characters could not adequately describe the highly inflected Japanese language, by the 8th century AD, the Japanese were using *Kanji* as phonetic symbols [9]. Each syllable was represented by one Chinese character. In the 9th century, two native phonetic syllabaries (*Kana*), *Hiragana* and *Katakana*, were developed by abbreviating these characters. Although the native phonetic syllabaries were developed, the Japanese retained *Kanji* in the writing system.

In Japanese, there are five vowels, which are Romanized as a, i, u, e, o, and 19 consonants [10]. These consonants are subdivided into:

- Seion: k, s, t, n, h, m, y, r, and w.
- Dakuon: g, z, d, and b.
- Han-dakuon: p.
- Others: sh, ch, f, and ts.

Kana consists of 48 basic characters which correspond to vowels and *seion*. Figure 6 shows a 5 by 10 matrix representation of *Kana* called the Fifty-sound Table. For example, the sound of the top character in the "h" column is *ha*. Adding a *dakuten* diacritical mark (") to the *ha* character, shown as the top character in the "b" column, changes the sound to *ba*. Similarly, adding a *han-dakuten* diacritical mark (°) to the charcter *ha*, shown as the top character in the last column, changes the sound to *pa*. Several *Kana* have smaller versions which are combined with other *Kana* to write the remaining sounds.

	Seion k s t n h m y r w	Dakuon g z d b	Han-dakuon p
a	あかさたなはまやらわ	がざだば	ぱ
i	いきしちにひみ　りゐ	きじぢび	ぴ
u	うくすつぬふむゆる	ぐずづぶ	ぷ
e	えけせてねへめ　れゑ	げぜでべ	ぺ
o	おけそとのほもよろを	ごぞどぼ	ぽ
n	ん		

Fig. 6. Fifty-sound table.

Hiragana are used to write particles, copula, the inflected endings of verbs and adjectives, and some nouns. Formerly, *Katakana* and *Kanji* were used in official documents and other publications. Nowadays, *Katakana* are mainly used to write words of foreign origin.

The Japanese not only developed *kana* but also generated new *Kanji*, which are identified as *kokuji*. Many of them were borrowed by the Chinese [11].

The large number of *Kanji*, with their complicated shapes and multiple pronunciations, were a great burden to the Japanese. To deal with these problems, the system of *furigana* (also called *Ruby* characters) was developed. Small versions of *Kana* were placed along the right-hand side of *Kanji* characters in the vertical-set or above *Kanji* in horizontal-set text to show their correct pronunciation. Before the end of World War II, virtually all *Kanji* in newspapers and popular magazines and books had furigana [9]. Nowadays, furigana is less frequently used.

Simplification of the Japanese writing system was proposed as early as the *Edo period* (1603–1868). In 1866, the exclusive use of *Kana* was proposed by the scholar Maejima Raisuke [12]. During the *Meiji Period* (1868–1912), the exclusive use of the Roman alphabet was also proposed. Several ways to Romanize Japanese were defined, such as *Hyojun-shiki, Hebon-shiki*, and *Nihon-shiki*. However, these methods did not catch on.

To restrict the number of *Kanji* used in writing, the government published a list of 1850 *Kanji* called *Toyo-Kanji* for general use in 1946. The government also simplified 600 in the set. The traditional shapes are called *Kyu-jitai,* and the simplified shapes are called *Shin-jitai.*

Although the pronunciation of many words had changed since around 1000 AD, their *Kana* spellings were to remain the same until 1946. In 1947, the new *Kana* spelling (*Shin-Kana-Zukai*), which is based on the modern pronunciation, was defined and gradually replaced the historical *Kana* spelling. Because of the historical *Kana* spelling and the traditional *Kanji* shapes, the recognition of documents printed before 1946 introduces additional challenges.

In 1981, the *Toyo-Kanji* was replaced by a new set called *Joyo-Kanji* consisting of 1,945 characters. Of this set, only 1,006 characters called *Gakushu Kanji* are formally taught in Japanese elementary schools. In addition, the government also designated a set of characters called *Jinmei-yo Kanji* for writing personal names. Currently, this set consists of 266 characters.

Modern Japanese text typically consists of 30% *Kanji*, 60% *Hiragana*, 10% *Katakana* [11]. No space is used to separate words in Japanese writing, hence word extraction needs syntactic and lexical analysis [13]. Because of multiple pronunciations, the correct pronunciation of a proper name may be difficult without "furigana". The spelling of a foreign word or name in *Katakana* is often not unique. The same problem exists in English. For example, a Russian composer Tchaikovsky also appeared as Tchaikovskyan and Tchaikovskian (also Tschaikovsky) [13]. This problem affects lexical analysis and information retrieval of Japanese text.

4. Encoding of Chinese Characters

The output of all OCR systems is a code string that represents the sequence of symbols on a page and, optionally, the typographic and layout information. At one time, several different six-bit, seven-bit, and eight-bit character encoding methods, some using zone-bits derived from the Hollerith card code, were used in the United States to represent the letters of the alphabet. The seven-bit ASCII code for 94 upper and lower case letters, numerals, punctuation and a few other common graphic symbols, and 34 control characters, is now in universal use, but mathematicians, chemists, accountants and chess players need many additional symbols. Even now, the encoding of many of these special symbols is not standardized. This lack of standardization has proved a handicap to OCR, both with respect to generating word-processor compatible OCR output, and to benchmarking OCR devices by comparing their output with labeled test data.

Similar difficulties, but on a larger scale, hamper Chinese OCR. Even for the same subset of characters, the collating order varies among the different coding systems, precluding systematic comparison and sorting [14]. Gradually, however, two dominant encoding systems have emerged: the GuoBiao (GB) code in Mainland China, and the Big-5 code in Taiwan and Hong Kong. Several releases of each of these codes that contain different character sets are in current use. In addition, each code can be modified for either internal (eight-bit) or external (seven-bit) use.

The correspondence between characters and code words is defined by a two or three-dimensional integer coordinate system. In the two-dimensional (matrix) encoding, row and column coordinates are used. In the three-dimensional system, each matrix is considered a plane. The mapping between the characters and the code points is implementation independent, but the actual codes assigned to each coding point depend on the specific implementation.

The current version of the GB code ("Code of Chinese Graphic Character Set for Information Interchange —Primary Set") is GB2311-80. It contains 3,755 Level 1 (common) and 3008 Level 2 simplified Chinese characters, 682 graphic symbols and alphanumeric characters (including upper and lower case roman, cyrillic, greek, and Japanese *Hiragana* and *Katakana*), and 1291 reserved or user-defined positions. The Level 1 characters are arranged by pronunciation (*Pin Yin*), and the Level 2 characters by radical and stroke count. There is a simple formula for translating any position in the 94×94 GB matrix into a two-byte code. The high-order bit of each byte is used to differentiate the ASCII subset, and certain cells are left blank to avoid conflict with the ASCII control characters. A new GB standard, GB-13000, based on the Han portion of the ISO 10646 Unicode, was announced in December 1993.

Big-5 is the most popular encoding method for the traditional characters normally used in Taiwan and Hong Kong. The character set contains 5,401 Level 1 and 7,652 Level 2 *Hanzi*, 470 roman characters and miscellaneous symbols, and 942 user-defined symbols, arranged in a 94×157 matrix. Different software vendors use somewhat different versions of the table for both seven-bit and eight-bit codes.

There are also two less widely-used popular coding methods in Taiwan and Southeast Asia. One, the Standard Interchange Code for Generally Used Chinese Characters (CNS), was established as the national standard in Taiwan in 1986. It is based on the ISO 2022 character set. The current version, CNS 11643-1986, contains 13,051 *Hanzi*, on two 94 × 94 planes. The characters are arranged by stroke count and radical. The seven-bit version of the codes uses an escape sequence to distinguish planes. A recently announced expansion, CNS 11643-1992, has more than 60,000 characters on sixteen planes. It includes rarely used and variant characters.

The other code, the Chinese Character Code for Information Interchange (CCCII), was developed by the Chinese Character Analysis Group in Taiwan. It was designed to accommodate the needs of all Chinese, Japanese, and Korean information interchange. The variant forms of each character are in the same row and column positions, but in a different plane. Successive versions, beginning in 1980, contained 4,807, 33,544, and 53,940 characters, respectively. CCCII is based on the ISO 646 seven-bit communication coding standard, and uses a 94 × 94 × 94 coding space for three bytes. Some subsets of the code are available in two-byte versions.

There is a fair chance that the 16-bit ISO 10466 codebook, developed jointly by the International Standards Organization and the Unicode Consortium, will eventually supersede all these codes. Unicode supports a set of about 21,000 Chinese (both traditional and simplified), Japanese, and Korean (CJK) characters, in addition to most of the world's alphabets.

It is clear that the size of the character sets to which codes have already been assigned far exceeds the size of the sets that can be currently recognized by any OCR system. Because software for interconversion among the various systems already exists, the existence of test data in different formats is less of a handicap to OCR development than the incompatibility of many Chinese computing environments. Nevertheless, as seen elsewhere in this volume, rapid progress is taking place on all fronts.

5. Encoding of Japanese Characters

In general, words in dictionaries and names in directories are first arranged by their pronunciations according to the ordering of *Kana* in the "50 sound table". Words with the same pronunciation are then ordered by the stroke counts of the corresponding *Kanji*. *Kanji* in *Kanji* dictionaries are organized by their radicals then the number of strokes.

There are two national character set standards containing *Kanji*: JIS X 0208-1990 and JIS X 0212-1990. JIS X 0208-1990 evolved from predecessors, JIS C 6226-1978 and JIS X 0208-1983, by adding a few characters. This set contains 6,355 *Kanji* and 524 non-*Kanji*. These *Kanji* are divided into the most common 2,965 characters as the JIS Level 1 *Kanji* and the less common 3,390 characters as the JIS Level 2 *Kanji*. The characters in the JIS Level 1 *Kanji* are ordered according to their pronunciations. On the other hand, the characters in the JIS Level 2 *Kanji* are ordered according to their radicals then stroke counts. JIS X 0212-1990 contains 6,067 supplemental characters. The 5,801 *Kanji* in the set are organized by their radicals then stroke counts. Other character sets defined by individual companies are known as *Japanese corporate character set standards*. Some of these sets contain characters that are not in JIS X 0208-1990.

In Japanese character standards, the characters are indexed by the *kuten* method that is independent of any encoding methods. Characters are arranged in a 94-row (or *ku* in Japanese) by 94-cell (or *ten* in Japanese) matrix. Each character is assigned a decimal four-digit code; the first two digits correspond to the row (*ku*) position and the second two digits correspond to the cell (*ten*) position.

The following three encoding schemes are widely used: JIS, Shift-JIS, and EUC (Extended UNIX Code). EUC is also known as *UNIXized JIS* (UJIS). The JIS encoding scheme uses 7-bit codes. Escape sequences are used to switch between one-byte characters, such as ASCII/JIS-Roman characters, and two-byte characters, such as *Kana* and *Kanji*. This method is primarily used for exchanging information electronically.

The shift-JIS (also called SJIS) was developed by Microsoft Corporation. This method also uses both one-byte codes and two-byte codes. Since the first byte of each two-byte character is outside of the range of one-byte characters, it does not require any escape sequence.

EUC, defined in ISO 2022-1993, was developed to process multiple character sets within a text [11]. The following two implementations for Japanese are commonly used: *EUC Packed Format* using one- and two-byte-per-character modes and *EUC Complete Two-byte Format* using two-byte fixed-width representation. EUC is widely used in workstation environments.

Information encoded by any one of these three methods can be translated into another form algorithmically, but mapping tables are required to translate them into Unicode [11]. Although Unicode, in the form of ISO/IEC 10646, is being adopted as JIS X 0221-19xx [15], Unicode has not been widely accepted in Japan.

6. Chinese Typography

Any serious work on Chinese OCR requires some knowledge of Chinese typographic conventions as well as of the syntax and semantics of the language itself. The ideographic script requires new approaches to both the recognition of individual glyphs and the analysis of layout. The problems are aggravated by western influences that have resulted in an essentially hybrid notation for technical material.

By the second century AD the classics were printed in China. Moveable metal type was used in Korea before it was rediscovered in Europe. Chinese seals and color prints enjoyed wide popularity. Type design and typography were respected crafts and are only now beginning to be automated. Nevertheless, because of the large character set, the number of different typefaces is relatively small, and the block-structure of the characters favors simple layouts. The upper-lower case distinction does not exist.

Character segmentation is greatly facilitated by the uniform pitch (mono spacing) of most printed matter. Blanks, however, are used sparingly, and roman characters are often interspersed with *Hanzi*.

There are four commonly used families of type: *Song, Fang Song, Hei,* and *Kai* (Fig.

(a) 宋體

(b) 仿宋體

(c) 楷體

(d) 黑體

Fig. 7. Chinese typefaces: (a) Song, (b) Fang Song, (c) Kai, and (d) Hei.

7). *Song* is named after the dynasty that gave it birth (but is called *Ming* by the Japanese). It is used as the body font in books, newspapers and magazines.

The *Fang Song* (imitation or quasi-Song), the horizontal and vertical strokes have almost the same thickness. The characters are tall, and there is a slight slant to the horizontal strokes. In mainland China, it is the required typeface for government documents. The *Hei* (Black) font, bold and very dark, is used for titles and eye-catching announcements. The *Kai* (Writing) font, rather ornate, imitates brush strokes. It is used as an associate font in newspapers and magazines. Other more artistic typefaces are used when it is desired to evoke calligraphy

Type sizes are not standardized to printers' points; instead, numbers are used as shown in Table 3. Computer systems based on the GB codes use, however, a system based on grades (12 grade = 0.25 mm) for printed output, although it is often converted to point size for the user interface.

Table 3. Chinese font sizes.

Chinese font size	Points	mm
No. 0	42.00	14.70
Small 0	36.00	12.60
No. 1	27.50	9.62
No. 2	21.00	7.35
Small 2	18.00	6.30
No. 3	16.00	5.60
No. 4	13.75	4.81
Small 4	12.00	4.20
No. 5	10.50	3.68
Small 5	9.00	3.15
No. 6	8.00	2.80
Small 6	6.88	2.42
No. 7	6.00	2.10

Typographic devices used for variety and emphasis include size changes, vertically and horizontally compressed styles, outline, shadow, underlined and tilted characters (Fig. 8). Instead of boldface, Hei type is used. Because of the detailed nature of the distinction

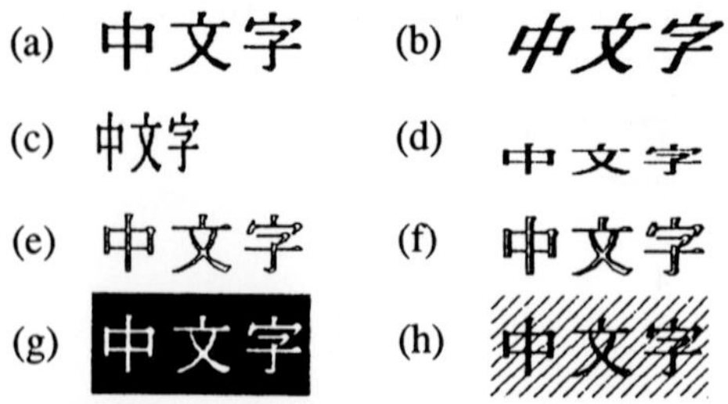

Fig. 8. Chinese typestyles: (a) normal, (b) tilted, (c) horizontally-compressed, (d) vertically compressed, (e) outline, (f) 3-D, (g) reverse video, and (h) shaded background.

among many Chinese characters, most material intended for OCR must be scanned at least at 400 dpi. While not really a typesetting characteristic, it must be noted that thin, translucent paper seems popular in China and may give rise to strike-through noise. Additional comments on digital fonts appear in the next section.

7. Japanese Typography

Traditionally, Japanese text is typeset vertically, and text-lines are placed from right to left on a page. The western style horizontal text flow is also acceptable. Thus, recognition of reading order consists of two problems: ordering of text zones and detection of text flow in a zone. The direction also affects the shapes or positions of a few punctuation marks and small *Kana*.

In general, text is printed with mono spacing. Long Arabic numbers and words made of alphanumeric characters can misalign characters. Punctuation marks, whose character boxes are either half height for vertical-set or half width for horizontal-set, are occasionally used to avoid bad line breaks. These *hankaku* (half size) punctuation marks also misalign characters.

In the western world, a font consists of three attributes: type size, typeface, and type style. Similarly, a Japanese font typically consists of three attributes: type size, typeface, and weight. Photo-typesetters also allow shape modifiers, such as condensed, tall, and oblique. Since Japanese text usually contains both *Kana* and *Kanji*, a *Kana* font and a *Kanji* font can be changed independently of each other.

In addition to *point*, a type size can be expressed in units *gou* and *kyu* [16]. (The unit *kyu* is used in photo-typesetting.) Table 4 shows their approximate relationships.

The two typefaces *Mincho* and *Gothic* are most widely used. *Mincho* is serif and is used for main body text. *Gothic* is sans serif and is used mainly for headings. Other popular typefaces, *Kaisho, Shincho*, and *Maru-Gothic*, are used for business cards and announcements (see Fig. 9).

A Japanese typeface consists of thousands of characters; therefore, only a limited number of typefaces was available before digital typography became reality. Since several

Table 4. The relationship among three units of type size.

Gou	Kyu	Points
1	38	24
2	32	21
3	24	16
4	20	14
5	15	10.5

(a) (b) (c) (d) (e)

Fig. 9. Japanese typefaces: (a) Mincho, (b) Gothic, (c) Kaisho, (d) Shincho, and (e) Maru-Gothic.

font editors are now commercially available, the number of typefaces for Japanese is expected to increase rapidly.

Kana words are Japanese ligature combining two or more *Katakana* characters in a single character box. They are used to abbreviate common units of measures and currencies and are often used in newspaper articles. *Kana* words are not in the JIS character sets but are supported by some Japanese corporate character sets. In the Unicode specification [5], these characters are called *squared words*, and common squared words are included. Examples are shown in Fig. 10.

Other special characters common in Japanese text are circled numerals and pictorial symbols, such as Japanese postal code and telephone number. Parenthesized *Kanji* characters, which fit in a character box, are used to abbreviate the description of company types, such as "Incorporated" and "Limited Liability Company"

In English, words are emphasized using bold letters, italics, underline, or quotation marks. Similar methods are also used to emphasize Japanese words (or phrases). In Japa-

(a) ㍉ ㌢ ㍍

(b) ① ② ③

(c) ㈱ ㈲ ㈹

(d) 「 」 『 』 【 】

Fig. 10. Japanese special symbols: (a) Kana (squared) words, (b) circled numerals, (c) parenthesized Kanji characters, and (d) Japanese parentheses.

Fig. 11. Artistic newspapers headings

nese, *Gothic* characters work as **bold** letters; however, no italicized characters are used. Instead of underlines, small dots are often placed along the right-hand side of characters in vertical-set or above characters in horizontal-set text. Japanese parentheses (single and double *Kagi-kakko*) and guillemots, « », are also commonly used to emphasize words. Japanese word processors also support other ways to emphasize words, such as the special effects shown in Fig. 8. In newspapers, headings often receive artistic treatment as shown in Fig. 11. Thus it is difficult to extract characters from such headings.

8. Chinese Input Methods

The goal of OCR and DIA is to provide an economical alternative to human document entry. As mentioned earlier, Chinese data entry methods are so slow and clumsy that there is a great deal of incentive for automation. However, familiarity with existing input and output methods, and a comprehensive Chinese operating environment, are necessary simply to conduct meaningful research on Chinese OCR.

The large Chinese keyboards that are still used in Chinese offices actuate the print head directly for the more common characters, while rare character slugs are individually inserted by the operator. Touch-typing takes on a new meaning on a keyboard with several thousand keys. Although large computer keyboards have been developed for ideographic applications [1], these devices are not available to most researchers.

At the other end of the scale is voice input, either with a specially trained device for a single speaker, or with omni-speaker capability. Voice input is not yet in common use, but may eventually provide the fastest (and noisiest) interactive input for Chinese text.

There are three types of data-entry systems now in use. By far the most common is the ASCII-terminal keyboard input in a dialog mode with screen display. The second method uses a pointer device with a display: one either points directly at the screen (touch sensor or light-pen), or drives a cursor using a mouse, joystick or trackball. The third method, on-line character recognition with a tablet or electronic ink, is discussed by H. J. Lee in Chapter 13 of this Handbook.[a] The following discussion focuses on the keyboard dialog.

More than 500 different methods for Chinese input have been invented, and standardized methods of evaluating them have been adopted [1]. For our purposes, we group the

a See also the chapter by J. Y. Kim and B. Sin.

ASCII keyboard methods into three categories: input by structure, by pronunciation, and by code. All of these methods are hierarchical in the sense that if an ambiguous string is entered, alternative characters are displayed for the final choice. Most of the current methods are derived from existing dictionary organizations, because finding a word in a Chinese dictionary presents essentially the same selection problem.

Structure methods are based on radicals, or strokes, or graphic corner configurations. The latter is currently gaining popularity in PRC. The character set is divided into a tree with about as many branches as there are keys, and each branch is assigned to a key. The resulting organization is a trade-off between ease of use and efficiency. Ease of use means "meaningful", easy-to-remember key assignments, while efficiency implies a Huffman code that minimizes the average number of keystrokes. Unfortunately, none of the three structural characteristics exhibits the desirable probabilistic properties for obtaining optimal key-stroke sequences.

With the phonetic entry methods one can type *Pin Yin* or *Zhu Yin* pronunciations directly. For common syllables, single-key equivalents exist. But because of the many homonyms, a screen display and additional key-strokes are necessary. Phrase input exploits the correlation between successive symbols. Typing phonetic codes for several *Hanzi* at once reduces the number of homonymic alternatives and, consequently, the number of keystrokes required for selection. Since these methods are based on *Mandarin*, they are less accessible to speakers of other dialects.

The coding methods requires typing in a four or five digit code. One of the commonly used codes is the telegraph code introduced in 1911. The original four digit code was based on a hundred-page code book with a 10×10 array on each page. Methods using GB or Big-5 also take advantage of the intrinsic co-ordinate structure of these coding schemes. As in English, code input is available in all systems for rare or user-defined characters.

Some user interfaces allow for combining several encoding methods. With constant practice it is possible to memorize enough codes in all of these systems to achieve some measure of proficiency. However, performance falls off rapidly with casual use.

9. Japanese Input Methods

The process of entering Japanese text consists of one to three steps depending on the keyboard used. *Kanji* tablets typically consist of 2,160 keys corresponding to 1860 *Kanji* and 300 non-*Kanji*. These tablets allow direct keyboard input like a western keyboard for English. Because of their size and the number keys, these tablets are not popular.

Kana keyboards are two keyboards in one. A Roman/*Kana* key allows a user to select the QWERTY layout for entering alphanumeric characters and a *Kana* layout for entering *Kana* characters directly. Two standardized *Kana* layouts are available: the JIS keyboard array (JIS X 6002-1985) and the new-JIS keyboard array (JIS X 6004-1986). Japanese sentences are entered according to their pronunciation using *Kana*. A single *Kanji*, a *Kanji* compound (*Jukugo*), or a *Kanji* phrase is converted from a string of *Kana*. Since many *Kanji* share the same pronunciation, the user selects the character from a list of candidates.

Western keyboards, such as QWERTY, can be used to enter Japanese text. A user enters Romanized Japanese strings. The software automatically and quickly converts the input string into a *Kana* string and displays it. The *Kana-Kanji* conversion process is the same as the *Kana* keyboard method.

Most input software also allows *Kanji* to be entered by radical, by stroke count, and by encoded value. It takes two steps to enter a *Kanji* using a radical method. First, a radical is selected from a radical table or by its name. Then the character is chosen from a list of candidates. Similarly, a stroke count method allows for selecting a character from a list of candidates whose stroke count matches with the specified stroke count. To enter a character using its corresponding encoded value, *Kanji*-encoded value dictionaries can be used to find the corresponding encoded value.

10. Oriental Output Methods

The earlier output devices for alphabetic characters, like chain printers and other impact printers using fully-formed type slugs, could not be easily adopted to ideographs. However, 24-pin dot-matrix printers, laser printers, and inkjet printers need no modification. An array of at least 15×16 (as opposed to 7×9 for roman characters) is required for *Hanzi*, and 24×24 is preferable. 300 dpi laser printers produce easily legible characters at normal point sizes.

The major considerations for Chinese output are the availability of digital fonts for a given character set, and the large amount of storage required to store a complete font. Ideographic fonts are stored using the same methods as alphabetic fonts: bitmap arrays, vector fonts, outline fonts based on mathematical curves, and hinted fonts. Scaling rules are even more important for quality Chinese output than for roman characters. Fonts in all of these formats are available for many of the character sets specified by the different encoding methods. However, no widely adopted standard exists, so many applications include their own font files.

PC processor and storage resources are still only barely sufficient to produce pleasing ideographic displays and printouts at a reasonable speed. Plug-in cards with read-only font storage are available. As in alphabetic languages, the availability of digital fonts and of font-design software may greatly increase the diversity of shapes that OCR devices must recognize. This may be partially compensated by improved paper, printing, and copy quality.

11. Conclusion

The large number of symbols to be recognized is certainly considered the major difference, from the point of view of OCR, between alphabetic and ideographic writing systems. Current Chinese and Japanese OCR systems typically recognize 3000–4000 symbols, generally so called Level-1 sets of characters. In addition, most recognize Latin letters and Arabic numerals, and European punctuation. Many Chinese systems accept either traditional or simplified shapes, though antique printed texts foil them as they do Western OCR. We are not aware of large-scale, automated and systematic tests of Chinese and Japanese OCR systems for printed text.

It is expected that the size of the character sets that can be recognized will double or triple in the next decade. This will bear the greatest benefit if additional standardization of the encoding of second-level character sets takes place. Unicode is one candidate for widespread adoption, but it still faces some technical and political hurdles.

With the advent of computerized page composition and typesetting, the number of different typefaces may approach the Western profusion of styles. However, for the time being official publications in the People's Republic of China remain restricted to a small number of fonts.

Aside from the large number of symbols and more complex shapes, the major difference between alphabetic and ideographic systems lies in the use of context. Syllables and words take on an entirely different meaning, and the distinction between character n-gram and lexicon-based disambiguation disappears. The inherently larger error rate, due to many more classes and similar shapes, puts an additional premium on contextual discrimination.

The presence of both horizontal and vertical lines of text, sometimes on the same page, has not proved a major obstacle to automated format analysis. The square shape of the Hanzi and Kanji ideographs facilitates character-level segmentation. The prevalence of raster displays and printers has put Western and Eastern development environments on a nearly equal footing, but the many systems of ideographic key entry still constitute a handicap in comparison to the almost complete standardization of English-language keyboards. The intrinsic difficulty of ideographic key entry can be expected to continue to spur further research and development of automated data entry systems for Chinese and Japanese characters.

Acknowledgments

George Nagy gratefully acknowledges the support of the Central Research Laboratory, Hitachi, Ltd. and of the Northern-Telecom/BNR Education and Research Networking Program. This work has been conducted in the New York State Center for Advanced Technology (CAT) in Manufacturing, Automation and Robotics. The CAT is partially funded by a block grant from the New York State Science and Technology Foundation.

Much of the material in this paper is drawn from Yucheng Liu's UNLV Computer Science Department MS thesis, *Chinese Information Processing* [18], where many of the topics are treated in greater detail.

References

[1] J. K. T. Huang and T. D. Huang, *An Introduction to Chinese, Japanese and Korean Computing*, Series in Computer Science, Vol. 12 (World Scientific, 1989).

[2] T. Hsia, *China's Language Reforms* (The Institute for Far Eastern Languages, Yale University, 1956).

[3] —, Customs form, *Sky Magazine*, Vol. 23, No. 3 (Halsey Publishing Co., 1995) 154–155.

[4] *Chinese Character Code for Information Interchange, and Variant Form*, Vol. III (1987), referenced in [1], p. 51.

[5] The unicode consortium, *The Unicode Standard, Version 1.0,* Vol. 1 and 2 (Addison-Wesley, 1991).

[6] J. Bettels and F.A. Bishop, Unicode: a universal character code, *Digital Technical J.,* Vol. 5, 3 (1993) 21–31.

[7] Y. R. Chao and L.S. Yang, *Concise Dictionary of Spoken Chinese* (Harvard University Press, 1945).

[8] G. Nagy, The dimensions of shape and form, in *Visual Form,* eds. C. Arcelli, L.P. Cordella, and G. Sanniti di Baja (Plenum Press, New York, 1992).

[9] R. A. Miller, *The Japanese Language* (The University of Chicago Press, 1967).

[10] N. H. Dickey, ed., *Funk & Wagnalls New Encyclopedia,* Vol. 15 (Funk & Wagnalls L. P., 1990).

[11] K. Lunde, *Understanding Japanese Information Processing* (O'Reilly & Associates, Inc, 1993).

[12] H. I. Chaplin and S. E. Martin, *A Manual of Japanese Writing, Book 1* (Yale University Press, 1967).

[13] H. Fujisawa, Full-text search and document recognition of Japanese text, *Proc. of Fourth Annual Symp. on Document Analysis and Information Retrieval* (University of Nevada, Las Vegas, 1995) 55–80.

[14] M.M.T. Yau, Supporting the Chinese, Japanese, and Korean languages in the OpenVMS operating system, *Digital Technical J.,* Vol. 5, 3 (1993) 63–79.

[15] "What does unicode mean for Japanese codes? Will it replace EUC, JIS and Shift-JIS?", http://www.stonehand.com/unicode/faq/cjk/japanese.html#q1 (1994).

[16] M. Sonobe, *Henshu Nyumon*, 4th ed. (in Japanese) (David Inc., 1987).

[17] M. F. Lee, *General Knowledge of Writing Common Chinese Characters* (Era Book Inc. 1982).

[18] Y. Liu, Chinese information processing, M.S. Thesis, University of Nevada, Las Vegas, 1995.

Handbook of Character Recognition and Document Image Analysis, pp. 305–329
Eds. H. Bunke and P. S. P. Wang

CHAPTER 11

MACHINE PRINTED CHINESE CHARACTER RECOGNITION

XIAOQING DING
Department of Electronic Engineering, Tsinghua University
Beijing, 100084, P.R.China

This chapter describes the principles, methods and practices of printed Chinese character recognition. A stroke structural feature based statistical recognition method, feature extraction and classifier design are described for solving the especially difficult problem faced by Chinese OCR systems. In addition, some preprocessing (such as binarization, normalization and segmentation) and postprocessing are introduced for improving recognition performance. Lastly, some exciting developments in practical printed Chinese character recognition system are described.

Keywords: Chinese character; Chinese character recognition; Chinese and English character recognition; Feature extraction; Pattern recognition; Pattern matching.

1. Introduction

Chinese character recognition is a technology used for entering Chinese characters into computer has been under research since the 1960s. The goal is to automatically convert letters, newspapers, magazines, etc. into a coded representation of the input characters. These inputted codes can be used for many computer processing tasks, such as database storage, information retrieval, text edition, machine translation, etc.

There are more than four hundred Chinese character coding schemes for keyboard typing of Chinese characters, such as the Pinyin inputting scheme which is a sound-based coding method, the five-stroke inputting scheme that is a stroke-based coding method, etc. However keyboard typing of Chinese characters is time-consuming and a nuisance. Chinese character entering is still the bottle-neck of information processing for the Chinese language.

Practical demands on inputting huge amounts of Chinese documents into a computer stimulate the need for research and development of machine printed Chinese character recognition. Exciting progress in research and development of commercial products was made in recent years.

Character recognition techniques include on-line handwritten character recognition, machine printed character recognition, off-line hand-printed character recognition and off-line handwritten character recognition with increasing difficulty. Now some commercial products of on-line handwritten Chinese character recognition systems and machine printed Chinese character recognition systems have appeared on the market. This paper describes the principles, methods and practice of machine printed Chinese character recognition and the recent advances.

2. Printed Chinese Characters

The Chinese character is a glorious and brilliant prestige symbol of Chinese culture. It is the oldest ideographical character set in the world and the only one still widely used now. Therefore, the Chinese character is closely related to the development of culture.

Before discussing Chinese character recognition, some properties of Chinese characters and some problems in recognition are introduced, which determine the requirements for a Chinese character recognition system.

CHARACTER STRUCTURE: Chinese ideographs are formed in a two-dimensional stroke based structure. Form, sound and meaning are the three essential elements of Chinese ideographs. Chinese ideographs vary in complexity from just one horizontal stroke to 36 or more strokes, and all are arranged in the same square area. There are four levels of structure in a Chinese character, i.e. character, radicals, strokes and stroke segments. A Chinese character consists of special radicals. About 200 to 300 kinds of radicals appear in Chinese ideographs. Radicals consist of certain strokes. There are about 20 to 30 kinds of strokes used in Chinese characters. The strokes consist of five fundamental stroke segments, i.e. horizontal（-）、vertical（|）、left-falling （/）、right-falling （\） and point（、） stroke segment.

SIMPLIFIED AND TRADITIONAL CHARACTERS: Simplified Chinese character forms ("Jian-Ti Zi") were published as the standard in 1964 and are used in the PRC now, but in most overseas publications Chinese traditional character forms ("Fan-Ti Zi") are still used as a standard.

FONT: In printed books, magazines and newspapers, etc., four major categories of font styles are mainly used in China. They are Song, Fang-Song, Black (or Hei) and Kai fonts. These are only the basic categories of font styles. There are some derivatives from one font style. For example, there is Book-Song, Title-Song and Newspaper-Song, etc., all derived from the Song font. Song fonts are the most frequently used font styles

清华文通印刷汉字识别系统　JianTi-Song
清华文通印刷汉字识别系统　JianTi-Fang Song
清华文通印刷汉字识别系统　JianTi-Black(or Hei)
清华文通印刷汉字识别系统　JianTi-Kai
清华文通印刷汉字识别系统　JianTi-Li Shu
清华文通印刷汉字识别系统　JianTi-Yuan
清華文通印刷漢字識別系統　FanTi-Kai
清華文通印刷漢字識別系統　FanTi-Hei

Fig. 1. Different fonts of printed Chinese characters.

for body text and headlines. Fang-Song fonts are similar to the Song font, but they have slightly slanting upward horizontal strokes which are often used for subheadings as well as for body text. Black fonts have equal thicknesses for Horizontals and Verticals, and are mostly employed for headlines or specially marked portions of text. The Kai font is modeled on handwritten Chinese characters, and is used in prefaces, subheadings, footnotes and to mark the names of persons inside the Song font's body text.

Besides the four major categories of font styles, there are other font styles used in China, such as Xiao-Yao, Wei-Bei, Yuan and Li-Shu, etc., especially for headlines and some body text. Different font styles have considerable variations of character instances and often appear in the same document. This greatly increases the difficulty of printed Chinese character recognition. Figure 1 shows some examples of simplified and traditional printed Chinese characters in different fonts.

SIZE: The size of machine printed Chinese characters is measured in pounds. One pound is equal to 0.35 mm. The sizes are presented by size numbers from a special large to size number 7 in decreasing order of sizes, see Fig. 2. For example, the size number 6 is 7.875 pounds, but the special large size number is 56 pounds, almost 7 by 7 times larger than size number 6. In Chinese texts, different types with different sizes are employed for headlines, subheadings, body text and footnotes, which often appear on the same document. For example, size 5, size 4, and size 3, are the most frequently used size numbers for body text, and size 1 and 2 for headlines. Size normalization is very important for recognition, because recognition features are extracted from the same square character area.

	清	清	清	清	清	清	清	清	清
Size number	Special large	Primary	1	2	3	4	5	6	7
Pound	56	35	28	21	15.75	14	10.5	7.9	6

Fig. 2. Machine printed Chinese character size.

CODE: The recognition of Chinese characters is concerned with transforming a Chinese character image into the corresponding code, which computers can manipulate. Chinese characters are usually represented by a two-byte code in computers for storage, transmission and internal processing. Standardized codes are necessary for information transmission.

The national standard of code set of the P. R. of China is the GB2312-80, which is "Code of Chinese Graphic Character Set for Information Interchange Primary Set" published in May, 1981. The GB2312 contains 6763 simplified Chinese characters. It is divided into two levels, Level 1 with 3755 frequently used characters and Level 2 with 3008 more rarely used characters. This code set of simplified Chinese characters is the standard and officially used in mainland China and Singapore, which is simplified from traditional Chinese characters by reducing their complexity.

BIG-5 Code is another code set that represents traditional Chinese characters. It was brought out in Taiwan and contains 5401 traditional, frequently used Chinese characters. This code set of traditional Chinese character forms is still the standard for overseas Chinese.

Here a very important problem for the OCR system is how many Chinese characters are chosen as a relevant set in recognition. It would seem at first that a larger set of characters which are able to be recognized should be better. However, many Chinese characters rarely appear in common use. In other words, about three thousand Chinese characters are nearly 99.9% of frequently used characters in China. Since the recognition accuracy that could be reached today is only about 98% to 99%, too many Chinese characters in the recognition set are unnecessary, since this greatly increases the computing complexity of recognition. Thus only about four thousand to seven thousand Chinese characters, i.e. 3755, 4000, 5401, 6763, etc., are chosen for the most recent recognition systems.

Chinese characters are large in number, different in printed style and type, and quite complex in structure for both the simplified as well as the traditional character forms. All these factors make serious difficulties in automatic Chinese ideograph recognition.

3. Machine Printed Chinese Character Recognition Methods

Chinese character recognition has been under research for nearly thirty years since 1966, when Casey and Nagy first started it. Since then, significant progress has been achieved and many methods have been proposed and tested. Since the 1980s, more researchers paid attention to printed Chinese character recognition. Some attractively machine printed Chinese character recognition systems were developed in China and appeared on the world market.

The researches of printed Chinese character recognition are still in progress: from theory and algorithm research to real-live applicable recognition system development, from single-font to multi-font, from simplified or traditional only to simplified and traditional Chinese character recognition, and from single Chinese ideograph recognition to Chinese-English bi-linguistic character recognition. Now improvement of the performance of recognition is pursued, especially in strengthening the robustness and lowering the error recognition rate.

3.1. *Structure Analysis and Statistical Pattern Recognition*

Two major kinds of approaches, the structure analysis and the statistical classification, are used in Chinese character recognition.

In the early days, some researchers emphasized the structure analysis method for printed Chinese character recognition, which was extracting the strokes of every Chinese character and deciding the attributes of the strokes and the relationship among them. Then the recognition was using the structure analysis or syntactic grammar matching of two-dimensional graphic syntax to identify Chinese characters.

More tests and researches have found that there are a few existing barriers. The most important is that the structural elements are hardly extracted correctly from a character image. If one stroke element extraction fails, a heavy recognition error has taken place. Another problem concerning two-dimensional structural relational graph matching or syntactic grammar method is its difficulty in dealing with a large number of structural elements of a character and large set of Chinese characters.

Structure analysis methods also have special difficulty in handling the effects of various noises, which cause rather complicated variations of structure elements and relations. So the structure analysis method often fails in practical application of character recognition.

Other researchers use straightforward pattern matching or template matching approaches. One is to directly compare the input character image array with reference character image matrix. The others are transform feature matching which are directly based on the two-dimensional character image or projections of the character image such as the Fourier transform and the Walsh transform, etc., that are similar to the direct image matching approach.

These kinds of template matching methods may get good recognition results only for a single font and for good quality prints. However it is almost impossible to solve the problems in practical applications, such as in variations of stroke width, different fonts and serious noises, etc. In such situations the character image varies resulting in making the recognition rate fall down sharply.

3.2. *Statistical Pattern Recognition based on Structure Feature for Printed Chinese Character Recognition*

In order to overcome these problems, we must combine the stroke structure features of Chinese character with the statistical recognition method, the structural features based statistical pattern recognition approaches have been developed for printed Chinese character recognition.

The reason is that the two-dimensional stroke structure contains the key information of the Chinese character and the statistical classification approaches are more robust with the noise met in practical applications. The approaches combining both statistical and structural merits are more successful in printed Chinese character recognition by now.

In printed Chinese character recognition, the most important requirement is high recognition accuracy under the condition of recognizing thousands of multi-font and omni-size Chinese characters and even in poor quality prints.

Thus in the Chinese character recognition systems 12 to 13 bits per character Information System Entropy *H(E)* exists, which represents the uncertainty of the recognition system and needs to be removed in the recognition process, i.e.

$$H(E) = -\sum_{i=1}^{n} P(\omega_i) Log P(\omega_i) \tag{3.1}$$

where ω_i is the i-th character category and $\omega_i \subset \Omega_n$,
Ω_n is the character set of all *n* categories in the system.

In English character recognition systems, only 7 bits per character Information System Entropy *H(E)* that exist need to be removed. The number of Chinese characters is far larger than the number of English characters in the recognition system, which means that the Chinese character recognition needs much more information to remove the uncertainty of the system than what a western character recognition needs.

Feature extraction and pattern classification are the fundamentals for the Chinese character recognition system. The feature extraction decides the potential optimum, that is the minimum of recognition error rate, and optimal classifier design should approximate this optimum as closely as possible. If Chinese character recognition is to be used in practical applications, more attention must be paid to the fact, that Chinese characters are often printed on thin, yellowish dark paper with an uneven structure, and poor ink and variations of character graph.

4. **Feature Extraction of Machine Printed Chinese Character Recognition**

4.1. ***Information Entropy Theory in Pattern Recognition***

Extracting more relevant information from the original character image into a smaller set of features is most important for automatic recognition.

This set of features is represented by a higher *N*-dimensional statistical feature vector $X = (x_1, x_2, \cdots, x_N)^T$ that is obtained as the description of every Chinese character sample. The effects of various fonts, sizes and noises are presented by the random variation of feature vector *X* with probability density function *p(X)* and conditional probability density function $p(X|\omega_k)$ when the random feature vector belongs to the category ω_k.

The conditional probability density function $p(X|\omega_k)$ and *a priori* probability $P(\omega_k)$ of category ω_k can be estimated from a learning procedure with a training set. For instance, when there is a Gaussian probability density function for feature vector *X*, then,

$$p(X|\omega_k) = \frac{1}{2\pi^{N/2}|\Sigma_k|^{1/2}} \exp\left\{-\frac{1}{2}(X - M_k)^T \Sigma_k^{-1}(X - M_k)\right\}, \quad (4.1)$$

where $$\Sigma_k = E\left\{(X - M_k)(X - M_k)^T\right\} \quad (4.2)$$

and $$M_k = E\{X|\omega_k\}. \quad (4.3)$$

The mean feature vector M_k of a category ω_k is used as a reference vector of the same category. It can be estimated by averaging the vectors of the known samples in a training set.

Classification is to compare the distances of the feature vector X of an unknown sample with the reference vectors M_k. Then the category of the reference vector yielding the minimum distance is assigned to the unknown input character pattern.

What determines the recognition accuracy of pattern recognition? What are the most effective features in certain recognition problems? In other words, what kind of feature is the best one for it? These are the most important problems concerned with feature selection and extraction and recognition system design.

From the information theory point of view, pattern recognition is a reducing entropy procedure due to canceling or minimizing the uncertainty when there is an unknown sample in the recognition system. The entropy is reduced when the information is taken from the extracted features of an unknown sample. The system entropy is reduced from the original the System Entropy *H(E)* to remain *a posteriori* Entropy *H(E|F)*. The System Entropy *H(E)* describes the uncertainty of an unknown sample in the recognition system, and *a posteriori* Entropy *H(E|F)* describes the minimized and left-over uncertainty of the recognition system after the recognition has taken place.

The recognition error probability is determined by *a posteriori* Information Entropy,

$$H(E|F) = -\sum_i \int p(X,\omega_i) \log P(\omega_i|X) dX. \quad (4.4)$$

It is a lower bound of classification error probability P_e of optimum Bayes classifier. That is because

$$P_e \leq \frac{1}{2} H(E|F) \quad (4.5)$$

has been proved, (Ref.[3]).

We can prove that the *a posteriori* Information Entropy *H(E|F)* satisfy

$$H(E|F) = I(E,F) - H(E). \quad (4.6)$$

It has been proved theoretically and experimentally that recognition property depends on the feature extraction, i.e. it depends on the Effective Information Entropy of the feature vectors in the recognition system, which is, (Refs.[1], [2])

$$I(E,F) = -\sum_{i=1}^{n} \int_{R^N} p(X,\omega_i) \log \frac{p(X/\omega_i)}{p(X)} dX \quad (4.7)$$

or $$I(E,F) = -\sum_{i=1}^{n} \int_{R^N} p(X,\omega_i) \log \frac{P(\omega_i|X)}{P(\omega_i)} dX \quad (4.8)$$

From that clearly, the error probability P_e is determined by $I(E,F)$ for a certain recognition system entropy $H(E)$. That is to say, for a certain pattern recognition problem the bigger the Mutual Information Entropy $I(E,F)$, the better are the features. That is to say, the feature extraction determines the potential optimal error minimum P_e through $H(E|F)$ of the recognition system.

According to maximizing the Mutual Information Entropy $I(E,F)$ and minimizing *a posteriori* Information Entropy $H(E|F)$, some feature selection principles were proposed.

We can also prove, that is, (Refs. [1], [2])

$$I(E,F) = H(F) - H(F|E), \tag{4.9}$$

where

$$H(F) = -\int_{R^N} p(X)\log p(X)dX \tag{4.10}$$

is the Feature Information Entropy, and

$$H(F|E) = -\sum_i \int_{R^N} p(X,\omega_i)\log p(X|\omega_i)dX \tag{4.11}$$

is the Conditional Feature Entropy of features in the system.

If the features have been extracted, its Feature Information Entropy $H(F)$ and Conditional Feature Entropy $H(F|E)$ determine the Effective Information Entropy $I(E,F)$ of the features, and also the lower bound of classification error probability P_e of the optimal Bayes classifier.

Therefore in order to get a high accuracy of recognition, two ways can be adapted. One is increasing the $H(F)$, while the other is decreasing the $H(F|E)$.

Increasing the $H(F)$ is to capture more information from the character image, i.e. the features contain more information in them. For instance, we can increase the dimension number of independent feature vectors for it.

Conditional Feature Entropy $H(F|E)$ describes the feature dispersion within a category. If the features are stable in a category, or dispersiveness is smaller, the $H(F|E)$ should be smaller too.

For instance, when the probability density of feature vector is a Gaussian distribution function, as in Eq. (4.1), we can get,

$$H(F) = E_X\{\log P(X)\}$$

$$\therefore H(F) = \frac{1}{2\ln 2}\left[N + \ln|\Sigma_X| + N\ln(2\pi)\right], \tag{4.12}$$

and

$$H(F|E) = \frac{1}{2\ln 2}\left[N + N\ln 2\pi + \sum_{i=1}^{n} P(\omega_i)\ln|\Sigma_i|\right], \tag{4.13}$$

$$I(E,F) = \frac{1}{2\ln 2}\left[\ln|\Sigma_X| - \sum_{i=1}^{n} P(\omega_i)|\Sigma_i|\right]. \tag{4.14}$$

when there *a priori* probability $P(\omega_i)$ is equal for all i (i= 1, 2,..., n), then $P(\omega_i) = \frac{1}{n}$, and if the covariance matrix Σ_i of every category is the same as Σ, i.e., $|\Sigma_i| = |\Sigma|$, for all i (i=1, 2,...., n), so that the Effective Information Entropy of the feature vectors should be

$$I(E,F) = \frac{1}{2\ln 2}\ln\frac{|\Sigma_X|}{|\Sigma|}. \tag{4.15}$$

When the feature vector X is N-dimensional independent, its covariance matrix should be diagonal. Therefore,

$$|\Sigma_X| = \prod_{j=1}^{N}\sigma_{jX}^2, \tag{4.16}$$

and

$$|\Sigma| = \prod_{j=1}^{N}\sigma_j^2. \tag{4.17}$$

From there, the Effective Information Entropy should be

$$I(E,F) = \frac{1}{\ln 2}\sum_{j=1}^{N}\ln\frac{\sigma_{jX}}{\sigma_j}, \tag{4.18}$$

$$H(F) = \frac{1}{2\ln 2}\left[N + N\ln 2\pi + 2\sum_{j=1}^{N}\ln\sigma_{jX}\right], \tag{4.19}$$

and

$$H(F|E) = \frac{1}{2\ln 2}\left[N + N\ln 2\pi + 2\sum_{j=1}^{N}\ln\sigma_j\right]. \tag{4.20}$$

From there, very important conclusions about feature extraction can be obtained:

(1) The ratio σ_{jX}/σ_j can be chosen as the superiority measurement of every feature, that is to say, the higher the ratio of global feature dispersion to a certain category's, the better the feature is.

(2) Every effective feature extracted must possess bigger variance σ_{jX} of global feature and smaller variance σ_j of a certain category. Especially, the smaller σ_j within every category for every feature should be steady even those of a poor-print quality and various types. This stability of features chosen in the recognition system determines the robustness of the recognition system.

(3) The separateness of the pattern recognition system depends not only on the superiority measurement of every feature, but also the dimensional number N of the independent feature vector, that means more independent features will provide more information to removing the unknown System Entropy $H(E)$.

From the above statement, the variance of the features in a category determine the complexity of the recognition problem, for example, from a single-font to a multi-font, from good print quality to poor print quality, even from machine printed to hand printed character recognition, all are increasing the variance of the features in a category and increasing the complexity of recognition. In many practical applications serious noises make the dispersion of features in a category increase heavily, that causes severe reduction of the recognition accuracy. Therefore, decreasing the variance of the features in the category is an important way to improving recognition performance.

4.2. *Effective Features for Printed Chinese Character Recognition*

4.2.1. *Structure based feature vector*

The feature extraction decides the potential optimum, the minimum of recognition error, and the better features have the smaller $H(F|E)$. In order to find the better features, the researchers have put in a lot of efforts. The features which are based on the character image are closely dependent on the variations in the fonts, shapes and noises of the character image. The strokes are hardly extracted correctly from character image and the strokes extracted often get mistakes because of printing defects and noises on the character image, therefore, it is impossible to get good recognition results by the structure recognition method or syntactic recognition method.

Some special features closely related to the character stroke structures, such as peripheral features, stroke density features, and gross meshed features, etc. see Figs. 3, 4, 5 and 6, can be extracted effectively, and these kinds of features based on the stroke structure are weakly related to the variation of fonts, sizes and noises. Therefore, their Conditional Feature Entropy $H(F|E)$ will be smaller, and therefore better than the features based on the character image. Also we can select much more features which are independent of each other to compose a large dimension of feature vector, so that enough available Feature Entropy $H(F)$ can be obtained. Thus the Effective Mutual Information Entropy $I(E,F)$ will be increased and large enough, because of

$$I(E,F) = H(F) - H(F|E). \tag{4.21}$$

It has been proved in practice that the statistical pattern recognition algorithms based on the stroke structure features are the most effective method for machine printed Chinese character recognition.

4.2.2. *Special key-points or peepholes logical matching*

Some special key-points or peepholes in the strokes as well as in the background represented by their positions and logical values are also superior features. Logical matching, which is to match the special key-points with the corresponding character image points, is applied in the recognition system (Ref. [7]). The key-points are a few endpoints, corner points and cross points, etc. see Fig. 5. The most important is how to choose and extract the key-points and how to position the key-points in strokes and in background. These key-points, chosen as the representations of the character, have enough information to identify unknown characters and have good robustness for printing noises. Special key-point or peephole logical matching is also a successful Chinese character recognition method.

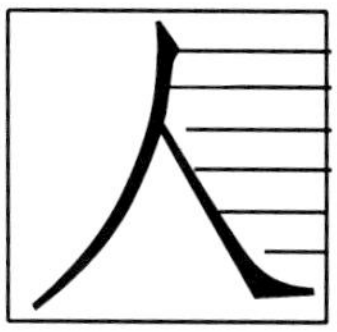

Fig. 3. Peripheral features.

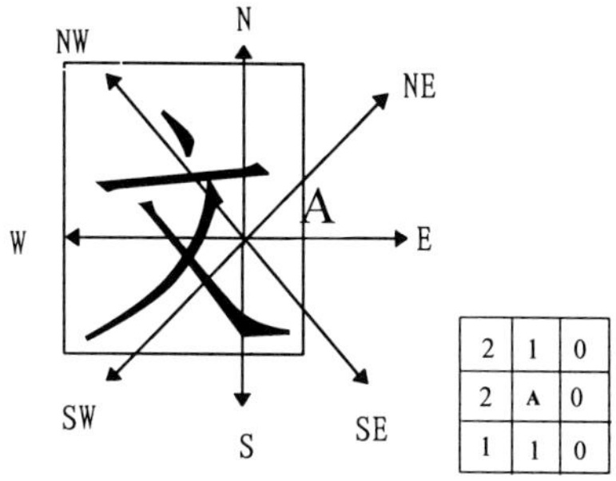

Fig. 4. Stroke density features.

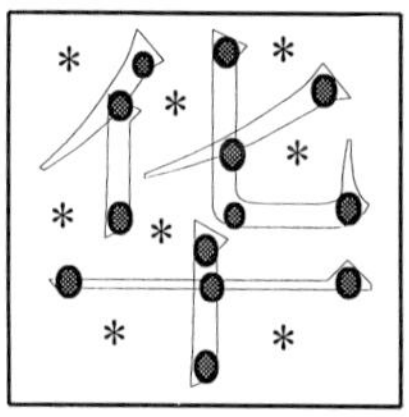

Fig. 5. Special Key-point features.

Fig. 6. Gross meshed features.

4.2.3. *Comprehensive recognition method for combinative optimal features*

The multiple features or multiple classifiers combined can be used to increase the *H(F)* and the Effective Information Entropy $I(F,E)$.

Suppose, the feature spaces are denoted by $F_1, F_2, \cdots, F_k$, the combined feature space is *F*, then

$$F = F_1 \cup F_2 \cup \cdots \cup F_K = \bigcup_{k=1}^{K} F_k\,, \tag{4.22}$$

If $$F_i \cap F_j = 0, \qquad i \neq j\,,$$

then $$H(F) = \sum_{k=1}^{K} H(F_k)$$

and $$H(F|E) = \sum_{k=1}^{K} H(F_k|E)\,,$$

because $$I(F_k, E) = H(F_k) - H(F_k|E)\,,$$

therefore $$I(F,E) = \sum_{k=1}^{K} I(F_k, E)\,. \tag{4.23}$$

So that, some comprehensive methods to combine multiple independent features can increase the Effective Information Entropy *I(E,F)*, and reduce the recognition error (Ref.

[13]). It has been successfully used in TH-OCR Chinese and English bi-linguistic character recognition system to get the highest recognition rate in various noisy environments (Ref.[6]).

5. Classification

5.1. *Minimum Distance Classifier*

The statistical pattern recognition approach is to divide the feature vector space into a mutually exclusive region in order to identify a single character category or a group of character classes. This aim is pursued by classifier designing, training or learning. The partition is usually performed through discriminate function $d_K(X)$ with the property when

$$d_k(X) \le d_i(X) \qquad \text{for all } i \neq k, \tag{5.1}$$

then $\omega(X) = \omega_k$,

it means the feature vector X belongs to the class ω_k.

The optimal discriminate function $d_k(X)$ in a certain feature vector space is a Bayes classifier on which *a posteriori* probability $P(\omega_j|X)$ is maximized, i.e.,

when

$$P(\omega_k|X) \ge P(\omega_j|X) \qquad \text{for all } i \neq k \tag{5.2}$$

then $\omega(X) = \omega_k$.

The Bayes classifier is an optimal classifier with a minimum error probability P_e. If the probability density function of the feature vector X is a Gaussian distribution, and the covariance matrix Σ_i of the feature vector X is the same for all categories, i.e., $\Sigma_i = \Sigma$ for all i (i = 1, 2,..., n), the Bayes classifier becomes a linear classifier. When the covariance matrix $\sum$ is diagonal matrix, i.e. the feature vector is independent of each other, the Bayes optimal classifier becomes a minimum distance classifier, i.e., when

$$d(X, X_k) \le d(X, X_i) \qquad \text{for all } i \neq k, \tag{5.3}$$

then $\omega(X) = \omega_k$,

here X is the N-dimensional feature vector of an unknown sample and X_k is the reference feature vector of category ω_k. A "Minimum distance criterion" is that the character category yielding the minimum distance is assigned to an unknown input character pattern.

The distance $d(X, X_k)$ between two feature vectors X and X_k is a measurement of similarity. A number of distance measures have been proposed, but the simplest one is the City-block distance, that is computed as

$$d(X, X_k) = \sum_{j=1}^{N} |x_j - x_j^k| \tag{5.4}$$

where $X_k = \left(x_1^k, x_2^k, \cdots, x_N^k\right)^T$.

The advantage of this distance measure is that it involves only adding and subtracting, and thus is high speed efficient in computing. It speeds up the Chinese character recognition.

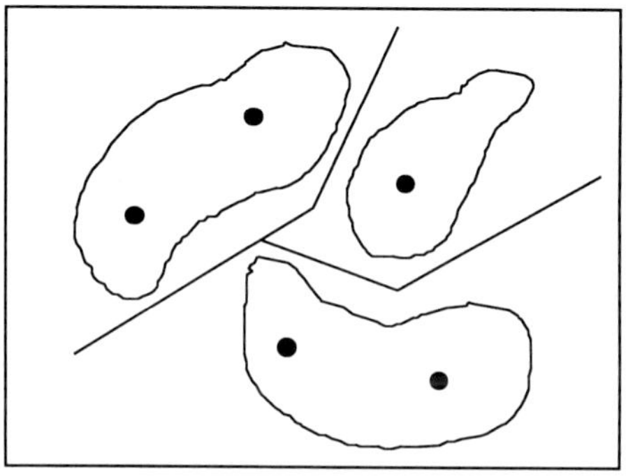

Fig. 7. Minimum distance linear classifier.

The classifier design is processed by a learning or training procedure. The goal is to establish the reference feature vectors $X_k = \left(x_1^k, x_2^k, \cdots x_N^k\right)^T$ $k = 1,2,\ldots,n$ of every category in the feature vector space, upon which the minimum distance principle can be used for every category pattern recognition. The reference feature vector X_k comes from statistical meaning feature vectors of known samples in the learning procedure. The more known the training samples, the better the reference feature vector that can be obtained. However, there are some problems in actual practice. The covariance is not the same in different categories, and sometimes the variation of known training samples is bigger than usual, so that the optimal classifier is non-linear, refer to Fig. 7. For these problems, more than one reference feature vectors for one category are established in the training procedure and the minimum distance linear classifiers can be used for approximating the non-linear optimal classifier.

5.2. *Tree Classification Design*

The statistical pattern recognition method, such as the minimum distance classification method, is to search for a minimum distance between unknown sample feature vector $X = (x_1, x_2, \cdots x_N)^T$ and the reference feature vectors of every known category representation $X_k = \left(x_1^k, x_2^k, \cdots x_N^k\right)^T$. The minimum distance principle decides the category of the unknown sample, see Eq. (5.3).

Here, one character recognition process requires computing and comparing all distances with the reference feature vectors of all categories in the system. The computing complexity for one distance calculation is about (*2N-1*) times the adding. There are n distance computings needed for a total number n of reference feature vectors, therefore, the computing complexity for one character recognition is $(2N-1)\bullet n$ times the adding and (n-1) times the comparing. The global decision of a single stage classifier is very complex and time-consuming for large set of character recognition, such as the Chinese character recognition.

Decision tree or hierarchical classification has been applied to deal with these problems. In a tree classifier, the global complicate recognition decision can be divided into a sequence of simple local decisions at different levels of the tree classifier. A sequence of subdecisions is used to assign an unknown sample into a pattern class. A tree classifier makes a global category decision much more quickly than the single stage classifier. For instance, for an L level balanced, equalized and non-overlapping tree classifier, in which every level possesses the equal number B of branches and the leaf node number is $n = B^L$. Thus the computing complexity will be reduced to $L \bullet \sqrt[L]{n}$ times of the distance computing and comparing. In other words, the tree classifier reduces the computing complexity from n times of the distance computing to $L \bullet \sqrt[L]{n}$ times. Therefore, the tree classifier greatly reduces the computing complexity of a recognition problem and it is always used in the Chinese character recognition for speeding up the recognition process.

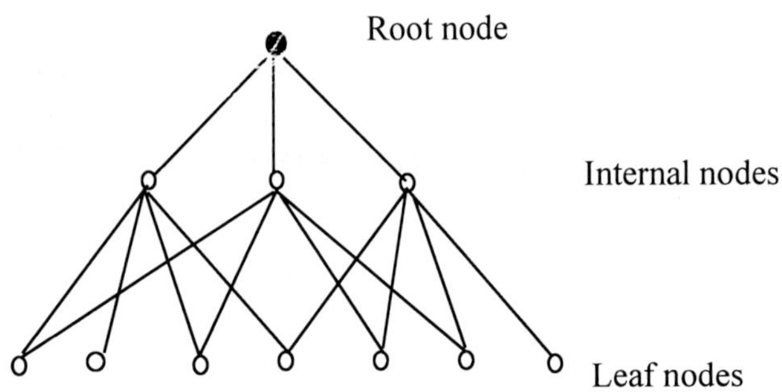

Fig. 8. Tree classifier.

The tree classifier is helpful in reducing the computing complexity, the optimal number of levels of tree would be $\ln n$ for the fastest response. For this kind of tree classifiers there would be much more levels of tree classifier for a large set of Chinese character recognition. Nevertheless the error accumulation in every level of the tree classifier makes a heavily increasing error. Suppose the L levels of the tree classifier, and the classification error probability of every level is the same as ε, thus the classification error probability of a global decision is equal to $-(1-\varepsilon)^L \approx L\varepsilon$ because an error decision of the up-stage is affecting the decision of the next stage. So that compared with a single stage classifier, L times the error increasing will take place in the tree classifier. This error accumulation is quite a serious problem for Chinese character recognition. Pursuing a higher correct recognition rate is more important than recognition speed up in a practical recognition system. A compromise is therefore made. Usually only two levels of a decision tree are used in a large set of Chinese character recognition. The first level is the pre-classification stage, from which some candidates have been found, and the second level is a precise decision stage, which determines a single category of an unknown sample as the result of recognition.

In a tree classifier, the features of different levels can be the same or different from each other.

Another important method to overcome the error accumulation of the tree classifier is overlapping. That is the middle nodes expand its leaf nodes, so that different middle nodes may have the same number of leaf nodes, see Fig. 8. The overlapping can reduce the error accumulation because the error decisions of the up-stage can be rectified and only have little affect on the decision of the next stage. In other words, the middle level division is not exclusive, that the leaf nodes may belong to more than one middle node. If the degree of overlapping is described by the overlapping coefficient $\eta \geq 1$, it is roughly estimated that the overlapping decreases the error accumulation from $L\varepsilon$ to $\varepsilon + (L-1)\varepsilon/\eta$.

It should be noted that the overlapping improves the recognition accuracy of the tree classifier at the cost of increasing the computing complexity and time consumption. The more the overlap, the more the computing complexity is increased. The overlap is the same as increasing the leaf node number from n to $\eta^{L-1} \bullet n$, upon which some trade-off between speed and accuracy should be carefully made in the tree classifier design.

Using the clustering technique for a tree classifier design, a single-stage classification problem can be converted into multistage small ones. The cluster procedure is to cluster leaf nodes which have closer feature vectors in a feature vector space, and forms the father-nodes of the tree classifier by these clustered nodes.

The leaf nodes of the classifier are established by the reference vectors of known sample feature vectors which are obtained from the learning or training procedure.

In the tree classifier design, the first step is to determine the topological structure of the tree classifier, that is the level number L of the classifier, the branch number B of subclasses in the middle stage, and the overlapping coefficient $\eta \geq 1$, etc. Then the second step is using the cluster technique to determine every middle node of B middle level nodes. The third step is to determine the leaf nodes belonging to every middle node according to the overlapping coefficient η. Finally, the tests and experiments will be needed to inspect the tree classifier design.

6. Chinese-English Bi-linguistic Character Recognition Method

Recently, Eastern and Western cultures interacting with each other in the world is on the increase. Chinese characters and English words often appear in printed materials simultaneously. The strong demands for Chinese-English characters document recognition make it necessary to quickly solve this problem.

Although the usual Chinese character recognition system can recognize the isolated English characters, high performance is impossible. The difficulty comes from the difference in their segmentation and recognition between Chinese ideographs and the English character. The segmentation of the English text, that often includes touched characters, is different from the Chinese character in a regular square area. It is very difficult to get high recognition accuracy for English characters which mixed in a large set of Chinese characters. The reasons being that English characters are composed of simple cursive lettering and lack structural information. Therefore, a bi-linguistic recognition scheme is an effective and efficient method, in which different segmentation algorithms, different features extraction and different classification methods are applied to deal with the Chinese character or English character recognition separately even when both appear in the same document and even on the same line.

For the bi-linguistic recognition scheme (Ref. [6]), see Fig. 9, the first problem to be solved is to correctly distinguish whether it is an English character or a Chinese character. After that, the Chinese character recognition method and the English recognition method are applied separately to deal with the corresponding characters.

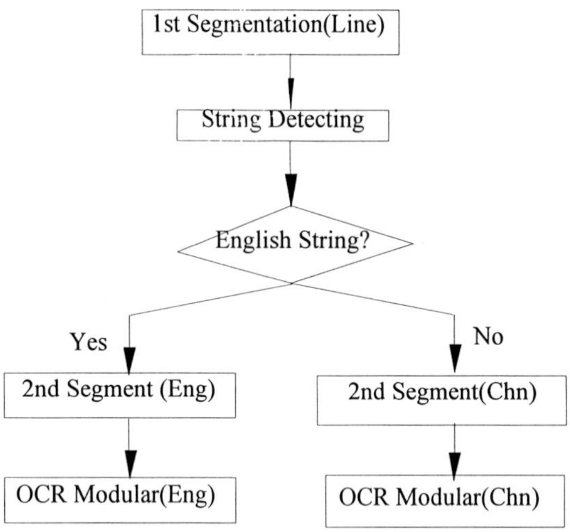

Fig. 9. Chinese-English bi-linguistic recognition scheme.

Distinguishing Chinese and English characters is based on the difference of their contour or shape, such as height, width, position and distance between neighboring characters. Chinese characters are block ideographs, with nearly uniform width, uniform height and uniform distance between neighboring characters. But, English has different widths, heights and are closer or even touching each other. This method sometimes has barriers, because there are a few separable Chinese characters, like "八", "北", "川" and "别", etc., which are composed of vertical disjuncted parts. The width of these parts is similar to the English character. Vice versa, some capital English characters have the same width and height as the Chinese character. So that some compensating methods such as English character segmentation and recognition methods have to be used and are incorporated for distinguishing the English characters. The correctness of the identified English character should be confirmed by character recognition, that is dependent on the recognition confidential distance. If the confidential distance is lower than a threshold, the "character" will be thought as a part of the Chinese character. Otherwise it will be an English character. Combining the parts of the Chinese character, it should be recognized again by Chinese character recognition algorithms.

It is unnecessary to confirm the correctness of the identified Chinese character, because Chinese and isolated English characters can be recognized by the convenient Chinese character recognition system, (Ref. [5]).

7. Printed Chinese Character Recognition System

A recognition system is typically composed of the following parts:

- scan inputting document pages,
- preprocessing: binarization, document analysis, segmentation, normalization, feature extraction, etc.
- classification: preclassification, recognition
- postprocessing
- outputting of results into an application

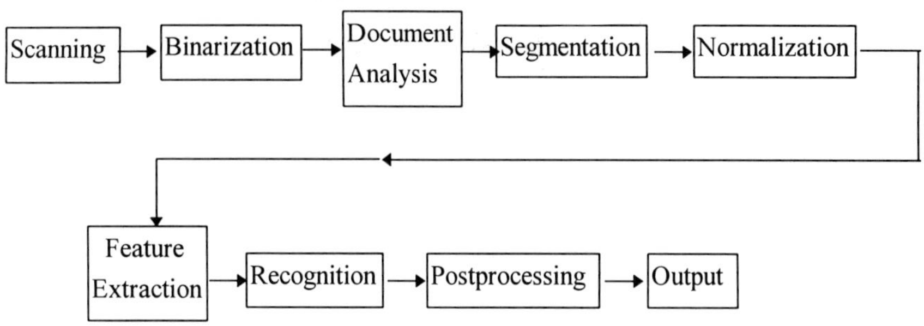

Fig. 10. Printed Chinese character recognition system.

7.1. Binarization of Scanned Image

The document image is the electronic form of a printed document created by the scanner producing a gray scale image, then the binary document image can be obtained through the binarization of this scanned image with a certain binarization threshold.

The threshold affects the quality of the binary image and, therefore, the recognition results seriously. If the selected threshold is too high, the image will be too white and many character strokes will be broken, refer to Fig. 10(c), if the selected threshold is too low, the image will be too dark and many character strokes will be merging and touching, refer to Fig. 10(b). Both situations decrease the recognition accuracy sharply. Careful selection of binarization threshold or adaptive modulation of binarization threshold is necessary.

There are two kinds of threshold selection algorithms. One is global, in which the same threshold is used in a whole page. The other is local, i.e. different thresholds adapting to the different local image such as different backgrounds, different font texts, etc., are used. It is clear that the adaptive local threshold can get better binarized results than a global one. Because of the limitations of computer memory and speed, generally the global threshold binarizing algorithms are used in some commercial Chinese character recognition systems, such as TH-OCR system. Considering the histogram of scanned gray image, a lot of gray image segmentation algorithms, such as maximal entropy method, some algorithms of maximizing separability between black and white, etc., can be used for automatic binary threshold selection. Otherwise sometimes the threshold selection is made by manual adjustment. In general, a suitable threshold, see Fig. 11(a), is necessary for the recognition to succeed.

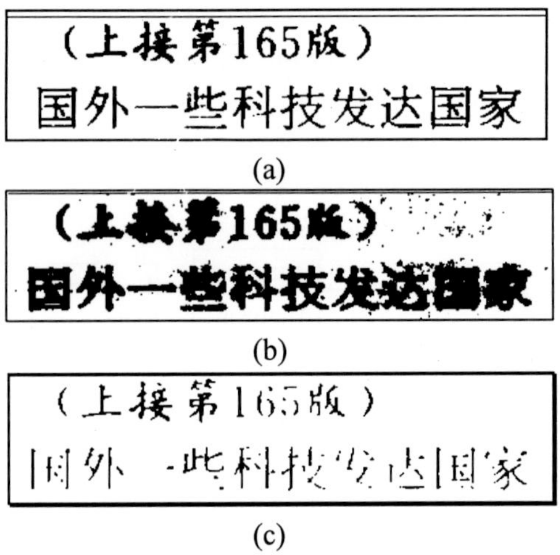

Fig. 11. Binarization with various thresholds. (a).Suited threshold. (b). Threshold too low. (c). Threshold too high.

7.2. Document Layout Analysis

A document image is the visual representation of a printed page such as a journal article page, a technical paper, an office letter, newspaper, etc. Typically, it consists of blocks of text that are composed of some columns with horizontal text lines or some rows with vertical text lines and are interspersed with tables, figures and headings in special fonts.

The document decomposition and structural analysis can be divided into three phases: The first is area segmentation that the document is decomposed into several rectangular blocks or some special areas. The approaches for segmenting document image components can be either top-down or bottom-up. The top-down techniques divide the document into major regions which are further divided into subregions based on the knowledge of the layout structure of the document. The bottom-up method progressively refines the data by layered grouping operations.

The second is area classification. The aim is to get an assignment of attribution (title, regular vertical text, regular horizontal text, picture, table, etc.) to all the blocks or special areas, using projection properties of individual blocks as well as some spatial layout knowledge, for instance, the texts have horizontal or vertical periodic projections, etc.

The third is logical grouping and ordering of blocks. For the OCR it is necessary to group and order text blocks to get whole recognized text files.

Document layout analysis and understanding are very important for automatic document recognition or document reading. However, it is not easy because of the flexibility of the document layout. Document layout is very complex to analyze automatically, especially for the newspaper, since every block may be too closely neighbored to separate, and sometimes a group of areas is untidily arranged, so that is difficult to separate, groupe and order. The most important thing in understanding the document layout is to use human knowledge, so that the research and development still goes on.

7.3. Character Segmentation

Text segmentation is composed of line segmentation and character segmentation. The line segmentation is to segment the text body into separated text lines, based on the line projection information usually. The character segmentation is a procedure to separate the text line into individual characters for recognition. The target of segmentation is to get an individual character and decide if it is an English character in a Chinese-English bi-linguistic character recognition system.

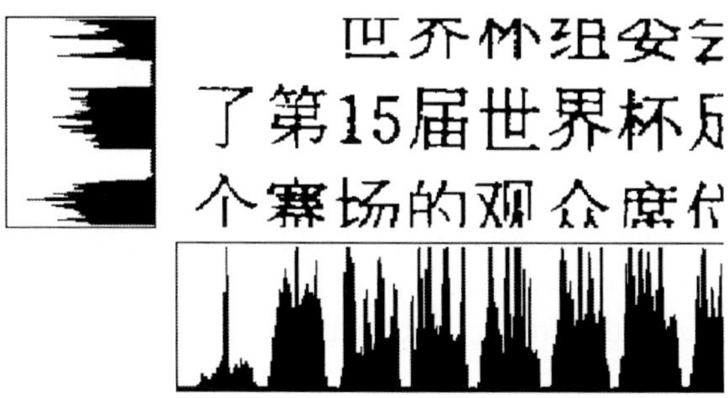

Fig. 12. Vertical and horizontal projections for character segmentation.

The correctness of character segmentation has a great impact on recognition accuracy. One segmentation error will cause at least one character recognition error. The emphasis on the accuracy of segmentation is necessary. The similarity of the width and height between an English and separable Chinese character components make segmentation difficult. The twice-segmented algorithm proposed, which includes pre-segmentation and second character segmentation is targeted to solve these problems.

Pre-segmentation is to segment every separated line text into character string groups based on the vertical projection information of characters and to detect the attribute of these groups, such as width, height, space, distance, etc. These will be used to determine if the group is English.

The second character segmentation has two branches, which is represented in Fig. 9. One is English string segmentation, segmenting touched strings into individual English characters using English segmentation methods to deal with these touched English strings, which are a special connected domain based method and a recognition method. Another is Chinese character string segmentation. Segmenting touched Chinese characters and combining separated parts of a Chinese character into an individual Chinese character are based on regular width and height knowledge of Chinese characters (Ref. [5]). The segmentation results of a Chinese and English mixed layout document are shown in Fig. 13.

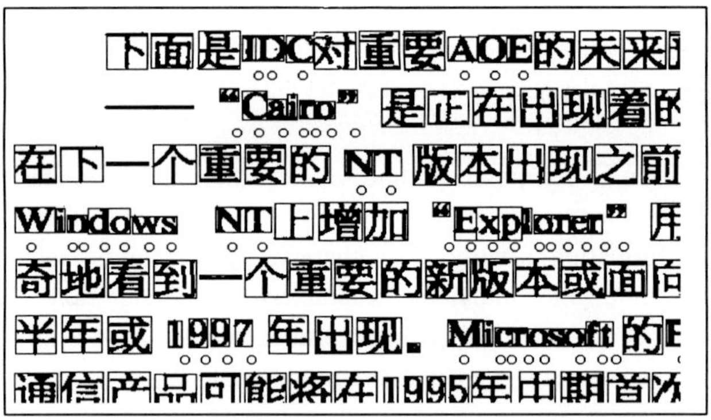

Fig. 13. Segmentation result of Chinese-English document.

The special problem is caused by Chinese characters in vertical text lines, in which the line and character segmentation is somewhat different with a horizontal text line. The difficulty is from some vertically separable characters, such as "二", "三" and "吉" etc., which are subject to being segmented incorrectly. The same method as that of the horizontal line text has been taken, except for line segmentation with vertical projection information and character segmentation with regular height information.

Experiments show that the twice-segmented algorithm is effective and efficient for Chinese or Chinese-English bi-linguistic character recognition system (Refs. [6], [10]).

7.4. Character Normalization (position and size normalization)

Successful machine printed character recognition methods are mainly structural feature based statistical pattern recognition methods, in which the character normalization affects the recognition result greatly, because features are extracted from normalized character image array. Normalizing every unknown segmented character image must be implemented carefully before feature extraction and classification. There are both position and size normalizations.

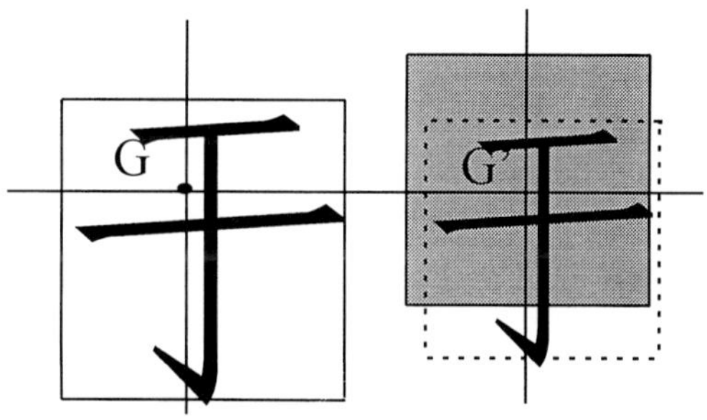

Fig. 14. Chinese character normalization (gravity center and size normalization).

There are two kinds of position normalization methods. One method is to use the outline contour center as the center of a normalized image array, that is sensitive to the noises affecting the outline contour. The other is to use the gravity center as the center of the normalized image array, which is insensitive to the noise, but difficult to reserve the whole body of the character in a certain block area (see Fig. 13). Some compromises between the two kinds of position normalization should be made, and some non-linear size normalization methods may be needed.

7.5. Post Processing of Recognized Text

What we discussed above is confined to isolated Chinese character recognition taken out of context. While Chinese characters predominantly occur as parts of words, phrases and sentences. A poorly formed or degraded Chinese character in isolation may be unrecognizable or incorrectly recognized, while, taking the context into consideration, recognition may be possible.

Contextual postprocessing can use context knowledge at the word level, at the sentence level or at a higher linguistic level in Chinese document recognition. An obvious method for representing the word knowledge would be to store the list of legal words or a lexicon. Another method is to use Markov models that capture first and higher order character transitional probabilities.

Contextual postprocessing is necessary to correct the recognition results with contextual knowledge to select the expected words. For example, the Chinese character recognition technique may not be able to distinguish reliably between character “革” and “草”. But if the next character is “新”, postprocessing will determine that “革” is correct, since “革新” is found in the Chinese lexicon.

A statistical method can be implemented using the hidden Markov methodology at word level. It is assumed that the context relation of a Chinese character string is described by a first order Markov transitional probability model. Upon the word-frequency-statistics of the Chinese language, the mean word length is about two Chinese characters. Therefore, only the directly neighbored two Chinese characters are processed. Longer words can be taken as a composition of several words of two characters. For example, the four character string “改革开放” can be decomposed as “改革” + “革开” + “开放” as two Chinese character words.

The transitional probabilities in the Chinese lexicon are estimated with a large Chinese corpus. For character-based word recognition, decisions are dependent on maximizing *a posteriori* transitional probability. It is to compare *a posteriori* transitional probabilities among the recognized characters, which recognized characters considered are in the context lexicon and in the ordering recognized results according to the distance with an unknown sample character.

The context postprocessing, the same as the character frequency treatment may sometimes have a positive result, but sometimes may cause negative effects. From the statistical point of view, if positive effects exceed negative ones, the context postprocessing will improve recognition accuracy.

In a Chinese character recognition system, many important factors affect the recognition accuracy, such as scan image resolution (enough higher scan resolution, such as above 300 dpi in general), binarization threshold, image skew (deskew is needed), character segmentation, character normalization, feature extraction, classification recognition and post-processing. If one of them is degraded, perhaps the recognition has failed or has degraded, so that better efforts must be put at every step in the printed Chinese recognition procedure.

8. Chinese Character Recognition System and Application

Chinese character recognition research in China can be roughly divided into two periods. The first period was in the 1980s, in which about 10 printed Chinese character recognition systems were developed. The second period was in the 1990s, in which a few of them were continually developed as applicable systems and appeared in the market as commercial products. For example, the first is TsingHua OCR (TH-OCR; TWReader for Windows) (Refs. [4], [6]), it is a Chinese-English bi-linguistic multifont printed character recognition system developed by Tsinghua University and now applied worldwide. The others are BI-OCR(Ref. [7]), developed by the Beijing Institute of Information Engineering and SY-OCR(Ref. [8]) developed by Sengyang Automatic Research Institute. They are also marketed in the China.

8.1. *Performance of Printed Chinese Character Recognition System*

For a practical printed Chinese character recognition system, the main specifications are:

(1)Recognizable character set

- Character categories: more than 4000 simplified Chinese characters or 5401 traditional Chinese characters, and English characters, punctuation marks, etc.
- Multi-fonts and various types of characters
- Omni-sizes of characters

(2)Recognition speed

(3)Recognition accuracy

The performance of the recognition system can be represented by the recognition rate P_C, rejection rate P_J, error recognition rate P_E (substitution rate) and reliability rate P_R, the relations among them are

$$P_C = 1-(P_E + P_J) \tag{8.1}$$

$$P_R = P_C/(P_C + P_E) \tag{8.2}$$

The requirements for recognition accuracy include not only a higher recognition rate P_C, but also a lower error recognition rate P_E. Sometimes the lower error rate P_E is much more important than the higher P_C. Therefore, some comprehensive methods of combining multiple classifiers have been developed, the most simple one is the voting method of multiple classifier decisions, which can reduce error rate P_E drastically and achieve a higher recognition reliability P_R than a single classifier decision at the cost of increasing computing complexity and reducing the recognition rate P_C. For example, the NI-OCR, an integrated machine printed Chinese character recognition system (Ref. [13]), combines three Chinese character recognition systems, i.e. TH-OCR, BI-OCR and SY-OCR by voting method, and obtains the lowest recognition error rate less than 0.2% and recognition rate above 98%.

(4) Recognition Robustness

The robustness is the most important for a practical recognition system, which represents the anti-noise ability of the recognition system facing various practical situations. This is an critically important point, that the higher recognition rate still remain in a poor application situation.

(5) Friendly user-interface and automatic operation

8.2. *A Practical Printed Chinese Character Recognition System*

In order to introduce the recent progress and advances in machine printed Chinese character recognition, THOCR-94, a multifont omni-size Chinese and English bi-linguistic printed character recognition system (Refs. [6], [10]), popularly used in China and worldwide, is described as follows.

THOCR developed by the Deptment of Electronic Engineering, Tsinghua University in 1994, is the first Chinese-English bi-linguistic printed character recognition system in the world. It can recognize four thousand simplified Chinese characters or 5401 traditional Chinese characters with multi-fonts (such as Song, Fang-Song, Black or Hei, Kai, WeiBei and XiaoYao and so on), omni-size and different types. It also can recognize more than 30 kinds of upright and slanted English character fonts and a lot of punctuation marks simultaneously.

It obtains the highest correct recognition rate of 98.6% tested by 240,000 Chinese characters in a practical document. For more than 30 kinds of upright and slanted English character fonts, its correct recognition rate is between 97% and 99%. For numerical recognition only, the rate can be as high as 99.5%.

The most distinguishing aspect is that THOCR has more robust strength than any other system, and can especially deal with the degraded printed documents and variations of printed fonts and character types, whether computer print-outs or photocopies and also faxed printed documents, a higher recognition performance is still maintained.

Chinese character recognition systems are now more often used in office automation, documents storing, re-editing, retrieving for Chinese information computer entering (electronization) automatically.

Recognition systems are also used for Form Recognition and Database automatic entering and Chinese-English Business Card Reader, etc. (Ref. [12]).

9. *Summary*

The theory, method and practice on printed Chinese character recognition are introduced in this chapter.

First, we introduce the problems faced by Chinese character recognition, especially for a large set of Chinese characters and the variations of Chinese types. Second, we discuss recognition methods and feature selection or extraction, Third, describe the structure features based statistic recognition methods which are effective for machine printed Chinese character recognition. Based on the Information Entropy Theory in Pattern Recognition we explain why feature extraction is so important and how it decides the recognition performance. Some effective stroke structure based features are introduced. Fourth, minimum error classifier, tree classifier design and some compromises between recognition speed and accuracy are discussed. Fifth, we discuss the compositions of printed Chinese character recognition system. Finally, the most popular used practical printed Chinese character recognition system, TH-OCR, is introduced.

Great progress in printed Chinese character recognition has been made, but problems still exist, such as reducing error rate, strengthening robustness, etc. so that continual research and development is needed and the pursuit of new techniques continue for the development of the applications of Chinese character recognition.

References

[1] Y. S. Wu and X. Q. Ding, *Chinese Character Recognition: Theory, Method and Practice* (High Educational Press, 1991, Beijing,.China).

[2] X. Q. Ding, Information entropy theory in pattern recognition, *Acta Electronica Sinica*, .**21**, 8 (1992) 2- 8.

[3] P. A. Devijver, On a new class of bound on Bayes risk in muitihypothesis pattern recognition, *IEEE Trans. Comput.* **C-23**, 1 (1974) 70- 80.

[4] X. Q. Ding, Y. S. Wu, Recognition of multi-font printed chinese characters by structure analysis, .*CCIPP/CLCS,* 1988, Toronto, Canada.

[5] Z. Zhang, X.Q. Ding, Twice-segmentation algorithm on mixed printed Chinese and English text, *Proc. ICCC94*, Singapore, June, 1994, 396- 402.

[6] H. Guo, X. Q. Ding, The development of high performance Chinese/English bi-linguistic OCR system, *Proc. CMIN'95*, Beijing, China, March, 1995, 248-253.

[7] X. Z. Zhang, *et al.*, Printed Chinese document recognition system, *Proc. 4th National Conf. Chinese character and speech recognition,* China, 1992, 168-173.

[8] L. Yang and Y. P. Wu, Multi-functional realizable Chinese character recognition system, *Proc. 4th National Conf. Chinese character and speech recognition*, China, 1992, 174-179.

[9] X. Q. Ding and F. X. Guo, Advances in recognition of printed Chinese character, *Proc. the 1st China-Korea Joint Symp. Machine Translati*on, CKJSMT'94, 1994, China, 129-134.

[10] H. Guo, X. Q. Ding, Realization of high-performance bilingual Chinese-English OCR system, *Proc. 3rd ICDAR'95*, Montreal, Canada, 1995, 978-981.

[12] J. H. Liu, X. Q. Ding and Y. S. Wu, Description and recognition of form and automated form data entry, *Proc. 3rd ICDAR'95*, Montreal, Canada, 1995, 579-582

[13] Y. H. Zhang, C. P. Liu and G. J. Li, Integrated printed Chinese character recognition system, *Proc. ICCC94*, Singapore, 1994, 107-113.

[14]. Y. X. Gu, Q. R. Wang, C. Y. Suen, Application of multilayer Decision tree in computer recognition of Chinese characters, *IEEE Trans. PAMI* **5**, 1(1983) 83-97.

[15]. P. P. Wang, R. C. Shiau, Machine recognition of printed Chinese characters via transformation algorithm, *Pattern Recognition* **5** (1973) 303-321.

[16] S. Mori, C. Y. Suen, Historical review of OCR research and development, *Proc.IEEE* **80**, 7 (1992) 1029-1057.

[17]H. Bunke and R. Liviero, Classification and postprocessing of documents using an error-correcting parser, *Proc. 3rd ICDAR'95*, Montreal, Canada, 1995, 222-226.

Handbook of Character Recognition and Document Image Analysis, pp. 331–355
Eds. H. Bunke and P. S. P. Wang

CHAPTER 12

CHINESE CHARACTER RECOGNITION IN TAIWAN

HSI-JIAN LEE
Department of Computer Science and Information Engineering
National Chiao Tung University
Hsinchu, Taiwan, 30050, R.O.C.

Research interest in Chinese character recognition in Taiwan in recent years has been intense, due in part to cultural considerations, and in part to advances in computer hardware development. This chapter addresses coarse character classification, candidate selection, statistical character recognition, recognition based on structural character primitives such as line segments, strokes and radicals, as well as postprocessing and model development.

Coarse character classification and candidate selection are used to reduce matching complexity; statistical methods of character recognition are shown to be effective; feature-matching which shows good performance is reported; and, structural-based methods able to distinguish between similar characters are investigated thoroughly. Since no temporal information is available for off-line recognition systems, the character test base is still limited. Methods used to extract structural primitives are also investigated.

Language models based on syntactical or semantic considerations are used to select the most probable characters from sets of candidates, and are applied in postprocessing in input sentence images. These models generally employ the dynamic programming methods. To increase identification capacity, various ways of grouping Chinese words into a reasonable number of classes are also proposed.

Keywords: Chinese character recognition; Candidate selection; Statistical character recognition; Structural character recognition; Language models.

1. Introduction

The development of Optical Character Recognition (OCR) systems is motivated by the need to cope with the enormous flood of documents such as bank checks, commercial forms, government records, credit card imprints and posted letters. Cultural considerations and the rapid development of computer hardware capacities have focused the attention of many researchers in Taiwan on Chinese character recognition. Researchers from universities such as National Chiao Tung University, Cheng Kung University and Central University, Academia Sinica, as well as research institutes like Telecommunication Laboratories (TL) of the Ministry of Transportation, Computer and Communication Research Laboratory (CCL) of the Industrial Technology Research Institute (ITRI), have concentrated on studies of document analysis and Chinese character recognition.

Characters typed or written on documents are scanned and digitized with optical scanners to produce digitized images. OCR systems then locate regions (usually guided by marks pre-printed on the input document) in which data have been printed or written on the input documents. Once these regions have been located, the data blocks are then segmented into character images. Instead of keeping the images in gray scale, it is a common practice to convert them into binary matrices to save memory space and computational effort. Noise elimination and normalization of size, orientation, or position may then be performed to facilitate extraction of distinctive features. Once the characteristics of the cleaned characters have been extracted, they are classified by comparing them with a list of reference characters in a knowledge base generated during the learning process.

Many different techniques are used for character recognition; they may be grouped into three general approaches: statistical methods, structural methods, and hybrid methods [1]. Statistical methods use a set of characteristic measurements usually called "global features" extracted from the characters to identify characters by partitioning the feature space. However, since in Chinese the characters under study are very complex and the number of classes is very large, the number of features required for recognition also becomes very large. Several strategies have been proposed to handle these problems and are described in Sec. 2. Structural methods express characters as compositions of structural primitives such as lines, curves, and loops, and then identify the characters by matching representations of its primitives with those of a reference character, or by parsing the representations according to sets of syntactic rules. These methods are more tolerant of irregularities in handwritten Chinese characters than statistical methods. However, characters cannot be rigidly described using grammar rules because of the existence of noise and numerical attributes. Since the advantages of one approach can remedy the shortcomings of another approach, hybrid methods that remove certain disadvantages have also been proposed.

Figure 1 outlines the general structures of an optical character recognition system. The modules in the system include preclassification and coarse classification, statistical and structural OCR, and postprocessing. Most systems proposed in the literature may be taken as a subset of this general system. In the following discussions, the processing flow indicated in Fig. 1 is followed.

The description of a character involves features and/or the rules that characterize the character. As mentioned above, the relevant studies are grouped into two major families: statistic-based methods and structure-based methods. From the literature, we find that the researchers of Chinese character recognition systems have put more emphasis on the structure-based methods, and have proposed many algorithms. To clarify the discussions, this family of methods is further divided into three subfamilies according to the structural primitives used. They are line segments, strokes, and radicals. These methods are discussed in Secs. 3 and 4; most of them were proposed recently by researchers in Taiwan.

We need to tolerate structural differences in input patterns of the same character, and to differentiate between distinct descriptions of different characters. If the character features are properly described, the algorithm should perform well.

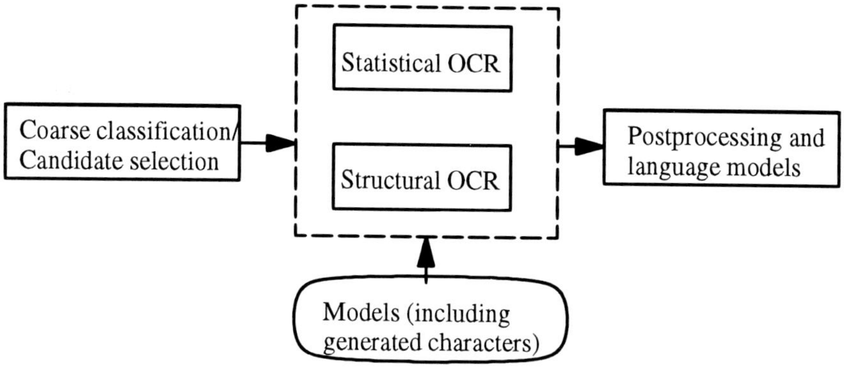

Fig. 1. The overall structure of an optical recognition system.

Anyway, the requirement is sometimes a contradiction. If we need to tolerate variations in a certain character, different characters with similar features may also be misrecognized as the same character. To solve this problem, postprocessing modules using the contextual information are proposed. They will be discussed in more detail in Sec. 5.

2. Coarse Classification and Candidate Selection

Character recognition systems match input characters with reference characters in a database. In Taiwan, the generally accepted standard for recognition system performance is the ability to recognize 5401 commonly-used characters (although several commercial products claim to have character sets exceeding 12,000 characters). However, since the Chinese language uses many thousands of characters, matching may be a lengthy process depending on the size of the database used, so time is a serious concern in Chinese character recognition system research. Two approaches have been proposed to address this concern: *coarse classification* algorithms and *hierarchical systems.*

2.1. *Coarse Classification*

These algorithms first use training samples to classify or cluster input characters into a relatively small number of classes. Chang and Wang [2] employed peripheral shape coding to pre-classify Chinese characters according to 37 fundamental stroke patterns, called radicals, which are then further partitioned into 25 categories. After they obtained the skeleton and feature points of an input character, they extracted four radicals from the four corners to form a codeword. Here they used end points, corner points, three-fork points and four-fork point as features. Based on the codewords, 5401 frequently-used handwritten characters can be classified into 2144 classes with 94% accuracy; each class contains 2.5 characters on average.

Cheng *et al.* [3] defined stroke substructures as a set of connected line segments such that any two strokes intersect or are L-connected. They defined 64 substructures to identify the 5401 characters. To separate the substructures in an input character, they first identified a salient line segment. Then, the nearby line segments were checked to see whether a possible substructure existed according to the built-up knowledge about the substructure. To pre-classify a character, all substructures extracted were arranged in a 1-D string according to a predefined order. The spatial relationships between two consecutive substructures were also inserted into the 1-D string. Characters with the same 1-D string were then grouped into the same clusters.

2.2. *Hierarchical Systems*

Hierarchical systems are constructed to speed up the matching process. These systems generally divide the recognition process into two stages. The first stage, which is called the *candidate selection module*, aims to reduce the candidate character set with respect to an unknown input character to a smaller set, called the *candidate set.* The precision of this module may be very high, for example, over 99%. The candidate selection problem has been studied by many researchers. Some well-known candidate selection techniques use features such as orthogonal expansion, stroke distribution, stroke analysis, background feature distribution and combinations of these. The algorithms are usually simple and fast to execute but are not powerful enough to distinguish between similar patterns. To identify uniquely the input character, more detailed structural information must be derived and elaborate algorithms have been designed to accomplish this. The second stage, generally called the *matching module*, aims to choose the best candidate from the candidate set. Note that if there is another postprocessing module, then several candidates will be chosen with respect to an input character.

A matching module with a candidate selection module is faster than one without, since the computation cost of a candidate selection module is much lower than that of a matching module. When a system is decomposed into a sequence of cascaded subsystems, its computation cost is the sum of the computation costs of the subsystems. The computation cost c_i of a subsystem i is proportional to the number of characters N_i in the database multiplied by the number of features f_i used in the subsystem; that is, $c_i = \alpha_i \cdot N_i \cdot f_i$, where α_i is a weight factor. The computation cost of a character recognition system without a candidate selection module is $\alpha \cdot N \cdot f$, where $N = 5401$ (the number of frequently-used characters) and f may range from 100 to 600. The computation cost of a recognition system with a candidate selection module is $\alpha_1 \cdot N_1 \cdot f_1 + \alpha_2 \cdot N_2 \cdot f_2$, where subscripts 1 and 2 denote the candidate selection and matching modules, respectively. $N_1 = N$, $N_2 \ll N_1$, f_1 is much smaller than f, and f_2 may be equal to f. Usually, $\alpha_1 \cdot N_1 \cdot f_1 + \alpha_2 \cdot N_2 \cdot f_2$ is smaller than $\alpha \cdot N \cdot f$.

A candidate selection module for handwritten Chinese characters may still entail high computation costs for processing highly dimensional feature vectors and handling a large database. Kumamoto *et al.* [4] proposed a system that speeds up the recognition process by adding a fast, high-precision candidate pre-selection module. The computation cost of the whole system is the summation of the three subsystems; that is, $\sum_{i=1}^{3} \alpha_i \cdot N_i \cdot f_i$, where $i = 1$, 2 and 3 represent the candidate pre-selection, candidate selection, and matching modules, $N_3 \ll N_2 \ll N_1$, and f_1=1. Tung *et al.* [5] proposed a model with $s+2$ stages of subsystems to minimize the computation cost $\sum_{i=1}^{s+2} \alpha_i \cdot N_i \cdot f_i$. The number of stages, s, in the candidate pre-selection module is determined by estimating the expected execution time required to process a set of training characters. At each stage, a specific feature of the input pattern is measured. This feature value is then compared with that of each character remaining in the candidate set. If the difference between the features is too large, then the character in the candidate set is removed. The comparison process is repeated stage by stage, and the number of characters in the candidate set decreases gradually. The characters remaining in the final stage are fed into the candidate selection module for further processing.

Lin and Fan [6] proposed a structure-feature based method to classify on-line handwritten characters into six structure types. They are one-element, left-right, up-down, up-left, left-up-right, and left-down. Since temporal information is included with on-line characters, the coordinates of the on-line data points can be represented as a function of time, and denoted as $X(t)$ and $Y(t)$. The maximum x and y coordinates from time 0 to time t, denoted as Max $X(t)$ and Max $Y(t)$ can be computed. According to these values, the structure types can be determined. For example, if Max $Y(t)$ reaches $HalfCharSize$ earlier than Max $X(t)$, then the character has a left-structure, where $HalfCharSize$ is half of the length of the character-bounding rectangle. After the structure type of an input character is determined, the character can be divided into separate parts accordingly. The insufficiency of this proposed method is that there are probably more than two components in most characters. Further decomposition of the components is sometimes necessary.

3. Statistical Character Recognition

The study of statistical character recognition has a long history. Some commercial Chinese character recognition systems which are now available use this approach with some modifications. Although there have been some problems, these systems are effective and can be implemented easily. The features used in this approach can be divided into two categories: local and global features.

The 2-D character image is the most direct representation for a character. Template matching, which directly compares the pixels of the unknown character with those in reference character images, is applied after some normalizations are performed on the unknown character. Linear normalization has been universally applied, but several non-linear normalization methods have been proposed for handwritten Chinese character recognition.

Point distribution features can be detected more easily, but they are less immune to noise and local distortions. A better way to describe characters is to adopt *local* properties, also called *geometrical* and *topological* features, including hook points, end points, cross points, corner points, T-shapes, and loops. These types of features have been selected to reflect structural information about characters. If the relations between these local properties are not specified, then the recognition scheme can be either a decision tree or a statistical matching method such as a Bayes classifier. Note that decision trees are generally used in the preclassification stage. As for the statistical method, the number of features required for Chinese character recognition can become very large and it is ineffective or computationally infeasible to solve this kind of problem.

The three features, peripheral background area (PBA), contour line length (CLL), and crossing counts (CNT), are frequently referred to by local researchers [7]. The first of these is defined as the average position of the first black pixel horizontally or vertically. The last is the average of the number of changes from white pixels to black pixels in a row or a column. They can be computed efficiently and have demonstrated good recognition abilities.

Transformation type features are designed for global use. They require large storage and long computation. Orthogonal transformations such as Walsh, Fourier, 2-D moment and Karhunen–Loeve are used to transform 2-D character images into new domains. Huang and Chung [8] used the Walsh transformation to separate similar complex Chinese characters into classes. They first performed coarse classification to partition the commonly-used characters by adopting 4C and 4P codes, where the 4C code is defined by encoding each of the four corner zones of a character into two levels and the 4P code is defined by encoding each of the four peripheral rectangular zones into 32 levels according to the number of points having some particular runs of black and white pixels. By using these codes, they partitioned 5401 Ming-font characters into 4096 groups, with 1–6 characters in each group. They then extracted the central portion of the input image and performed the Walsh transform. Since each group had at most six characters, they selected 2–5 Walsh coefficients with the most separability power to classify the given characters.

Tu and Ma [9] proposed the Sectionalgram, which used the layers obtained by dividing 2-D character images into several blocks, to compute the cumulative occurrence distributions. Character recognition was then conducted by a Markov dynamic programming procedure.

Li and Yu [10] utilized the Bayes rule to classify 5401 characters. An input character is first thinned to reduce written thickness variations. Then the character is normalized into a 60×60 image, and divided into a 10×10 grid frame according to the marginal densities of black pixels. The numbers of black pixels in four directions (horizontal, vertical and two diagonal) of each grid are counted and divided by the total number of black pixels to represent the probability distribution of the character. An 88.65% recognition rate has been claimed by using the Bayes decision rule for character classification. Four subsidiary features are designed to increase the rate to 93.43%.

Jeng [11] described a system built on a personal computer with three extension boards: a preprocessor board, a feature-extraction board, and a matching board. The features he used are called accumulated stroke features. He computed the lengths of the strokes going through each pixel along the horizontal, vertical, and two diagonal directions. Then he reduced the four direction matrices into 8×8 matrices by summing all elements in each block. A byte was used to represent the major stroke type in each block. That is, bits 0 to 4 represented no stroke, vertical stroke, stroke along 135° direction, horizontal stroke and stroke along 45° direction, respectively. In this way, an 8×8 feature matrix can be used to represent a character, in which each element needs a single byte. If a learning-set character has n samples, then the n 8×8 feature matrices can be logically OR-ed together. To recognize an input character, an AND operation can then be performed.

4. Structural Character Recognition

Human beings usually view characters as hierarchical organizations. From the lowest level to the highest level, the hierarchical organization can be classified into segment, stroke, radical, and character. Recognition based on the organization is called *structural character recognition*, and recognition algorithms can be divided into three categories: segment-based, stroke-based and radical-based.

In on-line character recognition, temporal information about the writing can be captured and the order of the strokes can be used to transform the unknown character into a 1-D stroke sequence. The problems encountered in on-line character recognition include stroke-order variations and stroke-number variations. In off-line character recognition, no dynamic information can be used. Different combinations of 1-D stroke sequences organized as a tree or a graph have been presented.

4.1. *Methods Based on Line Segments*

Chinese characters writing primitives are strokes that can be approximated by sequences of line segments. In the structural approach, the strokes of an input character are first transformed into unit-width thin lines. In the thinning process, T-junctions or cross junctions in those strokes are usually very difficult to process. A well-known result is that a cross junction is thinned into two Y-junctions, as shown in Fig. 2.

The strokes of a handwritten character can be written with different lengths and different directions by different writers or by the same writer at different times. If we can segment long strokes into short line segments, then most line segments can still be identified. If two strokes are incorrectly connected or touch as the two circles in Fig. 3 do, using short line segments can also solve partial recognition problems. Lee and Chen [12] adopted a dynamic programming method to calculate the similarity between a short line segment of an unknown character and a segment

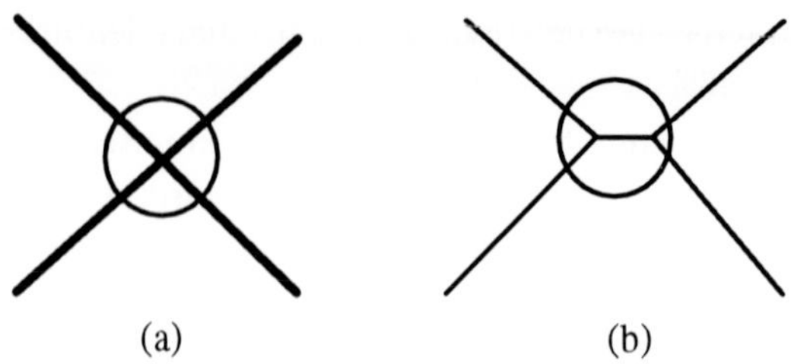

Fig. 2. The thinning of a cross junction. (a) The possible input. (b) The possible thinning result. There are two Y-junctions in the circle.

of a reference character provided that the neighboring segments were sorted by angles. The problem with this approach is that it is time-consuming because a large number of primitives must be presented in the input and reference characters.

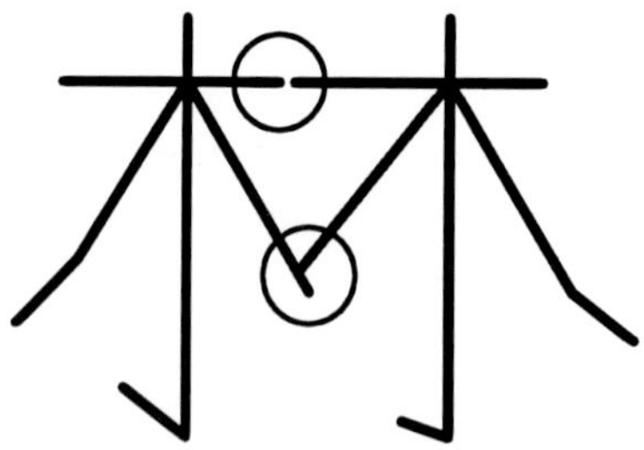

Fig. 3. A Chinese character. The two circles may be considered connected strokes.

If a reference character is represented as short line segments along with a neighboring segment list, then the *relaxation technique* can be applied to match an unknown character with a reference character [13]. The relaxation technique propagates local information via iterative processing to relax the ambiguities between objects and labels.

In on-line character recognition, a reference pattern can usually be represented as a sequence of line segments. String matching techniques can then be used to recognize the unknown character. Hsieh *et al.* [14] proposed stroke-order and stroke-number free bipartite weighted matching for on-line handwritten Chinese character recognition. The similarity measure between line segments of input and reference characters is defined by length, coordinates, and orientation. Since some line segments in the input character may not appear in the reference character, and vice versa, they introduced unmatched line segments associated with penalties. Their matching goal becomes that of finding a match such that the sum of the weights of matching edges and the penalties of unmatched vertices is minimized. The matching result is generated by the Hungarian method, which depends on the given costs and penalties. In order to obtain a good match, the distance between line segments and the penalties of line segments are given carefully. They further proposed a greedy algorithm based on the Hungarian method to restrict optimal match that satisfies the constraints of geometric relation. For each iteration in the greedy algorithm, a matched pair is deleted if their relation to their neighbors does not match. A new match is then found by repeatedly applying the Hungarian method until they find a stable match that preserves the geometric relation has been found.

In handwritten characters, the angle of a segment or a stroke may vary widely. Usually, the angle can be assigned to one of four or eight ranges, which can also be uniformly divided or non-uniformly divided. To increase segment classification robustness, Chou *et al.* [15] divided the angles into four ranges: $A = \{(295°, 340°)$, $(255°, 285°), (185°, 245°)$, and $(350°, 70°)\}$. A unique code was assigned to segments in each of these ranges. If a segment had angles of $(245°, 255°)$, or $(285°, 295°)$, or $(340°, 350°)$, it was assigned to the two neighboring ranges in A.

Chou *et al.* [15] tried to recognize connected-stroke characters on-line. They argued that if the additive segments can be identified, then the method used to recognize regular-style characters can be used to recognize connected-stroke style characters. They assigned each segment to one of these three categories: real, virtual, and null. The real segment is the one existing in regular style characters; the virtual and null segments result from connected strokes. Since segments with left-up direction $(70°, 185°)$ are usually unstable segments, they are considered useless and are called null segments. Chou *et al.* first deleted the null segments from the input stroke sequence. Then they identified the linked strokes with a finite-state recognizer from the remaining segment sequence, in which all segments of linked strokes belong to real segments. If the subsequence was not recognized, then the segments in the subsequence were taken as real and virtual segments in turn.

They used an adaptive search range of segment numbers in the first coarse classification stage. The possible stroke-number deviation of each radical was built into a knowledge base. The existence of virtual segments in the radical was used to modify the possible number of strokes. According to the modified stroke number, candidate radicals could be determined. In handwriting, radicals are written one by one. The second coarse classification selects the first and the last candidate radicals by traversing the first radical deviation tree forward and the last radical deviation tree backward. In the character-matching phase, a matching tree is constructed. The evaluation of the similarity function is implemented by using the A* algorithm. After this phase, some confusion sets may still exist. A universal recognition mechanism is used to select the most likely candidates. They used relative geometric positions between segments by computing the direction angles among the matching pairs of stable segments. The candidate with the smallest accumulated distortion distance from the input character was taken as the final result.

Chou and Tsai [16] proposed an iterative relaxation scheme to match stroke segments in the input character to those in reference characters. In this iterative scheme, at most one segment in the input character can be matched with a segment in the reference character, and vice versa. This constraint generates compatible and incompatible matches involving the best and second-best matches for a given stroke. If the second-best match competes with the best match, the strength of the best match is reduced; if the second-best match does not compete, then the strength of the best match is increased. By repeating this process, the system will eventually achieve a global stable matching. Once the matching is completed, a similarity measure is evaluated, and the best match is chosen.

The disadvantages of matching segments include the following: (1) The basic primitives of a character are strokes. In on-line recognition of regular handwritten characters, segments are seldom used. (2) The chances of having a one-to-one correspondence between reference segments and input segments are lower than that between reference and input strokes. (3) The large number of segments involved in matching causes a large amount of computation. On the other hand, to model a cursive (connected-stroke) handwritten character in an on-line recognition system, line segments are taken as more natural primitives than strokes because it is very difficult to predict the possible connections in consecutive strokes. This can be taken as an advantage of using segments as matching primitives.

4.2. *Methods Based on Strokes*

A stroke in handwriting on paper is the writing primitive made during the pen-down to pen-up interval. In on-line character recognition, a stroke can be identified naturally. Nevertheless, an on-line stoke in connected-stroke style handwriting may correspond to several regular handwritten strokes. In off-line character recognition, there is no temporal information available, so extracting the strokes from a character is a difficult task. Many methods related to the extraction of strokes are reported. Figure 4 shows the primitive strokes generally used. This section first discusses the methods for on-line and off-line character recognition and then clarifies the methods for stroke extraction.

Lin *et al.* [18] proposed an on-line character recognition approach involving a deviation-expansion model for representing reference character stroke order and stroke number variations. They are built on knowledge of the reference character. For example, if a character has four strokes, the stroke-order deviation is (-1, 1) and the stroke-number deviation is 0. This means there are four different stroke order sequences: 1243, 1324, 2134, and 2143. Pattern matching is done by a matching graph constructed according to the deviation-expansion model of the reference character and the stroke sequence of the unknown input character. Using the graph, a similarity function was evaluated by utilizing a dynamic programming matching method. The cumulative classification rate of choosing the ten most similar characters is claimed to be 98%. The drawback of this approach is that construction of the deviation knowledge is very subjective and very time-consuming.

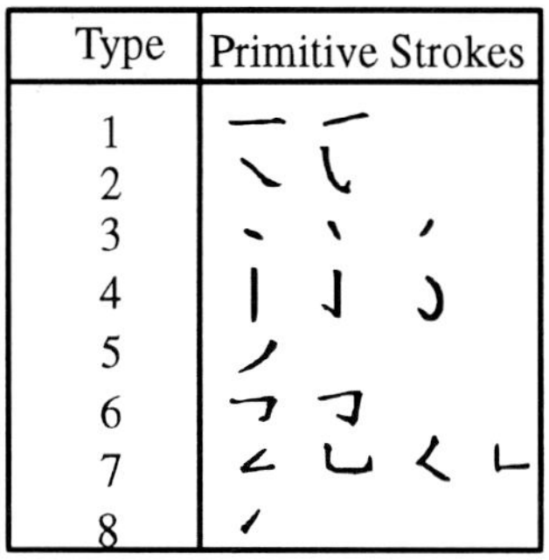

Fig. 4. The primitive strokes.

On-line recognition by Chen *et al.* [19] utilized the input stroke sequence to represent an unknown character as a 1-D string. Matching was then conducted, which is an easier approach than off-line character recognition. In off-line character recognition, strokes are not input one by one. Thus, angles, positions, and stroke properties must be adopted to rearrange the extracted strokes into a 1-D string. A difficulty for these methods is that the stroke order may not be unique if the character is written with rotation, translation, or scaling. A better method for transforming the 2-D strokes into a 1-D representation is still needed.

The most elaborate algorithms for off-line handwritten Chinese character recognition use strokes as primitives for character description. They can be grouped according to the degree of stroke relationship usage. The simplest method uses no relationships. Character recognition merely finds the stroke correspondence. Cheng *et al.* [20] developed a fuzzy set approach to recognizing handwritten Chinese characters. They described a stroke by its type (including slope and direction) and location. Two fuzzy set membership functions are defined for location and type of strokes. A function of fuzzy entropy was used to measure the stroke similarity. All similarities between corresponding strokes were obtained by solving the assignment problem using the fuzzy entropy cost function, and then averaging to derive the similarity between two Chinese characters. Note that they applied the Hungarian method to solve the assignment problem.

Wang *et al.* [21] used relaxation methods to recognize handwritten characters. They found strokes by linear fitting and quadratic curve-fitting algorithms after thinning. Stroke features included center point coordinates, length, slope angle, crossing points, and the angles between any two strokes. After stroke extraction and size normalization, redundant strokes were detected and deleted, which greatly reduced the matching time. A redundant stroke was defined as a short line segment that forms an acute angle with another longer line segment at one end point and has no connecting line segment at the other end point. After relaxation matching, certain rules designed to solve the following problems were applied. (1) The corner point between two connecting segments being replaced by short line segment. (2) Two line segments accidentally connected at their end points. These situations were determined by detecting whether a small matching probability in a stroke existed and if the stroke length, angle, type and so on satisfied a constraint. Some of these distorted characters were reconstructed and their feature vectors were modified to reflect the change. They were then sent back through the feedback relaxation matching process once again.

From the syntactic recognition point of view, a graph is the most direct representation of a 2-D character. Chen and Lieh [22] constructed two-layer attribute graphs to represent characters by using radicals and strokes as primitives. In the first-layer graph, the primitives were the character radicals and the relations between them were their relative locations. In the second-layer graph, the primitives were the strokes of radicals and the relations between them were their relative orientations. The relaxation matching technique was used in the learning stage to synthesize different attribute graphs into a 2-layer random graph. The relaxation matching method was also applied in the recognition stage to match the attribute graph of an input character with the random graph of each reference character. A similarity measure based on whether or not the attribute graph is a possible outcome of the random graph was used for recognition. Chan and Cheung [23] extended the attribute graph to handle fuzzy attributes for Chinese character recognition. They proposed a measure for the degree of matching of fuzzy-attribute graph monomorphism, which is NP-complete and time-consuming.

Hsieh and Lee [24, 25] utilized on-line character models for off-line Chinese character recognition. They described Chinese characters according to a manually-built 1-D string of interleaved stroke relationships, called an *on-line model*, and based on stroke types and writing sequences. Eight stroke types were defined by regular expressions with angles as the attributes. Twelve definite relations and nine indefinite relations between any two consecutive strokes were also defined. To model a character, the on-line stroke writing sequence, which includes stroke number and stroke order, can be used. However, strokes are not available for off-line character recognition. A stroke-finder was required to pick out all the correct strokes in an unknown character. This stroke extraction method must be invariant to scale, translation, and rotation transformations to recognize large numbers of handwritten characters. The matching process was formulated as a tree-searching algorithm guided by the relationships in the on-line model. The matching was successful if a feasible path was found and the number of missing strokes was fewer than a given threshold value.

Since there are significant variations in the way the same character is handwritten, numerical information is incorporated to recognize noisy characters [25]. Using an on-line model, Hsieh and Lee transformed a 2-D unknown character into a 1-D stroke sequence which could be represented as a multi-stage graph. The Viterbi algorithm which could handle stroke insertion, deletion, splitting, and merging, was used to compute similarities between unknown characters and reference characters. Unknown characters were matched against all reference characters and recognition was based on the one with the highest degree of similarity.

Dynamic programming matching by split-and-merge was proposed by Tsay and Tsai [26]. They also introduced pseudo primitives, i.e., segments added between consecutive strokes of regular written characters that aid in the recognition of connected-stroke writing. Assuming that the input script is written in correct stroke order, a pseudo stroke in the input script always corresponds to a pseudo primitive in the reference character. In the candidate selection phase, they first selected candidate characters by matching the pseudo string of the input with that of the reference character. In the detail matching phase, they matched each input stroke with corresponding strokes in each candidate character, where each stroke was represented as a substring of pseudo and real primitives. The pseudo primitive in the substring of a candidate reference character was used to split the substring of an input stroke if the input stroke matched more than one stroke in the candidate reference character. This algorithm can only resolve connected-stroke handwriting questions but cannot handle the problem of stroke-order variation. By combining the split and merge operations and the conventional edit operations (substitutions, deletions, and insertions), the distinction or similarity value between the input character and each reference pattern can be obtained.

Since stroke-based methods are very popular for character recognition, methods of *stroke extraction* from characters have been studied by many researchers. Strokes are usually extracted by following these steps: thinning, edge-linking, and line or curve approximation. Since thinning-based methods entail noise and time-consumption problems, non-thinning algorithms have also been proposed by researchers.

Lin and Chen [27] proposed a non-pixel-based thinning algorithm. From the run-length coding of an image, they constructed a graph. In the graph, each node was associated with one of three attributes: vertical lines, horizontal lines, and points, which were determined by the nodes above and below and the run length stored in that node. They eventually constructed the skeleton of a character by line-fitting using the attributes of and relationships between nodes in the character graph. For example, they used the center points of vertical runs for horizontal line fitting.

Chen [28] presented a method that extracts strokes from a run-length encoded image. Since the width of a stroke in a written character is nearly constant, characters are partitioned into two kinds of segments: horizontal and vertical. A picture graph is built to represent the relationships among the segments. Based on the graph, a stroke extractor is used to extract linear and curved strokes.

Tung *et al.* [29] also investigated the possibility of replacing thinning with a two-pass vectorization algorithm based on Pavalidis's algorithm. The algorithm uses the fact that Chinese characters are composed largely of vertical and horizontal line segments. After both types of line segments have been extracted, an additional process is required to merge the extracted results. Experiments were performed to compare the performance of the algorithm with conventional thinning-based algorithms. When the recognition rate and processing speed were measured, it was found that the conventional thinning-based stroke extractor performed better than the vectorization-based algorithm. The processing speed was faster than the conventional ones, but only slightly faster since handwritten characters are usually thin enough.

Liao and Huang [30] proposed a method for extracting strokes from thinned images. Since a cross intersection is usually split into two connected Y-junctions, they proposed a method to recover it. For every fork point f, they found the largest circle within the character centered at f. If more than two circles intersected, they were merged and a new center is defined as the average of the centers of the intersecting circles. The new center is connected to the strokes outside the intersecting circle. All possible pairs of stroke segments connecting at the same fork point are considered, and the Bernstein–Bezier curve is used to fit each pair, smooth the data, and find its trend. From the result of this curve-fitting, they decided which pair belonged to the same stroke. Inflection points with very small curvature radii were found, and the stroke segmentation was carried out based on these inflection points.

Tseng and Chung [31] proposed a knowledge-based stroke extraction algorithm for Chinese characters printed in a variety of fonts. They analyzed the Chinese character structures and summarized them as stroke knowledge. This knowledge was coded as 20 parameters and incorporated into the stroke-finding algorithm. Instead of using any preprocessing like thinning or other transformation, they scanned binary images row by row. The knowledge about horizontal strokes, for instance, includes the width of the strokes being greater than the height, the stroke beginning where the length of the dark area in a row changes from short to long, and the stroke ending where the length of the dark area in a row changes from long to short.

4.3. *Methods Based on Radicals*

Most Chinese people view Chinese characters as compositions of radicals. To look up a character in a dictionary, we use a radical of the character as an index. To model the variations of the character, we have to consider combinations of radical variations. Suppose that a character is composed of two radicals, the ith radical has n_i variations, and there are M characters and N radicals. To model the variations, we have to consider $\sum_M n_i \times n_j$ variations if we take whole characters into consideration, and $\sum_N n_i$ variations if we describe only the radicals. However, in the latter case, the combination methods also have to be considered. In other words, if we model a character as a graph, we have to identify subgraphs corresponding to radicals from the graph. This is generally very difficult in the recognition process. Since modelling all the variations of handwritten characters is still impossible, radical-based recognition methods have been considered by many researchers as a natural and practical approach to solving the recognition problem. Another approach to coping with this problem is to adopt a robust recognition method which can tolerate writing variations.

The number of reference characters stored in a database can be greatly reduced if characters are recognized according to radicals. About 250 radicals are needed to make up all Chinese characters. To identify radicals in a character, some people use background data to separate the radicals without identifying them, while others detect connected components. However, radicals may touch each other or be inherently connected; radicals are not easily separated. Therefore, this method is not suitable for off-line handwritten Chinese character recognition because the gap models between radicals vary in unpredictable ways.

Cheng and Hsu [32] separated radicals according to the heuristics of stroke connections. Long strokes, sub-long strokes, and midpoint distributions of the strokes were used to separate radicals. Almost 75% of the Chinese characters can be divided into two parts by means of this dividing strategy. The remaining 25% must be processed according to other special rules.

A background-thinning approach to dividing a character was also proposed by Cheng and Hsu [33]. A curved dividing path was found by thinning the background. However, this method is based on two assumptions. One is that the gap between two radicals is continuous in a certain direction when the character is divisible and its radicals are not joined by any strokes. The other is that if the radicals are joined with any strokes, then the disturbance or unnatural connection will appear at the point joined.

Radical identification is much more difficult than radical separation. Liao and Huang [34] proposed a least-square-error estimation method that identifies radicals by matching an unknown character with each reference radical. The matching algorithm finds transformation parameters from reference line segments to all combinations of line segments in the unknown character. Here the transformations considered include scaling, rotation and translation. The evaluation of the difference between the radical C_1 and a part of a character C_2, named a substructure, is defined as the sum of the squares of errors. The radical with the minimum transformation error is the identified radical. Since the number of substructures C_2 extracted is very large, geometric relationships such as angle, relative positions and stroke intersection are used to prune incorrect subcharacters. Since the number of substructures may still be very large, the computation time may be extensive. They used a tablet to input 114 characters and defined 62 radicals to test the recognition system. The identification rate was claimed to be above 99%.

Hsieh [35] further refined the radicals in the Da-Yi input method into 209 radicals suitable for radical extraction. These radicals are represented by second-order online models. Based on the models, a Viterbi algorithm is applied to extract the possible radicals from the input characters. Then all extracted radicals are arranged as a connected graph, in which an edge links two radical nodes, provided that the two radicals have no common strokes. By finding the maximum clique, the most possible radicals can be identified. Since the radicals in the Da-Yi input method are known with respect to the refined radicals, Hsieh mapped the extracted radicals to the Da-Yi radicals and then found the character identity from the radical-character mapping table of the Da-Yi input system. The relationships among strokes, Da-Yi radicals, refined radicals and characters are shown in Fig. 5. Figure 6 shows that six possible radicals are identified from the input character and the three radicals forming a maximum clique are the components of the input character.

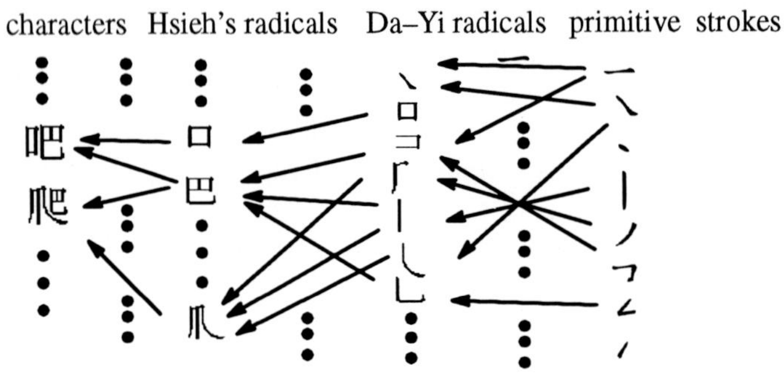

Fig. 5. The relationship among strokes, Da Yi radicals, Hsieh's radicals, and characters.

Wang *et al.* [36] proposed a hierarchical relaxation-based matching scheme for radical recognition of input characters. Their system consists of three phases: partial matching, knowledge database matching, and whole-character matching. They extracted blocks from the input character according to the 19 possible positions of the radical types. They defined 32 radicals. In the partial matching phase, an input character is decomposed into 19 components according to the positions of the 19 radical types. The standard radicals are stored in the database, and transformations are developed to represent a radical appearing in 19 different positions inside a square character. The matching of radicals is performed by the modified relaxation method under constraints of the number of strokes and the composing position of each radical. The second phase is matching with the knowledge database, which contains the radicals and radical types, i.e., placement, of each character. In this phase, all possible combinations of the recognized radicals are matched with the knowledge database. If an input character is recognized as more than one candidate character or some radicals are not recognized, the candidate characters are matched with the input character by the relaxation algorithm. They found the constitutent radicals and associated radical type of each candidate character from the knowledge database. Then the feature vector of the whole candidate character was formed by combining the feature vectors of the constitutent radicals. The candidate character with the minimum distance is accepted as the recognized result.

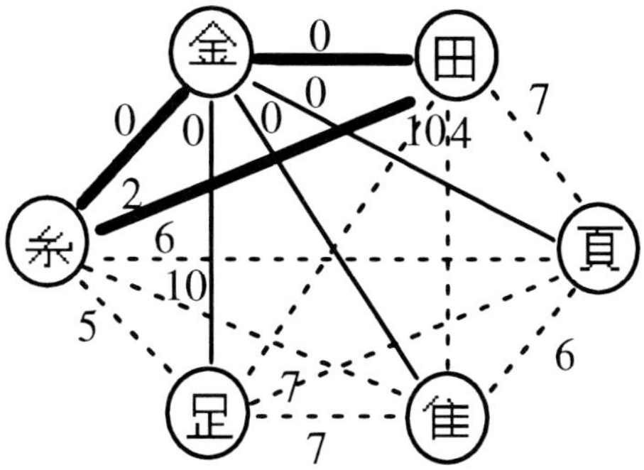

Fig. 6. The maximum clique corresponding to the components of an input character.

5. Postprocessing and Language Models

When the input to an OCR system is a text file instead of a set of characters, we can use contextual information either to select the most possible candidate or to correct recognition errors. We usually call the system a *text* recognition system, which is composed of two stages. The first stage is a character-matching module, which generally selects a fixed number of candidates for each input character. The second stage is a *language model*, which selects the most promising sentence from the candidate sets generated by the first stage. Because a natural language can be regarded as a Markov source, a contextual postprocessor is usually implemented using a Markov language process [37].

Different language models have been proposed. In the following we introduce a general method proposed by Lee and Chang-Chien [37]. They used the bigram word model for postprocessing. Since the number of *words* is very large in Chinese, each word in an input sentence is first segmented and then mapped to a class. Note that a word is the basic primitive with complete syntactic and semantic attributes. They used the POS (part-of-speech)-based language model, which includes 30 part-of-speech to represent the groups or classes. Some significant results were shown when the character recognition rate was not high enough.

To apply the language models, some contextual information has to be found from the training corpus. Since the POS model has shown high performance on word segmentation, the bigram POS language model is used to segment the sentences in the corpus. Let the words in a segmented sentence be $w_1, w_2, \cdots, w_N$. Let $G(w_i)$ represent the group in which the word has been mapped. The segmented sentence is transformed into $G(w_1), G(w_2), \cdots, G(w_k)$ to train the transition probability, that is, the contextual information $P(G(w_i)|G(w_{i-1}))$. The conditional probability of the word w_i, $P(w_i|G(w_i))$, is trained similarly.

If the top N candidates are selected for each input character, a transition graph can be constructed for each input sentence image. After a word transition graph is constructed, a language model is applied for contextual postprocessing. Below, we describe the general operation of the bigram Markov language model used as a contextual postprocessor. After a multistage transition graph for all candidate characters of the sentence has been constructed, a dynamic programming method is applied to find the most promising sentence hypothesis. An example of a multistage part-of-speech transition graph is shown in Fig. 7. There are two candidate characters for each input character. All possible character strings are generated from these candidate characters and the lexicon is searched to find all words among the generated strings. In this example, since both characters in the first candidate set are one-syllable words, the first stage of the multistage part-of-speech transition graph contains two nodes, each of which includes the character, word length, and the part-of-speech. The other nodes in Fig. 7 are generated in the same way. Note that the matching score of a word is omitted in the example.

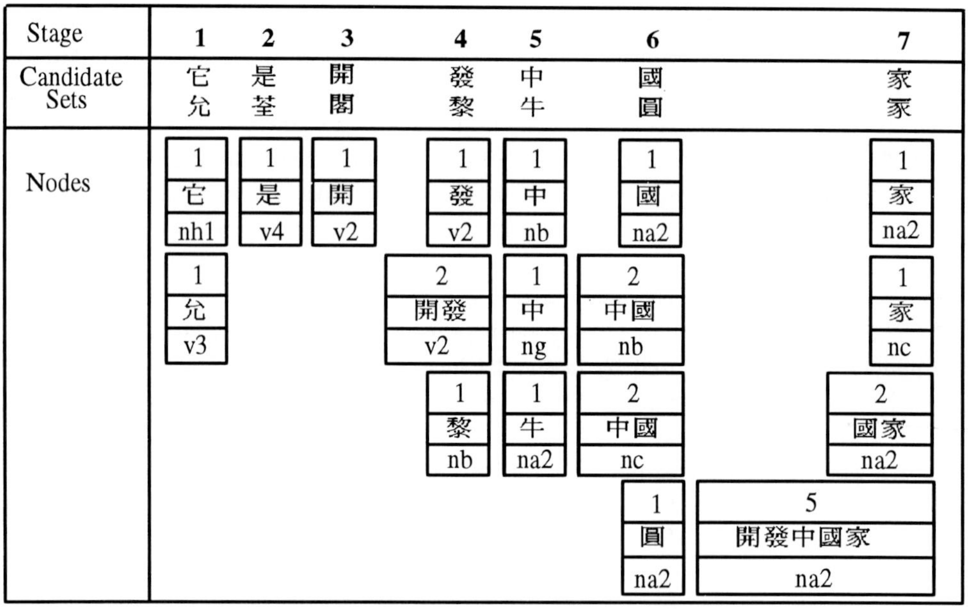

Fig. 7. An example of a multistage part-of-speech transition graph.

Let $I = I_1 I_2 \cdots I_L$ be a sequence of character images, where I_i is the ith character image in the input sentence I, and L is the length of the sentence. Each character image I_i is recognized as M candidate characters $c_{i1} c_{i2} \cdots, c_{iM}$. Each candidate character c_{ik} has a matching score MS_{ik}. In the candidate character sets, there are M^L sentence hypotheses. The goal of the language model is to determine a sentence hypothesis $\hat{S} = c_{1i} c_{2j} \cdots, c_{L,k}$ that has the maximum likelihood among all sentence hypotheses S. The occurrence likelihood of a sentence hypothesis $S = c_1 c_2 \cdots, c_L$ is given by $P(S|I)$, where c_i is one of the candidate characters in the ith candidate set. The goal can be represented as

$$P(\hat{S}|I) = \max_{S} P(S|I).$$

The bigram contextual probability at the word level, $P(w_i|w_{i-1})$, can be modified as

$$P(w_i|w_{i-1}) \approx P(G(w_i)|G(w_{i-1}))P(w_i|G(w_i)).$$

The probability $P(S|I)$ can be computed as

$$P(S|I) \approx \prod_{i-1}^{N} P(G(w_i)|G(w_{i-1}))P(w_i|G(w_i)) \prod_{k} P(c_k|I_k).$$

The term $P(c_k|I_k)$ is the matching score of word w_i, which is the product of the matching scores of the constituent characters c_k.

The recognition result for an input character in the character recognition system can be correct or incorrect. When a single recognition module is used, it is difficult to determine the correctness of the matching result. Accordingly, the categories of all input characters must be confirmed in the following stage. This is a heavy burden for the language model. To relieve this problem, Tung and Lee [38] presented a general approach to increasing character recognition accuracy by detecting and correcting erroneously-identified characters. A high-accuracy two-stage recognition system for recognizing 5401 handwritten Chinese characters was demonstrated. In the first stage of the system, two matching modules recognize an input character simultaneously. The first matching module uses a direction feature and the second a generalized feature. The input image is divided into nonuniform image blocks and a 4-dimension direction-feature vector is extracted from each image block. Figure 8 shows several Chinese characters segmented nonuniformly according to the number of black pixels. The features of nonuniformly segmented blocks are more stable than those of uniformly segmented blocks when the form of the handwritten characters in a category varies widely.

A character is rejected at the first stage when the matching results of the two modules are not the same. Because their system recognizes most of the input characters correctly and outputs a small number of candidates for each rejected character, a bigram Markov language model in the second stage can choose a candidate with high recognition accuracy for each rejected character according to contextual information. Experiments were performed on sentences consisting of characters extracted from the CCL/HCCR1 database. In the first stage, the rejection rate for input characters was 13% and the recognition rate for accepted characters is 95.9%. In the second stage, the recognition rate for the characters rejected by the first stage was 91.2%. Thus, the overall recognition rate for the input handwritten text was 95.9% × 0.87 + 91.2% × 0.13 (=95.2%).

Figure 9 shows an example of using the language model. In Fig. 9(a), the images of a sentence are given. In Fig. 9(b), the recognition results of the first stage are processed by the language model and the final results are obtained. Since most of

Fig. 8. Examples of segmenting Chinese characters nonuniformly.

the characters recognized in the first stage have only one candidate, the contextual ambiguity for the language model decreases significantly. That is, the rejected characters can be processed very quickly even when the size of the candidate set for each rejected character is large, e.g., more than 20 characters.

In a language model, if the number of parameters used to describe contextual information is small, the ability to correct recognition errors in the character recognition stage will be insignificant due to insufficient information. For example, if the words in a dictionary are clustered into about 30 parts-of-speech, only a few recognition errors can be corrected using contextual information. By contrast, if the number of parameters used to describe the contextual information is very large, the parameter training process will be difficult, and the memory required will make the execution of the language model impractical. For instance, if a Chinese language model adopts a word bigram to describe contextual information, the language model may consume all available system memory.

Lee and Tung [39] presented a method for clustering the words in a dictionary into word groups in a Chinese character recognition system. The Chinese synonym dictionary *Tong2yi4ci2ci2lin2*, which has semantic features, is used to train the semantic attribute weights of the word classes. These weights were then updated according to the words in the Behavior dictionary, which has a rather complete word set. Then, the updated word classes were clustered into m groups according to semantic measurement by a greedy method. The words in the Behavior dictionary were finally assigned to the m groups. The parameter space required for bigram contextual information by the character recognition system is m^2. Experimental results show the recognition system with the proposed model performed better than a character-based bigram language model.

Instead of using two character recognition systems to detect possible recognition errors, Wang *et al.* [40] found the characters most likely to be misrecognized with a single recognition system. They used character n-gram models to segment sentence hypotheses and select the most likely characters for these misrecognized characters.

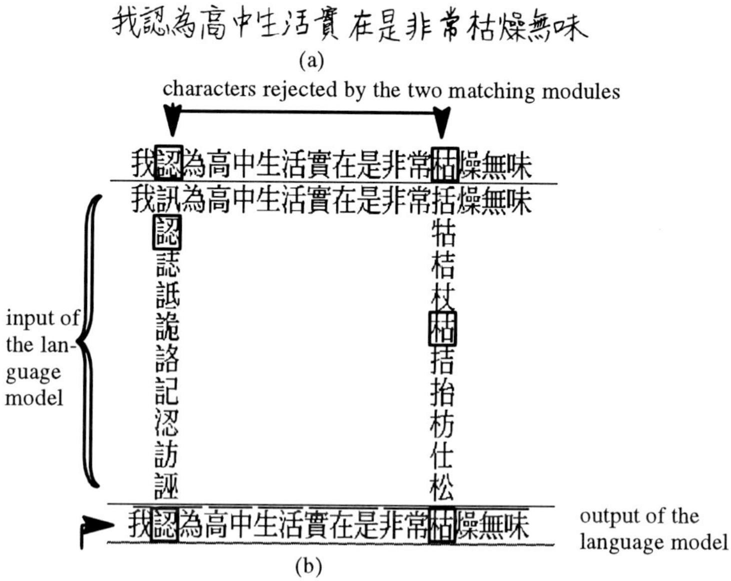

Fig. 9. (a) Character images of a sentence. (b) The candidate sets of a sentence used as the input to the language model and the final result.

They also proposed a method for selecting correct characters that do not appear in the set of candidate characters.

If the outputs for an input character in a recognition system are not unique, the set of outputs can be referred to as a *confusion set*. There are several on-going projects concerning analysis of confusion sets. The set must be analyzed by various recognition methods. Language models are just one method, and other models have already appeared. Fan *et al.* [41] presented a system for recognizing characters in a confusion set by using the knowledge about Chinese characters in the system. For the pattern pair matching made on an input character, the matching process is made by using problem reduction strategy, goal-driven inference like matching, and AND-OR tree searching algorithm.

6. Models and Character Generation

Typically, a large number of character images are required to test the performance of an OCR system. The CCL/HCCR1 database developed by CCL/ITRI is commonly used by local researchers. This database contains 5401 frequently-used Chinese characters and there are about 200 samples of each character. These handwritten characters were written by more than 2,600 people, including junior high school and college students, and employees of CCL/ITRI. The 200 samples of each handwritten character were arranged according to writing quality. Figure 10 shows some characters in the database.

Fig. 10. Some characters in the CCL/GCCR1 database.

Collecting a large database of character images is time-consuming. Instead of using a large testing database in evaluating the performance of character recognition systems, it would be more convenient to use an artificial character generator that can generate character images with a wide range of variations. Tung *et al.* [29] proposed a handwritten Chinese character generator with two stages. In the first stage, each radical of a character is generated stroke by stroke. The strokes generated must satisfy the relative stroke relations defined in the reference radical. The radicals generated are then combined into a line-vector character such that their positions satisfy the relative radical relations specified in the reference character. In the second stage, a thickening procedure is applied to thicken each stroke in the line-vector character. A character image is then generated from the line-vector character.

For the character recognition problem, it is known that a priori knowledge of the pattern structures can be utilized as a guide to obtain an accurate match efficiently. Naturally, the strokes play an important role in guiding the system toward correct recognition. Based on this high-level structural information, a *hierarchical deformation model* was proposed to describe the deformation of on-line cursive Chinese characters [42]. The character recognition approach consists of two levels of matching processes. First, the attribute string-editing algorithm matches two sequences of turning points extracted from the input and reference characters to determine the stroke matches. Next, constrained parabola transformation is used to reduce the difference between the matched strokes appropriately.

Chou and Chen [43] also introduced a stochastic deformation model. A character is represented by a stochastic process that forms a sequence of piecewise stochastic curves, called *stochastic cubic Bezier curves*, and that includes some random noise.

The parameters of this stochastic representation are estimated by real observing a great many actual characters. They offered a much more precise description of character deformations. Thus, the difficulty of distinguishing between similar characters can be reduced.

7. Conclusion

In this chapter, we have presented many algorithms for recognizing Chinese characters. Statistical and structural approaches have been discussed. Language models proposed for post-processing of input characters in a body of text have also been given in detail.

There are many possible applications in the field of character recognition. More robust and effective recognition algorithms have to be developed. Different applications with different requirements and operational environments will induce different recognition algorithms. Some applications may require especially high recognition rates and some may require fast processing. The design philosophies will be different. In general, we think that a system with preclassification, detailed matching and post-processing modules will be the general architecture used for Chinese character recognition. In the detailed matching modules, the multi-expert approach which adopts statistically-based recognition algorithms may be the right choice. For characters in confusion sets, we can apply the structurally-based recognition algorithms. The final remaining uncertain candidates can be checked by a Markov language model.

The advances in computer systems, tablets and scanners have inspired advances in research on character recognition. Many researchers think it is time to apply the character recognition technologies to different document processing systems, to name a few, business card organizers, newspaper organizers, personal digital assistants (PDAs), check recognition systems, letter address readers, map and geographic information system (GIS) input systems, and enhanced desktop publication systems. Several prototype systems are under development. It is expected that practical systems will appear in the market in the near future.

References

[1] W. H. Tsai and K. S. Fu, Attributed grammar — a tool for combining syntactic and statistical approaches to pattern recognition, *IEEE Trans. Systems, Man, and Cybernetics* **SMC-10** (1980) 73–885.

[2] H. D. Chang and J. F. Wang, Preclassification for handwritten Chinese character recognition by a peripheral shape coding method, *Pattern Recog.* **26** (1993) 711–719.

[3] R. H. Cheng, C. W. Lee, and Z. Chen, Preclassification of handwritten Chinese characters based on basic stroke substructures, *Proc. 4th Int. Workshop on Frontiers in Handwriting Recognition,* Taipei, Taiwan, 1994, 176–184.

[4] T. Kumamoto, K. Toraichi, T. Horiuchiz, K. Yamamoto and H. Yamada, On speeding candidate selection in handprinted Chinese character recognition, *Pattern Recog.* **24** (1991) 793–799.

[5] C. H. Tung, H. J. Lee and J. Y. Tsai, Multistage pre-candidate selection in handwritten Chinese character recognition systems, *Pattern Recog.* **27** (1994) 1093–1102.

[6] T. Z. Lin and K. C. Fan, Coarse classification of on-line Chinese characters via structure feature-based method, *Pattern Recog.* **27** (1994) 1365–1378.

[7] L.-T. Tu, *et al.*, Recognition of handprinted Chinese characters by feature matching, *Proc. 1st National Workshop on Character Recognition*, Taiwan, 1991, 166–175.

[8] J. S. Huang and M. L. Chung, Separating similar complex Chinese characters by Walsh transform, *Pattern Recog.* **20** (1987) 425–428.

[9] T. Y. Tu and Y. L. Ma, Character recognition by stochastic sectionalgram approach, *Pattern Recog.* **21** (1988) 593–601.

[10] T. F. Li and S. S. Yu, Handprinted Chinese character recognition using the probability distribution feature, *Int. J. Pattern Recog. Artif. Intell.* **8** (1994) 1241–1258.

[11] B. S. Jeng, Optical Chinese character recognition using accumulated stroke features, *Optical Engineering* **28** (1989) 793–799.

[12] H. J. Lee and B. Chen, Recognition of handwritten Chinese characters via short line segments, *Pattern Recog.* **25** (1992)543–552.

[13] F. H. Cheng, W. H. Hsu, M. C. Kuo, Recognition of handprinted Chinese character via stroke relaxation, *Pattern Recog.* **26** (1993) 579–593.

[14] A. J. Hsieh, K. C. Fan, and T. I. Fan, Bipartite weighted matching for on-line handwritten Chinese character recognition, *Pattern Recog.* **28** (1995) 143–151.

[15] K. S. Chou, K. C. Fan, T. I. Fan, C. K. Lin, and B. S. Jeng, Knowledge model based approach in recognition of on-line Chinese characters, *IEEE J. Selected Areas in Communications* **12** (1994) 1566–1575.

[16] S. L. Chou and W. H. Tsai, Recognition handwritten Chiense character by stroke-segment matching using an iteration scheme, *Int. J. Pattern Recog. Artif. Intell.* **5** (1991) 175–197.

[17] F. H. Cheng, W. H. Hsu, and M. Y. Chen, Recognition of handwritten Chinese characters by modified Hough transform techniques, *IEEE Trans. Pattern Anal. Machine Intell.* **PAMI-6** (1989) 386–405.

[18] C. K. Lin, K. C. Fan, and F. T. P. Lee, On-line recognition by deviation-expansion model and dynamic programming matching, *Pattern Recog.* **26** (1993)259–268.

[19] K. J. Chen, K. C. Li, and Y. L. Chang, A system for on-line recognition of Chinese characters, *Computer Processing of Chinese and Oriental Languages* **3** (1988) 309–318.

[20] F. H. Cheng, W. H. Hsu, and C. A. Chen, Fuzzy approach to solve the recognition problem of handwritten Chinese characters, *Pattern Recog.* **33** (1990) 475–484.

[21] A. B. Wang, J. S. Huang, and K. C. Fan, Recognition of handwritten Chinese characters by modified relaxation methods, *Image and Vision Computing* **12** (1994) 509–522.

[22] L. H. Chen and J. R. Lieh, Handwritten character recognition using a 2-layer random graph model by relation matching, *Pattern Recog.* **23** (1990) 1189–1205.

[23] K. P. Chan and Y. S. Cheung, Fuzzy-attribute graph and its application to Chinese character recognition, *Computer Processing of Chinese and Oriental Languages* **4** (1989) 85–98.

[24] C. C. Hsieh and H. J. Lee, Off-line recognition of handwritten Chinese characters by on-line model-guided matching, *Pattern Recog.* **25** (1992) 1337–1352.

[25] C. C. Hsieh and H. J. Lee, A probabilistic stroke-based Viterbi algorithm for handwritten Chinese characters recognition *Int. J. Pattern Recog. Artif. Intell.* **7** (1993) 329–352.

[26] Y. T. Tsay and W. H. Tsai, Attributed string matching by split-and-merge for on-line Chinese character recogntion, *IEEE Trans. Pattern Anal. Machine Intell.* **15** (1993) 180–185.

[27] J. Y. Lin and Z. Chen, A Chinese character thinning algorithm based on global features

and contour information, *Pattern Recog.* **28** (1995) 493–512.

[28] L. H. Chen, A new approach for handwritten character stroke extraction, *Computer Processing of Chinese and Oriental Languages* **6** (1992) 1–17.

[29] C. H. Tung, Y. J. Chen and H. J. Lee, Performance analysis of an OCR system via an artificial handwritten Chinese character generator, *Pattern Recog.* **27** (1994) 221–232.

[30] C. W. Liao and J. S. Huang, Stroke segmentation by Bernstein–Bezier curve fitting, *Pattern Recog.* **23**, (1990)1167–1188.

[31] L. Y. Tseng and C. T. Chuang, An efficient knowledge-based stroke extraction method for multi-font Chinese characters, *Pattern Recog.* **25** (1992) 1445–1458.

[32] F. H. Cheng and W. H. Hsu, Radical extraction from handwritten Chinese characters by background thinning method, *Trans. IECE* **E71** (1988) 88–98.

[33] F. H. Cheng and W. H. Hsu, Radical extraction by background thinning method for handwritten Chinese characters, *Proc. 1987 Int. Conf. on Chinese Computing*, 1987, 175–182.

[34] C. W. Liao and J. S. Huang, A transformed invariant matching algorithm for handwritten Chinese character recognition, *Pattern Recog.* **23** (1990) 1167–1188.

[35] C. C. Hsieh, Model-guided recognition of handwritten Chinese characters, Ph.D. Thesis, National Chiao Tung University, 1992.

[36] A. B. Wang, K. C. Fan and J. S. Huang, Optical recognition of handwritten Chinese characters by hierarchical matching, *Computer Processing of Chinese and Oriental Languages* **8** (1994) 193–210.

[37] H. J. Lee and C. H. C. Chien, A Markov language model in handwritten Chinese text recognition, *Proc. 2nd Int. Conf. Document Analysis and Recog.*, Tsukuba, Japan, 1993, 72–75.

[38] C. H. Tung and H. J. Lee, Increasing character recognition accuracy by detection and correction of erroneously-identified characters, *Pattern Recog.* **27** (1994) 1259–1266.

[39] H. J. Lee and C. H. Tung, A language model based on semantically clustered words in a Chinese character recognition system, *Proc. 3rd Int. Conf. Document Analysis and Recog.*, Montreal, Canada, 1995.

[40] J. F. Wang, H. S. Shiau and H. M. Suen, A linguistic decoder for postprocessing of Chinese character recognition, *Proc. 4th Int. Workshop on Frontiers in Handwriting Recog.*, Taipei, Taiwan, Dec. 1994, 402–409.

[41] K. C. Fan, C. K. Lin and K. S. Chou, Confusion set recognition of on-line Chinese characters by artificial intelligence technique, *Pattern Recog.* **27** (1995) 303–313.

[42] W. T. Chen and T. R. Chou, A hierarchical deformation model for on-line cursive script recognition, *Pattern Recog.* **27** (1994) 205–219.

[43] T. R. Chou and W. T. Chen, A stochastic deformation model for the on-line recognition of cursively similar characters, *Proc. 4th Int. Workshop on Frontiers in Handwriting Recog.*, Taipei, Taiwan, 1994, 206–215.

Handbook of Character Recognition and Document Image Analysis, pp. 357–380
Eds. H. Bunke and P. S. P. Wang

CHAPTER 13

RESEARCH IN JAPANESE OCR

SARGUR N. SRIHARI, GEETHA SRIKANTAN, TAO HONG and STEPHEN W. LAM
CEDAR, SUNY at Buffalo, 520 Lee Entrance
Amherst, New York 14228-2567, USA

Recognition of Japanese machine-printed documents poses several challenges. The variation of document layout styles, vertical and horizontal text alignment, mixed pitch characters, and the large character set size (over 3000 in everyday use), contribute to the complexity of Japanese OCR system design. Variations in font styles and the structurally complex character set are other contributing factors in design. After a brief overview of previous Japanese OCR research, we outline the directions of research in Japanese OCR. To illustrate these issues we present details in the design and performance of a Japanese OCR system developed at CEDAR.

Keywords: Japanese OCR; Pattern recognition; Large character set; Japanese document recognition system.

1. Introduction

Research in Japanese optical character recognition (OCR) began in Japan in the late 1950s. Character recognition is one of the earliest topics of pattern recognition research in Japan [1, 2]. This coincides with the early stages of pattern recognition research activities in North America and Europe.

The Japanese character set is a combination of several scripts: Kanji (Chinese characters), Kana (Japanese consonant and syllabary characters), Roman and Arabic numerals. Kanji are Chinese ideographs which were introduced into the Japanese language a long time ago [3, 4]. In the electronic standard, JIS X 0208-1990, which is considered to be a definitive description of the Japanese character set [5] (see Fig. 1), there are 6,879 characters. Among them, 6,355 are Kanji, divided into two distinct sections — JIS Level I and Level II; where Level I spans 2,965 of the most frequently used Kanji characters. For more details on the Japanese writing system, character set standards and other aspects of Japanese information processing, please refer to [5, 6].

Research and development in OCR has been pursued actively in Japan. This overview of Japanese OCR is extracted from the research papers from Japan which were published in English in the last two decades. However, we realize that many important papers written in Japanese have not been translated into English as yet. Further, as Kanji recognition is one of the most challenging aspects of the Japanese OCR task, many studies related to Japanese OCR can be found under the subject

Miscellaneous symbols	、。，．・：；？！ … ∬Å‰♯♭♪†‡¶◯
Numerals	0 1 2 3 4 5 6 7 8 9
Roman characters	A B C D E F G H I J … q r s t u v w x y z
Hiragana	ぁあぃいぅうぇえぉお … りるれろゎわゐゑをん
Katakana	ァアィイゥウェエォオ … ロヮワヰヱヲンヴヵヶ
Greek characters	Α Β Γ Δ Ε Ζ Η Θ Ι Κ … ο π ρ σ τ υ φ χ ψ ω
Cyrillic characters	А Б В Г Д Е Ё Ж З И … ц ч ш щ ъ ы ь э ю я
Line-drawing elements	─│┌┐┘└├┬┤┴ … ┠┯┨┷┿┝┰┥┸╂
JIS Level 1 kanji	亜唖娃阿哀愛挨始逢葵 … 瓦亘鰐詫藁蕨椀湾碗腕
JIS Level 2 kanji	弌丐丕个丱丶丼丿乂乖 … 齟齧齬齪齷齲齶龕龜龠
Additional kanji	堯槇遙瑤凜熙

Fig. 1. Example characters in JIS table (from [5]).

of *Chinese Character recognition* or *Chinese OCR* [7, 8, 9].*

The rigorous efforts in Japanese optical character recognition research is propelled by strong social and economic demands. The automatic recognition of document data can significantly lower operation costs and increase productivity. This need is pre-eminent in banking, insurance, government and postal services.

Besides machine-printed document data, a significant portion of document data is generated by hand-printing. Therefore, automatic recognition of hand-printed characters, which include Kanji and Kana, has become a crucial area in Japanese character recognition research.

To meet these economic demands, the goal of Japanese character recognition research is to develop a highly accurate recognition method for hand-printed characters, as well as for machine-printed characters. Although many character readers are commercially available, intensive research is still being conducted to improve the accuracy and data throughput.

A brief summary of Japanese OCR in general is presented in the rest of this section. In Sect. 2, we describe the new CEDAR Japanese character image database. In Sect. 3 we present a discussion of a working Japanese OCR system to illustrate the various aspects of this research area.

1.1. *State-of-the-Art in Character Recognition*

Several issues need to be considered when defining the state-of-the-art in Japanese OCR, including character classes, speed, accuracy, image quality and related conditions. However, recognition accuracy is the most important and representative index. In Japan, the research and development in Japanese OCR is carried out by many research laboratories (such as NTT and ETL), universities (such as Tokyo,

*See also the chapters on Chinese OCR in this book.

Table 1. Commercial products of JOCR (Most information is from [10]).

Company	Model	Date	Character set	Speed (char /sec.)	Accuracy	Price (million yen)
Toshiba		1977	2000 single font Kanji	100	98.4%	
Toshiba	V595	1984	2000 (H) Kanji + ANKS	50		>25
NTT	OCR50	1985	4000 multiple font Kanji + ANKS	25	99%	35
Sanyo	CLL-2000	1985	3000 (H) Kanji + ANKS	5	93%	1.98
Sanyo	CLL-2000D	1986	3000 (H) Kanji + ANKS	5 Kanji 10 ANKS	93%	1.98
Toshiba	V-3050	1986	2200 (H) Kanji + ANKS	150		8.5
Fujitsu	FACOM 6678A	1986	3200 (H) Kanji + ANKS	40		>7
NTT	OCR60	1986	3176 (H) Kanji + ANKS	20 Kanji 100 ANKS	98%	16.5
Sanyo	CLL-3300K	1986	3000 (M) Kanji + ANKS	5	>99%	1.98
Matsushita		1986	2000 (M & H) Kanji + ANKS	10	>95%(sm) >99%(sf)	
Ricoh	CR106	1986	2000 (M & H) Kanji + ANKS	8	>98%(h) >99%(m)	
Mitshubishi	M6560	1987	2416 (H) Kanji + ANKS	6.6	95% 99%(PP)	51

M – machine-printed;

H – handwritten;

ANKS – alphabeta+numeral+kana+symbol;

PP – postprocessing

Osaka and Nagoya Universities) and a few large companies. Table 1 lists some commercial products which are available on the Japanese market.

1.2. *Design of Japanese Document Recognition Systems*

Like systems for English document recognition, a Japanese document recognition system usually has three major components: layout analysis, character segmentation/recognition, and postprocessing. Given an image of a document page, layout analysis will locate text blocks and further segment them into text lines. In a Japanese document, a text line can be either vertical or horizontal. Layout analysis techniques for Japanese documents [11, 12] and multi-lingual documents [13, 14] have been developed.

In Japanese text images, a text line is segmented directly into a sequence of characters as word boundaries are not easily distinguishable. Information from connected-component analysis and projection profile analysis can be used to assist character segmentation [12, 15, 16, 17]. Feedback from a character recognizer can also be used to resolve ambiguities and correct errors in segmentation. Techniques in Japanese character segmentation and recognition will be discussed in more detail in Sect. 3.

Postprocessing is typically intended to improve accuracy by detection and correction of OCR errors. Linguistic knowledge sources such as a dictionary and transition probability between characters, and domain knowledge are exploited in postprocessing [18].

There is a trend, also, to integrate document recognition system with other information processing systems such as information retrieval. Based on a Japanese OCR system, a retrieval system for index-free full-text search has been designed [19].

1.3. *Recognition Methods for Japanese Characters*

The major challenges in developing a Japanese OCR system are as follows [20]:

(i) Large variety of print layouts.
(ii) Vertical and horizontal alignment of text.
(iii) Large number of character categories.
(iv) Structural complexity of each character pattern.
(v) Existence of character patterns that have similar shape and structure.
(vi) Wide variety of character shapes due to different typefaces and image quality in machine-printed documents, or different writing styles in handwriting.

Character recognition methods can be broadly classified into two paradigms: *structural analysis* and *pattern matching.* In the structural analysis methods, feature extraction is carried out from the viewpoint of character pattern primitives. Many character features have been devised to satisfy design criteria of the researcher/engineer. It is now commonly believed that local-geometrical-contour features are most useful, and that they are naturally implemented in decision tree, finite state automata relaxation matching, and other classification methodologies [21]. These methods have been applied to the recognition of hand-printed characters, other than Chinese ideographs, or Japanese Kanji.

In the pattern matching methods, global and uniform features are used to simplify the recognition scheme [21]. The feature extraction methods studied in this field can be categorized as based on 1) orthogonal expansions, 2) stroke distribution, 3) stroke analysis, and 4) background feature distribution. The Karhunen-Loeve (KL) expansion has been used very successfully for both machine-printed and hand-printed Kanji recognition [20]. Stroke distribution methods, such as the local direction contribution (LDC) and the stroke density, are very useful [12]. Others such as cellular features, surrounding area features, complexity index, histogram, and gradient features, are also useful [20, 22].

Similarity or distance measures have been proposed to achieve more reliable and accurate performance in the discrimination process. The combination of stroke direction features and the multiple similarity method [23] or the modified quadratic discriminant functions (MQDF) [24] have been applied to the recognition of hand-printed Kanji, as well as omni-font machine-printed characters.

Neural network models have been widely applied to character recognition. However, most of these networks have been used to recognize a small number of classes, such as alphabets, numerals and Kana. It is difficult to extend these network models to Kanji recognition, as there are over 3000 Kanji classes. For Kanji recognition, it is desirable to have neural network training rules that result in fast convergence. Further, the architecture of such networks would have to be designed such that the computational requirements are feasible. A large scale neural network model, known as "multiple modified LVQ neural network", has been developed to achieve high performance on machine-printed Kanji recognition [25].

1.4. *Multiple-Stage Classification and Postprocessing*

Kanji characters are complex in their stroke structure and several characters are structurally similar. Feature descriptors which can discriminate different character classes, typically have large dimensionality. Correspondingly computation time requirements can be high, when these features are used. Efficient multi-stage recognition methods have been developed [12], where coarse or approximate classification stages are followed by finer and more accurate classification stage. The key to the success of the multi-stage recognition method is the reliability of the coarse classification stage, that is, the correct class must be present in the choices output from the coarse classifiers. Classification is error-prone when the input pattern is degraded.

Contextual information, such as character n-gram, dictionary, semantic knowledge and domain knowledge have to be applied [19, 26, 18], to detect and correct errors. Linguistic error correction disambiguates recognition results, particularly in identifying a correct Kanji character among candidates which are structurally similar.

1.5. *Commercial Applications*

There are several applications of current optical character readers, particularly in office and industrial automation.

Office automation applications are predominant, particularly, applications in business form data entry, account for more than half of the total market. Here, OCR systems are used for processing bills, sales slips, and remittance forms. Insurance companies and banks are now obtaining their own OCR system to increase total office productivity. In these applications, recognition accuracy is crucial for the success of the OCR systems. Automation of mail sorting is another important application, where large volumes of documents need to be recognized reliably at real-time speed.

A newer application is in text data entry combined with machine translation, word processing, database building, and reprinting. In text data entry applications, OCR performance requirements are not as stringent as in the business form data entry applications. Higher error and rejection rates of the OCRs are tolerated for the following reasons:

(i) Text reading is performed first, followed by subsequent processes.
(ii) Complete recognition is not possible for documents printed in unknown type faces or complex mixed text/image format.
(iii) Correction of text is inevitable, prior to advancing to the next process.
(iv) Correction and text reading can be accomplished concurrently with the same machine.
(v) Sophisticated spell-checking software helps the user find OCR errors semi-automatically.

1.6. *Current Research Areas*

The present state of the Japanese character recognition technology has been reviewed here in terms of recognition method, implementation, and practical applications. As a result of extensive research efforts conducted for more than a quarter of a century, great progress has been made in this field. However, several issues remain open – generalized OCR machines for hand and machine-printed data are sought, higher accuracy and fast performance are also sought. OCR systems for handling degraded and poor quality images are in constant demand. In summary, the main challenges ahead are:

(i) Realization of ultra high accuracy recognition for machine-printed characters.
(ii) Considerable recognition rate improvement for low quality handwritten characters.
(iii) Development of a unified recognition method suitable for multi-lingual character patterns, such as Roman, Greek, and Chinese language symbols.

2. CEDAR Japanese Character Database

At CEDAR, we have created a Japanese character image database, which is available on CD-ROM. This database has two components: (a) document images for the development of document analyzer and segmenter modules, and (b) a character image database for recognizer development. Japanese documents were digitized at 400 ppi on a flatbed scanner. These documents are from varied sources — facsimile, photocopies, newspaper, book, journal and magazines — and span diverse document layouts and print qualities. A total of 264 pages have been scanned into the document database. Characters extracted from these images have been tagged with the truth value in the JIS code. Approximately 180,000 character images are available for recognizer development in this database (see Fig. 2 for samples of character images from the database). The data has been partitioned into train and test sets. Both the document and character databases along with accessing software and documentation are available on CD-ROM. In addition to this CEDAR CD-ROM, ETL and the University of Washington have made available databases, for OCR system development.[†]

[†] See also the chapter by I. Guyon, R. Haralick, J.J. Hull and I. Phillips in this book.

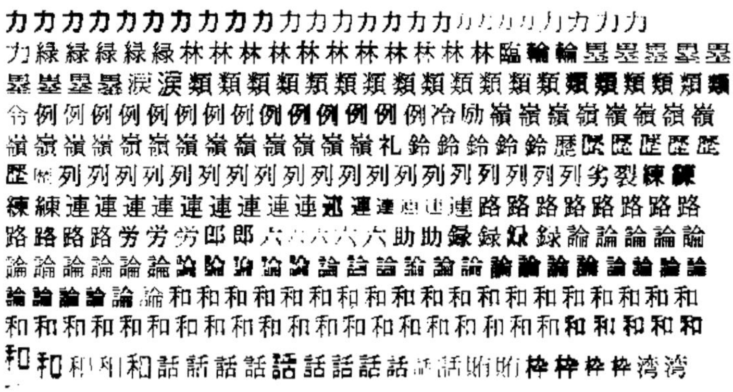

Fig. 2. Samples of character images from CEDAR dataset.

3. The Design of a Japanese OCR System

We now describe a Japanese OCR system for omnifont machine-printed text developed at CEDAR over the past few years. This system is designed to lift text automatically from scanned documents and segment text into character units. The main challenges are the wide variations in data quality (facsimile, photocopy, newspaper etc.), low scan resolution (200ppi), large character set (JIS Level 1 & 2) and the variety of font and point sizes. Further, the system is expected to perform without knowledge of context. One of the main issues dealt with in this project is the creation of a large database of Japanese document images to overcome a lack of representative data for system training. The system comprises a document analyzer, page and text segmenters as well as character recognizers and postprocessors. A user interface was also developed for this system. We describe each module in detail in the following subsections. The results of each module based on the CEDAR Japanese document and character image database are also presented.

3.1. *Document Analyzer*

This module is designed to detect and correct skew in document images. A fast directional profile analysis is used in the detection of skew angles between ±30. A simple affine transformation is used for skew correction. The document analyzer is required to perform fast skew detection on images containing graphics, tables and figures as well as text. An outline of the skew detection algorithm follows; refer to [27] for a complete description.

3.1.1. *Skew detection*

This algorithm is an extension of a technique by Ishitani [28]. The document image is divided into local regions. In each region a directional profile analysis is performed to detect skew. Overall document skew is estimated as a consensus among the local region estimates. Local regions of the entire document can consist of text, graphics and figures, see Fig. 3. This division into multiple regions increases the text hit rate, a factor that influences further processing.

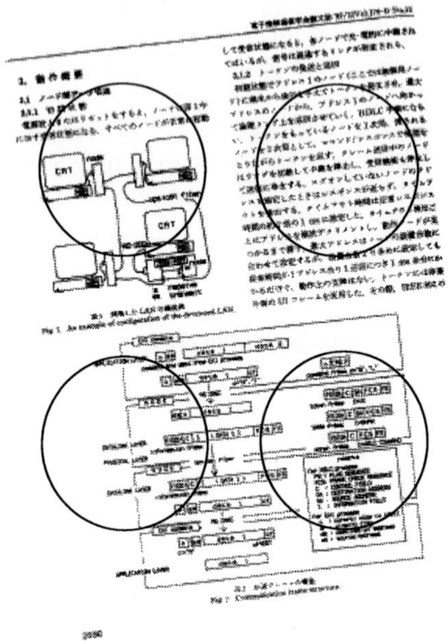

Fig. 3. Local regions are extracted for directional profile analysis from the document image.

Local region complexity variance is profiled over a range of orientations. Complexity is measured by computing black pixels on scan lines oriented at some angle in the range. For a particular orientation θ, the variance in complexity $V(\theta)$ is:

$$V(\theta) = \frac{1}{n} \sum (N_i - M)^2,$$

where n is the number of scan lines, N_i is the complexity of the scan line i, and $M = \frac{1}{n} \sum N_i$. Notice that the variance increases as θ approaches the actual skew angle.

Complexity variance measurements from a range of orientations define a directional profile, see Fig. 4. The angle which maximizes the profile is reported as the skew angle of this local region. Observe that the mix of graphics, figures and text in local regions can cause directional profile analysis to yield misleading estimates of the skew angle. However, it is likely that skew estimates on several text regions cluster near a particular value. Hence the document skew angle is estimated as the weighted average of clustered skew estimates.

3.1.2. *Skew detector results*

The skew detector was tested with 467 artificially skewed images, with varying

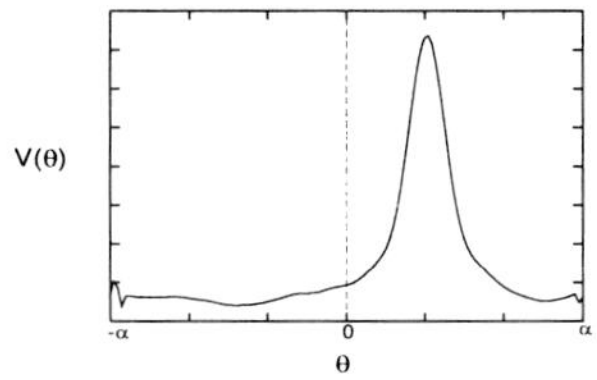

Fig. 4. Directional profile obtained by taking complexity variation measurements in several directions.

amounts of noise, fragmentation, black pixel density and ratio of text to graphics. Of these, skew was detected within 0.5% accuracy of actual skew in 71% of the images, within 1.0% of actual skew in 92% of the images and within 2.0% of the actual skew in 98% of the images. Results are also improved with high quality images when compared with poor quality document images. The document analyzer achieves 97% accuracy within ±1% of the actual skew on good quality document images, and 86% on noisy or complex documents. The average time required for skew detection is 2 cpu seconds on a SPARC-10.

3.2. *Document Segmenter*

This module incorporates a local-to-global approach for discrimination between text/non-text regions as well as segmentation of non-rectangular blocks, and is robust in the presence of noise.

The objective here is to decompose a scanned document image into regions, which contain homogeneous entities, such as text, graphics and half-tones. Further, this decomposition has to be performed across a wide variety of documents with no prior knowledge of document types nor constraints on the document layout. Other significant aspects of this module include robustness with respect to low quality data, capability of segmenting non-rectangular as well as rectangular regions, and applicability to multi-lingual document images.

Rather than use a purely bottom-up or purely top-down approach this document segmentation module incorporates a hypothesize-and-test strategy that is also local-to-global. The main limitation of bottom-up page segmentation is that distance metrics among components as well as run-length estimation are unreliable on noisy and degraded images. Many top-down approaches do not perform satisfactorily in complex layouts (example, large white streams) and can only detect rectangular regions. An outline of this algorithm is presented next, for details refer to [29].

3.2.1. *Local to global page segmentation algorithm*

This algorithm is based on two steps: (a) hypothesis generation and (b) hypothesis testing.

A segmentation hypothesis is generated by recursively partitioning the document into smaller regions and locating plausible partition points from local regions.

Refer to Fig. 5 for initial regions generated by the recursive partitioning and Fig. 6 for examination of plausible partition points. A quad-tree of the image regions is formed while allowing for partial segmentation on low-quality/non-rectangular regions. Partitioning stops when either a high confidence partial segmentation is attained or the region is too small for further subdivision. Partial segmentation is assisted by profile analysis on local regions and statistical analysis on connected components.

Segmentation hypothesis testing involves the creation of region candidates based on a neighbor constraint satisfaction approach; for an illustration see Fig. 7. Regions are then classified based on profile and run-length analysis. Creation of region candidates is accomplished by combining partial segmentations of four neighboring regions. Profile and component information is used to terminate propagation and block classification.

Fig. 5. Recursive partitioning of document image into smaller regions.

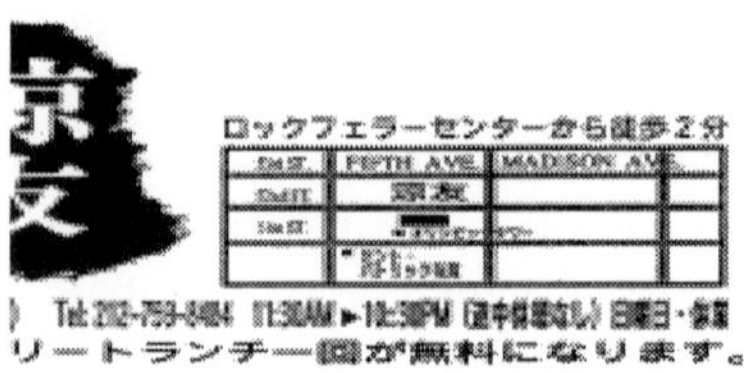

Fig. 6. Locate plausible partitions from local regions.

3.2.2. *Document segmenter results*

The document segmentation module was tested with 200 Japanese and English documents. The performance of the document segmenter ranges between 92%

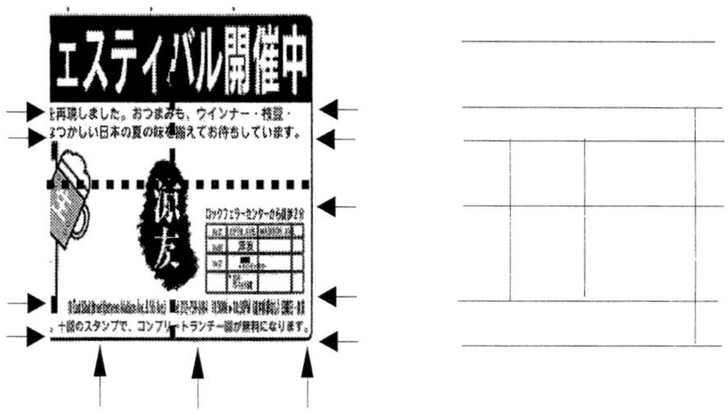

Fig. 7. Test hypothesized region candidates via constraint satisfaction.

completely correct segmentations and 94% partially correct segmentations. On an average this module requires 27 cpu seconds on a SPARC-10. The processing time of this module is directly proportional to the size and quality of the image as well as the layout complexity. Hence, on book images it takes only about 17 seconds.

3.3. *Text Segmenter*

This text segmenter module incorporates a divide-and-conquer approach. The segmentation task is divided into smaller ones and each task is then handled individually.

The text segmenter module was designed with the objective of segmenting multi-line text blocks into character units, and performing very reliably on high quality images and reasonably well with noisy images. Among the challenges in developing such a system is the diversity in document image quality. Print styles vary from clean and uniformly separated text units to noisy and degraded. Font styles vary and inter-symbol spacing varies between closely spaced and widely spaced. When Japanese and English characters appear together, there is a large difference in size of each character type. It is often hard to distinguish noise from characters and segment characters of mixed pitch.

This module segments a text block into lines and characters via six subprocesses — image quality estimation, character size estimation, text alignment detection, text-alignment detection, line and character segmentation. An overview of this module is shown in Fig. 8. Each of the subprocesses is described briefly here, for details refer to [30].

3.3.1. *Image quality estimation*

This stage determines those aspects of image quality, particularly relevant to text segmentation. Several features are used in characterizing image quality, including:

(i) periodicity of profile pulse size (uniform/other)
(ii) abnormal size of a pulse (noisy/other)
(iii) symmetricity of pulse (font-varying/noisy)

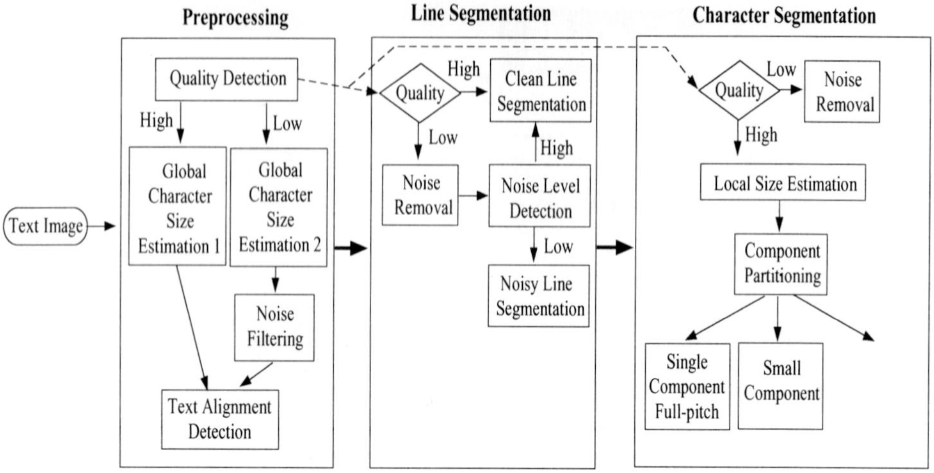

Fig. 8. Overview of text segmentation.

(iv) ratio of pulse-width/height (noisy/other)

(v) number of connected components in image (small/large)

3.3.2. *Character size estimation*

Character size is estimated globally based on histogram plots of connected components. In clean images, the highest peak in the component histogram corresponds to the width of full pitch characters. In noisy images, this component histogram is smoothed, and the mean width of the histogram is evaluated. The highest peak in the histogram which is greater than the mean is chosen as the width of full pitch characters; for an illustration see Fig. 9.

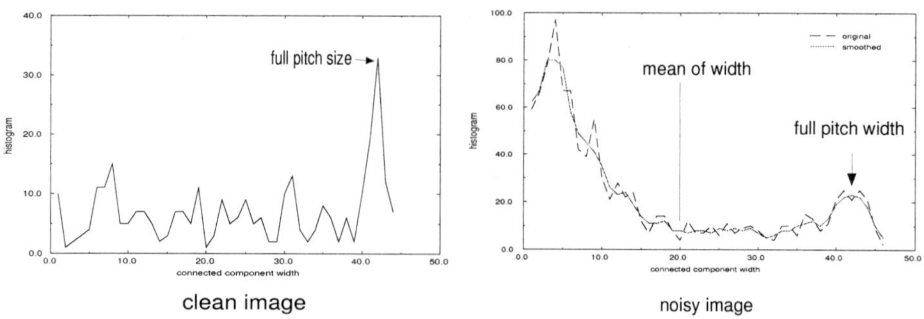

Fig. 9. Global character size estimation.

3.3.3. *Text alignment detection*

Two approaches to determine whether the text is vertically or horizontally aligned are described here. The first approach is based on a minimum spanning tree of components, while the second is based on direct projection profile analysis.

A minimum spanning tree is built with the connected components forming nodes in this tree. Each connected component is represented by its center and the minimum spanning tree is constructed with these points (this is similar to the technique of Ittner and Baird [13]). Text alignment is identified as the most frequently occurring edge orientation. In another approach, the variance in horizontal and vertical projection profiles is analyzed. The direction of largest variance is the direction of the alignment; see Fig. 10.

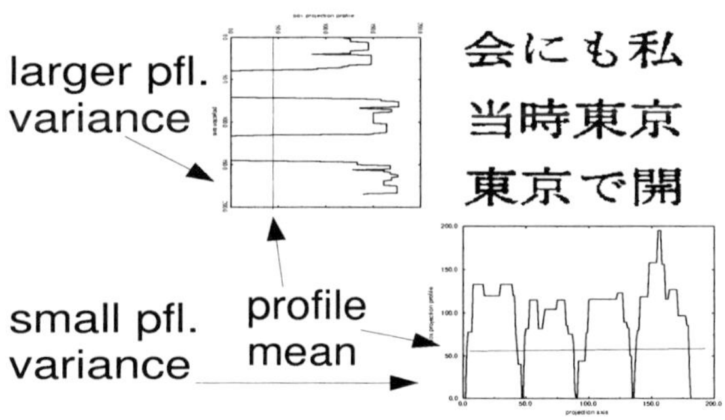

Fig. 10. Projection profile variance based estimation of text alignment.

3.3.4. *Line segmentation*

In clean and high quality documents, separation of text blocks into lines is relatively simple. It is accomplished by estimating line breaks from valleys in the projection profiles. For noisy and degraded images, line segmentation is not as simple, as the projection profiles are not as sharply defined. Our approach is to filter out noise components during the estimation of line breaks. The projection profiles of text components are then smoothed and a recursive threshold is applied until there is only a small deviation in the main pulse widths of the projection profile. At this stage, the approach used for estimating line breaks of clean images can be applied; see Fig. 11.

3.3.5. *Character segmentation*

Initially character components are located by performing a projection profile analysis of black pixels and identifying white spaces as component separators. See Fig. 12 for an illustration of this method. Local character size estimation is then performed for accuracy, based on a mean size estimate and line width. A rule-based approach is then used to partition components into three groups. These groups identify single component characters, partial character components and touching characters (or multi-character components). The sub-character components are

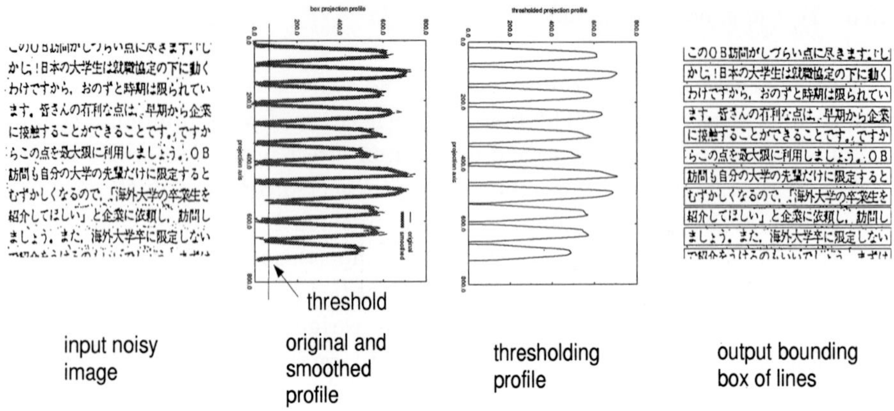

Fig. 11. Line segmentation after profile processing.

merged based on a few simple rules to form complete characters. Touching characters are split based on further analysis and heuristics. At this stage a character recognizer has also been used to validate proposed splitting points. Fig. 13 indicates how components may be merged or split under the rule-based scheme.

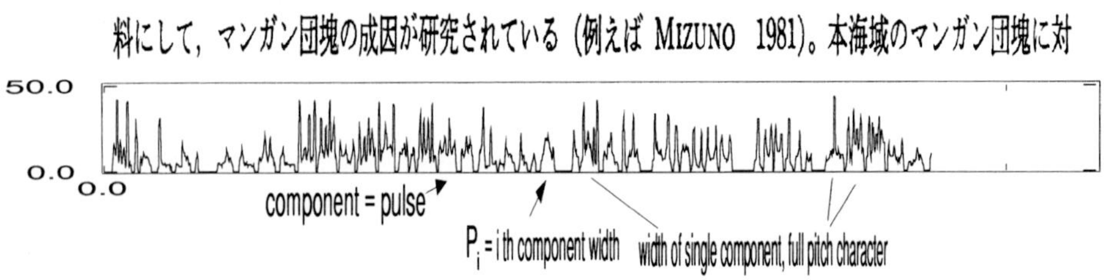

Fig. 12. Character segmentation from projection profiles.

3.3.6. *Text segmenter results*

The text and character segmenter was evaluated with respect to several criteria, as shown in Tables 2 and 3. Text segmenter performance varies between 98.5% for good quality images and 94% for degraded images. Simple connected component analysis is used for high quality printed text, while more complex analysis is applied for low quality document images.

3.4. *Character Recognizers*

Two independent character recognition modules have been developed to handle 3300 classes of Hiragana, Katakana and JIS level 1 Kanji, alphanumeric and symbol characters. Hence the main challenges are to design recognizers that are fast as well as accurate in handling complex character shapes. The recognizers are based on

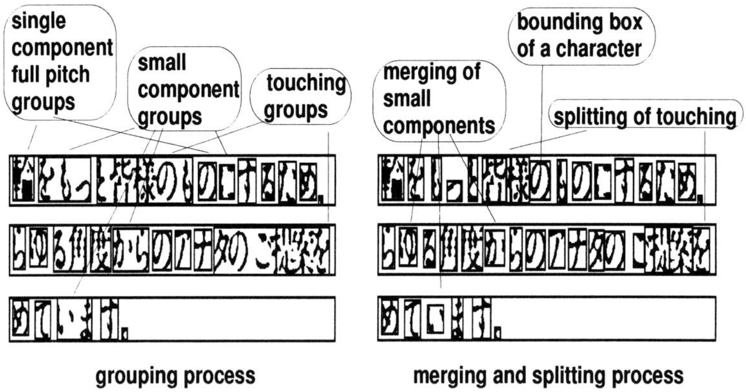

Fig. 13. Character segmentation — splitting and merging of components.

Table 2. Correct rate of character and line segmentation subsystems: CharsegC and LinesegC correspond to performance with high quality images; charsegN and LinesegN correspond to performance with noisy images.

Text Segmenter Performance - I					
Quality	Alignment	CharsegC	LinesegC	CharsegN	LinesegN
98.2%	100%	98.5%	100%	94.2%	98.6%

quadratic discriminant and fast nearest-neighbor classifiers. After preliminary experiments using several kinds of feature sets, two of the best performing feature sets were chosen. These are the Local Stroke Direction (LSD) and Gradient, Structural and Concavity (GSC) feature sets. These feature sets are described next, followed by a discussion of the minimal error subspace classifier and the nearest neighbor classifier. An outline of the postprocessing method is also presented.

3.4.1. *Local stroke direction features*

When given a character image, its LSD feature vector can be computed as follows [21] (see Fig. 14 for a detailed example): first, for each black pixel, compute its directional run-length for each of the four directions and normalize as a ratio to the total run-length in all directions; second, partition the image into $n \times n$ areas and compute the directional run-length of each area as an average of the pixels in the area. Here, we choose n equal to 8. Therefore, the size of LSD feature vector is $4 \times 8 \times 8 = 256$. In Fig. 14 (c), the values in the LSD feature vector are scaled

Table 3. Timing of character and line segmentation subsystems: CharsegC and LinesegC correspond to performance with high quality images; charsegN and LinesegN correspond to performance with noisy images.

Text Segmenter Performance - II					
Quality	Alignment	CharsegC	LinesegC	CharsegN	LinesegN
> 100 img/s	> 1000 img/s	41.5 c/s	> 1000 img/s	14.3 c/s	6.2 img/s

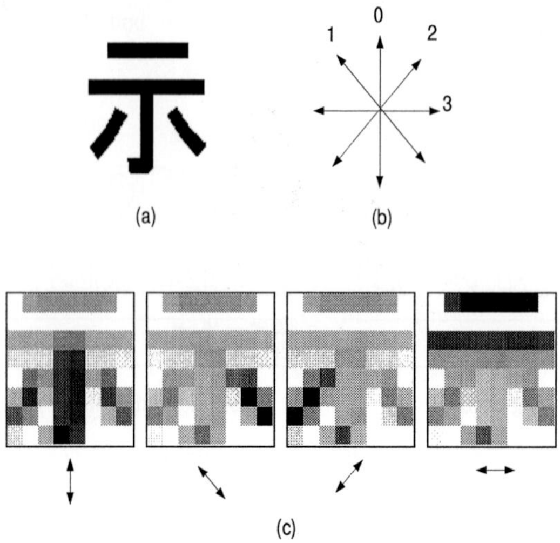

Fig. 14. A Kanji character image and its local stroke direction (LSD) feature vector. (a) Character image; (b) Scan directions; and (c) Local stroke direction feature (size: $4 \times 8 \times 8$).

to integers in the range of 0 and 255 so that they can be visualized in a grayscale image.

3.4.2. *Gradient, structural and concavity features*

The gradient and structural features encode local structure, while the concavity features are global descriptors extracted from binarized images.

A gradient map is constructed from the normalized digit image, by estimating gradient value and direction at each pixel. Fig. 15 contains the gradient map of the character image. Histograms of directional features are recorded in each region – indicating the presence (or absence) of a small number of oriented ranges of gradients in the region. Directional histograms from each region are concatenated into a fixed-length *gradient feature* vector. *Structural features* are computed from the gradient map by examining similarities in gradient direction in a local neighborhood of each pixel. These features record curvature in an approximate fashion. Curvature histograms indicating presence of changing orientation of character contours are estimated in each region. These histograms are concatenated from each region in a fixed-length structural feature vector. Gradient and structural features have been described in more detail in [31]. *Concavity features* are coarse global descriptors and are of three kinds: pixel density, large stroke, and true concavity. The large stroke features encode horizontal and vertical strokes in the image. Concave regions enclosed with a character are also recorded. Concavity features have been described in detail in [32].

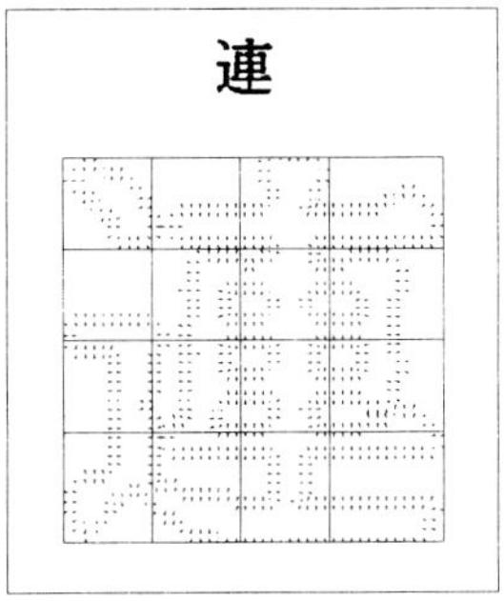

Fig. 15. Gradient and structural features extracted from gradient map of character images.

3.4.3. *Minimum error subspace classifier*

Subspace methods have been a major field of study in pattern recognition [33, 34], particularly for classification and feature subset selection. The minimum error subspace classifier is a discriminant function derived from the Karhunen-Loeve expansion. It is given by:

$$g_j(X) = |X - M_j|^2 - \sum_{i=1,k} \{\phi_{i_j}^T (X - M_j)\}^2 .$$

Here, $g_j(X)$ defines the discriminant classifier for the jth class, given an input measurement/feature vector X; M_j is the mean vector and ϕ_{ij}^T is the transpose of the ith eigenvector of the covariance matrix for the jth class; k is chosen as the dimension of the subspace defined by the dominant k eigenvectors. An unknown test feature vector is evaluated with these discriminant functions for each class, that is, $g_j()$, for $j = 1, 2, ..., C$, where C is the number of classes. The class of X is determined as that of the discriminant function for which the residual error is least, that is: $g_J(X) \leq g_j(X)$, $\forall j \in \{1, 2, .., C\}, j \neq J$.

Projections of an unknown test vector on the subspaces defined by the dominant eigenvectors of each class are subtracted from the projections on the entire eigenspace. The subspace of the class which best represents the unknown feature vector, results in the least residual error (and results in the least difference in projections between the whole space and the subspace of the principal eigenvectors). The class corresponding to the subspace with the least residual error is chosen as the class-identity of the unknown test vector.

This classifier can also be derived from the modified quadratic discriminant function [35], and can be interpreted as a quadratic discriminant classifier.

The training phase of this classifier requires the computation of the covariance matrices and their eigen-decomposition. For each class, the eigenvectors are then sorted in decreasing order of the corresponding eigenvalues. The mean vector M_j is evaluated for each class. Only the dominant eigenvectors (k, in this case) are used in the subsequent processing. During testing, the unknown or test feature vector

Table 4. Character recognition results.

ME Subspace classifier results		
Top 1	Top 5	Top 10
93.48%	97.86%	98.45%

is projected onto the subspace defined by these dominant vectors, for each class. The term $|X - M_j|^2$ defines the projection of the test vector onto the whole eigen-space. Hence, the class of an unknown vector is determined by the best subspace projection, given by the least residual error.

The classifier had been trained and tested on the CEDAR dataset, which contains 175,988 character images from 3,354 categories. A training set consisting of 118,417 samples, was generated by randomly collecting two-thirds of the samples from each category. The test set consists of the remaining 57,571 samples. Although the samples were extracted from document pages scanned at 400 ppi, they were downsampled to 200 ppi by applying a multi-rate method [36] to simulate degradation effect and used in our experiments. The performance of the ME subspace classifier, based on the LSD feature set is listed in Table 4.

3.4.4. *A nearest-neighbor classifier*

A NN classifier is designed for Japanese character recognition [37]. New algorithms for prototype reduction, hierarchical prototype organization and fast NN search are implemented in the classifier. In these algorithms, we use k-nearest and k-farthest neighbor lists to estimate the distribution of samples in the feature space of a given sample set. Given a set of samples, $P = \{p_0, p_1, ..., p_{n-1}\}$, for each sample p_i, its k-nearest neighbor list, $N_i = \{n_{io}, n_{i1}, ..., n_{i,k-1}\}$, and its k-farthest neighbor list, $F_i = \{f_{io}, f_{i1}, ..., f_{i,k-1}\}$, can be computed by comparing the sample with the rest of the samples.

Given a training set, the step of prototype reduction is to select a subset of samples which can represent the whole training set. Fig. 16 illustrates the process of prototype selection. The subset can be generated by deleting those redundant samples from the training set. By checking the pre-computed nearest neighbor list of each training sample, the prototype reduction algorithm decides whether a training sample can be deleted without negative effect on correct classification of itself and other samples if the sample is removed. Previous methods, such as Hart's *condensed nearest neighbor rule* (CNN) and Gates' *reduced nearest neighbor rule* (RNN), have to call the procedure of classification iteratively to test whether the deletion(or adding) of a prototype affects the correct classification of the other prototypes [38, 39]. The advantage of the method here is to avoid such an iterative process.

A brute-force NN algorithm suffers from its computational complexity because every prototype has to be matched with the input pattern to see whether it is close to the input pattern. To speed up NN classification, we proposed a fast NN search

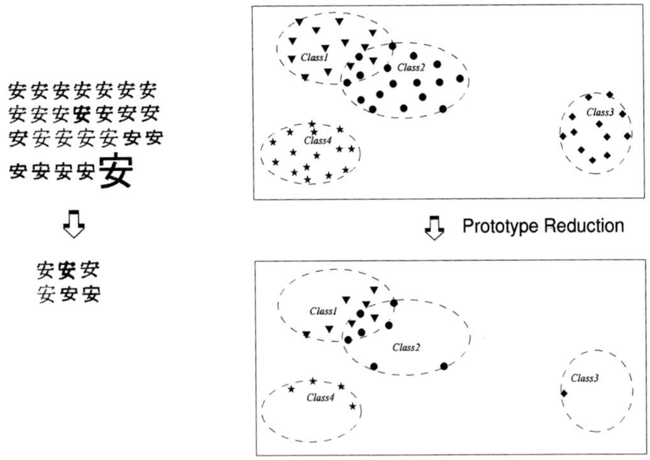

Fig. 16. Prototype reduction.

algorithm, FNN. The basic idea is to avoid unnecessary comparisons. Suppose there is a test sample X to be classified (see the triangle in Fig. 17). It has to be compared with prototypes from the prototype set. After the comparison between the input pattern X and prototype p_i, we know that the distance between them is very small. For those prototypes in p_i's k-farthest neighbor list, such as f_{i1},f_{i2} and f_{i3} shown in Fig. 17, without going further to compare each of them with the input sample X, we know the distance will be very large and therefore they cannot be in X's k-NN list. Similarly, after the comparison between X and prototype p_j, we know that the distance between them is very large. For those prototypes in p_j's k-nearest neighbor list, such as n_{j1},n_{j2},n_{j3} and n_{j4} shown in Fig. 17, without going further to compare each of them with the input sample X, we know the distance will be very large and therefore they cannot be in X's k-NN list.

In order to further speed up the search process, the prototype set can be organized into a hierarchical representation. The prototype set is divided into two levels. The first level is the so-called "centroid set". For a centroid, there may be a set of prototypes which are stored in the second level. The set for a centroid is called as the centroid's "package". A centroid can approximately represent those prototypes in its package. The process to create such a hierarchy is to pack prototypes for selected centroids. Two versions of the algorithm are designed: one is *farthest-like-neighbor-based*; another is *nearest-unlike-neighbor-based*.

After packing prototypes, the k-nearest neighbor list and the k-farthest neighbor list of each prototype in the centroid set C will be computed. The NN classifier can be described as a two-step procedure. Given a test sample X, its approximate k-nearest neighbors $\overline{N}_X$ in the centroid set C can be calculated using the search algorithm described above. Then the prototypes in the packages of those centroids

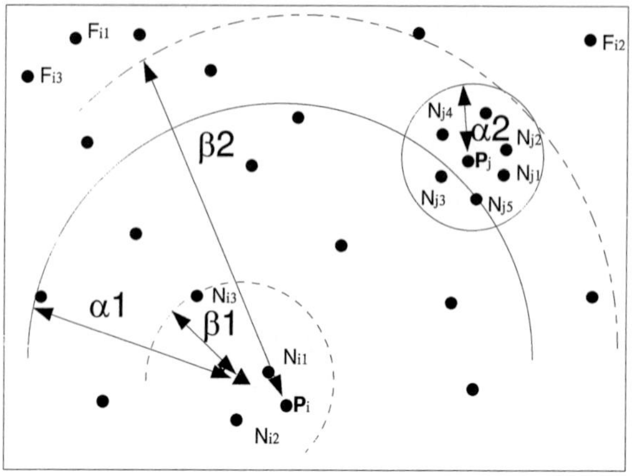

Note: ● denotes prototype; ▲ is an input pattern x.

Fig. 17. Speed up nearest neighbor search by avoiding redundant comparisons.

in $\overline{N}_X$ will be compared with X to generate the final k-nearest neighbors N_X for the sample.

The NN classifier using LSD features and city-block distance measure was trained and tested on the ETL machine-printed dataset. This dataset, contains 26002 train and 26002 test images. Prototype reduction resulted in 5117 of the 26002 prototypes to be retained for classification. Performance on the independent test set, based on these reduced prototypes resulted in 97.9% accuracy. On an average, each test image was compared with about 56.8% of the prototypes by the fast algorithm presented above.

The NN classifier was also integrated with the *ME Subspace* method to achieve higher accuracy [40]. On 200 ppi testing samples from CEDAR's dataset, its top 1 correct rate is 95.83%.

3.5. *Visual Similarity Analysis of Kanji for Postprocessing*

A Kanji can be composed of several elements, such as radicals and other simple components, based on some lexicographical principles. Although radicals are well-defined lexicographically, it is time-consuming to manually locate all possible character radicals and describe the structure of each Kanji character according to the radical set. Given a set of Kanji character images, we try to automatically determine which character images are partially similar based on feature matching [41]. We use *LSD* features (see Fig. 14) here. By analyzing the difference between two feature vectors, we can determine whether they are partially similar. For example, in Fig. 18(a), two Kanji characters are shown. Fig. 18(b) shows the distance distribution of their difference. We can find out that distance scores from the bottom part are significantly smaller than that from the top part. The result suggests that the bottom parts of those two images are very similar. The similar region of two LSD features can be approximately represented by a mask (Fig. 18(c) for the mask derived from Fig. 18(b)). A mask defines a relation between character images.

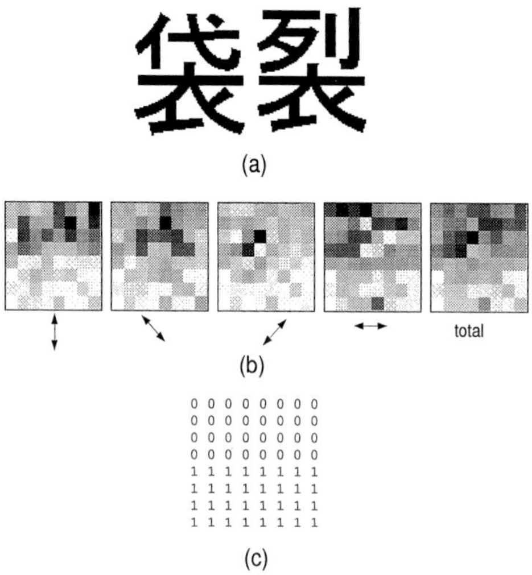

Fig. 18. Two Kanji images and their difference in LSD. (a) Character image pair; (b) Difference in LSD feature; and (c) Mask obtained.

By applying feature-based visual similarity analysis to character images from a specific typeface, we can compute a similarity matrix which records possible visual inter-category relations.

For a character image with several candidates(see Fig. 19 for the top 5 candidate lists of three Kanji characters), visual similarity between character images and character categories can provide useful information to select a proper choice. Given a text page, character images can be extracted after the stage of character segmentation. Visual inter-character relations between character images can be computed. For example, for images a, b and c in Fig. 19(a), two relations, $R4(a,b)$ and $L3(b,c)$, are derived (see Fig. 19(b)). Here, $R4(a,b)$ means that images a and b have the similar right part; $L3(b,c)$ means images b and c have the similar left part. Correct choices for images a, b and c should keep those relations in the similarity matrix. The result of this analysis is shown in Fig. 19(c). Considering the candidates of image c in Fig. 19(c), we know that all candidates have the same left part by checking a pre-computed similarity matrix. To distinguish those candidates, we have to focus on the information from their right parts. By matching the right part of the image with the right parts of prototypes for categories that those candidates belong to, those candidates can be re-ranked by their new confidence scores which are based on the information from the right part. The first candidate will be selected as the decision for the image. For the example above, its final result is shown in Fig. 19(d).

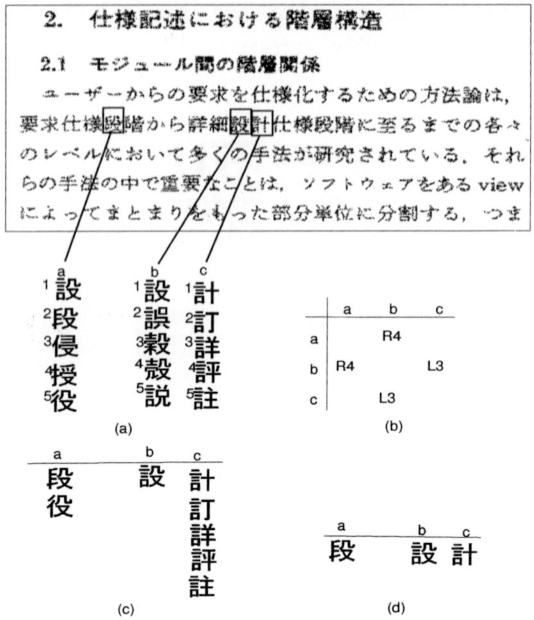

Fig. 19. Kanji candidate selection using visual similarity. (a) Top 5 Candidates provided by OCR; (b) Visual similarity between Kanji images; (c) Candidate reduction after using visual relations betwee images; and (d) After re-testing the candidates by partial matching.

4. Conclusion

We have presented an overview of some of the main directions of research in the automated reading of Japanese documents. The challenges in page layout analysis and character recognition for this class of documents have been described. The Japanese OCR system developed at CEDAR has been described in detail to illustrate many of the design challenges. A Japanese character image database available from CEDAR was also presented.

Ongoing research is oriented towards higher performance — accuracy and speed — for good quality document images, improved recognizers for low quality images (such as facsimile and photocopies) and systems for multilingual text recognition. Application of OCR technology in specific office and industrial automation tasks also present areas for future work.

Acknowledgements

The significant contribution of Dr. Jonathan Hull in the direction of the early part of this research is greatly appreciated. This research would not have been possible without the contributions of several students at the State University of New York at Buffalo. In particular, we would like to thank Victor Zandy, Qun Feng Liao, Chi Fang, Brian Grom, and Zhou Hong.

References

[1] T. Sakai, A history and evolution of document information processing, in *Proc. of the Second Int. Conf. on Document Analysis and Recognition*, Oct. 1993, 377-384.

[2] S. Mori and C. Y. Suen and K.Yamamoto, Historical review of OCR research and development, *Proc. of the IEEE* **80,7**, 1992, 1029-1058.

[3] J. D. Becker, Typing Chinese, Japanese, and Korea, *IEEE Computer*, **18**, **1**, 1985, 27-34.

[4] J. K. Huang, The input and output of Chinese and Japanese characters, *IEEE Computer*, **18,1**, 1985, 18-24.

[5] K. Lunde, *Understanding Japanese Information Processing* (O'Reilly and Associates, Inc.), 1993.

[6] R. Matsuda, Processing information in Japanese, *IEEE Computer*, **18**, **1**, 1985, 37-45.

[7] G. Nagy, Chinese character recognition: a twenty-five year retrospective, in *Proc. of the 9th Int. Conf. on Pattern Recognition*, 1988, 163-167.

[8] R. Suchenwirth and J. Guo and I. Hartmann and G. Hincha and M. Krause and Z. Zhang, *Optical Recognition of Chinese Characters* (Vieweg), 1989.

[9] T. H. Hildebrant and W. Liu, Optical recognition of handwritten Chinese characters: advances since 1980, *Pattern Recognition J.*, **26,2**, 1993, 205-226.

[10] X. Z. Zhang, *Techniques for Chinese Character Recognition* (Qinghua University, Beijing, China), 1992 (in Chinese).

[11] Q. Luo and T. Watanabe and N. Sugie, A structure recognition method for Japanese newspapers, in *First Symp. on Document Analysis and Information Retrieval*, 1992, 217 - 234.

[12] T. Akiyama and N. Hagita, Automated entry system for printed documents, in *Pattern Recognition*, **23**, 1990, 1141-1154.

[13] D. J. Ittner and H. S. Baird, Language-free layout analysis, in *Proc. of the Second Int. Conf. on Document Analysis and Recognition*, 1993, 336-340.

[14] S. Tsujimoto and H. Asada, Major components of a complete text reading system, *Proc. of the IEEE*, **80**, 1992, 1133-1149.

[15] S. Ariyoshi, A character segmentation method for Japanese printed documents coping with touching character problems, *Proc. of the 11th Int. Conf. on Pattern Recognition*, 1992, 313-316.

[16] Y. Kobayashi and K. Yamada and J. Tsukumo, A segmentation method for handwritten Japanese character lines based on transitional information, in *Proc. of the 11th Int. Conf. on Pattern Recognition*, 1992, 487-491.

[17] Y. Maeda and F. Yoda and K. Matsuura and H. Nambu, Character segmentation in Japanese hand-written document images, in *Proc. of the 8th Int. Conf. on Pattern Recognition*, 1986, 769-772.

[18] A. Konno and Y. Hongo, Postprocessing algorithm based on the probabilistic and semantic method for Japanese OCR, in *Proc. of the Second Int. Conf. on Document Analysis and Recognition*, 1993, 646-649.

[19] H. Fujisawa and K. Marukawa, Full-text search and document recognition of Japanese text, in *Fourth Symp. on Document Analysis and Information Retrieval*, Apr. 1995, 55-80.

[20] K. Sakai and S. Hirai and T. Kawada and S. Amano and K. Mori, An optical Chinese character reader, in *Proc. of the 3rd Int. Conf. on Pattern Recognition*, 1976, 122-126.

[21] S. Mori and K. Yamamoto and M. Yasuda, Research on machine recognition of handprinted characters, *IEEE Transactions on Pattern Analysis and Machine Intelligence*, **6**, 1984, 386-405.

[22] R. Oka, Handwritten Chinese character recognition by using cellular feature, in *Proc. of the 6th Int. Joint Conf. on Pattern Recognition*, 1982, 783-785.

[23] T. Ijima and H. Genchi and K. Mori, A theory of character recognition by pattern

matching method, in *Proc. of the First Int. Joint Conf. on Pattern Recognition*, 1973, 587-594.

[24] F. Kimura and K. Takashina and S. Tsuruoka and Y. Miyake, Modified quadratic discriminant functions and the application to Chinese character recognition, in *IEEE Transactions on Pattern Analysis and Machine Intelligence*, **9**, 1987, 149-153.

[25] K. Miyahara and F. Yoda, Printed Japanese character recognition based on multiple modified LVQ neural network, in *Proc. of the Second Int. Conf. on Document Analysis and Recognition*, Oct. 1993, 250-253.

[26] K. Kigo, Improving speed of Japanese OCR through linguistic preprocessing, in *Proc. of the Second Int. Conf. on Document Analysis and Recognition*, 1993, 214-217.

[27] S. W. Lam and V. C. Zandy, Skew detection using directional profile analysis, in *Proc. of Machine Vision Applications*, 1994, 95-98.

[28] Y. Ishitani, Document skew detection based on local region complexity, in *Proc. of Second Int. Conf. on Document Analysis and Recognition*, 1993, 49-52.

[29] S. W. Lam, A local-to-global approach to complex document layout analysis, in *Proc. of Machine Vision Applications*, 1994, 431-434.

[30] S. W. Lam and Q. F. Liao and S. N. Srihari, A divide-and-conquer approach to Japanese text segmentation, in *Proc. of the Conf. on Document Recognition, 1995 SPIE Symp.(SPIE95)*, 1995, 216-227.

[31] G. Srikantan and S. W. Lam and S. N. Srihari, Gradient-based contour encoding for character recognition, *Pattern Recognition J.*, to appear.

[32] J. T. Favata and G. Srikantan and S. N. Srihari, Handprinted character/digit recognition using a multiple feature/resolution philosophy, in *Int. Workshop on the Frontiers of Handwriting Recognition, IWFHR - 4*, 1994, 57-66.

[33] E. Oja, *Subspace Methods in Pattern Recognition* (John Wiley & Sons), 1983.

[34] K. Fukunaga, *Introduction to Statistical Pattern Recognition* (Academic Press), 1990.

[35] F. Kimura and M. Shridhar and Y. Miyake, Relationship among quadratic discriminant functions for pattern recognition, in *Int. Workshop on the Frontiers of Handwriting Recognition, IWFHR - 4*, 1994, 418-422.

[36] G. Srikantan and D. S. Lee and J. T. Favata, Comparison of normalization methods for character recognition, in *Proc. of the Third Int. Conf. on Document Analysis and Recognition*, Aug. 1995, 719-722.

[37] T. Hong and S. W. Lam and J. J. Hull and S. N. Srihari, The design of a nearest-neighbor classifier and its Use for Japanese character recognition, in *Proc. of the Third Int. Conf. on Document Analysis and Recognition*, Aug. 1995, 270-273.

[38] P. E. Hart, The condensed nearest neighbor rule, *IEEE Transactions on Information Theory*, **IT-14, 3**, 1967, 515-516.

[39] G. W. Gates, The reduced nearest neighbor rule, in *IEEE Transactions on Information Theory*, **IT-18, 3**, 1972, 431-433.

[40] T. Hong and G. Srikantan and V. C. Zandy and C. Fang and S. N. Srihari, Character recognition in a Japanese text recognition system, in *Proc. of the Conf. on Document Recognition, 1996 SPIE Symp.(SPIE96)*, 1996.

[41] T. Hong and S. W. Lam and J. J. Hull and S. N. Srihari, Visual similarity analysis of Chinese characters and its uses in Japanese OCR, in *Proc. of the Conf. on Document Recognition, 1995 SPIE Symp.(SPIE95)*, 1995, 245-253.

Handbook of Character Recognition and Document Image Analysis, pp. 381–396
Eds. H. Bunke and P. S. P. Wang

CHAPTER 14

ONLINE RECOGNITION OF KOREAN HANGUL CHARACTERS

JIN H. KIM and BONGKEE SIN†
Computer Science Department, Korea Advanced Institute of Science and Technology
Yusung-ku, Taejon, 305-701, Korea

This paper describes an HMM network-based approach to online recognition of Korean Hangul characters. The primary concern of the approach is hidden Markov modeling of letters and explicit modeling of inter-letter patterns or ligatures appearing in cursive script. By using ligature HMMs the variability of inter-letter patterns is resolved. By alternate concatenation of the two kinds of HMMs, a network model for all legal Korean characters has been designed. Given the network, the recognition problem is formulated as that of finding the most likely path from the start to the end node. A DP-based search for optimal input-network alignment gives simultaneous letter segmentation and character recognition — up to 93.3% correct on unconstrained samples.

Keywords: Online character recognition; Cursive character; Korean Hangul; Ligature; Hidden Markov model; Finite state network; Viterbi search; Letter segmentation; Perturbation smoothing.

1. Introduction

Each piece of handwriting is highly variable and ambiguous. Different persons write in different styles, and their writings may also vary at different times (Fig. 1(a)). This characteristic has been termed as the variability in pattern recognition. It is not easy to find a framework that is flexible enough to model this characteristic and based firmly on theory so that we can devise character recognizers systematically. The other dimension of difficulty in handwriting recognition is the ambiguity in character shape, which is, in a sense, orthogonal to the variability. The problem is due to the fact that different characters are often written very similar to each other. Typical examples are shown in Fig. 1(b). The primary cause of ambiguity lies in fast and sloppy writing characteristic, which is usually a writer's style affected by the motor program learned from long practice since childhood. The ambiguity implies that we cannot recognize characters entirely based on shape analysis and that effective use of contextual and a priori knowledge is necessary to

†The current address is Multimedia Development Team, SW Research Labs, Korea Telecom, 17 Woomyeon-dong Seocho-ku, Seoul 137-792 Korea

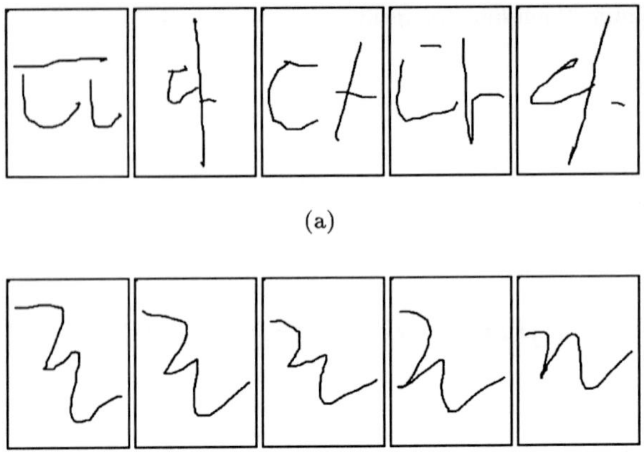

Fig. 1. (a) Variable shape of a Hangul character, and (b) different but ambiguous characters.

achieve human performance.

Because of the two characteristics of handwriting, character recognition by machine is, if not impossible, a highly difficult problem that is not tractable with simplistic methods based on human insight. But as humans read most script without difficulty, even correcting misspellings, it is believed that human performance is ultimately achievable by machine. This fact has been the rationale of our study of handwritten character recognition.

For the Korean Hangul character, there are several attempts to recognize online cursive characters [1]. However, they suffer from the variability problem. It is primarily due to the lack of solutions that are able to deal with cursive strokes. Moreover, they resort to external segmentation prior to recognition, which obviously limits the performance of the recognizers.

In this paper, a network-based approach to the recognition of online Korean Hangul characters is described. The approach is based on modeling all and only the legal characters with a finite state network using a set of hidden Markov models. An individual hidden Markov model (henceforth HMM) models either a letter or an inter-letter pattern (called *ligature*). Then just like the order of writing letters in a character, the overall network model for all characters is designed into a sequential concatenation of letter HMMs and inter-letter pattern HMMs.

The proposed approach contrasts with others in that we explicitly model each ligature as a separate entity, and that we describe handwriting homogeneously as a network. By allowing the ligature HMMs to accept both pen-down ligatures and pen-up moves, the character network can model both cursive and discrete patterns. The recognizer based on the network performs recognition by using any of the well-known graph-searching or dynamic programming algorithms. Another feature is that the recognizer performs recognition and letter segmentation simultaneously, by performing a search based on the criterion of input-network alignment.

2. Recognizer Model

The MAP (maximum a posteriori) rule for pattern recognition states that a system makes a decision based on the maximization of a posteriori probability $Pr(W|X)$, or, equivalently by way of Bayes' rule,

$$Pr(\hat{W}, X) = \max_{W} Pr(X|W)Pr(W). \tag{2.1}$$

$Pr(X|W)$ is the probability that the script writer will generate output script X from character W. In order to estimate $Pr(X|W)$, we will design in this section a handwritten character network based on HMMs that accounts for a script's shape and its variability. $Pr(W)$ is the a priori probability of a character W or the probability that the text generator will produce the character. The model for $Pr(W)$ is termed language model and is often separately modeled as a Markov source

$$Pr(W) = \prod_{k=1}^{K} Pr(w_k|w_1 \cdots w_{k-1}), \tag{2.2}$$

where $W = w_1 \cdots w_K$. To reduce model parameters, a simplifying assumption, N-th order Markov process, is made so that

$$Pr(W) = \prod_{k=1}^{K} Pr(w_k|w_{k-N} \cdots w_{k-1}). \tag{2.3}$$

2.1. *Modeling Unit*

In Korean Hangul, a letter[a] is considered the most basic and natural unit of describing characters. The choice is made from a compromise between the difficulty of modeling and the trainability. For each letter, we design an HMM. The models have a left-right transition structure (Fig. 2(a)) that constrains the temporal characteristic of online handwriting signals [2]. There is a unique initial state and a unique final state. The number of states is determined by the length and structure of letter patterns. There are a few states for simple letters and well over ten for long and complex letters.

In a handwritten script, ligatures occurring at letter boundaries are the dominant source of shape variability. The conventional notion of ligature involves only the pen-down linking patterns between strokes. In general, however, the sequence of strokes are written smoothly leaving behind characteristic traces at the strokes'

[a]This is not precisely the same as the alphabet; there are composite letters consisting of two basic letters.

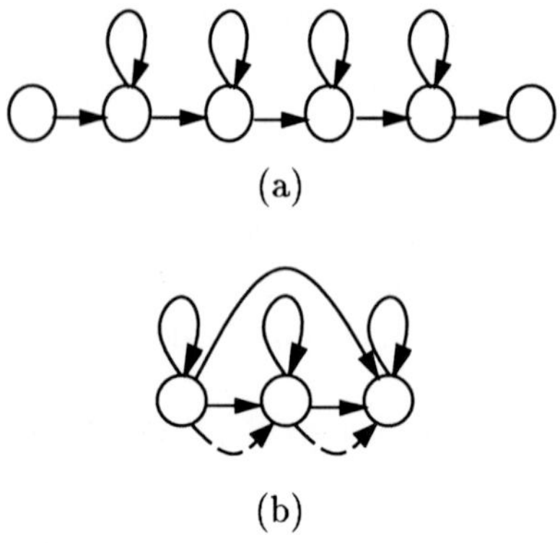

Fig. 2. (a) Letter HMM, and (b) ligature HMM with null-transitions shown in broken arrows.

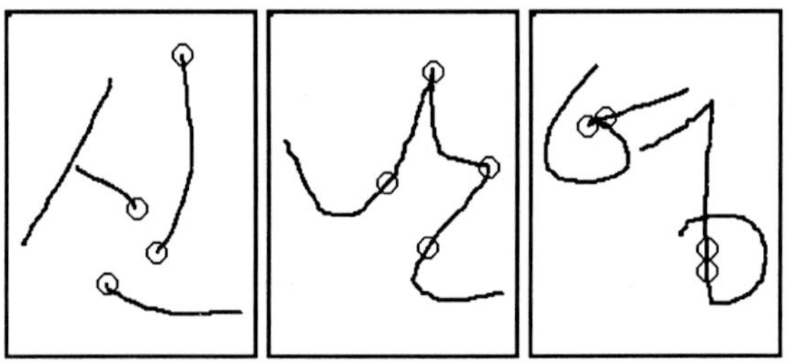

Fig. 3. Ligatures: (left to right) pen-up moves, pen-down drags, and ligature deletion.

ends, which is attributed to physical constraints and psychological needs. These patterns convey information about the stylus motion and the position of the next letter just like conventional ligatures.

In order to represent the stylus locus faithfully and consistently, we extend the conventional definition of ligature to pen-up moves which are imaginary or invisible and assumed straight. In addition the ligature will also include deletion of linking patterns that can arise where the end of the preceding stroke and the start of the following stroke are coincident. See the left and right box, respectively, of Fig. 3.

The structure of ligatures in general is simple with a small variation. They are approximately linear with similar directions. Therefore a simple topology as shown in Fig. 2(b) will suffice. The broken arrows in the figure represent null transitions that produce no observations. They are introduced to model the ligature deletion, in case ligatures are absent and preceding and following strokes are directly linked.

2.2. Character Model

A Korean character consists of either two or three letter types: initial consonant C, middle vowel V, and final consonant Z. Each type can appear only at a specific position within a character, and written in the order of C, V and then Z, if it exists. Ligatures appearing between letters in handwriting are denoted as L_{cv} for those between C and V, and L_{vz} for those between V and Z. Then a handwritten character is defined as an augmented sequence

$$\text{Handwritten Character} := C\ L_{cv}\ V \mid C\ L_{cv}\ V\ L_{vz}\ Z.$$

On the other hand the Korean character is not a simple left-to-right concatenation of letters but a spatial arrangement of them as shown in Fig. 3. We have to incorporate this knowledge in some way.

What ligature means to Hangul character modeling is the character's topology or the spatial arrangement of component letters inside an imaginary bounding box. The information therein has long been noted, both in online and offline character recognition [1]. Sample statistics have shown us that ligature appearances are dependent on the way the previous stroke ends and the way the following stroke starts relative to the previous ends. Based on this observation, we have clustered letter models. Figure 4 shows the result for C letters clustered by letter-ending patterns.

C_1	ㄱㄲ
C_2	ㄴㄷㄸㄹㅁㅂㅃㅌㅍ
C_3	ㅅㅆㅈㅉㅊ
C_4	ㅇㅎ
C_5	ㅋ

Fig. 4. Five clusters of initial consonant letters C.

Similar clusterings are made for vowels V and final consonants Z. Refer to the arc labels in Fig. 5. Then we have determined a set of ligature classes by Cartesian products as of (C, V)'s and (V, Z)'s, each of which is modeled by a single HMM. The above classification enables us to reduce the number of ligature models. The resulting set of models are similar to the *generalized triphones* [3] in continuous speech recognition.

Once the set of ligature models have been designed, the construction of character networks is straightforward. With the introduction of a dummy node corresponding to each letter cluster, and by sharing common letters and ligatures among character networks, we can obtain a single character network as shown in Fig. 5.

In the figure the leftmost node is the start node and the rightmost one the end node. Each node in the second column corresponds to a cluster of C letters with a

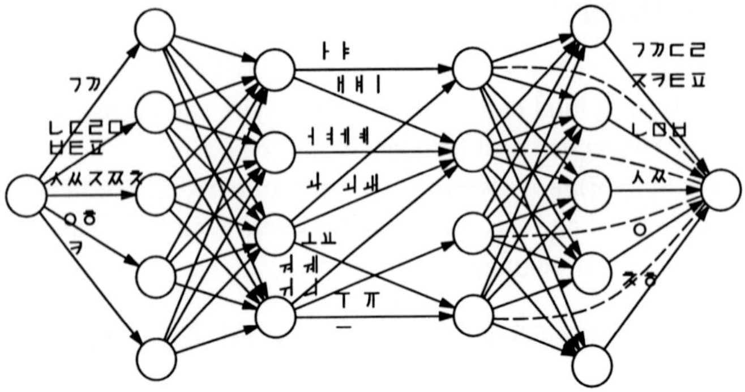

Fig. 5. Hangul character network model *BongNet*; null arcs shown in broken lines model the case of final consonant (Z) deletion.

similar ending pattern. The third column nodes correspond to clusters of V letters by the starting tip pattern. The mesh of arcs between the two columns represent the ligature classes. The network has in total 40 ligature models, half L_{cv}'s and half L_{vz}'s.

3. Data Coding

Data acquisition in online handwriting is a time-sequential registration of the locus of stylus on a digitizer tablet, the result of which is given as a sequence of two-dimensional coordinates (x, y)'s. Such a raw handwriting cannot be directly analyzed per se by most recognizers. Therefore several kinds of preprocessing techniques are applied first. Following these, the data is converted to a form that a recognizer can process.

3.1. *Preprocessing*

Handwriting usually contains several types of noise that comes from both electromechanical defects and imperfect hand motor control. Preprocessing involves a set of signal processing techniques that reduce the noise in the tablet data. Examples often employed are smoothing, filtering, wild point correction, dehooking, and dot reduction [4]. Since, however, any kind of preprocessing invariably entails loss of information through shape distortion, the extent of that effect should be kept minimal. In our experiments, only the techniques of reducing wild points that appear randomly, and removing obvious hooks that appear at stroke ends have been used.

3.2. *Chain Coding*

Given a preprocessed handwriting it is required to convert it to a form suitable for HMM evaluation. We use Freeman code [5] that effectively encodes the local

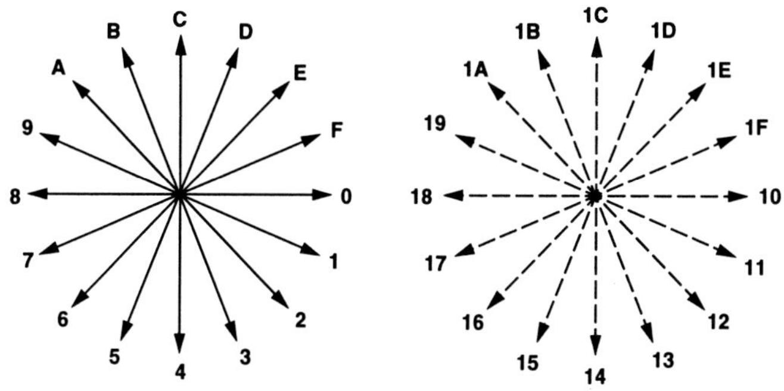

Fig. 6. Direction codes for pen-down stroke (left) and pen-up move (right).

feature of stylus locus. Each code in the chain represents the local direction information in the corresponding position in the script.

In online handwriting, the direction of pen-up moves provides the information as to where the next stroke or letter starts. We describe a pen-up move as a stroke spanning from the end point of the previous stroke to the start point of the following stroke. Then we explicitly encode the pen-up stroke with an alternate coding scheme. The following is the procedure for converting a piece of handwriting into a chain code.

(1) We insert a pen-up stroke between each pair of strokes. The stroke is assumed straight and represented as a pair of points, coincident with the end point of the preceding stroke and the start point of the following stroke respectively. Its pen-up status is retained. In this way the complete locus of stylus for a character is described by an *augmented* sequence of strokes.

(2) All the strokes are resampled spatially into a sequence of equidistant points. It is done by following the locus of the strokes in order, and generating a point by linear interpolation for each quantum step. This applies to both the pen-down strokes and the pen-up strokes. The resampling step size is determined as a fraction of the size of the character.

(3) Each pair of consecutive points is converted into a direction code. For pen-up strokes we use another set of direction codes (Fig. 6). The complete chain code for a character is then fed to the BongNet, see Fig. 5.

We have tested two schemes of direction coding: eight directions and sixteen directions. In this encoding stage, there is another loss of information owing to quantizing direction angles. The loss will be greater in eight direction coding than in sixteen direction coding. In spite of the loss, however, the character model of the preceding section works well as will be shown through a detailed comparison of these coding schemes in Sec. 5.2.

4. Recognition

Dynamic programming-based Viterbi search in a finite state network (FSN) is widespread and has been shown to be successful in speech recognition [6]. Based on the principle of optimality [7], it is a technique of finding a path that best aligns to an input code sequence.

Character recognition, given the network of Fig. 5, is based on the same technique. Each complete path $W = w_1 w_2 \cdots w_5$ from the start node to the end node in the network determines a unique letter sequence corresponding to a character $C = cvz$ where $c = w_1$, $v = w_3$ and $z = w_5$, and w_2 and w_4 are ligatures. Given an input sequence $X = x_1 x_2 \cdots x_T$ we can obtain the best alignment to the best path $\hat{W}$ by applying the Viterbi algorithm that maximizes the $P(X|W)$. Each alignment of X to a path W defines a partition

$$X(\tau_1),\ X(\tau_2),\ \ldots,\ X(\tau_5)$$

where $X(\tau_k) = x_{\tau_{k1}+1} \cdots x_{\tau_k}$ corresponds to w_k and satisfies $0 = \tau_0 \le \tau_1 \le \tau_2 \le \cdots \le \tau_K = T$. Let us denote such a partition as $\tau = (\tau_1, \tau_2, \ldots, \tau_K)$. Then we can rewrite the first component of the right-hand side of Eq. (2.1) as

$$Pr(X|W) = Pr(X(\tau_1) \cdots X(\tau_5)|w_1 \cdots w_5)\,. \tag{4.1}$$

Here the odd index terms correspond to letters and the even index terms to ligatures between letters.

To avoid the exploration of all combinatorial possibilities, we apply first-order Markovian assumption and conditional independence assumption among the sequence of segments $X(\tau_1), \ldots X(\tau_5)$. Although it is not valid in cursive script recognition in general, the assumption is justified provided that each letter and ligature model describes the variability of the corresponding segmental patterns well. Then we can rewrite Eq. (4.1) as

$$Pr(X|W) = \max_{\tau} \prod_{k=1}^{5} Pr(X(\tau_k)|w_k)\,. \tag{4.2}$$

stating the best choice among the segmentation possibilities. The above equation states that the overall recognition problem can be decomposed into a sequence of subproblems of letter segmentation and matching.

The letter/ligature recognition given the subsequence $X(\tau_k)$ is computed using the Viterbi algorithm. Let us assume that a letter/ligature HMM is given as an N-state, left-right type of model with state indices given in order $s_1, s_2, \ldots, s_N$ as shown in Fig. 7.

The forward computation of the Viterbi algorithm in state s_j of an HMM at time t is given by the recursion

$$\delta_t(j) = \max_{1 \le i \le j} \{\delta_{t-1}(i) a_{ij} b_{ij}(x_t)\},\ j = 1, \ldots, N \tag{4.3}$$

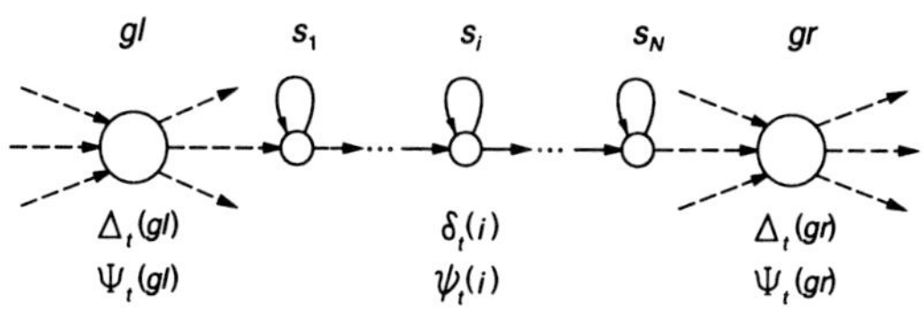

Fig. 7. Viterbi computation in an FSN.

where the $\delta_t(j)$ is the joint likelihood of the partial symbol sequence thus far observed and the best partial state sequence for reaching the state s_j. The argument which maximizes Eq. (4.3) is kept in $\psi_t(j)$ to retrieve the best state sequence later.

For the initial state s_1 of an HMM, there is a null inward transition from the preceding node gl. In this case we have

$$\delta_t(1) = \max\{\delta_{t-1}(1)a_{1,1}b_{1,1}(x_t),\ \Delta_t(gl)\} \tag{4.4}$$

where $\Delta_t(gl)$ denotes the joint probability of the partial observation sequence $X_1^t = x_1 \cdots x_t$ and the best partial path arriving at the node gl from the start node of the network.

The computation done at dummy nodes between HMMs of the network is similar to the maximization done at HMM-internal states. For each time t and node gr, the Viterbi maximization is performed over all inward arcs labeled by HMMs m as

$$\Delta_t(gr) = \max_m \delta_t^m(N_m)\,. \tag{4.5}$$

Here $\delta_t^m(N_m)$ is the likelihood of the final state s_{N_m} of the HMM m at time t corresponding to $Pr(X(\tau_k)|w_k)$ in Eq. (4.2). Again, the backtrack information is maintained in $\Psi_t(gr)$ pointing to the final state of the model m that maximizes Eq. (4.5). When the forward pass of the algorithm is finished, we can retrieve the best sequence of models (or states) and the best segmentation of X. The label sequence of the result becomes the recognition result.

5. Experiments

5.1. *Experimental Context*

For model training, unconstrained samples of two texts H and K have been collected from 37 and 21 subjects, respectively, using a digitizer connected to a workstation. The total set of H samples amounts to 45,000 characters. The subjects were high school students. The set K consists of about 7,000 characters written by those who have received college education or above and write more cursively.

The test data include two other sets, L and M, written by ten other high school students and eight other college graduates, respectively. They amount to 8,100 and 1,700 characters, respectively.

To train letter and ligature models given character samples, it is required to specify the position of each letter comprising the characters. This is done manually by a teacher who can tell the likely points. Although the points are usually highly ambiguous in cursive script because of the smooth connection of strokes, they can be chosen and the variability can be modeled in statistical terms. Finally, for each set of classified segments, the corresponding HMM was trained with at most fifty iterations of the forward/backward algorithm [8].

5.2. *Character Recognition*

The first test concerns the performance of the two coding schemes with eight and sixteen directions, respectively. For this we have created two models Λ_8 and Λ_{16} trained with eight and sixteen direction codes, respectively, based on set H. Their performance on test set L and M is shown in Table 1.

Table 1. Performance with eight and sixteen direction codes (in % correct recognition).

Models(train set)	L	M
$\Lambda_8(H)$	84.24	84.99
$\Lambda_{16}(H)$	87.82	83.27
$\Lambda_8(H+K)$	86.16	88.55
$\Lambda_{16}(H+K)$	88.10	87.43

As shown in the table, the sixteen direction coding model performs better than the eight direction coding models for test set L. The result meets our expectation that is based on the degree of quantization error of chain codes.

However, in case of test set M, the result is reversed. As a means of pinpointing the cause we have added set K to the training set producing two models $\Lambda_8(H+K)$ and $\Lambda_{16}(H+K)$. In Table 1 their performance measured on test set M shows $\Lambda_{16}(H+K)$ to be still inferior by 1.12% in the recognition rate. (But notice that it was 1.72% for $\Lambda_{16}(H)$.) We can attribute this result partially to under-training of HMMs or more precisely to the data characteristic as remarked previously, that is, general stylistic difference between the subject groups. In a further test with models trained with data set $H+2K$ that contains K twice, the recognition rates of $\Lambda_8(H+2K)$ and $\Lambda_{16}(H+2K)$ are almost equal to each other, 88.32% and 88.26% respectively, thus supporting our hypothesis.

Finally, the performance of the above models without character language model is shown in Table 2. By using a simple bigram language model we have witnessed error reduction of about 40% which argues for a good language model.

Table 2. The effect of language model with test set M.

Model(train set)	no language	language	% error reduction
$\Lambda_8(H+K)$	81.55	88.55	37.94
$\Lambda_{16}(H+K)$	77.28	87.43	44.67

5.3. *Output Parameter Smoothing*

HMM requires a large amount of training samples for reliable estimation of its parameters. Owing to the lack of data, however, it is often not possible to satisfy this requirement in practice. In such cases the HMM parameters will be overfitted to the given training set. This is especially the case with output probability distributions (PDs) having a majority of the parameters. As a means of alleviating the problem, a simple Gaussian smoothing technique which we call *perturbation smoothing* has been applied to all HMM output PDs after the forward-backward algorithm. It is a form of nonlinear smoothing (blurring) with a Gaussian filter [9]. For each HMM whose parameters have been estimated, we perturb the observation PDs so that, for a symbol x, the output probability $b_{ij}(x)$ of the transition from s_i to s_j is reestimated as a weighted sum of $b_{ij}(y)$ for all y in the neighborhood of x as

$$\hat{b}_{ij}(x) = \sum_{y \in NGB(x) \cup \{x\}} w(y;x)\, b_{ij}(y)$$

where $NGB(x)$ denotes the set of neighbor codes of x within a given perturbing range and $w(y;x)$ is a weight function satisfying

$$\sum_{y \in NGB(x) \cup \{x\}} w(y;x) = 1.$$

With the perturbing distance limited to immediate neighbors, that is, $c \pm 1$ for code c, and with a varying weight function $w(x;c)$

$$w(x;c) = \begin{cases} \frac{n}{n+2}, & \text{if } x = c \\ \frac{1}{n+2}, & \text{otherwise,} \end{cases}$$

for $n > 0$, we smooth the output distribution parameters of all HMMs. See Fig. 8 for a sample result of an output PD.

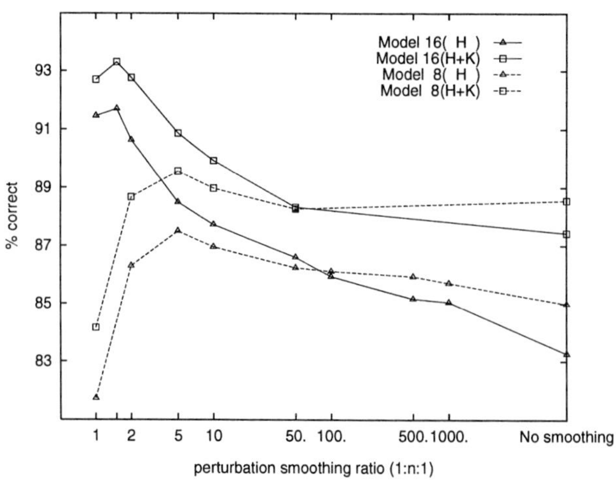

Fig. 9. Perturbation effect on a model performance with varying ratios.

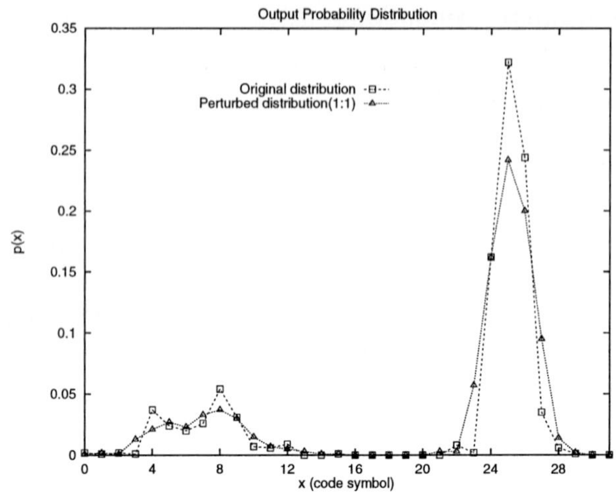

Fig. 8. Perturbation result of an output PD with $n = 1$.

The four basic models in the previous section underwent the smoothing procedure with varying perturbation parameters. The trend of performance change is shown in Fig. 9. The figure says that performance improves in general with more perturbation smoothing (as it goes from right to left in figure) to a critical point, $n = 1.5$ or weight ratio $w(c-1;c) : w(c;c) : w(c+1;c) = 1 : 1.5 : 1$ for sixteen direction code models (up to 93.3%) and $n = 5$ for eight direction code models. The performance increase by parameter smoothing is not so conspicuous for eight direction code as for sixteen direction code, but the recognition rate precipitates implying over-smoothing at the left end of the figure. This suggests that there is under-training in the sixteen direction code but not in the eight direction code.

If we perturb the smoothed PDs again, the effect will be similar to the perturbation smoothing with a greater amplitude or neighborhood size. The test result of the re-perturbation showed no further significant improvement nor degradation. Still more smoothing, however, resulted in gradual degradation.

In summary, even with the above results, the question still remains as to whether the smoothing effect is really due to under-training or the performance increase actually resulted from the effect of leveling the discrimination powers among HMMs. More extensive experiments are required in order to arrive at a convincing answer.

5.4. *Utility of Ligature Model and Networking*

The third set of experiments concerns the utility of ligature models and of model clustering. Since it is not easy to make a fair comparison between the ligature-based approach and other conventional ligature-free approaches, we have resorted to an indirect way of checking the utility. It is based on the idea of controlling the role of ligatures and HMM networking. For this we have created three additional character networks. Their structures are so designed as to nullify or de-emphasize the role of ligature models. The networks are:

Random Ligature Single Path (RLSP) is a network with a single ligature HMM for each of the two ligature models L_{cv} and L_{vz} as shown in Fig. 10(a). Their parameters are random or not trained. Therefore the network has essentially no ligature modeling nor an elaborate network design with model clustering.

Trained Ligature Single Path (TLSP) is a network with the same topology as RLSP, but has trained ligature HMMs, L_{cv} and L_{vz}.

Multiple Ligature Single Path (MLSP) is a network with multiple trained ligature HMMs, but has only one network arc between letter HMMs C and V, and between V and Z. Each of the arcs is multiply labeled by L_{cv} HMMs and L_{vz} HMMs, respectively, as shown in Fig. 10(b). Compared with the network in Fig. 5 and described below as MLMP, this is its degenerate version: all the nodes (large circles vertically aligned) at each stage have been merged, thus removing the effect of networking.

Multiple Ligature Multiple Path (MLMP) is a network having multiple ligature paths each labeled by a distinct ligature HMM. BongNet as shown in Fig. 5 is the network of MLMP.

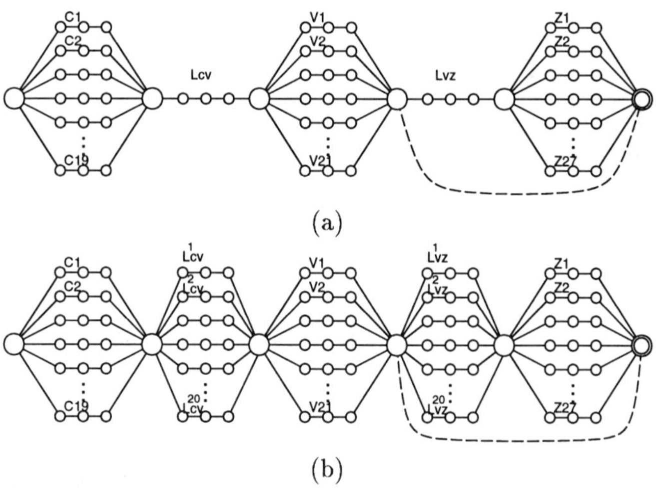

Fig. 10. Degenerate character networks de-emphasizing the effect of ligature modeling and networking.

Table 3. The effect of ligature modeling and HMM networking with test on set *M*.

Model(train set)	RLSP	TLSP	MLSP	MLMP
$\Lambda_{16}(H)$	40.57	73.13	76.16	83.27
$\Lambda_{16}(H+K)$	44.37	80.31	78.41	87.43

The experiment was carried on the test set M and the result is shown in Table 3. According to the table the performance of the RLSP character model lingers around $40 \sim 45\%$ recognition, significantly below the product of the letter recognition rates, that is, $90.12(\text{for } C) \times 86.52(\text{for } V) \times 94.87(\text{for } Z) = 73.97\%$. The phenomenon can be construed as a result of the absence of ligature modeling and an appropriate letter segmentation mechanism. Since the random ligature model can match any part of the strokes either pen-up or pen-down, the chance of missegmentation is great, and in such cases good letter HMMs alone will seldom lead to a high recognition rate. Thus it is our assertion that the above result strongly suggests the need for explicit ligature modeling.

The models TLSP and MLSP, on the other hand, have single and multiple trained ligature HMMs, respectively, and show roughly the expected rates of recognition as in the above calculation. In these models, the ligature models are context-independent since any HMM (letter) can follow any HMM (letter) regardless of the intermediate ligature HMM. Then, depending on the handwriting patterns, there is a good chance of finding the individually most likely models (or path segments), which does not agree with the optimality criterion of network searching for complete input-path alignments, as described in Sec. 4. This can be seen in Table 3 where the HMM network of MLMP reduces errors by 30 to 40%. Here we have used BongNet for the MLMP. Similarly we consider that this advocates the utility of elaborate design of HMM networks.

5.5. *Analysis of Decoding*

As a result of network searching, an optimal path or a sequence of letter labels and the corresponding input-to-network alignment are determined simultaneously. The recognition result is the character that is defined by the letter sequence retrieved by back-tracking, which is triggered at the final node. In case more flexibility is desirable, more than one hypotheses can be produced by computing the required number of local hypotheses all along the way (i.e., at all nodes and HMM states) at each time.

On the other hand, optimal letter boundaries can be obtained from the alignment. Along the back-tracked path, the time at which the start state of each HMM is met for the first time during the Viterbi forward pass is the segmentation point (in time frame) between the current letter and the previous letter. Figure 11 shows some examples of segmentation results mapped to the raw handwriting. The segmentation results as shown are highly natural and similar to those indicated by a human. Although the points in Fig. 11(a) are easy to detect by analyzing the stroke shape, once discreteness is known, many of the points lying in the middle of locally straight regions of strokes (Fig. 11(b)) are unprecedented in other conventional approaches based on shape analysis. As there are no guiding features around such points, it is very difficult to find them unless the recognizer is equipped with a

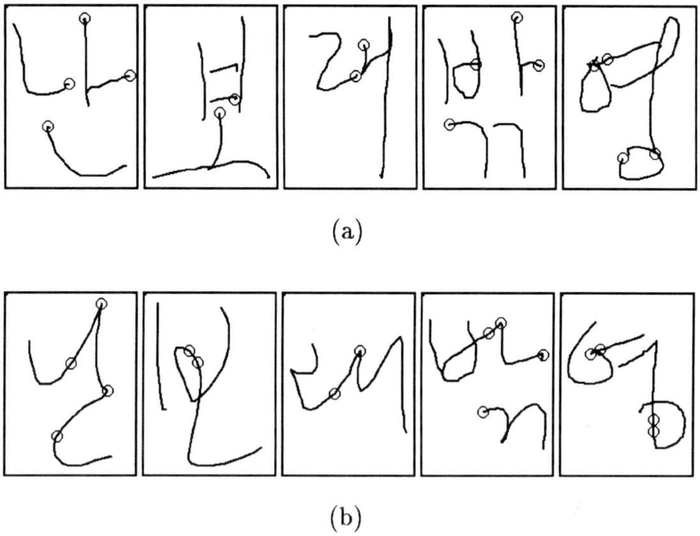

(a)

(b)

Fig. 11. Character segmentation points indicated by small circles; (a) easy and (b) difficult cases.

priori character-specific knowledge. In the proposed model of recognition, however, it is straightforward and those points are determined in simple probabilistic terms.

6. Conclusion

An HMM-based approach to the online recognition of Korean Hangul characters has been described. It has been shown to work with good performance on unconstrained handwriting. The modeling technique for time sequential patterns is sufficiently general so that it can be applied to other alphabet-based languages or even multiple languages, and possibly to word sequence recognition with some minor extensions.

Our extended definition and explicit modeling of ligatures enable us to model the primary source of variability in character shape. The local variability in time and space has been resolved with the tool of HMM. By modeling letters and ligatures using HMMs, and by the method of organizing them into a network, it has become possible to have a well-defined and consistent framework, for both representation and computation.

Experimental results have confirmed that, through the explicit use of ligature models, the letter segmentation problem can be elegantly solved. Although the task of manual segmentation in preparing training samples is labor-intensive, it is not considered a serious problem since it can be solved by using the technique of bootstrapping.

References

[1] H. D. Lee, T. Aagui and M. Nakajima, On-line recognition of cursive Korean characters by descriptions of basic character patterns and their connected patterns, *Proc. SPIE Conf. on Visual Communications and Image Processing '88*, Massachusetts, USA, 1988, 1001–1111.

[2] L. R. Rabiner, A tutorial on hidden Markov models and selected applications in speech recognition, *Proc. IEEE* **77**, 2 (1989) 257–285.

[3] K. F. Lee, Context-dependent phonetic HMMs for speaker-independent continuous speech recognition, *IEEE Trans. ASSP* **38**, 4 (1990) 599–609.

[4] C. C. Tappert, C. Y. Suen and T. Wakahara, The state-of-the-art in on-line handwriting recognition, *IEEE Trans. PAMI* **12** (1990) 787–808.

[5] H. Freeman, On the digital computer classification of geometric line patterns, *Proc. National Electronics Conf.*, vol.18, 1962, 312–324.

[6] C. H. Lee and L. R. Rabiner, A frame synchronous network search algorithm for connected word recognition, *IEEE Trans. ASSP* **37**, 11 (1989) 1649–1658.

[7] R. Bellman, *Dynamic Programming* (Princeton Univ. Press, Princeton, 1957).

[8] L. E. Baum and J. A. Egon, An inequality with applications to statistical estimation for probabilistic functions of Markov process and to a model for ecology, *Bull. Am. Meteorol. Soc.* **73** (1967) 360–363.

[9] R. C. Gonzalez and R. E. Woods, *Digital Image Processing* (Addison-Wesley, New York, 1992).

Handbook of Character Recognition and Document Image Analysis, pp. 397–420
Eds. H. Bunke and P. S. P. Wang

CHAPTER 15

ARABIC CHARACTER RECOGNITION

ADNAN AMIN
School of Computer Science and Engineering, University of New South Wales, Sydney, New South Wales 2052, Australia

Machine simulation of human reading has been the subject of intensive research for almost three decades. A large number of research papers and reports have already been published on Latin, Chinese and Japanese characters. However, little work has been conducted on the automatic recognition of Arabic characters because of the complexity of printed and handwritten text, and this problem is still an open research field. The main objective of this chapter is to present the state of Arabic character recognition research throughout the last 15 years for both on-line and off-line recognition.

Keywords: Arabic characters; On-line and off-line recognition; Handwriting recognition; Segmentation; Feature extraction; Character recognition.

1. Introduction

For the past three decades there has been increasing interest among researchers in problems related to machine simulation of the human reading process. Intensive research has been carried out in this area with a large number of technical papers and reports in the literature devoted to character recognition. This subject has attracted immense research interest not only because of the very challenging nature of the problem, but also because it provides the means for automatic processing of large volumes of data in postal code reading [1, 2], office automation [3, 4], and other business and scientific applications [5–7].

Much more difficult, and hence more interesting to researchers, is the ability to automatically recognize handwritten characters [8–10]. The complexity of the problem is greatly increased by the noise problem and by the almost infinite variability of handwriting as a result of the mood of the writer and the nature of the writing. Analysing cursive script requires the segmentation of characters within the word and the detection of individual features. This is not a problem unique to computers; even human beings, who possess the most efficient optical reading device (eyes), have difficulty in recognizing some cursive scripts and have an error rate of about 4% in reading tasks in the absence of context [11].

The different approaches covered under the general term character recognition fall into either the on-line or off-line category, each having its own hardware and recognition algorithms.

In on-line character recognition systems, the computer recognizes the symbols as they are drawn [12–16]. The most common writing surface is the digitizing tablet, which typically has a resolution of 200 points per inch and a sampling rate of 100 points per second, and deals with one-dimensional data.

Off-line recognition is performed after the writing or printing is completed. Optical Character Recognition, OCR [17–21], deals with the recognition of optically processed characters rather than magnetically processed ones. In a typical OCR system, input characters are read and digitized by an optical scanner. Each character is then located and segmented and the resulting matrix is fed into a preprocessor for smoothing, noisereduction, and size normalization. Off-line recognition can be considered the most general case: no special device is required for writing and signal interpretation is independent of signal generation, as in human recognition.

Many papers have been concerned with the recognition of Latin, Chinese and Japanese characters. However, although Arabic characters are used in several widespread languages, little research has been conducted toward the automatic recognition of Arabic characters because of the strong cursive nature of its writing rules.

Many researchers have been working on cursive script recognition for more than three decades. Nevertheless, the field remains one of the most challenging problems in pattern recognition and all the existing systems are still limited to restricted applications.

There are two strategies which have been applied to cursive script recognition. According to Lecolinet and Baret [10], they can be categorized as follows:

(i) Holistic strategies in which the recognition is globally performed on the whole representation of words and where there is no attempt to identify characters individually. These strategies were originally introduced for speech recognition and can fall into two categories:

   (a) Methods based on distance measurements (Edit Distance, Dynamic Programming).

   (b) Methods based on a probabilistic framework (Markov Chains or Hidden Markov Models).

(ii) Analytical strategies in which words are not considered as a whole, but as sequences of small size units and the recognition is not directly performed at word level but at an intermediate level dealing with these units, which can be graphemes, segments, pseudo-letters, etc. Analytical strategies fall into two main categories:

   (a) Explicit segmentation, in which words are explicitly segmented into letters (or pseudo-letters) which are then recognized individually. Contextual high-level knowledge (lexical, syntactic or semantic knowledge) is then used to ensure the proper word identification.

(b) Implicit segmentation, in which segmentation and recognition take place at the same time. That is, words are segmented into letters and so recognition-based segmentation is performed. Letter segmentation is then a by-product of letter recognition.

More details about cursive script recognition can be found in various other chapters of this book.

The objectives of this chapter are to identify the problems related to handwritten Arabic characters, and describe different methods for the recognition of handwritten Arabic characters for both on-line and off-line recognition. The remainder of this chapter is organized as follows: Section 2 reviews some of the basic characteristics of Arabic writing. Section 3 covers different approaches for segmentation and feature extraction, and presents various methods adopted for recognition in both on-line and off-line systems. Finally, concluding remarks are given in Sec. 4.

2. General Characteristics of the Arabic Writing System

A comparison of the various characteristics of Arabic, Latin, Hebrew and Hindi scripts are outlined in Table 1. Arabic, like Hebrew, is written from right to left. Arabic text (machine printed or handwritten) is cursive in general and Arabic letters are normally connected on the base line. This feature of connectivity will be shown to be important in the segmentation process. Some machine printed and handwritten texts are not cursive, but most Arabic texts are, and thus it is not surprising that the recognition rate of Arabic characters is lower than that of disconnected characters such as printed English.

Table 1. Comparison of various scripts.

Characteristics	Arabic	Latin	Hebrew	Hindi
Justification	R-to-L	L-to-R	R-to-L	L-to-R
Cursive	Yes	No	No	Yes
Diacritics	Yes	No	No	Yes
Number of vowels	2	5	11	–
Letters shapes	1 – 4	2	1	1
Number of letters	28	26	22	40
Complementary characters	3	–	–	–

Arabic writing is similar to English in that it uses letters (which consist of 28 basic letters), numerals, punctuation marks, as well as spaces and special symbols. It differs from English, however, in its representation of vowels since Arabic utilizes various diacritical markings. The presence and absence of vowel diacritics indicates different meanings in what would otherwise be the same word. For example, مدرسة is the Arabic word for both "school" and "teacher". If the word is isolated, diacritics are essential to distinguish between the two possible meanings. If it occurs in a sentence, contextual information inherent in the sentence can be used to infer the appropriate meaning. In this chapter, the issue of vowel diacritics is not treated, since it is more common for Arabic writing not to employ these diacritics. Diacritics are only found in old manuscripts or in very confined areas.

The Arabic alphabet is represented numerically by a standard communication interchange code approved by the Arab Standard and Metrology Organization (ASMO). Similar to the American Standard Code for Information Interchange (ASCII), each character in the ASMO code is represented by one byte. An English letter has two possible shapes, capital and small. The ASCII code provides separate representations for both of these shapes, whereas an Arabic letter has only one representation in the ASMO table. This is not to say, however, that the Arabic letter has only one shape. On the contrary, an Arabic letter might have up to four different shapes, depending on its relative position in the text. For instance, the letter (ع A'in) has four different shapes: at the beginning of the word (preceded by a space), in the middle of the word (no space around it), at the end of the word (followed by a space), and in isolation (preceded by an unconnected letter and followed by a space). These four possibilities are exemplified in Fig.1.

(a) (b) (c) (d)

Fig. 1. Different shapes of the Arabic letter 'A'in (a) beginning (b) middle (c) end (d) isolated.

In addition, different Arabic characters may have exactly the same shape, and are distinguished from each other only by the addition of a complementary character (see Appendix). Complementary characters are positioned differently, for instance, above, below or within the confines of the character. Figure 2 depicts two sets of characters, the first set having five characters and the other set three characters. Clearly, each set contains characters which differ only by the position and/or the number of dots associated with it. It is worth noting that any erosion or deletion of these complementary characters results in a misrepresentation of the character. Hence, any thinning algorithm needs to efficiently deal with these dots so as not to change the identity of the character.

Arabic writing is cursive and is such that words are separated by spaces. However, a word can be divided into smaller units called subwords (see Appendix). Some Arabic characters are not connectable with the succeeding character. Therefore, if one of these characters exists in a word, it divides that word into two subwords. These characters appear only at the tail of a subword, and the succeeding character forms the head of the next subword. Figure 3 shows three Arabic words with one, two, and three subwords. The first word consists of one subword which has nine letters; the second has two subwords with three and one letter, respectively. The last word contains three subwords, each consisting of only one letter.

Fig. 2. Arabic characters differing only with regard to the position and number of associated dots.

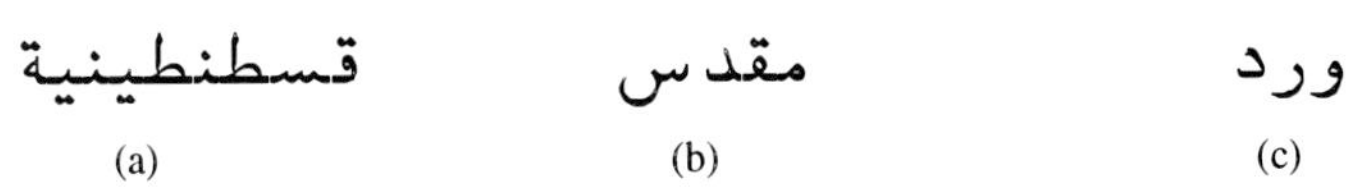

Fig. 3. Arabic words with constituent subwords.

Arabic writing is similar to Latin in that it contains many fonts and writing styles. The letters are overlaid in some of these fonts and styles. As a result, word segmentation using the baseline, which is the line on which all Arabic characters are connected, is not possible. Furthermore, characters of the same font have different sizes (i.e. characters may have different widths even though the two characters have the same font and point size). Hence, word segmentation based on a fixed size width cannot be applied to Arabic.

3. Recognition of Arabic Characters

This section covers different techniques used for the recognition of Arabic characters in both on-line and off-line systems.

3.1. On-Line Recognition Systems

An on-line character recognition system typically uses a magnetic graphic tablet as the main input device. Such a tablet operates through a special pen in contact with the surface of the tablet which emits the coordinates of the plotted points at a constant frequency. Breaking contact prompts the transmission of a special character. Thus, recording on the tablet produces strings of coordinates separated by signs indicating when the pen has ceased to touch the tablet surface.

On-line recognition has several interesting characteristics. First, recognition is performed on one-dimensional data rather than two-dimensional images as in the case of off-line recognition. The writing line is represented by a sequence of dots whose location is a function of time. This has several important consequences:

- The writing order is available and can be used by the recognition process.

- The writing line has no width (Fig. 4).
- Temporal information, like velocity can also be taken into consideration.
- Additionally, penlifts can be useful in the recognition process.

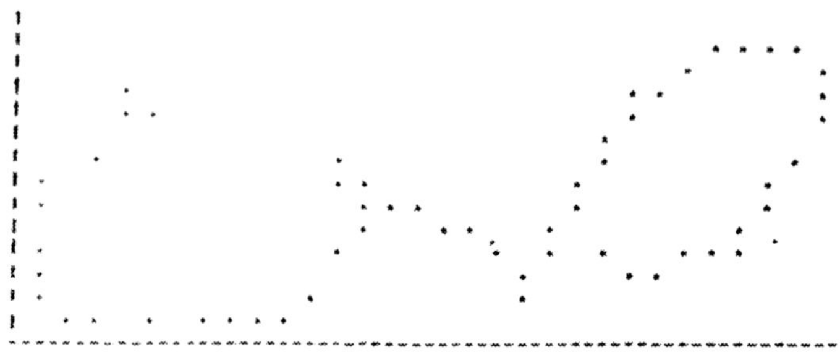

Fig. 4. Acquisition of Arabic character ص

The IRAC (Interactive Recognition of Arabic Character) system [22] adopted a structural classification method for recognizing on-line, handwritten isolated Arabic characters. The system consists of three major steps. First, preprocessing using an algorithm inspired by Berthod and Jancenne [23] which accomplished two objectives: the reduction of the number of points and the correction of any deformation (Fig. 5). Features such as the shape of the main stroke, the number of strokes, the characteristics relative to the group of dots and the presence of an eventual "zigzag" (which can be either isolated, e.g. ك or attached, e.g. س) are then extracted from the character. If these features do not permit the recognition of a character by a simple dichotomic consultation of the dictionary, backtracking is performed in order to correct the shape of the main stroke. Finally, secondary features of the main stroke such as frame size, the start point and the curvature are extracted using a distance function in order to remove the ambiguity and provide an exact match.

Amin [24] proposed a system for on-line Arabic word recognition. The hand drawing is directly segmented into characters on the basis of certain heuristic criteria. For every word, the characters are connected to each other by horizontal segments from right to left. Statistically, these connections appear:

(i) almost always after an intersection point,

(ii) often after a cusp point,

(iii) sometimes simply after a change of curvature.

These points serve remarkably as separators in the segmentation process and directly permit one to obtain a list of the characters of the word component. However, since the separators are determined by statistical considerations, the segmentation has to proceed through successive essays (i.e. tentative tries), and the character recognition module is responsible for validating each essay. Hence, the two modules (segmentation and recognition) are highly interactive.

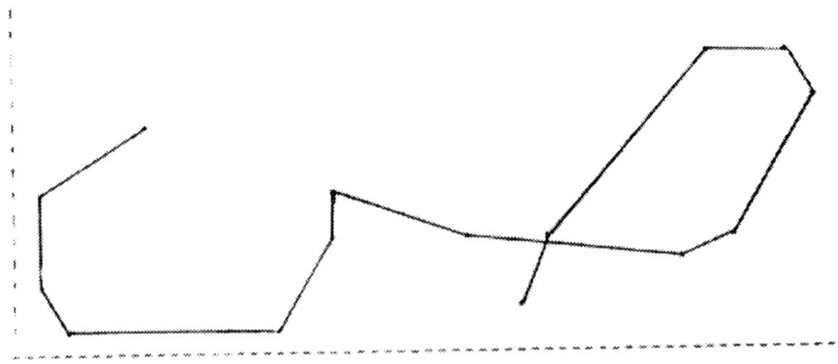

Fig. 5. Sampling and smoothing of the character ص given in Fig. 4.

Each class of separator is affected by a priority. The intersection point has the highest priority and the cusp point has the lowest priority. An Arabic handwritten word is segmented into characters by using the separators according to their priority and their appearance in the word. With the help of each separator, portions of the handwriting are extracted and transmitted to the character recognition module for identification. If no character has been recognized, the corresponding segmentation is cancelled and a new tentative try is carried out with the separator of the next lower priority. For example, assume E is the end of the last character identified. If the portion of handwriting between E and the first intersection point is not identified as a character, then the second essay will be carried out on the portion between E and the next cusp point.

The character recognition module is similar to the system which recognizes isolated characters; this system is in fact a further development of the method for the recognition of isolated Arabic characters, described in [22].

Finally, three hypotheses at most, i.e. the three best score candidates provided by the character recognition module, are associated with each of the characters which are extracted from the handwritten Arabic word by the segmentation module. The set of hypotheses form a lattice (Fig. 6). Identification of the word consists of traversing the lattice to find the path of the best score corresponding to a word in the dictionary. Binary diagrams [25] are used for resolving ambiguities and for eliminating all candidates which are not present in the dictionary of known words.

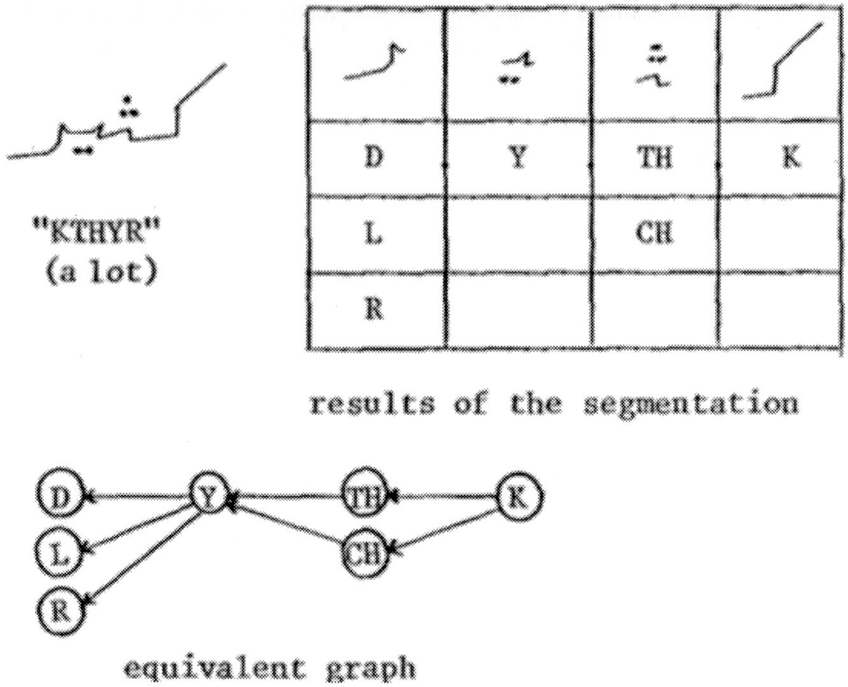

Fig. 6. An example of segmentaing the Arabic word كثير and the equivalent graph

Figure 7 illustrates two examples of the segmentation into characters and the recognition of two Arabic words حديقة ، كثير.

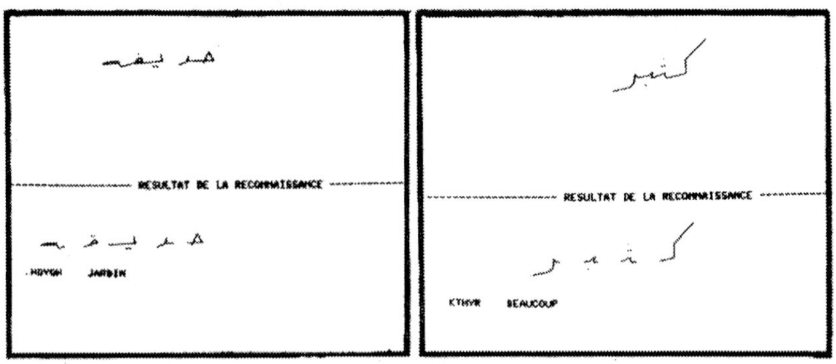

Fig. 7. Two examples of the recognition of Arabic words. The part above of each frame represents the handwriting after sampling, while the part below shows the segmentation result into characters and their translation into French.

In [26] two methods were presented for recognizing cursive Arabic words. The first is a syntactical method based on the segmentation of words into primitives. These primitives are:

(i) open curves

∪ ∩ ⊂ ⊃

a b c d

(ii) closed curves

e

f

(iii) vertical/horizontal stroke

g

h

(iv) cusp point

i

(v) inflection point

j

(vi) group of dots

k l m

(vii) shape of the secondary stroke

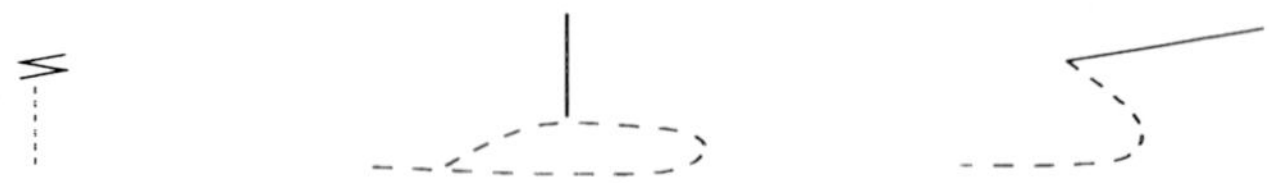

zigzag vertical bar horizontal bar

(viii) pencil lift

The syntactic representations of the characters are merged into a tree structure whose arcs are labelled with primitive names (Fig. 8). Recognition consists of finding a path through the tree, using the primitive classes of the description of the character to be recognized as instructions for choosing the arcs to take. The terminal node of the path gives the name of the recognized character. The recognition of words works from the list of primitives provided by the segmentation process. It consists of finding all combinations of successive characters utilizing binary diagrams to eliminate all ineligible combinations of letters.

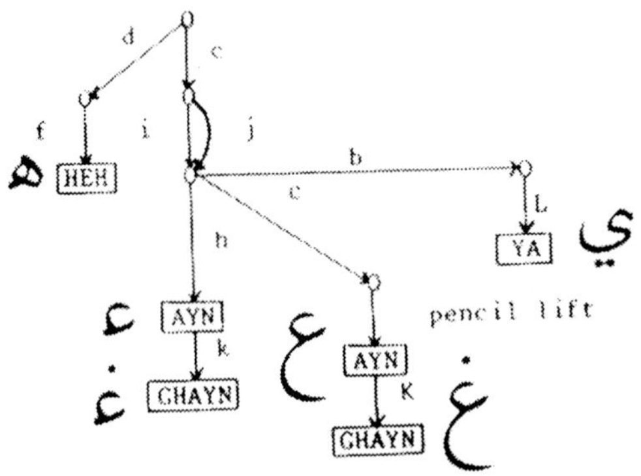

Fig. 8. A piece of the tree structure grouping the representation of Arabic characters.

The second method of [26] is global without segmentation into characters. It uses some properties of Arabic writing, in which seven characters cannot have a left connection. These characters are:

ا د ذ ر ز و لا

This property causes words to be decomposed into pieces (Fig. 3 (c)) whenever one of these characters occurs.

This method uses the notion of stroke instead of character. A vector defining the main parameters of a word (number of secondary strokes, size and position of groups of dots, number of intersection points, number of cusp points) makes it possible to recognize a word and each of its constituent strokes. Whenever the same vector yields several possibilities, they are classified according to a score computed from secondary parameters of each stroke (the start point, form and angular variation of the main stroke).

Furthermore, to enhance the recognition rate, a syntactical and semantical analyzer that verifies the grammatical structure and the meaning of the Arabic sentence is used [27]. The syntactic analysis module is inspired by the work of Wood [28] on ATN's (Augmented Transition Network) [28] and adopted the same notation. The structure of a sentence is given by a graph whose arcs are labeled with symbolic names like <verb>, <nominal-group>, etc. Some action and conditions (compatibility of the adjectives, verbs, nouns, etc.) may be associated with an arc. The syntactic analysis is carried out by using the syntactic class of each successive word of the input to determine which arcs to take in the network. When an arc is chosen, all its conditions should have been satisfied and, in this case, the actions associated with it are executed. If the conditions are not satisfied, the analyzer discards the path and a backtrack is triggered. The parser works on a word lattice which contains the three best-scored candidates provided by the segmentation. The system adopted a best-first strategy. The analyzer builds a syntactical tree describing the structure of the sentence.

The semantic analysis consists of verifying that the combination of the words associated with the syntactic classes at the leaves of the tree is meaningful. It should be noted that a leaf can be associated with various reference words belonging to the same syntactic class. The semantic analyzer has two tasks:

(i) to verify the compatibility of the subject and the object in accordance with the action implied by the verb,

(ii) to complete the syntactical tree in order to remove the ambiguities in the parsing process.

Each word is associated with classes regarding its meaning and its implications in a sentence. For example, a wolf is a wild animal eating meat but a turkey is a domestic animal eating corn. The compatibility between the semantic classes of the subject, the verb and the object in building a sentence is given by the second set of classes for the object. The verification of compatibility in the sentence to be analyzed is expressed in terms of conditions attached to the arcs of the syntactic network. This method is no longer efficient when we stray outside the realm of a restricted vocabulary.

Further on-line recognition systems are now described. El-Sheikh and El-Taweel [29] proposed a system for recognizing properly segmented handwritten Arabic characters. Four groups of characters are identified depending on the position of the character within a word (isolated, beginning, middle or at the end). Moreover, each group is further classified into four subgroups depending on the number of strokes (one, two, three, or four) in the character. The classification of characters within each subgroup is done by means of some features, such as the existence of a maximum or a minimum on either the horizontal or vertical direction of the main stroke, the ratio between length and width, the type of secondary stoke and other features.

Al-Emami and Usher [30] presented a system for on-line recognition of handwritten Arabic words. Words were entered via a graphic tablet and segmented into strokes based on the method proposed by Belaid et al. [31, 32]. In the preliminary learning process, specifications of the strokes of each character are fed to the system, while in the recognition process, the parameters of each stroke are found and special rules are applied to select the collection of strokes that best match the features of one of the stored characters. However, few words were used in the learning and testing processes, which makes the performance of the system questionable.

3.2. *Off-Line Recognition Systems*

Off-line character recognition systems typically use a scanner as the main input device. Off-line recognition can be considered as the most general case: no special device is required for writing and signal interpretation is independent of signal generation, as in human recognition.

3.2.1. Character segmentation

The segmentation phase is a necessary step in recognizing printed Arabic text. Any error in segmenting the basic shape of Arabic characters will produce a different representation of the character component.

Two techniques have been applied for segmenting machine printed and handwritten Arabic words into individual characters: implicit and explicit segmentations.

(i) Implicit segmentation (straight segmentation): In this technique, words are segmented directly into letters. This type of segmentation is usually designed with rules that attempt to identify all the character's segmentation points.

(ii) Explicit segmentation: In this case, words are externally segmented into pseudo-letters which are then recognized individually. This approach is usually more expensive due to the increased complexity of finding optimum word hypotheses.

In all Arabic characters, the width at a connection point is much less than the width of the beginning character. This property is essential in applying the baseline segmentation technique [33, 34]. The baseline is a medium line in the Arabic word in which all the connections between the successive characters take place. If a vertical projection of bi-level pixels is performed on the word (Eq. (3.1)),

$$v(j) = \sum_i w(i,j) \tag{3.1}$$

where $w(i, j)$ is either zero or one and i, j index the rows and columns, respectively, the connectivity point will have a sum less than the average value (AV) (Eq. 3.2)

$$AV = (1 / Nc) \sum_{j=1}^{Nc} Xj \tag{3.2}$$

and where Nc is the number of columns and Xj is the number of black pixels of the jth column.

Hence, each part with a sum value much less than AV should be a boundary between different characters. However if the histogram produced from the vertical projection does not follow the condition of Eq. (3.3), the character remains unsegmented, as illustrated in Fig. 9.

By examining Arabic characters , it is found that the distance between successive peaks does not exceed one third the width of the Arabic character. That is

$$|d_k| < d_l / 3 \tag{3.3}$$

where d_k is the distance between kth peak and peak k+1, and d_l is the total width of the character.

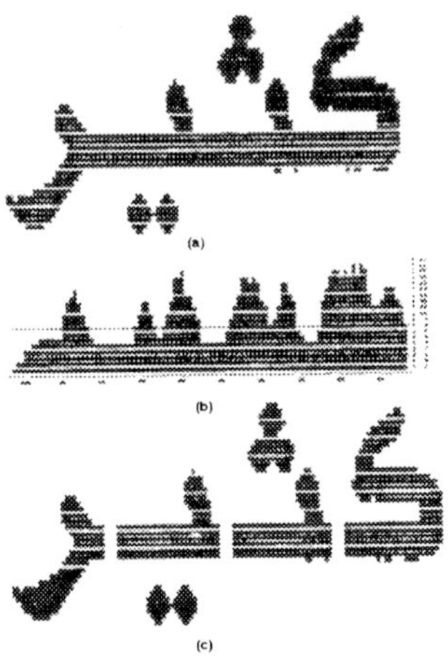

Fig. 9. An example of segmentation of the Arabic word كثير into characters (a) Arabic word (b) histogram (c) word segmented into characters.

Moreover, at the end of a word or a subword, Eq. (3.4) is also to hold.

$$L_{k+1} > 1.5 * L_k \tag{3.4}$$

where L_k is the kth peak in the histogram. This rule is brought to bear because of the inter-connectivity of Arabic characters and their shapes at the end of a word.

An example of a histogram of a word that should be segmented into five parts based on the average value appears in Fig. 10. However, when the last rule is applied only the first four segmentation locations should be chosen, hence reducing the number of characters to four since both conditions are to hold simultaneously.

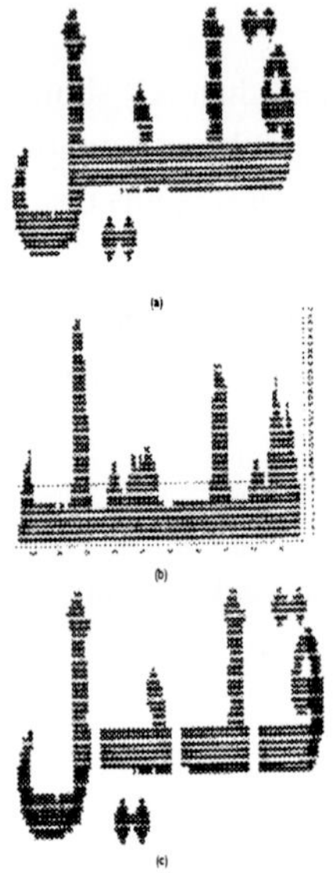

Fig. 10. An example of the Arabic word قليل and its segmentation into characters (a) Arabic word (b) histogram. (c) word segmented into characters.

This approach depends heavily on a predefined threshold value related to the character width. Moreover, this approach will not work effectively for skewed images.

Almuallim and Yamaguchi [35] proposed a structural recognition technique for Arabic handwritten words. Their system consists of four phases. The first is preprocessing, in which the word is thinned and the midline of the word is detected. Since it is difficult to segment a cursive word into letters, words are segmented into separate strokes and classified as complementary characters, strokes with a loop and strokes without a loop. These strokes are then further classified using their geometrical and topological properties. Finally, the relative positions of the classified strokes are examined, and the strokes are combined in several steps into the string of characters that represents the recognized word. System failures in most cases were due to incorrect segmentation of words.

Segmentation is also achieved by tracing the outer contour [36] of a given word and calculating the distance between the extreme points of intersection of the contour with a vertical line. The segmentation is based on a horizontal scan from right to left of the closed contour using a window of adjustable width w. For each position of the window, the average vertical distance h_{av} is calculated across the window. At the boundary between two characters, the following conditions should be met:

(i) $h_{av} < T$. In this case, a silence region is detected, which means that the average vertical distance over the window should be less than a certain preset threshold T.

(ii) Detected boundaries should lie on the same horizontal line (the base line).

(iii) No complementary characters should be located (above or below the base line) at a silence region.

Readjustment of parameters w and T as well as backtracking may occur if segmentation leads to a rejected character shape. Figure 11 illustrates some examples of this method.

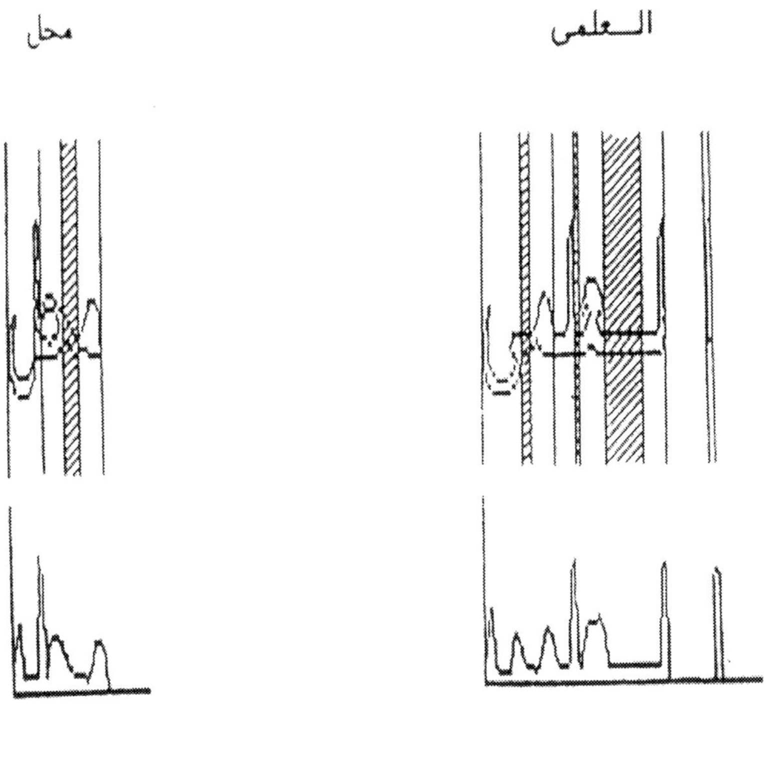

Fig. 11. Segmented Arabic words and the corresponding contour heights.

El-Khaly and Sid-Ahmed [37] segment a thinned word into characters by following the average baseline of the word and detecting when the pixels start to go higher or lower than it.

Abdelazim and Hashish [38] use the technique of traversing an energy curve (similar to that used in speech recognition, to discriminate the spoken utterance from the silence background), which shows the number of black pixels in each column of the digitized word, to segment the word into characters. This curve is traversed and a threshold value is used to select significant primitives leaving out silent zones.

Shoukry [39] used a sequential algorithm based on the input-time tracing principle which depends on the connectivity properties of the acquired text in the binary image domain. This algorithm bears some resemblance to an algorithm devised by Wakayama [40] for the skeletonization of binary pictures.

The SARAT system [41] used outer contours to segment an Arabic word into characters. The word is divided into a series of curves by determining the start and end points of the word. Whenever the outer contour changes sign (from a positive to a negative curvature) a character is segmented as illustrated in Fig. 12.

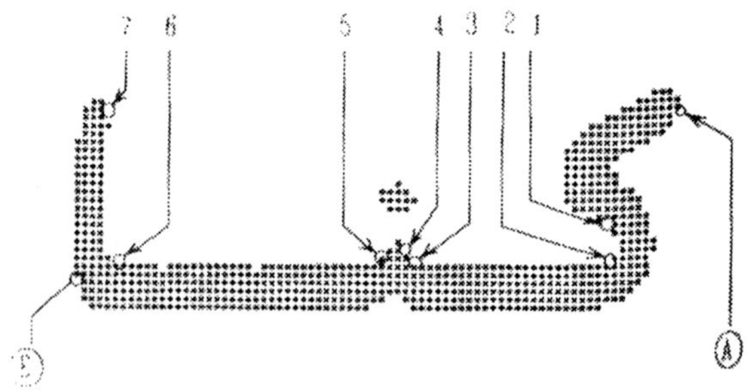

Fig. 12. An example of a segmented subword, with start point A, end point E, and horizontal lines 2–3 and 5–6.

Kurdy and Joukhadar [42] use the upper distance function of the subword, which is the set of the highest points in each column. They assign to each point of the function a token name by comparing the point's height to the height and token name of the point on its right. Using a grammar, they then parse the sequence of tokens of a subword to find the connection points.

Finally, Amin and Al-Sadoun [43, 44] adopted a new technique for segmenting Arabic text. The algorithm can be applied to any font and it permits the overlay of characters. There are two major problems with the traditional segmentation method which depends on the baseline:

(i) Overlapping of adjacent Arabic characters occurs naturally, see Fig. 13 (a). Hence, no baseline exists. This phenomenon is common in both typed and handwritten Arabic text.

(ii) The connection between two characters is often short. Therefore, placing the segmentation points is a difficult task. In many cases, the potential segmentation points will be placed within a character rather than between characters.

The word in Fig. 13 (a) was segmented utilizing a baseline technique. Figure 13 (b) shows the proper segmentation and the result of the new segmentation method is shown in Fig. 13 (c).

The new technique can be divided into four major steps. First is the digitization step in which the original image is transformed into a binary image utilising a scanner (300 dpi). Second, there is a preprocessing step in which the Arabic word is thinned using a parallel thinning algorithm. Third, the skeleton of the image is traced from right to left using a 3 * 3 window and a binary tree is constructed. The Freeman code [45] is used to describe the skeleton shape. Finally, the binary tree is segmented into subtrees such that each subtree describes a character in the image.

Fig.13. Example of an Arabic word محمد and different techniques of the segmentation.

3.2.2. Feature extraction and recognition

It is known that features represent the smallest set that can be used for discrimination purposes and for a unique identification for each character. Features can be classified into two categories:

(i) Local features which are usually *geometric* (e.g. concave/convex parts, type of junctions: intersections/T-junctions/endpoints etc.).

(ii) Global features which are usually *topological* (connectivity, number of connected components, number of holes, etc.) or *statistical* (Fourier transform, invariant moments, etc.).

Nouh et al. [46] suggested a standard Arabic character set to facilitate computer processing of Arabic characters. In this work, thirteen features, or radicals, which represent parts of characters are selected by inspection. The recognition is based on a decision tree and a strong correlation measurement. The disadvantage of the proposed system is the assumption that the incoming characters are generated according to specified standard rules.

Parhami and Taraghi [47] presented a technique for the automatic recognition of machine printed Farsi text (which is similar to Arabic text). The authors first segment the subword into characters by identifying a series of potential connection points on the baseline at which line thickness changes from or to the thickness of the baseline. Although they also have some rules to keep characters at the end of a subword intact, they segment some of the wider characters (e.g. س) into up to three segments. Then they select twenty features based on certain geometric properties of the Farsi symbols to construct a 24 bit vector that is compared with entries of a table where an exact match is checked first. The system is heavily font dependent, and the segmentation process is expected to give incorrect results in some cases.

Table lookup is used for the recognition of isolated handwritten Arabic characters [48]. In this approach, the character is placed in a frame which is divided into six rectangles and a contour tracing algorithm is used for coding the contour as a set of directional vectors by using a Freeman code. However, this information is not sufficient to determine Arabic characters, therefore extra information related to the number of dots and their position is added. If there is no match, the system will add the feature vector to the table and consider that character as a new entry.

Amin and Masini [33] adopted a structural approach for recognizing printed Arabic text. Words and subwords are segmented into characters using the base-line technique. Features such as vertical and horizontal bars are then extracted from the character using horizontal and vertical projections (Fig. 14). Four decision trees, chosen according to the position of the character within the word which was computed by the segmentation process, have been used. The structure of four decision trees allows a rapid search for the appropriate character. Furthermore, trees are utilized in distinguishing characters that have the same shape but appear in different positions within a word.

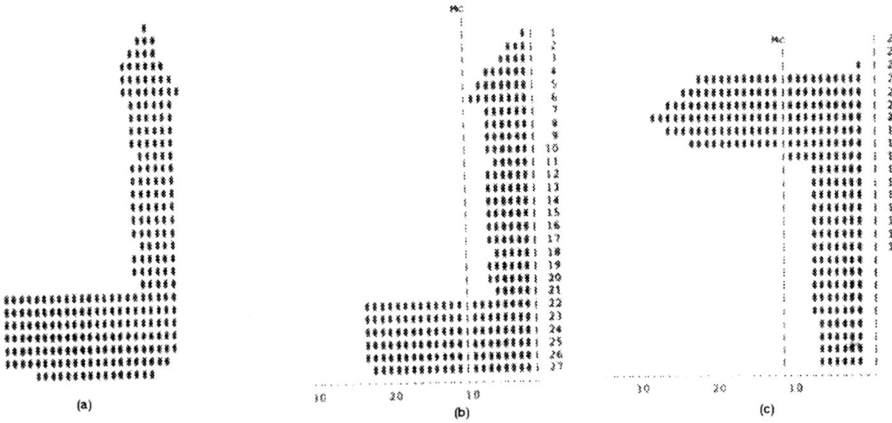

Fg. 14. Vertical and horizontal scanning of the character ل a) character ل (b) horizontal scanning (c) vertical scanning.

Amin and Mari [34] proposed a new technique for multifont Arabic text which includes character and word recognition. A character is divided into many segments by a horizontal scan process (Fig. 15). In this way, segments are connected to form the basic shape of the character. Segments not connected with any other segment are considered to be complementary characters. By using the Freeman code [45], the contour detection process is applied to these segments to trace the basic shape of the character and generate a directional vector through a 2 * 2 window. A decision tree is then used for the recognition of the characters. Finally, a Viterbi algorithm [49] is used for Arabic word recognition to enhance the recognition rate. The main advantage of this technique is to allow an automatic learning process to be used.

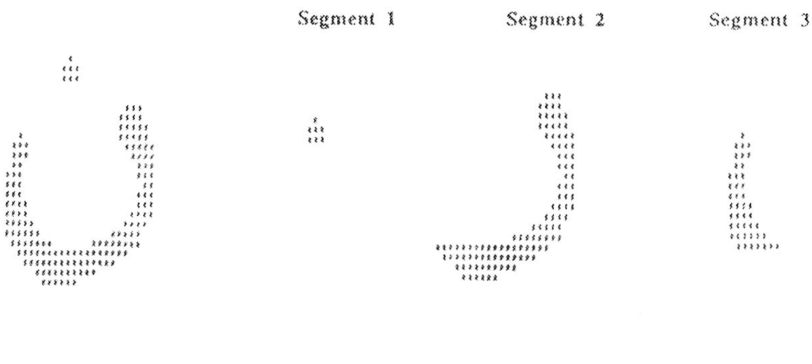

Fig. 15. Major segments of character ن

The study reported in [37, 50, 51] utilizes moment invariant descriptors to recognize the characters. Other techniques include a set of Fourier descriptors from the coordinate sequences of the outer contour which is used for the recognition [36]. Also, in [52] each character is assigned a logical function where characters are preclassified into four groups depending on the existence of certain pixels in a specified location of the image.

In [53] table lookup is adopted for the recognition of isolated Arabic characters. In this approach, the character is placed in the frame window and divided into small windows to extract some features. These features include end points, intersection points, corners, and the relationship between the length and width of the window frame (Fig. 16). Characters are identified by an association between feature points and their locations within the window frame. The recognition is achieved by finding a match between unknown characters and entries in a lookup table.

To enhance the recognition rate of an OCR system, some characteristic morphological properties of the Arabic language can be used. Amin and Al-Fedaghi [6] describe a method for spell correction of Arabic words. They correct spelling errors and complete words that have some unrecognized characters using an algorithm that depends on the frequencies of roots and patterns in Arabic.

Amin and Al-Sadoun [54] proposed a structural approach for recognizing handwritten Arabic characters. The binary image of the character is first thinned using a parallel thinning algorithm and then the skeleton of the image is traced from right to left using 3 * 3 window in order to build a graph to represent the character. Features like straight lines, curves and loops are then extracted from the graph. Finally, a hierarchical classification (similar to a decision tree) is used for the recognition of the characters.

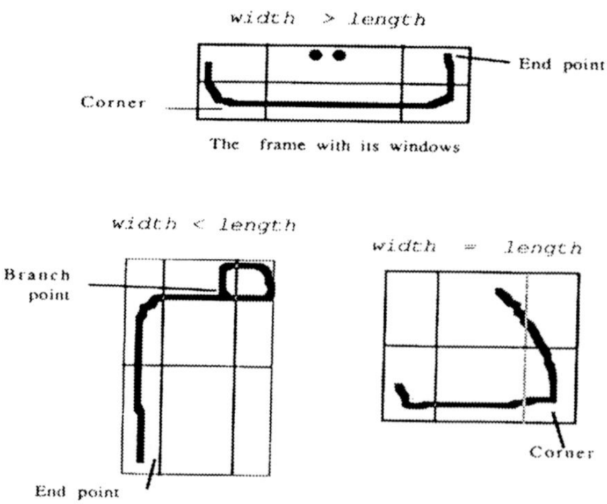

Fig. 16. The three different classes characters.

Finally, Al-Badr and Haralick [55] proposed a system to recognize machine printed Arabic words without prior segmentation by applying mathematical morphology operations on the whole page to find the locations where shape primitives are present. They then combine those primitives into characters and print out the character identities and their location on the page.

4. Conclusion

This chapter presented the problems related to printed and handwritten Arabic characters, and much of the important research work was briefly described in an attempt to present the current status of Arabic character recognition research. This is still an open research area and there is no commercial Arabic OCR system available yet. This is because of the segmentation problem, which is in fact similar to the segmentation of cursive script in many languages, and because of the complexity of Arabic characters. Moreover, all the algorithms presented in this chapter deal with unvocalized text and the recognition of vowel diacritics is an extremely important research area in the Arabic language.

As stated previously, no vital computational techniques in this area have yet been fully explored. As such, this field is of importance for future research.

References

[1] L. D. Harmon, Automatic recognition of printed and script, *Proc. IEEE,* **60**, 10 (1972) 1165–1177.

[2] A. A. Spanjersberg, Experiments with automatic input of handwritten numerical data into a large administrative system, *IEEE Trans. Man Cybern.* **8**, 4 (1978) 286–288.

[3] L. R. Focht and A. Burger, A numeric script recognition processor for postal zip code application, *Int. Conf. Cybernetics and Society*, 1976, 486–492.

[4] J. Schuermann, Reading Machines, *6th Int. Conf on Pattern Recognition,* Munich 1982, 1031–1044.

[5] R. Plamondon and R. Baron, On-line recognition of handprint schematic pseudocode for automatic Fortran code generator, *8th Int. Conf on Pattern Recognition*, Paris, 1986, 741–745.

[6] A. Amin and S. Al-Fedaghi, Machine recognition of printed Arabic text utilising a natural language morphology, *Int. J. of Man-Machine Studies* **35**, 6 (1991) 769–788.

[7] D. Guillevic and C. Y. Suen, Cursive script recognition: A fast reader scheme, *2nd Int. Conf. on Document Analysis and Recognition,* Japan, 1993, 311–314.

[8] M. K. Brown and S. Ganapathy, Preprocessing technique for cursive script word recognition, *Pattern Recogn.* **19**, 1 (1983) 1–12.

[9] R. H. Davis and J. Lyall, Recognition of handwritten characters a review, *Image and Vision Computing* **4**, 4 (1986) 208–218.

[10] E. Lecolinet and O. Baret, *Cursive Word Recognition: Methods and Strategies, Fundamentals in Handwriting Recognition,* ed. S. Impedovo 1994, 235–263.

[11] C. Y. Suen, R. Shingal and C. C. Kwan, Dispersion factor: A quantitative measurement of the quality of handprinted characters, *Int. Conf. of Cybernetics and Society,* 1977, pp. 681–685.

[12] D. J. Burr, Designing a handwritten reader, *5th Int. Conf. on Pattern Recognition,* Miami, USA, 1980, 715–722.

[13] A. Shoukry and A. Amin, Topological and statistical analysis of line drawing, *Pattern Recogn. Lett.* **1** (1983), 365–374.

[14] J. Kim and C. C. Tappert, Handwriting recognition accuracy versus tablet resolution and sampling rate, *7th Int. Conf. on Pattern Recognition,* Montreal, 1984, 917–918.

[15] J. R. Ward and T. Kuklinski, A model for variability effects in handprinted with implication for the design of handwritten character recognition system, *IEEE Trans. Man Cybern.* **18** (1988) 438–451.

[16] F. Nouboud and R. Plamondon, On-line recognition of handprinted characters: Survey and beta tests, *Pattern Recogn.* **25**, 9 (1990) 1031–1044.

[17] C. Y. Suen, M. Berthod and S. Mori, Automatic recognition of handprinted characters, the state of the art, *Proc. IEEE* **68**, 4 (1980) 469–483.

[18] J. R. Ullmann, Advance in character recognition,*Application of Pattern Recognition,* ed. K. S. Fu, 1982, 197–236.

[19] V. K. Govindan and A. P. Shivaprasad, Character recognition A–review, *Pattern Recogn.* **23**, 7 (1990) 671–683.

[20] S. Impedovo, L. Ottaviano and S. Occhinegro, Optical character recognition– A survey, *Int. J. of Pattern Recogn. and Artif. Intell.* **5**, 1&2 (1991) 1–24.

[21] S. Srihari, From pixel to paragraphs: the use of models in text recognition, *Second Annual Symp on Document Analysis and Information Retrieval*, Las Vegas, USA, 1993, 47–64.

[22] A. Amin, A. Kaced, J. P. Haton and R. Mohr, Handwritten Arabic characters recognition by the IRAC system, *5th Int. Conf. on Pattern Recognition,* Miami, USA, 1980, 729–731.

[23] M. Berthod and P. Jancenn, Le pretraitement des traces manuscrits sur une tablette graphique, *2eme, congres AFCET-INRIA,* Toulouse, France, 1979, 195–209.

[24] A. Amin, Machine recognition of handwritten Arabic word by the IRAC II system,*6th Int. Conf. on Pattern Recognition,* Munich, 1982, 34–36.

[25] E. M. Riseman and R. W. Ehrich, Contextual word recognition usig binary diagrams, *IEEE Trans. Computer* **c–20**, 4 (1971) 397–403.

[26] A. Amin, G. Masini and J. P. Haton, Recognition of handwritten Arabic words and sentences, *7th Int. Conf. on Pattern Recognition,* Montreal, 1984, 1055–1057.

[27] A. Amin, IRAC: Recognition and understanding systems,*Applied Arabic Linguistic and Signal and Information Processing*, ed. R. Descout, Hemisphere, New York, 1987, 159–170.

[28] W. A. Wood, Transition network grammars for natural language analysis, **CAMC-13**, 10 (1970) 591–602.

[29] T. S. El-Sheikh and S. G. El-Taweel, Real-time Arabic handwritten character recognition, *Pattern Recogn.* **23**, 12 (1990) 1323–1332.

[30] S. Al-Emami and M. Usher, On-line recognition of handwritten Arabic characters, *IEEE Trans. Pattern Anal Machine Intell.* **PAMI-12** (1990) 704–710.

[31] A. Belaid and G. Masini, Segmentation of line drawings for recognition and interpretation, *Technology and Sc.Informatics* **1**, 2 (1983) 121–134.

[32] A. Belaid and J. P. Haton, A syntactic approach for handwritten mathematical formula recognition, *IEEE Trans. Pattern Anal. Machine Intell.* **PAMI-6** (1984) 105–111.

[33] A. Amin and G. Masini, Machine recognition of muti-fonts printed Arabic texts, *8th Int. Conf. on Pattern Recognition,* Paris, 1986, 392–395.

[34] A. Amin and J. F. Mari, Machine recognition and correction of printed Arabic text, *IEEE Trans. Man Cybern.* **9**, 1 (1989) 1300–1306.

[35] H. Almuallim and S. Yamaguchi, A method of recognition of Arabic cursive handwriting, *IEEE, Trans. Pattern Anal. and Machine Intell.* **PAMI-9** (1987) 715–722.

[36] T. El-Sheikh and R. Guindi, Computer recognition of Arabic cursive script, *Pattern Recogn.* **21**, 4 (1988) 293–302.

[37] F. El-Khaly and M. Sid-Ahmed, Machine recognition of optically captured machine printed Arabic text, *Pattern Recog.* **23**, 11 (1990) 1207–1214.

[38] H. Abdelazim and M. Hashish, Arabic reading machine, *10th National Computer Conf.* Riyadh, Saudi Arabia, 1988, 733–740.

[39] A. Shoukry, A sequential algorithm for the segmentation of typewritten Arabic digitized text, *Arabian J. Sc. and Eng.* **16**, 4 (1991) 543–556.

[40] T. Wakayama, A core-line tracing algorithm based on maximal square moving, *IEEE Trans. Pattern Analysis and Machine Intell.* **PAMI-4** (1982) 68–74

[41] V. Margner, SARAT- A system for the recognition of Arabic printed text, *11th Int. Conf. on Pattern Recognition*, 1992, 561–564.

[42] B. M. Kurdy and A. Joukhadar, Multifont recognition system for Arabic characters, *3rd Int. Conf. and Exhibition on Multi-lingual Computing (Arabic and Roman Script)*, U.K, 1992, 731–739.

[43] A. Amin and H. Al-Sadoun, A segmentation technique of Arabic text, *11th Int. Conf. on Pattern Recognition,* 1992, 441–445.

[44] H. B. Al-Sadoun and A. Amin, A new structural technique for recognizing printed Arabic text, *Int. J. of Pattern Recognition and Artif. Intell.* **9**, 1 (1995) 101–125.

[45] H. Freeman, On the encoding of arbitrary geometric configuration, *IEEE. Trans. Electronic Comp.* **EC-10** (1968) 260–268.

[46] A. Nouh, A. Sultan and R. Tulba, An approach for Arabic character recognition, *J. Eng. Sc.* **6**, 2 (1980) 185–191.

[47] B. Parhami and M. Taraghi, Automatic recognition of printed Farsi texts, *Pattern Recog.* **14**, 6 (1981) 395–403.

[48] S. Saadallah and S. Yacu, Design of an Arabic character reading machine, *Proc. of computer Processing of the Arabic language,* Kuwait, 1985.

[49] D. Forney, The Viterbi algorithm, *Proc. IEEE* **61**, 3 (1973) 268–278.

[50] S. El-Dabi, R. Ramsis and A. Kamel, Arabic charcter recognition system: Statistical approach for recognizing cursive typewritten text, *Pattern Recogn.* **23**, 5 (1990) 485–495.

[51] H. Al-Yousefi and S. S. Udpa, Recognition of Arabic characters, *IEEE Trans. Pattern Anal. and Machine Intell.* **PAMI-14** (1992) 853–857.

[52] A. Nouh, A. Ula and A. Sharaf-Edin, Boolean recognition technique for typewritten Arabic character set, *Proc. 1st King Saud Univ. Symp. on Computer Arabization*, Riyadh, 1987, 90–97.

[53] K. Jambi, Arabic charcter recognition: Many approaches and one decade, *Arabian J. Sc. Eng.* **16**, 4 (1991) 499–509.

[54] A. Amin and H. Al-Sadoun, Handprinted Arabic character recognition system, *12th Int. Conf. on Pattern Recognition*, 1994, 536–539.

[55] B. Al-Badr and R. Haralick, Segmentation-free word recognition with application to Arabic, *3rd Int. Conf. on Document Analysis and Recognition*, Montreal, 1995, 355–359.

Appendix:

The following are definitions of terms used throughout this paper.

Subword: A potion of a word including one or more connected characters.

Complementary character: A portion of a character that is needed to complement an Arabic character. These are normally a dot, a group of dots or a zigzag (hamza). This may appear on, above, or below the base line (Fig. 17).

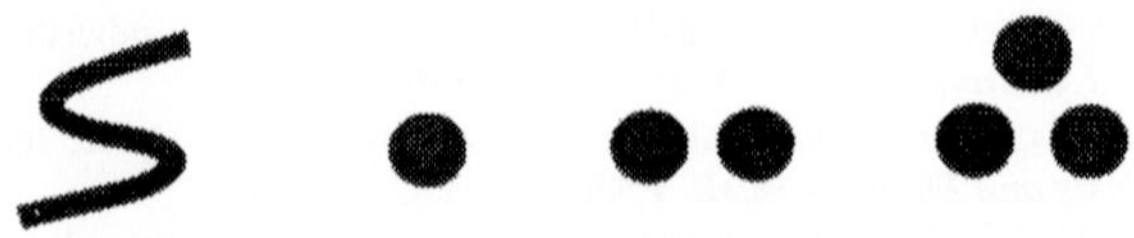

Fig. 17. Complementary characters.

Cusp point: a positive (negative) curve intersecting another positive (negative) curve with an intermediate angle greater than 150 degrees.

Group: A consecutive non-broken run of horizontally aligned pixels is termed a group. One group is separated from another by at least two white pixels.

Inflection point: a positive (negative) curve intersecting another negative (positive) curve with an intermediate angle less than 40 degrees.

Stroke: is a portion of a word including one or more characters between two successive penlifts.

Segment: One or more groups are said to constitute a segment if they have one or more consecutive black pixels in common, or it is not connected to any other group.

ANALYSIS OF STRUCTURED DOCUMENTS

Handbook of Character Recognition and Document Image Analysis, pp. 421–456
Eds. H. Bunke and P. S. P. Wang

CHAPTER 16

THE REPRESENTATION OF DOCUMENT STRUCTURE: A GENERIC OBJECT-PROCESS ANALYSIS

DOV DORI
Faculty of Industrial Engineering and Management
Technion, Haifa 32000, Israel

DAVID DOERMANN, CHRISTIAN SHIN
Center for Automation Research, University of Maryland at College Park,
College Park, Maryland 20742-3275, USA

ROBERT HARALICK
Intelligent Systems Laboratory, University of Washington,
Seattle, Washington 98195, USA

IHSIN PHILLIPS
Seattle University
Seattle, Washington 98122, USA

MITCHELL BUCHMAN
Dept. of Electrical Engineering, University of Maryland,
Baltimore County, Maryland 21228, USA

DAVID ROSS
RAF Technologies, Inc.,
Redmond, Washington 98052, USA

Document understanding has attained a level of maturity that requires migration from ad-hoc experimental systems, each of which employs its own set of assumptions and terms, into a solid, standard frame of reference, with generic definitions that are agreed upon by the document understanding community.

The logical structure of a document conveys semantic information that is beyond the document's character string contents. To capture this additional semantics, document understanding must relate the document's physical layout to its logical structure. This work provides a formal definition of the logical structure of text-intensive documents. A generic framework using a hierarchy of textons is described for the interpretation of any text-intensive document's logical structure. The recursive definition of textons provides a powerful and flexible tool that is not restricted by the size or complexity of the document. Frames are analogously used as recursive constructs for the physical structure description.

To facilitate the reverse engineering process which is required to derive the logical structure, we describe DAFS, a Document Attribute Format Specification, and demonstrate how our framework can serve as a conceptual framework for enhancements of DAFS.

Keywords: Document understanding; Document layout analysis; Physical structure; Logical structure; Document format standards.

1. Introduction

The field of document image understanding encompasses the technology required to allow the information contained in paper documents to be accessed electronically. Although the task may seem easily definable, the general expressibility of a document suggests that document image understanding must involve more than simply recognizing a string of characters on a page and putting them into the format of a word processing system.

Historically, Optical Character Recognition (OCR) has been more intensively researched. Consequently, it has attained a considerable level of maturity. Document page layout analysis constitutes a subject of intensive research and it, too, has reached a certain level of maturity. Physical layout analysis, combined with OCR, provides for complete reproduction of documents.

Logical document structure is a hierarchy that conveys the semantics of the document. The same logical document structure can be formatted in a variety of physical layouts by changing such variables as page and font size, spacings between paragraphs and between sections, number of columns, etc. In all of these layouts, the semantics of the document remains unaltered. Logical structure analysis determines the document's semantic structure and provides data appropriate for information retrieval involving more than string matching. For example, one would like to query all the abstracts of all papers in a database which have some keyword combination in a title or section heading and were written within a certain time period. To do this, the resulting data structure(s) should be precisely defined and agreed upon by the OCR and document understanding community.

The analysis of a document image involves both the physical decomposition of the page and the derivation of the logical or semantic meanings of the salient fields or regions defined by the decomposition. For example, if we are given a newspaper page, the physical analysis involves extraction of blocks of text, graphics and halftones as well as the identification of attributes such as font size and style. The structural analysis, on the other hand, may involve using the layout clues to identify headlines, locate bylines, group paragraphs from different columns which belong to the same article, or associate a picture with the article which references it and the photographer who took it. In general, the analysis involves the extraction and use of attributes and structural relationships in the document to label document components within the contextual rules dictated by the document class or type (memo, letter, journal article, newspaper, etc.).

Document understanding is complicated by the fact that relevant information can be expressed in a non-textual medium, such as mathematics, drawings and graphics. Hence, understanding art line drawings, engineering line drawings, perspective projections, graphs, and special kinds of documents like complex mathematical formulae and music scores, are all part of the document image understanding task.

The fact that hardcopy documents convey a great deal of semantic information via their *logical structure*, expressed as the two-dimensional arrangement of the text and non-text elements on the page, suggests that structural information is an essential part of understanding. Although there are many ways to store and display the physical appearance of the same document, the document's logical structure should remain the same, because it reflects semantics that is part of the author's intention, but beyond the ASCII stream of characters. Determining the logical structure of a document, therefore, constitutes an important aspect of document image understanding.

In this chapter, we will address the problem of providing a standard, formal framework for the management of physical and logical descriptive information extracted from document images, and address issues of intermediate representation which arise during the analysis process. Although the techniques used in the analysis of document images differ widely from those used in the document creation process, it is useful to examine the fundamental representations used by the more developed document processing community.

Many standards exist for representing the logical structure of a document. In publishing applications, documents are "encoded" via standards such as SGML in preparation for the actual printing process. In document understanding and page decomposition, however, we perform "reverse encoding", seeking to reverse-engineer the meaning from an image of the printed page. The greatest difference between encoding for document creation and reverse encoding of document images is that during the reverse encoding process, there may be varying levels of uncertainty in the interpretation of aspects of the document. In the document creation process, this ambiguity is not present, since the document is usually encoded by the same person who created the representation of the document. A data format for document reverse encoding must have a mechanism for representing these ambiguities.

In document image reverse encoding, it is important to move back and forth between the interpretation of the physical structure of the document and the semantic structure of the document. Although they rely heavily on one another and may share common aspects, they are still two different ways of perceiving the structure of a document. The fact that typically there is not a one-to-one mapping between the physical and logical structure complicates this requirement.

While document creation proceeds in a serial manner, document image decomposition usually traces through a document hierarchically rather than serially. This is a result of the way reverse engineering processes are usually applied to documents. An example of this is that discrimination between text and non-text regions is often performed for an entire document before any character recognition is performed.

Throughout this process, the ultimate goal is to use physical attributes to obtain a consistent and valid interpretation of the semantic attributes. For example, the text blocks should be semantically ordered by the "reading order" (sequencing) of the text units. Further, each text unit should be assigned a semantic label. In a business letter, the sender's address, receiver's address, date, opening salutation, body, closing, and signature can generally be inferred by their relative locations on the page. Likewise, in a technical article, the title, author(s), abstract, keywords, sections, displayed equations, tables, graphs, illustrations, footnotes, page numbers, reference list, and other logical components can be deduced by their locations and/or sequencing, as well as the fonts, styles and sizes of the characters that make them up. The resulting recognized text strings are formatted such that their two-dimensional layout, deduced from the 2-D layout analysis, are recorded along with the text itself.

The resulting complex data structure, if constructed correctly, captures the entire semantics of the original document. However, it is in a much more condensed form, providing for both data compression and noise removal. This data structure enables one to retrieve information through querying and to reproduce the original page document with practically no noise.

In Sec. 2 of this chapter we provide an overview of the state-of-the-art in both geometric and logical interpretation. In Sec. 3, we propose a working framework for the logical structure of text-intensive documents. A key definition is that of a *texton*,[a] which provides for recursion and a quantitative definition of document complexity. Being a complex system, analysis of document structure and layout requires a sound methodology. We employ the object-process analysis (OPA) methodology [13,12] and object-process diagrams (OPDs), which are the graphic tool of OPA, to express both the structure and behavior of a document analysis system within a coherent, unified frame of reference.

In Sec. 4, we describe a new and powerful document attribute format specification, called DAFS, which provides mechanisms for representing and maintaining both physical and logical information during the reverse encoding process. We also show how the logical framework of Sec. 3 can be implemented directly using DAFS.

2. Literature Survey

Papers covering all areas of document image analysis can be found in the Proceedings of the 1991, 1993 and 1995 *International Conferences on Document Analysis and Recognition* [ICDAR]; the 1992, 1993, 1994 and 1995 Annual Symposia on Document Analysis and Information Retrieval, sponsored by the University of Nevada, Las Vegas; and the International Conferences on Pattern Recognition [ICPR]. The brevity of this survey necessitates that many papers from these conferences are not mentioned. For surveys of document image analysis and document image understanding see [6] and [38].

2.1. *Geometric Layout Analysis*

A geometric page layout of a document image page is a hierarchical specification of the geometry of different kinds of maximal homogeneous regions. A region is homogeneous if all its area is of one type, e.g., a character, a line of characters, a text paragraph, or a figure. Formally, a *geometric page layout* Φ is an ordered pair $(\mathcal{R}, \mathcal{S})$, where $\mathcal{R}$ is a set of regions and $\mathcal{S}$ is a labeled hierarchical spatial relation on $\mathcal{R}$. A

[a]Not to be confused with the earlier use of the term by Julesz in connection with visual texture perception [25].

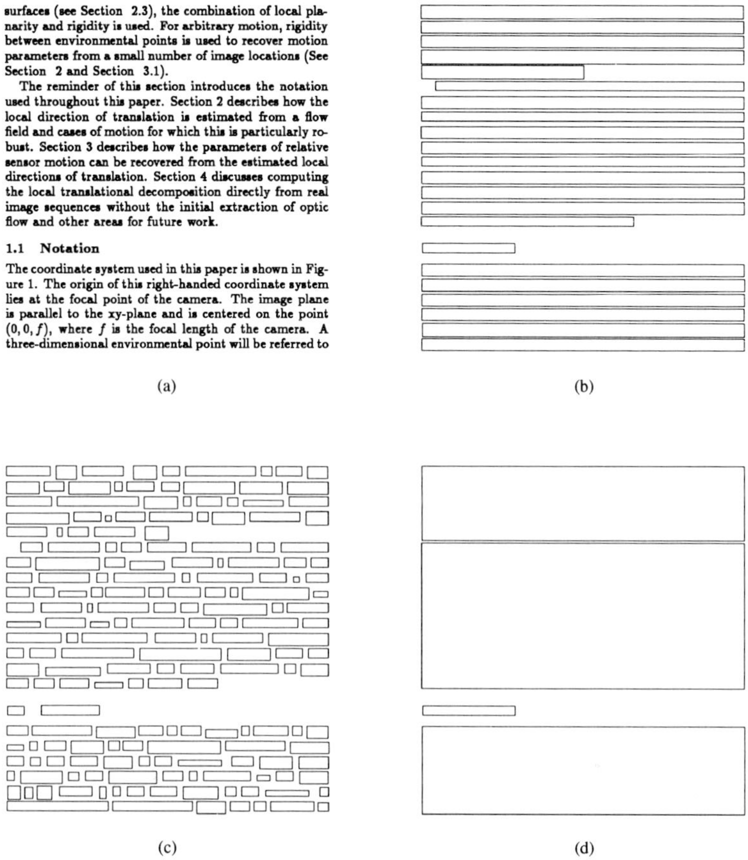

Fig. 1. Example of document image decomposition: (a) a document image written in English, (b) text line bounding boxes, (c) word bounding boxes, (d) text block bounding boxes.

region R is described by an ordered pair (T, θ), where T defines the type of the region and θ is the parameter vector of values for the region. The parameter vector may also include uncertainties for any of the parameters. Uncertainty can be a parameter standard deviation, or a tolerance interval, represented as a probability pair. For the entire parameter vector, the uncertainty can be specified by a covariance matrix. Figure 1 illustrates the character, line, and paragraph hierarchy for a small text sample.

Many of the algorithms for determining geometric layout employ the operations of mathematical morphology.[b] Although the original algorithm developers generally did not describe their algorithms in terms of mathematical morphology and were probably not even aware that their algorithms could be described in such terms, our descriptions will utilize the operations of mathematical morphology. This will

[b] An introduction to mathematical morphology is given in the chapter by Ha and Bunke in this book.

allow us to be brief and precise.

Early work on page segmentation was done by Wahl *et al.* [90], using a technique called the constrained run length smoothing algorithm. It essentially consists of a morphological closing of the document image with a horizontal structuring element of specified length (they used 300) intersected with a morphological closing of the document image with a vertical structuring element of specified length (they used 500). The intersection is then morphologically closed with a horizontal structuring element of specified length (they used 30). The bounding rectangle of the connected components of the resulting image constitute the block segments. Features of a block include its area, height, and width, the number of black pixels in the segment on the original document image, and the mean horizontal black run lengths of the original image within the segment. Text areas are then classified into *text, horizontal solid black lines, graphic and half-tone images*, and *vertical solid black lines.* No measure of performance is given.

Nagy and Seth [30] and Nagy *et al.* [31] employ an X–Y tree as the representation of a page layout. The root node of an X–Y tree is the bounding rectangle of the full page. Each node in the tree represents a rectangle on the page. The children of a node are obtained by subdividing the rectangle of the parent node either horizontally or vertically, with horizontal and vertical cuts being alternately employed at successive levels in the tree. Hao *et al.* [18] describe a variation on this technique.

Fisher *et al.* [16] sample a 300 dpi document image by a factor of 4 and use a run length smoothing algorithm. They then compute the connected components of the run length smoothed image. The connected components and their bounding boxes constitute the blocks of the geometric page layout. They extract connected component features such as component height, width, aspect ratio, density, perimeter, and area for classifying each block as text or non-text.

Lebourgeois *et al.* [29] sample a document image by a factor of 8 vertically and 3 horizontally. Each pixel in the sampled image corresponds to an 8×3 window in the original image. If any pixel in this 8×3 window is a binary 1, then the sampled image has a binary 1 in the corresponding pixel position. The sampled image is then dilated by a horizontal structuring element to effectively smear adjacent characters into one another. Each connected component is then characterized by its bounding rectangle and the mean horizontal length of its black runs. Connected components having vertical height within given bounds and mean horizontal run length within given bounds are then labeled as text. Lines outside the given bounds are labeled as non-text lines. Components labeled as text regions are then vertically merged into larger blocks using rules that take alignment into account. Blocks are also subdivided to separate them at horizontal peninsulas. No measure of performance was given but it was indicated that the method needs improvement.

Bloomberg [5] uses morphological operations on a document image at various resolutions to identify font style for each word. Class labels include *bold, italic*, and *normal.* The method employs a small vertical dilation, followed by a close-open sequence to remove noise, followed by a hit or miss transform to identify seed points of characters in the italic class or bold class. The words which are in italic or bold can then be delineated by conditionally dilating the seed with a pre-calculated word segmentation mask. No accuracy performance results are given.

Saitoh and Pavlidis [31] sample a document along eight vertical and four horizontal lines and then extracting connected components. They then classify each component into *text, text or noise, diagram or table, half-tone image, horizontal separator*, or *vertical separator*, using block attributes such as height, height to width ratio, connectivity features of the line adjacency graph, and whether there are vertical or horizontal rulings. Page rotation skew is estimated from a least squares line fit to the center points of blobs belonging to the same block. Blocks are subdivided based on the height of the lines in a block and the vertical distance between. The technique was tried on 52 Japanese documents and 21 English documents. No quantitative measure of performance was given.

Hinds *et al.* [20] sample a 300 dpi document image by a factor of 4 and compute from it a burst image. This image is obtained from the distance transform or the erosion transform of the document image using a two-pixel vertical (horizontal) structuring element for portrait (landscape) mode images. The burst image selects only the column (row) relative maximum pixels of the erosion transform. They then compute the Hough transform of the burst image, incrementing each Hough bin by the value of the pixel in the burst image, provided this value is less than 25. The rotation skew of the image is then determined by searching the Hough parameter space for the bin having the largest accumulated value; its angle is the rotation skew angle. They tested the technique on 13 document images, and correctly determined the rotation angle on all the images. The inter-line spacing on one document image was not correctly determined and the landscape/portrait mode was incorrectly determined on five document images.

Pavlidis and Zhou [33] determine geometric page layout by analyzing the white areas of a page by looking for long white intervals on the vertical projection. The column intervals are then converted into column blocks, merging small blocks into larger blocks. Blocks are clustered according to their alignments and the rotation angle is estimated for each cluster. The column blocks are then outlined. Finally, each block is labeled as text or non-text, using features such as the ratio of the mean length of black intervals to the mean length of white intervals, the number of black intervals over a certain length, and the total number of intervals. No performance results are given.

Baird [3] discusses a computational geometry technique for geometric page layout which finds the maximal rectangles covering the white areas of the page. The rectangular regions not covered by these white rectangles can then be classified as text or non-text.

Amamoto *et al.* [2] determine geometric page layout by operating on the white space of the sampled document image. They open this white space with a long horizontal structuring element and open it again with a long vertical structuring element. The union of these two openings constitutes the white space of the blocks, which are then extracted from this white space. They decide that a block is a text block if the length of the longest black run length in the vertical or horizontal direction is smaller than a given threshold. A decision is made as to whether the writing is horizontal or vertical based on the number N_H of blocks whose widths are greater than twice their heights and the number N_V of blocks whose heights are greater than twice their widths. If $N_H > N_V$, the decision is horizontal writing; otherwise, vertical writing. Each block is then assigned a class label from the set *text, figure, image, table*, and *separation line.* No performance results are given.

O'Gorman [32] presents what he calls the docstrum technique for determining geometric page layout. This technique involves computing the k nearest neighbors for each of the black connected components of the page. Each pair of nearest neighbors has an associated distance and angle. By clustering the components using the distance and angle features, the geometric regions of a page layout can be determined.

Ishitani [23] determines the rotation skew angle as that direction in which the variance of the complexity of the white-black transitions is greatest. Specifically, a set of lines is defined for each angle. The difference between successive angles is $0.01°$. Each line's complexity is then measured, where complexity is defined as the number of white-to-black transitions along the line. The variance of the white-to-black transition counts is then determined. The angle which maximizes this variance is the estimated rotation skew angle. It is reported that this measure does not have difficulties with document pages which have large areas of non-text. The method was tested on 40 300 dpi document images taken from magazines, newspapers, manuals and scientific journals. It was reported that the rotation angle was measured to within an accuracy of $0.12°$.

Chen and Haralick [7] use an algorithm based on the recursively computed morphological opening and closing transforms. First the image is subsampled to a 100 dpi resolution. The recursive closing transform is computed, a histogram is generated from the transformed image, and a threshold is determined using a regression tree function of the histogram values, where the regression tree was previously determined off-line. The closing transform is then thresholded to produce a closing of the subsampled image that fills inter-character gaps. Then, on this closed image, a recursive opening transform is computed, and, in a similar manner, a threshold is determined from the histogram of the values of the opening transform. The opening transform is then thresholded to produce an opened image in which the character ascenders and descenders have been removed. Then the connected components of the opened image are computed and line fits to each connected component are obtained. Using the estimated orientation of each of the fitted lines and the residual fitting error, the skew angle of the document page image is obtained from a robust estimation of the fitted line orientations. To determine the performance of the algorithm they used the UW-I document image database [34] having 1147 distinct document images each associated with a ground truth skew angle. This set of images was then rotated through eleven rotation angles: $0°, \pm1°, \pm2°, \pm3°, \pm4°$, and $\pm5°$, to yield a set of $1147 \times 11 = 12617$ images. The absolute difference between the estimated skew angle and the ground truth skew angle was less than 0.5° for more than 99% of the images and less than 1° for virtually all of the images.

Hirayama [21] develops a technique for determining the geometric layout structure of a document which begins by merging character strings into text groups. Border lines of blocks are determined by linking edges of text groups. Then blocks which are over-segmented are merged and a projection profile method is applied to the resulting blocks to differentiate text areas from figure areas. Hirayama reports that on a data set of 61 pages of Japanese technical papers and magazines 93.3% of the text areas and 93.2% of the figure areas were correctly detected.

Ittner and Baird [24] determine geometric layout by doing skew and shear angle corrections, partitioning the page into blocks of text, inferring the text line orientation within each block, partitioning each block into text lines, isolating symbols within each text line, and finally merging the symbols into words. The rotation skew angle is determined by taking the projections of the centers of the connected components of the black pixels on the page at a given angle. The angle is iteratively updated to optimize the alignment without having to compute the projection over each possible angle. After rotating the image, shear is corrected by a similar technique. They report an accuracy of less than three minutes of arc, and indicate that the method fails on perhaps one in 1000 images. Blocks are determined by the white space covering technique of Baird [3]. They report that on 100 English document image pages from 13 publishers and in 22 styles, 94% of the layouts were correctly determined. The orientation of the text lines in a block is determined from the minimum spanning tree of the connected components of the black pixels. The mode of the histogram of the directions of the edges in the minimum spanning tree is the orientation of the text lines. Symbols in a text line are determined by taking the projection in the direction orthogonal to the text line. The projection profile is checked for a dominant frequency and the segmentation into characters is done from the projection profile using the knowledge of the dominant frequency. To determine the words in a text line, they determine a scalable word-space threshold for each text block separately. Then each text line is independently segmented to distinguish between the inter-character spacing and the inter-word spacing.

Ankindele and Belaid [1] determine a geometric page layout that permits blocks to be polygonal as well as rectangular. They determine the elongated white spaces in the document image and then find intersections of these white spaces. Points of intersection are candidate vertices for polygonal blocks. The polygonal blocks are then extracted from the geometry of the intersection points. No performance results are given.

2.2. *Logical Layout Analysis*

The emphasis in the work on logical layout analysis is on developing processes (algorithms) to carry out various tasks related to logical segmentation.

Tsujimoto and Asada [39] assume that each block of the geometric page layout contains exactly one logical class. They organize the geometric page layout as a tree. Each new article in a document such as a newspaper begins with a headline which is in the head block. They find the paragraphs which belong to the head block by rules relating to the order of the geometric page layout tree and are able to assign logical structure labels of *title, abstract, sub-title, paragraph, header, footer, page number,* and *caption.* They worked on 106 document images and correctly determined the logical structure for 94 of them.

Fisher [15], an extension of Fisher *et al.* [16], describes a rule-based system to identify the geometrical and logical structure of document images. Ingold and Armangil [22] describe a formal top-down method for determining logical structure. Each document class has a formal description that includes composition rules and presentation rules. The technique has been tested on legal documents.

Chenevoy and Belaid [8] use a blackboard system in a top-down method of logical structure analysis of a document image. The system is defined in a Lisp formalism and has a hypothesis management component using probabilities.

Kreich *et al.* [28] describe a knowledge-based method for determining the logical structure of a document image. To obtain the blocks they search for the largest text blocks because these are the most characteristic elements in the document layout. The search consists of grouping together the connected components which are close enough to each other. Once text blocks are determined, lines are found within each of the text blocks and words within the lines. The determination of document layout structure is based on interpreting documents and their parts as instances of hierarchically organized classes. They have defined over 300 classes of document images and their parts. No performance results are given.

Derrien-Peden [11] describes a frame-based system for the determination of structure in a scientific and technical document image. The basis of this system is a macro-typographical analysis. The idea is that in scientific and technical documents, changes of character size or thickness of type, white separating spaces, indentation, etc., are used to make visual searching for information easier. The technique therefore searches for such typographical indications in the document and recovers the document's logical organization without any interpretation of its semantic content. The first step is the determination of the geometric page layout, keeping a *part of* relationship between blocks. The logical structure determination removes running heads and footnotes and searches for the text reading order. Text blocks are then compared to logical models of classes and each text block is assigned a class. No performance results are given.

Yamashita *et al.* [41] use a model-based method. Character strings, lines, and half-tone images are extracted from the document image. Vertical and horizontal field separators (long white areas or black lines) are detected based on the extracted elements; then appropriate labels are assigned to character strings by a relaxation method. Label classes include *header, title, author, affiliation, abstract, body, page number, column, footnote, block* and *figure.* The technique was applied to 77 front pages of Japanese patent applications. They reported that the logical structure for 59 of these documents was determined perfectly.

Dengel [9] discusses a technique for automatically determining the logical structure of business letters. He reports that on a test set of 100 letters, the recipient and the letter body could be correctly determined. Saitoh *et al.* (1993) determine logical layout based on text block labels of *body, header, footer*, and *caption.* They tested their technique on 393 images of mainly Japanese, but including some English, documents. To characterize performance they measured the average number of times per image an operator has to correct the results of the automatically produced layout. They report that on the average 2.17 times per image, areas not suitable for output have to be discarded; 0.01 times per image, mis-classified areas have to be correctly labeled; and 1.09 times per image, a text area has to be reset. With respect to text ordering they report that it required moving connections 0.47 times per image on the average, making new connections 11 times per image, and re-assigning type of text 0.36 times per image.

3. Logical and Physical Document Description

In order to describe a document's physical and logical characteristics consistently, it is advantageous to first distinguish between the document's content and its structure.

3.1. *Document Content versus Document Structure*

The *content* of a document is the information contained in the document, which is not bound to a particular representation format. At the lowest level, this information may include a stream of characters, and these characters make up successively higher-order objects built on top of each other such as words, sentences, paragraphs, etc., up to the complete document level.

The *physical structure* of a document is how the document's content is laid out on the physical medium. The same content can be organized in a variety of ways and therefore can have many physical layouts, which stem from different values of the attributes of the physical components (point size, line spacing, page size, etc.). The medium is traditionally paper, but may be any visual host, such as a computer screen or photographic film, which emulates the layout on the page. Bearing this extension in mind, we refer to paper documents as the classical representatives.

The *logical structure* of a document's content is how the content is organized prior to the enforcement of a particular physical structure. A text-intensive document, for example, typically consists of sentences, words and characters, and possibly higher-order constructs such as sections and chapters for an article, or address block, body and signature block for a business letter.

It is essential to note that the **same** content can be viewed with respect to both physical structure and logical structure. A major goal of this chapter is to describe the two types of structure and how they can interact.

3.2. *Generic Document Structures*

Text-intensive documents can be classified into many types including books (textbooks, edited books,...) technical papers, correspondence (business letters, memos,...), newspaper articles, and conference proceedings. When attempting to describe the structural relationships between document components, it is essential to provide a level of abstraction, so as not to be caught up in the terminology associated with a specific type of document. To this end, we define terms that are generic at both the logical and physical levels. The *generic document structure* terminology allows us to define type-independent relationships among document components.

Following this rationale, we distinguish between the generic (both logical and physical) document structure and the instantiated generic document structure. An instantiation of a generic structure allows us to begin discussing a particular class of document, and the relationships between known entities. Such a distinction is in accordance with the basic concepts of object-oriented analysis (OOA), where objects are instances of their respective classes.

Terms such as column, line and symbol in the context of physical structure, or chapter, section, and subsection in the context of logical structure, are instances of the generic terms we define below. Some of them, such as the physical term "line" and the logical term "sentence", are applicable to a large variety of documents, while others, such as the logical term "session", are document-type-specific.

Finally, a *specific structure* (physical or logical) refers to the structure of a single given document, and describes the structure of that instance of the document.

3.3. *Realizing Document Structures*

The progression from generic to specific instances of a document is fairly straightforward, but the ability to reverse the process can be an important component of any analysis system. At the first level, the object class "Document" is instantiated as to its type, both physical and logical, according to the document content. The structure of each instance of the generic document is thus dependent on the document's type.

By making such distinctions, it becomes possible to prune the generic document tree to narrow the analysis to a given document type. For example, there is no use talking about a document's physical component called "volume" or the logical component "chapter" when the document under consideration is of the type, "business letter". Such distinctions can be made at an early point.

A second level of instantiation involves instantiating the object class of a particular document type, such as "Proceedings", to its specific structure, such as the "Proceedings of Document Analysis Systems '94 Conference" [10].

We do not attempt to provide a full classification of all the text-intensive document types, let alone the particular logical or physical structure of each type. In the remainder of this section, however, we do provide a set of definitions and tools that enable this task to be done consistently and coherently, along with a few examples.

3.4. *Generic Physical Structure*

The following are the definitions related to the generic physical structure.

Frame – an area within a page of a document, which may consist of a collection of (lower level) frames and/or blocks.

Root Frame – a frame which encompasses the entire physical document.

Page – a frame which occupies a rectangular region of a page, on which the document, or part thereof, is physically recorded. A page is the basic physical document object.

Page Set – a frame which consists of multiple pages. An instance of a page set might be a single volume or chapter, for example.

Block or **Simple Frame** – a terminal, lowest-level frame, which, at the given level of granularity, need not or cannot be further decomposed.

A frame is defined to be a recursive component, as it may itself consist of one or more frames. A block is the terminal frame, which defines a region on the page and has content. Depending on the desired level of granularity, a block's content may correspond to a column, line, word, or character, for example.

Each type of frame has an associated set of attributes. Information about a frame's location, position on the page, justification, etc. is defined by its attributes. For example, a character is characterized by such attributes as font, boldness, point size, inclination; a line, by length; etc.

Being the leaves of the tree, only blocks have content, while items higher in the hierarchy, which are compound frames, are lists of pointers to lower-level frames. The block's granularity must be at least as fine as the smallest logical component to be described. Thus, for example, if the lowest-level logical component is a word, then the block cannot be a line—it must be at least a word, and it may be a character. This data structure enables the reconstruction of the structure of a frame at each level.

Consider, for example, a paragraph which is split over two pages, or a word which is split over two columns. In both cases, a single logical component (paragraph or word) is a combination of two frames, which should be concatenated to yield the entire logical component.

If, on the other hand, a single physical component is found to correspond to more than one logical object, such as the string "092596", which corresponds to a date with three logical sub-components (month, day, year), then the physical representation may be subdivided to reflect the finer granularity. This is quite rare, however, as the physical structure is normally aimed at reflecting the logical one.

An instance of a text-intensive document may have blocks and frames corresponding to a:

Character – a block containing the image of a symbol in the document's language.

Word – a frame containing a group of one or more aligned characters, separated by white space. A word is both a logical object in a document (to be defined below), which conveys a certain meaning, and a physical object—the frame (page area) containing the union of its constituent characters. If word is a block, then its content is the image of the word; otherwise, it is a frame consisting of characters (each of which is a block), with each character having its image. The specializations of word are as follows:

Subword – a frame containing a group of one or more aligned characters, which make up the first part or the last part of a word.

Preword – a subword containing the first part of a word. A preword is located at the end of a line and ends with a hyphen.

Postword – a subword containing the last part of a word. A postword is located at the beginning of a line and completes its preceding preword to a whole word.

Line – a frame containing (1) a collection of one or more aligned words and (2) at most one preword and at most one postword.

Stack – a frame containing a collection of one or more lines stacked on top of each other and possibly separated by a non-empty white space, such that its logical content is one paragraph[c] at most.

Column – a frame containing a collection of one or more stacks on top of each other. A page may contain one or more columns. In structured documents, this number is generally fixed throughout the document.

3.5. *Generic Logical Structure*

Like the physical structure, which is represented as a hierarchy of frames, the generic logical structure is similarly viewed as a tree, in which the leaves are typically the characters, or symbols, and the root is the entire document. This structure is depicted in Fig. 2.

To be able to describe the logical document hierarchy generically, and not be restricted by a particular number of levels and associated level names, such as "section" and "chapter", we define the term "texton" as the logical analog of "frame".

[c] Paragraph is a logical term defined below.

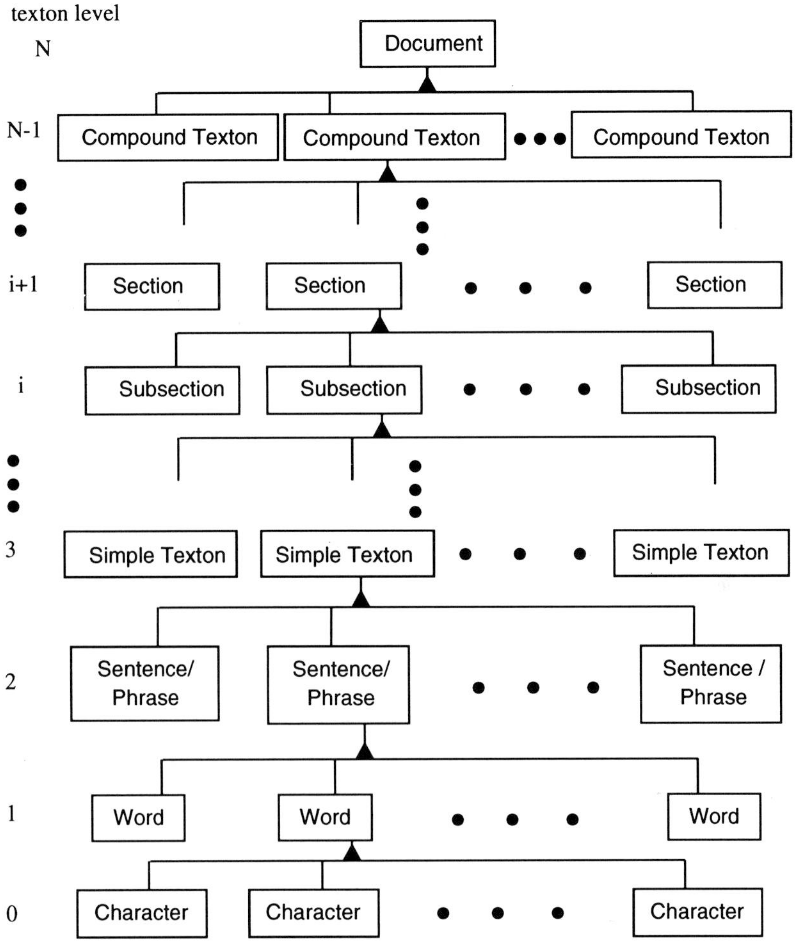

Fig. 2. The logical structure of a document described as a tree.

Texton – a logical component of a text-intensive document, which consists of one or more (lower level) textons or simple textons.

Root Texton – a texton which is the entire document.

Examples of root textons include book, encyclopedia, concordance, dictionary, journal, newspaper, magazine, report, scientific paper, cover letter and business letter.

Simple Texton – a logical component of a text-intensive document, which is not further divided.

Instances of a simple texton are paragraph, sentence, phrase, word and character. If, for example, word is the simple texton in a particular document, then any subcomponent such as a character is a *primitive texton* document.

Compound Texton – a texton consisting of a distinct header, body, and optional trailer.

An example of a compound texton is a section of a document, which has a header (the section head), a body (the set of paragraphs), and no trailer. Another, less obvious example of a compound texton is a signature block in a letter, which contains a header (the closing), a body (the signature), and a trailer (the printed name).

As shown in Fig. 2, scanning the logical structure from the top down, the entire document (encyclopedia, book, article, business letter, etc.) is the root texton—the root of the tree. Below it is a varying number of levels of textons. The black triangle along the paths connecting a whole to its parts in Fig. 2 is the aggregation symbol [12].

Texton is the logical analog of the physical frame. Like frame, the definition of texton is recursive, and the halting condition is that the constituent texton is a simple texton, i.e., the base logical unit, typically a character. The recursive definition of texton encompasses the entire spectrum of logical levels in any text-intensive document, just as the root frame encompasses all the physical levels.

Instances of a texton, in a text-intensive document, include:

Character – a texton which is a symbol in the document's language. Normally, character is a basic texton.

Word – a texton containing a sequence of one or more characters, which has some meaning in the document's language.

As we have noted, both character and word have logical as well as physical definitions. The difference between a logical character and a physical character is that a logical character is the symbol itself, while a physical character is the image representation of the logical character. Likewise, the difference between a logical word and a physical word is that a logical word is a semantic-conveying object, while a physical word can be considered either as the image representation of the logical word or as an ordered collection of its comprising physical characters. Note that there is no logical analog to the physical term subword, whose existence stems from spatial arrangement considerations.

We continue with the definitions of higher-level textons.

Phrase – a texton which is a meaningful collection of one or more words that do not necessarily form a complete grammatical sentence.

Examples of phrases include a title of a document or part thereof, a name of a person or an organization, an address, or a (possibly nested) itemized list of such entities.

Sentence – a meaningful collection of one or more phrases which correspond to a valid grammatical sentence, complete with punctuation.

Paragraphon – a texton which is a generalization of a paragraph. It consists of a group of one or more sentences and/or phrases.

Each one of the items above is an example of a paragraphon, where the title is a phrase, and it is followed by another phrase and optional sentence(s).

Unlike character and word, higher-level textons have different names than the corresponding frames, because the physical structure departs from the logical one, and the correspondence becomes more and more fuzzy as we climb up the two hierarchies. Thus, above word at the physical level is the frame called line, while the corresponding textons at the logical level are phrase and sentence. However, it is obviously unlikely that a single sentence occupies exactly one line. At the next level up, paragraphon is analogous to stack, but again, the correspondence is only partial, because a paragraphon may stretch across more than one stack, if it starts at the end of a column and ends at the first stack of the next column, or even across several whole columns and pages.

At yet higher levels, the relationships between textons and frames are type-dependent. For example, the texton chapter in a textbook normally starts at a new page, as does a paper in a proceedings.

A document may contain logical elements which are referenced from multiple independent points within the document itself. They are often self contained logical units (i.e., textons) and should be treated as such. To handle such components, we define a *referenced* texton.

Referenced Texton – a graphic or textual texton which is referenced from the document.

Examples of referenced components include figures, appendices, footnotes, citations, continuation text bodies (e.g. in newspapers) and even complete documents. The header of a texton is a label or identifier which is "referenced" by a pointer.

Pointer – a referencing texton pointing from the main text of the document to the referenced texton. This pointer is typically a phrase or a sentence, such as "continued on page 5, column 3", "see Figure X", or "(Author, 1995)".

Graphon – a referenced texton in a document whose nature is mainly non-textual, and whose function is to illustrate, explain or demonstrate the text. Examples of graphons are line drawings (engineering drawings or art-line), half-tones, photographic (black and white or color) images, maps, diagrams, charts, tables, etc.[d]

In graphons, if text exists, it supplements or enhances the graphic. A graphon has graphic (imagery or geometric) contents and an optional *caption*, which itself is a texton, and consists of a mandatory *caption header* (the graphon identifier), and an optional caption body (the textual title or explanation of the graphon). The caption header is mandatory, because it serves as a reference and is pointed to by the text.

Since a graphon, like the figures in this document, normally occupies a considerable portion of the page area, the physical location of a graphon is frequently allowed to *float* in the neighborhood of where it is referred to in the main text for the first time.

Finally, in addition to the hierarchical structure given by the recursive definition of the texton, the logical structure must also preserve the reading order.

Reading order – the order in which the characters or symbols in a text-intensive document must be traversed for the document to be correctly understood.

The reading order corresponds to a depth-first visit of the document's logical structure. Reading order normally makes sense only within and between the text-intensive components of a document. In the case of referenced textons, the texton appears physically only once, but may appear logically at many locations. A pointer denotes the logical appearance, so the reading order follows a round-trip "visit" to the referenced texton.

Graphons tend to "float" and can be referenced from multiple locations in the main text. Hence, like any referenced texton, the read order is preserved by requiring a visit from the reference pointer (typically a phrase) to the graphon and back.

3.5.1. *Document complexity*

We have seen that a text-intensive document has a hierarchy whose textons depend partially on the document's logical type and may represent chapters, sections, subsections, parts, etc. Hence, the number of texton levels in a document is finite, normally not greater than 10. This number depends on the nature of the document

[d] A table is a boundary case between text and graphon. We classify it as a graphon, because even though it contains text, the text normally does not have a definite linear reading order and it is normally enclosed within graphics—the lines that separate rows and columns.

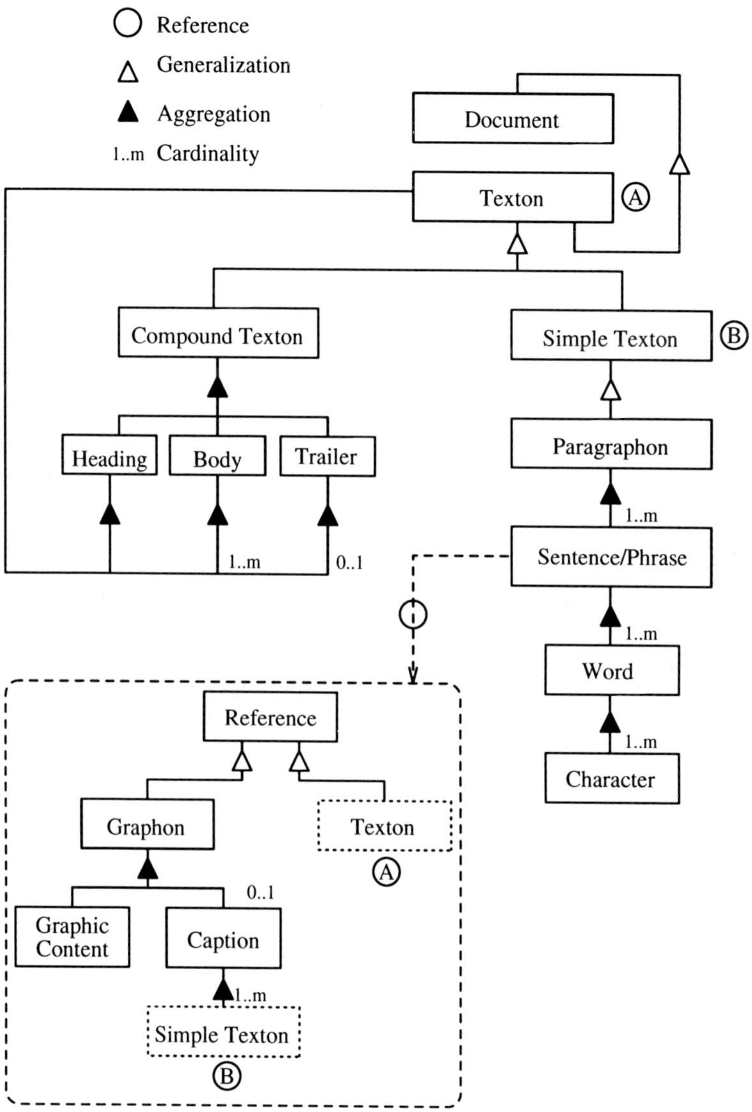

Fig. 3. The object-process diagram of the generic logical structure of a text-intensive document.

and indicates its structural complexity. The numbering of the levels is bottom up, with zero assigned to the character level.

The *logical complexity* of a document is the level number of the document's root texton.

Consider, for example, a journal paper, whose body consists of sections. The body of each section is a paragraphon. Assigning the level numbers 0, 1, and 2 to the character, word, and sentence levels, respectively, a paragraphon is a level 3 texton, and the entire document is a level 4 texton. Hence the complexity of this document is 4. If at least one of the sections is divided into subsections, and no subsection is divided into sub-subsections, then the document complexity is 5.

Although usually there is a relation between the document's size and its complexity, these two terms should not be confused. The size can be measured by the number of pages, words or characters. A dictionary, for example, may be a very large document, but its complexity is not necessarily high. Similarly, an outline may be relatively small, but may have much higher logical complexity.

3.5.2. *Simple and compound textons*

Having defined textons and their roles in the document logical structure, we turn to a more abstract and comprehensive description of logical document layout than the one given in Fig. 2. Figure 3 is an object-process diagram, or OPD [12,13], which describes the structure of a document.

The object Document is a specialization of a Texton, which is the root of the structure. This is denoted by the generalization symbol—the blank triangle going from Texton to Document. Texton is a generalization of Compound Texton and Simple Texton. This is denoted by the blank triangle from Texton to both Compound Texton and Simple Texton in Fig. 3. A *simple texton* is a generalization of a paragraphon.

A character is defined to be a *level 0 texton. A word* is a level 1 texton, as it consists of one or more characters, and a sentence is a level 2 texton. A simple texton in the main text of the document is therefore a level 3 texton. Below it in the main text reside the sentence or phrase (level 2 texton), the word (level 1 texton), and the character (level 0 texton). As we show below, these level numbers may vary for side text, such as the table of contents in a book.

Although in the simplest form, one may conceive of a primitive document consisting of a single character, perhaps conveying a coded message, a single-word document, a single-sentence/phrase document, or a single-paragraph document, we consider the simplest document to be a document which is a compound texton. Therefore, the minimal complexity of any document is 4. A simple document, such as a standard business letter, is an example of a level-4 document. It has a header (sender and recipient identification and subject), a body (one or more paragraphs: level 3 textons), and a trailer (salutation, signature, etc.).

The black triangle between Compound Texton on one hand, and Header, Body, and Trailer on the other hand, is an aggregation (whole-part) relation, expressing the fact that a compound texton consists of these three parts. The default cardinality (participation constraint) of the aggregation symbol is (1..1):(1..1), i.e., exactly one (minimum 1 and maximum 1) part for exactly one whole.

Consider a texton of level n. The cardinality of the header of this texton is 1, i.e., there is exactly one texton of level $n-1$ which is the header of the level n texton. The cardinality of the texton's body is $1..m$, meaning that there are between 1 and many textons of level $n-1$ in the body of the level n texton. Finally, the cardinality of the (optional) texton's trailer is 0..1, i.e., there is at most one texton of level $n-1$ functioning as the trailer of the level n texton. The "0..1" next to Trailer indicates that Trailer is optional. In other words, a texton has either two or three parts and must have exactly one Header, one Body and at most one Trailer. In summary, Header has exactly one texton, Body has a number of textons between 1 and many (denoted "$1..m$" in Fig. 3) textons, and Trailer, if it exists, has one texton.

For example, a section in a paper is a compound texton. It has a header (the section title); a body, consisting of one or more paragraphons; and no trailer. As another example, a textbook is a compound texton, whose header is everything from the beginning of the book to the beginning of the first chapter. The body of the book consists of a number of chapters and its trailer is everything from the end of the last chapter to the end of the book (appendices, glossary, index, etc.).

3.5.3. *The recursion in texton definition; the body path*

Since Compound Texton is Texton, and Compound Texton has Header, Body and Trailer, each having at least one Texton, we get a recursive definition. As in any recursion, to avoid infinite looping, a halting condition must exist. The halting condition, as expressed in the object-process diagram of Fig. 3, occurs when the textons of Header, Body and Trailer of the Compound Texton are all Simple Textons. When a texton is simple, the recursion stops, because from this level downward, we descend through the phrase level and the word level down to the character level. In the case of a referenced texton, the pointer—a Simple Texton in the main text—links it to preserve the reading order, while the referenced texton itself is a Compound Texton, whose header is the identifier the pointer points to.

Body Path is the path in the tree structure going from the root node—the entire document—through successively decreasing levels of compound textons, all the way down to the simple texton (the paragraphon level), such that the path always visits the body of each texton. Since by definition any compound texton has a body, such a path is guaranteed to exist, and it is unique.

The level of a character, which is the last node—the leaf—along the body path, is defined to be zero. This implies that along the body path the level number of a word is 1, the sentence/phrase level is 2, and the level of the paragraphon—the simple texton—is 3. Note that these numbers are not necessarily the same for characters, words, sentences and paragraphs which are not nodes along the body path. As we show in the example below, the level numbers may be higher or lower than the ones along the body path, depending on whether the path from the top texton (the document level) is longer or shorter than the length of the body path. The fact that of the three compound texton parts only two are mandatory gives rise to a 2-3 tree structure, as we demonstrate in the example in the next section.

3.5.4. *The DAS94 proceedings—A case in point*

To demonstrate the use of the concepts and terms presented above, and to show how document complexity is defined, consider the document *Proceedings of DAS94* [10]. The structure of the document is described in Fig. 4. The structure is detailed down to the Simple Texton level. Since a compound texton may have either three parts (Header, Body, and Trailer) or two parts (Header and Body), the resulting structure in Fig. 4 is a 2-3 tree.

As indicated in the legend of Fig. 4, the body path is marked by thick line segments. The level numbers are written in parentheses next to the corresponding textons along the path. The body path visits the nodes "Proceedings of DAS94", "Session", "Paper", "Section", "Subsection", and "Paragraph", in that order. Assigning the number 3 to the paragraph level and counting up we find that "Subsection" is at level 4, "Section" is at level 5, "Paper" is at level 6, "Session" is at level 7, and the entire document, "Proceedings of DAS94", is at level 8. Hence the complexity of this document is 8.

As a compound texton, "Proceedings of DAS94" has a header, a body and a trailer. The header is a level 7 texton, which, in turn, consists of three level 6 textons: a header—"Front Page" and "Copyright note", a body—"Chairmen's Message", and a trailer—"Table of Contents". Chairmen's Message consists of a level 5 header—the title "Chairmen's Message," a level 5 body, consisting of seven level 4 paragraphons, and a level 5 trailer, containing two level 4 paragraphons. The first phrase is "Kasierslautern, October 1994," and the second is the names of the two document editors. As we see here, both the paragraphs and the phrases, which are basic textons, are at level 4 rather than 3. The reason is that the path traversed here is not the body path. As already noted, a basic texton is guaranteed to be at level 3 only when it is on the body path. In other paths it may be more (as here) or less than 3. The path that ends with "Author", "Affiliation", and "Address", for example, is the longest one. It is longer by two edges than the body path. Therefore "Address", which is a simple texton, is a level 1 texton in this case, as shown at the bottom of Fig. 4. Table of Contents is a level 6 texton, consisting of a header—the title "Table of Contents," a body, and no trailer. The body of the Table of Contents consists of eight items. Each item is a level 5 texton called Session Contents. It has a header—session number and name, a body—a phrase (itemized list) of three level 4 items, each called Paper Details, and no trailer. Each Paper Details item is a level 3 texton. It consists of three paragraphons, each containing a single phrase. The first phrase is the paper name, the second phrase is the author name, and the third phrase is the page number.

In most of the papers, Section consists directly of paragraphs, but several papers

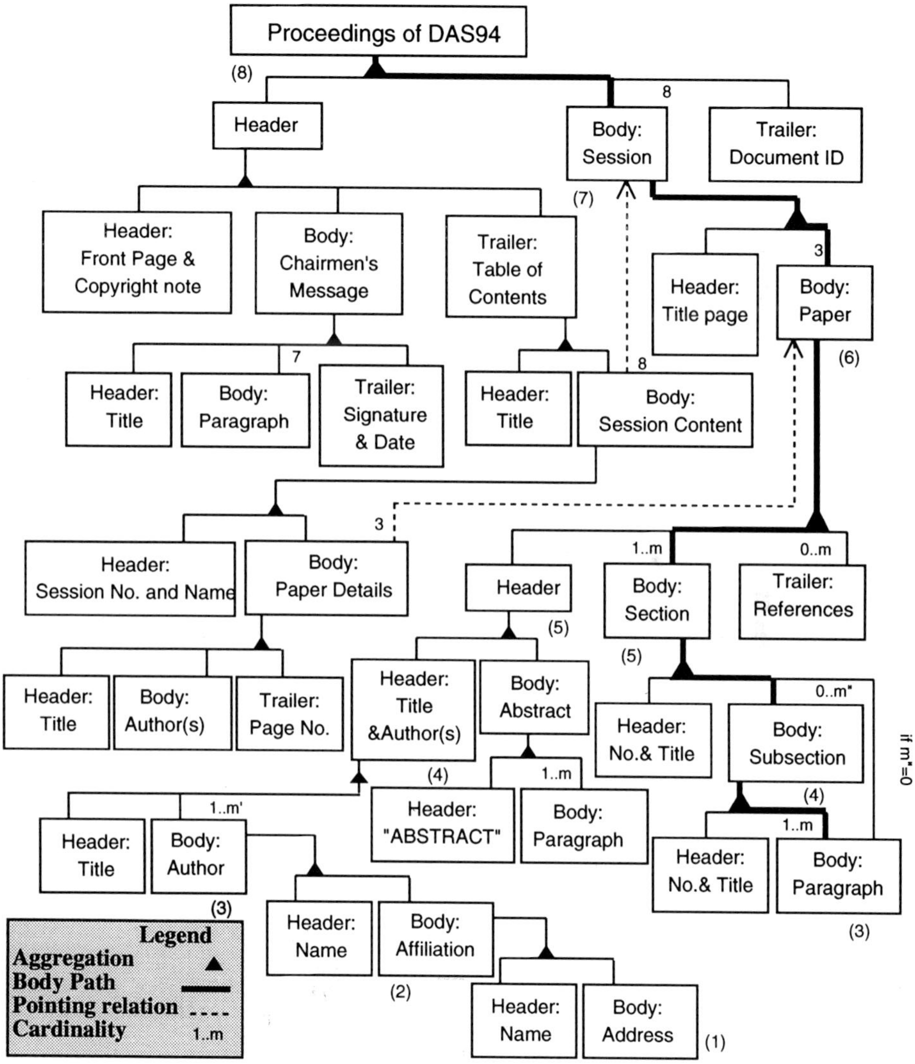

Fig. 4. An OPD describing an instance of a generic logical structure—the Proceedings of DAS'94 [10].

have subsections (see for example page 139 in the document). To accommodate this variability, we add the condition "if $m'' = 0$" along the aggregation link from Section to Paragraph in Fig. 4, where m'' is the number of subsections in a section. This means that if there are no subsections in the section, then Section consists directly of paragraphs.

3.6. *Relating Physical and Logical Structure*

Having defined generic terminology for both logical and physical structure, it is straightforward to relate the two at the content level, in most cases. Figure 5 shows the structure of a simple document, a multiple page chapter. The chapter is a compound texton, with a header (title) and two body components (one abstract and one section). The abstract is a simple texton and the section is a compound texton, consisting of two simple textons (paragraphons).

The physical structure subdivides the document into rectangular blocks. The content is shared in both structures. Note that the structure allows logical components (i.e., the abstract) to be split over two pages. For most documents, the logical complexity will be higher.

4. The Document Attribute Format Specification—DAFS

Integrating the physical structure with the logical one is a difficult problem. We now return to the physical layout as described by DAFS—Document Attribute Format Specification.

While many formats exist for composing a document from electronic storage onto paper, no satisfactory standard exists for the reverse process. DAFS is a file format specification for documents with a variety of uses. It was developed under the Document Image Understanding (DIMUND) project funded by ARPA and is meant to be the file format for all documents whose content has been examined either manually or automatically and which form parts of DIMUND databases. In addition, DAFS-formatted documents are used in the Illuminator project, where they are employed for training and testing document image understanding tools.

DAFS is intended to be a standard for the representation of document images and their partial interpretations during document decomposition. It is hoped that DAFS will prove to be general enough to enjoy widespread use, particularly in applications such as OCR and document image understanding. Several standards have been developed which address the creation or composition of documents, but none of these standards is well suited to the problem of document decomposition. There are many applications which would require some form of document decomposition, including character recognition and document image understanding. DAFS is a new format, designed explicitly for representing reverse encoding—the encoding of decomposed documents. As such, DAFS is designed to allow representation of both the physical and semantic information contained within a document image, but it is desired that this format have many applications beyond these specific ones. With this as a goal, the principle was established that the format should be extensible to meet the needs of a broad base of users, and that the format should not impose unnecessary assumptions on potential users.

4.1. *DAFS Design*

This section describes the philosophy which was used in the development of DAFS.

4.1.1. *Object-oriented design*

In the process of reverse encoding a document image, it becomes clear that it would be convenient to be able to handle portions of the image as discrete objects. An object can be any part of a document that may be defined in a stable form. Under DAFS, an object is called an "entity", and is essentially one or more rectangular pieces of the document image. Note that a DAFS "entity" is analogous to a SGML element and has nothing to do with the SGML concept of entity. Examples of useful entities are "paragraph", "character", and "document". Each may be part of a document, but needs to have a sufficiently stable form to be unambiguously defined. By implementing this specification in an object-oriented fashion, each entity may have any number of properties associated with it, allowing information about the entity to be entered. (See "DAFS Entities" for more information.) Another advantage of an object-oriented format is that an object may also contain other objects, which is the key to a hierarchical structure, and that attributes can be extended to specializing objects.

4.1.2. *Hierarchical design*

Often the objects we wish to define in a document fit into a hierarchical relationship. For example, a character may be contained within a word; that word may then be contained within a line of text; and this in turn may be within a paragraph, within a page, all of which may be part of a document. DAFS provides the ability to create "parent", "child" and "sibling" relationships between entities, forming the basis for specifying any hierarchy desired.

The use of a hierarchical structure for describing objects within a document can make the description of the document more compact. Users can leave out the layers or details that they do not need. For example, an OCR application might use the structure document-paragraph-line-word-character in processing a document image. A pitch and phase detector, working with the same image, might use the structure document-line and have no need for word and character information.

4.1.3. *Extensibility*

As with any large software system, we do not expect that any primary design will be general enough to accommodate all potential applications of DAFS. Hence, an important design goal of DAFS is to give it the greatest potential for extension. In addressing such future needs, the current specification is designed to provide a practical degree of extensibility, and to permit asynchronous revisions of data and applications to coexist.

The use of discrete objects to describe document structure permits easy extension to the set of objects available for this purpose. In another venue, the availability of an unlimited number of properties for each object permits user extensions to document descriptions. The user may classify, modify, or record information about a pre-existing document's content by creating and using new properties. (See "Properties" for additional information.)

In order to preserve the extensibility of DAFS and to maintain compliance with this specification, a process that conforms to DAFS must either interpret code values as specified, or pass these values through and not interpret them at all. A process must not change a code that it cannot interpret.

4.2. *Primary DAFS Requirements*

DAFS must meet a number of requirements in order to fulfill its purpose.

In general, DAFS must be powerful enough to serve as the format for database and tools document interchange. It was developed in the hope that at some point, the specification will develop into a standard for data interchange in the document understanding community.

Secondarily, DAFS should offer expandability, allow alternatives/possibilities (e.g., allow a supposed character's content to be labeled 'm' or 'rn' or 'iii'), allow confidence values for each alternative, so that a measured choice can be made among them, support many human languages, help tools process images rapidly, and be human readable (to allow editing on non-DAFS tools), among others.

4.3. *DAFS and Existing Standards*

The definition of DAFS has been influenced by a number of current standards. Some of these are from private companies, while others are international standards. Among the private formats are IBM's RFT:DCA, Microsoft's Rich Text Format (RTF), and Digital Equipment Corporation's CDA. International standards have been implemented by the International Organization for Standardization (ISO) and the Comité Consultatif International Télégraphique et Téléphonique (CCITT) to assist in the open exchange of documents. These standards include the Open Document Architecture (ODA), ISO 8613, and the Standard Generalized Markup Language (SGML), ISO 8879. The three of these which most influenced DAFS are SGML for overall structure and text handling, CCITT group IV for images, and ISO 10646 for Unicode.

These formats have been made available to the public and are implemented by numerous vendors. Each was created to provide a common interchange format for moving documents among heterogeneous document formatting and text publishing systems. Since they were originally conceived for encoding documents for publishing applications, none of these formats adequately addresses the problems of document decomposition and reverse encoding.

4.3.1. *DAFS and SGML*

The advantages of SGML are so strong it was decided to use it as a basis for DAFS. Some of these advantages include:

1. SGML is well-designed. Much thought has gone into what makes up a document; how to handle natural hierarchies and nesting of hierarchies; how to link document content and markup, yet be able to distinguish between the two.
2. SGML is designed to be general and to allow modification. For example, the character set can readily be changed, enabling SGML to handle documents in many foreign language scripts.
3. While by no means perfect in handling non-English text, SGML offers several reasonable alternatives. These are discussed in "DAFS-U Storage Format".
4. SGML was designed to be easy to parse. It is possible to start anywhere in an SGML document and be able to discover where you are without scanning from the beginning. Elements not recognized by the current application are easily ignored.
5. The user does not need to know everything about a document to begin using SGML. It is not necessary to know the total number of paragraphs, for example.
6. SGML is a well-known and widely used standard.

DAFS is being implemented as a special SGML application with a set Document Type Definition (DTD), but with some built-in features providing the extensibility and flexibility to cover a wide range of applications. The three main disadvantages of SGML and their solutions are discussed below.

1. SGML discourages encoding of physical characteristics. It was decided to overlook SGML inhibitions about encoding physical attributes. A DTD was created which encodes essential attributes, including physical ones like "bold" and "point size", and which allows users to add their own.
2. SGML does not handle images easily. Document decomposition applications will often require both image and text (for example, the image of a page and the corresponding OCR'd text). SGML was not designed with this need in mind. The difficulty arises from the need to distinguish intentional SGML tags from chance sequences of image bytes which, by chance, read the same. To overcome this, DAFS could pursue one of the following options:

- Escape the images. This method of handling the problem makes reading and writing tedious.
- Write the image after the final tag of the SGML data. Unfortunately, the end of an SGML file is not well defined, and not all SGML parsers would necessarily interpret this in the same way.
- Write the image in an associated external file which is referenced by the text file. Maintaining multiple files for one document can be inconvenient and is not aesthetic, but there is no question regarding which bytes are image and which are text.

The decision was to use references to external image files, arguing that the negative aspects of storing a single document across more than one file are offset by the certain knowledge of which bytes are text and which are image. Including image with text via external files is not without precedent. The CALS (Computer-aided Acquisition and Logistics Support) standards of the US Department of Defense call for just such handling of mixed text and image. Under CALS, text is to be converted to SGML and image to another format, and the two information types are stored in separate but linked files.

3. SGML applications are not necessarily portable from one DTD to another, yet different applications will require different DTD's. We are alleviating this problem by carefully constructing the DAFS DTD to be as generally applicable as possible, and by building in a limited expandability through DAFS "properties". These properties provide a way for users to create new categories of information about a document's entities, and are discussed further under "Properties".

4.4. *DAFS Tags*

The following suggestions regarding tags have guided the creation of the DAFS tagset and DTD:

1. Because DAFS aims to support many languages and scripts, DAFS will incorporate the Unicode character set. SGML (and by extension, DAFS) allows any character to be used in a tag, and the idea of specifying a defined set of tags violates the complete freedom of SGML. Nevertheless, it is highly recommended that all tag characters be the Unicode equivalent of ASCII. Limiting tag characters to ASCII helps maintain their human-readability, and eases the interconversion of DAFS-U and DAFS-A—two DAFS file types, discussed in "DAFS Storage Formats".
2. It is anticipated that new tags will be developed by users, but in the interest of portability of the resulting documents, the DAFS-defined tags should be used as much as possible.
3. High-frequency entity names should be kept short.

4.5. *DAFS Storage Formats*

DAFS has three storage formats, each designed for a different purpose, but all three are wholly interconvertible. These are the compact binary format DAFS-B (BINARY), and the "human-readable" DAFS-A (ASCII) and DAFS-U (Unicode). In DAFS-B, image and text (if any) are stored in a binary format in one file. DAFS-A is a direct application of SGML. DAFS-U is similar to DAFS-A, but modified to allow Unicode characters as content. In DAFS-A and DAFS-U, images (if any) are included in external linked files. Software libraries which enable reading and writing of all three versions have been developed and made publicly available.

4.6. *DAFS Entities*

DAFS entities are conveniently defined objects within a document such as a paragraph or word. In essence, an entity is one area of an image which is usually defined by a bounding box (though it need not be). An entity can have content, which might be the text it encompasses, properties, such as bounding box, font and point size, and hierarchical relationships with other entities, allowing specification of read orders and page layouts.

Prospective users have requested that a list of document elements and characteristics be definable under DAFS and DAFS has been structured to make each of the following DAFS entity types available.

Doc	The document as a whole.
Page	A given page.
Column	A column of text.
Paragraph	A delimited block of text comprising a paragraph.
Line	A line of text.
Newline	A newline, or carriage return.
Word	A word in the text.
Glyph	A single character in the text. "Glyph" rather than "character" is used because we are referring to an area of image which is meant to be a character, but which may not actually be correctly segmented.
Space	A Glyph whose textual content is the space character.

4.6.1. *Hierarchical relationships*

DAFS permits the creation of parent, child and sibling relationships between entities, providing easy representation of the hierarchical structures of a document. As an example, consider a paragraph entity made up of words which are in turn composed of glyphs or characters. The component glyphs are the child entities of each word, while the paragraph is the parent of each word. The other words in the paragraph are a given word's siblings.

4.6.2. *Properties*

An entity may have various attributes or "properties" associated with it. A few predefined properties are listed below.

bounding box	Rectangular box delimiting the entity or portions of it.
font class	Information on the character font (e.g. Courier or Helvetica).
point size	Size of the printed characters.
bold	Characters printed with thicker lines for emphasis.
italic	Characters printed with slanting lines for emphasis.

DAFS permits the creation of an unlimited number of user-defined properties. A property is used to describe or classify an entity and its contents, and exists only in association with the entity to which it refers. In SGML applications, such attributes are generally predefined in the DTD. DAFS introduces user-defined properties as a way for users to create their own entity categories and descriptions, without the need to alter the underlying DAFS DTD. It provides flexibility for handling a large variety of applications, yet protects the ability to share tools and data.

4.6.3. *Confidences*

Since DAFS is meant for use with document decomposition, and since there is always some uncertainty or ambiguity associated with determining exactly what the content of an entity is, DAFS must be able to assign confidence values to all its entities.

conf	Contains the confidence in the value of an entity's contents or its properties. It defaults to a single unsigned byte 0–255.
alternative set	This is a list of alternatives suggested or allowed as the content of an entity. Alternatives within an alternative set are meant to be read as either the first one or the second, and so on.

4.6.4. *Alternative sets and property ranges*

A related idea is the allowed range of values for properties. We anticipate the existence of tools using DAFS which test the effectiveness of automatic character recognizers, page decomposers, and other image understanding tools. For example, an automatic page decomposition tool might put a bounding box around a paragraph, different from the one the human creating the test set had assigned. The testing tool must determine whether the machine-set bounding box is close enough to the human-created 'ideal'. Since exact matches are not required for this type of application, the exact values of some properties may be uncertain. DAFS accommodates this need with entities which set the allowed range of properties. The property "bold" is another example. It may have an allowed range of 100–200 for font class 1. If the "boldness" of a glyph is measured as 132, this is within the range and it will be concluded that the glyph is in fact bold. For consistency, all other bold glyphs like it from font class 1 should also have a boldness of 132.

4.6.5. *Alternative contents: or and borrow*

DAFS must be able to handle alternative values for entity content. Any attempt to decompose a document will engender areas of uncertainty. A classic OCR uncertainty, for example, involves distinguishing 'I' (capital I), '1', and 'l' (lower case L). If just one of the three is selected as "most likely", the fact that the other two were very nearly as likely is lost. DAFS provides easy means of preserving and presenting sets of such alternatives. The use of alternatives is available not only for sets of characters, but for any other kind of entity as well.

The key to DAFS alternatives involves the concepts of child type and entity borrowing. An entity's children may be of "And" type, such as the component glyphs of a word, which are all meant to be presented together. "Or" type children, on the other hand, are alternatives of one another; only one of the set can be present at one time. A glyph may have "Or" children 'I', '1', and 'l', allowing a variety of techniques to be tried in selecting the best of the possibilities.

Entity borrowing is another useful device which permits easy data sharing among entities. As an example, consider a document that has a read order different from the physical order of the entities on the page. The Document could be an "Or" type entity with two child entities. The first child would arrange the Paragraphs, Words, etc. to represent the read order. The second would arrange them to represent page layout, borrowing the same images used by the read-order child, but arranging them differently. The Borrow concept allows the data to appear only once, but to be arranged and used in multiple ways. Through Borrowing, DAFS files can be more compact than would otherwise be possible.

4.6.6. *Alternative read orders and page layouts*

Documents can have multiple allowed read orders and page layouts, and it will be desirable to encode them into the document itself for the applications which use them. Automatic testers might use this information when evaluating page decomposition systems. Alternative read orders and page layouts rely on the entity borrowing capability discussed above, so that the same entities from the image of the document can be linked together in different orders.

5. Representing Textons and Specific Structure in DAFS

During the document understanding process, we must analyze and infer the specific instances of the logical and physical structures of the document. DAFS provides mechanisms which allow us to represent both the logical and physical structures, as well as alternative structures and associated confidences. Using "or" children, multiple hypotheses can be carried along in the reverse encoding process.

Recall that DAFS provides a basic entity structure, which may have a property list and data, as well as child, image and borrowed entities. The property list has both system-level and user-defined properties for information such as the entity name, relationships among children, and bounding boxes for physical entities.

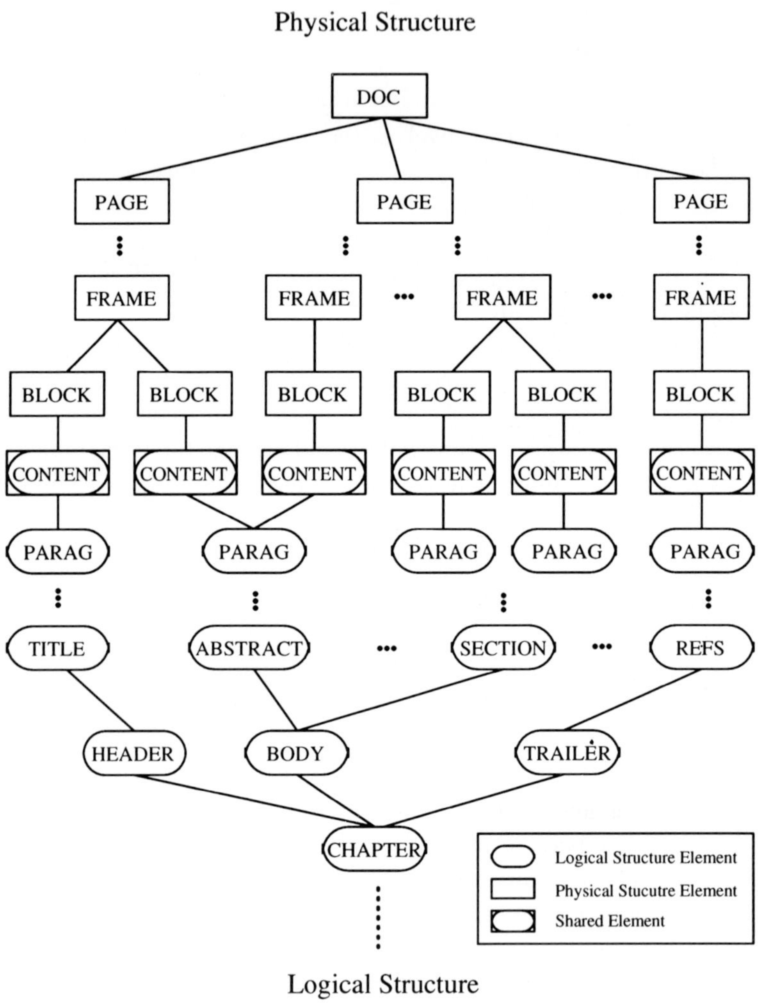

Fig. 5. A sparse schematic representation of the physical and logical structure of the content of a document "chapter".

The encoding of physical information is straightforward and was highlighted previously in the description of DAFS (Sec. 4). Each physical "object" in the document is represented by a single DAFS entity. From the generic description, each physical entity will correspond to a component of the generic physical description—Page set, Page, Frame, or Block. Frames will have content which corresponds to a list of frames and/or blocks, and blocks will have content which corresponds to a physical region on the page.

In the DAFS representation, a physical entity's *type* will correspond to a physical document component, such as column, paragraph, line, word or character. Which physical components are valid, however, is clearly dependent on the document's type.

Logical entities will similarly be organized in a hierarchical manner. Entities will be used to represent logical components and are, once again, document class dependent. For a journal article the types may correspond to a page, section, paragraph, etc., whereas for a business letter, the entity types may include the address block, greeting, and salutation, for example. Each logical entity is related to other entities with parent, child, and sibling relationships, inferred from the hierarchy. The concept of a referenced texton is implemented in DAFS by a user defined *reference* property which provides the ID of a logical (or physical) entity which it references.

There are also several special-purpose entities which we define to aid in organizing the physical and logical entity trees. The *Document Root* represents the root of the document and has two children—the *physical root*, pointing to the physical hierarchy, and the *logical root*, pointing to the logical hierarchy. The children of the physical root are simply the largest physical components of the document, e.g. pages in a journal article, page sets in a multi-volume document, etc. The children of the logical root are the maximal logical components. One component is likely to be the main body, and the remaining components are often referenced textons (or subdocuments) such as figures. Distinguishing between logical components at such a high level is necessary because of the fact that although figures are embedded in the physical document, they are disjoint (and referenced) from the main body of the document.

The terminal components for both hierarchies are characters and graphic blocks. Both hierarchies descend independently toward the leaf nodes, where they share the character and graphic components via the DAFS borrow construct. From this representation, given a logical component, one can obtain relevant physical information and vice versa.

6. Summary

The structure of a document conveys semantic information that is beyond its character string contents. To capture this additional semantics, document understanding must perform a reverse encoding of the document and relate the physical layout to the logical structure. This work has proposed a formal generic framework for the definition and interpretation of any text-intensive document's physical and logical structure, that is not restricted by the size or complexity of the document. The physical document is described by a hierarchy of frames and the logical structure of text-intensive documents is described as a hierarchy of textons. The definition of textons provides a powerful and flexible tool for document logical structure analysis. We also propose a method for determining quantitatively, in an objective, reproducible, and unbiased way, the complexity of such documents.

We have also presented a description of a new and powerful document attribute format specification, DAFS, which provides mechanisms for representing and maintaining both physical and logical information during the reverse encoding process, and have shown how it can be used to relate logical and physical structure at the content level.

Acknowledgments

The partial support of this research by the Technion VPR Fund and by the Advanced Research Projects Agency is gratefully acknowledged, as is the help of Azriel Rosenfeld in editing this chapter.

The current versions of the DAFS document, DAFS Library and Illuminator Software are available at http://documents.cfar.umd.edu.

References

[1] O.T. Akindele and A. Belaid, Page segmentation by segment tracing, *Proc. 2nd ICDAR*, Tsukuba, 1993, 341–344.

[2] N. Amamoto, S. Torigoe, and Y. Hirogaki, block segmentation and text area extraction of vertically/horizontally written document, *Proc. 2nd ICDAR*, Tsukuba, 1993, 739–742.

[3] H.S. Baird, Background structure in document images, *in Advances in Structural and Syntactic Pattern Recognition*, ed. H. Bunke (World Scientific, Singapore, 1992) 253–269.

[4] H.S. Baird, Document image defect models and their uses, *Proc. 2nd ICDAR*, Tsukuba, 1993, 62–67.

[5] D.S. Bloomberg, Multiresolution morphological approach to document image analysis, *Proc. 1st ICDAR*, Saint-Malo, 1991, 963–971.

[6] R.G. Casey and G. Nagy, Document analysis—A broader view, *Proc. 1st ICDAR*, Saint-Malo, 1991, 839–849.

[7] S. Chen and R. Haralick, An automatic algorithm for text skew estimation in document images using recursive morphological transforms, *Proc. ICIP*, Austin, TX, 1994, 139–143.

[8] Y. Chenevoy and A. Belaid, Hypothesis management for structured document recognition, *Proc. 1st ICDAR*, Saint-Malo, 1991, 121–129.

[9] A. Dengel, Initial learning of document structure, *Proc. 2nd ICDAR*, Tsukuba, 1993, 86–90.

[10] A. Dengel and L. Spitz, eds. *Proc. DAS94—Document Analysis Systems*, Kaiserslautern, Germany, 1994.

[11] D. Derrien-Peden, Frame-based system for macro-typographical structure analysis in scientific papers, *Proc. 1st ICDAR*, Saint-Malo, 1991, 311–319.

[12] D. Dori, Object-orocess analysis: Maintaining the balance between system structure and behavior," *Journ. Logic Comput.*, **5** (1995) 1–23.

[13] D. Dori, I. Phillips and R.M. Haralick, Incorporating documentation and inspection into computer integrated manufacturing: An object-process approach, in *Applications of Object-Oriented Technology in Manufacturing*, ed. S. Adiga (Chapman & Hall, London, 1995).

[14] F. Esposito, D. Malerba, G. Semeraro, An experimental page layout recognition system for office document automatic classification: An integrated approach for inductive generalization, *Proc. 10th ICPR*, Atlantic City, 1990, 557–562.

[15] J.L. Fisher, Logical structure descriptions of segmented document images, *Proc. 1st ICDAR*, Saint-Malo, 1991, 302–310.

[16] J.L. Fisher, S.C. Hinds, and D.P. D'Amato, A rule-based system for document image-segmentation, *Proc. 10th ICPR*, Atlantic City, 1990, 567–572.

[17] H. Fujisawa and Y. Nakano, A top-down approach for the analysis of documents, *Proc. IAPR Workshop on Syntactic and Structural Pattern Recognition*, Murray Hill, NJ, 1990, 113–122.

[18] X. Hao, J.T.L. Wang, and P.A. Ng, Nested segmentation: An approach for layout analysis in document classification, *Proc. 2nd ICDAR*, Tsukuba, 1993, 319–322.

[19] J. Higashino, H. Fujisawa, Y. Nakano, and M. Ejiri, A knowledge based segmentation method for document understanding, *Proc. 8th ICPR*, Paris, 1986, 745–748.

[20] S.C. Hinds, J.L. Fisher and D.P. D'Amato, A document skew detection method using run-length encoding and the hough transform, *Proc. 10th ICPR*, Atlantic City, 1990, 464–468.

[21] Y. Hirayama, A block segmentation method for document images with complicated column structures, *Proc. 2nd ICDAR*, Tsukuba, 1993, 91–94.

[22] R. Ingold and D. Armangil, A top-down document analysis method for logical structure recognition, *Proc. 1st ICDAR*, Saint-Malo, 1991, 41–49.

[23] Y. Ishitani, Document skew detection based on local region complexity, *Proc. 2nd ICDAR*, Tsukuba, 1993, 49–52.

[24] D.J. Ittner and H.S. Baird, Language-free layout analysis, *Proc. 2nd ICDAR*, Tsukuba, 1993, 336–340.

[25] B. Julesz and J.R. Bergen, Textons, the fundamental elements in preattentive vision and perception of textures, *Bell Syst. Tech. J. 62* (1983) 1619–1645.

[26] J. Kanai, T.A. Naratker, S.V. Rice, and G. Nagy, Performance metrics for document understanding systems, *Proc. 2nd ICDAR*, Tsukuba, 1993, 424–427.

[27] T. Kanungo, R.M. Haralick and I.T. Phillips, Global and local document degradation models, *Proc. 2nd ICDAR*, Tsukuba, 1993, 730–734.

[28] J. Kreich, A. Luhn, and G. Maderlechner, An experimental environment for model based document analysis, *Proc. 1st ICDAR*, Saint-Malo, 1991, 50–58.

[29] F. Lebourgeois, Z. Bublinski, and H. Emptoz, A fast and efficient method for extracting text paragraphs and graphics from unconstrained documents, *Proc. 11th ICPR*, The Hague, 1992, 272–276.

[30] G. Nagy and S.C. Seth, Hierarchical representation of optically scanned documents, *Proc. 7th ICPR*, Montreal, 1984, 347–349.

[31] G. Nagy, S.C. Seth, and S.D. Stoddard, Document analysis with an expert system, in *Pattern Recognition in Practice II*, eds. E.S. Gelsema and L.N. Kanal, (North Holland, Amsterdam, 1986) 149–159.

[32] L. O'Gorman, The Document spectrum for bottom-up page layout analysis, in *Advances In Structural and Syntactic Pattern Recognition*, ed. H. Bunke (World Scientific, Singapore, 1992) 270–279.

[33] T. Pavlidis and J. Zhou, Page segmentation by white streams, *Proc. 1st ICDAR*, Saint-

Malo, 1991, 945–953.

[34] I.T. Phillips, S. Chen, and R.M. Haralick, CD-ROM document database standard, *Proc. 2nd ICDAR*, Tsukuba, 1993, 478–483.

[35] I.T. Phillips, J. Ha, R.M. Haralick, and D. Dori, The implementation methodology for a CD-ROM english document database, *Proc. 2nd ICDAR*, Tsukuba, 1993, 484–487.

[36] T. Saitoh and T. Pavlidis, Page segmentation without rectangle assumption, *Proc. 11th ICPR*, The Hague, 1992, 277–280.

[37] T. Saitoh, M. Tachikawa, and T. Yamaai, Document image segmentation and text area ordering, *Proc. 2nd ICDAR*, Tsukuba, 1993, 323–329.

[38] Y.Y. Tang, C.Y.Suen, C.D. Yan, and M. Cheriet, Document analysis and understanding: A brief survey, *Proc. 1st ICDAR*, Saint-Malo, 1991, 17–31.

[39] S. Tsujimoto and H. Asada, Understanding multi-articled documents, *Proc. 10th ICPR*, Atlantic City, 1990, 551–556.

[40] F.M. Wahl, K.Y. Wong, and R.G. Casey, Block segmentation and text extraction in mixed text/image documents, *Computer Graphics Image Proces. 20* (1982) 375–390.

[41] A. Yamashita, T. Amasno, H. Takahashi, and K. Toyokawa, A model based layout understanding method for the document recognition system, *Proc. 1st ICDAR*, Saint-Malo, 1991, 130–138.

[42] C.L. Yu, Y.Y. Tang and C.Y. Suen, Document architecture language (DAL) approach to document processing, *Proc. 2nd ICDAR*, Tsukuba, 1993, 103–106.

Handbook of Character Recognition and Document Image Analysis, pp. 457–484
Eds. H. Bunke and P. S. P. Wang

CHAPTER 17

INTERPRETATION OF ENGINEERING DRAWINGS

KARL TOMBRE
INRIA Lorraine & CRIN/CNRS, Bâtiment LORIA, 615 rue du jardin botanique
54602 Villers-lès-Nancy CEDEX, France

and

DOV DORI
Faculty of Industrial Engineering and Management, Technion
Haifa 32000, Israel

Technical documents are documents of some science or engineering domain, in which graphics provide the basis for the information conveyed by the document. Engineering drawings in general and mechanical engineering drawings in particular are a typical example. The interpretation of such drawings entails converting them into some kind of CAD representation, which has a high-level meaning, including 3D structure. In this chapter, we review the different methods and techniques used to achieve this goal, which is still an active research topic. After presenting the general framework in which such a system must work, we introduce the three levels of interpretation: lexical, syntactic and semantic. We give an overview of the low- and intermediate-level tools used in this context: vectorization, recognition of arcs, hatching and dashed lines, etc. At the syntactic level, we describe the task of recognizing annotations in general and dimensioning in particular. We also elaborate on the functional analysis that is possible on each view. Finally, we describe the available techniques and the open problems for 3D reconstruction from several views.

Keywords: Engineering drawings interpretation; Vectorization; Raster-to-vector conversion; Dimensioning; Annotations; CAD; Lexical analysis; Syntactic analysis; Functional analysis; 3D reconstruction.

1. Needs and Motivations

1.1. *General Context*

An engineering drawing is a graphic product definition. Prior to the introduction of CAD/CAM systems, an engineering paper drawing model was the major means of design; numerous mechanical engineering drawings of operational products still exist only in this form, and are needed for maintenance of existing systems and/or as a basis for designing the next generation of the product the model of which they describe. Paper drawings resulting from subcontractors' CAD systems which are not used by the organization that ordered them are also very common.

Automated understanding of engineering drawings within a project design and documentation environment provides for the possibility of converting both paper and electronic drawings into a CAD representation. Engineering drawings are the main means of communication among professionals involved in the project execution process. An object-oriented analysis of the design and documentation system, based on a large scale project case study, underlines the central role played by drawings and the duality of paper and electronic engineering drawings [1].

In CAD representations, some meaning is originally attached to each graphic primitive; thus they can be directly used for CAM. On the other hand, to be meaningful, drawings in both paper and electronic media must be correctly interpreted by a professional or an intelligent system. Drawing interpretation is proposed to eliminate most of the tedious human labor associated with digitizing and assigning meaning to drawings. If a paper drawing is to be converted into CAD, a total rework of the existing design must be carried out. Although commercial products in this field alleviate the burden of low level processing, such as vectorization, this is still a time-consuming, error-prone and unproductive process, which nevertheless requires skilled personnel. Since this interactive, human-intensive labor can only be done by costly trained professionals, it is frequently avoided and is done very selectively. This, in effect, amounts to ignoring many valuable design man-years put on paper of already "debugged" products.

As pointed out in [2], in many cases, stores of paper drawings are still growing faster than those of CAD model files, and the conversion problem will be topical for at least the next decade. Moreover, most experts agree that the "paperless society" dream of the eighties is becoming more remote with each newly developed type of printer and plotter and with each decline in their prices. Humans find paper just too convenient to give it up for a "soft" screen image which cannot be easily folded and carried along. This reality is equally true for textual and technical documents. However, the significant recent progress made in scanning and high-volume secondary storage technology has made electronic archiving of paper drawings an economically viable option. This new possibility underscores the absence of an automatic capability to intelligently process these digital images of drawings so that they can be incorporated into a CAD database.

Mechanical design and manufacturing information for 3D solid objects has been effectively conveyed through sets of engineering drawings — annotated orthographic projections and optional cross sections. The inverse problem involves the 3D reconstruction of given line drawings of multiple views of an object. The line drawings are represented as scanned, binary images, and the objects themselves can have surfaces that are either planar, spherical or cylindrical.

Until the late eighties, this problem was largely considered as a geometry problem. More recently, it has begun to be addressed more realistically as a computer vision problem or, more specifically, as a document understanding problem, where the line drawings are represented as noisy, scanned images rather than symbolically. Here, the vertices and lines are not represented explicitly. In fact, some of them might be missing, while extraneous ones might be randomly present due to noise, caused by folding of the paper, stains, photocopier noise, etc. Some initial image processing for noise cleaning and enhancement is typically performed, followed by primitive extraction. These primitives are then combined to constitute larger patterns which, in turn form the multiple views that are combined together to form a final interpretation of the 3D object. The ultimate goal is to be able to represent the interpreted 3D solid in a 3D CAD format of a particular CAD system or a neutral CAD file format, such as IGES.

There are several reasons for the growing interest in the problem. Companies which now use CAD systems may have large archives of paper drawings. An object oriented analysis of a case study involving a power plant construction project is described in [1]. If these archives are to be retrieved for inclusion in the company's information system, the paper documents must somehow be converted to the appropriate CAD representation. Once a drawing is in CAD format, all the advantages of database storage, retrieval and query become available. The evolution of CAD systems from low-level, basic geometric representations to higher-level models also requires document analysis capabilities, even when no paper drawings are involved. Companies which have been working with computer-aided drafting systems, or with CAD systems that only handle low-level primitives, would like to convert their data to higher level representations, in order to incorporate them into modern, feature-based type CAD systems. For this purpose, a CAD conversion system should also be able to recognize machinable manufacturing features from some low-level representation.

To put technical drawing interpretation in an economic perspective, we quote some excerpts from [3]:

> *There are approximately 3.5 billion engineering drawings of various types in the United States and Canada [alone], with about 26 million new ones added each year. The annual cost of filing, copying, accessing, and preparing these drawings for distribution exceeds one billion dollars. [...] A major advantage of CAD systems is that [...] the time needed to modify a drawing is typically 13% to 33% of the time needed to accomplish the same revision using paper-and-pencil techniques. [...] Only 13% (!) of the existing, active drawings are available in CAD form. In 1990, about 20% of the drawings produced in the U.S. were created in CAD form and about 25% were CAD revisions of older drawings. The remaining 55% were done [on the drawing board, using] traditional paper-and-pencil drafting techniques. Cost-benefit analyses show that if a drawing is expected to be modified several times, it is advantageous to convert [it] from hard copy to electronic format.*

1.2. *Commercial Systems and Their Limitations*

Several commercial products have tried to meet the demand for automating engineering drawings (see for instance the overview in [4]). A number of those systems have achieved satisfactory results for low-level tasks, such as vectorization, text/graphics separation (when no text touches the graphics), and symbol recognition through template matching techniques. However, their performance is still limited for various reasons:

- The vectorization techniques, usually based on supervised skeletonizing or some other medial-axis approach (see Sec. 3.3), perform fairly well in general, but they are disturbed by noise and by inherent limitations stemming from discretization and the lack of "intelligence" in the process.
- Most existing text/graphics separation methods are strongly affected by text strings touching the graphics, as they usually rely on connected components analysis.
- Template-matching methods are not sufficient for the recognition of complex symbols, which may vary in pose and scale and which sometimes are labeled with text strings.

Due to these limitations, many systems include a vector editor for manual correction of the results of the raster-to-vector conversion. This manual editing has often been deemed too time-consuming by customers of commercial systems. To overcome this, another class of commercial solutions has been proposed. These are semi-automated systems, where the user, guided by the scanned raster image of the document and by a set of simple tools, such as line tracking, inputs the CAD model by himself/herself. The amount of domain-dependent knowledge in these systems is at best very limited. They usually stop short of applying such knowledge, which could have raised them beyond the basic lexical, or primitive recognition level.

1.3. *Is High-level CAD Conversion Achievable?*

As we have already noted, drawings in both paper and electronic (i.e. raster images of paper drawings) media are contrasted with CAD representations, in that the latter attach semantic meaning to graphic entities, enabling them to be manipulated, redesigned and to serve as a basis for Computer Aided Manufacturing (CAM). Non CAD-based drawings have no such inherent semantics; to be converted into CAD, they must be correctly understood and interpreted, either by a human or by a machine. In addition, complete CAD systems know about, and are able to handle, 3D objects and not only 2D views; true CAD conversion from paper should therefore also include the reconstruction of 3D models from 2D views.

The optimal situation in this arena would be when a roll of paper containing a complex design of some part to be manufactured can be considered equivalent to a file in some standard CAD format. The drawing usually conforms to ISO, ANSI, or some other standard. It may also be a mixture of standards and contain "shortcuts" that can be easily understood by humans but challenges any "reasonable" automated system. By scanning an acceptable quality drawing of this kind and running the resulting raster file through the set of drawing understanding algorithms, it should be ultimately possible to come up with a complete interpretation. At least theoretically, if humans can do it, so too can computers. In the area of OCR, such a situation is not as remote as it may be in the area of engineering drawing understanding. Two basic differences between drawing understanding and OCR are the following:

1. The input to OCR is a printed page, which is normally structured in the sense that text is expected to be aligned along lines and columns, with equal spacing and some expected font size and shape. Drawings, on the other hand, contain a complex web of graphics—geometry and annotation—interleaved with text at various locations and orientations. Therefore, the structure of printed documents cannot be assumed for drawings.
2. While in OCR only lexical level recognition is required, for drawings we demand also syntactic and semantic understanding. In other words, OCR is required to convert a printed page into an equivalent computer (e.g. ASCII) representation, but not to understand the content of the text or even to check its syntactic correctness. For drawings we demand that what people understand from the drawing, namely a mental picture of the 3D object, be reconstructed by the computer-based system. Such requirement is equivalent to demanding that OCR be able to perform natural language understanding and perhaps even logical inference from the printed text. This underscores the big gap between the two seemingly similar technologies and may hint at the underlying difficulties facing engineering document understanding.

The situation as described here may make us wonder whether the ultimate 3D drawing understanding is at all an achievable goal. In our opinion, the answer to this question is that although we recognize the difficulties associated with full and complete understanding of complex drawings, we do have to keep in mind this remote goal and approach stepwise towards achieving it. Without such ultimate goals, we are doomed to remain not far above the level of pixel-based operations that characterize current commercial systems.

1.4. *About This Chapter*

In this chapter, we argue that it is possible to build interpretation systems which are capable of converting the original paper drawings into high-level CAD databases. After having presented in this section the general framework in which such a system must work, we give an overview of the low- and intermediate-level tools used in this context: vectorization, recognition of arcs, hatching and dashed lines, etc. Section 4 describes the important task of recognizing annotations in the drawing, emphasizing especially dimensioning, which is the core of annotation. Section 5 discusses the type of analysis that can be applied to individual views, and Sec. 6 describes the available techniques and the open problems for 3D reconstruction from several views.

This chapter focuses on mechanical engineering drawings. Other visually-oriented technical documents, including cadastral maps, plant layouts, facilities, gas, telephone, electricity and cable TV utilities, flow charts, electric and electronic diagrams, and civil engineering charts (of roads, highways, bridges...), have distinct syntax and semantics that differ from one domain to another. The degree of domain specialization increases along with the height of the level of document understanding. However, many of the underlying principles presented in this chapter, when appropriately modified as needed, are applicable to other domains of technical document understanding.

2. The Phases of Drawing Understanding

A complete drawing understanding process can be roughly broken up into three phases. Any drawing understanding system consists of one or more of the following three phases:

1. **Lexical Phase**: This early phase starts with noise reduction and is mainly concerned with primitive recognition. Here, the basic structural constituents of the engineering drawing are recognized. We describe the basic tools used for this phase in Sec. 3.
2. **Syntactic Phase**: Here, drafting rules and standards are embedded in a grammar or a rule-based system and are used to check the syntactic correctness of the drawing. Correctness is checked with respect to syntax, but not with respect to feasibility of the drawn object. A typical problem may be that although the dimensioning follows the correct syntax, it may not make sense, but this will not be discovered until the semantic phase. Syntactic analysis of mechanical engineering drawings focuses on the annotations layer, which we describe in Sec. 4.
3. **Semantic Phase**: Here 2D and 3D understanding of the object(s) described in the drawing takes place. Among other things, the process verifies that the drawing represents a feasible object and that the dimensioning defines the object unambiguously. Kinematic analysis may also be applied at this stage. An example of kinematic and of functional analysis is described in Sec. 5.3.

3. Lexical Analysis

3.1. *Noise Sources and Noise Filtering*

Images of real engineering drawings are characterized by the following features:

- They are in raster format resulting from scanning. Hence, information about the vertices, lines and faces is not explicit.
- Scanned images are inherently noisy. Lines in the image may be broken and random dots and lines may appear throughout the image plane due to accidental folding, wrinkles, stains, repeated photocopying of blueprints, aging of the paper and ink, and other reasons.

There is a variety of sources of noise, including the following:

- *Degradation of the paper medium*: Originals and photocopies tend to fade over time. Their usage makes them stained, worn and torn and they may be glued by tape. The noise introduced may be in the form of straight lines (folding, tapes, etc.), arcs (coffee mugs), and other irregular shapes (e.g., grease stains). The straight line noise requires a different treatment than the irregular shape noise. While the latter can be treated by adaptive thresholding, wires (lines and arcs) are hard to remove automatically. They can either be removed by a human interacting with the system at a preprocessing phase or automatically, at a later stage (no earlier than the syntactic phase). Only at this stage can the system infer that those wires are redundant, as they can be part of neither the geometry nor the annotation, and even then it is quite a risky operation, as we rely on the system to make decisions that require intelligence.

- *Degradation due to photocopying and scanning*: The photocopying and scanning processes degrade the document further. Photocopying introduces such noise as blurring and speckles, while the scanning process introduces quantization noise, wrong thresholding noise, skew, and random pixel noise.
- *"Logical" noise*: Geometry wires, dimension-sets, or parts thereof may be broken, or at least partially missing in the drawing. They can be easily restored mentally and graphically by humans, but this operation, like that of removing extraneous redundant wires, discussed above, is hard for machines to perform, due to the high-level syntactic and semantic considerations that are required.

3.2. *The Basic Primitives*

Having dealt with noise and reduced it as much as possible, the first phase in drawing understanding is the lexical analysis phase. Analogous to the operation of a compiler for high-level programming languages, the aim of this phase is to segment tokens, referred to as *primitives*, from the raster image. Mechanical engineering drawings contain four basic primitives: bars (straight line segments with non-zero width), arcs, arrowheads, and textboxes. Bars and arcs have many attributes in common, and are therefore generically called *wires*. A description of the method used to detect wires and arrowheads within the Machine Drawing Understanding System can be found in [5], while [6] describes the extraction of textboxes from engineering drawings.

Primary detection of wires is domain independent, but it may be aided by higher level phases. For example, if an arrowhead is found, the dimensioning grammar tells us that it must point at a wire. Hence, we should search for a wire through which the tip of the detected arrowhead passes. If no such wire is found, then either the segmented arrowhead is a false alarm (something that may just look like an arrowhead), or more relaxed parameters (such as the distance between the computed arrowhead tip and the wire's medial axis) should be employed. Likewise, if a textbox is found, then if it is a dimensioning textbox, it should pass next to a wire which is the tail of a leader (an arrow). These two examples demonstrate the inter-dependency of the relative location of the primitives with respect to each other. Formulated as constraints, these relations may be helpful in determining whether or not certain primitives have been correctly identified. This implies that the recognition and understanding process cannot be linear and unidirectional. Rather, iterations and backtracking from advanced to earlier phases should be planned to enhance the recognition rate.

3.3. *Vectorization and Bar Detection*

Bars are the most prominent constituents in most engineering drawings. They appear in the raster image as elongated rectangles of black pixels, usually with noisy edges. *Vectorization*, or recognition of bars (straight line segments) in engineering drawings is a preliminary stage for higher level interpretation. It is the main problem in low-level processing. The task calls for massive processing of voluminous scanned raster files. This is a heavy undertaking if each pixel is to be addressed at least once, as required by the Hough Transform or thinning-based methods. Nevertheless, many vectorization algorithms are designed to operate on a skeleton of pixels. This implies that some thinning algorithm must be applied to the raster image before vectorization can take place. Since thinning is normally a multi-pass process that examines every pixel at least once, it is by nature a computationally intensive process. To do the segmentation, still further processing is needed.

Most segmentation algorithms get a skeleton as an input, and find a subset of the skeleton's pixels. *Polygonal approximation* is a simple and convenient way for the representation of digital curves. Its advantages over other curve representation schemes, like higher order splines and other functional representations, are its computational simplicity, coding efficiency, good preservation of local properties, and the low complexity of feature extraction.

There are three basic algorithmic approaches to polygonal approximation of digital curves, yielding polygonal vertices which are subsets of the curve points: Merge, Split, and Split and Merge [7]. Merge algorithms are based on linear scanning of the digital curve, where at each point of the curve a decision is made whether the point should be merged with the previous set of points and belong to the same vector, or be the first point in a new vector. Most of the algorithms use this approach, including [8–12]. The subset of pixels from the digital curve is defined such that the polygonal approximation to the path formed by connecting each pair of consecutive pixels is not farther than ϵ from any of the skeleton's pixels. The definition of ϵ differs from one segmentation algorithm to another, leading to different segmentations of the same skeleton. Reumann and Witkam [13] propose a stripe algorithm, in which a segment is continued until a pixel is found to lie in a distance larger than a parameter δ from the centerline of the stripe. Sklansky and Gonzalez [9] define a uniform Euclidean norm ϵ and keep adding pixels to the string approximated by the curve as long as the maximal distance does not exceed ϵ. Wall and Danielsson [14] propose a fast algorithm based on the algebraic area between the curve and the approximating segment. Among the more advanced Merge algorithms that take global considerations into account are [15], which iteratively reduces the number of vertices and converges to a local minimum, and [16], which finds a polygon with a minimal number of vertices using dynamic programming.

Dunham [17] also defines ϵ and finds the minimal number of segments that approximate a given skeleton using a recursive function. Yuan and Suen [18] devise an algorithm for detecting straight lines from chain codes based on a quantization scheme. Around pixels, a passing area is constructed so that a line formed by these pixels must pass through this area if it is straight.

In real engineering drawings lines are usually more than one pixel wide. This is due to the relatively high resolution needed for the various recognition tasks. Many existing vectorization methods initially apply either morphological operations on the input raster file or some variant of the Hough transform. Kasturi's group [19], for example, has employed the thinning algorithm specified in [20] as a first step in line extraction, and a Hough-based method for locating text strings. Other vectorization techniques involve a combination of thinning with non-thinning methods or run length codes [21]. Line tracking after preprocessing and thinning is employed by Fahn *et al.* [22] and by Nagasamy and Langrana [23]. Other techniques avoid thinning by doing window following [24], by subsampling the raster image into meshes [25, 26], by using gray level data, by analysis of the strokes parallel to the line direction, etc. Most of these operations are computationally intensive, because each pixel in the image must be inspected at least once.

The Hough Transform, which also addresses each pixel, needs extra memory and processing to handle the multi-dimensional accumulator array and to determine the edges and widths of the detected lines. Width is an important parameter in engineering drawings, because drafting standards require that geometry lines be thicker than annotation ones, hence it may provide an important criterion for separating the geometry from the annotation in the drawing.

The Machine Drawing Understanding System (MDUS) applies a specially designed algorithm, called Orthogonal Zig-Zag, or OZZ for short, to carry out the bar detection task [27]. The underlying idea of OZZ is inspired by a light beam conducted by an optic fiber: a one-pixel-wide "ray" travels through a black pixel area designating a bar, as if it were a conducting pipe. The ray's trajectory is parallel to the drawing axes, and its course zig-zags orthogonally, changing direction by 90° each time a white area is encountered. Accumulated statistics about the two sets of black run-lengths gathered along the way provide data for deciding about the presence of a bar, its endpoints and its width, and enable skipping junctions.

We have found both theoretically and empirically that the sparse-pixel approach of OZZ is better for extracting bars from engineering drawings than the Hough Transform (HT). Experiments resulted in about twenty-fold reduction in both space and time complexity compared to HT, while the recognition quality was found to be about 40% higher.

3.4. *Arc Segmentation*

Existing methods for arc segmentation can be divided into two groups. The first group includes methods to detect circular objects that are based on the Hough Transform, while the second group is motivated by the need to recognize objects in a scene. Algorithms in the first group attempt to locate circles in (possibly noisy) images by a certain transformation. Algorithms in the second group are primarily concerned with the extraction of meaningful features from objects by estimating their edge curvature. The features extracted from the object contour can be used for object classification and recognition.

HT is normally used for arc segmentation in the case of isolated points that potentially lie on circles or circular arcs. A circle can be described by $(x-a)^2+(y-b)^2=r^2$, where (a, b) is the circle center and r is its radius. HT uses this equation to map each (x, y) image point into all parameter points which lie on the surface of an inverted right angled cone, whose apex is at $(x, y, 0)$. The circle parameters a, b and r are identified by the intersection of many conic surfaces. This is done by examining the peaks in a 3D accumulator array. This 3D accumulator peak detection is in general too costly to be applicable. To reduce the dimensionality of the problem from 3 to 2, the Adaptive Hough Transform (AHT) method [28] was developed.

To locate the circle center, AHT incorporates the constraint that the vectors, which are normal to the circle boundary, must all intersect at the circle center (a, b). The intersection of many of these lines indicates the circle center in the image. The radius of the circle can then be found by histogramming $r^2 = (x - a)^2 + (y - b)^2$ and locating the largest peak in the r^2 histogram.

Although AHT converts circle detection into 2D peak detection, it still requires pixel-by-pixel image operations and demands a large memory space to accumulate the frequency of each circle parameter. Locating peaks in accumulators is also a heavy task. To make HT more applicable, other works, including [29–31], have attempted to present optimized variations. However, since they are derived from the original HT, they also require considerable time and space. For large engineering drawings, this is prohibitively slow and expensive.

Other arc segmentation methods belong to the curvature estimation group. Motivated by object recognition, the aim of these algorithms is to extract meaningful features from objects by estimating their edge curvature. Whereas Hough-based methods can handle isolated points, the common requirement of curvature estimation algorithms is that the input be a digital curve, i.e., a one-pixel-wide line. Asada and Brady [32] propose the curvature primal sketch to represent significant changes in curvature along the bounding contour of a planar shape. They define a set of parametrized primitive curvature discontinuities and derive expressions for their convolution with the first and second derivatives of a Gaussian. They interpret the significant changes in curvature at various scales. The input is a bounding contour of the object to be recognized, while the output may be a semantic network or a filtered response graph.

Rosin and West [33] describe a method of segmenting curves in images into a combination of circular arcs and straight lines. It finds the best fit of a combination of arcs and straight lines to the data. The input to the algorithm is a digital line, hence the image must be preprocessed by an edge detector to extract connected pixels or use boundary descriptions from chain-coded binary images.

O'Gorman [34] proposes to carry out feature extraction by estimating the curvature along digital lines and determining the features from the curvature plots. For the difference of slopes (DOS) approach, curvature at a point is estimated as the angular difference between the slopes of two line segments fit to the data before and after the point. After finding the curvature of each point in the data, the curvature plot is usually smoothed to reduce noise. The specialized difference of slopes method DOS+ proposes to use a non-zero, small positive gap between the two segments that estimate the lines between which the angle is measured. DOS+ has been shown to be effective for estimating curvature in terms of signal detectability.

On top of their computational complexity, arc segmentation algorithms are not suitable for detecting arcs with small angles, because the peaks that arcs with small angles produce are not high enough to distinguish them from noise. Detected arcs require postprocessing to define their endpoints, because Hough-based algorithms define only the arc center and radius. Finally, the width of each detected arc has to be somehow related to the width of the signal in the transform domain. This does not seem to be a trivial problem, and it is relevant only if the input to the Hough-based algorithm is not a result of a contour extraction preprocessing (thinning or edge detection) which causes all contours and lines to lose their original width and become one-pixel wide.

Curvature estimation methods assume that the input is a digital line, implying that the image must undergo some contour extraction preprocessing. Not only does this prerequisite put a heavy computational burden on a system, but it also causes line drawings to lose important information about their original line widths. Since curvature estimation methods are initially driven by the need to extract features that emerge from the combination of straight lines and circular arcs, they are not even concerned with the extraction of the more basic arc parameters center and radius, which Hough-based methods attempt to determine. In summary, existing arc segmentation methods do not provide adequate means for extracting arcs from engineering drawings. First, none of the methods detects line width. Second, their main focus is either on detecting circles or wide-angle arcs from noisy images or extract features from curvature estimation for object recognition.

The motivation of the arc segmentation implemented by Dori's group in MDUS is to explicitly define where arcs exist in the drawing and determine their parameters, including angle (which may vary from several to 360 degrees), radius, center, endpoints, and, last but not least, line width. Line width is very important in higher-level drawing understanding, because geometry lines are required by drafting standards (ISO and ANSI) to be twice as thick as annotation lines. Rather than using one of the existing methods, MDUS takes advantage of the knowledge about the bars detected in the previous Bar Detection phase of MDUS. Some of these bars, which are linear approximations of arcs, provide the basis for an arc segmentation algorithm—Perpendicular Bisector Tracing (PBT)—that is both effective and efficient [35].

As noted, bars are detected by the OZZ vectorization algorithm. Some of the detected bars are linear approximations of arcs. As such, they provide the basis for arc segmentation. An arc is detected by finding a chain of bars and a triplet of points along the chain. The arc center is first approximated as the center of mass of the triangle formed by the intersection of the three perpendicular bisectors of the chords these three points define. The location of the center is refined by recursively finding more such triplets and converging to within no more than a few pixels from the actual arc center after two or three iterations. The high performance of the algorithm is due to the fact that it avoids both raster-to-vector and massive pixel-level operations, as well as any space transformations.

3.5. *Arrowheads, Textboxes, and Other Graphic Elements*

Arrowheads are important anchoring elements in mechanical engineering drawings. There is a variety of arrowhead shapes, but the most frequently used are the filled triangles. The two parameters—height and base length of these triangles—may vary from one drawing to another, but they may be assumed to be about uniform for dimension-sets within the same drawing. Larger triangles are normally used as part of the cross section symbol. In MDUS, arrowheads are detected using the Supervised Arrowhead Recognition algorithm [5]. The basic idea of this algorithm is to first learn the arrowhead parameters by searching arrowheads in slots located at the edges of bars. Once enough arrowheads have been found to determine the parameter values with confidence, a more thorough search is carried out within slots located at the edges of all the wires throughout the drawing plane. The size of the slot is set using the parameter values found in the parameter learning phase.

Several higher-level graphic entities are composed of the extracted primitives, mainly bars. The two most important ones are dashed wires and hatched areas. Dashed wires are dashed lines and arcs. There are two main types of dashed lines: the first consists of basically equal dashes while the second is a dash-dotted line. The equal dash line denotes hidden lines, while the dash-dotted one is used to mark symmetry axes. In manually prepared drawings, the dashes of the dash-dotted lines are often very long and few, and the "dots" are shorter dashes. Such lines may contain as few as two dashes and one "dot." This explains why it is sometimes difficult to detect these lines. On the other hand, their recognition may be instrumental, as it adds knowledge about symmetry. This knowledge can be helpful in the syntactic phase for determining and/or validating the existence of primitives around symmetry axes in the corresponding 2D views (see Sec. 5.1).

Dotted circles are also of the same two types. Equal dash circles denote hidden circles, while dash-dotted circles denote distribution circles, i.e., circles on which centers of holes are located, such that the holes are equally located around the circle. Recognition of such circles may be quite difficult, since it entails recognizing small angle arcs that are very similar to short bars.

Hatched areas denote cross sections in matter and can be traced by locating bars that have about the same slope and are about equally spaced.

4. Annotations and Dimension-Sets

Having found the sets of the different primitives in the drawing, the next major phase is *annotation*, which can be divided into two major categories: dimensioning annotation and non-dimensioning annotation.

Dimensioning comprises of a set of dimension-sets, each of which denotes the measure (length or angle) between two sites (geometry wires) in the object. Understanding dimensioning annotation has been addressed by Dori *et al.* [5], Pao *et al.* [36], Collin [37], and Lai and Kasturi [38]. Due to the proper dimensioning requirement, the number of dimension-sets is proportional to the complexity of the object, as expressed by the number of its faces. The more complex the object is, the more dimension-sets are needed to determine it. Since the white space on the drawing paper (and screen, for that matter) is limited, each additional dimension-set complicates the drawing in an increasing marginal fashion, since it interferes with the ones already drawn and cannot overlap or clutter any of the existing ones.

Any dimension-set can be characterized by a set of seven attributes: Standard, Group, Type, referencing, leading, pointing, and texting. Formally, the dimension-set is described by a *descriptor*—a seven parameter function

$$descriptor(Standard, Group, Type, referencing, leading, pointing, texting),$$

symbolized as $\delta(\Sigma, \Gamma, \mathcal{T}, \rho, \lambda, \pi, \xi)$.

Each one of these seven attributes pertains to a specific feature of the dimension-set and has an enumerated domain of legal values. The attributes, along with their legal values, are listed as follows:

1. **Standard**, Σ, specifies the Dimensioning and Tolerancing standard according to which the dimension-set (and the whole drawing) is drawn. Its domain is $\Sigma = \{A, I\}$ for ANSI and ISO, respectively. We restrict the discussion to ANSI Standard, hence the standard attribute value is set to A.

2. **Group**, Γ, specifies the group of dimension-set types (see next attribute) of the dimensioned feature. Its domain is $\Gamma = \{C, R, S\}$ for *Cardinal*, *Revolution*, and *Special*.

3. **Type**, $\mathcal{T}$, reflects the nature of the feature being dimensioned. There are nine dimension-set types, classified into three groups:

   (a) linear **L** and angular **N** – the cardinal group,

   (b) cylinder **C**, diameter **D**, and radius **R** – the revolution group, and

   (c) chamfer **H**, ordinate **O**, point **P**, and arc **A** – the special group.

4. **Referencing** specifies the combination of the reference component (the object pointed to by the arrowhead) in the dimension-set. Its domain is $\rho = \{s, c, m\}$, standing respectively for *simple*, when the reference consists of only a geometry, *compound*, when the reference consists of a witness and a geometry, and *mixed*, when any one of the two references may be either simple or compound.

5. **Leading** specifies the number of leaders in the dimension-set. Its domain is $\lambda = \{b, u\}$ for *binary* (two leaders) and *unary* (one leader), respectively.

6. **Pointing** specifies the pointing direction of the leaders relative to themselves. Its domain is $\pi = \{t, a, p\}$, standing respectively for *towards* the dimension-set center (the point in the middle between the arrowhead tips), *away* from the dimension-set center, and *perpendicular* to the direction from the arrowhead tip to the dimension-set center.

7. **Texting** specifies the location of the text relative to the leader pair. Its domain is $\xi = \{g, d\}$, standing respectively for *along*, when the text is located along the leader pair, and *outside*, when the text is located outside of the leader pair.

The first three attributes, Standard, Group, and Type, have to do with the *meaning* of the dimension-set, and are therefore called *semantic* attributes. The rest of the attributes—referencing, leading, pointing, and texting—deal with the dimension-set's *form*, and are therefore called *syntactic* attributes.

The number of legal values of the domain, $|\delta|$, is the Cartesian product of the cardinalities of the seven individual attributes:

$$|\delta| = |\Sigma| \cdot |\Gamma| \cdot |\mathcal{T}| \cdot |\rho| \cdot |\lambda| \cdot |\pi| \cdot |\xi| = 1944$$

This big number is fortunately only a theoretical upper limit. Restricting our discussion to the ANSI Dimensioning and Tolerancing standard, out of 1944/2=972 possible combinations in the upper limit, the dimensioning language, as described by the standard, allows

Table 1. The 28 legal combinations of ANSI dimension-set attributes.

<table>
<tr><th>Attribute Symbol
V_1=Attribute $Value_1$;
V_2=Attribute $Value_2$;</th><th colspan="28">Attribute Value</th></tr>
<tr><td>Standard Σ
A=ANSI;
I=ISO</td><td colspan="28">A</td></tr>
<tr><td>Group Γ
R=Revolution;
C=Cardinal; S=Special</td><td colspan="10">C</td><td colspan="14">R</td><td colspan="4">S</td></tr>
<tr><td>Type $\mathcal{T}$
L=Linear; N=Angular;
D=Diameter; C=Cylinder;
R=Radius; O=Ordinate;
H=Chamfer; P=Point; A=Arc</td><td colspan="6">L</td><td colspan="4">N</td><td colspan="7">D</td><td colspan="4">C</td><td colspan="3">R</td><td>O</td><td>H</td><td>P</td><td>A</td></tr>
<tr><td>Referencing ρ
c=compound;
s=simple; m=mixed</td><td colspan="4">m</td><td colspan="2">c</td><td colspan="4">c</td><td colspan="3">s</td><td colspan="4">c</td><td colspan="4">c</td><td colspan="3">s</td><td>s</td><td>c</td><td>s</td><td>c</td></tr>
<tr><td>Leading λ
b=binary;
u=unary</td><td colspan="4">b</td><td colspan="2">u</td><td colspan="4">b</td><td>u</td><td colspan="2">b</td><td colspan="4">b</td><td colspan="4">b</td><td colspan="3">u</td><td>b</td><td>u</td><td>u</td><td>b</td></tr>
<tr><td>Pointing π
t=towards; a=away;
p=perpendicular</td><td colspan="2">a</td><td colspan="2">t</td><td colspan="2">a</td><td colspan="2">a</td><td colspan="2">t</td><td>t</td><td>t</td><td>a</td><td colspan="2">a</td><td colspan="2">t</td><td colspan="2">a</td><td colspan="2">t</td><td colspan="2">a</td><td>t</td><td>p</td><td>t</td><td>t</td><td>a</td></tr>
<tr><td>Texting ξ
g=along;
d=outside</td><td>g</td><td>d</td><td>g</td><td>d</td><td>g</td><td>d</td><td>g</td><td>d</td><td>g</td><td>d</td><td>d</td><td>d</td><td>d</td><td>g</td><td>d</td><td>g</td><td>d</td><td>g</td><td>d</td><td>g</td><td>d</td><td>g</td><td>d</td><td>d</td><td>d</td><td>d</td><td>d</td><td>g</td></tr>
</table>

only 28 combinations enumerated in Table 1. This is just 2.88% of this upper limit of 972 for the ANSI standard.

Non-dimensioning annotation includes symbols for surface quality, welding, threading, bearing, textual manufacturing and finishing instructions, table of hole center coordinates for drilling, etc. A system that attempts to tackle real-life drawings should address these annotations and understand them at the highest level possible. This type of annotation is superimposed alongside the dimensioning annotation and should not interfere with it.

5. 2D Analysis

A human examining an engineering drawing, even with no prior knowledge about draftsmanship, would normally recognize elements such as line segments of different width, circular arcs, annotation text, hatched areas, etc. These elements are in fact the most basic structures in a technical drawing; it is therefore not surprising that they are also the first objects sought by automated systems, as discussed in the previous sections. But technical draftsmanship entails much more than just putting together some vectors and other graphical primitives; it is based on the mathematical properties of orthographic projections and on rigorous standards for representing non-ambiguously all the information necessary for the whole sequence of people working with it, from engineers in the design department to the machine-tool operators at the shop floor. Even when a single view is concerned, an engineer "sees" much more than just graphical primitives. We refer to 2D analysis of engineering drawings as the interpretation process which can be applied to a single 2D view in the drawing.

5.1. *Structural and Syntactic Analysis*

Once basic structures, or *primitives*, have been extracted from the document, a straightforward idea is to *combine* them into increasingly more complex objects in order to recognize higher-level structures. For that purpose, it is quite natural to use structural and syntactic pattern recognition methods, as technical documents usually follow strict representation rules, which provide for a standardized way of drawing the various elements. These standards can thus be represented by a grammar or by a set of production rules, which enable the recognition of complex structures from the most basic ones. The "knowledge engineering" process concerns the question: "When reading a drawing, how do you decompose it into structures that are more and more elementary?" Conversely, the question can be posed as "What are the representation rules followed by the draftsman (who applies the drafting standard) for creating higher-level objects from basic elements?"

Several systems developed in the field of technical document analysis apply this approach. Some of these systems are described below.

Cappellini *et al.* [39] propose a system which identifies primitives on hand-drawn drafting. The basic idea is to consider entities in engineering drawings as special symbols and to recognize them by a hybrid approach, combining graph matching and classification. However, the level of semantics reachable by such a system remains quite low, as higher-level entities are seldom storable as model symbols.

Lu and Ohsawa [40] use a knowledge based system, which vectorizes the drawing by matching opposite line borders, and recognizes various entities specific to technical drawings, especially the components of dimension-sets, such as arrowheads. Once again, there is no higher-level analysis of the drawing.

The ANON system developed at the University of Sheffield [2] is based on a structural description of engineering drawings, using frames to represent components such as lines, curves, dimension-sets, etc. and the relations between these components. The interpretation itself follows strategy rules written in the `yacc` syntax; the parsing allows the recognition of entities such as dimensions or broken lines. The representation of the a priori knowledge using frames yields more abstraction power than what is available through a flat set of grammar rules.

WIZ, a system designed at the university of Hamburg (Germany), takes an artificial intelligence approach to the recognition of entities in engineering drawings [41]. The aim in this case is to design a generic system, where symbols and other entities are described as basic graphical objects, with constraints defined between such objects [42].

Kasturi's group at Pennsylvania State University has been very active in this field. In addition to developing robust algorithms for recognizing geometric entities, such as dashed lines, hatched areas or dimensioning, they have designed a system for recognizing more complex graphics entities [19] and have applied this know-how to the analysis of various drawings of telephone equipment [43,44] and more generally to the analysis of engineering drawings [45].

The Machine Drawing Understanding System, MDUS, currently under development at the Technion, Israel Institute of Technology [5], is designed to automate the process of converting mechanical engineering drawings into an accepted standard for exchange of graphic information among CAD/CAM systems, such as IGES (Initial Graphic Exchange Specification), or DXF.

Another typical example of analyzing the document in terms of higher-level structures, based on domain knowledge, is that of the minimum closed block drawn in thick lines, which is used in the Celesstin system [46] as the basic structural component of mechanical engineering drawings and the basic "building stone" for interpretation phases. The assembly of these blocks along an axis leads to the recognition of entities such as screws, shafts, and — from them — ball bearings, gears, etc. A set of simple assembly rules allows the progressive assembly of larger objects from this block entity. For instance, it is possible to group blocks into symmetrical entities when following a dot-dashed symmetry axis [47]. Figure 1 illustrates the recognition of shafts.

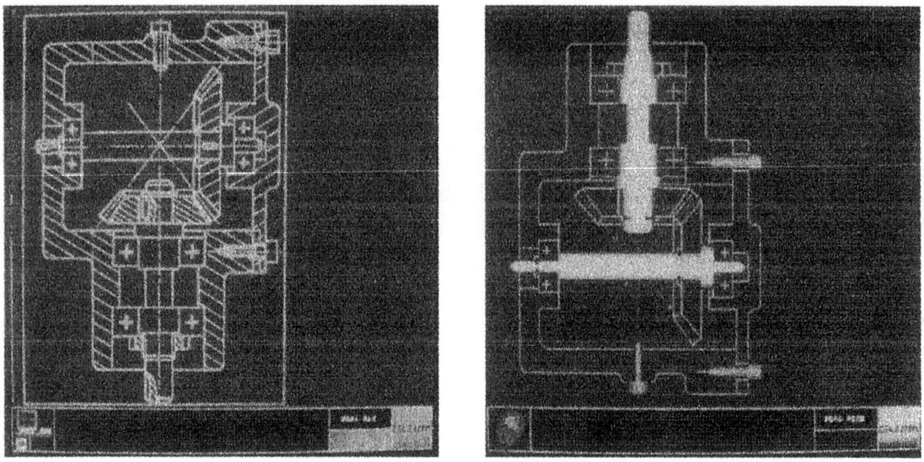

Fig. 1. A scanned drawing and the result of recognition of shafts and screws in Celesstin.

5.2. *Contextual Layer Separation*

The segmentation techniques used in document analysis, such as text–graphics separation for instance, rely on the analysis of connected components. However, the *physical* layers extracted by these methods are far too crude to be readily useful for higher-level interpretation, for the following reasons:

- Lines, which are physically connected, can belong to different *logical* layers. Basically, we distinguish two logical layers: the *geometry* layer and the *annotation* layer.

The geometry layer contains real geometry lines, that correspond to the edges of an object. The annotation layer contains everything else: hatching lines, dimensioning lines, manufacturing instructions and symbols, etc. We need good techniques for vectorizing these different types of lines, knowing in addition that the requirements are not the same for each type: precise localization is most important for the lines in the geometry layer, as they represent edges of the object described in the drawing. However, the exact location of hatching lines, for instance, is much less important.

- As the segmentation procedures do not use a priori knowledge about the meaning of the extracted strings, they tend to be more or less "blind" to the difference between a string of hyphens and a dotted line, or between a small graphical symbol and an isolated character. There should be a feedback from character recognition to graphical segmentation.
- Although the rules state that text should not intersect or even touch the graphics, in practice this often occurs. Thus, text–graphics segmentation will find the characters of a string which are connected components on their own, but will miss those which touch a line. Retrieval of the whole string then requires some non-trivial post-processing phase. This is an important point, where the introduction of context may improve drastically the quality of vectorization.

Actually, all these items show that there is still a long way to go before having a good, *context-based* vectorization and segmentation process for technical drawings. Typically, such a system may first perform "crude" vectorization and character separation, and then progressively extract as separate layers the dimensioning, the textual annotations, the symbolic lines (e.g. hatching), the edges, etc. Partial re-runs of vectorization and of character recognition may be necessary at locations where text and graphics — or hatching and edges — intersect, to improve the overall quality and reduce the number of recognition errors.

5.3. *Functional Analysis*

From 1989 to 1991, the group of Tombre has worked on Celesstin, an integrated, blackboard-based prototype system which converts drawings into a CAD description [46]. The first versions of this system were essentially based on structure and syntax to recognize entities such as shafts, screws, ball bearings or gears on a single view of a mechanical device. The system decomposes the vectorized document into a set of blocks having contextual attributes (hatching, threading, etc.) and analyzes these blocks by focusing on technical elements located along the axis lines. In the last version, Celesstin IV, we experimented with semantic knowledge rules. By focusing on a specific area of mechanical engineering, we were able to show that it is possible to analyze a single view of a drawing at the level of technological functionalities. We designed two "experts", one focusing on *disassembling*, based on the assumption that it *must* be possible to disassemble a mechanical setup, and the other dealing with the *kinematics* of the whole setup. The kinematics expert determines the functionalities of various entities from their behavior when a rotation motion is applied around the identified axes in the drawing [47,48].

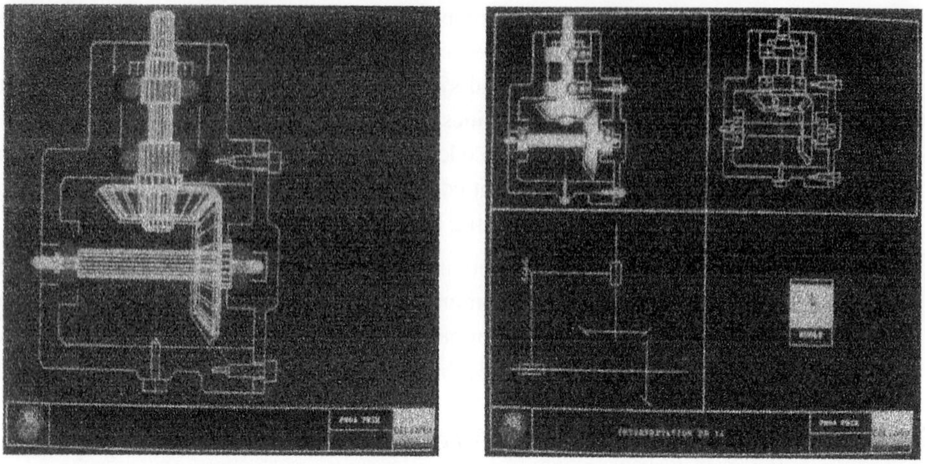

Fig. 2. Result of kinematics analysis and of functional interpretation in Celesstin.

Figure 2 illustrates the results of this functional analysis. Although we are aware of the fact that even in the area of mechanical engineering, our prototype is far from covering all possible functional interpretations, we believe that this work suggests a possible methodology for extracting functional information from technical drawings.

Similar needs also increasingly arise in the CAD world itself, even when no paper drawings are involved. Companies which have been working with computer-aided drafting systems or with CAD systems operating only with low-level primitives now wish to convert their data to higher-level representations, in order to incorporate them into a *feature-based CAD* system. This leads to the development of systems for converting 2D geometric representations to a higher-level model by implementing various schemes for automating the recognition of machinable manufacturing features [49, 50]. A number of recent papers on this topic can be found in Ref. [51].

6. 3D Reconstruction

The evolution of CAD is both towards functional representation and 3D models, complete with geometrical and semantic/functional attributes. Functional interpretation alone is therefore not sufficient. We should also be able to reconstruct a 3D model from different views. This need arises not only when starting from a paper drawing, but also in the numerous cases where the available data are a set of geometrical 2D views constructed by some computer-based drafting system, that needs to be converted into a real 3D CAD model.

Many purely geometric methods have been designed for combining several orthogonal views into a 3D model. The basic idea is to look for sets of consistent matching hypotheses between features extracted from three orthogonal projection views of an object [52].

A first family of algorithms, usually known as the "fleshing out projections" concept, formalized by Wesley and Markowsky [53] and Preiss [54], is based on the reconstruction of a *wire-frame* model by matching vertices and edges. The real faces of the object are then found by propagating constraints on this wire-frame. Many researchers have implemented variants of this method, including Lequette [55]; Yan *et al.* [56], who take advantage of the additional constraints yielded by the fact that some lines are dashed and some are solid, to improve the efficiency of the wireframe construction; Lysak and Kasturi [57], who propose a bottom-up approach going from edges to faces and then on to volumes, which works both for polyhedral and for non-polyhedral objects; Martí *et al.* [58], and many others. One of the problems with the "fleshing out projections" approach is that these methods have to generate "pseudo-elements" from matchings which do not correspond to real faces or vertices of the object. The complexity of the process which eliminates these pseudo-elements tends to be high, although some efficient methods have been recently proposed, including the decision-chaining algorithm by Marston and Kuo [59].

Another family of algorithms is *volume-oriented.* The idea is to first find 3D subparts and then combine them to build the complete object [60]. This is the basic approach of several works. Chen and Perng [61] decompose each view into several predefined types of subviews before reconstructing the corresponding subparts. Grabowski *et al.* [62] interpret every 2D projection line as a surface with material direction, i.e., a halfspace, and generate surface elements which, in turn, are combined to yield the 3D reconstruction. Kitajima and Yoshida [63] and Tomiyama *et al.* [64] recognize volume-oriented primitives and use them to construct a CSG tree.

All these methods have in common two limitations which narrow their domain of applicability:

- The data on which they work have to be perfect, i.e. the methods assume that a clean, idealized set of projections is provided, uncluttered by annotation and free of noise or uncertainties. This is of course impossible, due to the noise introduced by the digitization and vectorization of the drawing, as discussed above.
- These methods only apply to the geometric part of the drawing; however, as discussed, a large part of the drawing pertains to the annotation, or the "linguistic" dimension, and to a large amount of symbolic information.

The first limitation is usually not a problem when the input data comes from another CAD system, but in the case of scanned paper drawings, the geometry of the drawings must be corrected before fleshing out the projections. Such corrections are possible if we perform dimensioning analysis, as seen (Sec. 4). A step towards incorporating dimensioning information in the 3D interpretation process has been made by Weiss and Dori [65], who apply a variational geometry rule base to validate the dimensioning of recognized 2D views. Another approach to solve this problem has been explored by Iwama *et al.* [66], who use probabilistic relaxation matching in order to get matches in the presence of imprecise data.

The second limitation has only been dealt with in some experimental systems. Kanai *et al.* [67] add topological constraints to their reconstruction system, as they look for curvilinear surfaces by matching hand-drawn curves and reconstructing a Bezier surface. Kim *et al.* [68] make the contextual assumption that the system deals with sheet-metal objects.

There is still a lot of research to be carried out on this aspect of mixing geometry with semantics in the interpretation process. It may even be nearly impossible to reach a comprehensive solution, as the symbolic/semantic part of the information is in essence strongly context-dependent.

6.1. *An Example of the Implementation of Fleshing-out Projections*

As an example of the way geometric matching works, we present here an implementation based on the "fleshing-out projections" paradigm [69]. The algorithm is based on previous work by R. Lequette [55], who, in turn, was inspired by Wesley and Markowsky [53] and by Sakurai [70]. The reconstructed solids are represented using a B-Rep (Boundary Representation) scheme.

Data: In this implementation, we start with an ASCII input file describing the geometry of the three views, typically representing the result of a complete vectorization and dimensioning analysis process.

Preprocessing: The first step of the algorithm consists of locating all the intersections between lines, and specific nodes, i.e. outline nodes (points where curves have tangents parallel to the projection axes) and tangency nodes (points where two curves have a first order tangency connection). A second preprocessing step splits up the lines, starting from the previously found nodes, into lines which connect only two nodes at their extremities.

Search for vertices: The search is based on the fact that the three views share some coordinates. At this point we must also take into consideration five main vertex types, as illustrated by Fig. 3.

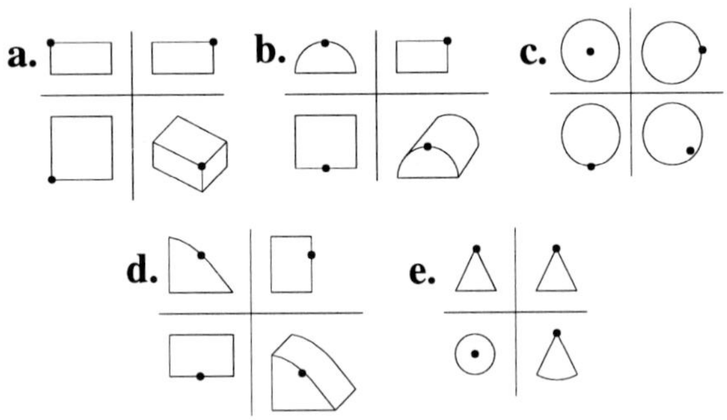

Fig. 3. A plain vertex (a), an outline vertex (b), a double-outline vertex (c), a tangency vertex (d) and a cone vertex (e).

Search for edges: A wire-frame of the object which is represented by the views is reconstructed. The projections of the edges of this wireframe are the lines found on the three views. As for the vertices, there are several kinds of edges: *axial edges*, which are perpendicular to one of the projection planes; *semi-axial edges*, which are not axial but parallel to one of the projection planes; *oblique edges*, which are in general position; *outline edges* which can be found on solids of revolution and are occluding contours, which do not correspond to an actual 3D edge; and *tangency edges*, which separate two faces whose support surfaces have a first order connection along the edge.

Search for surfaces: We now find the surfaces which contain the solid's faces. This step is based on the fact that a surface is defined by two edges sharing a common vertex. The type of these two edges and the relationships which exist between them determine the type and the features of the generated surface. The three most common cases are:

1. segment + segment $\Rightarrow$
   - *plane* if the segments are collinear
   - nothing otherwise
2. segment + circular arc $\Rightarrow$
   - *cylinder* if segment $\perp$ circle plane
   - *plane* if segment $\subset$ circle plane
   - *cone* otherwise
3. circular arc + circular arc $\Rightarrow$
   - *sphere* if axes of arcs intersect
   - *plane* if arcs are coplanar
   - *torus* otherwise

After this stage is completed, we have all the support surfaces for the object's faces and all the solid's edges contained in these surfaces.

Search for faces: For each surface found by the previous stage, we arrange the different edges. The first thing to do is to find the circuits on each surface. This is done by sorting the edges in trigonometric order with respect to the normal to the surface at each vertex, and by traversing the resulting graph of oriented edges. As a face can be determined by several circuits, we must also find out which circuits have to be associated with each other to make a face. This is achieved by classifying the circuits into finite and infinite circuits. Finite circuits are those which were found by traversing the oriented graph in trigonometric order with respect to the normal to the surface, whereas infinite circuits are those corresponding to the reverse trigonometric order. A face is then created for each finite circuit, while each infinite circuit is interpreted as a hole in the smallest face in which it is contained.

Search for solids: Only a subset of the faces found in the previous step actually belong to the solid's border. The others are *pseudo-faces*, as previously mentioned. The challenge in this last step of the algorithm is to find a group of faces which make up a real solid. The algorithm for achieving this is based on the *Möbius principle* [71], which states that when a *real edge* is on the border of two real faces, the traversals of these two faces follow the edge in opposite directions. The search for a set of consistent faces is implemented by traversal of the whole graph of possible face configurations, with backtracking when necessary, to find all possible solids. There is no guarantee that this algorithm always finds a single solution.

Figure 4 shows some of our first results. The input data are represented graphically as three views in each case, and the result of the reconstruction is a *Gnuplot* output, where each real face is separated to show how the program works.

Fig. 4. Examples of reconstructed solids, from [72].

We are currently investigating the possibility to add symbolic information to this process [69].

7. Conclusion and Perspectives

It is not expected that systems in the near future will be fully automatic. A certain degree of human involvement is necessary to resolve ambiguities and support the system's proper execution. The vast majority of the routine conversion work, though, is expected to be automated, or else the system will not be cost-effective. We are constantly working to improve our systems' performance and robustness. For example, the lexical phase of MDUS is currently being revised using object-process analysis (OPA) [73] and existing C code is wrapped by C++ in preparation for coding the syntactic and semantic phases of drawing interpretation.

Understanding engineering drawings is both a challenging open research area and a pressing issue, calling for practical solutions. Robust drawing recognition systems with increasingly higher intelligence level should be constructed in the near future if we wish the vision of an integrated view of hard and soft version of technical documentation to become a reality. The need for solving this problem is not just a topic for a relatively short transition period. Rather, it is supposed to stay with us for a long time. First, the stores of paper documentation are not showing any signs of decline. Even if and when it does, transition from early to advanced CAD systems would still render CAD conversion a viable application domain.

Future applications may include cooperation among humans and robots, where both need to refer to the same drawing. Since no one suggests that people are going to be able to interpret coded files representing drawings, robots must be equipped with the capability of understanding drawings the way people do.

The magnitude and complexity of the automated drawing understanding problem implies that making a significant progress within a few years requires the concentrated and coordinated efforts of interested researchers and manufacturing organizations worldwide. We envision that once CAD conversion attains a certain level of maturity, it will be incorporated as an integral, standard component of a Computer Integrated Manufacturing (CIM) environment.

References

[1] D. Dori, Automated understanding of engineering drawings: An object-oriented analysis, *J. Object Oriented Programming* **7** (1994) 35–43.

[2] S. H. Joseph and T. P. Pridmore, Knowledge-directed interpretation of mechanical engineering drawings, *IEEE Trans. PAMI* **14** (1992) 928–940.

[3] A. J. Filipski and R. Flandrena, Automated conversion of engineering drawings to CAD form, *Proc. IEEE* **80** (1992) 1195–1209.

[4] L. S. Wolfe and J. de Wyze, An update on drawing conversion, *Computer Aided Design Rep.* **8** (1988) 1–11.

[5] D. Dori, Y. Liang, J. Dowell and I. Chai, Sparse-pixel recognition of primitives in engineering drawings, *Machine Vision and Appl.* **6** (1993) 69–82.

[6] D. Dori and I. Chai, Extraction of text boxes from engineering drawings, *Proc. SPIE/IS&T Symp. on Electronic Imaging Science and Technology, Conf. on Character Recognition and Digitizer Technologies*, San Jose, CA, 1992, 38–49.

[7] T. Pavlidis, *Structural Pattern Recognition* (Springer-Verlag, New York, 1980).

[8] J. Sklansky, R. L. Chazin and B. J. Hansen, Minimum perimeter polygons of digitized silhouettes, *IEEE Trans. Comput.* **21** (1972) 260–268.

[9] J. Sklansky and V. Gonzalez, Fast polygonal approximation of digitized curves, *Pattern Recog.* **12** (1980) 327–331.

[10] I. Tomek, Two algorithms for piecewise continuous approximation of functions of one variable, *IEEE Trans. Comput.* **23** (1974) 445–448.

[11] Y. Kurozomi and W. A. Davis, Polygonal approximation by minmax method, *Computer Graphics and Image Process.* **19** (1982) 248–264.

[12] C. M. Williams, An efficient algorithm for the piecewise linear approximation of planar curves, *Computer Graphics and Image Process.* **8** (1978) 286–293.

[13] K. Reumann and A. P. M. Witkam, Optimizing curve segmentation in computer graphics, *Int. Computer Symp.*, 1974, 467–472.

[14] K. Wall and P. Danielsson, A fast sequential method for polygonal approximation of digitized curves, *Computer Vision, Graphics and Image Process.* **28** (1984) 220–227.

[15] U. Montanari, A note on minimal length polygonal approximation to a digitized contour, *Commun. ACM* **13** (1970) 41–47.

[16] J. G. Dunham, Optimum piecewise linear approximation of planar curves, *IEEE Trans. PAMI* **2** (1980) 327–331.

[17] J. G. Dunham, Optimum uniform piecewise linear approximation of planar curves, *IEEE Trans. PAMI* **8** (1986) 67–75.

[18] J. Yuan and C. Y. Suen, An optimal algorithm for detecting straight lines in chain codes, *Proc. 11th ICPR*, Den Haag, Netherlands, 1992, 692–695.

[19] R. Kasturi, S. T. Bow, W. El-Masri, J. Shah, J. R. Gattiker and U. B. Mokate, A system for interpretation of line drawings, *IEEE Trans. PAMI* **12** (1990) 978–992.

[20] J. F. Harris, J. Kittler, B. Llwellyn and G. Preston, A modular system for interpreting binary pixel representations of line-structured data, in *Pattern Recognition Theory and Applications*, eds. L. F. Pau, J. Kittler and K. S. Fu (D. Reidel Publishing Company, 1982) 311–351.

[21] M. Furuta, N. Kase and S. Emori, Segmentation and recognition of symbols for handwritten piping and instrument diagram, *Proc. 7th ICPR*, Montreal, Canada, 1984, 612–614.

[22] C. S. Fahn, J. F. Wang and J. Y. Lee, A topology-based component extractor for understanding electronic circuit diagrams, *Computer Vision, Graphics and Image Process.* **44** (1988) 119–138.

[23] V. Nagasamy and N. A. Langrana, Engineering drawing processing and vectorization system, *Computer Vision, Graphics and Image Process.* **49** (1990) 379–397.

[24] T. Sato and A. Tojo, Recognition and understanding of hand-drawn diagrams, *Proc. 6th ICPR*, Munich, Germany, 1982, 674–677.

[25] X. Lin, S. Shimotsuji, M. Minoh and T. Sakai, Efficient diagram understanding with characteristic pattern detection, *Computer Vision, Graphics and Image Process.* **30** (1985) 84–106.

[26] P. Vaxivière and K. Tombre, Subsampling: A structural approach to technical document vectorization, in *Shape, Structure and Pattern Recognition*, eds. D. Dori and A. Bruckstein (World Scientific, 1995) 323–332.

[27] I. Chai and D. Dori, Orthogonal zig-zag: An efficient method for extracting lines from engineering drawings, in *Visual Form*, eds. C. Arcelli, L. P. Cordella and G. Sanniti di Baja (Plenum Press, New York, 1992) 127–136.

[28] J. Illingworth and J. Kittler, The adaptative Hough transform, *IEEE Trans. PAMI* **9** (1987) 690–698.

[29] R. S. Conker, A dual plane variation of the Hough transform for detecting non-concentric circles of different radii, *Computer Vision, Graphics and Image Process.* **43** (1988) 115–132.

[30] C. Kimme, D. H. Ballard and J. Sklansky, Finding circles by an array of accumulators, *Commun. ACM* **18** (1975) 120–122.

[31] L. O'Gorman and A. C. Sanderson, The converging squares algorithm: An efficient method for locating peaks in multidimensions, *IEEE Trans. PAMI* **6** (1984) 280–288.

[32] H. Asada and M. Brady, The curvature primal sketch, *IEEE Trans. PAMI* **8** (1986) 2–14.

[33] P. L. Rosin and G. A. West, Segmentation of edges into lines and arcs, *Image and Vision Comput.* **7** (May 1989) 109–114.

[34] L. O'Gorman, An analysis of feature detectability from curvature estimation, *Proc. IEEE Conf. on CVPR*, Ann Arbor, 1988, 235–240.

[35] D. Dori, Vector-based arc segmentation in the machine drawing understanding system environment, *IEEE Trans. PAMI* **17** (1995) 1057–1068.

[36] D. Pao, H. F. Li and R. Jayakumar, Graphic features extraction for automatic conversion of engineering line drawings, *Proc. First ICDAR*, Saint-Malo, France, 1991, 533–541.

[37] S. Collin, Syntactical analysis of technical drawing dimensions, in *Advances in Structural and Syntactic Pattern Recognition, Proc. Int. Workshop on Structural and Syntactic Pattern Recognition*, ed. H. Bunke (World Scientific, 1992) 280–289.

[38] C. P. Lai and R. Kasturi, Detection of dimension sets in engineering drawings, *IEEE Trans. PAMI* **16** (Aug. 1994) 848–855.

[39] V. Cappellini, F. Flamigni and A. Mecocci, A system for automatic drafting encoding, *Proc. IAPR Workshop on Computer Vision*, Tokyo, Japan, 1988, 173–178.

[40] W. Lu, Y. Ohsawa and M. Sakauchi, A database capture system for mechanical drawings using an efficient multi-dimensional graphical data structure, *Proc. 9th ICPR*, Rome, Italy, 1988, 266–269.

[41] B. Pasternak and B. Neumann, Adaptable drawing interpretation using object-oriented and constraint-based graphic specification, *Proc. 2nd ICDAR*, Tsukuba, Japan, 1993, 359–364.

[42] B. Pasternak, Processing imprecise and structural distorted line drawings by an adaptable drawing interpretation kernel, *Proc. IAPR Workshop on Document Analysis Systems*, Kaiserslautern, Germany, Oct. 1994, 349–365.

[43] J. F. Arias, C. P. Lai, R. Kasturi and A. Chhabra, Interpretation of telephone system manhole drawings, *Proc. 2nd ICDAR*, Tsukuba, Japan, 1993, 365–368.

[44] J. F. Arias, A. Prasad, R. Kasturi and A. Chhabra, Interpretation of telephone company central office equipment drawings, *Proc. 12th ICPR*, Jerusalem, Oct. 1994, 310–314.

[45] C. P. Lai and R. Kasturi, Knowledge-based understanding of engineering drawings, Department of Computer Science and Engineering Technical Report CSE-93-005, The Pennsylvania State University, University Park, Pennsylvania 16802, Sept. 1993.

[46] P. Vaxivière and K. Tombre, Celesstin: CAD conversion of mechanical drawings, *IEEE Computer Mag.* **25** (1992) 46–54.

[47] P. Vaxivière and K. Tombre, Knowledge organization and interpretation process in engineering drawing interpretation, *Proc. IAPR Workshop on Document Analysis Systems*, Kaiserslautern, Germany, Oct. 1994, 313–321.

[48] P. Vaxivière, Interprétation de dessins techniques mécaniques, Thèse de doctorat, Institut National Polytechnique de Lorraine, Vandœuvre-lès-Nancy, Feb. 1995.

[49] S. Meeran and M. J. Pratt, Automated feature recognition from 2D drawings, *CAD* **25** (1993) 7–17.

[50] J. H. Vandenbrande and A. A. G. Requicha, Spatial reasoning for the automatic recognition of machinable features in solid models, *IEEE Trans. PAMI* **15** (1993) 1269–1285.

[51] *Proc. IFIP Int. Conf. on Feature Modeling and Recognition in Advanced CAD/CAM Systems*, Valenciennes, France, Nov. 1994.

[52] R. H. Haralick and D. Queeney, Understanding engineering drawings, *Computer Graphics and Image Process.* **20** (1982) 244–258.

[53] M. A. Wesley and G. Markowsky, Fleshing out projections, *IBM J. Res. Dev.* **25** (1981) 934–954.

[54] K. Preiss, Constructing the solid representation from engineering projections, *Computers and Graphics* **8** (1984) 381–389.

[55] R. Lequette, Construction automatique de solides à partir de vues orthogonales de type dessin industriel, Thèse de doctorat, Université de Grenoble, 1987.

[56] Q.-W. Yan, C. L. Philip Chen and Z. Tang, Efficient algorithm for the reconstruction of 3D objects from orthographic projections, *CAD* **26** (1994) 699–717.

[57] D. B. Lysak and R. Kasturi, Interpretation of engineering drawings of polyhedral and non-polyhedral objects, *Proc. First ICDAR*, Saint-Malo, France, 1991, 79–87.

[58] E. Martí, J. Regincós, J. Lopez-Krahe and J. J. Villanueva, A system for interpretation of hand line drawings as three-dimensional scene for CAD input, *Proc. First ICDAR*, Saint-Malo, France, 1991, 472–480.

[59] R. E. Marston and M.H. Kuo, Reconstruction of 3D objects from three orthographic projections using a decision-chaining method, *Proc. IAPR Workshop on Machine Vision Applications*, Kawasaki, Japan, Dec. 1994, 423–426.

[60] H. Yoshiura, K. Fujimura and T. L. Kunii, Top-down construction of 3-D mechanical object shapes from engineering drawings, *IEEE Computer Mag.* **17** (1984) 32–40.

[61] Z. Chen and D.-B. Perng, Automatic reconstruction of 3D solid objects from 2D orthographic views, *Pattern Recog.* **21** (1988) 439–449.

[62] H. Grabowski, S. Rude and B.-Y. Liu, Automatic recognition of product models from technical drawings, *Proc. IFIP Int. Conf. on Feature Modeling and Recognition in Advanced CAD/CAM Systems*, Valenciennes, France, 1994, 627–644.

[63] K. Kitajima and M. Yoshida, Reconstruction of CSG solid from a set of orthographic three views, *Proc. IEEE Int. Conf. Systems Engineering*, Kobe, Japan, Sept. 1992, 220–224.

[64] K. Tomiyama, T. Nakamura and T. Koezuka, Auxiliary lines in three view drawings for 3D shape reconstruction using CSG method, *Proc. IEEE Int. Conf. Systems Engineering*, Kobe, Japan, Sept. 1992, 250–256.

[65] M. Weiss and D. Dori, A scheme for 3D object reconstruction from dimensioned orthographic views, *Proc. Third ICDAR*, Montréal, Aug. 1995, 335–338.

[66] T. Iwama, T. Horiuchi, K. Toraichi, H. Yamada and K. Yamamoto, An algorithm to restore a curved object from three orthographic views by using probabilistic relaxation matching method, *Proc. IAPR Workshop on Machine Vision Applications*, Kawasaki, Japan, 1994, 439–442.

[67] S. Kanai, S. Furushima and H. Takahashi, Generation of free-form surface models by understanding geometric and topological constraints on rough sketches, *Proc. IEEE Int. Conf. Systems Engineering*, Kobe, Japan, Sept. 1992, 246–249.

[68] C. Kim, N. Tsuchida, M. Inoue and S. Nishihara, Understanding three-view drawings of mechanical parts with curved shapes, *Proc. IEEE Int. Conf. Systems Engineering*, Kobe, Japan, Sept. 1992, 238–241.

[69] C. Ah-Soon and K. Tombre, A step towards reconstruction of 3-D CAD models from engineering drawings, *Proc. Third ICDAR*, Montréal, Aug. 1995, 331–334.

[70] H. Sakurai, Solid model input through orthographic views, *Computer Graphics* **17** (1983) 243–247.

[71] C. M. Hoffmann and J. E. H. Hopcroft, Geometric ambiguities in boundary representations, *CAD* **19** (1987) 141–147.

[72] C. Ah-Soon, Analyse de dessins industriels : reconstruction et interprétation, Rapport de DEA, Université Henri Poincaré Nancy I, Vandœuvre-lès-Nancy, Sep. 1994.

[73] D. Dori, Object-process analysis: Maintaining the balance between system structure and behaviour, *J. Logic and Computation* **5** (1995) 1–23.

Handbook of Character Recognition and Document Image Analysis, pp. 485–502
Eds. H. Bunke and P. S. P. Wang

CHAPTER 18

ANALYSIS OF PRINTED FORMS

DEBASHISH NIYOGI, SARGUR N. SRIHARI and VENU GOVINDARAJU
CEDAR, SUNY at Buffalo, 520 Lee Entrance, Suite 202
Amherst, New York 14228-2567, USA

Automatic analysis of images of printed forms is a problem of both practical and theoretical interest, due to its importance in office automation, and due to the conceptual challenges posed for document image analysis. The automatic reading of optically scanned forms consists of two major components. The first is the extraction of the data image from the form; the second is the interpretation of the image as coded alphanumerics, and is commonly referred to as optical character recognition (OCR). The individual steps involved in forms analysis include image pre-processing, forms identification, field extraction, data interpretation, and contextual post-processing. In this chapter we outline the issues and current state-of-the-art in the analysis of printed forms. Forms analysis poses several challenges, given the enormous variety of current form layouts and contents, and some of these research issues are explored. In particular, two current forms analysis systems developed at CEDAR are described.

Keywords: Forms analysis; Forms editing; Data entry; Image pre-processing; Forms processing applications.

1. Introduction

A paper form is any preprinted document designed to elicit information. The problem of automatically interpreting an optically scanned and digitized image of a filled-out paper form containing typed and handwritten entries is considered in this chapter. The goal of forms analysis is to segment the form into the basic components of text (characters and words), line segments and boxes, and then analyze the components to extract relevant information from the form.

Forms analysis has grown to be a major component of document image understanding. Imaging-based technology, as well as several innovative forms analysis techniques, have made it much easier, quicker and cheaper to process the hundreds of different types of forms that typically need to be processed.

Most common documents that are processed are forms — even bank checks are nothing but small, simple forms. All kinds of organizations use forms: finance, education, retail, health care, shipping and transportation. Forms hasten market research, order entry, payment and delivery. Governments, especially, deal in a vast number of forms.

As industry and government interact with larger numbers of companies and people, the number of forms multiplies. Forms processing lets the user handle more data with the same number of people, and imaging to databases and data entry.

Instead of having many people keying in data off paper forms, forms processing extracts the data from images and uses it to populate databases.

In forms processing, one handles completed forms that arrive via fax or scanned documents. After cleanup and processing, the form images can be routed on a network or stored. After extracting the data from a form and verifying its accuracy, one does not need to keep the form. (One part, the signature, might need to be archived, but there is no need to record indefinitely whether someone filled in a box with a checkmark, an X, or a solid circle.) Once the correctly extracted data enters the database, the data becomes the important resource to safeguard and process, not the form from which it came.

1.1. *Characteristics of forms*

Forms are more complex and harder to work with than other imaging documents, due to the following [1]:

- *They contain more than just text.*
  With documents, one can OCR the raw text (usually machine-readable, and created by typewriters, laser or dot matrix printers) and put it in a full-text database. One can also retrieve the images using index or keyword searches. Flexible text search engines allow the user to find text even if it contains many OCR errors. But forms can contain more than machine-readable text — they often come with checkmarks, handprint and signatures, all of which the system must recognize and process.
  Even barcodes take on new weight with forms. While some systems use barcodes for auto-indexing images, form packages use them to identify the form or the provider of the form. Some users remove a peel-off barcode from an inventory item and stick it onto a form. The forms processing application converts the code to a number and uses it to modify the inventory database.
- *They make data extraction more difficult.*
  Text-oriented OCR works with text in paragraph format. Such OCR systems do not typically look for text inside one-letter boxes, between lines, placed in different positions on the page, etc. Forms processing systems, on the other hand, need to OCR text in discrete chunks surrounded by graphic elements, such as lines and boxes and "combs". Sometimes the text overlaps. Thus, OCR for forms packages have to recognize a character, even with lines dissecting it. With the line in place, the character is marred. With the line removed, the character is broken. Either way, recognition becomes difficult for OCR.
- *They compress more slowly.*
  Forms contain lines, fancy type, logos and guide text (instructions that help users to fill the form). These elements hinder compression. Most compression algorithms, optimized for text and unbroken runs of white pixels, slow down when they have to process forms.The difference is similar to that between how

fast one can fax a simple cover page and a busy page with a lot of text and graphics.

- *They compress less efficiently.*
  Because it contains a lot of extra lines, fonts, logos and guide text, a compressed form can be four to 10 times the size of the same data without the form. Users can save space by dropping out the form and just compressing the data. A blank form can be stored at the user workstation and recombined with the data to display or print it. Doing this can boost performance and reduce storage needs by a factor of 10.
- *They may contain data needed to process them.*
  In standard imaging systems, one can scan documents, index them, run them through OCR for full-text retrieval, then store them in one place. But with forms, one might perform different OCR operations on them to extract different types of data based on preliminary data that is read from the form (e.g., type of form).
- *They involve handprint.*
  Forms are designed to be filled in by hand. However, handprinting is more difficult to read than machine-generated text, since there is more variation. Fortunately, forms present blocks or fields to constrain the handprint. Without these, standard OCR systems would have a difficult time reading a paragraph or a full page of handprint well. In forms analysis, context-based text reading becomes more important than simple OCR.

2. Background

Current research in forms analysis is ongoing at several universities and research institutions. Recent research papers in the area of forms processing describe various approaches towards the solution of different components of the overall forms processing problem.

Casey and Ferguson [2] describe a method developed at IBM for entry of a wide variety of forms that contain machine-printed data and that are often produced in business environments. The function, called Intelligent Forms Processing (IFP), accepts conventional forms that call for information to be printed in designated blank areas, but in which the information may exceed boundaries due to poor registration when printing. The IFP system uses a setup phase to create a model of each form that is to be read. Scanned forms containing data are compared against the matching form model. Special algorithms are employed to extract data fields while removing background printing (e.g., form lines) intersecting the data. The extracted data images are interpreted by an OCR process that reads typical monospace fonts. New fonts may be added easily in a separate design mode. If the data are alphabetic, a lexicon may be assembled to define the possible entries.

Taylor *et al.* [3] describe techniques for manipulating electronic images of forms in preparation of data interpretation. A combination feature extraction/model-based approach is used for forms identification, registration, and field extraction. Forms identification is implemented with a neural network. The system is demonstrated on Internal Revenue Service forms.

Watanabe *et al.* [4] propose a new method to recognize the layout structure of table-form documents analytically. Their method identifies the line segments directly and interprets the layout structure with the extracted line segments. Traditional approaches to extract characters from table-form documents erase all line segments in order to utilize the character recognition techniques or apply the OCR techniques. In this approach, however, the mutual relationships among line segments is very important for the structure analysis because the line segments specify the document layout structure. In addition, the individual blocks partitioned by the line segments define the particular domains of meaningful items.

Yan *et al.* [5] propose a knowledge-based form understanding system for automatic data entry. It includes two parts, the design and implementation of a Form Description Language (FDL), and the understanding program. This system can process various kinds of forms and produce images of different items in the form in any desired order. Moreover, this system has learning capability. It can update the description according to its analysis so that it can understand the form more quickly and accurately afterwards.

Yuan *et al.* [6] introduce a new method for extracting the filled items from the pre-printed parts of a form by using an unfilled form as a model object, and a filled form as a moved version of the model. The motion is detected by a new technique called GRHT (generalized randomized Hough transform), and then the two forms are registered so that the filled items can be extracted by some simple operations.

Chandran and Kasturi [7] describe a system for the extraction of structural information of a table from its image. Following the initial binarization and deskewing operations, the image is analyzed to extract all horizontal and vertical lines that are present. The table's dimensions are estimated based on these lines. Unlike other systems, this procedure does not depend on the sole existence of lines to mark the item blocks. White streams are recognized in both the horizontal and vertical directions as substitutes for any missing demarcation lines. A structure interpretation procedure uses the extracted demarcation information to identify each of the item blocks in the table. Subsequently, the inter-relations of these item blocks is used to recognize the structure of the tabulated data.

Monger *et al.* [8] describe the design and implementation of a user-friendly graphical package for the interactive layout and content description of document images. The specific problems addressed are those associated with predefined forms, such as proposal forms used in the insurance business, which have been completed by hand. The document image description allows position and content of the handwritten fields to be defined and labeled, and their inter-relationships specified. This information assists the automatic processing of handwritten forms using OCR techniques by enabling appropriate OCR algorithms and local syntax constraints to be applied to each field to produce the best recognition performance.

2.1. *Forms Processing Research at CEDAR*

Active research on forms processing has been ongoing at CEDAR (Center of Excellence for Document Analysis and Recognition). Research techniques developed at CEDAR have been implemented in current delivered products. The forms processing techniques developed at CEDAR that have been published so far include the following:

Wang and Srihari [9] describe an approach to the extraction of text, both typed and handwritten, from scanned and digitized images of filled-out forms. In decomposing a filled-out form into three basic components of boxes, line segments and the remainder (handwritten and typed characters, words, and logos), the method does not use a priori knowledge of form structure. Characters broken by line removal are rejoined using a character patching method.

Lam *et al.* [10] present a robust form reader with high adaptability and trainability. The form reader consists of two modules: field registration and data recognition modules. The field registration module acquires knowledge about the forms of interest, and the data recognition module recognizes text data on filled forms using the acquired knowledge. The capability of the reader increases progressively through supervised learning. The form reader has been trained to read a large variety of forms with machine-printed data. The adaptability and trainability of the system has been demonstrated through the experiments.

3. CEDAR Forms Analysis

Current research and development in forms analysis at CEDAR includes the use of forms processing techniques in several ongoing projects, including the Name and Address Block Reader project for reading IRS tax forms, and a Forms Analysis project for extracting information from business reply cards.

3.1. *Name and Address Block Reader*

In the Name and Address Block Reader (NABR) project, sponsored by IRS & Grumman Data Systems Corp., address block labels are extracted from IRS tax forms. Only the address block is received from the IRS tax forms, and data is extracted from these address blocks.

All but one of the forms have "drop-out" guidelines, which allow a person to fill out the form within the boundaries marked, but also allow the form to be processed electronically without the guidelines being present to interfere with the recognition of the characters. The only exception to the above is the IRS Form 1040EZ, which does not have drop-off guidelines. Therefore, for this form, the guidelines have to be first removed before any further processing.

Fig. 1 shows some examples of name-and-address blocks extracted from various IRS forms. The forms represented are:

(a) 1040EZ (non-dropout)
(b) 1040EZ label

(c) 941 non-label (dropout)
(d) IRPS non-label (dropout)

On receiving an image, this system first attempts to find out whether it is a label image or not. It is important to get the anchor points for proper registration of the 1040EZ non-label image owing to the non-dropout guide lines which interfere with the segmentation and recognition processes. A set of significant features such as shape and location of instructions found within the input image is used for the anchor information. Then the guide lines are removed, and the printed or handwritten information is extracted and sent to a segmentation process. The segmented image is further processed with a character recognition algorithm that identifies the individual characters in the label and then performs contextual post-processing to correct any character recognition error.

It is not necessary to apply the anchor information for the non-label images in the other forms because of the use of dropout guide lines. Information of name, address, city, state and ZIP Code is extracted from the areas predefined according to the various forms.

For the label images in all forms, anchor information is also useful for proper process. Common anchor points are the Social Security Number with check digits or the identification numbers (Employer Identification Number, Payer's Identification Number, or Recipient's Identification Number). Extracted information is sent to a different character recognition algorithm designed for OCR-A font.

More details on this system can be found in [11].

3.2. *Analysis of Reply Cards*

The objective of this project is to develop a Forms Analysis system to analyze certain types of forms, e.g., postal reply cards. The system is designed to do the following:

(i) Acquire the images of the information-bearing side of the reply cards.
(ii) Determine the recipient's ID to determine the required service(s) and to retrieve relevant a priori information regarding that particular reply card.
(iii) Extract the information contents (both address and non-address) automatically.
(iv) Forward the image to a site for further manual processing, if automated processing fails.
(v) Organize the extracted information in a format specified by the recipient.
(vi) Aggregate the processing results of the reply cards destined for the same recipient.

The system has the capability of processing both handwritten and machine-printed reply cards. Processing the reply cards involves reading both the address and non-address information. The speed of processing the addresses is 2.2 images per second for handwritten addresses and 2.8 images per second for machine-printed addresses. Fig. 2 shows the overall system functional design.

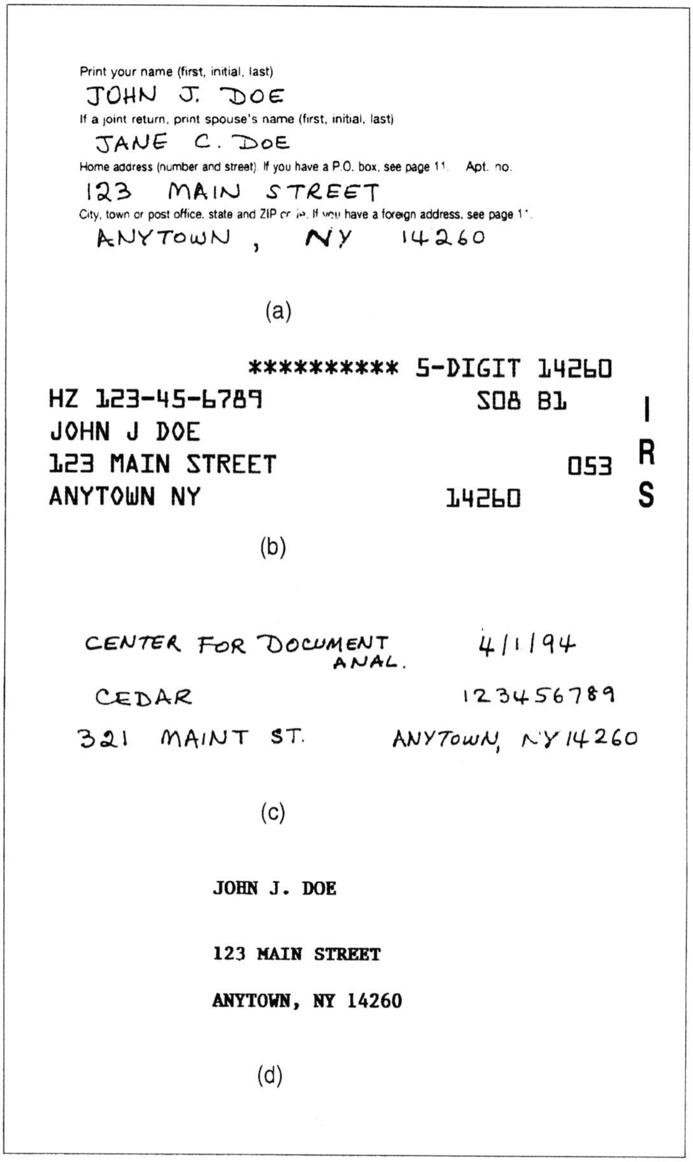

Fig. 1. Name and address blocks in IRS forms.

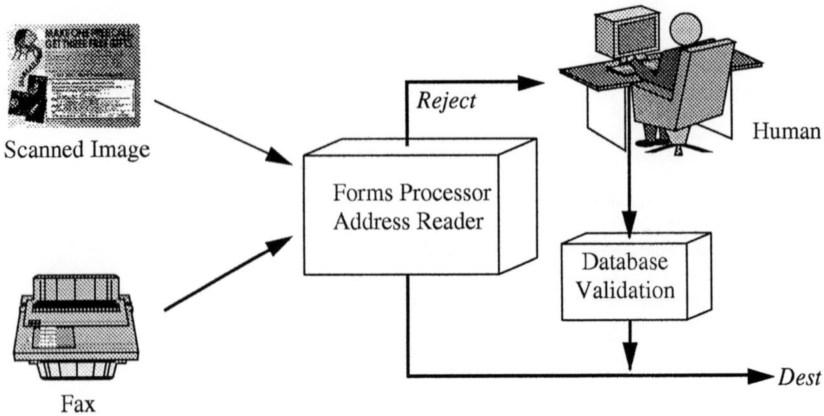

Fig. 2. Functional design of the CEDAR forms analysis system.

The demonstration system includes three different modules: (i) off-the-shelf forms processing product that can correct for skew, remove form guide lines, and extract sub-fields, (ii) OCR engine to read every sub-field in both the address and non-address sections of the reply card, and (iii) a reject re-entry software for manual processing of (some) reply cards.

A survey of forms processing software was performed as part of this project, and a forms processing package is part of the overall system. The forms processing software selected for this project needed to be reliable, require little end-user training, and be able to be interfaced in the overall design. Specifically, there were several tasks called for by the functional design. They were:

Form Definition — A user should be able to easily define zones to be extracted from a form through a comfortable graphical user interface.

Form Identification — The software should be able to identify a filled-in form reliably using a set of previously scanned blank forms.

Form Removal — The package should perform accurate form removal, even on low quality or skewed images.

Preprocessing — It is desirable, although not required, to have a package that performs some line and noise removal on the extracted zones.

Reject Re-entry — Images which cannot be accurately recognized by the system's OCR should be displayed and highlighted to prompt the operator for corrective values if necessary.

The above tasks needed to be executed in real time as images are output from the scanner with the exception of the offline reject re-entry processing. This means that the forms processing software must be able to directly interface with the control structure written in C code.

3.2.1. *Survey of commercial form readers*

Most forms processing packages operate in a series of steps:

Zone Design — The user scans a blank form and defines all zones in which data to be recognized is contained. Zones are drawn with a mouse for easy and efficient definition. Attributes for all zones are compiled for use during recognition.

Scanning — Forms are scanned individually or in batches using a number of commercially available scanners suited to the application requirements. Variable data from the defined zones are stored for recognition.

Recognition — Images of the variable data are converted to ASCII and written to mass storage. All characters which cannot be accurately recognized are marked for post processing.

Post Processing — This paperless step may be performed when the user desires. Each unrecognized character and all others within a zone are magnified and displayed in their original image form on the computer screen. The operator is then prompted to enter the correct character which is then placed in the output stream.

Data Transfer — Data is edited into the desired output format for transfer to other applications.

Several commercially available forms processing and OCR packages were included in the survey. They were: AEG, ComCom Systems Inc., Recognition Research Inc. (RRI), Recognition Inc., Nestor Inc., TiS America, Symbus Technology, and Sequoia Data Systems. Of these, the following products were evaluated in greater detail.

AEG 6160 ICR:

The AEG system is able to recognize characters at great speeds, but it is limited to machine-printed and upper-case-only hand-printed characters. It is currently not available for the Sun platforms and cannot be used for mail processing applications.

Nestor NestorReader ICR:

Nestor's character recognition system is comprised of an all-software package. Hand and machine-printed alphanumeric character recognition is possible, with an advertised recognition rate of 99.8% on the NIST test set. Recognition can be done at a speed of 15 characters per second on a Sun SPARCstation. Segmentation of unconstrained touching characters can be done, and Nestor claims to handle sloppy writing and low resolution (Fax) degraded images.

TiS America FormOut!:

TiS America's FormOut! contains only form identification and removal routines that can be interfaced with a C program. The software development kit programmer's manual was used in detail to evaluate the system since it included examples of features available through the software. Noise, line, and skew removal is performed on the whole image during form removal, although the quality of these features appeared to be limited. Form definition was not very robust and required more effort than other packages.

RRI FIP2000:

RRI offers a complete forms processing package, from form identification, through form removal, and also reject re-entry. A demonstration copy of the system was obtained from RRI and helped greatly in the evaluation of the product. The task of defining a blank form is performed fairly easily by a user with little training. Attributes can be assigned to zones for use by the OCR downstream, and all image can be accessed through C functions. Reject re-entry provides two modes of operation. Individual character mode allows an operator to quickly process a continuous stream of rejected characters without regard to context. In full context mode, the rejected character is displayed with surrounding image data. RRI claims their research shows that allowing an operator to switch from one mode of work to the other actually increases overall productivity. Currently, the form definition and reject re-entry programs are not available for Sun SPARCstations, but are available for MS-DOS.

Based on the evaluation of the products mentioned above, RRI's FIP2000 form removal package along with their reject re-entry program was found to best suit the needs of the Forms Analysis system. Feature qualities found to exceed those from other products include:

- Ease of blank form definition. Their FIPEdit program provided a comfortable graphical user interface for selecting zones for removal.
- Many C functions available in the FIP2000 library to allow all necessary data extraction required by the control structure.
- Fast forms processing. Processing time, including form identification and removal, was under 1 second for all test images.
- Moderate cost.

3.2.2. *CEDAR forms processing software*

The major tasks in the forms processing component of the system are Forms Registration, Forms Identification, and Forms Reconstruction. A forms editor is required to create the definition of each type of form that the system is required to process. A forms library is also necessary, so as to store the different form definitions.

RRI's forms processing package has been replaced by a package developed at CEDAR. This task involved creating a forms editor, designing form definition files, and creating C library routines that can be linked by the programmer(s) of custom forms processing software. The library functions are used by the programmer(s) to write software for form identification, form removal and form reconstruction. The CEDAR forms processing package will serve this project as well as other forms processing projects.

The current status of the CEDAR forms processing software is as follows:

Forms Editor:

A forms editor must be easy for the user to use, yet be complex enough to handle the many different forms that it may come across. The CEDAR forms editor has a user friendly X Windows-Motif interface. There are menus and tool bar buttons that give the user access to editor features.

Menu options and pop-up windows enable users to specify control definitions and various user preferences. The user is able to create zones, fields, tags, lines and registration components to be saved to the form definition file. This form can be re-read by the editor for further definition or for visual feedback to the user. Processes allow the user to scroll on the image, zoom on the image and rotate the image. The user is also able to select form definition components. Once the form definition component is selected it can then be resized, moved or deleted. Also the editor allows the user to set attributes for fields and checkboxes. The editor has been integrated with the library so that form definitions produced with the editor are used by the library as input to the form processing.

Forms Library:

The forms library provides the user with a collection of functions that are important to forms processing. These library functions act on the images created by the forms editor. The high level processes are Forms Identification, Forms Removal and Reconstruction. A full specification has been developed for these functions.

Low level design work has provided a set of low level (internal) functions to be used by the Forms Processing Library. This helps keep the library relatively small. A high level specification finalizes the presentation of the library to the user. All main features of the library (with the exception for reconstruction) were integrated and tested to see a form through from registration to field snippets. This includes the library loading form definitions into memory. The library has been wrapped into high level user-callable functions according to the specifications.

Alternate forms of registration have been added to handle images that do not correspond to the final Reply Card specification. The library has undergone extensive testing with different combinations of user-available library functions being called. This quality assurance process ensures that the library (along with the editor) are in a state to be released.

3.3. *CEDAR Data Entry*

CEDAR has developed a field re-entry package on the Sun Sparc platform. We have used the experience gained here along with ideas learned from our analysis of several commercial data entry packages to develop a quality data entry system precisely suited for this application.

3.3.1. *Data entry software*

Recently a survey of data entry software was conducted as part of the ongoing Forms Analysis project. This survey compared the reject re-entry and manual entry features of various commercial forms processing packages, as well as the associated costs. Table 1 shows the comparison of the data entry features for these packages.

3.3.2. *Design considerations*

As mentioned above, CEDAR has surveyed over 20 forms processing and data entry companies in its endeavor to identify a commercial package that would fit the needs of the Forms Analysis project, and be easy to install and integrate with the rest of the system. We found that although several companies offer powerful data entry application development functionality, they require considerable effort to design applications. Some packages offered only simple re-entry of characters within a snippet field image, while most could not support importing of recognition data into their proprietary file format. Furthermore, none could directly make use of information stored in the form definition files.

After reviewing these packages, features were identified that would be attractive to users:

- A comfortable development environment which is simple to learn and use.
- MS windows based.
- Quick application creation.
- Drag and drop control of fields.
- Provide many validations for field data.
- All options and controls available through pop-up windows.
- Allow flexible image and data entry window placement.
- Use of standard input and output file formats.
- Allow entry of data for a whole image.
- Allow entry of data for selected snippet of the whole image.
- Allow re-entry of recognition results with highlighting option of low confidence characters.
- Allow linking of fields to each other.
- Tracking of operator keystroke statistics.

Table 1. Data entry software review

Company	Reject Re-entry	Manual Entry	Notes
Autodata	X		Focus is auto data entry.
Cardiff	X		Re-entry of their OCR'd data.
ComCom	X		Re-entry of their OCR'd data.
Computer Systems Co.			No data entry.
DataCap	X	X	Manual data entry possible with programming language provided. Applications cannot be easily created.
DataLex		X	Full feature Windows data entry, easy to integrate, no reject re-entry currently.
Diamond Head			Powerful auto data entry system using Visual BASIC interface. Focus on image scanning and display. No data entry.
GTESS	X		Auto data entry and validation.
Imagination			No data entry.
Kofax	X	X	Recommended TextWare.
Ligature	X		OCR'd text pages. Simple re-entry of page data.
Maxsoft			No data entry.
Microsystems Tech.			No response from company.
Mitek	X		Re-entry of their OCR'd data.
Principia	X		Mainly mark sense recognition with re-entry.
Recognition International	X	X	They just redistribute TextWare's DOS software. More powerful systems available.
RRI	X		Comfortable character re-entry tool. No snippet re-entry.
SCS	X	X	OS/2 full featured image data entry. Difficult programmer interface. They suggest SCS setup for clients.
Solution Technology			No data entry, only image viewing.
SPSS	X	X	Auto data entry with re-entry. No image based manual entry.
Tangent Systems			No data entry.
TextWare	X	X	Simple DOS based image data entry. Validation checks to handle most situations. Not very flexible.
TiS America			No data entry, only forms processing.
TurboPower Software			Borland C++ functions to do simple image data entry and validation.

3.3.3. *Full data entry from images*

No single commercially available data entry software package has all of the features needed by the Forms Analysis system. The CEDAR data entry package incorporates the desirable functionality described above and also introduces features that specifically benefit the system. These missing features include:

- Providing data entry and reject re-entry under one development environment in MS Windows.
- Use of form definition file attributes and image coordinates for data entry application design.
- Marking poor quality images for re-get and retrieval requests.

The data entry application development process is shown in Fig. 3. Users can create the forms definition manually using the specification of a reply card, or automatically using a form definition file for the card as defaults for field set up. The Screen Definition interface creates a definition file that is loaded into the Data Entry Processor interface along with a batch file describing which cards are to be processed (see Fig. 4). Several operators can be running Data Entry Processors on separate PCs to perform data entry in parallel.

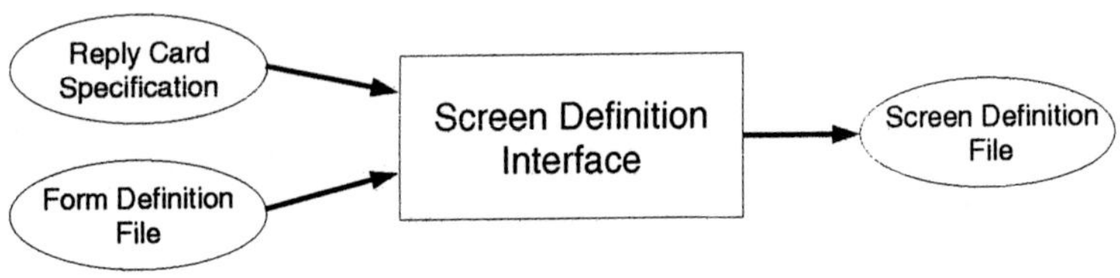

Fig. 3. Screen definition process.

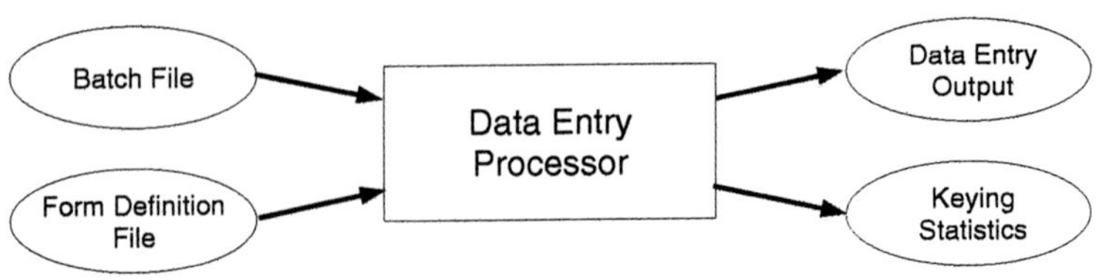

Fig. 4. Data entry process.

3.3.4. *Screen definition*

As mentioned previously, the data entry package provides a means to create data entry screens in one of two ways. The easiest way for screen development is to let the system create the screen definition based on attributes and image coordinates found in a corresponding form definition file. Using a form definition file, many of the field validations can be initialized. The screen developer only has to verify that these validation attributes are correct and enter any additional attributes that may be needed for data entry, but are not needed for recognition. Using form definition files to initialize the screen definition will leave the field order as it appeared in the form definition. After a screen has been defined using a form definition, the developer will have the ability to modify the screen definition in any manner to customize the screen for data entry.

If a form definition file is not to be used to initialize a screen definition, the data entry package gives the developer the ability to define all parts of the data entry screen. A screen window is drawn with the mouse by the developer to store a data entry window and image data. The data entry window holds the fields that the operator uses to enter the card data. The data entry window size and placement can be defined by drawing it with the mouse and it can be moved to any location within the screen window. Within the data entry window, fields are defined by dropping them with the mouse in the window. Default values are assigned to the field attributes initially, but can be changed through a popup attribute window.

The attributes available for a field include, but are not limited to:

- field name
- title string
- data display length data type
- print type
- classes
- entry order number
- output field number
- image number
- field image location (top, left, bottom, right)
- check digit algorithm
- double key flag
- lexicon
- range
- required field
- field mask
- number of characters
- case
- default value
- confidence threshold (for character highlighting)
- highlight color
- equation (to combine values from other fields)
- checkbox group (to link checkboxes together)

The data entry order defaults to the order in which the fields were created, but this order can be changed. The screen developer can link snippet field images or their location within a full image to a data entry field. This information is stored in the associated field attributes.

3.3.5. *Data entry processor*

As shown in Fig. 4, operators run the Data Entry Processor interface to perform the actual manual data entry functions. The Data Entry Processor reads a batch file which specifies the reply card files that are to be processed and the screen definition file that should be used for the data entry interface. The Data Entry Processor screen is initialized using the screen definition file, and all attributes for each field would be applied as the operator goes from field to field. The Data Entry Processor executes in an efficient manner such that the operator will not be significantly impacted by the program's processing overhead.

Operator keying statistics are output for each batch session. These statistics include items of interest common to typical data entry environments.

3.3.6. *Data entry file interface*

The Data Entry Processor is able to read standard TIFF format files as input. It is able to use full, removed, and snippet image data, along with field and character image coordinates. Recognition data is used to highlight low confidence characters for snippet verification and reject re-entry.

The Data Entry Processor writes the entered data to the Manual_Results section of the input TIFF file. This data can then be converted to a format that is able to be imported to the clients database using standard utilities available through the current system.

3.3.7. *Reject re-entry*

In addition to the full data entry of reply card images and snippets, a reject re-entry program has been developed which allows an operator to verify or correct recognition results at the character level. The user is able to use this program integrated with the data entry program, or separately as a stand alone program. As shown in Fig. 5, the Character Re-entry Processor operates on form images specified in an input batch file and, given a confidence threshold, displays those character images that have recognition confidence values below the threshold.

The character images are displayed in a manner that allows efficient re-entry. Many character images are displayed on the screen at one time to provide the operator with a look-ahead buffer. If the value cannot be determined from the character image, the snippet field is displayed as a contextual aid.

If the operator believes that the form image is of poor quality, options are available through the reject re-entry interface allowing the operator to make the appropriate requests to the system. Keying statistics are tracked as in the data entry interface.

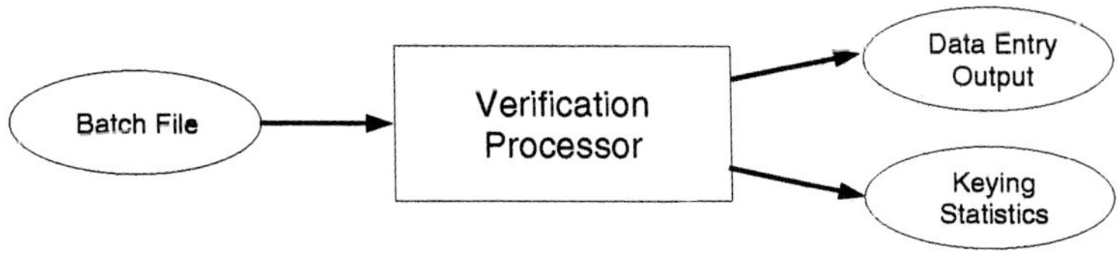

Fig. 5. Reject re-entry process.

3.3.8. *Reject re-entry file interface*

The Character Re-entry Processor is able to read standard TIFF format files as input. It uses the character coordinates and image data from snippet data in the TIFF file to display the individual characters, and the recognition result confidence values are used to determine which characters are to be displayed.

The Character Re-entry Processor combines the entered data with the already recognized data and writes this to the Manual_Results section of the input TIFF file. This data can then be converted to a format that is able to be imported to the clients database using standard utilities available through the current system.

3.3.9. *Commercial data entry package*

In order to provide data entry and reject re-entry capabilities in the system in the short term, CEDAR has chosen two commercial packages based on our review of packages mentioned previously. This software closely satisfies the requirements of the system.

CEDAR has created interface routines that allow the commercial packages to "plug into" the Forms Analysis platform with little or no changes to existing software. This involves converting the TIFF files containing images and recognition data into a format that is readable by the commercial software. Additionally, interface routines have been written to convert the commercial packages' output file format back to the TIFF format.

One key benefit of the CEDAR data entry software is its ability to read form definition files directly when creating data entry screens. All commercial packages create their own format data entry screen configuration files through a manual interface, allowing different validation checks to be specified for fields. CEDAR is investigating converting the form definition files to the commercial proprietary formats to be used in creating data entry interfaces.

4. Conclusions

We have given an overview of different concepts involved in the automatic analysis of printed forms. We have described the problem of forms analysis as the task of extracting accurate information using some knowledge of the structure of the form. Two systems developed at CEDAR, namely, the Name & Address Block

Recognition system and the CEDAR Forms Analysis system, have been described. In particular, the data entry subsystem of the latter has been described in detail. The capabilities of some commercial forms processing packages have also been highlighted. Forms analysis is one of the most active research areas in the field of document analysis, and much of the future research progress is expected to quickly lead to viable commercial products.

Acknowledgements

We are grateful to David Bartnik for providing much of the information on the CEDAR Forms Analysis system, including the data entry software comparison table. Thanks are also due to Dr. Yong-Chul Shin and Vemulapati Ramanaprasad, who provided details on the NABR project, and to Sridhar Parthasarathy for information on the CEDAR Forms Processing package.

References

[1] Lee Mantelman, Up to 98% labor savings with automated forms processing, *Imaging Mag.* **3**, 9 (1994) 14–30.
[2] R.G. Casey and D.R. Ferguson, Intelligent forms processing, *IBM Syst. J.* **29**, 3 (1990) 435–450.
[3] S.L. Taylor, R. Fritzson, and J.A. Pastor, Extraction of data from preprinted forms, *Machine Vision and Appl.* **5** (1992) 211–222.
[4] T. Watanabe, H. Naruse, Q. Luo, and N. Sugie, Structure analysis of table-form documents on the basis of the recognition of vertical and horizontal line segments, in *Proc. ICDAR-91*, 1991, 638–646.
[5] C.D. Yan, Y.Y. Tang, and C.Y. Suen, Form understanding system based on form description language, in *Proc. ICDAR-91*, 1991, 283–293.
[6] J. Yuan, L. Xu, and C.Y. Suen, Form items extraction by model matching, in *Proc. ICDAR-91*, 1991, 210–218.
[7] S. Chandran and R. Kasturi, Structural recognition of tabulated data, in *Proc. ICDAR-93*, 1993, 516–519.
[8] D. Monger, S. Leedham, and A. Downton, An interactive document image description for OCR of handwritten forms, in *Proc. ICDAR-93*, 1993, 524–527.
[9] D. Wang and S.N. Srihari, Analysis of form images, in *Proc. ICDAR-91*, 1991, 181–191.
[10] S. Lam, L. Javanbakht, and S.N. Srihari, Anatomy of a form reader, in *Proc. ICDAR-93*, 1993, 506–509.
[11] S.N. Srihari, Y-C Shin, V. Ramanaprasad, and D-S Lee, Name and address block reader system for tax form processing, in *Proc. ICDAR-95*, 1995, 5–10.

Handbook of Character Recognition and Document Image Analysis, pp. 503–528
Eds. H. Bunke and P. S. P. Wang

CHAPTER 19

PAPER-BASED MAP PROCESSING

HIROMITSU YAMADA *
Image Understanding Section, Electrotechnical Laboratory
Tsukuba, Ibaraki, 305 Japan

Feature extraction and symbol/character recognition from a paper-based map is discussed. The objective is a conversion from a scanned image to a computer internal representation using an appropriate data structure; roughly, we describe a raster-to-vector conversion. A method using the MAP concept (Multiple-Angled (directional) feature planes with Parallel operation and matching) is proposed. The MAP Operation Method extracts geometric features using directional erosion-dilation operations, and the MAP Matching Method recognizes fixed-shaped symbols by overall directional template matching. In both of these methods, feature representation on multiple directional planes and parallel operations are consistently used. During this raster-to-vector conversion, the following points are stressed: raster rather than vector representation, parallelism and directionality, independent extraction of each feature, and simultaneous segmentation and recognition. Through this example, problems of conventional methods, limits of the proposed approach, and future problems are discussed.

Keywords: Paper-based map; Raster-to-vector conversion; Parallel processing; Directional feature; Mathematical morphology; Generalized Hough transform; Multi-Angled Parallelism.

1. Introduction

1.1. *GIS: Geographic Information System*

Geographic Information System (GIS) is a computerized system which handles map information [1, 2]. GIS allows users to collect, edit, analyze and manage large volumes of spatially referenced data. It is organized as a four-part system consisting of input, storage, analysis and output (see Fig.1).

The sources of data for the GIS are (1) alphanumeric information, such as census and crop reports, field notes, and tabulations of historical meteorological data, (2) pictorial or graphic information, such as photographs and maps, and (3) remotely sensed data in digital form, such as that obtained from the LANDSAT satellite.

These data are grouped into two types of information: geometric and textual. The characteristic feature of GIS is that the data are spatially indexed, allowing the geometric information and the textual information to be combined and displayed in a variety of formats and media. This means that the spatial and textual data have to be interlinked in a consistent way in the database. GIS is a typical realization of a multi-media database system.

*yamada@etl.go.jp

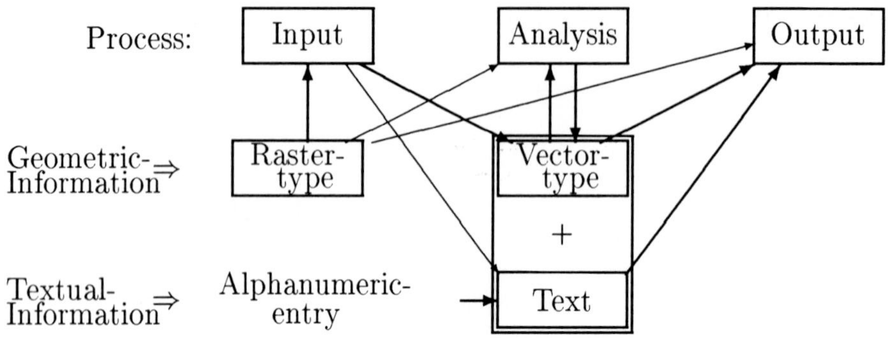

Fig. 1. Geographic Information System.

The textual information is further divided into two groups: (1) nominal information such as names of cities or rivers, and (2) scalar information such as elevation values. The textual information can be input by conventional alphanumeric entry equipment.

Geometric entries are classified into points, lines and regions. A point feature seldom has meaning by itself, and is usually merged into other features, as part of a dotted line, a symbol or a character. A datum point of elevation information in a Japanese map may be an exception since it is a nominal symbol and has meaning on its own (see Fig.12). However, a point as an internal entry is important to specify the location of textual information. It is also important as a feature point like a terminal point, crossing point or refracting point found on linear features.

A region can be expressed by a sequence of border lines. In that sense, it can be a higher-ordered feature of line segments. As another higher-ordered feature, the relation between the primitive component features has to be expressed in some way. In addition, layer representation where each layer (image) expresses different kinds of information such as contour lines, land usage, population, etc., is another way of describing information.

When we consider these kinds of information in the database, we have to assume that each geometric entry has some structure associated with it, that is, (1) a point will have a corresponding nominal attribute and they are linked together, and (2) a set of vectors could be linked together that specifies a region. On the other hand, geometric data in the external world has no structure, like a photograph which is input by an optical scanner. Input functions — the accumulation and organized representation of data — remain a major impediment to the adoption of GIS technology.

1.2. *Geometric Data: Raster and Vector*

(1) Most elemental representations of two-dimensional information such as photographs and graphics use a *pixel array*, where external two-dimensional information is expressed by an array of gray-levels, colors, or multi-spectral data. An external

two-dimensional plane is scanned raster by raster like a television image. That is the reason why this type of representation is called raster-type. (2) In order to compress the data, *run length* encoding is used, especially for binary images. This is also a kind of raster representation. However, it is not suited for two-dimensional processing. Only limited processing can be applied directly on a run length format. (3) In the *chain code* representation, the border of a region of a binary image is expressed by a coordinate of the starting point and a sequence of directions to a neighboring cell. This type of expression is closely related to the vector expression which will be discussed later. However, in the sense that all information can be recovered, we will include it here as a raster-type, although it can also be said to be an intermediate expression between raster-type and vector-type. (4) In a *pyramid* data structure, the same original image is expressed by multiple arrays of different resolutions. Typically, when the upper layer has twice the resolution of the lower layer, it is called a quad-tree data structure [3, 4, 5]. This type of data structure is used not only for raster-type data but also for vector-type data.

Raster representation has a rather flat structure. In order to realize linkages between geometric data and textual data, the data must have some structure. (1) The most fundamental structure of geometric information is a collection of *polygons*. A polygon is expressed by a sequence of directed line segments, i.e., vectors, and the data type is called vector representation. This is realized by an approximation of chain codes by line segments. By using this representation, a textual attribute at a vertex can be linked from the start-point or end-point of the vector. One region can be picked up instantly from a large database. However, since neighboring regions have different vectors for the same border, it is difficult to update the data.

(2) In an *arc-node structure*, a point is the fundamental element of the lowest level of representation and an arc is represented by a sequence of points. A node is an endpoint of an arc. By using this representation, links can be given from both a node and an arc. Neighboring regions can share the same border line.

(3) When the relations between nodes, arcs and textual entries are stressed, *relational structure* is often used. In this structure, the fundamental elements are nodes and arcs, but the topological information and attributes of the elements are separated. Thus there are more tables but it becomes easier to update attributes and the various linkages between attributes.

1.3. *Objective: Raster to Vector*

The objective here is the conversion from an external image data to an organized internal data representation, i.e., raster-to-vector conversion. First we review typical operations related to raster-to-vector conversion.

The first step of the conversion is vectorization, which is performed typically by first thinning and then approximating by line segments [6, 7]. Vectorization in character recognition is reviewed in [8].

Thinning warps shape information, especially when applied to wide lines. In order to make the features more precise, information along contours, or both thin lines and contours, are used [9, 10, 11, 12]. However, when we use contour information instead of thin lines, an extra process to connect corresponding lines at the crossings is needed. Naturally, finer details require greater computation. More complex feature extraction to refine line approximation will be discussed later.

Usually, raster-to-vector conversion involves more than just vectorization. The next stage is separation of text and graphics. A standard technique for the separation is detection of connectedness, in which large or long components are viewed as non-character entries and components of suitable size are viewed as candidates for characters [7, 13, 14, 15]. In order to extract a sequence of characters, an arrangement of extracted blocks are checked, e.g. by using a Hough transform [16]. When there is contact between a character and its background, simple information from connective components is not enough for the segmentation.

Complex geometric features like roads drawn by parallel lines or hatching regions are extracted afterwards [9, 10, 17]. This extracted information is used for higher level applications, such as extraction of features by question and answering [18] or automatic placement of names [19].

1.4. *Representation: Raster than Vector*

As stated earlier, vector representation can be useful to organize the geometric data of geographic information. However, vector representation is not a unique solution for the final representation. Here, the minimum requirement for the organization is, for example, to indicate the peak of a mountain to link to its name or the information of elevation. In that sense, if we can extract a feature point, interpret and locate it from original data, the raster representation can also fulfill this primal requirement. At the same time, the raster representation is also important for a faithful and natural display of the source data. Some GIS in effect have representation of both rasters and vectors, and uses them when necessary as the need arises.

Moreover, to determine which representation is better during the raster-to-vector conversion is another problem. Conventionally, data have been converted to vectors too early, in my opinion. Vector representation loses two-dimensional proximity. The resultant representation is very different depending on whether an object touches others or not. It is difficult to recover continuous two-dimensional information from this very discrete description.

Usually, character recognition is difficult in vector representation, so some kind of separation or recognition of characters is done before vectorization. However some problems are left between linear features and characters or symbols, e.g., touching characters. It is often said that this kind of problem should be solved afterwards by using higher-level knowledge, meaning, result of process, etc. However, vector representation makes it difficult to solve. As a result, such unresolved problems have a tendency to remain unresolved.

A database system (information system) for design and manufacturing is a CAD/CAM system. A paper map in GIS corresponds to an engineering drawing in a CAD/CAM system, so they have the same problems with the input of the data to the database system. Although a map is a kind of drawing, it exhibits some differences from an engineering drawing. The first difference is that because we are trying to represent natural data in GIS, the spatial distribution of the components is fixed whereas in an engineer drawing there is freedom to rearrange the components to aid our understanding. The second difference is that we want to condense as much information as possible in a map. In order to overlap the information, we use colors and different kinds of lines like dotted lines, bold lines, and parallel lines. For this kind of natural data of GIS, raster representation is more required than the artificial CAD/CAM data.

1.5. *Parallelism: as Architecture and as Independence*

We stress the use of raster representation during the raster-to-vector conversion process as long as possible. However, the use of raster representation affects not only storage but also computation. We have to access more redundant data than in the case of vector representation. It has been argued that the common iterative methods are of no use in a cost-effective OCR system without cheap, highly parallel processing power [8]. In that sense, pixel-wise parallelism inevitably becomes necessary. If we develop a pixel-wise parallel algorithm, we can use dedicated hardware for image processing and the realization becomes concrete. Thus, we emphasize the importance of pixel-wise parallelism in this chapter.

In addition to the pixel-wise parallelism, independent processes may be parallelized. A typical example is graphic and text processing, in other words, segmentation and recognition of text. Usually, a concatenate process is taken; linear feature extraction, then segmentation of a character box, and then identification of the separated character. Sometimes sequence changes such as character segmentation (and recognition) and then vectorization of the rest as graphic data. In any case, recognition of each character is tried after the segmentation process. However, if a character touches its background, this kind of process becomes difficult. In order to make the system robust, two processes of character segmentation and recognition should be independent, in other words, simultaneously processed. If two processes are independent, we may have a chance to combine the two results to obtain better results.

1.6. *MAP: Multi-Angled Parallelism*

Given these considerations, a method using pixel-based representation and pixel-wise parallel operations is proposed [20, 21, 22].

First, we use mathematical morphology for a base of parallel feature extraction. In order to enforce the performance, we introduce directionality, i.e., non-isotropic (directional) operations on multiple directional planes (images). We call this the MAP (Multi-Angled Parallel) Operation Method.

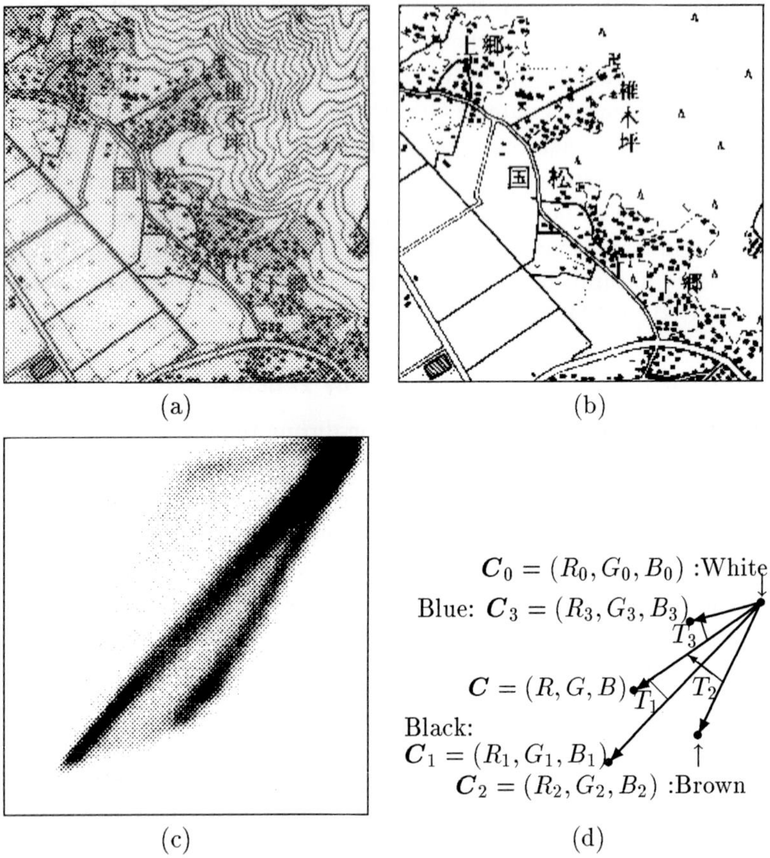

Fig. 2. 1/25000 topographic map of the Tsukuba area and color separation. (a) Three-color image (b) Black color image b (c) Color histogram (x=red, y=green) (d) Representative colors; C_0, C_1, C_2, C_3 are representative colors of white, black, brown, and blue, respectively.

At the same time, the MAP concept, i.e., Multiple-Angled feature planes with Parallel operations, is used for symbol recognition. This is called the MAP Matching Method, which is a kind of overall template matching method. It has the following strengths: (1) it uses edge strength and edge direction instead of gray-levels as a matching elements, and (2) it enables pixel-based parallel computation.

In this chapter, these methods are proposed as a basis for discussion. The following points are stressed: raster representation rather than vector representation, parallelism and directionality, independent extraction of each feature, and simultaneous processing of segmentation and recognition. Through an example, problems of conventional approaches, limits of the proposed approach, and future problems are discussed.

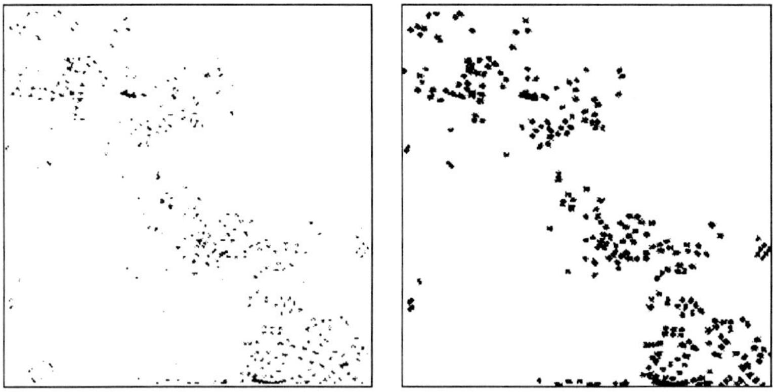

(a) (b)

Fig. 3. Results of Erosion (a) and Opening (b). (a) Eroded image $E^1_{n8}E^1_{n4}b$ (b) Building region f^{block}.

2. Color Separation

Before examining feature extraction, we will touch on color separation. In our experiment we used a 1/25000 Japanese standard topographic map printed in three colors, and scanned at a resolution of 300 dpi (see Fig.2(a)). In some cases such as a thematic map that a color is used to fill regions [23], however in our case, it is used to distinguish the species of lines.

Fig. 2(c) and (d) show a color histogram of scanned image where the x- and y-axes correspond to red and green. Note that this histogram has three linear distribution clusters, corresponding to black, brown and blue, respectively. It is essential that each distribution be linear in 2D-space or cylindrical in 3D(RGB)-space. By using this property, i.e. the excluding property of the color hue, the color separation is performed using the ratio of the distances to the center lines of the cylinders.

Suppose $\boldsymbol{C}_k = (R_k, G_k, B_k)$, $k = 0, 1, 2, 3$ are representative colors of white (background), black, brown and blue, respectively. These representative colors are set manually in advance of the experiments. Then, from the value $\boldsymbol{C} = (R, G, B)$ at each pixel, the value e_k of each color separated image with respect to $k = 1, 2, 3$ is obtained as follows (see Figs.2(c) and (d)).

$$
\begin{aligned}
e_k &= b_k \cdot h_k, && \text{for } k = 1, 2, 3, \\
b_k &= \max\{\ 0,\ \min\{\ 1,\ 1 - \ \left|\frac{|\boldsymbol{C} - \boldsymbol{C}_0| \cos T_k}{|\boldsymbol{C}_k - \boldsymbol{C}_0|} - 1\ \right|\ \}\}, \\
h_k &= \max_{k' \neq k} \frac{\sin T_{k'}}{\sin T_k + sinT_{k'}},
\end{aligned}
\tag{2.1}
$$

where T_k is the angle between $\boldsymbol{C}_k - \boldsymbol{C}_0$ and $\boldsymbol{C} - \boldsymbol{C}_0$ in RGB-space, b_k is the strength of color k in terms of brightness, and h_k is the strength of color k in terms of hue. More precisely, b_k is the difference of brightness from the representative point $\boldsymbol{C}_k$ along the representative color axis $\boldsymbol{C}_k - \boldsymbol{C}_0$, and h_k is the relative distance to the center lines of the representative colors. Thus, the inherent characteristics of the objects must be considered to refine the system.

3. Feature Extraction: MAP Operation

3.1. *Mathematical Morphology*

As stated earlier, we use pixel-based representation as much as possible. Consequently, pixel-based parallelism has to be kept in mind to make the approach workable by using a special purpose image processor. Mathematical morphology is a well-known technique used in applications such as printed circuit board inspection [24]. It has also been investigated from a mathematical viewpoint and can form an important component of dedicated image processors [25, 26].

Mathematical morphology is a set theoretic formalization of neighborhood operations of digital (in our case binary) images. Erosion and dilation are starting (basic) operations of this system. Erosion, $Ef(= E_{n8}f)$ in the first row in Table 1, is a logical AND operation with 3 by 3 neighbors with respect to every pixel on the image f. This is nothing more than a 3 by 3 neighbor operator and is suitable for any dedicated image processors using the pixel-based parallel concept. Erosion removes pixels touching other pixels that are part of the background (value 0). This removes a layer of pixels from the periphery causing some shrinking. Dilation $D_{n8}f$ is a complementary operation to add a layer of pixels to the periphery.

In a mathematical treatment, 3 by 3 neighbors are considered as small image S called a structuring element, and a set operation is defined between an input image f and the structuring element S. In Fig.4, assuming that S consists of four points, each of which is at a distance $\boldsymbol{v}$ from its origin, the erosion and the dilation can be expressed by the following equations,

$$\begin{aligned} E_S f &= \bigcap_{v \in S} f[-\boldsymbol{v}], \\ D_S f &= \bigcup_{v \in S} f[-\boldsymbol{v}], \end{aligned} \tag{3.2}$$

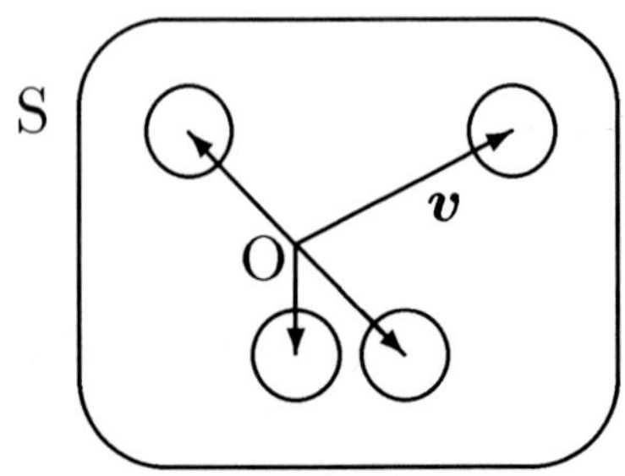

Fig. 4. Structuring element. In this case, the structuring element has four points.

and the opening and closing as follows:

$$Open_S f = \bigcup_{v \in S} (E_s f)[\boldsymbol{v}], \qquad (3.3)$$
$$Close_S f = \bigcap_{v \in S} (D_s f)[\boldsymbol{v}].$$

Here, the notation $[\boldsymbol{v}]$ means a translation by the vector $\boldsymbol{v}$. In Table 1, the notation $[d]$ represents a bit shift in direction d, which is defined by $0 \sim 7$ clockwise from the right horizontal, and the addition and the subtraction is followed by modulo 8 to obtain the direction.

Erosion E and dilation D, and *Open* and *Close* have duality. That is, corresponding operators have complementary results for the complement of an input image f, i.e., $Ef = \overline{D\overline{f}}$. By changing the operation $\cap$ to min and $\cup$ to max, the system can be applied to gray-level images. Furthermore, it can also be extended to continuous input waveforms.

The basic applications of mathematical morphology, erosion and dilation operations with isotropic neighbors, have been used in map processing. In [11], this process was utilized to separate solid objects and thin entities. In [27], a more complex procedure was used to segment the region of a thematic map.

Fig.3(b) is a result of erosion and then dilation operations (shown in Table 2) to the original black image b of Fig.2(b). This f^{block} almost completely corresponds to the building region in the map.

However, conventional operators extract only solid objects, i.e., building regions in the map. In order to enhance the performance, we introduce directional operators which enable a distinction between line and point features.

3.2. *Contour instead of Thin Line*

In the extraction of geometric features in a map, two approaches are taken: thinning and contour following. The former obtains thin (medial-axis, skeleton) lines, and the latter obtains border chains. When the object is drawn by relatively uniform thin lines, the thinning approach is much more compact and easily obtains the topology of lines like terminal points or crossings. However, some loss of information occurs, such as loss of line-width, and shape warping at crossings or corners. Sometimes, it is difficult to judge whether a short line segment is a signal or merely junk. When there are solid objects, the warping becomes more pronounced.

In order to refine the system, this lost information has to be recovered as the need arises. In [28], the original data is labeled by the line width, and then thinning is performed on each labeled image. In [11], thin entities are expressed by their core lines and thick objects by their boundaries. In [12], both thinning and contour following are used to separate touched texts from graphics.

From these considerations, when we extend morphological operators by introducing directionality, we implement the following procedure. First solid objects are excluded by conventional erosion-dilation operations with isotropic neighbors.

Table 1. Multi-angled parallel (MAP) operation.

Direction	Mask	Operation
Erosion		
No	No	$Ef = E_{n8}f = f \cap (\bigcap_{d=0}^{7} f[d])$
No	With	$E_{:g}f = E(f \cup g)$
Single	No	$E_{=d}f = f \cap f[d]$
Fan	No	$E_{>d}f = f \cap (\bigcup_{\delta=d-1}^{d+1} f[\delta])$
Dilation		
No	No	$D_{n4}f = f \cup (\bigcup_{d=0,2,4,6} f[d])$
No	With	$D_{:g}f = g \cap Df$
Single	No	$D_{=d}f = f \cup f[d+4]$
Fan	No	$D_{>d}f = f \cup (\bigcup_{\delta=d+3}^{d+5} f[\delta])$
Fan	With	$D_{>*d:g}f = (D_{>d-2:g} \cup D_{>d+2:g})f$
Macro Operation		
Open		$Open_{>*d:g}^{m}f = D_{>*d:g}^{m}E_{>*d}^{m}f$
Close		$Close_{>*d:g}^{m}f = E_{>*d}^{m}D_{>*d:g}^{m}f$
Terminal		$End_{d}f = f \cap \overline{E_{>d}f}$
Convex hull		$Fill_{odd}\ f = \bigcup_{d=1,3,5,7}(D_{=d-1} \cap D_{=d+1})f$
Diagonal hull		$Fill_{even}\ f = \bigcup_{d=0,2,4,6}(D_{=d-1} \cap D_{=d+1})f$
Isolate pixel		$Zero\ f = f \cup (\bigcup_{d=0}^{7} f[d])$
Shorten		$Short\ f = E_{>*0}f \cup E_{>*2}f$
Other Examples		
Iteration		$E^2D^2f = E(E(D(Df)))$
Composition		$(D_{=1} \cap D_{=3})f = (D_{=1}f) \cap (D_{=3}f)$
Direction-0		$f[0](i,j) = f(i+1,j)$
Direction-1		$f[1](i,j) = f(i+1,j+1)$
All directions		$f = \bigcup_{d=0}^{7} f_d$
Two directions		$E_{>*d}f = (E_{>d-2}f \cap E_{>d+2})f$
Two directions		$D_{>*d:g}f = (D_{>d-2:g} \cup D_{>d+2:g})f$
Mask by itself		$Open_{:*}^{m}f = D_{:f}^{m}E^{m}f$

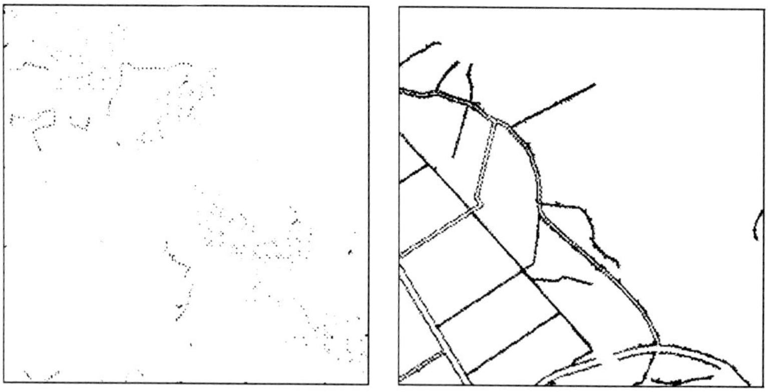

(a) (b)

Fig. 5. Results for dots and lines. (a) Dot feature f^{dot} (b) Line feature f^{lineb}.

Then we aim to operate on the contour (edge) rather than the inner region of the line because: (1) our objects are line-wise except for building regions, (2) as stated earlier, contour features are finer than line features, and (3) thinning is complex when implemented by morphological operators.

Now, let us define directional operators. Erosion-dilation basic operators are shown in Table 1, where erosion operators are represented by E, dilation operators by D and several macro operations formed by the combination of these two are presented. The superscript for an operator is the number of repeated operations, while the leading half of the subscript (such as $n4$, $n8$, $=d$, $=*d$, $>d$ or $>*d$) represents the direction of propagation, and the image information g following ":" denotes a mask image. Note here that the directional neighbors for erosion and dilation are opposite, and erosion and dilation have no property of duality. This definition is adopted in order to ensure that the erosion or dilation phenomenon at the end point in direction d is appeared at the same end.

The feature extraction procedures are determined by the combination of erosion-dilation, order of macro operators, and AND/OR operations. Each elemental operator within them are determined by defining the number of repetitions, the direction of propagation and the mask image.

From an original binary black image b (Fig.2(b)), eight directional planes depending upon the direction of edge are first made. These eight groups represent directional feature planes of different directions, and are called collectively a directional feature field. The creation of edges f_d^{contr} and f_d^{edge} ($d = 0, ..., 7$) are given by the equations in Table 2, where f is an image obtained by a certain operation, and f_d is a plane in the direction d. f^{contr} is a contour feature of an edge and f^{edge} is a basic edge feature.

Fundamentally, the small point feature f^{dot} (see Fig.5(a)) is extracted as a non-linear feature where all linear information disappears during a directional opening operation. Thus, thick solid objects are excluded by conventional isotropic mor-

Table 2. Feature extraction by MAP operation.

Feature	Procedure
f_d^{contr}	$b \cap \overline{b[d+4]}, \qquad d = 0,2,4,6$
f_d^{contr}	$f_{d-1}^{contr} \cup f_{d+1}^{contr}, \qquad d = 1,3,5,7$
f_d^{edge}	$f_d^{contr} \cap (f_d^{contr}[d-2] \cup f_d^{contr}[d+2]), \qquad d = 0,2,4,6$
f_d^{edge}	$\{f_{d-1}^{contr} \cap (f_{d-1}^{contr}[d-2] \cup f_{d-1}^{contr}[d+2])\}$ $\cup\{f_{d+1}^{contr} \cap (f_{d+1}^{contr}[d-2] \cup f_{d+1}^{contr}[d+2])\}, \qquad d = 1,3,5,7$
f_d^{short}	$Open^2_{>*d:*}\{f_d^{edge} \cup E^4_{>*d}(f_d^{edge} \cup D^2_{>d-2:f_d^{contr}} End_{d-2} f_d^{edge}$ $\cup D^2_{>d+2:f_d^{contr}} End_{d+2} f_d^{edge})\}$
f^{block}	$D^3_{:b} E^1_{n8} E^1_{n4} b$
f^{buil}	$D_{:f^{block}}\{f^{block} \cap \overline{\bigcup_{d=0}^{7} D^5_{>d} D^5_{>*d:f_d^{short}} Open^1_{>*d:*}(f_d^{contr} \cap \overline{f^{block}})}\}$
f^{dot}	$b \cap \overline{D^4_{:b} \bigcup_{d=0}^{7} Open^4_{>*d:*} b}$
f_d^{middle}	$Open^5_{>*d:*}\{f_d^{short} \cup E^4_{>*d}(f_d^{short} \cup D_{=d-2:b} D_{>d-2:b} End_{d-2} f_d^{short}$ $\cup D_{=d+2:b} D_{>d+2:b} End_{d+2} f_d^{short})\}$
f_d^{longb}	$Open^{17}_{>*d:*} f_d^{middle}$
f^{lineb}	$D^2_{:b} f^{longb}$
f_d^{longw}	$Open^{17}_{>*d:*}\{f_d^{middle} \cup E^8_{>*d}(f_d^{middle} \cup D^3_{=d-2} D_{>d-2} End_{d-2} f_d^{middle}$ $\cup D^3_{=d+2} D_{>d+2} End_{d+2} f_d^{middle})\}$
f^{cbase}	$D^2_{:b}\{(D^{100}_{:b} f^{lineb}) \cap \bigcup_{d=0}^{7}(f_d^{middle} \cup D_{>*d:f_d^{contr}} f_d^{short})\}$
f^{curve}	$D^2_{n8:b}\{\bigcup_{d=0}^{7} f_d^{short} \cap \{\bigcup_{d=0}^{7} f_d^{longb} \cup (D^2_{n8:b} \bigcup_{d=0}^{7} f^{cbase}$ $\cap \overline{D^1_{n8:b} f^{buil}})\}\}$
f_d^{hcore}	$D_{>d+4:\bar{d}} D_{=d+4:\bar{b}}\{(D_{:f_d^{contr} \cap \overline{f_{d+2}^{short}}} D^7_{>d-1:f^{contr}} End_d f_{d+2}^{short})$ $\cap (D_{:f_d^{contr} \cap \overline{f_{d-2}^{short}}} D^7_{>d+1:f^{contr}} End_d f_{d-2}^{short})\}, \qquad d = 1,\ 5$
$f^{honline}$	$D^{30}_{:f_d^{short}}(f_{d-2}^{hcore} \cup f_{d+2}^{hcore}), \qquad d = 3,\ 7$
f^{hatch}	$Open^2_{=*1}\{(D^5_{>7:\bar{b}} \cap D^5_{>3:\bar{b}}) \cup (D^4_{>3:b} \cap D^4_{>7:b})\} f^{honline}$
f_d^{rand}	$D^6_{>d+4:\bar{b}}\{(f^{curve} \cup f^{longw}) \cap f_d^{middle} \cap \overline{f^{hatch}} \cap \overline{f^{tall}} \cap \overline{E f^{buil}}\}$
f_d^{road}	$f_d^{rand} \cap f_{d+4}^{rand}$
$f_d^{chshort}$	$Open^1_{>*d:*}\{f_d^{edge} \cup E^4_{>*d}(f_d^{edge} \cup D^2_{=d-2} D^4_{>d-2:f_d^{contr}}$ $End_{d-2} f_d^{edge} \cup D^2_{=d+2} D^4_{>d+2:f_d^{contr}} End_{d+2} f_d^{edge})\}$
f^{chbase}	$b \cap \overline{f^{lineb}} \cap \overline{f^{dot}} \cap \overline{f^{buil}} \cap \overline{D^{10}_{n8:b} f^{hatch}} \cap \overline{D^{10}_{n8:b} f^{tall}}$
f^{chcore}	$D_{=d+4:\bar{b}}(f^{chbase} \cap f_d^{chshort})$
f^{char}	$D^3_{n8} E^8_{n8}(Fill_{odd} \cap Fill_{even})^{16} Close^6_{n8} D^{15}_{n8:f^{chbase}} D^2_{n8} Open^6_{n8}$ $(Fill_{odd} \cap Fill_{even})^{16}(f^{chbase} \cup Close^3_{n8})$ $Open^1_{n8} Close^1_{n8}[\bigcup_{d=0}^{7}\{f_d^{chcore} \cap (\bigcup_{d' \neq d, d\pm 1} f_{d'}^{chcore})\}]$

phological process, and by introducing the MAP operator, point features can be distinguished from others. From now on, the extraction of linear features becomes most important as a basic feature in a map.

3.3. *Linear Feature: Directionality than Connectivity*

In many cases, connectivity is a important cue in extracting linear features. However, it is a very discrete property: if a break exists in a line, the description becomes totally different. Moreover, since connectivity is a global property, it does not fit the local parallel operations we are considering now.

There is another type of information useful in extracting linear features: directionality, i.e., some kind of local straightness of the distribution of pixels. In some operations this property has been used to reinforce the local information as a linear feature. In [29], line crossing was detected by radial search in a non-thinning image. In [30], in order to separate overlapped linear features, directional distance transformation was used. In [12], both thinning and contour following are used. Linear segments whose attributes include both directionality and connectivity with their neighbors are extracted, and characters that touch backgrounds are separated. In all of these operations, some kind of directionality was used to extract linear features.

We can summarize the process of extracting linear features using the MAP Operation as follows (see Table 2).

The MAP Operation Method extracts the linear feature by using the straightness in each directional plane. First, directionality is very loosely defined where the directional planes are set by contour points (f_d^{contr}), i.e., a contour point in a diagonal direction is also set in the horizontal and/or vertical plane(s). Then the directionality is cleaned up by checking the 3×3 arrangement in the directional plane and f_d^{edge} is obtained. However, this feature is known to be easily disturbed by noise along the contour, so the noise is absorbed by connecting contour points held between the connected components in f_d^{edge}, which creates f_d^{short}.

Furthermore, the longer line f_d^{longb} is extracted from collinear short line segments f_d^{short} by connecting whose terminal points are separated by black pixels and not white pixels (see Fig.6). This process of connecting black pixels is effective at the crossing of lines (see Fig.5(b)).

Thus, when using straightness for the extraction of linear features in each directional plane, a small "core" of the linear feature is first extracted which is grown by the arrangement of the neighboring cores along the contour and then further grown on the condition that the feature can be connected not only on the contour but also on the black pixel. This procedure whereby indefinite features are merged with the advance of the globality of the feature enables stable feature extraction.

3.4. *Compromise with Connectivity*

We also must consider connectedness in order to obtain more "natural" lines. Linear features are extracted using the straightness obtained in the previous proce-

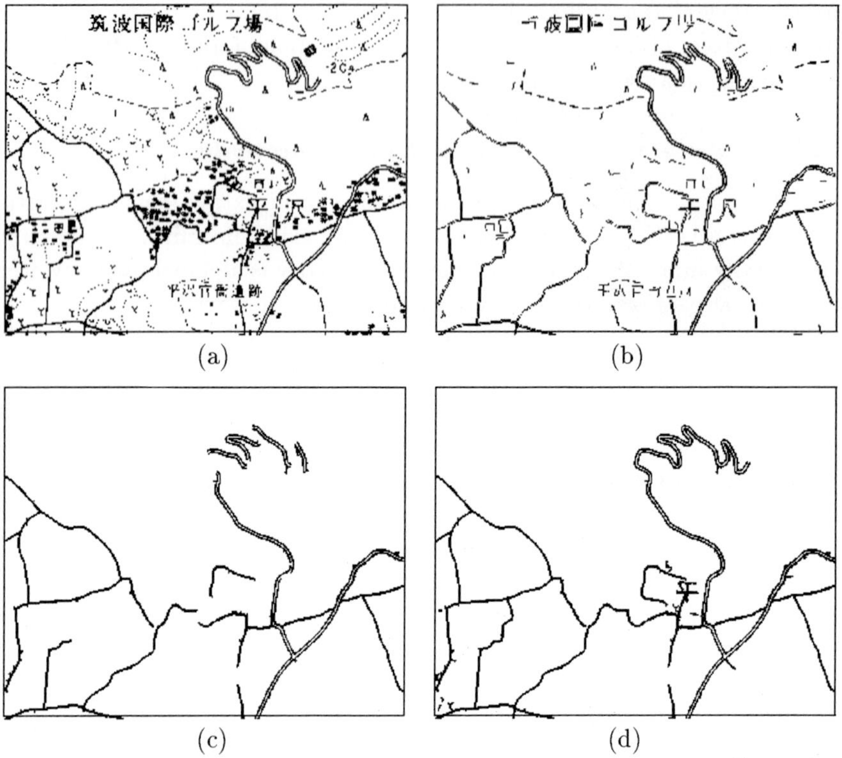

Fig. 6. Linear feature extraction. (a) Black image b (b) Middle line $f^{middle} = \bigcup_{d=0}^{7} f_d^{middle}$ (c) Long line f^{lineb} (d) Curve f^{curve}.

dure. However, features are sometimes lost where the curve is steep if we only use straightness as in Fig.6(c). This suggests that we should introduce "connectedness" in order to extract linear features of high curvature. For this purpose, information is exchanged between directional planes, the connectedness between segments on different planes is obtained, and the curved line (f^{curve}) is recovered (see Fig.6(d)). In the extraction of (f^{curve}), 100 iterations are used. This means we try to extract information nearly equivalent to connective components.

Here, the image processor attached to a conventional workstation which we used has a characteristic that the operation between directional planes is slower than that within the directional plane, so that we decided to operate in the planes as much as possible and then exchange the information between planes only when absolutely necessary. The operation within a plane and the operation between planes should be carried out together. To sum up, connectivity and directionality should be unified in a process of linear feature extraction.

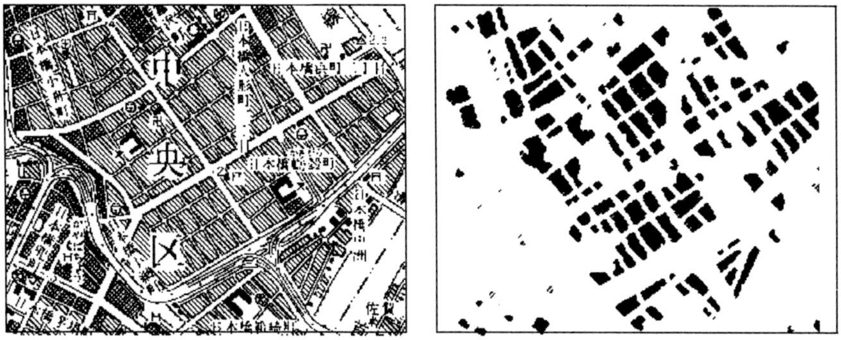

(a) (b)

Fig. 7. Hatched region extraction. (a) Black image b (b) Hatched region f^{hatch}.

3.5. *Extraction of Complex Features*

The fundamental reason for the introduction of directional operators on directional planes is the distinction between small points and lines. By combining these processes, we can extract more complex features like hatched areas and roads drawn by parallel lines.

A hatched area consists of a dense set of lines with a certain alignment

$$f^{hatch'} = D^4_{n8} E^6_{n8} D^2_{n8} (f_3^{short} \cup f_7^{short}). \tag{3.4}$$

The first (right) part extracts a parallel aligned feature, and the remaining (left) part fills the region. However, this simple procedure extracts a road as a hatched region if the road has the same orientation as the diagonal lines. The improved procedure in Table 2 excludes such cases by using the property that the lines of the hatching are connected to each other at both (or a single) end(s) (see Fig.7).

Character area is also extracted as a region where many lines of a certain length are concentrated. Note here that the "length" of lines ($f^{chshort}$ in Table 2) to detect the character area is not the same with the length of lines (f^{middle} or f^{short}) to extract, say, a road or river. More generally, the property as a "line" is not always the same. That is the reason we stress the parallelism, i.e., independence of procedures for each "line".

4. Symbol Recognition: MAP Matching

4.1. *Segmentation by Recognition*

In paper-based map processing, the separation between text and graphics is a major sub-goal of the total process. In many operations, a connective component is used to extract the text, i.e. first segmentation then recognition is performed. However, if a character touches neighboring graphics, this process becomes difficult. In that case, segmentation and recognition have to be processed simultaneously, i.e., segmentation by recognition.

In this section, we pursue this idea by activating parallel computations as much as possible.

4.2. *Hough Transformation and Template Matching*

When we do recognition without segmentation, exhaustive matching, i.e., matching with all categories with all locations, becomes necessary. The Hough transform is a well-known method of exhaustive matching.

The proposed method is related to the generalized Hough transformation method [31]. In the Hough transform, *voting* is performed using an *accumulator*. The Hough transform has a voting error, because the voting location differs from the location of the input feature. The error of the voting is proportional to the product of the error of input direction and the distance to the reference point. Thus, the error depends on the position of the reference point. By comparison, the proposed MAP Matching Method has no error with respect to the voting location because the voting place itself moves "onto" the input feature. The error of the input feature only decreases the counter value and has no positional effect. Furthermore, the voting in the Hough transform complicates parallel operation.

The MAP Matching Method is also related to the dynamic programming (DP) matching method [32]. When the MAP Matching Method is extended to an elastic model, it is a parallel computation of the DP matching method. An important difference is that DP matching evaluates the angle difference on all of the pixels along the model segments whereas MAP Matching evaluates the angle difference on restricted pixels only, i.e., at the evaluation points. Thus, MAP Matching can be viewed as a simplified, parallel version of DP matching. But MAP Matching Method is unique: it can express multiple directional strength at each pixel because of its representation of multiple directional planes.

4.3. *MAP Matching Method*

The reference shape model for recognition is expressed as a series of M evaluation points sampled from the contour points of a binary image in the following way:

$$R = \{(r_m, p_m, q_m), m = 1, 2, ..., M; r_m = 0 \sim 7\}. \tag{4.5}$$

where r_m specifies the directional feature plane at the evaluation point, and (p_m, q_m) is displacement from the prior evaluation point (see Fig.8). If $m = 1$, (p_1, q_m) represents the displacement from the end evaluation point M. The reference shape model for a factory is shown in Fig.9(d), where " $\vdash$ " represents the direction. For instance, " $\perp$ " stands for the direction 6. In the actual reference shape model, data such as displacement and the referred directional feature plane for each evaluation point are given in the form of a list. Fig.9(c) shows an image for creating a reference shape model which contains images of symbols to be recognized.

When performing matching through the use of such a reference model, directional feature planes are defined for the input image. That is, from input black

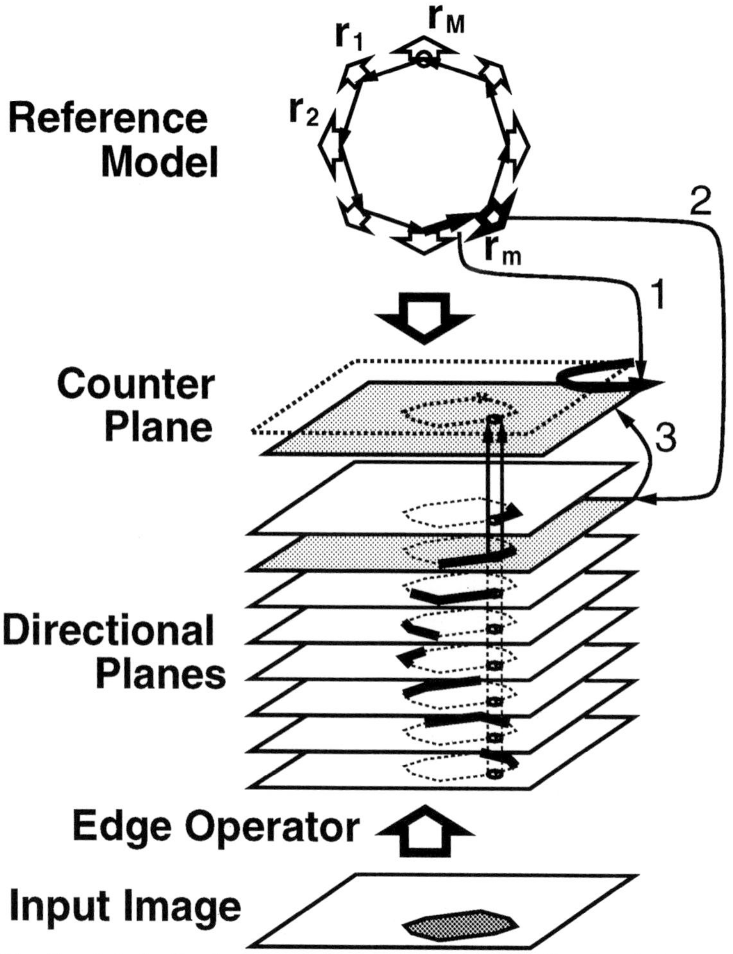

Fig. 8. Directional derivative from input image is set into each directional plane. The counter plane is cleared first. The following two operations of translation and renewal are iterated from $m = 1$ to $m = M$. (1) Translation of the counter plane by the displacement of r_m, and then (2) Renewal of the counter plane by the addition of (2.1) the maximum of neighbors in old counter plane and (2.2) the directional plane specified by the direction of r_m.

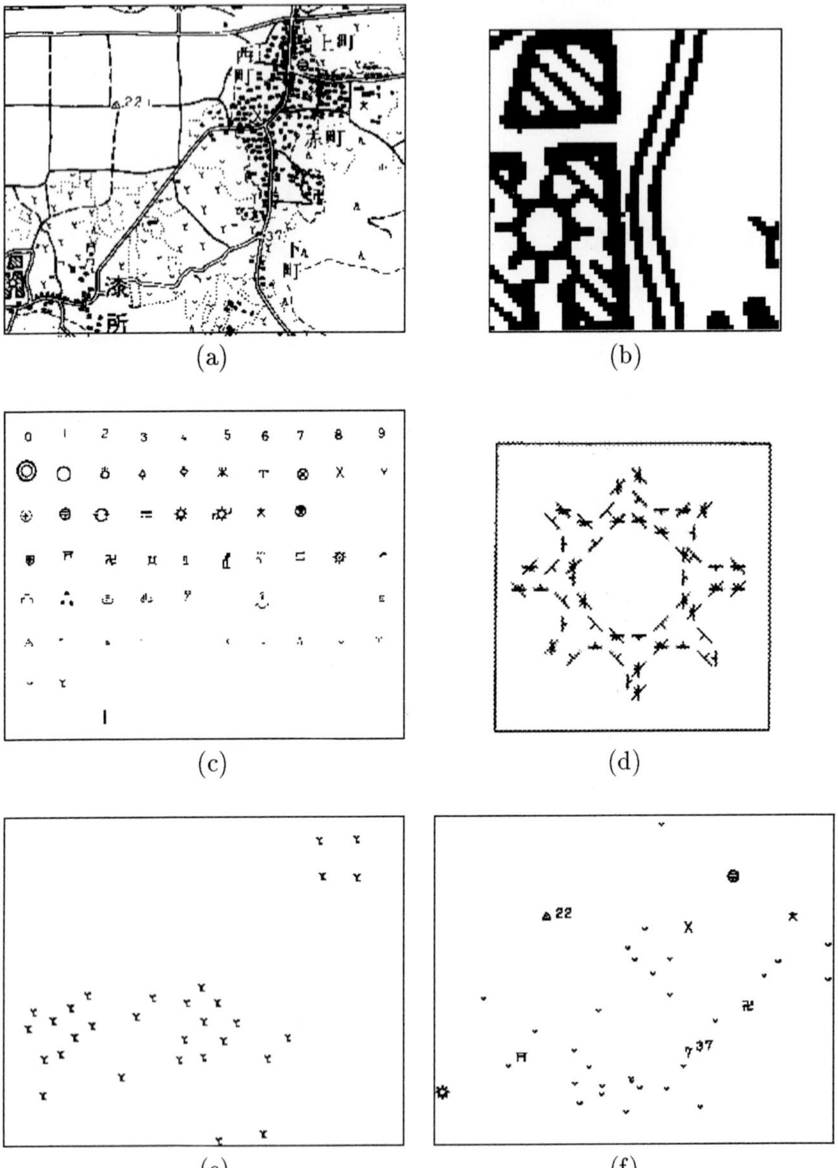

Fig. 9. Symbol recognition from the black image (a) of a topographic map. The MAP Matching Method extracts even when there are many contacts with background such as (b) which is enlarged at the lower left corner of the original image, where (c) is a legend image for model, and (d) is the model of factory. Many symbols are extracted simultaneously by parallel computation of the method, as in the case of the mulberry field (e). (f) is the results for other symbols.

image $b(i,j)$, eight directional planes f_d^{edge} are made by the MAP Operation procedure in Table 2. Each directional plane holds the edge information of that direction.

In the course of matching, a counter plane $c(i,j)$ is translated by a vector specified in the reference model, and the directional feature plane specified by the evaluation point is added in parallel to the counter plane.

$$c(i,j) = c(i - p_m, j - q_m) + \max_{(\epsilon_i,\epsilon_j)} \{ f_{r_m}^{edge}(i + \epsilon_i + \sum_{m'=1}^{m} p_{m'}, j + \epsilon_j + \sum_{m'=1}^{m} q_{m'}) \}. \quad (4.6)$$

When $c(i,j)$ and f_d^{edge} are expressed by C and F_d as whole images, and the notation $[x,y]$ denotes the translation of an image by (x,y), the equation can be rewritten as follows (see also Fig.8).

$$C = C[p_m, q_m] + \max_{(\epsilon_i,\epsilon_j)} \{ F_{r_m}[\epsilon_i, \epsilon_j] \}. \quad (4.7)$$

If the procedures of displacement and parallel addition are executed for all the evaluation points in accordance with the reference model, the similarity with the model at each point of the input image is stored at the counter plane $c(i,j)$ in a 1-to-1 correspondence. If the input image contains a symbol to be recognized, the similarity on the counter plane is highest at the center of the symbol and declines toward its periphery. If a point having a local maximum similarity above a certain level is found on the counter plane, it corresponds to the center of the symbol to be recognized. This technique is implemented by the repetitive use of "parallel translation" and "summation", an algorithm which can be processed at high speed by an image processor.

4.4. *Compromise between Segmentation and Recognition*

So far in this section, we have utilized exhaustive matching. This enables us to extract, for example, the factory symbol shown in Fig.9(b) where the symbol is touching with the background. We can also extract many mulberry field symbols as in Fig.9(c) at the same time with the use of parallel computation.

However, some segmentation tasks should be performed prior to the recognition, because the computation would otherwise become too large. The next section provides such an example of compromise between segmentation by recognition and segmentation prior to recognition.

5. Kanji Character Recognition

Theoretically, Kanji characters (Chinese-Japanese characters) can also be read by the MAP Matching Method; however, this is impractical since the sequence model of the MAP Matching Method takes too long to process the large number of categories and complex morphology of Kanji characters.

Thus, Kanji character recognition is performed by exhaustive pattern matching to the restricted region [33]. From the original black-colored image Fig.10(a), a

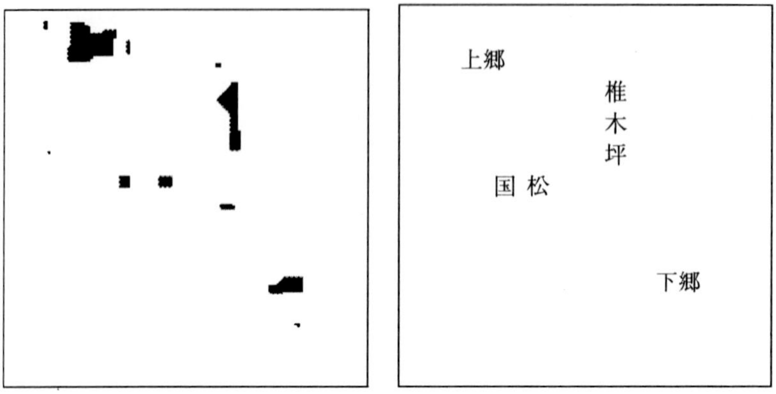

(a) (b)

Fig. 10. Result of Kanji character recognition. (a) Character region f^{char} (b) Recognition result.

candidate region for characters is extracted as f^{char} (see Table 2 and Fig.10(b)). Matching by correlation using directional patterns is performed at every location whose center is in the candidate region of f^{char}.

The directional pattern $G(i, j, \nu), 1 \leq \nu \leq 4$, is created as follows. First, the direction is defined on the contour point of a binary input pattern, and quantized into four directions (vertical, horizontal, and diagonal directions) and four directional patterns are made. Then these 16 × 16 patterns are blurred by a Gaussian operator and sampled into 4 × 4 patterns. A dictionary is made from 3543 categories × 5 samples. Similarity is defined as a normarized correlation. Normalization of the size is not performed, i.e., a single-font Gothic type and Ming type dictionary is used. The introduction of a candidate region reduces the computation to 1/20. The correct recognition rate was found experimentally to be 43/44 for Gothic type characters, and 92/100 for Ming type characters.

6. Recognition of Elevation Value

6.1. *Elevation Value*

Until now, we have investigated the method of separating characters and features from a single input image. In this section, within a framework of parallel computation, we show that these methods can be unified to accomplish a concrete objective, in this case, recognition of elevation information.

The description of elevation information in a topographic map consists of numeric text and one of three symbols: a triangulation point "△", a bench mark "□ · " or a datum point " · " (see Fig.11). The numeric value is not subjected to rotation with respect to the topographic map, and the character string is always held horizontally. Both triangulation points and bench marks have one subdecimal digit, while datum points have either none or one subdecimal digit. Numerals for

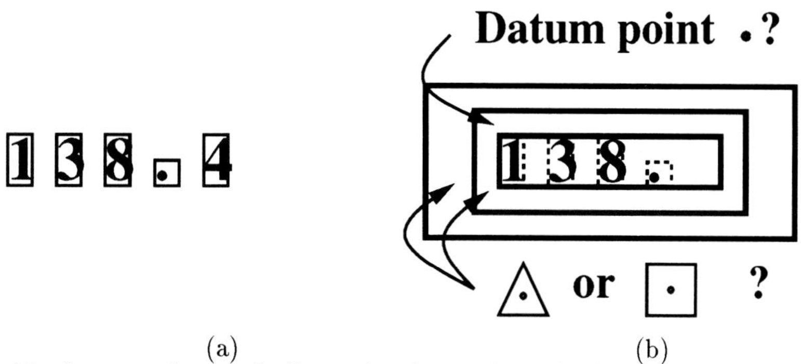

Fig. 11. Sequence of numerals. Numerals, trianguration point, bench mark, and datum point are first extracted independently, then the sequence is checked.

the integer part of the elevation value are larger than those in the subdecimal part. However, we are only concerned with recognizing the integer part.

The basic procedure for recognizing the elevation value is to recognize numerals and the aforementioned three symbols by MAP Matching. However, recognition of a shape by MAP Matching is inevitably limited. A typical example is the detection of the symbols " | " and " · ". These two very simple symbols are encountered much more frequently than might be expected. This is not a problem peculiar to the recognition of elevation values, but a common difficulty when extracting and recognizing known shapes out of complicated backgrounds, since " | " and " · " are the simplest and most basic geometric form, i.e., lines and points, respectively.

In this paper, advanced recognition is attempted through a combination of the MAP Matching with the MAP Operation. For " | ", lines are screened with respect to direction and length by the MAP Operation, and the remaining candidates ultimately confirmed by MAP Matching. The symbol " · " is extracted by the use of geometric characteristics having no line feature in any direction nor thickness seen in the building area. Moreover, datum point candidates are distinguished from decimal point candidates by counting the number of pixels contained in point-like features.

6.2. *Experiments*

Using the procedure described above, a recognition experiment was performed with 50 samples (Data Nos. tsukuba 50 ~ tsukuba 99) taken out of subimages of size 640 × 512 of a 1/25,000 scale topographic map of Tsukuba, scanned at a resolution of 300dpi. The results are shown in Table 3.

Of 75 elevation values, 68 were recognized correctly (recognition rate: 90.6%). None of the extracted elevation values were misrecognized as incorrect numerals or symbols ("mis-read"). Seven elevation values were not extracted ("un-read"). Four of these were not recognized because their datum points were in contact with other images (Fig.12(b), left and center). For the remaining 3 un-read datum points,

Table 3. Results of elevation recognition.

Data number	No. of elevation values	Correct read	Error-read: Mis-read	Error-read: Un-read	Over-read
Tsuku50∼59	14	12	0	2	2
Tsuku60∼69	16	16	0	0	2
Tsuku70∼79	16	13	0	3	2
Tsuku80∼89	13	13	0	0	1
Tsuku90∼99	16	14	0	2	2
Total	75	68	0	7	9

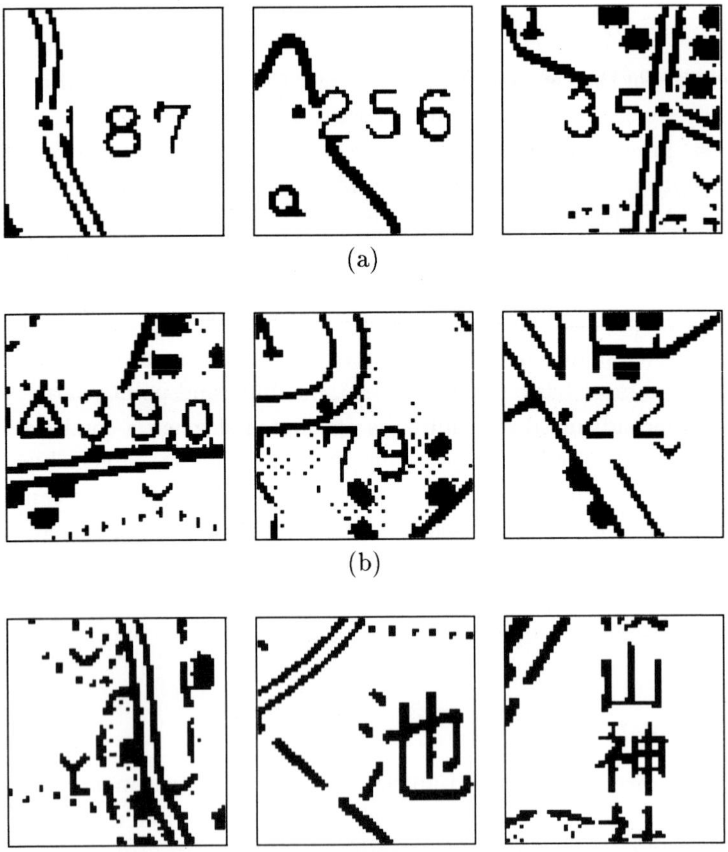

Figure 12: Result of elevation extraction. (a) Example of correct read. (b) Example of un-read where all data and decimal points are touching. (c) Example of over-read where the last example was read as a bench mark with 1m, while the others were read as a datum point with 1m.

elevation values were correctly recognized, but they were not recognized as datum points because their decimal points were in contact with other features (Fig.12(b) right). In all cases of recognition failure, point features could not be extracted owing to contact with other objects. "Over-reading", that is, extraction of an elevation value from a location where none existed, occurred in nine cases. Examples of mis-reading datum points due to the extraction of numerals or symbols out of the Kanji areas and point features in peripheral areas were: "· 1" in four cases, "· 6" in one case and "□1" in one case. Moreover, in three cases, "· 1" was recognized as an elevation value through the combination of a dashed line and point features. In Fig.12(c), a misrecognized datum point " · " is surrounded by a circle, and a misrecognized "1" by a rectangle.

The mean recognition time for images including 2 elevation values was about 15 minutes, timed from the start of directional feature field preparation on a "black image" using a SUN/3 and an image processor VITec.

6.3. *Discussion*

This chapter has presented an example of unification consisting of feature extraction by the MAP Operation and symbol recognition by MAP Matching, both of which were implemented with parallel operations in the directional feature field. It should be noted that model patterns involve different levels of shape complexity. For patterns with a high level of complexity, the primary task is the distinguishing of similar shapes, while in simple patterns such as points and line segments the primary task is not the recognition of rigid-shaped patterns, but rather the discrimination of information extracted from patterns of geographic features. In order to achieve these subgoals, it is necessary not only to analyze the geometric information of the object itself, but also to analyze the role of the object in relation to its surroundings, that is, its "meaning". In the example of elevation interpretation, the rule of the alignment between elevation symbol and numeral sequence performs this role. By using this rule, the meaning of " | " was settled.

Future work will include separating contiguous point-like features, recognizing the "· 1" formed by dashed line and point-like features, and improving the accuracy of edge feature extraction in the MAP methods.

7. Conclusion

This chapter discussed feature extraction and symbol/character recognition. Important features of this process are: raster representation rather than vector representation, parallelism and directionality, independent extraction of each feature, and simultaneous processing of segmentation and recognition.

Parallelism and Directionality: These features are exploited using the MAP concept which stands for Multiple-Angled feature planes with Parallel operations. The MAP Operation Method extracts geometric features using directional erosion-dilation operations, and the MAP Matching Method recognizes fixed-shaped symbols by overall directional template matching. Both of these methods consistently use parallel operations and feature representation on multiple directional planes.

Raster representation instead of Vector representation: By using pixel-based parallel computation, the possibility of using pixel-based representation in the recognition process was pursued. By these processes in this chapter, vectorization has not been done yet. As stated earlier, the vectorization is indespensable to unify geometric data with text data. However, this may be a rather easy job for each separated layer.

Independent extraction of each feature: Independence of the process was stressed; e.g., between line detection and point detection, between lines for characters and lines for roads. This will enable the combination of both features at higher stages in the future.

Simultaneous processing of segmentation and recognition: This is also a kind of independence between the processes, i.e., text and graphics.

All of these are directed toward making the system more complex, and the proposed methods are still incomplete. In order to make the system more robust and stable, we have to clarify the limits of each component technology. In that sense, independent development is important. In order to unify at a higher level, high quality component technology must be developed.

Acknowledgements

The author would like to express his sincere gratitude to his colleagues at the Image Understanding Section, Machine Understanding Division, Electrotechnical Laboratory for their invaluable discussions and suggestions.

References

[1] G. Nagy and S. Wagle, Geographic data processing, *Computing Surveys* **11** (1979) 139–181.

[2] R. Kasturi, R. Fernandez, M. L. Amlani, and W.-C. Feng, Map data processing in geographic information systems. *Computer* (1989) 10–20.

[3] H. Samet, C. A. Shaffer, R. C. Nelson, Y. G. Huang, K. Fujimura, and A. Rosenfeld, Recent developments in linear quadtree-based geographic information systems, *Image and Vision Computing* **5** (1987) 187–197.

[4] T. Matsuyama, L. V. Hao, and M. Nagao, A file organization for geographic information systems based on spatial proximity, *Comput. Vision Graph. Image Proc.* **26** (1984) 303–318.

[5] M. Sakauchi and Y. Ohsawa, The AI-MUDAMS: the drawing processor based on the multidimensional pattern data structure, *Computer Architechture for Pattern Analysis and Image Database Management* (IEEE Publ. 1986) 154–161.

[6] M. T. Musavi, M. V. Shirvaikar, E. Ramanathan, and A. R. Nekovei, A vision based method to automate map processing, *Pattern Recognition* **21** (1988) 319–326.

[7] L. Boatto, V. Consorti, M. D. Buono, S. D. Zenzo, V. Erano, A. Esposito, F. Melcarne, M. Meucci, A. Morelli, M. Mosciatti, S. Scarci, and M. Tucci, An interpretation system for land register maps, *Computer* **25** (1992) 25–33.

[8] R. W. Smith, Computer processing of line images: a survey, *Pattern Recognition* **20** (1987) 7–15.
[9] L. Heutte, J. M. Ogier, Y. Lecourtier, C. Olivier, Two aspects of automatic map treatment; road and texture extractions, *Proc. 11th ICPR* (1992) 109–112.
[10] M. Ejiri, S. Kakumoto, T. Miyatake, S. Shimada, and H. Matsushima, Automatic recognition of drawings and maps, *Proc. 7ICPR* (1984) 1296–1305.
[11] R. Kasturi, S. T. Bow, W. El-Masri, J. Shah, J. Gattiker, and U. Mokate, A system for interpretation of line drawings, *IEEE Trans. Pattern Anal. Machine Intell.* **12** (1990) 978–992.
[12] S. Shimotsuji, O. Hori, M. Asano, K. Suzuki, F. Hoshino and T. Ishii, A robust recognition system for a drawing superimposed on a map *Computer* **25** (1992) 56–59.
[13] H. Bley, Segmentation and preprocessing of electrical schematics using picture graphs, *Comput. Vision, Graphics, Image Processing* **28** (1984) 271–288.
[14] C. P. Lai and R. Kasturi, Detection of dimension sets in engineering drawings *IEEE Trans. Pattern Anal. Machine Intell.* **16** (1994) 848–855.
[15] A. J. Filipski and R. Flandrena Automated conversion of engineering drawings to CAD form, *Proc. IEEE* **80** (1992) 1195–1209.
[16] L. A. Fletcher and R. Kasturi, A robust algorithm for text string separation from mixed text/graphics images, *IEEE Trans. Pattern Anal. Machine Intell.* **10** (1988) 910–918.
[17] T. Nagao, T. Agui, and M. Nakajima, An automatic road vector extraction method from maps, *Proc. ICPR* (1988) 585–587.
[18] R. Kasturi and J. Alemany, Information extraction from images of paper-based maps, *IEEE Trans. Soft. Eng.* **14** (1988) 671–675.
[19] H. Freeman, An expert system for the automatic placement of names on a geographic map, *Information Sciences* **45** (1988) 367–378.
[20] H. Yamada, K. Yamamoto, T. Saito, and S. Matsui, MAP: Multi-Angled Parallelism for feature extraction from topographical maps, *Pattern Recognition* **24** (1991) 479–488.
[21] H. Yamada, K. Yamamoto, T. Saito, and K. Hosokawa, Directional mathematical morphology and reformalized Hough transformation for the analysis of topographic maps, *IEEE Trans. Pattern Anal. Machine Intell.* **15** (1993) 380–387.
[22] H. Yamada, K. Yamamoto, T. Saito, and K. Hosokawa, Recognition of elevation value in topographic maps by Multi-Angled Parallelism, *Int. J. Pattern Recognition and Artificial Intelligence* **8** (1994) 1149–1170.
[23] R. Espelid, N. A. Alvsvaag, J. Eileng, and I. Skauge, Automatic Digitizing of the colour-layer of thematic maps, *IAPR Workshop on Machine Vision Applications* (1990) 299–302.
[24] M. Ejiri, T. Uno, M. Mese, and S. Ikeda, A process for detecting defects in complicated patters, *Comput. Graphics Image Process.* **2** (1973) 326–339.
[25] J.Serra, *Image Analysis and Mathematical Morphology* (Academic Press, London, 1982).
[26] R. M. Haralick, S. R. Sternberg, and X. Zhuang, Image analysis using mathematical morphology, *IEEE Trans. Pattern Anal. Machine Intell.* **PAMI-9** (1987) 532–550.
[27] M. M. Ansoult and P. J. Soille, Mathematical morphology: a tool for automated GIS data acquisition from scanned thematic maps, *Photogrammetric Engineering and Remote Sensing* **56** (1990) 1263–1271.
[28] S. Suzuki and T. Yamada, MARIS: map recognition input system, *Pattern Recognition* **23** (1990) 919–933.
[29] M. Ejiri, S. Kakumoto, T. Miyatake, S. Shimada, and I. Iwamura, Automatic recognition of engineering drawings and maps, in *Image Analysis Applications*, eds. R.Kasturi and M.M.Trivedi (Marcel Dekker, New York, 1990).
[30] T. Kaneko, Line structure extraction from line-drawing images *Pattern Recognition* **25**

(1992) 963–973.

[31] D. H. Ballard, Generalizing the Hough transform to detect arbitrary shapes, *Pattern Recognition* **13** (1981) 111–122.

[32] H. Yamada, C. Merritt, and T. Kasvand, Recognition of kidney glomerulus by two-dimensional dynamic programming matching method, *IEEE Trans. Pattern Anal. Mach. Intelligence* **10** (1988) 731–737.

[33] T. Saito, H. Yamada and K. Yamamoto, Character recognition of topographical map by exhaustive pattern matching, *Tech. Report of IEICE Japan* **PRU90-37** (1990) in Japanese.

Handbook of Character Recognition and Document Image Analysis, pp. 529–555
Eds. H. Bunke and P. S. P. Wang

CHAPTER 20

INTERPRETATION OF MAPS: FROM BOTTOM-UP TO MODEL-BASED

RIK D.T. JANSSEN
Delft University of Technology, Faculty of Applied Physics, Pattern Recognition Group, Lorentzweg 1, 2628 CJ Delft, The Netherlands

This chapter describes aspects of methods for map interpretation. Maps are made according to rules, but this only slightly simplifies the design process of interpretation systems. This is because the same map can be drawn by different people having different styles. Also, updates on the same map can be months apart, causing aging effects. Sometimes, the rules are more obeyed, sometimes, they are less obeyed. A robust map interpretation system has to take this all into account.

In this chapter two systems for cadastral map interpretation are compared: a bottom-up system and a model-based system. The model-based system uses expectations of reality to correct and improve the interpretation. Apart from a better performance, a number of problems faced in the bottom-up system can be solved more easily in a model-based system. Also, parts of the model-based system can be re-used in an interpretation system for other types of maps.

Keywords: Model-based interpretation; Bottom-up interpretation; Cadastral map interpretation; Vectorization.

1. Introduction

Automatic map interpretation is a relatively new field in image processing [1–9]. Until recently, maps like cadastral maps were mainly drawn by hand on paper or film, and then stored in large cabinets. The disadvantage of this compared to computer accessible maps is the large cost for drawing, storing and maintaining them, and their fragility since paper maps are easily torn. After some updates a completely new map must be made to incorporate these changes.

By contrast, computer stored maps are readily accessible, changes are easy to incorporate, and they can be accessed from different locations using graphic terminals. Information can be processed using e.g. Geographic Information Systems (GIS) [10, 11].

Nowadays, new maps are often made with the computer. Existing maps (drawn on paper or film) can be acquired by scanning them and storing them on disk or tape. However, for meaningful queries some kind of interpretation is necessary. This can be done in several ways, of which human interpretation is the most obvious, and, because of the time required, also the most expensive.

In Sec. 3 (based on [6]), a bottom-up system for cadastral map interpretation is presented.

The basic knowledge of rules cartographers use for drawing is incorporated in the system. This high level knowledge enables us to anticipate the objects that are to be recognized. Problems associated with a good vectorization are discussed. Several "clean up" rules to improve the vectorization are proposed. This is different from the approach followed in e.g. the map recognition system of [1], in which a human operator is required for improvement of the vectorization generated by the computer.

In Sec. 5 (based on chapter 5 in [9]), a different, model-based interpretation method is presented. It appears that a number of problems encountered with the bottom-up system can be solved relatively easily in the model-based system. To give an introduction to model-based image interpretation, a brief literature review is given in Sec. 4.

We have chosen to interpret maps rather than other kinds of line drawings (e.g. schematic diagrams, CAD drawings, etc.). This was done because cadastral maps are relatively simple compared to other line drawings. Also, cadastral maps are made with a model (the rules used when drawing the map). For the Dutch cadastre, these rules are even laid down in a regulation by the Dutch Government [12] (in English: the Minister of Public Housing, Zoning, and Environment).

Another reason maps were chosen is because map interpretation is an economically interesting problem. Still a large number of maps only exist on paper or film, and do not have a computerized representation (in spite of the accessibility and the simplicity of incorporating changes in the latter representation). The reason for this is that the most straightforward method for converting from paper to the computerized representation is by redrawing the map (with a mouse on a digitizing table) and interpretation by hand, but this is cumbersome and laborious. Since computers have not succeeded very well (yet) in the automatic interpretation of maps it is also a very interesting research problem.

2. Cadastral Maps

Examples of cadastral maps are shown in Figs. 1 and 2. On cadastral maps *parcels* are drawn. A parcel consists of one or more polygons. Each parcel has one *parcel number* (no more, no less). The cadastral map specifies where the area represented by the parcel can be found in the real world. The number, which is usually in the middle of one of the polygons of the parcels, is used for reference. Different polygons which belong to the same parcel are connected using *connection signs* (small dotted S-like shapes). Examples are shown in Fig. 3.

A parcel number is sometimes written outside the polygon(s) of the parcel: if the polygons are too small such that they cannot contain the parcel number, the parcel number is written next to the parcel, enclosed with dots, and a dotted line is made from this dotted circle to the inside of one of the polygons of the parcel, see e.g. parcel 3568 in the top rightmost corner of map "leiden" in Fig. 2.

Sometimes, *building hatches* are found on cadastral maps (Fig. 4). Map "geo" has some of them. These are short parallel lines pointing to the inside of the buildings on a parcel. In the remainder of this chapter, a reference to a cadastral map is to be read as a reference to a Dutch cadastral map.

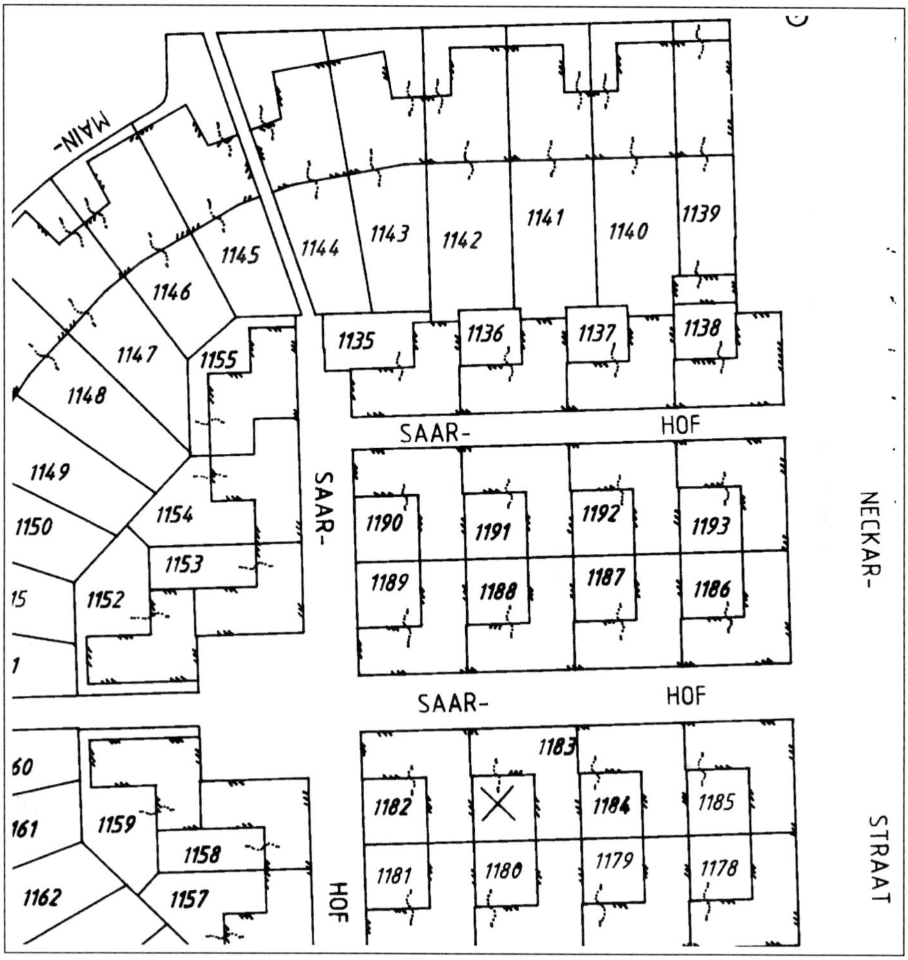

Figure 1: Section (1970 × 2000 pixels) of map "geo".

Figure 2: Section (2000 × 2000 pixels) of map "leiden".

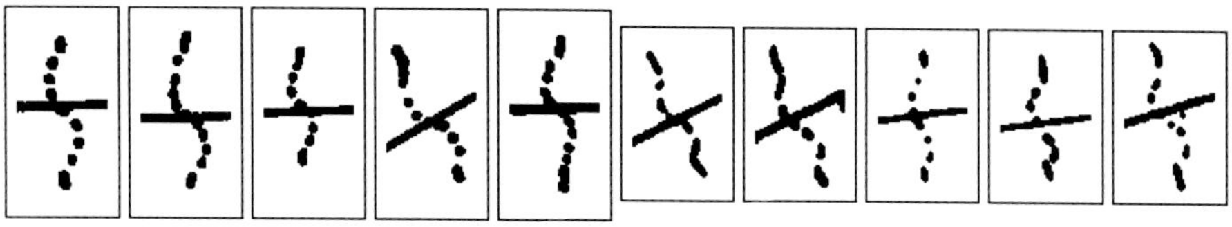

Figure 3: Examples of connection signs. The five on the left are from map "geo", the five on the right from map "leiden".

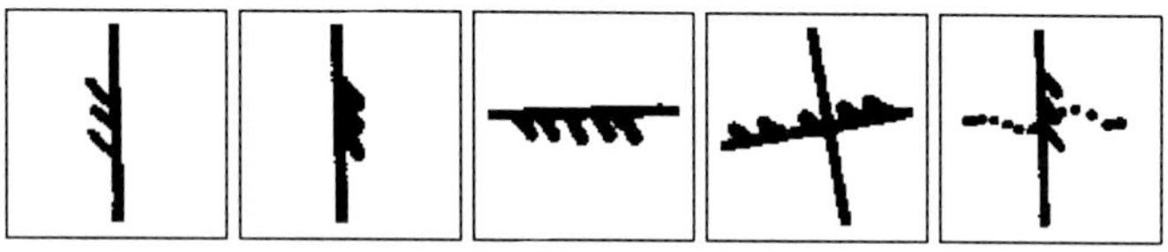

Figure 4: Examples of building hatches in map "geo". Building hatches may have crossing lines or can be drawn through connection signs as shown.

3. Bottom-Up Interpretation of Cadastral Maps

In this section, building blocks for a bottom-up cadastral map interpretation system are described. The emphasis is on the problems encountered with building such an interpretation system. Possible solutions are presented. The complete map interpretation system, including evaluation results and further improvements, is described in both [6] and in chapter 2 of [9] (the latter is more extensive).

A common bottom-up interpretation strategy is to first remove noise in the input image, followed by skeletonization. This is required for vectorization. Finally, detection of connection signs, parcel numbers, and parcels is performed.

3.1. *Noise Removal*

The problem

Often, there are small irregularities in the bitmap of the scanned image. Depending on the resolution, the quality of the scan and the background of the original drawing, various preprocessing steps may be required. Examples of "noise" are given in Fig. 5. A distinction is made between small irregularities in the bitmap caused by e.g. pixels having "wrong" values (Fig. 5(a)–(c)), between irregularities caused by a non-continuous flow of ink from the drawing pen (Fig. 5(d)), and the remainder, for example resulting from aging of the original drawing or (coffee) stains, or a case as in Fig. 5(e).

All these can influence the following processing operations (vectorization, interpretation) in a negative way.

A solution

In the case of pixels having "wrong" values, often a median filter suffices to adjust these. Sometimes, regular image processing operations may be useful, such as opening or closing.

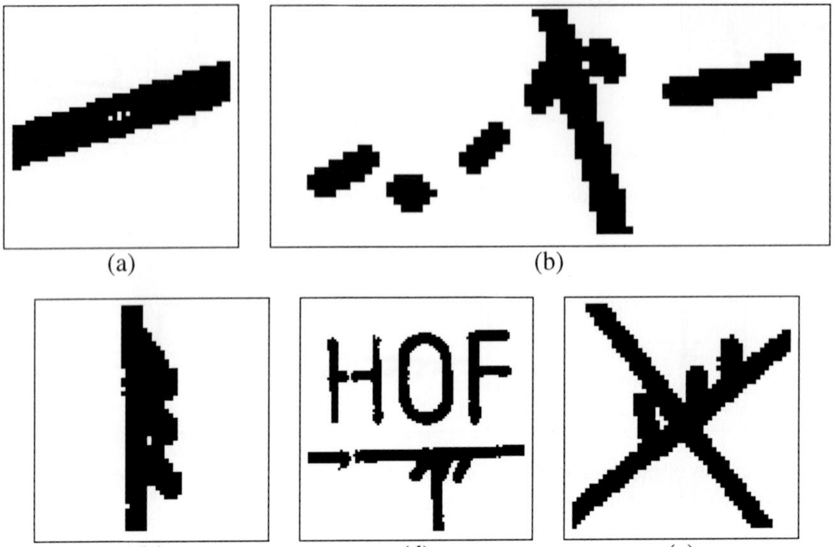

(a) (b) (c) (d) (e)

Figure 5: Examples of fractions of maps with pixels having "wrong" values. In the last example, the pixels do not really have a wrong value but a gap is introduced because the building hatches and the lines are intersecting.

One can also choose to (partially) postpone the correction to a later phase when more is known about the image or the objects in the image.

3.2. *Skeletonization*

The problem

Most vectorization approaches for line drawings require lines of one pixel thickness in the input image. To prevent discontinuities, the thickness of the lines after scanning should be at least one pixel. However, in the scanned image lines with a larger thickness can appear since lines in the original image may be of varying thickness.

To apply the vectorization algorithms to this type of drawings, the lines in the scanned image have to be reduced to one pixel thickness. This is usually done with an eight-connected skeletonization algorithm. However, it has several disadvantages: corners and T-junctions are often distorted because they may consist of very short spurious or unnecessary lines. Apart from that, skeletonization is sensitive to rotation.

A solution

To solve these problems there are two approaches: a sensible choice of skeletonization algorithm, or just to ignore them and to postpone correction until after vectorization. In the latter case small edges may be observed in the vectorization.

In the former case, one can choose between a number of skeletonization algorithms [13–18]. We have used a pseudo Euclidean skeleton [15]. This is an improvement of the Hilditch skeleton [13] because a better (almost Euclidean) metric is used. It ensures that the skeletonization is less sensitive to rotations when compared to the Hilditch skeleton. Another advantage is that the skeleton is more "straight", especially in the neighborhood of junctions. For an example see Fig. 6. Additionally, we have used "clean up" rules to correct the vectorization if necessary.

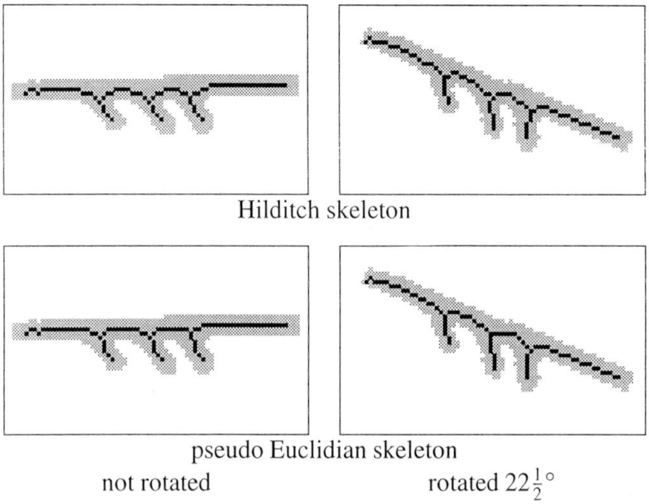

Figure 6: Differences between the Hilditch and the pseudo Euclidian skeleton. Significant differences can be observed in the neighborhood of the junctions.

3.3. *Vectorization*

The problem

Following skeletonization, the resulting image is vectorized. Again, there are numerous algorithms which can be used (see e.g. [19–23]). For an overview see Janssen and Vossepoel [23].

For a simple and fast vectorization algorithm producing good results the Douglas–Peucker algorithm [19] can be used. This method uses a maximum allowed perpendicular distance in pixels t_{dp} from the vectorization to the curve to approximate as a criterion. The algorithm starts with two dominant points of the curve as nodes of the vectorization. When the maximum perpendicular distance from this vectorization to the curve is larger than t_{dp}, a new node is added to the vectorization at the place of the maximum deviation. This process is repeated recursively. See Fig. 7.

Due to the use of skeletons, the vectorization may have short, unnecessary or spurious edges. We have designed different rules (*clean up* rules) to refine the vectorization.

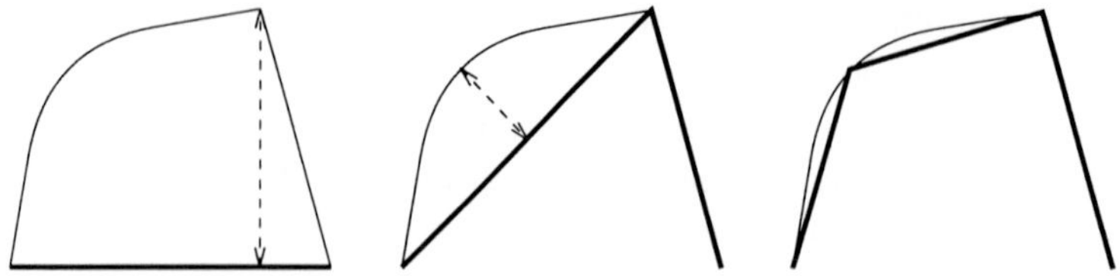

Figure 7: The Douglas–Peucker algorithm in subsequent iterations. The thick lines are the edges from the vectorization; the thin lines are from the original image. The dashed arrow (perpendicular to the approximating edge) gives the maximum (perpendicular) distance. New nodes are inserted until the length of the dashed arrow is less than or equal to t_{dp}.

A solution: "clean up" rules for the vectorization

There are various possibilities for creating clean up rules for the vectorization. We have used the following rules:

Rule 1. A rule to close gaps in the vectorization. This is done by finding two end nodes that have a small distance. If their edges make a small bend angle (e.g. $|\alpha| \leq 5°$), the two nodes are replaced by a new node in their middle. For an example, see Fig. 8. The nodes (i) and (ii) are removed and replaced by (iii).

Rule 2. A rule to remove link nodes whose two edges make an angle of almost 180°. This can be done if the bend angle between the edges is e.g. $|\alpha| \leq 5°$. This is shown in Fig. 9. Node (i) is removed in the process.

Rule 3. In corners in the vectorization, often a short unnecessary edge is generated as shown in Fig. 10. By setting constraints on the length of the edges and their angles, the short edge can be removed. The new node is the intersection of the base of the two long edges (i) and (iii). It replaces edge (ii) as shown in the figure.

Rule 4. T-junctions usually consist of three long edges and sometimes spurious short edges. An example is given in Fig. 11 with three short edges ((iv), (v) and (vi)), but it can also happen that there are only one or two short edges. The short edges can be removed by setting constraints on the angles with other edges and constraints on their lengths.

Rule 5. X-junctions are sometimes vectorized as two separate T-junctions as shown in an example in Fig. 12. Then, the two T-junctions are connected with a very short line. The rule to correct X-junctions uses the length of the edge between nodes (i) and (ii). If this length is short, the two nodes are snapped to a new node in between the two old ones.

The results of these refinements are a reduction in the number of nodes and edges, at the cost of a small increase in vectorization error (this has been shown in [6]).

3.4. *Detection of Connection Signs*

The problem

Connection signs are drawn in many different ways (cf. Fig. 3). This is because maps are drawn by hand by different people at different moments (e.g. at creation of the drawing or at

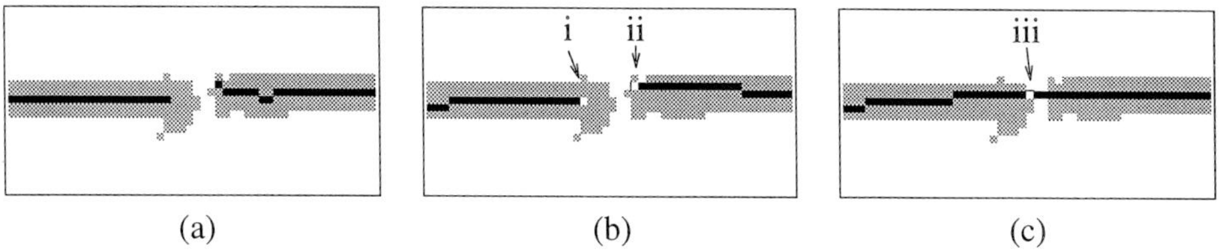

Figure 8: Example of the rule to close gaps in the vectorization. (a) shows the skeletonization plotted on the bitmap, (b) the raw vectorization from the skeleton, and (c) the resulting refined vectorization.

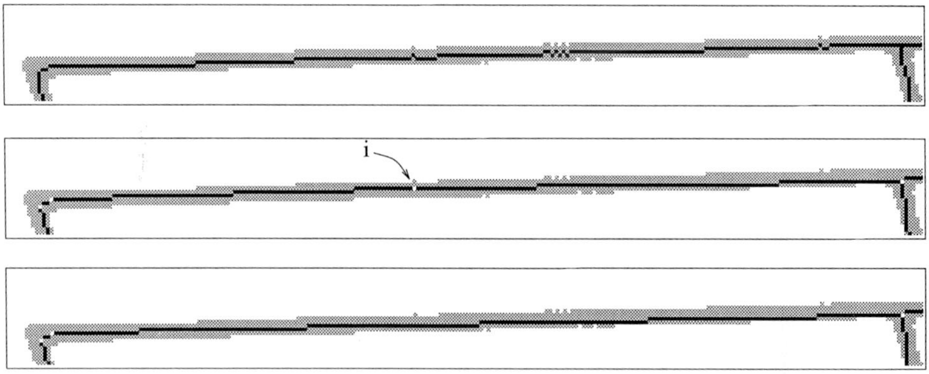

Figure 9: Example of the rule to remove certain link nodes. The top figure shows the skeletonization plotted on the bitmap, the middle figure the raw vectorization from the skeleton, and the bottom figure the resulting refined vectorization.

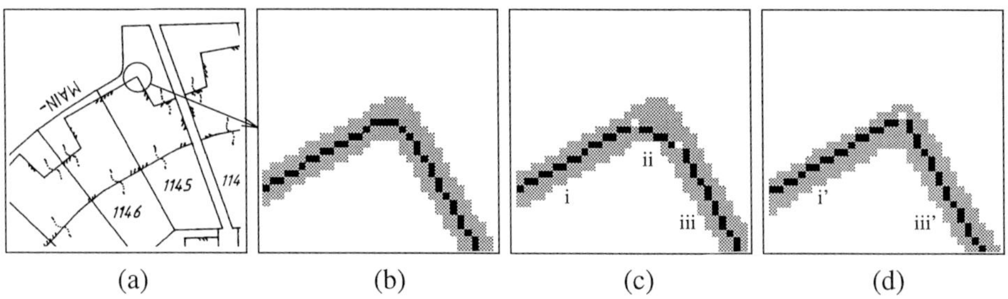

Figure 10: Example of the corner clean up rule. (a) a small part of map "geo", (b) enlargement of the skeletonization of (a) plotted in black on it, (c) the raw vectorization from the skeleton. Nodes are white, edges black. Edge (ii) is unnecessary because it is short. Using constraints on the length of the edges involved and their angles, the vectorization can be corrected as shown in (d).

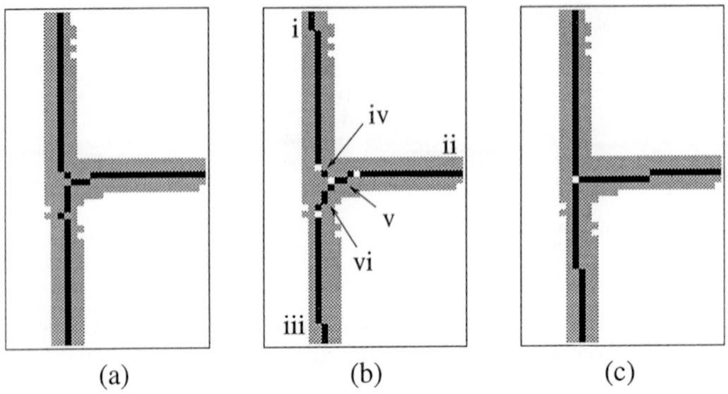

Figure 11: Example of the rule to correct T-junctions. (a) shows the skeletonization plotted on the bitmap, (b) the raw vectorization from the skeleton, and (c) the resulting refined vectorization.

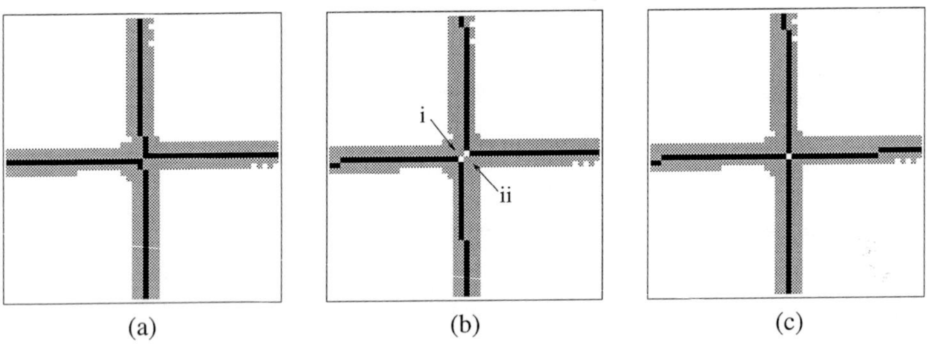

Figure 12: Example of the rule to correct X-junctions. (a) shows the skeletonization plotted on the bitmap, (b) the raw vectorization from the skeleton, and (c) the resulting refined vectorization.

a later update). Usually, there are dots separated from each other in a serpentine-like shape, but it also happens that dots are connected to each other or that they are connected to the line they cross. In extreme cases only the serpentine-like shape can be observed, and all the dots are connected to each other and to the line they cross.

A solution

In the simple case (connection signs consisting of small separate blobs located near each other), connection signs can be found with standard image processing procedures. If the original image is dilated, the connection signs will become connected to the other lines, and after skeletonization the short lines are part of the connection signs. Next, they can be isolated. This is illustrated with an example in Fig. 13.

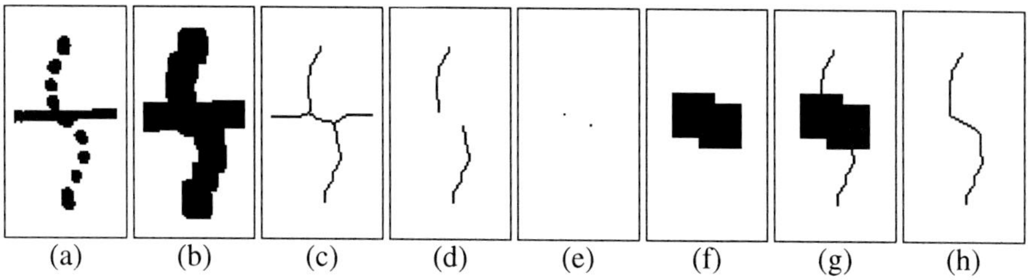

Figure 13: Example how connection signs are found. (a) the original connection sign (excerpt from a larger image), (b) the dilated connection sign, (c) the pseudo Euclidian skeleton, (d) only the branches from the skeleton, (e) the branch points from the skeleton in (c) that touch (d), (f) dilation of the branch points, (g) OR of (d) and (f), (h) pseudo Euclidian skeleton from (g).

Other solutions

The solution given in the previous paragraph does not work for every connection sign. A more elaborate model-based detection mechanism is described in Sec. 5.3. In the case that connection signs are connected to lines, also other knowledge has to be used as described in Sec. 5.5.

3.5. *Detection of Parcel Numbers*

The problem

Just as connection signs, parcel numbers are also drawn in many different ways. Fortunately, map drawers are trained to have some kind of even handwriting, or sometimes templates are used. Thus, one may argue that recognizing numerals and strings is easier in maps than in freehand writings. The problem, however, is that writing can appear in any orientation, and that symbols —contrary to the drawing rules— sometimes touch lines or touch each other.

A solution

In the simple case (numerals separated from each other and not touching lines) parcel numbers are found as follows. Small objects in the bitmap (i.e. numbers and characters) are AND-ed with the inverse of the street image to remove the small objects that are in the street (i.e. the street names). If the parcel numbers and the street names are put in small, separate images, they can be sent to a character recognition system.

Other solutions

The solution given in the previous paragraph does not work for every parcel number. A more elaborate model-based detection mechanism is described in Sec. 5.4. In the case that parcel numbers are connected to lines, other knowledge also has to be used as described in Sec. 5.5.

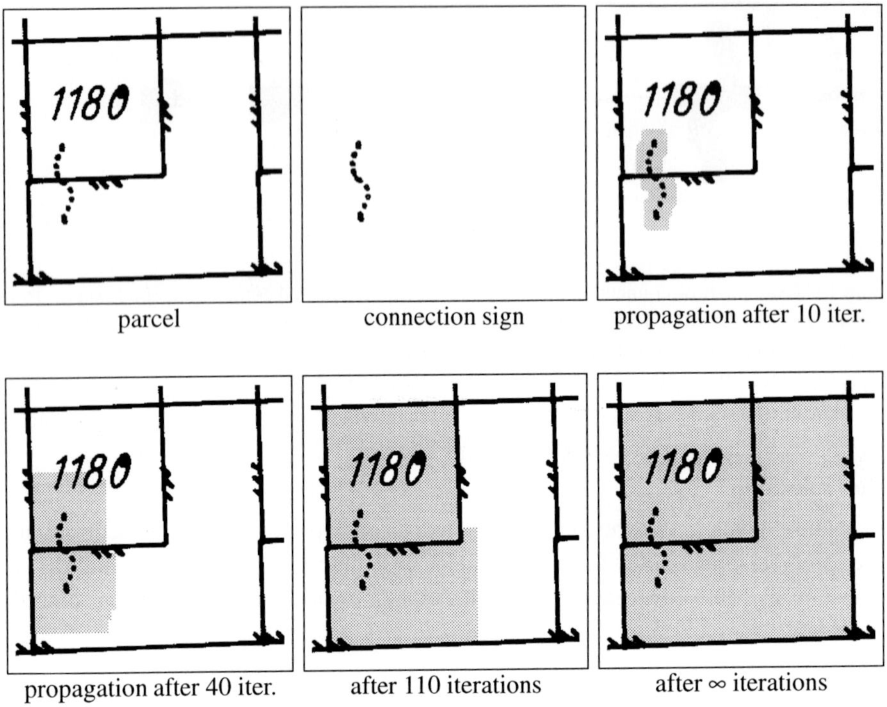

Figure 14: Propagation of the connection sign in its parcel can be used to find parcels. The different stadia are shown in subsequent figures. Propagation "grows" a region either a specified number of iterations or until it cannot be expanded any more (this can be called: fill using ∞ iterations).

3.6. *Detection of Parcels*

A solution

Parcels can be detected with a propagation of the connection signs in the original bitmap. For an example see Fig. 14. Propagation is a process for extracting connected components [24, 25]. The term "propagation" was used for example by Goetcherian [26] and Preston and Duff [27].

Other solutions

With the method described in the previous paragraph, not every parcel can be found. For instance, in the case that connection signs have not been found, parcels cannot be found because their detection depends on connection signs. Or, parcels consisting of only one polygon do not have connection signs. These cannot be found at all. A more elaborate model-based mechanism using knowledge of the possible errors which can be made when interpreting cadastral maps is described in Sec. 5.5.

4. Literature Review on Model-Based Interpretation

This section reviews the published approaches of several authors to the problem of model-based interpretation. The emphasis is on describing their notion of model-based. Because there is no commonly used definition, a subdivision has been made into authors who describe the objects to interpret with a model, authors who base the interpretation method on a model, and authors who combine these two.

4.1. *Object Description Based on a Model*

Several authors use the term "model-based interpretation" to refer to an algorithm in which the description of the objects to interpret is based on a model. This is done because the form or other features of the objects can be difficult to describe with fixed parameters. Thus, a possible reason to use a model-based description of objects is that the variance within the same class is large. Another reason is that the objects to represent are too detailed and that a simplification is desired.

Cooper, Bryson, and Taylor [28] describe a system to locate boundaries of overlapping and touching chromosomes. The chromosomes are described by a shape model which captures their size and shape, and the variation along the boundary. This model is obtained in a learning process in which example chromosomes are presented to a user, who draws an axis. From this axis the boundary of the chromosome is derived automatically. Any errors made are corrected interactively, after which statistics are updated.

Rohr [29] describes a system for recognizing movements of pedestrians. He uses an approach which models the pedestrian by simplifying him to 14 cylinders with elliptic cross sections (for the head, torso, arms, and legs) (this originates from the cylinder model with circular cross sections from Marr and Nishihara [30]). The parameters necessary to describe the sizes of the cylinders and their movements with respect to each other are initially set using averaged anatomical data. Next, these parameters are refined by estimations from both single and consecutive images.

4.2. *Interpretation Method Based on a Model*

Other authors use the term "model-based interpretation" to refer to an algorithm in which the interpretation method is based on a model. Often, this is done because it structures the interpretation algorithm. The system consists of different levels, usually arranged as one low level, zero or more intermediate levels, and one high level. These refer to the "basicness" of the objects and the operations at each level. Low level objects and low level interpretation refer most often to objects as pixels or objects directly derivable from the pixels, and to operations processing these kinds of objects. High level objects and operations refer to the representation of the most abstract objects in the application, and to the reasoning with these kinds of objects. Their representation is completely dependent of the application. Most often, each level has its own knowledge repository, which can sometimes be accessed from higher levels.

Mayer *et al.* [5, 7] present an interpretation system for a Bavarian cadastral map. Their model is organized in four levels (from high to low): semantic objects, graphics and text, image graph, and image. It is represented with a semantic network inspired by ERNEST [31].

The semantic object level contains all objects from the map legend. In the graphics and text level, mainly neighbor relations and operations analyzing the thickness, position, etc. of objects are used. The image graph level consists of separated attribute graphs, each corresponding to a connected component in the image, and the image level contains the (binary) raster input image.

Puliti and Tascini [32] present an interpretation system to support diagnostic biomedical image analysis (support the physician's clinical diagnosis of diabetes or systemic sclerosis diseases). It consists of three levels (from high to low): high level biomedical object interpretation, low level object interpretation, and image segmentation. A distinction is made between generic world knowledge (knowledge about visual perception, necessary for determining some meaning about the initial tokens supplied by the segmentation process), and specific problem-domain knowledge (knowledge about nail fold capillaries of the human finger, to determine the final interpretation and to solve possible ambiguities in the token detection).

The segmentation module generates an initial segmentation, which is refined by the low-level interpretation module to give more meaningful objects. In turn, it interacts with the segmentation module to e.g. compute necessary feature values. The recognition module attempts to recognize objects or parts of objects with a goal-driven approach. After the generation of a hypothesis, attempts are made to falsify it. If that is not possible, it is accepted. The recognition module may influence the segmentation and the low-level interpretation by modifying control parameters.

4.3. *Both Object Description and Interpretation Method Based on a Model*

The approaches from the previous two sections can be combined. Then, the description of the objects to interpret is based on a model, and the interpretation method is based on a model (not always consisting of different levels). This is done because e.g. a structured interpretation is desired, and the variability of the objects is large. Several combinations can be made: the model of the object can consist of several layers and the model of the interpretation is not layered [33], the model of the interpretation can be layered and the model of the object not [34], and both can be layered [9, 35].

Some authors define a model-based method as consisting of a separate learning stage and matching stage (see e.g. Chang and Leou [36]). This is seen as a special form of the interpretation method. In the learning stage, a model is constructed from each object to interpret. The model should be able to accurately describe the object, including the variation in which it can appear in the "real world". Robust detection methods are provided. Then, in the matching stage, the objects are interpreted. If objects are rejected they may be saved for later processing, e.g. by consideration of an operator or by an automatic procedure. This may be followed by changes in the model or in the values of the parameters. In this way, the model "learns" to adapt on specific properties of the data.

Denisov and Dudkin [33] describe a method for the recognition of chromosomes. They use a model of a chromosome consisting of several layers, from bottom to top on the lowest layer the size, shape and brightness, then on a higher layer constraints on these values (a priori information), next a layer considering the density profile, and finally a layer with predicates allowing for e.g. direct access to information about band position and brightness intensities. Their interpretation method is hypothesis construction/verification.

Kise *et al.* [34] describe a frame based system for analyzing office document images (a frame describes a collection of attributes that a given object normally possesses and thus allow us to reason with "commonsense knowledge"). Its aim is separation of layout and contents (no interpretation is done), and it is based on their analysis system for business (visiting) cards [37]. They define a (hierarchical) model that represents the layout structure and the logical structure of the target documents. The model is constructed automatically by generalizing instances of sample documents, and incrementally refined by error feedback. Two kinds of frames are used: layout frames (to describe features of a distinct layout object) and similarity frames (to describe spatial similarity, difference or relation between layout objects).

In den Hartog *et al.* [35] an interpretation system for utility maps is described. They have an explicit description of the knowledge necessary for recognition of the different objects (houses, pipelines, arrows, etc.) and the relations between these objects (an arrow is next to a pipeline, etc.), in a specially designed high level language. Then, a semantic network is constructed from this description. This network consists of nodes describing the geometrical properties of the objects, and links describing the spatial relationships between these objects. The interpretation flow is guided by the semantic network because priorities are attached to the links. These links generate the search actions, and they determine the execution order of the search actions. Thus, this is a system with a layered object model and layered interpretation model because both the objects on the map and the interpretation flow are based on the same semantic network, and a semantic network is a multi-level structure.

Janssen [9] describes a model-based interpretation structure which is used in a system to interpret cadastral maps and large scale base maps. The object description is based on a model because then the objects can be described in a modular way and variation between objects of the same class can be described more easily. Moreover, the interpretation method is based on a model because it structures the interpretation algorithm and because objects found on a higher level can influence the construction of objects on lower levels.

Three models are used. The first model is the *model of the interpretation flow*, describing the flow of the interpretation process. It describes what modules are present, how these are connected, and what the information flow is during the interpretation process. It does not describe the implementation of the modules, but provides only a description of their functionality. The second model is the *model of the map*, specifying which objects can be found in the map, the relation between the objects, where they must be found, and what can go wrong in this process. It also specifies which structures must be found (e.g. for cadastral maps: parcels), how they must be found, and what can go wrong. Methods are provided to solve problems. The last model is the *model of the object*, describing how an object can be found. This model is closely intertwined with the model of the map. Every type of object has its own model of the object.

The model of the interpretation flow can be independent of the application, while the model of the map and the models of the objects are very dependent on the application. However, it is very well possible that parts of the model of the map can be used for different models of the map. In this way re-usage is possible.

4.4. *Discussion*

The basic idea of model-based image interpretation is to approach interpretation problems in a different way. One of the usual interpretation methods is a bottom-up strategy: starting with the pixels on the lowest level, objects are formed, which are further grouped on higher levels, until a satisfactory interpretation has been found. However, this does not offer possibilities to use a priori knowledge extensively, nor does it allow recognized parts on higher levels to solve detection problems on lower levels. Also, it can be very difficult to change an interpretation algorithm for one application to another, slightly different type of application.

The model-based approach is a combination of bottom-up and top-down strategies which can be intertwined. A priori knowledge about the images or the objects in the images ("expectations of reality") can be used at various places and on various levels: on a low level to group pixels (e.g. into blobs from connection signs), to guide the segmentation, or to correct the vectorization, and on higher levels to guide, improve, correct, or simplify the interpretation (e.g. to construct parcels from connection signs and parcel numbers).

If errors are detected (e.g. a parcel does not have a number) the top-down link allows us to influence low level processes, for example by starting a new search for objects which must be present at some place (in each polygon of the parcel a search is done to find a parcel number, e.g. by changing parameters locally or hypothesizing that the number is connected to a line).

Ideally, the a priori knowledge must be specified explicitly in some way. It can be represented as parameters input to the program and as statements in C and C++ code, but another method could be to define some kind of high level language allowing us to specify that e.g. connection signs consist of several blobs in a serpentine like shape that cross a line.

Another point of model-based interpretation is that when changing to a similar application, models should be easy to change, and it should be possible to re-use parts (this may also be true for bottom-up or top-down interpretation).

5. Model-Based Interpretation of Cadastral Maps

In this section, building blocks are described for a model-based interpretation system for cadastral maps. The emphasis is on model-based solutions for problems encountered with the bottom-up system. Also some other aspects of this interpretation methodology are discussed. The system presented in this chapter is taken from chapter 5 of [9]. It was designed for model-based interpretation of cadastral maps, and extended to model-based interpretation of large-scale base maps. The modifications were fairly straightforward and the evaluation results were good, showing that model-based interpretation is a flexible interpretation methodology.

One of the interesting engineering problems in map interpretation is to find out what can go wrong (we considered this part of the learning procedure), and trying to make modules which correct for this. Ideally, the interpretation process should not crash if the designer did not anticipate something, e.g. parcel numbers connected to the lines of the parcel, which is not allowed according to the drawing rules. The system described below is reasonably robust because it can correct for drawing mistakes.

We have chosen to represent the scanned map in *vector objects* (objects which can be represented with vectors) and in *pixel objects* (a group of connected pixels, a connected component). This distinction has been made because it may be advantageous to be able to describe objects as either pixel objects or vector objects. For instance, the lines in a binary image can be described using both representations. An advantage of using the pixel representation is that pixel based operations such as dilation and erosion can be used, which provide e.g. a simple method for determining the distance to other pixel objects. Likewise, an advantage of using the vector representation is that vector based operations can be used. This is very useful in for instance determining the intersection of two lines.

Pixel objects can be "converted" to vector objects by determining their minimum bounding rectangle or by determining their convex hull (although this may not always be very accurate). In this way vector-based operations can be used. Vector objects can be "converted" to pixel objects by drawing the minimum bounding rectangle or the convex hull of a vector object in an image. In this way pixel-based operations can be applied. Since these types of "conversion" operations result in approximations, they should only be applied if necessary.

The drawing rules for cadastral maps are guidelines for the construction of the map model. Each parcel has exactly one number, and polygons of one parcel are connected with connection signs. The parcel number must be written in the middle of one of its polygons, and numerals may not touch the side of the polygon or each other [12].

The remainder of this section describes a model-based method for finding parcels in cadastral maps. It has the same structure as Sec. 3.

5.1. *Noise Removal and Skeletonization*

"Noise" and small irregularities in the input image are removed from the binary input image as described in Sec. 3.1. For skeletonization, the pseudo Euclidean skeletonization algorithm has been used as described in Sec. 3.2.

5.2. *Vectorization*

Vectorization is done using the adaptive vectorization method described in Janssen and Vossepoel [23]. This method is based on a sequence of a standard vectorization algorithm (the Douglas–Peucker algorithm as described in Sec. 3.3) and maximum threshold morphology, which can be iterated until a fitting criterion is met.

This method works as follows. First, "anchor points" (nodes capturing the gross structure of the lines in the image) are found using a coarse vectorization. The position of each anchor point is corrected by means of morphological operations, after which it is fixed. Next, the vectorization is refined between the anchor points. The method is adaptive because the original input image is used for correcting and improving the vectorization.

Before vectorization, the image is enclosed in a rectangle, so that polygons can be constructed. This is done because polygons are the "basic units" in cadastral maps (if parcels have one or more connection signs, they consist in fact of several polygons "connected" with these connection signs).

5.3. *Detection of Connection Signs*

Instead of using global operations as dilation and skeletonization as described in Sec. 3.4 to construct connection signs, in the model-based detection algorithm connection signs are constructed from a sequence of blobs. These blobs have to satisfy several criteria based on size and distance parameters (see Fig. 15).

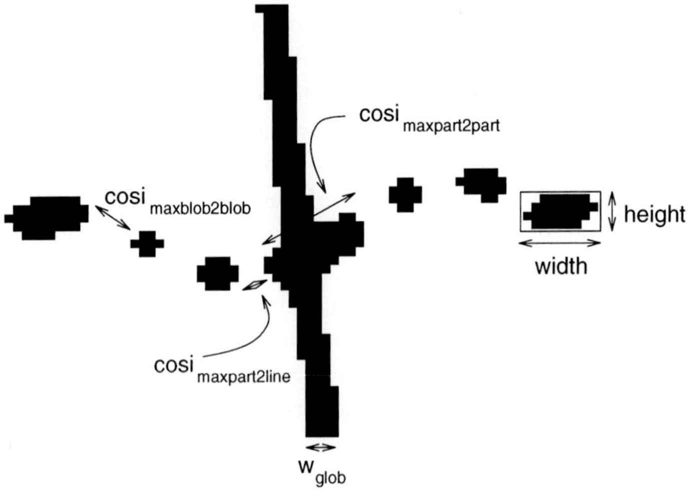

Figure 15: Parameters used in finding connection signs. w_{glob} is the average line thickness.

First, the small blobs of the connection sign are isolated based on the width and the height of the minimum bounding rectangle of each separate blob. Next, these small blobs are grouped (to locate all the blobs of a connection sign at the same side of a line) by finding from each blob the two closest blobs (possibly on both sides of the first blob). Grouping is determined based on the distance: the other blobs may be at most $cosi_{maxblob2blob}$ pixels away from the first blob. This result is called *part of a connection sign*.

Finally, two parts of a connection sign are grouped to one *full connection sign* (or connection sign if no confusion arises) if the distance from one part to the nearest line < $cosi_{maxpart2line}$ pixels, and the distance from one part to the closest other part through a line is < $cosi_{maxpart2part}$.

This method does not detect connection signs completely connected to the line separating the two parts (i.e. all the blobs are connected to each other and to the line). Detecting these connection signs requires more knowledge, and is therefore postponed to the detection of parcels (Sec. 5.5).

5.4. *Detection of Parcel Numbers*

Parcel numbers are constructed from grouped characters and numbers. Thus, first separate characters and numbers are found based on the width and the height of the minimum bounding rectangle enclosing the character, and next these separate characters are grouped to strings if they are close enough. See Fig. 16.

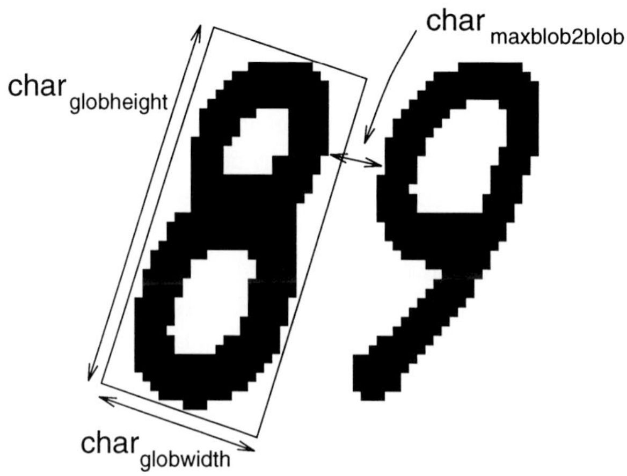

Figure 16: Parameters used in finding parcel numbers.

This method does not detect parcel numbers touching the boundary of the parcel. Detecting these requires more knowledge, and is therefore postponed to the detection of parcels (Sec. 5.5).

5.5. *Detection of Parcels*

Parcels are formed from polygons, connection signs, and/or parcel numbers. This is done by determining which full connection signs intersect two or more polygons, and constructing a candidate parcel from them. Next, for each set of numerals their enclosing polygon is determined. If such a polygon is part of a candidate parcel formed in the previous step, the set of numerals is incorporated in that candidate parcel. Otherwise, a new candidate parcel is constructed from the polygon and the set of numerals. Finally, for the remaining polygons a candidate parcel is constructed. These remaining polygons can be characterized as neither intersecting with a connection sign, nor enclosing a set of numerals. This is done to aid the interpretation process with a "remaining" category. To make a distinction between the different kinds of candidate parcels, they are labeled:

FULL parcel: (parcel if no confusion arises) a candidate parcel having one parcel number and zero or more connection signs, i.e. a parcel conforming to the definition of "parcel";

BOUNDARY parcel: a candidate parcel of which one or more of its polygons touch or intersect the boundary of the image. For this parcel objects necessary for the interpretation may be on the other side of the boundary, so a complete interpretation is not always possible;

UNKNOWN parcel: a candidate parcel not satisfying one of the labels above (a "remaining" category).

Check against the model of the cadastral map

The check against the model tests whether every candidate parcel conforms to the definition of FULL parcel or not. In the latter case, an analysis is made on what can be done to make it conform to the definition. The labels assigned to the candidate parcels are used in this process.

Candidate parcels having the label FULL parcel are conforming to the definition, and are not considered any more. BOUNDARY parcels may conflict with the definition, but are not considered because information necessary for the interpretation may be on the other side of the boundary. However, whether this is true or not cannot be known in advance.

The remaining candidate parcels (with the label UNKNOWN parcel) are investigated to see if the interpretation can be improved. The cases in Table 1 can be distinguished. The table can be seen as a "truth table" for checking the candidate parcels. The numbers give the number of objects of that type necessary for choosing that case. Hypotheses are generated from left to right. If both case 1 and case 2 do not apply, the candidate parcel cannot be relabeled.

Table 1: "Truth table" for checking the candidate parcels. The abbreviations used are "cosi" for connection sign, and "parnr" for parcel number.

pixel object	parcel	case 1: missing cosi or missing parnr	case 2: cosi incor. labeled as parnr	no solution can be found
parnr	1	0	$[2, \infty >$	$[0, \infty >$
cosi	$[0, \infty >$	$[0, \infty >$	$[0, \infty >$	$[0, \infty >$
label	FULL parcel	UNKNOWN parcel	UNKNOWN parcel	BOUNDARY parcel, UNKNOWN parcel

Case 1. **Observation:** there are 0 parcel numbers.

Hypothesis 1: the algorithm missed a connection sign. Try to find one.

Action 1: this is the situation in Fig. 17. Basically, for each polygon of the candidate parcel, each edge is considered to find an unused part of a connection sign. If such a part is found within $\leq cosi_{maxpart2midline}$ pixels distance from the edge of the polygon (Fig. 18), on the other side of the edge a search region is created, in which (hopefully) the other part can be found, but connected to the edge itself. The aim is to find a cut line in this search region to separate the connected part.

If such a line can be found, a new connection sign is constructed, and used for the construction of a new parcel. Next, the interpretation continues with another problem.

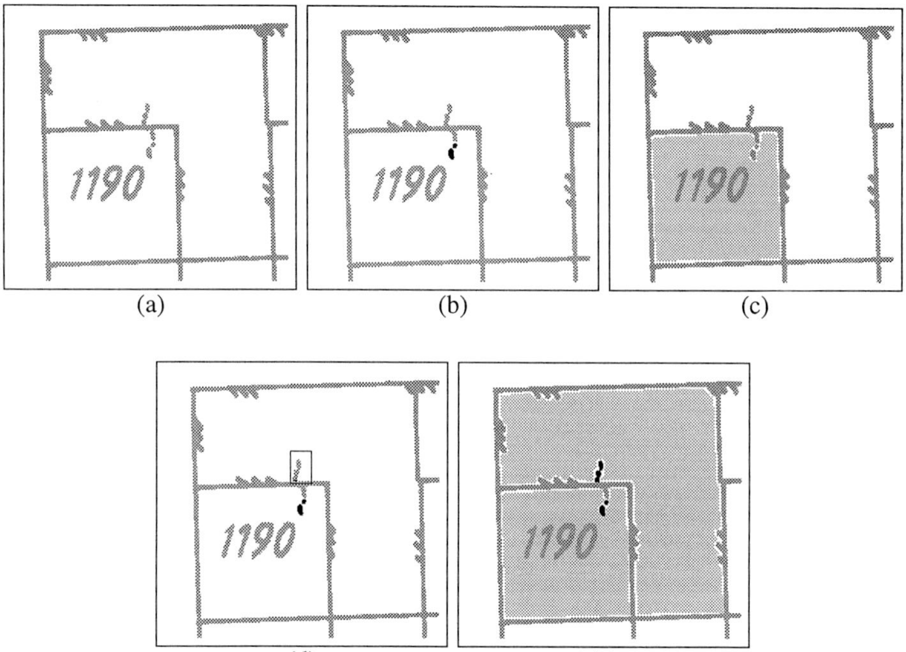

(a) (b) (c)

(d) (e)

Figure 17: An example of using model knowledge to find missing connection signs. In (a) the original image is given. The only part of a connection sign that can be found is shown in black in (b) (the other part cannot be found because it is connected to the lines of the image). Since it consists of only one part, it is not used, and a parcel consisting of one polygon is constructed, shown in light shading in (c) (this satisfies the definition of "parcel having one number and no connection signs"). In the "check against model" part, the not labeled polygon is detected. Since it is not a parcel, something must have gone wrong (a connection sign may have been missed). From this polygon a search is done for a part of a connection sign (the model specifies that the other part may be connected to the lines of the image). The search region generated (a black rectangle) is shown in (d). Indeed, a part of a connection sign is found, and a valid connection sign can be constructed. The new parcel is shown in (e).

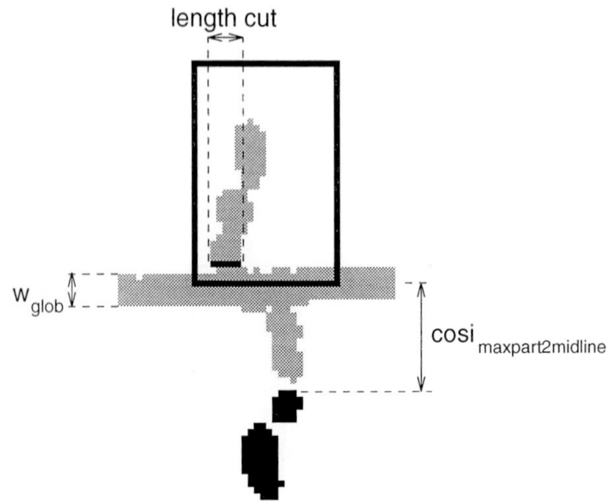

Figure 18: Parameters used for finding connection signs connected to lines. This is an enlargement of Fig. 17(d). w_{glob} is the average line thickness.

Observation: there are 0 parcel numbers.
Hypothesis 2: the algorithm missed a parcel number. Try to find one.
Action 1: the first action is to determine if the height of the bounding rectangle of an unused pixel object satisfies the global height of a character. If that is true, and if such a pixel object is inside one of the polygons of a candidate parcel, it is assumed that it is a string of characters connected to each other not detected before because its width was too large. A new parcel is constructed. Next, the interpretation continues with another problem.

Action 2: this is the situation in Fig. 19. Basically, for each polygon in the candidate parcel each node in the vectorization having more than two branches is checked. A search region is constructed around that node pointing to the inside of the polygon (Fig. 20). This search region is used to determine if a parcel number is connected to the lines of the image.
If in this search region a cut line can be found which separates the (possible) parcel number and the lines of the image, a new parcel number is constructed and used for the construction of a new parcel. Next, the interpretation continues with another problem.

Case 2. Observation: there are ≥ 2 parcel numbers.
Hypothesis: the algorithm labeled part of a connection sign as a parcel number. Try to find out if that is true, and if so, relabel.
Action: sometimes, parts of connection signs are mislabeled as parcel numbers (specifically, they look like a "1"). This can be checked with the procedure to detect

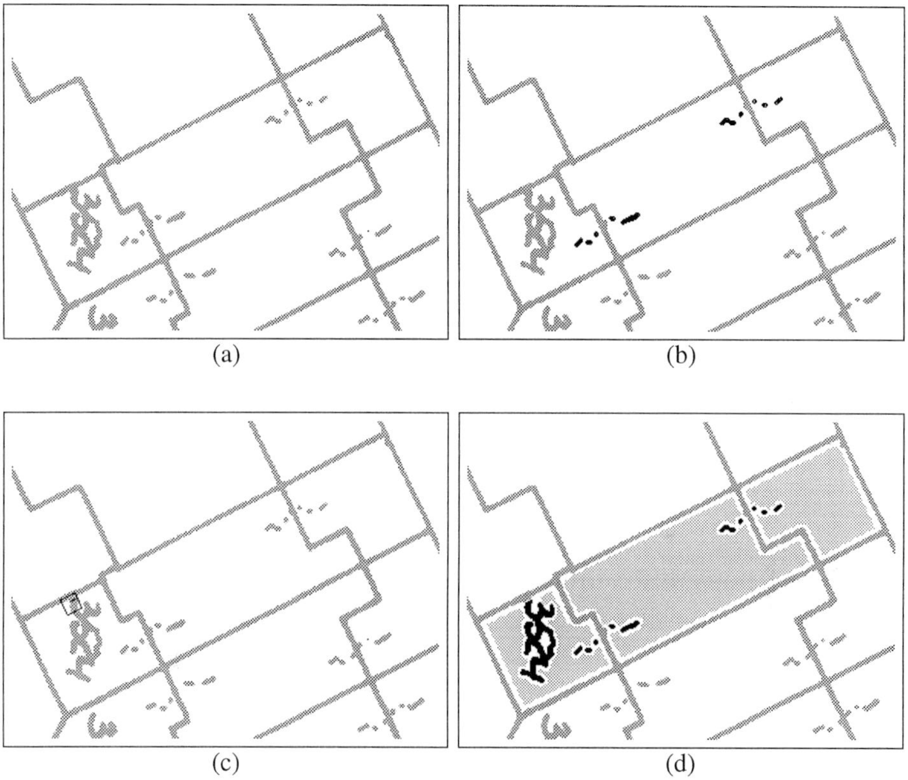

(a) (b)

(c) (d)

Figure 19: An example of using model knowledge to find missing parcel numbers. In (a) the original image is given. The connection signs found are shown in black in (b). However, a parcel cannot be constructed because a parcel number is missing (it is connected to the lines of the image, therefore it could not be found, and the model requires that a parcel has a parcel number). To find a parcel number, in each polygon of the candidate parcel a search is done for a parcel number. The search is in a region around each branch point of the vectorization (the model specifies that a parcel number may be connected to the lines of the image, and this will influence the vectorization by inserting an extra node). In (c), a search region in the leftmost polygon is shown. In the black rectangle, it shows a connection between the lines of the image and the parcel number. With this parcel number and the connection signs a parcel can be constructed as shown in light shading in (d).

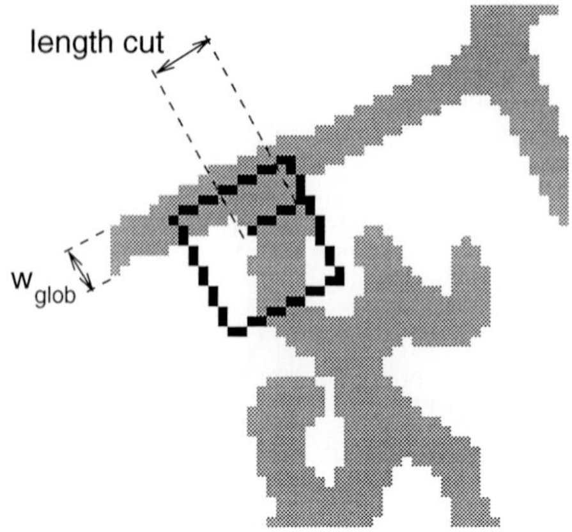

Figure 20: Parameters used for finding parcel numbers connected to lines. This is an enlargement of Fig. 19(c). w_{glob} is the average line thickness.

parcel numbers, but now with a global character width of the average line thickness (the expected width of a "1"). If such objects can be found, a check is done to find out whether or not a connection sign can be constructed, and if so, a new parcel is constructed. Next, the interpretation continues with another problem.

6. Conclusions and Summary

This chapter described two methods for the interpretation of cadastral maps: a bottom-up method and a model-based method. Both methods are used to illustrate problems and solutions encountered in map interpretation.

The advantage of using a model-based approach compared to a bottom-up approach is that a priori knowledge can be used more effectively. This can be seen because consistency checks can be incorporated much more easily in the model-based system. In case of an inconsistency, the vector and/or pixel representation of an object permits us to correct for this. The changes are fed back to the lower levels of the system and the interpretation is redone for the inconsistent objects. Another advantage of the model-based methodology is that modules can be re-used in other interpretation systems.

Apart from that, the performance of the model-based system is better than the performance of the bottom-up system. This is because connection signs connected to the lines of the image, connection signs labeled incorrectly, or parcel numbers connected to the lines of the image are handled more adequately by the former than by the latter approach.

Still, the model-based map interpretation system for cadastral maps can be improved by an even more accurate search for parcel numbers and connection signs. Sometimes, parts of connection signs were not found because they were confused with building hatches. Since this did not happen in the learning set, this could not be anticipated. Parcels are sometimes not found because a line of the parcel boundary is broken next to a junction. A more elaborate reconnection mechanism may solve this (possibly in connection with the adaptive vectorization algorithm).

A very interesting research issue is the acquisition of the a priori knowledge and the representation of that knowledge. In the current systems, parameters "representing knowledge" were chosen by hand and used as a priori knowledge in the algorithms. The model-based interpretation algorithm may be improved by using some predefined values (for example derived from prototype objects) to detect the various objects in the image, and refining these values automatically during the interpretation process.

Acknowledgements

This research was supported by the Foundation for Computer Science in the Netherlands (SION) and the Netherlands Organization for Scientific Research (NWO). The maps in this chapter were reprinted with permission of the Dutch Cadastre. Research has been conducted at the Delft University of Technology and the TNO Institute of Applied Physics, both in Delft, the Netherlands.

References

[1] M. Ejiri, S. Kakumoto, T. Miyatake, S. Shimada and K. Iwamura, Automatic recognition of engineering drawings and maps, in *Image analysis applications*, eds. R. Kasturi and M. M. Trivedi, chap. 3, (Marcel Dekker Inc., 1990), 73–126.

[2] S. Suzuki and T. Yamada, MARIS: map recognition input system, *Pattern Recogn.* **23**, 8, (1990), 919–933.

[3] D. Antoine, CIPLAN, a model-based system with original features for understanding French plats, in *Proc. 1st Int. Conf. on Document Analysis and Recognition*, Saint-Malo, France, Sept. – Oct. 1991, 647–655.

[4] L. Boatto, V. Consorti, M. Del Buone, S. Di Zenzo, V. Eramo, A. Esposito, F. Melcarne, M. Meucci, A. Morelli, M. Mosciatti, S. Scarci and M. Tucci, An interpretation system for land register maps, *Computer* **25**, 7, (1992), 25–33.

[5] H. Mayer, C. Heipke and G. Maderlechner, Knowledge-based interpretation of scanned large-scale maps using multi-level modelling, in *Int. Archives of Photogrammetry and Remote Sensing, Proc. 17th ISPRS Congress*, Washington, U.S.A., Aug. 1992, 578–585.

[6] R.D.T. Janssen, R.P.W. Duin and A.M. Vossepoel, Evaluation method for an automatic map interpretation system for cadastral maps, in *Proc. 2nd IAPR Int. Conf. on Document Analysis and Recognition*, Tsukuba Science City, Japan, Oct. 1993, 125–128.

[7] G. Maderlechner and H. Mayer, Automated acquisition of geographic information from scanned maps for GIS using frames and semantic networks, in *Proc. 12th IAPR Int. Conf. Pattern Recognition*, Jerusalem, Israel, Oct. 1994, 361–363.

[8] N. Ebi, B. Lauterbach and W. Anheier, An image analysis system for automatic data acquisition from colored scanned maps, *Machine Vision and Appl.* **7**, (1994), 148–164.

[9] R.D.T. Janssen, The application of model-based image processing to the interpretation of maps, Ph.D. thesis, University of Technology Delft, ISBN: 90–9008408–8, 1995.

[10] B.W. Hickin, D.J. Maguire and A.J. Strachan, *Introduction to GIS: The ARC/INFO Method* (Midlands Regional Research Laboratory, Leicester, 1991).

[11] ESRI, Redlands, *Understanding GIS: The ARC/INFO Method*, 6th edition, 1992.

[12] Minister van VROM, *Instructie kadaster*, Ministerie van VROM, Dienst van het kadaster en de openbare registers, 1989 (in Dutch).

[13] C.J. Hilditch, Linear skeletons from square cupboards, in *Machine Intelligence 4*, eds. B. Meltzer and D. Mitchie, chap. 22, (Univ. Press Edinburgh, 1969), 403–420.

[14] L. Dorst, Pseudo-euclidian skeletons, in *Proc. 8th IAPR Int. Conf. Pattern Recognition*, Paris, France, Oct. 1986, 286–288.

[15] B.J.H. Verwer, Improved metrics in image processing applied to the Hilditch skeleton, in *Proc. 9th IAPR Int. Conf. Pattern Recognition*, Rome, Italy, Nov. 1988, 137–142.

[16] J.W. Brandt and V.R. Algazi, Continuous skeleton computation by Voronoi diagram, *Computer Vision, Graphics and Image Process.: Image Understanding* **55**, 3, (1992), 329–338.

[17] C.Y. Suen and P.S.P. Wang, eds., Thinning methodologies for pattern recognition, *Special Issue of Int. J. of Pattern Recogn. and Artif. Intell.* **7**, 5 (1993).

[18] C.-C. Han and K.-C. Fan, Skeleton generation of engineering drawings via contour matching, *Pattern Recogn.* **27**, 2, (1994), 261–275.

[19] D.H. Douglas and T.K. Peucker, Algorithms for the reduction of the number of points required to represent a digitized line or its caricature, *The Canadian Cartographer* **10**, 2, (1973), 112–122.

[20] C.M. Williams, Bounded straight-line approximation of digitized planar curves and lines, *Computer Graphics and Image Process.* **16**, (1981), 370–381.

[21] Y. Kurozumi and W.A. Davis, Polygonal approximation by the minimax method, *Computer Graphics and Image Process.* **19**, (1982), 248–264.

[22] K. Wall and P.-E. Danielsson, A fast sequential method for polygonal approximation of digitized curves, *Computer Vision, Graphics and Image Process.* **28**, (1984), 220–227.

[23] R.D.T. Janssen and A.M. Vossepoel, Adaptive vectorization of line drawing images, to appear in *Computer Vision and Image Understanding*, 1996.

[24] J. Serra, *Image Analysis and Mathematical Morphology* (Academic Press, 1982).

[25] R.C. Gonzalez and R.E. Woods, *Digital Image Processing* (Addison Wesley, 1992).

[26] V. Goetcherian, From binary to grey tone image processing using fuzzy logic concepts, *Pattern Recogn.* **12**, 1, (1980), 7–15.

[27] K. Preston and M.J.B. Duff, *Modern Cellular Automata, Advanced Applications in Pattern Recognition* (Plenum Press, 1984).

[28] D.H. Cooper, N. Bryson and C.J. Taylor, An object location strategy using shape and grey-level models, *Image and Vision Computing* **7**, 1, (1989), 50–56.

[29] K. Rohr, Towards model-based recognition of human movements in image sequences, *Computer Vision, Graphics and Image Process.: Image Understanding* **59**, 1, (1994), 94–115.

[30] D. Marr and H.K. Nishihara, Representation and recognition of the spatial organization of three-dimensional shapes, *Proc. Royal Soc. London B* **200**, (1978), 269–294.

[31] H. Niemann, G. Sagerer, S. Schröder and F. Kummert, ERNEST: A semantic network system for pattern understanding, *IEEE Trans. Pattern Analysis and Machine Intell.* **12**, 9, (1990), 883–905.

[32] P. Puliti and G. Tascini, Knowledge-based approach to image interpretation, *Image and Vision Computing* **11**, 3, (1993), 122–128.

[33] D.A. Denisov and A.K. Dudkin, Model-based chromosome recognition via hypotheses construction/verification, *Pattern Recognition Lett.* **15**, 3, (1994), 299–307.

[34] K. Kise, M. Yamaoka, N. Babaguchi and Y. Tezuka, Model based system for analyzing document images, in *Proc. 11th IAPR Int. Conf. Pattern Recognition*, The Hague, The Netherlands, Aug. – Sept. 1992, 647–650.

[35] J.E. den Hartog, T.K. ten Kate and J.J. Gerbrands, Knowledge-based interpretation of utility maps, to appear in *Computer Vision and Image Understanding*, 1996.

[36] Y.-L. Chang and J.-J. Leou, A model-based approach to representation and matching of object shape patterns, *Pattern Recognition Lett.* **13**, (1992), 707–714.

[37] K. Kise, K. Momota, M. Yamaoka, J. Sugiyama, N. Babaguchi and Y. Tezuka, Model based understanding of document images, in *IAPR Workshop on Machine Vision Applications*, Tokyo, Japan, Nov. 1990, 471–474.

Handbook of Character Recognition and Document Image Analysis, pp. 557–582
Eds. H. Bunke and P. S. P. Wang

CHAPTER 21

RECOGNITION OF MATHEMATICAL NOTATION*

DOROTHEA BLOSTEIN† and ANN GRBAVEC
Department of Computing and Information Science
Queen's University, Kingston, Ontario, Canada K7L 3N6

Recognition of mathematical notation involves two main components: symbol recognition and symbol-arrangement analysis. Symbol-arrangement analysis is particularly difficult for mathematics, due to the subtle use of space in this notation. We begin with a general discussion of the mathematics-recognition problem. This is followed by a review of existing approaches to mathematics recognition, including syntactic methods, projection-profile cutting, graph rewriting, and procedurally-coded math syntax. A central problem in all recognition approaches is to find a convenient, expressive, and effective method for representing the notational conventions of mathematics.

Keywords: Computer recognition of mathematical notation; Notational conventions; Symbol recognition; Symbol-arrangement analysis; Syntactic methods; Projection-profile cutting; Graph rewriting.

1. Introduction

Over the centuries, people have developed a specialized two-dimensional notation for communicating with each other about mathematics. The notation is designed to represent ideas in a way that aids mathematical thinking and visualization. It is natural and convenient for people to communicate with computers using this same notation. This involves conversion between mathematical notation and internal computer representations. Under current technology, two-dimensional mathematical notation can be generated by computers, but recognition facilities are not widely available: the task of translating mathematics into a computer-processable form usually falls to a human user (Fig. 1). By relieving the user of the burden of translation, a mathematics recognition system enhances the usefulness of computers as a tool for mathematics and document-handling.

---

* This research is supported by Canada's Natural Sciences and Engineering Research Council.
† blostein@qucis.queensu.ca

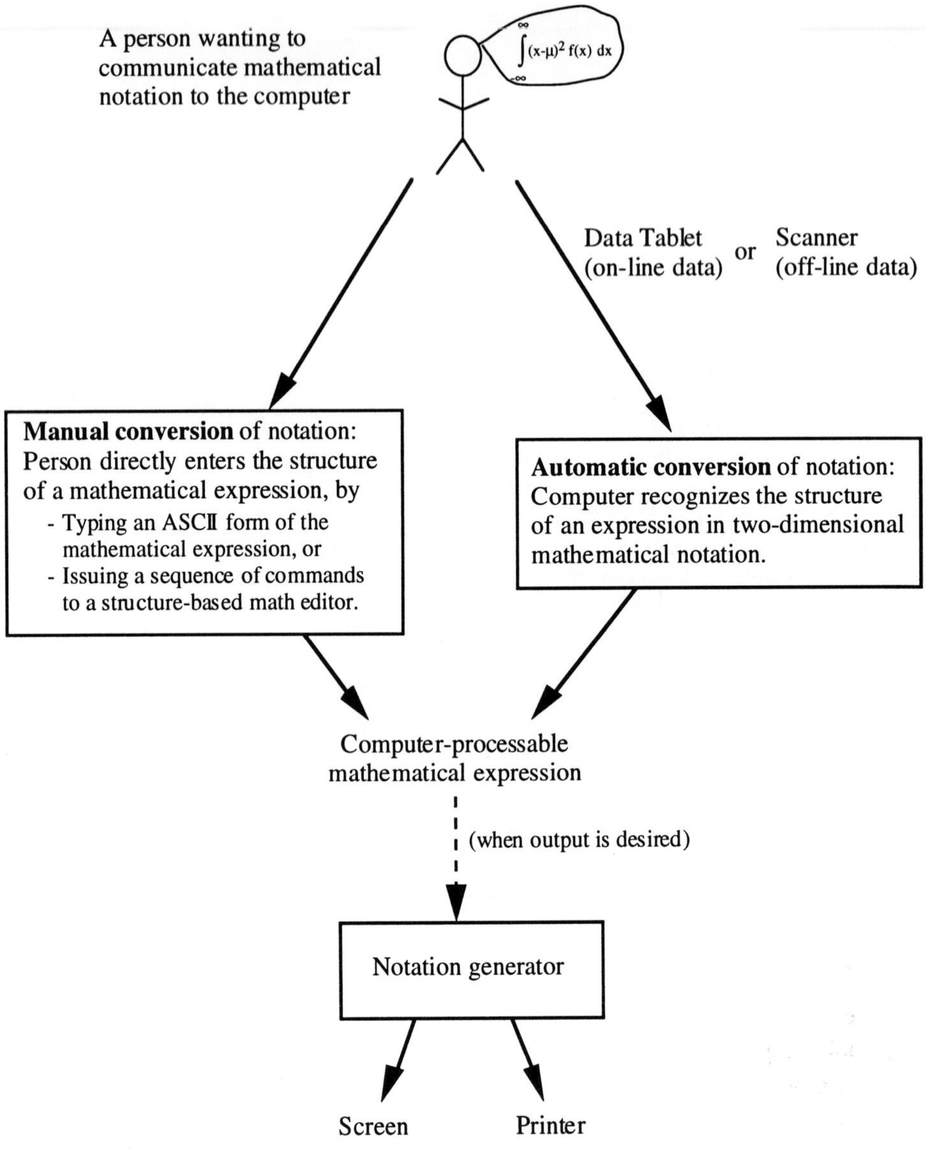

Fig. 1. Conversion between 2D mathematical notation (which people work with) and a computer-processable representation. Input of expressions is shown in the top part of the figure: the automatic-conversion pathway is desirable, but not yet widely available. Currently, most people who need to enter mathematical expressions use manual conversion (typing an ASCII form of the expression, or issuing editor commands to enter the structure of the expression [1]). In contrast, the technology for output of mathematical notation is mature and widely available (bottom part of the figure).

The requirements that must be met by a mathematics-recognition system depend on the application. Two main uses of recognition systems are as follows.

- On-line entry of mathematical notation, for a person creating a document for printing. A handwritten entry on a data tablet is one appropriate input means. A successful recognizer must be able to cope with the variability of handwriting; on-line timing information can be a great help.
- Off-line recognition of mathematical notation. Here, a previously typeset document is scanned; the mathematical expressions it contains must be located and then analyzed. This can be done for a variety of purposes. An example of small-scale use is a reading machine for the visually impaired. Large-scale use arises in the scanning and interpretation of a large collection of technical documents, for the creation of a database.

1.1. Six Processes for Mathematics Recognition

Mathematical notation uses two-dimensional arrangements of symbols to transmit information. Understanding a mathematical expression involves two (possibly concurrent) activities: symbol recognition and symbol-arrangement analysis. The former converts the input image into a set of symbols. The latter analyzes the spatial arrangement of this set of symbols (relative to the given notational conventions of the 2D language for expressing mathematics) to recover the information content of the given mathematical notation.

The two major recognition activities (symbol recognition and symbol-arrangement analysis) are performed by all mathematics-recognition systems. Subdividing further, the mathematics-recognition problem can be discussed in terms of six processes [2]:

Process	Activity
(1) early processing — noise reduction, de-skewing, etc	symbol recognition
(2) segmentation, to isolate symbols	
(3) recognition of symbols	
(4) identification of spatial relationships among symbols	symbol-arrangement analysis
(5) identification of logical relationships among symbols	
(6) construction of meaning	

These processes can be executed in series, or in parallel with later processes providing contextual feedback for the earlier processes. Any implementation must perform these recognition steps, although an implementation can mix steps together, so that the various processes are not clearly delineated in the code. The order of these recognition steps can vary somewhat. For example, partial identification of spatial and logical relationships can be performed prior to symbol recognition; this is done in projection-profile cutting (Sec. 3.2), and in Faure and Wang's structure recognition (Sec. 3.5).

For research purposes, it is possible to bypass the symbol-recognition step in order to concentrate on symbol-arrangement analysis. Test-data for the recognition system can be obtained by scanning a mathematical expression, and manually simulating the symbol-recognition step. This approach has been taken in [3] [4] [5] [6]. Here, perfect (error-free) symbol-recognition results are used to test the symbol-arrangement analysis method. To create a complete recognizer, one must: (1) supply an automatic symbol-recognizer, (2) modify the symbol-arrangement analysis so that it can handle errors and uncertainty from the symbol recognizer, and (3) possibly provide feedback from symbol-arrangement analysis to symbol recognition. For example, the work by Dimitriadis *et al.* [7] builds on Anderson's work [3] in this way.

1.2. Notational Conventions

Notational conventions play a central role in mathematics recognition. The notational conventions define the two-dimensional language, the mapping between the marks on the paper and the information that these marks represent. All six of the above recognition processes require knowledge of notational conventions. For example, noise reduction (1) needs knowledge of the notation in order to preserve small or thin symbols, such as the decimal point. Segmentation (2) needs information about how symbols can overlap. Symbol recognition (3) needs information about symbol appearance; perhaps a font defining fixed symbols, and structural descriptions for parameterized symbols such as the integrals and matrix brackets. The identification of spatial relationships (4) requires information about which spatial relationships are significant for encoding information. The identification of logical relations among symbols (5) can require extensive use of notational conventions; for example, large parts of an expression may need to be examined to determine whether a spatial relationship that looks like a subscript is coincidental, or whether it actually has the logical meaning *subscript*. Finally, the construction of meaning (6) relies heavily on the knowledge of notational conventions: this is the step in which the diagram is fully converted into the information it represents. Existing mathematics-recognition systems use a variety of methods to represent and apply notational conventions; these are summarized in Secs. 3 to 6.

1.3. Definition of Mathematical Notation

The first step in designing a mathematics recognition system is to define the recognition problem. We should begin with a definition of mathematical notation: a definition of the syntax and semantics of this two-dimensional language. Unfortunately, mathematical notation, like most diagrammatic notations, is not formally defined. It is only semi-standardized, allowing many variations and drawing styles. Here is a brief review of some sources of information about mathematical notation.

- **Written descriptions:** Various descriptions of mathematical notation have been published. Some presentations are oriented toward human readers who want to solve typesetting problems (e.g. [8] [9] [10]). Others are geared toward a computational treatment of typesetting problems (e.g. [11]). These descriptions are oriented toward generation of mathematical notation; they do not make explicit the knowledge of notational conventions that is needed to solve recognition problems.

- **Descriptions built into mathematics-recognition systems:** Existing mathematics-recognition systems use a variety of methods for representing notational conventions. These are reviewed below.
- **Descriptions built into mathematics-generation systems:** Generators of mathematical notation (which are usually bundled with editors) contain comprehensive, though often proprietary, descriptions of notational conventions.
- **Human experts:** Interviews with a highly-experienced user of mathematical notation can help in defining the notational conventions appropriate for a mathematics recognizer.

Several decades ago, Martin suggested that the first step in automating the processing mathematical notation is to make a study of the notation. He presents a brief list of the notational conventions found in use in technical publications ([12], p. 79).

To simplify the recognition problem, attention can be restricted to a certain style of mathematical notation. For example, during the development of Chou's system [13], testing was conducted with a large set of mathematical expressions, all drawn from a textbook which was known to be typeset in TROFF. This restriction to a certain style of mathematics typesetting increases the uniformity of notational conventions that the recognizer encounters. Clearly, recognition is easier if one knows the technology used to create the document being recognized.

Here we do not undertake a summary of the notational conventions of mathematics. Instead, we limit ourselves to a brief discussion of the factors that influence how mathematical symbols should be grouped during the recognition process. This grouping results from interpreting the appearance of a particular mathematical expression, relative to the notational conventions for mathematics. Relevant grouping factors include the following.

(1) Grouping factors defined for mathematical notation in general:

- **Operator range:** The *range* of an operator defines the possible spatial locations for operands. The range for operators is established by conventional usage, and can vary from one typesetting style to another. For example, the bounds of an integral can appear above and below the integral symbol, or they can appear to the top-right and bottom-right of the integral symbol. Computationally, operator range can be defined in various ways, including grammar rules [3] [13], division rules in a structure specification scheme [4], and rules for applying projection-profile cuts [14]. Details of the definition of *operator range* vary from one author to another.
- **Operator precedence:** Operator precedence is defined by a ranking of mathematical operators, to indicate the priority of evaluation of operations. Authors vary in their precise definition of operator precedence.

(2) Grouping factors arising within a particular mathematical expression:

- **Symbol identity:** The identity of a symbol restricts how it can be grouped with other symbols. In most cases, symbol identity determines whether the symbol is an operator or an operand. Once the identity of an operator symbol has been recognized, knowledge about the general grouping factors (operator range and operator precedence) is applied in the process of locating the operands.

- **Relative symbol placement:** Relative symbol placement is crucial to the identification of implicit mathematical operators. Whereas explicit operators are represented by a symbol (+ - Σ), implicit operators are indicated by the relative location of operands (implied multiplication, subscript, exponentiation). Eight basic spatial relationships are commonly used in mathematics-recognition systems: left, right, above, below, above right, above left, below right, below left (e.g. [15]). Identification of these relationships can be difficult; Sec. 5.1 discusses methods for distinguishing subscript, in-line, and superscript relations.
- **Relative symbol size and case:** Relative symbol size provides a clue for resolving ambiguous spatial relationships. This is particularly important for implicit operators. For example, a decreased font size provides evidence that the relation between symbols is subscript or superscript, rather than implied multiplication. Symbol size provides a cue for recognition, but not a hard constraint; this is particularly true for handwritten mathematical expressions, where symbol size can be erratic.

These factors (operator range, operator precedence, symbol identity, relative symbol placement, relative symbol size and case) interact in complex ways. Additional terminology can be useful in describing this interaction. For example, Chang's term *operator dominance* is defined for operators in the context of a particular expression. Operator A dominates over operator B if operator B's range nests completely within the range of operator A. Operator dominance provides a partial ordering on the evaluation of operations; operator precedence information is used to complete the ordering [4].

The grouping of symbols into subexpressions is a central problem in mathematics recognition. Below we review various computational approaches to solving this problem. Syntactic methods rely upon parsing to eventually determine the correct groupings (Secs. 3.1, 4.1). Projection-profile cutting exploits the existence of white space to efficiently extract (part of) the structure of an expression (Sec. 3.2). Graph-rewriting rules can be used to encode knowledge of operator precedence, operator range, and other symbol-grouping factors (Sec. 3.3). Procedurally-coded recognition rules can be used as well (Sec. 3.4). An interesting modularization of the mathematics-recognition problem is discussed in Sec. 3.5.

1.4. Problems in Recognizing Mathematical Notation

We now discuss problems that arise in the recognition of mathematical notation. It is interesting to compare the mathematics-recognition domain with other document-analysis domains. A few comparisons are mentioned below, but further work is needed to obtain a deeper understanding of the similarities and differences between mathematics-recognition and other document-recognition problems.

1.4.1. Noise versus small symbols

Various forms of preprocessing are common in document-image analysis. These serve to reduce noise, remove skew, and so on. In preprocessing mathematical notation, it is important to choose algorithms that will preserve small symbols, which are critical to the meaning. These include dots, commas, and symbol annotations such as A'.

1.4.2. Symbol segmentation and recognition

Symbol segmentation is relatively easy in printed mathematical notation, because there is little symbol overlap compared to other notations such as music ([16]; see also the chapter by Bainbridge and Carter in this book). Handwritten mathematical notation may contain many heavily-overlapping symbols; this can be very difficult to segment if only off-line data is available.

Symbol recognition in mathematical notation is difficult because there is a large character set (roman letters, greek letters, operator symbols) with a variety of typefaces (normal, bold, italic), and a range of font sizes (subscripts, superscripts, limit expressions). Certain symbols have an enormous range of possible scales (e.g. brackets, parentheses, Σ, Π, $\int$).

1.4.3. Ambiguity in the role of symbols

A characteristic of mathematical notation is that most symbols have a well-defined meaning. That is, "2" always means *two*, "+" always means *plus*, and so on. Compare this to the meaning that a line has in a road map; it could represent a road, a river, a fence, or a large number of other possibilities. However there are a few common symbols in mathematical notation that do feature such ambiguity in their role. A dot can represent a decimal point, a multiplication operator, a symbol annotation such as $\dot{x}$, or noise; a horizontal line can indicate a fraction line or a minus sign. The only way to determine the meaning of such symbols is through the contextual information provided by surrounding symbols. Thus, resolving ambiguity in role is a problem that must be addressed during symbol-arrangement analysis.

1.4.4. Identifying significant spatial relationships

Much of the meaning in mathematical notation is carried by the spatial relationships between the symbols. Although mathematics is a relatively standardized notation, considerable variation is permitted in relative symbol placement. In light of this variability, it is not clear how to define and identify meaningful spatial relationships among symbols in an expression. Spatial relationships are especially critical for the recognition of implicit mathematical operators; these are indicated by the spatial arrangement of operands, rather than by an explicit operator symbol. Superscripts, subscripts, implied multiplication, and matrix structure are indicated implicitly by the layout of operands, whereas addition, subtraction, and division use explicit operator symbols.

A difficult problem is to distinguish between configurations that represent horizontal adjacency and those that represent superscripts or subscripts. One source of difficulty is the continuum of positions between the horizontal adjacency and superscript (or subscript) configurations: 2x 2x 2x 2^x 2^x.

Mathematics interpretation is complicated by the fact that symbols in many different positions can potentially affect the meaning of a given symbol. There is a dense web of interdependencies among symbols. A large local neighborhood must be considered to interpret mathematical notation. From a given symbol, "adjacent" symbols may be relatively far away (say 5 to 10 times the symbol size, in every direction). This contrasts with music notation, where there are much stronger constraints on the location of relevant symbols.

1.4.5. Ambiguity of relative symbol placement

Mathematical notation uses subtle spatial relationships to indicate logical relationships: given a certain spatial relationship between two symbols, it can be quite difficult to determine the logical relationship between these symbols. We illustrate this with the subscript relation. (Keep in mind that we cannot locally deduce the baseline for a line of mathematical notation.) The configuration x_i could indicate that the symbol 'i' is a subscript of 'x' (as in the expression $x_i y_j$), or it could be a coincidental alignment (as in the expression a^{x_i}). Martin provides an interesting series of examples of ambiguity arising from subtleties of symbol placement ([12], p. 83). Additional examples can be found in [6]. The ambiguity of spatial relationships is greatly increased for handwritten mathematical expressions; people take a lot of freedom with the placement and alignment of symbols. Proper resolution of symbol-relationship ambiguities can require contextual information from the entire expression.

1.4.6. Little redundancy in the notation

Mathematical notation has fairly little redundancy, when compared with other notations such as music [2]. Redundant representation of information provides a recognition system with constraints and cross-checks, particularly useful for the interpretation of noisy images. The relative lack of redundancy in mathematical notation means that less information is available for resolving symbol-recognition ambiguities.

1.5. Scope of Mathematics-Recognition Systems

Comparisons among mathematics-recognition systems are complicated by the fact that each system defines the recognition problem in its own way, placing different limitations on the types of input that should be recognizable. Definition of the recognition problem includes the following considerations.

- The mathematical notation may be handwritten or typeset. In case of handwritten notation, the input data may be on-line or off-line.

- Isolated mathematical expressions may be given as input, or the system may have to first locate mathematical expressions within a text page.
- Recognizers accept different sets of notational constructs. For example, some authors use *mathematical notation* to mean arithmetic mathematical notation, whereas others include matrix notation as well.
- Recognizers make various assumptions about the layout style of the notation. Some systems expect a very regular layout, assuming, for example, that a summation-limit-expression must lie within the x-extent of the Σ symbol. Other systems allow great flexibility in symbol placement.
- Some recognizers begin with the raw image as input, whereas others begin with symbols as input. The latter systems can be tested by human simulation of a perfect symbol recognizer. This human simulation sometimes includes lexical identification as well (e.g. using integers and floating-point numbers as input tokens [3]).

This concludes our introduction of the mathematics-recognition problem. We now turn to a survey of existing work in the area, organized according to major sub-parts of the problem: finding mathematical expressions in a document, methods for symbol-arrangement analysis (syntactic methods, projection-profile cutting, graph rewriting, and procedurally-coded rules), handling noise, identifying subtle spatial relationships, and recognition of handwritten expressions. Various factors are traded off in the design of a recognition system; these include efficiency, simplicity, readability, compactness, and generality.

2. Finding the Mathematical Expressions in a Document

Mathematical expressions are typically embedded in text documents, either as offset expressions, or embedded directly into a line of text. Thus, the first step in mathematics recognition is to identify where expressions are located on the page. (This does not apply to on-line recognition systems, where input comes from a person writing on a data tablet. In this case, text and equations generally are not mixed.)

Most work we survey assumes that the recognition system begins with an isolated mathematical expression. An exception is Lee and Wang, who present a method for extracting both embedded and offset mathematical expressions in a text document [17]. Text lines are labeled as offset expressions based both on internal properties and on having increased white space above and below them. The remaining text lines consist of a mixture of pure text and text with embedded expressions. These lines are converted to a stream of tokens. Certain tokens are recognized as belonging to an embedded mathematical expression; no details of this process are given, except that it is done "according to some basic expressions forms".

3. Methods for Symbol-Arrangement Analysis

Existing recognition systems use a variety of approaches for representing the notational conventions of mathematics, and for using these to drive the recognition process. Here we review four approaches: syntactic methods, projection-profile cutting, graph-rewriting, and procedurally-coded rules.

3.1. Syntactic Methods

A common strategy for recognizing mathematical notation involves some form of two-dimensional grammar. Syntactic analysis is appealing for mathematical notation because the notation has an obvious division into primitives (i.e. symbols), a recursive structure, and a well-defined syntax (compared to other diagrammatic notations). Grammar rules are used to define the correct grouping of mathematical symbols, and to define the meaning of the resultant grouping.

3.1.1. Coordinate grammars

Anderson presents coordinate grammars for the recognition of commonly-used arithmetic and matrix mathematical notation [3]. This work provides an excellent example for illustrating the methods of syntactic pattern recognition. Since Anderson presents his complete grammars, a precise understanding of his system can be obtained, both at the abstract and detailed levels. Correctly recognized symbols and symbol-groups (integers and real numbers) are assumed as input.

A coordinate grammar operates on *sets* of symbols, rather than on the strings of symbols used in traditional formal language theory. Nonterminals and terminals of the grammar are attributed with bounding-box and center coordinates, as well as with a string m, an ASCII encoding of the meaning of the symbol-set. The final interpretation result is given by the m attribute of the grammar's start symbol.

Using a top-down parsing scheme, each grammar-rule application starts with a set of symbols and a syntactic goal (e.g. EXPRESSION, FACTOR, LIMIT). The grammar rule specifies how to subdivide the set of symbols into several subsets, each with its own syntactic subgoal. Partitioning conditions are used to describe spatial constraints. For example, a division rule (syntactic goal DIVTERM) subdivides the set of symbols into a horizontal bar, a set of symbols above the horizontal bar (syntactic goal EXPRESSION), and a set of symbols below the horizontal bar (syntactic goal EXPRESSION). If this partitioning fails (e.g. there is no horizontal bar, or some symbols are to the right or left of the horizontal bar) then the production rule reports failure, causing the parser to try another alternative. On the other hand, successful partitioning gives the parser two syntactic subgoals; if these succeed, then the division-rule's m value is constructed by appropriate combination of the subgoals' m values. Particular care is taken to find efficient partitioning strategies for the grammar rules describing implicit mathematical operations (implied multiplication, subscript, superscript): these rules do not have a terminal symbol (such as a horizontal bar or a plus sign) on the right-hand side, and must therefore try several partitioning alternatives.

In summary, a coordinate grammar provides a clear, well-structured approach to the recognition of mathematical notation. Slow parsing speeds can be a drawback. (Anderson points out that, in general, more efficient recognizers could be achieved by exploiting the special topological features of the notation and sacrificing some of the flexibility of a purely syntax-directed approach.) A major difficulty is to extend the syntactic approach to handle errors or uncertainty from symbol recognition; Sec. 4.1 describes Chou's work in this direction. Also, tests based on bounding-box coordinates provide a fairly crude recognition of spatial relationships. (For example, a limit expression under a Σ is required to have an x-extent that lies strictly within the Σ's x-extent.) Modifications or extensions of the coordinate-grammar formalism could be devised, to reduce restrictions on the symbol layouts permitted in recognizable expressions.

More recently, Belaid and Haton report on a recognition system using both a top-down and a bottom-up parser [18]. The coordinate grammar they use is simpler than Anderson's; a smaller subset of mathematical notation is analyzed, with emphasis on symbol recognition as well as symbol-arrangement analysis. In cases of uncertainty, the symbol recognizer provides lists of alternatives; contextual information applied during parsing can sometimes choose the correct alternative. For example, the parser can choose between "(" and "C". When the symbol alternatives can play similar syntactic roles (such as "S" and "5"), user interaction is necessary to solve the ambiguity.

3.1.2. Structure specification schemes

Chang uses *structure specification schemes* to recognize the structure of mathematical expressions [4]. This is a restriction of the syntactic approach, designed to be efficient in searching for the structure of an expression. Recognition time is $O(n^2)$, for an input expression consisting of n symbols. This time is for symbol-arrangement analysis: correctly-recognized symbols are assumed as input.

Structure specification schemes are based on the existence of operators, which divide the pattern into one or several sub-patterns. The scheme is specified as a set of division rules, one per operator, which indicate this spatial partitioning into subpatterns. Precedence numbers, given as part of a division rule, are used to resolve ordering of operators when neither operator dominates the other. Chang presents precise, detailed descriptions of his algorithms for applying division rules to find a well-formed structure in the given input. (A well-formed structure is equivalent to the parse tree of a syntactically-correct expression.)

A structure specification scheme has less expressive power than the picture processing grammars that Chang used previously. In other words, any structure specification scheme can be translated into a picture processing grammar that describes the same set of patterns; the inverse translation, from grammar to structure specification scheme, is not always possible. To offset this loss of descriptive power, there is the aforementioned increase in efficiency, as well as an increase in usability. (Chang states that a table of division rules is more concise, and easier to design and comprehend than a table of picture-grammar rules.)

Chang presents small-scale tests that illustrate the use of structure specification schemes for the recognition of mathematical notation, and for the recognition of document-layout structure (e.g. the reading order of paragraphs). Two thresholds, t and α, are used. Operators are classified as being on the same line if their centroids are within t. Also, an operator is regarded as being contained in a region if a certain fraction α of its total area (e.g. 70%) is contained in that region.

3.2. *Projection Profile Cutting*

Projection-profile cutting (also called *structural analysis*) is a technique that has been used in a variety of document-image analysis applications. In recognition of music notation, full-page projections are sometimes used to separate staves, with localized projections used to isolate particular music symbols [16]. Projection profiles have also been used for page-layout analysis, to divide a page into columns and paragraphs (see the chapter by Ha and Bunke in this book). Here we consider the use of projection-profile cutting to express the structure of a mathematical expression.

To analyze the structure of a mathematical expression, Okamoto *et al.* apply recursive projection-profile cutting [14] [19]. This method provides an analysis of symbol-layout structure, prior to the recognition of symbol identity, and without the use of an expensive parsing step. Cutting by the vertical projection profile is attempted first, followed by horizontal cuts for each resulting region. This process repeats recursively until no further cutting is possible. The resulting spatial relationships are represented by a tree structure. Special processing is used for square-roots, subscripts, superscripts; these cannot be handled by a projection profile cut. More extensive special processing is reported in subsequent work [6]. For example, they state that

> *There are many cases in which the symbols of a printed mathematical expression are printed very close to each other or there is no blank space in terms of vertical projection profile in the expression. For this reason the width of a blank space is calculated by deleting 10% of both sides of symbols, except in the case of horizontal bars of fractions or summation symbols which have an important role in the width of symbols ([6], pp. 435–436).*

Sec. 3.4 discusses this work further, in the context of procedurally-coded math syntax. For completeness, we give references to three mathematics-recognition papers written in Japanese, two of which originate from Okamoto's project [20] [21] [22].

Faure and Wang use X and Y projections in their processing of handwritten mathematical expressions [23]. They describe situations under which projection methods fail. These include square root symbols, closely-written symbols, and skew. They define mask-removal operations to address these situations (see Sec. 3.5).

J. Ha *et al.* propose another mathematics recognition system that uses a recursive X-Y decomposition [15]. Based on bounding-box attributes of primitives, an initial expression tree is constructed in a top-down fashion. In a bottom-up traversal, the initial tree is checked for syntax errors. Tree nodes are split or merged to fix syntax errors: this corrects for missing primitives and other errors. This work is in the design stage; an implementation is planned.

It appears that projection profile cutting is a simple and efficient technique, suitable for analyzing some aspects of mathematical-expression structure. A complete system must provide extensive additional processing to cover cases not amenable to projection analysis.

3.3. Graph Rewriting

Graph rewriting is a general computational technique, in which information is represented as an attributed graph and computation proceeds by updating the graph through the application of graph-rewriting rules. (A graph-rewriting rule $g_l ::= g_r$ specifies that a subgraph isomorphic to g_l can be replaced by graph g_r. Further information is associated with the graph-rewriting rule, to specify how attribute-values for g_r are computed, and to specify the creation of edges between g_r and the main graph.) Graph rewriting has been applied to mathematics recognition as follows [5]. The existence of a symbol-recognizer is assumed: the input graph contains one node to represent each symbol, with node attributes recording the spatial coordinates of the symbol. Given this initial graph, which contains no edges, graph-rewriting rules are applied to add edges representing potentially-meaningful spatial relationships. Further application of graph-rewriting rules is used to prune and modify these edges, identifying the logically-important relationships from the spatial relationships. Information about operator precedence, stored as edge attributes, helps to determine how symbols are to be grouped into subexpressions. When a subexpression is recognized, the corresponding subgraph is replaced by a single node that represents the subexpression. The final output for a successfully-recognized expression is a single node whose meaning attribute represents the high-level meaning of the input expression as a character string. Fig. 2 illustrates a sample graph-rewrite rule.

The recognition process in [5] is organized into four phases.

Build Add edges between symbols that are related by potentially-meaningful associations. Edges are labeled Above, Below, Left, Superscript, Subscript.

Constrain Apply knowledge of notational conventions to remove contradictions and resolve ambiguities. This includes:

- Remove contradictory associations; remove the more distant of two similar associations
- Disambiguate horizontal lines as fractions or minus signs
- Disambiguate dots as decimals, multiplication signs, or noise
- Disambiguate diagonal associations — exponents and subscripts

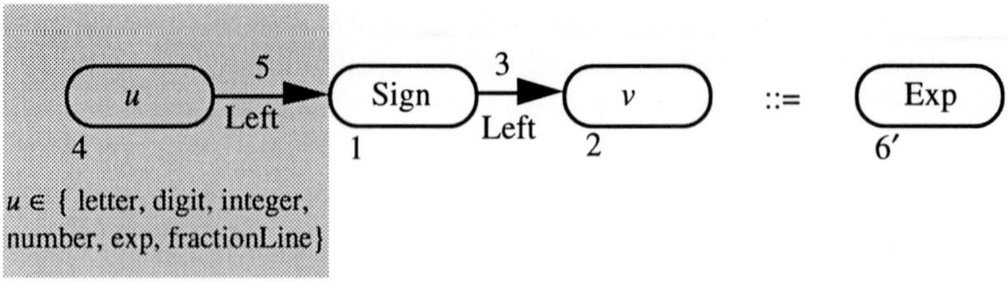

Application Condition ≡ ($v \in$ { Letter, Digit, Integer, Number, Exp }) & (prec_in(3)=1)

Embedding ≡ In_{Left} = {(1, 6')}, Out_{Left} = {(2, 6')},
Out_{Above} = {(1, 6'), (2, 6')}, Out_{Below} = {(1, 6'), (2, 6')},
In_{Super} = {(1, 6')}, Out_{Super} = {(2, 6')},
In_{Sub} = {(1, 6')}, Out_{Sub} = {(2, 6')},

Attribute Transfer ≡ {m(6') = m(1) (m(2)); reset_prec(6') }

Fig. 2. A graph-rewrite rule to interpret a signed item as a subexpression. The following steps are used to apply this graph-rewrite rule to a host graph.

1. Locate g_l^{host}, a subgraph of g that is isomorphic to the unshaded part of the left-hand side of the production rule. (If no isomorphic subgraph can be found, report failure of rule application.) This particular rule looks for a node with a Sign label that has a Left-labeled edge connecting it to some node with a variable label denoted v.
2. Test the application conditions. If they fail, report failure of rule application. This rule has both textual and graphical application conditions. The textual application conditions test that the label v is a member of the set {Letter, Digit, Integer, Number, Exp}, and that the precedence attribute prec_in of the Left edge has the value 1. The graphical application condition uses shaded region(s) to specify host-graph structure that must be *absent* for rule application to occur. In this case, rule-application is prohibited if there is an operand in front of the Sign node, since in that case the Sign should be interpreted as an addition or subtraction, rather than as a negation or explicit plus sign.
3. Remove g_l^{host} from the host graph; the remainder of the host graph is called RestGraph. Keep track of the pre-embedding edges (i.e. edges that connected g_l^{host} to RestGraph).
4. Create g_r^{host} by making a copy of the right-hand side of the production rule. In this case g_l^{host} consists of a single node with label Exp (short for Expression).
5. Embed g_r^{host} into RestGraph: use the embedding information to translate pre-embedding edges into post-embedding edges. Here, the embedding phrase In_{Left} = {(1, 6')} specifies that an incoming Left edge to node 1 causes creation of an incoming Left edge to node 6'. Other edge labels are treated analogously, e.g. transferring outgoing Above edges from nodes 1 and 2 to node 6'.
6. Compute attribute values for g_r^{host}, using the attribute transfer function. Here, the meaning of the Exp node (represented by the attribute m, a string) is computed by concatenating the meanings of nodes 1 and 2, with appropriate insertion of parentheses.

Rank Use information about operator precedence to group symbols into subexpressions.
Incorporate Interpret subexpressions

The use of Build and Constrain phases allows recognition of expressions with considerable flexibility in the relative placement of symbols. Build is allowed to form edges generously, assuring that all relevant associations are included. Any excess edges are removed by the Constrain phase.

This graph-rewriting system interprets an expression in a bottom-up manner, with no backtracking. The recursive nature of subexpressions is handled as follows. Graph-rewriting rules in the Rank phase encode information about operator precedence and operator range as edge attributes. An ordering is assigned to the edges around each node, indicating which operator has highest priority locally. Next, Incorporate rules group the highest-priority symbols into subexpressions, replacing them by a single node. When Incorporate cannot find further subexpressions, the Rank rules are re-applied to update edge attribute values. Incorporate and Rank phases alternate until only a single node remains. The meaning attribute indicates the meaning of the entire input expression.

The recognition system consists of approximately 60 graph-rewriting rules. Execution is slow due to the crude graph-rewriting environment. Good results are obtained on the small set of test expressions that have been tried; these test expressions include highly irregular symbol positioning, as could occur in handwritten expressions. The current work does not handle errors from symbol recognition; however, spurious dots and poorly-aligned symbols are interpreted correctly.

In summary, graph-rewriting is a promising approach for mathematics recognition. Graph rewriting offers a flexible formalism with a strong theoretical foundation for manipulating two-dimensional patterns. It has been shown to be a useful technique for high-level recognition of circuit diagrams and musical scores (references in [5]). The organization into Build, Constrain, Rank, and Incorporate phases facilitates the construction and integration of new rewriting rules. Future research must address the error-prone nature of symbol recognition, either allowing the graph-rewriting system to process input consisting of a number of possible interpretations for each symbol, or providing for feedback from the graph-rewriting system to a symbol recognizer.

3.4. Procedurally-Coded Math Syntax

H. J. Lee *et al.* have developed a mathematics recognition system using procedural code to embody the syntax of mathematical notation [17] [24] [25]. After symbol recognition, the mathematical expression is represented by a list of symbols in random order. To recognize appropriate symbol groups (i.e. subexpressions), procedural code is used. After a group of symbols is recognized as a subexpression, it is represented by a bounding box, and further grouping takes place. To indicate the flavor of these rules, a few sample rules are repeated here. A length threshold of 20 pixels is used to classify a horizontal line as a long bar or a short bar. If a long bar has symbols both above and below it, it is treated as a division. If there are no symbols above it, it is treated as boolean negation. If a short bar has no symbols above or below it, it is treated as a minus sign. If it has symbols above or below it, then combination characters (such as $=$ $\geq$ $\leq$ $\supseteq$) are formed. Various sources of errors are discussed in [25], including: scanning (noise and broken characters); thinning (e.g. confusion between "x" and "λ"); and expression formation (mainly due to incorrect recognition of subscript versus in-line relations). Other aspects of this system are discussed in Sec. 2 (locating mathematical expressions in text) and Sec. 4.4 (splitting of connected symbols). In summary, a benefit of this procedural approach is fast execution; the authors also find it convenient to code recursively in this way. A disadvantage is that this approach provides a rather ad hoc representation of notational conventions, making the procedural code difficult to maintain or scale up.

The recognition system by Okamoto *et al.* [6] [14] [19] combines projection-profile analysis with an extensive array of procedurally-coded recognition rules. The system has been tested on more than 100 expressions, taken from journals or generated by TeX. Their system is able to correctly recognize most of the expression structures. The variety and complexity of their successfully-recognized expressions is impressive. This system began with the use of projection-profile cutting and has evolved through the addition of specialized code segments to correct particular recognition errors arising in earlier versions of the system. Thus it is difficult to summarize the design of the system; the reader is referred to [6] for a description of the extensive list of processing steps and thresholds used. This paper provides a compelling illustration of the complexity of the mathematics-recognition task, of the many symbol configurations that must be considered. The authors state that syntactic approaches, using parsing, are untenable because the great variety of possible expressions makes it impossible to provide an a priori syntax definition for all possible expressions. This, combined with the computational complexity of parsing, motivates them to instead use a large collection of procedurally-coded recognition rules. Their comment about a priori syntax definition requires some clarification: any recognition method, including procedurally-coded rules, implicitly or explicitly defines the syntax of recognizable expressions. Apparently Okamoto *et al.* find it easier and more practical to provide an implicit syntax definition, coded as procedural rules, rather than an explicit syntax definition, coded as grammar rules.

3.5. Data-driven and Knowledge-driven Modules

Faure and Wang present an interesting, systematic description of the mathematics-recognition problem [23]. The emphasis is on extracting the structure of handwritten expressions, with symbol recognition performed only in restricted circumstances. (The authors plan to add modules for symbol recognition and full syntax recognition later.) Evidence is cited that humans recognize the structure of an expression prior to performing complete symbol recognition. For example, when a person is reading a handwritten expression aloud, syntactic structures such as a division bar are recognized correctly, even before noticing that certain symbols in the numerator or denominator are illegible.

Input data is provided by people writing on a digitizing tablet. The system works with freely-written material – the users are not restricted to particular styles of mathematical notation, and the expressions users write deviate from a horizontal writing line. The sample expressions shown in the paper [23] exhibit significant, non-uniform skew, and many instances of hasty writing (such as a division bar that is significantly shorter than the numerator, and a Σ symbol with a gap between the two strokes used to form the symbol).

The recognition system is organized as a collection of independent modules, which communicate through a shared memory. The shared memory contains the input data, as well as the representation tree, which is updated as recognition progresses. (This architecture is reminiscent of a blackboard architecture, although here the shared memory does not trigger module application, but a fixed order of module application is used. Document-image-analysis systems that use a blackboard architecture are reviewed in [2].) Currently four modules are used:

- Acquisition
- Data-driven segmentation. This module builds an initial relational tree. It does not take writing-order into account, and thus applies to off-line data as well. No contextual information is available to the procedures of this module: the information available to an operator is restricted to the expression-subimage to which it is applied. Projections onto the X and Y axis are used; when these fail (e.g. due to a square root symbol, closely-written symbols, or excessive skew), a mask-removal operation is attempted. The first step is to detect a mask: first, special routines look for square roots and fraction bars; if these fail then any long or tall stroke is sought. Next the mask is removed, and the remaining symbols are analyzed using X or Y projections.
- Knowledge-driven segmentation. This module corrects the relational tree produced by the previous module. It uses domain-specific knowledge, looking for a set of specific patterns, and updating the relational tree by pattern-dependent corrections. This module has knowledge about a subset of the lexicon, the syntactic rules related to these symbols, the writing order (e.g. the long stroke is written before the dot in "i" or "j"), and symbol shape (e.g. the strokes of "E", "F", and "Σ" may not be connected). The patterns that this module looks for are structured as a hierarchy of frames. Once relational-tree nodes have been identified for correction, a complex procedure is needed to update the remainder of the tree in accordance with the corrected nodes.
- Line labeling. This module labels the spatial relationships (subscript, superscript, same line) between components of a writing line. Statistical labeling is used, as discussed in Sec. 5.1.2.

Modules for symbol recognition and full syntax analysis will be added as system development continues. This system provides an interesting subdivision of the mathematics-interpretation problem into subproblems, and it demonstrates the ability to cope with challenging handwritten input. The organization of this system as a set of independent modules imposes a welcome structure on the software. A structured approach eases the task of designing, coding, and debugging the complex and diverse rules needed during mathematics recognition.

4. Noise, Distortions, and Connected Symbols

Recognition of mathematical notation is greatly complicated by noise and distortions in the input. Various computational approaches can be used to deal with this problem; these include stochastic grammars, deterministic grammars with confidence-values for tokens, and postprocessing error-correction rules.

4.1. Stochastic Grammars

Chou uses a stochastic grammar to recognize noisy typeset equations [13]. The most likely parse is taken as the correct interpretation of the input. Each production rule in a stochastic grammar has an associated probability. The probability of a particular parse tree is computed by multiplying together the probabilities of each production used in the parse. Two-dimensional patterns are described by allowing each production rule to use either vertical or horizontal concatenation (Sec. 5.1.1). Square-roots are not recognized by the grammar described in [13]. To add them, the square-root sign would have to be treated as two symbols, so that the necessary spatial relationships can be described by horizontal and vertical concatenation.

A dynamic programming algorithm is presented for finding the most likely parse; complexity of this algorithm is $O(n^3)$. An iterative algorithm is used to learn grammar-parameters (such as production-rule probabilities) from training data: parsing all the training expressions gives an estimate of the expected number of times each production rule is used.

Conceptually, the stochastic grammar goes all the way down to the pixel level. As a practical matter, this is implemented in two parts. First, symbol candidates (with associated probabilities) are generated by exhaustive template matching (match every symbol-template to every image location; see Sec. 4.4). Next, the stochastic grammar is used to find the most likely parse of the symbol candidates; the calculation of a parse-probability includes contributions from the symbol-probabilities of all symbols used in the derivation.

Testing reported in [13] is created by copying bitmaps from a previewer for the eqn/troff typesetting system; noise is added by flipping each image bit with probability 0.01. The resulting images are quite challenging, due both to added noise and to the previewer's low resolution (100 dots per inch). Subsequent testing (mentioned during

Chou's demonstration at SSPR90, Murray Hill, New Jersey) involved scanning all equations in a mathematics textbook.

In summary, Chou mentions the following advantages of stochastic grammars for document recognition. The approach treats the interpretation process by parsing from the grammar-start-symbol to the pixel level, with no hard decisions made at any intermediate stage. The result is statistically optimal, given the noise model and grammar model. Noise, ambiguity, and touching characters are handled well. Grammar-parameters can be trained to achieve optimal performance on particular printers, fonts, layouts and scanners. To offset this, the following disadvantages are mentioned: computational expense, skew sensitivity (preprocessing must include good skew correction), and difficulty in accounting for kerning.

More recently, Kopec and Chou are performing document-image analysis using methods adapted from communication theory [26]. Here a finite-state model of the image-creation-process is included as part of the recognition system. This interesting, highly-promising approach is illustrated on the interpretation of the phone book's yellow pages columns. Experiments with the interpretation of other notations, such as music notation, have been undertaken as well.

4.2. Error Correction with a Coordinate Grammar

Dimitriadis *et al.* report on on-going work that extends Anderson's coordinate grammar to add error detection and correction [7]. Unfortunately, this is a preliminary report with few details. Dimitriadis' Ph.D. thesis (in Spanish; Dept. of Automatic Control and Systems, University of Valladolid, cited as "expected 1991") may provide more information. On-line data from a data tablet is processed by grouping strokes and applying elastic matching to recognize symbols; each symbol is represented by a series of candidates, with confidence measures. Parsing starts by considering the candidates with highest confidences; if the resulting global confidence figure is rather low, other candidates are tried, and a globally optimum decision is made. At the time of their report, this parsing scheme was in the planning stage; they do not discuss the danger of a combinatorial explosion of candidate-combinations.

4.3. Postprocessing Error-correction Rules

Heuristic rules for error-correction are discussed in [17]. Examples of these rules are as follows. Knowledge of special function names can correct character confusions ("5in" is corrected to "sin"; "cOs*x*" is corrected to "cos*x*"). The operands of a binary operator normally appear in the same font type and font size; this can be used to correct a character confusion such as "$l<i<n$" to "$1<i<n$". Knowledge of legal subscripts can eliminate further confusions, such as correcting 1_2 to l_2 and 5_2 to S_2.

4.4. Connected Symbols

Connected symbols are common both in typeset and handwritten mathematical expressions. Separation of such symbols is a difficult problem, which has been extensively studied in OCR and other domains as well. Here we review a few approaches to symbol-separation that have been used in mathematics-recognition systems.

Anderson assumes correctly-recognized symbols as input [3]. A grammar rule (number A56) shrinks each symbol's bounding box down to a center point. This enables subsequent coordinate-tests to yield the desired results. For example, in the expression $\frac{y}{x}$ the modified bounding box of the "y" lies wholly above the division bar.

To recognize symbols, Lee and Lee begin with thinning followed by curve detection [25]. Connected symbols are separated using a dynamic programming algorithm. First, the curve list is matched to each of the reference symbols. If the maximum similarity value is lower than a threshold, the curve list is passed to the dynamic programming algorithm. Here the curve list undergoes curve splitting and loop merging, and is put into a canonical order. Then the reference symbols are matched to the sequence of curves at the start of the curve list.

Chou's stochastic grammar does not require explicit separation of connected symbols [13]. Instead, the lexical access stage matches all templates in the font dictionary to every position in the image. (The font dictionary has separate entries for different font sizes.) A match is retained if the hamming distance (number of differing pixels) between the template and the image is below the three-sigma threshold. (Using a noise model in which image bits are independently flipped with probability P_e, the three-sigma threshold includes all matches where the number of disagreeing bits is within three standard deviations of the mean number of disagreeing bits. The probability of a greater number of disagreeing bits is less than 0.13%.) The result is a list of all characters that have significant probability of being present in the image. These hypothesized characters are then processed by the stochastic grammar (Sec. 4.1).

Systems that accept on-line input face a rather different segmentation problem. Where off-line systems must separate symbols that overlap in the image, on-line systems must gather together the set of strokes that form a single symbol. As described in Sec. 6, Chen and Yin [27] do this by analyzing the bounding boxes of successive strokes.

5. Identifying Spatial Relationships

It is difficult to recognize the spatial and logical relationships among symbols in a mathematical expression (Secs. 1.3 and 1.4). Here we review techniques for recognizing subscript, in-line, and superscript relations, and also briefly discuss the recognition of matrix notation.

5.1. Distinguishing Subscript, In-line, Superscript

The identification of subscripts, in-line, and superscript relations is a complex and important problem. For example, the system reported in [6] makes recognition errors that are typically errors in subscript and superscript recognition; this is a mature system that is capable of correctly recognizing many complex expression structures.

5.1.1. Global thresholds

Some authors address this problem by choosing global threshold parameters to distinguish the relations (e.g. [3] [4] [27]). As another example, Chou's stochastic grammar [13] encodes horizontal and vertical concatenation criteria into each production rule. Two examples are as follows. An *<Expression>* and a *<Factor>* can be horizontally concatenated only if both have the same pointsize, and their baselines differ by at most one pixel. A *<Symbol>* and a *<Subscript>* can be horizontally concatenated only if the subscript has a smaller pointsize and a lower baseline.

5.1.2. Statistical labeling

Extensive discussion of subscript and superscript identification is provided by Wang and Faure [28]. This paper highlights the complexity of the task, and offers an interesting approach to its solution. Given a sequence of symbol-bounding-boxes that are part of the same writing line, the task is to label the relationships between pairs of symbols as exponent (E), in-line (L), or subscript (S). (Note that these are spatial relationships, not logical relationships. The symbol pairs in the expression a^2b^3 are labeled $E\ S\ E$; later processing must determine that the b is not logically a subscript of the 2.) Both adjacent and non-adjacent pairs of symbols must be labeled. For example, x^{a_i} and $x^a i$ have the same pairwise labelings (E followed by S); to distinguish these expressions, the x-to-i relationship must be labeled as either E or L. Processing takes place on bounding-box information alone; symbol identity is recognized later. Thus, constraints based on symbol identity (Sec. 5.1.3), cannot be applied here. Similarly, character-baseline information is unavailable. (Ambiguity arises in cases such as y_c versus bc; these have similarly-configured bounding boxes, but different spatial relations among the symbols.)

Training data (35 expressions produced by 7 writers) is used to construct a recognizer. Two features are measured from each pair of bounding boxes: the height-ratio and vertical offset. A modified Ho–Kashyap algorithm is used to place decision boundaries in this feature space. The space is divided into six regions (exponent, either exponent or in-line, in-line but on the exponent side, in-line but on the subscript side, either in-line or subscript, subscript). This is done separately for each of the three following zones: taller box followed by a shorter box; two boxes of approximately equal height; shorter box followed by a taller box.

During classification, the feature space is used to calculate an initial estimate of the probabilities of the E, L and S labels for the given symbol pair. The final labeling is found by searching for the most-probable labeling that also satisfied the contextual constraints. These constraints eliminate impossible situations such as the following: in the symbol-triple ab^c, if ab is labeled L and b^c is labeled E, then $a\ ^c$ must be labeled E. Testing results (using expressions produced by writers who did not participate in the production of training data) are promising. In a comparison with human labeling (based on a presentation of bounding boxes), the system had slightly higher error rates than the humans. The authors propose addressing this by introducing directionality into the recognition process.

This labeling process constitutes one recognition module in the complete recognition system [23] discussed in Sec. 3.5.

5.1.3. Constraints based on symbol identity

Another approach to sub- and superscript recognition is mentioned in [17]. Recognition is based on spatial coordinates as well as local context. The system makes use of a priori information regarding legal configurations for superscripts and subscripts: for example, a^x and x_n are legal, whereas B_* and $X^{\&}$ are not. Similar constraints are used in [6].

5.2. Recognizing Matrices

Matrix-recognition can be restricted to explicit matrices, or it can include linked vectors, in which a line connecting two matrix elements denotes an indeterminate number of intervening elements.

Anderson presents a coordinate grammar for matrix notation, which recognizes both explicit and linked matrices [3]. The matrix recognition is heavily dependent on the availability of three parameters:

mhmax	maximum separation between two horizontally adjacent matrix elements
mvmax	maximum separation between two vertically adjacent matrix elements
vmax	maximum vertical separation between adjacent characters belonging to the same matrix element

The coordinate grammar recognizes rows, columns, or diagonals one at a time. Therefore it never obtains an overall view of matrix structure. For example, suppose that the input is a "matrix" with a variable number of entries in each row. The grammar does not notice this; it parses the input as an EXPLICITARRAY and reports the number of matrix columns as equal to the number of entries in the last row of the matrix. It would be interesting to explore whether the coordinate-grammar formalism can produce a fuller analysis of matrix structure.

In [6], analysis of explicit matrices begins by finding a pair of delimiters of the same size and type. Next a horizontal projection profile is computed for the delimited region; if this profile exhibits periodicity (indicating several rows of similar height), then a vertical projection profile is used to separate the columns. Finally, each matrix entry is analyzed as an expression in its own right.

Another approach to matrix recognition is mentioned in [17]. First a pair of enclosure symbols are found. Symbols within this area are grouped based on proximity; no details of the proximity metric are given.

6. On-line Recognition of Handwritten Expressions

A data tablet offers the possibility of using timing information to assist with the recognition of a mathematical expression.

Chen and Yin describe a system for recognizing handwritten mathematical notation [27]. The emphasis of this work is on symbol segmentation (i.e. collecting the strokes that make up a single symbol) and symbol recognition. Symbol segmentation proceeds by surrounding each stroke with a frame (a bounding box), and testing consecutive frames for overlap in X and Y projections. Special tests are performed to unify two-part symbols such as i, j, $=$, $\leq$, $\geq$. Also, special tests are performed for the square-root symbol, which obscures gaps in the X and Y projections of symbols inside it. Symbol recognition is performed using a nearest-neighbor algorithm, with features coded from stroke types (line, clockwise curve, counterclockwise curve, circle). Multi-character classes are created for confusion-prone symbols, such as "1" versus "/", "7" versus "7", or "C" versus "(". After nearest-neighbor classification, specific contextual constraints are tested to disambiguate symbols in these multi-character classes. For example, if two operands are found on either side of a "1 or /" character, then this character is classified as a "/", otherwise as a "1". If the system rejects a character, or fails to disambiguate it, the user is asked for correction. Recognition results are presented for seven writers entering eight test expressions; the overall symbol-recognition rate is 94%.

On-line data is analyzed in [7]; few details are given, but they do mention the use of elastic matching (dynamic time warping).

The system by Faure and Wang [23] [28] is geared toward recognizing the structure of handwritten mathematical notation. On-line information is used as needed during the knowledge-driven segmentation module. Considerable variability in the handwriting can be tolerated. Sec. 3.5 provides further discussion of this work.

Recognition of handwritten expressions is treated in [18], as discussed in Sec. 3.1.1. Timing information (from the data tablet) is used in extracting strokes for symbol recognition, but is not used during syntactic analysis.

7. Summary and Conclusions

Mathematics recognition is a practically-important problem. Recognition is difficult because mathematical notation is only semi-standardized, conveys meaning through subtle use of spatial relationships, involves a large alphabet of symbols and a large range of font sizes, and contains little redundancy in its representation of information. Comparison among existing mathematics-recognition systems is complicated by the myriad ways in which the mathematics-recognition problem can be defined. Input may be typeset or handwritten, on-line or off-line, input expressions may be isolated or mixed with text, recognizable expression layouts may be restricted to a greater or lesser degree. In addition, some recognizers assume pixels as input, whereas others assume that symbol-recognition has already taken place. This great variation in problem-definition makes it difficult to compare the strengths and weaknesses of existing mathematics recognition systems. Research is needed to clarify the definition of mathematical notation and its dialects. Such a definition could provide a basis for more meaningful comparison among mathematics-recognition systems.

A variety of system architectures are used for mathematics recognition. Syntactic analysis involves parsing an expression using a two-dimensional grammar, such as a coordinate grammar or stochastic grammar, after symbol recognition has been performed. A grammar provides a clear, explicit representation of mathematical-notation syntax, although parsing can be computationally expensive. Projection-profile cutting determines the structure of an expression by recursively dividing the image into subregions (using minima in the horizontal and vertical projection profiles); this can be done before symbol recognition is performed. This simple and efficient technique is suitable for solving a restricted part of the mathematics-recognition problem. Graph rewriting represents symbols and symbol-relationships as a graph, rewriting parts of the graph as recognition proceeds. This offers a flexible means of applying extensive knowledge of notational conventions; however, current graph-rewriting environments execute quite slowly. Procedurally-coded rules provide step-by-step instructions for recognizing an expression. Unless a systematic organization of rules is provided, such a system can become ill-understood and unmanageable due to the unexpected interactions among the unstructured, often ad hoc recognition rules. The data-driven and knowledge-driven modules of Faure and Wang illustrate one approach for systematically organizing procedurally-coded recognition rules.

Whatever recognition approach is used, the structure of recognizable expressions must be characterized. Such a syntax for mathematics may be defined implicitly or explicitly. A major objective is to define the syntax cleanly, to provide a unifying framework for handling the myriad details and exceptions that arise during mathematics recognition.

References

[1] Y. Nakayama, Mathematical formula editor for CAI, *Proc. ACM SIGCHI Conf. on Human Factors in Computing Systems*, Austin, Texas, April 1989, 387–392.

[2] D. Blostein, General diagram-recognition methodologies, *Proc. Int. Workshop on Graphics Recognition*, University Park, Pennsylvania, Aug. 1995, 200–212.

[3] R. Anderson, Two-dimensional mathematical notation, in *Syntactic Pattern Recognition, Applications*, ed. K. S. Fu (Springer -Verlag,1977) 147–177.

[4] S. Chang, A method for the structural analysis of two-dimensional mathematical expressions, *Information Sciences* **2**, 3 (1970) 253–272.

[5] A. Grbavec and D. Blostein, Mathematics recognition using graph rewriting, *Third Int.Conf. on Document Analysis and Recognition*, Montreal, Canada, Aug. 1995, 417–421.

[6] H. Twaakyondo and M. Okamoto, Structure analysis and recognition of mathematical expressions, *Proc. Third Int. Conf. on Document Analysis and Recognition*, Montreal, Canada, Aug. 1995, 430–437.

[7] Y. Dimitriadis, J. Coronado, and C. de la Maza, A new interactive mathematical editor, using on-line handwritten symbol recognition, and error detection-correction with an attribute grammar, *Proc. First Int. Conf. on Document Analysis and Recognition*, Saint Malo, France, Sept. 1991, 242–250.

[8] T. Chaundy, P. Barrett, and C. Batey, *The Printing of Mathematics* (Oxford University Press, 1957).

[9] K. Wick, *Rules for Typesetting Mathematics*, translated by V. Boublik and M. Hejlova (The Hague, Mouton, 1965).

[10] N. Higham, *Handbook of Writing for the Mathematical Sciences* (SIAM, Philadelphia, 1993).

[11] D. Knuth, Mathematical typography, *Bulletin of the American Mathematical Soc.* **1**, 2 (1979).

[12] W. Martin, Computer input/output of mathematical expressions, *Proc. 2nd Symp. on Symbolic and Algebraic Manipulations*, New York, 1971, 78–87.

[13] P. Chou, Recognition of equations using a two-dimensional stochastic context-free grammar, *Proc. SPIE Visual Communications and Image Processing IV*, Philadelphia PA, Nov. 1989, 852–863.

[14] M. Okamoto and B. Miao, Recognition of mathematical expressions by using the layout structure of symbols, in *Proc. First Int. Conf. on Document Analysis and Recognition*, Saint Malo, France, Sept. 1991, 242–250.

[15] J. Ha, R. Haralick, and I. Phillips, Understanding mathematical expressions from document images, *Proc. Third Int. Conf. on Document Analysis and Recognition*, Montreal, Canada, Aug. 1995, 956–959.

[16] D. Blostein and H. Baird, A critical survey of music image analysis, in *Structured Document Image Analysis*, eds. H. Baird, H. Bunke, and K. Yamamoto, (Springer-Verlag, 1992) 405–434.

[17] H. Lee and J. Wang, Design of a mathematical expression recognition system, *Proc. Third Int. Conf. on Document Analysis and Recognition*, Montreal, Canada, Aug. 1995, 1084–1087.

[18] A. Belaid and J. Haton, A syntactic approach for handwritten mathematical formula recognition, *IEEE Trans. Pattern Analysis and Machine Intell.*, **6**, 1 (1984) 105–111.

[19] M. Okamoto and A. Miyazawa, An experimental implementation of document recognition system for papers containing mathematical expressions, in *Structured Document Image Analysis*, eds. H. Baird, H. Bunke, and K. Yamamoto (Springer-Verlag 1992) 36–53. (Earlier version in *Proc. IAPR Workshop on Statistical and Structural Pattern Recognition*, Murray Hill, New Jersey, June 1990, 335–350.)

[20] A. Murase, T. Satou, and M. Nakagawa, Prototyping of METAH, a recognition system for on-line handwritten mathematical expressions, *Information Processing Society of Japan SIG Notes*, **93**, 35 (1993) 25–32 (in Japanese).

[21] M. Okamoto and H. Higashi, Mathematical expression recognition by the layout of symbols, *Trans. IEIEC* **J78-D-II**, 3, (1995) 474–482 (in Japanese).

[22] M. Okamoto and H. Twaakyondo, Mathematical expression recognition by projection profile characteristics, *Trans. IEIEC* **J78-D-II**, 2 (1995) 366–370 (in Japanese).

[23] C. Faure and Z. Wang, Automatic perception of the structure of handwritten mathematical expressions, in *Computer Processing of Handwriting* (World Scientific, 1990) 337–361.

[24] H. Lee and M. Lee, Understanding mathematical expressions in a printed document, *Proc. Second Int. Conf. Document Analysis and Recognition*, Tsukuba, Japan, Oct. 1993, 502–505.

[25] H. Lee and M. Lee, Understanding mathematical expressions using procedure-oriented transformation, *Pattern Recognition* **27**, 3 (1994) 447–457.

[26] G. Kopec and P. Chou, Document image decoding using markov source models, *IEEE Trans. Pattern Analysis and Machine Intelligence* **16**, 6 (1994) 602–617.

[27] L. Chen and P. Yin, A system for on-line recognition of handwritten mathematical expressions, *Computer Processing of Chinese and Oriental Languages* **6**, 1 (1992) 19–39.

[28] Z. Wang and C. Faure, Structural analysis of handwritten mathematical expressions, *Proc. Ninth Int. Conf. on Pattern Recognition*, Rome, Italy, Nov. 1988, 32–34.

Handbook of Character Recognition and Document Image Analysis, pp. 583–603
Eds. H. Bunke and P. S. P. Wang

CHAPTER 22

AUTOMATIC READING OF MUSIC NOTATION

DAVID BAINBRIDGE*
Department of Computer Science, University of Waikato
Hamilton, New Zealand

and

NICHOLAS CARTER†
Department of Music, University of Surrey
Guildford, Surrey, GU2 5XH, United Kingdom

The aim of Optical Music Recognition (OMR) is to convert optically scanned pages of music into a machine-readable format. In this tutorial level discussion of the topic, an historical background of work is presented, followed by a detailed explanation of the four key stages to an OMR system: stave line identification, musical object location, symbol identification, and musical understanding. The chapter also shows how recent work has addressed the issues of touching and fragmented objects—objectives that must be solved in a practical OMR system. The report concludes by discussing remaining problems, including measuring accuracy.

Keywords: Optical music recognition; Primitive shape decomposition; Stave line identification; Musical object location; Symbol identification; Musical semantics.

1. Introduction

The aim of Optical Music Recognition (OMR) is to convert optically scanned pages of music into a versatile machine-readable format. Like other fields of Document Image Analysis (DIA), the task of recognizing such an image requires the comprehension of two-dimensional relationships. This distinguishes the problem from the extensively researched area of Optical Character Recognition (OCR), where—in its most basic form—only one-dimensional lines of text are processed. In fact, music typically includes text, so OCR is a sub-task of OMR. Music is further distinguished by the overlapping or super-imposing of shapes, such as notes on a stave, and is fundamental to the OMR problem.

Music notation is undoubtedly complex and, like any natural language, is con-

*d.bainbridge@cs.waikato.ac.nz
†n.carter@surrey.ac.uk

tinually evolving. There is no fixed and comprehensive definition of music notation but many books cover the subject and the widely-used core is referred to as Common Music Notation (CMN) [1–5]. Sometimes composers invent new notational symbols or systems in unusual situations, but in practical terms CMN represents the main body of printed music in the world today. Although the notation for CMN is standard, different terminology has arisen. Table 1 presents the equivalent terminology.

Table 1. Equivalent musical terminology.

Symbol	Equivalent terms
𝄞	G clef or treble clef
𝄡	C clef or alto/tenor clef
𝄢	F clef or bass clef

Symbol	American Term	English Term
𝅝	whole note	semibreve
𝅗𝅥	half note	minim
𝅘𝅥	quarter note	crotchet
𝅘𝅥𝅮	eighth note	quaver
𝅘𝅥𝅯	sixteenth note	semiquaver
𝄻	whole rest	semibreve rest
𝄼	half rest	minim rest
𝄽	quarter rest	crotchet rest
𝄾	eighth rest	quaver rest
𝄿	sixteenth rest	semiquaver rest

1.1. *Background*

The first published OMR work was carried out at Massachusetts Institute of Technology in the mid 1960s by Pruslin [6,7]. A subset of CMN that concentrated on musical notes was recognized. It lacked features such as clefs, time signatures, and rests, but permitted complex stem-sharing chords. Prerau [7–10], originally intended to extend Pruslin's work. Unfortunately he discovered that Pruslin's stave-removal distorted most musical features other than note clusters and beams, and so he had to develop an alternative approach.

Prerau observed that a musical feature could usually be identified by the dimensions of its bounding box. However, as the same musical feature could appear larger or smaller in a different piece of music, the idea could not be generalized directly. To make his observation independent of absolute size, the dimensions of the box were normalized with respect to the height of the stave. A survey of different styles of music was undertaken and a database built. This database was the main heuristic used to classify shapes. In contrast to Pruslin's work, Prerau's system recognized clefs, rests, certain time signatures, and accidentals, however it could not handle chords.

Early work was ultimately limited by technology, and merely acquiring image data was a significant part of these early projects. Prerau, for example, used a flying spot scanner to achieve a scan resolution of approximately 225 dots per inch (dpi). It was not until the early 1980s that the next significant development occurred. In a major undertaking by Waseda University (Japan), a robotic organist was built [11–13]. Demonstrated at the 1985 International Exposition, the robot would play the organ by voice request, or by reading sheet music using its electronic "eye" (a video camera). It would also play in time with someone singing. The real-time processing necessary was achieved by using dedicated hardware: 17 16-bit computers and 50 8-bit computers with fiber optic data-links. The project lasted $3\frac{1}{2}$ years, and involved four professors and some 50 students. Had commercial equipment and services not been donated, it is estimated that the project would have cost US $2,000,000 [13].

The OMR component of the Wabot-2 was the first system to recognize a comprehensive set of CMN, including chords. The real-time processing requirement led to a "brute force" approach for many OMR problems. Significant use was made of template matching, which was rendered practical by being realised in hardware. More recently the cost of hardware suitable for OMR has come within the budget of many research centers. A reasonable configuration would consist of a flatbed scanner with 200–400 dpi resolution, connected to a workstation with appropriate memory and disk space.[a] Beneficial to the growth of OMR work is the fact that no hardware item is peculiar to OMR. Consequently the necessary resources are frequently available within a department already. Since 1985 there have been over 20 projects, ranging from Honors to Ph.D. level and beyond.

Initially projects concentrated on static systems, where the recognition of musical shapes were "hard-wired" into the system. A more recent trend—and a sign of the maturity of the subject—is an emphasis on designing general systems capable of recognizing arbitrary music in an extensible manner. Unfortunately there is insufficient space in this chapter to describe all the implemented systems. Later in the chapter, when the key tasks for OMR are discussed, prominent work will be described. For a wider awareness of OMR approaches, the reader is directed towards "A critical survey of music image analysis" [14], and "Optical recognition of music notation: A survey of current work" [15].

Recently two commercial products have been released and a third exists as a demonstration version. MidiScan [15] runs on a PC and produces MIDI files for playback only. Due to its commercial orientation, little is known about the algorithms used, other than at a functional level. MidiScan first locates the staves

[a] An A4 page scanned at 300 dpi takes around 80–100 kilobytes when compressed.

automatically, and then enters an interactive phase where the user can correct any mistakes. Next the system classifies all the detected shapes and displays a split screen of the original image and the corresponding part of the "recognized page." Another interactive phase allows corrections to be made and once the user is satisfied with the result, a MIDI file is generated. Since MidiScan's application is audio, it need only recognize a restricted set of CMN: notes (including chords), accidentals, rests, ties, bar lines, time signatures, and clefs. It ignores all other shapes, such as accents, slurs, text, tempo, fingering, and volume markings.

The second product, NoteScan [15], requires a Macintosh computer and feeds its results into the Nightingale music notation editor [3]. Work by the company on a PC version is less developed and output is only suitable for audio. The Macintosh version is capable of recognizing notes (including chords), accidentals, bar lines, time signatures, and clefs. However it does not yet recognize accents, staccato, slurs, ties, text, and other "secondary" features of music.

Score Reader, by Yamaha, is a Windows application running on a PC. It is is similar to MidiScan in that it is intended for MIDI playback only.

Other OMR projects have concentrated on medieval music [16,17] and handwritten formats [18–20]. Working with such material is, understandably, harder since the clarity and regularity of the graphics is considerably reduced. For example, a handwritten note head is more likely to be physically separated from its note stem and will more typically consist of a short stroke rather than the neat oval found in engraved scores. Results are promising, but work on these topics lags behind the success of printed CMN in much the same way handwritten OCR research follows in the wake of printed OCR.

1.2. *Applications*

Consider some of the benefits of an OMR system: one could listen to a piece of printed music without any training in musical notation; a clarinetist could scan a melody written at pitch and have it transposed automatically; a soloist could have the computer provide an accompaniment for rehearsal purposes; an editor could read in an old edition, correct errors and generally "touch it up" using a music publishing program; or a publisher could arrange an orchestrated work as a piano part or chamber ensemble score and convert the piece to Braille at almost no extra cost.

Such examples illustrate that the level of musical detail required by an OMR system is dependent on its intended use. For audio playback, it is sufficient to recognize the musical features on and around the stave (the "core" of music). Items such as the title and fingering information are superfluous, and depending on how crude the playback can be, dynamic markings may also be ignored. Alternatively, an OMR system for editorial enhancement must recognize all musical information on the page. In this situation, typography is the primary goal, hence it is not strictly necessary for the OMR system to understand the musical significance of the graphical shapes, only what the primitive shapes are, and where they are located.

The two examples detailed above represent extreme situations. Other applications require a mixture of graphical recognition and musical understanding. The ultimate goal of OMR work is a system capable of graphically recognizing *and* musically understanding all the information present in a page of music.

2. Overview of the OMR Process

The problem of OMR is simplified by decomposing it into smaller tasks. There is general agreement on the sub-tasks that must be solved, although researchers are still investigating the suitability of different methods for solving these problems. Existing work, in general, can be broken down into four categories: stave line identification, musical object location, symbol identification and musical understanding. In theory each sub-task can be solved in isolation, although in practice implemented systems often overlap stages and solve some sub-tasks simultaneously. For the purposes of the chapter they will be discussed in turn.

2.1. *Stave Line Identification*

To recognize a page of music the staves must, naturally, be detected at some point. Performing this operation at the start of the OMR process is beneficial since it reveals the height of the staves, a useful metric for defining the dimensions of other musical objects in the image, such as note heads and treble clefs. Stave identification is usually achieved by detecting and grouping individual stave lines.

By far the most widely used method for detecting stave lines is *horizontal projection*. A horizontal projection maps a two-dimensional bitmap into one-dimension by recording the number of set pixels in each row of the bitmap in a one-dimensional array, where an entry in the array corresponds to the y coordinate of the row in the bitmap. Under such projections, stave lines appear as distinct peaks which can be easily detected. Numerous variations exist, but the basic technique remains the same. Figure 1 shows an example piece of music. In Fig. 2, its horizontal projection is shown.

It would be naïve to assume all music is scanned level, consequently calculating the horizontal projection for the entire page would reveal little. Instead, the algorithm is applied to proportionally smaller regions of the page. The located short segments of stave can then be logically connected and the angle of skew inferred. Most common is picking a segment on the left-hand side and right-hand side of a page, then connecting the lines, one for one, using $y = mx + c$. A variation on this was implemented by the Wabot-2 project [12], where a low-pass filter (equivalent to a short horizontal projection) located stave lines in contiguous regions across the page.

Competing non-horizontal projection methods are the line and stave fragment-finding with subsequent piecewise-linking process, used by Carter [21–27]; and the vertical scan lines with subsequent mutual similarity checking process, used by Glass [28].

Fig. 1. Example piece of music.

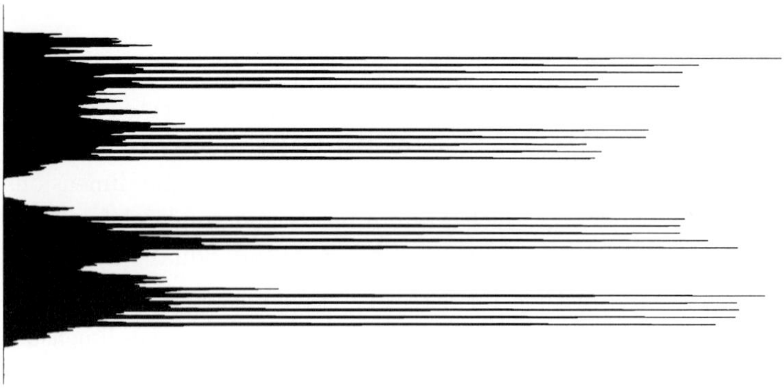

Fig. 2. Horizontal projection of example piece of music.

All the above algorithms naturally discover the angle of skew for the image, and if desired, the image can be enhanced by rotation or shearing. If such a step is carried out, it is advised that the standard image processing technique of performing the transform in reverse, be used [29]. That is to say, compute where each pixel comes from, rather than where each original pixel goes to. This reduces the number of "holes" that might appear as a result of the transform.

Skew is not the only problem caused by scanning. A page scanned from a tightly bound book will be distorted. Depending on the direction of the spine (vertical or horizontal) this can lead to bowing stave lines or foreshortening of musical shapes. Frequently such deformations harmlessly occur only in the margin of the work, so the number of affected scanned works is small by comparison to the number of "normal" works. Thus most OMR researchers do not consider deformation an essential problem, and most are content to design systems that are tolerant of deviations from the norm, but do not attempt to correct the distortion.

2.2. *Musical Object Location*

Musical object location is commonly achieved by removing the stave lines, thus graphically isolating the various shapes. Prerau removed stave lines completely, without regard for other musical features, thus dissecting all shapes that overlapped the stave. Assembling the remaining pieces was a substantial part of his work. A more popular technique can be illustrated by reference to the work of Clarke *et al.* [30], and keeps most musical features whole. In this algorithm, a stave line is removed piecemeal. Considering one vertical slice of a stave line at a time (working say from left to right), evidence of musical objects existing, either above or below the stave line is checked for. If no evidence is found, then the slither of stave line is removed. A common template when searching for evidence is shown in Fig. 3. If pixel B is set and at least one of X, Y, or Z is set, then the stave line remains.

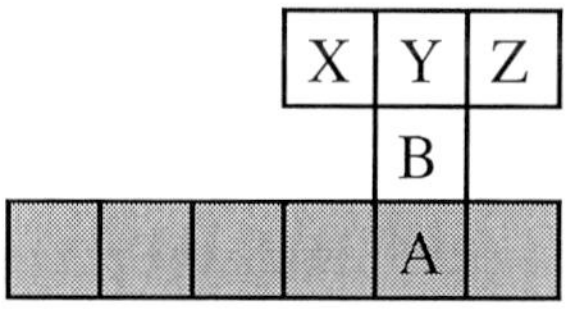

Fig. 3. The template proposed by Clarke *et al.* for removing stave lines.

Of course, stave lines do not faithfully follow a straight line. Even if all the stave lines were drawn perfectly straight and parallel, quantization causes aliasing effects. In other words, a slither of stave line may not be exactly where the formula predicts. Tolerances, therefore, need to be included in the algorithm.

Even with careful stave line removal, some objects still become fragmented. The problem arises when a thin part of a musical feature, typically a line, blends tangentially into a stave line. Commonly cited examples are minim note heads and bass clefs. Such cases of fragmentation are visible in Fig. 4, where the stave lines have been removed from the example piece of music, using the described algorithm.

Not all projects remove the stave lines to locate the objects. Since it is known where the stave lines are, another solution is to *ignore* stave lines. This approach has been successfully used by Glass, in NoteScan, and in the Wabot-2 project. Although ignoring stave lines avoids the problem of fragmentation explicitly, the problem must be solved implicitly to decide where one object stops and the next starts. The difficulty of this task for an OMR system increases in proportion to the complexity of the music.

2.2.1. *Fragmentation*

Fragmentation represents a serious problem to the latter recognition stages—none of the component parts of a fragmented object look like the original object.

Fig. 4. The example piece of music with stave lines removed.

There have been various attempts to address this problem.

Early work made use of musical context [30,31], to detect particular fragmentation cases. Such algorithms work well when the music layout is simple (with only one voice per staff, for example), however such ad hoc schemes become less attractive when more complex scores are considered.

A different approach is to consider the issue earlier on, and eliminate the possibility of fragmentation by improving the stave line removal algorithm. Such an algorithm is described by Martin and Bellissant [32]. The algorithm is similar to the basic stave line removal algorithm, except a more intricate computation is used to decide if an object is present.

Although Martin and Bellissant reported reduced fragmentation, it did not correctly preserve all cases. Additionally, some objects that should remain separate (and would have remained separate under the basic algorithm) became joined. The fact that objects become joined may detract from the attractiveness of this algorithm. This is not the case since touching objects occur in printed works anyway (see Sec. 2.2.2), so that problem must be faced at some point. However, even if it were possible to enhance this algorithm so that all objects were correctly isolated, eliminating fragmentation caused by stave line removal is, perhaps, a false goal. Noise from scanning, dirt on the scanned surface, or foreign bodies introduced whilst printing can all lead to fragmented objects in the starting image for OMR. Problems of fragmentation, therefore, must be faced later on in the OMR process. There would be no harm in reducing the number of cases of fragmentation by applying this algorithm, however the more sophisticated check greatly increases the computational expense of the algorithm. Independent work by Bainbridge [33], reported the sophisticated algorithm to be close to 15 times more expensive, in processor time, than its basic counterpart. It is therefore recommended that a general approach be sought to deal with all fragmentation issues. Solutions to general fragmentation are discussed in Sec. 2.2.3.

2.2.2. *Objects that are joined*

A distinction is made here between two classes of joined objects: objects that legitimately become joined, such as slurs and "hairpin" crescendos crossing bar lines; and objects that illegitimately touch, such as an accidental or the ends of a slurs being drawn too close to its corresponding note head. To aid clarity, the first class of joined objects will be called *super-imposing*, whilst the term *touching* will refer to the latter scenario.

Strictly speaking, touching objects should never arise. Texts on the topic of music notation layout [4,5] stress the importance of a clear, distinct layout when positioning objects such as accidentals and slurs, and examples of objects touching reflect poor craft-work. Traditionally when typesetting music, the spacing of staves and stave systems is decided first, followed by the division of measures. Finally the clefs, key signatures, time signatures, notes etc. are added. It is at this point that the typesetter may find the estimated space for notes insufficient, and thus pack objects tightly together causing objects to touch. If this were the only consideration, a typesetter could perhaps quickly sketch a draft copy, observe the bottlenecks and rearrange things slightly for the final copy. There are, however, other constraints the typesetter must consider, such as natural breaks in the music to turn the page, as well as the aesthetic look of each page. As the title, *The Art of Music Engraving* [4], suggests, music notation layout is not a mechanical process. It is a complex task.

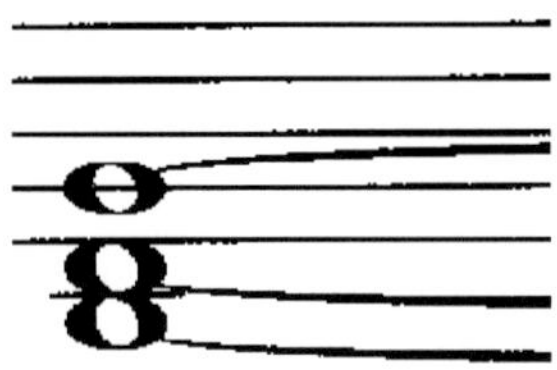

Fig. 5. Careless craft-work sometimes results in touching objects.

It could be argued that an OMR system should not have to consider touching objects. More care should be taken by the typesetter. Such an argument is naïve. There is an extensive body of existing music printed through the centuries that includes touching items, and experience in the field has shown that touching objects are common—even in modern works. In Fig. 5, a typical example is shown. A practical OMR system must address the complications of both super-imposed objects *and* touching objects.

Objects that are joined pose a problem to a matching algorithm as the conglomerate shape looks like neither shape. This description is reminiscent of the fragmentation problem. In fact though fragmentation and joining are opposites in dilemma, they are often resolved together in an OMR system.

2.2.3. *Coping with general fragmentation and joining*

As OMR systems attempt to recognize more "real-world" pages, issues of fragmentation and touching are becoming increasingly important. In this situation, the observation by Prerau that many musical shapes are distinguishable by their bounding box, which results in an efficient recognition process, can no longer be reliably used. That is not to say the technique should no longer be used, just that there needs to be a shift in emphasis away from bounding box checks towards a heavier reliance on graphical classification algorithms.

The issue of fragmentation is tackled by Bainbridge [34], in the primitive identification stage of the system. The system makes use of the shape information present in the primitive pattern recognition routines, in combination with the isolated objects to define regions to be searched for primitives, however the check for their existence can be performed back in the original image, rather then in the image with the isolated shapes. Such a system is effectively managing "tri-state" pixels for the bitmap: clear, set, and "clear but was originally set." Using this strategy and removing the detected primitives, then repeating the process deals with objects that touch. A fuller description of the system is presented in Sec. 2.3.1.

Work by Coüasnon *et al.* [35,36], utilizes a derivative of Definite Clause Grammars (DCG) [37]. Amongst other things, the DCG controls when to join and segment shapes. The process is reminiscent of Clarke's ad hoc musical context joining work, only more sophisticated (directed and targeted) since it is woven into the matching processes, unlike Clarke's work where it was done as a pre-processing step when little is known about the musical shapes.

A concern with the general fragmentation and touching solutions that have been discussed, is the lack of empirical evidence verifying that they will actually work. This will change over time. For the moment we will assume that issues of broken and joined objects are resolved correctly by the OMR system, and proceed to discuss the next important component of an OMR system: symbol identification.

2.3. *Symbol Identification*

The topic of pattern recognition covers a wide variety of problems [38], and there are numerous techniques that are applicable to detecting musical shapes. To aid our discussion of symbol identification strategies, let us consider some of the properties of musical features and compare this with the similarities and differences to other document image analysis tasks.

The graphical properties of music are significantly different to printed text. This point is illustrated in Fig. 6. For example, the same musical shape can have more than one graphical representation. In some circumstances the difference is a simple alteration such as stretching; more complex alterations include rotation and

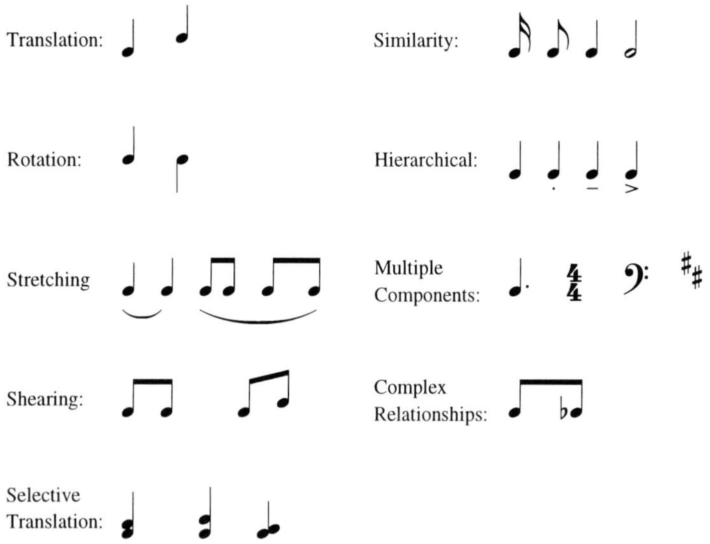

Fig. 6. Properties of music not found in printed text.

shearing. Another important difference between text and music is that each text item is intentionally graphically different. In music, however, shapes are sometimes graphically similar on purpose, where the minor differences convey additional information. This is also true for the case of hierarchy, which additionally serves to illustrate the frequent use, in music, of multiple components. This contrasts with text, where multiple components are relatively rare, and are normally processed using ad hoc methods. Finally, music makes use of more intricate spatial layouts than text. Two-dimensional spatial information, both within a shape and between shapes is used to convey musical information, unlike text which has a predominantly one-dimensional layout.

These examples serve to illustrate that OMR is substantially different to OCR. On the other hand, the nature of the music recognition problem is strongly structured and the topic of structured DIA is a rich source of techniques [39]. Techniques from this area that have been successfully adapted and applied to OMR are: Definite Clause Grammar [35,36], Graph Grammar [40,41] Musical Knowledge with Constraints [42], and a Decision Tree Classifier [43].

Additionally, unique aspects of recognizing music have led researchers to develop new methodologies. Fujinaga's system, for example [44–52], integrates a recognizer, editor and learner. The recognizer utilizes projection methods and a k-nearest neighbor classification algorithm [53]. The result is viewed using a music notation editor, where any corrections made by the user are passed on to the learning module. Here, the weights of the feature vectors used for classification are recomputed, aided by a genetic algorithm, to give an improved match. Even with the genetic algorithm, this is an expensive operation; consequently the system is designed to perform the learning operation during processor "idle-time." The system is naturally extensible. When a musical feature is encountered for the first time it will be mis-classified. This is corrected by the operator during the interactive phase. As more examples of the new shape are encountered and corrected, the system will gradually learn how to distinguish the shape automatically.

Fujinaga's system also serves to illustrate an important sub-division of the symbol identification stage of an OMR system. Some musical features are complex and variable in shape—beamed notes being one such example. In Fig. 7, two graphically different musical notes that are musically identical are shown. An additional step was employed in Fujinaga's system that broke down the larger objects. This process of decomposition is common in OMR systems.

Fig. 7. Musical features can be complex and variable in size.

2.3.1. *Decompose into primitives*

Music shapes, particularly those conveying pitch information, are graphically complex, making them difficult to recognize in a single pass. Instead—like other DIA problems—simpler geometric shapes, called *primitives*, are located. Examples of primitives are note heads, stems, tails, slurs, and beams. The decomposition into primitives is a natural step to take because even if it were possible to recognize shapes "straight off," it would still be necessary to extract the position of the note heads for the musical semantic phase of the system (see Sec. 2.4).

Some researchers have, in the past, selected just one technique for detecting primitives and used it exclusively, to see how appropriate the method is. Alternatively, researchers have used a selection of methods, whichever is the most appropriate for the primitive shape. Needless to say, there are numerous rival pattern recognition techniques, all suitable. Pattern recognition algorithms successfully applied to OMR are: projections [14], slicing techniques [30], connectivity analysis [31], nearest neighbor [53], template matching [54], neural networks [55,56], and the Hough transform [57].

Primitives can be broadly categorized into three groups:

- *Text sized shapes* which vary little in size: examples include accents, and stress marks. They are usually single components.
- *Distinctive sized shapes* which vary little in size: examples include treble clef, and rests.
- *Variable sized shapes*: examples include crescendo/decrescendo "hairpins," slurs,

and beams.

The groupings characterize the type of pattern recognition routines that are applicable. It was pointed out earlier that music includes text. A suitable OCR technique therefore, becomes part of the OMR pattern recognition framework. Conveniently, the group *text sized shapes* are also suitable for recognition using standard OCR matching algorithms.

Distinctive sized shapes already contain the information necessary for classification, i.e. the feature that makes them distinctive. The trick is to design a routine that exploits this. Such algorithms are normally less expensive than methodologies that classify general shapes. For example, the bounding box of a treble clef is distinctive and the shape is not prone to fragmentation, or other objects touching it. Its bounding box, therefore, would be sufficient to classify treble clefs. To be absolutely sure, perhaps a prudent step would be to additionally insist on a y-projection that matches the profile of the treble clef. This is still a significantly cheaper operation than, for example, using a general template matching solution.

Variable sized shapes are not so common in other pattern recognition problems. They require techniques that support parameterized feature descriptions, such as the Hough transform, slicing techniques and connectivity analysis.

In general, an OMR system is designed to be a framework where the most appropriate pattern recognition routine can be used to detect a particular primitive. In most OMR systems the selection of suitable pattern recognition routines, and the design of functions that recognize particular primitives are implemented with "hard-wired" code. Work by Bainbridge [31,33,34,58–60], describes one way this step can be generalized.

At the heart of this system is a language called PRIMELA,[b] which permits the specification of arbitrary primitive graphical shapes in a homogeneous manner. The definition of a primitive includes a graphical description as well as stating which matching technique to apply, selected from a collection of pattern recognition routines. The graphical description must be arbitrarily scalable, hence freeform curves [61] are particularly suitable for the role. The current set of pattern recognition routines includes: projections (vertical and horizontal), Hough transform, bounding box checks, template matching, slicing techniques, and connectivity analysis, although this set could be extended to include other pattern recognition techniques should they prove useful for OMR.

The specially designed language serves to abstract the description of primitives which are to be located, from the computational engine that is capable of doing the work. It is this that gives the project its extensibility.

The best methodology for symbol identification is still an open question in OMR.

2.3.2. *Assemble primitives into musical features*

Now that the isolated shapes have been decomposed into primitives, they need

[b]PRIMELA is an acronym for PRIMitive Expression LAnguage.

to be assembled into musical features. Essentially the task is that of expressing the valid taxonomy of musical features. Syntactic Pattern Recognition [62–64], is a rich source for approaches to solve this problem. In particular, formal grammars have already been applied with some success to a variety of two-dimensional recognition problems, including engineering drawings [65,66], circuit diagrams [67], and flow charts [68].

There is a hierarchy of grammar methodologies, based on the type of data-structure they generate. Commonly used data-structures are strings, arrays, trees, and graphs; increasing in expressive power as listed, but also increasing the complexity of the parser required. String grammars form the basic model and are frequently used to specify computer languages for compilers, however they are not particularly suited to two-dimensional problems. The use of data-structures that are themselves two-dimensional, more naturally model structured document problems. The most powerful structure is a graph grammar [69]. Comparable to string grammars, where restrictions in the type of grammar (context sensitive, context free, and so on) permit more efficient parsers, graph grammars are often restricted to sub-classes, since this reduces the degree of complexity in parsing such languages.

An attributed programmed graph grammar [70] forms the basis for the work by Fahmy and Blostein [40,41]. Traditionally a graph grammar translates an input graph into an output graph that reflects a higher level of understanding of the document. The work by Fahmy and Blostein differs slightly in that the starting point is a collection of isolated nodes that represent the primitive shapes detected in the image. The graph grammar processes these nodes, forming links that represent interactions between primitives, consequently generating a connected graph as output. Initially the system used a "Build-Weed-Incorporate" model to resolve conflicts that can arise when forming links. Later this was altered to a "Build-Constrain-Incorporate" model when the system was extended to cope with the more realistic scenario of uncertain data. Uncertainty was introduced into the system by tagging primitives with values reflecting how reliable the pattern recognition match was.

Work by Coüasnon *et al.* combines a grammar with programming logic to form a definite clause based grammar. The primary role of the grammar is to specify the taxonomy of musical features. It also controls joining and segmentation (briefly described earlier) as well as defining a sequential order to process objects on the staves—effectively specifying how to "read" music.

Put simply, a Definite Clause Grammar (DCG) is a Backus Naur Form (BNF) grammar conforming to slightly different notational conventions. In Fig. 8, a BNF grammar has been rewritten as a DCG. The motivation for this change in syntax comes from Prolog programming, where an automatic process can translate a DCG into Prolog clauses. Combining a DCG with the facilities offered by Prolog leads to a powerful platform, permitting the elegant implementation of parsers.

In the project by Coüasnon *et al.* [35,36], the notation for a DCG is augmented with two additional constructs: a Position Operator takes the form $A\ \boldsymbol{P}\ B$ and means A and, at position $\boldsymbol{P}$ in relation to A, we find B; and Factorization takes the

Backus-Naur Form	Definite Clause Grammar
`<s> ::= a b`	`s --> [a], [b].`
`<s> ::= a <s> b`	`s --> [a], s, [b].`

Fig. 8. A DCG is a BNF grammar with different syntax.

form $A\ (B : C)$ and means $A\ B$ and $A\ C$. The implemented translator converts a specified grammar into λProlog, a higher order dialect of Prolog, as this language more naturally supports the features of the extended DCG.

Figure 9 is based on an example presented by Coüasnon *et al.* [36]. The example is only intended to illustrate the principle, and thus has been simplified by omitting the productions for chords, dots, accents, and slurs. BNF is used to express the grammar for convenience, since it is trivial to change this to the extended DCG. The rule for <NoteGrU> can be interpreted as, "There is a <NoteGrU> if there is a stem, and close to the left of the down tip of the stem, there is a <NoteHead>."

```
<NoteGr>       ::= <NoteGrU> | <NoteGrD>
<NoteGrU>      ::= stem closel_downtip <NoteHead>
<NoteHead>     ::= <Head> closel_same_line [<Accidental>]?
<Head>         ::= black_note_head | white_note_head
<Accidental>   ::= natural | flat | sharp | double_flat | double_sharp
```

Fig. 9. A simplified excerpt of the grammar used by Coüasnon *et al.*

As a final comment, if an OMR system is to be extensible, not only must the recognition of primitive shapes be arbitrary, but so too must the technique for expressing the legal taxonomy of the primitive shapes.

2.4. *Semantics of Music Notation*

This stage of an OMR system converts the graphically recognized shapes into a musical data structure. Basically, it comes down to interpreting the spatial relationships found in the image. In OCR this is a simple step (almost trivial), however it is much more complex for music. Positional information is extremely important. The same graphical shape can mean different things in different situations. For instance, to determine if an object between two notes is a slur or a tie, the pitch of the two notes must be considered; a dot *above* a note head changes its performance character principally by reducing its duration whereas a dot *to the right of* a note head increases the note's duration by 50%. Another, more subtle, example is given by variations in alignment, where notes do not align in the score (i.e. do not occur at the same horizontal position) but are intended to occur concurrently.

Also, establishing the association of "free-floating" objects such as slurs with the appropriate notes is an important task in understanding music.

In the literature, how this is done is usually omitted, since the operations are specific to a particular implementation. In broad terms, this phase of an OMR system consists of (possibly) multiple passes over a graph-like data structure, creating links, deleting links, or modifying the attributes stored at nodes due to the effect of one musical feature—say a key signature—on other musical features, such as notes.

Both of the systems for musical taxonomy described in Sec. 2.3.2 in fact simultaneously deal with taxonomy and musical semantics. In the graph grammar approach the "Incorporate" stage of the process realises the musical semantics by using the formed links to store data as attributes to individual nodes, directed by the production rules. Such a description is similar to the general outline of the musical semantic process just given above. In the system developed by Coüasnon *et al.* it was previously mentioned that the DCG incorporates "how to read music." From this the musical interpretation of the piece can be generated "on the fly" using actions embedded in the grammar.

3. Experimental Results

How the accuracy of an OMR system should be measured is still an open question. Existing systems cope with different imperfections in the original image and have a variety of symbols in their vocabulary. Even if these factors were the same for a set of systems, musical features can be defined in different ways so as to alter the number of items and hence errors/substitutions on a page. For example, is an A major key signature (which consists of three sharp signs) one item or three? One way to approach this issue is to ask how many editing operations would be necessary to correct a particular error in the output of the recognition system. This would, however, tie the measurement to one particular music printing package where the editing would take place. It is also difficult to calculate exactly how accurate a system is when it comes to recognizing symbols such as slurs and beams where exact positioning and thickness may not be reproduced in the reconstruction. As was alluded to in Sec. 1.2, if a system is aimed at typography, then such details are important. If the aim is simply to transfer the musical information that "there is a slur encompassing this group of notes" then the exact position, thickness and curvature of the symbol are less important and may be set using the musical knowledge built into the reconstruction software.

In relating the accuracy of an OMR system to the notation editing software used for reconstruction, another factor emerges. This is the closeness of the integration between the recognition engine and the end-use application. Indeed, the amount of user-interaction may vary from little or none to being a significant part of the process. Drawing a comparison with commercial OCR software, the direction taken seems to be one where the recognition system and the text editor, spell checker and formatter used to work with the results are becoming more closely integrated and with user-definable degrees of interactivity.

Work is currently proceeding on a standard file format for music notation, called Notation Interchange File Format (NIFF), which may assist in providing a means to compare accuracy of systems. Assuming that NIFF is adopted widely by the music notation software authors (which, at the time of writing, seems likely) then the providers of commercial OMR software will probably follow suit and start writing NIFF files. Hopefully, direct side-by-side comparisons using the same input image would then be possible and at least provide relative accuracy figures.

In contrast to the potentially diverse "end-points" of an OMR system—with the complications this entails—the parameters involved with the initial stages of an OMR system are more tightly defined. This, combined with the natural development of an OMR system which starts with the early stages, has led to a better understanding of this part of the OMR problem. This is reflected by the strong similarity in design of the initial steps of different OMR systems. There is still room, however, for refinement in certain areas. For example, although shearing a page to correct for skew is known to be faster than rotating, it does introduce errors (in vertical alignment) into the corrected image. Experimentation is required to determine whether the error is acceptable. Also some experimentation is needed to clarify the significance of distortion in a scanned image.

4. Remaining Problems

Undoubtedly, there is room for further research work in the field of automatic recognition of music notation. There is such a vast range of printed music in existence that it is an enormous task to develop a recognition system which can cope with a wide scope of input images. Not only does printed music vary in size and font but also in print quality and complexity. As has been discussed, symbols touch, markings vary in size whilst retaining the same classification and the super-imposition of symbols is fundamental to the notation system.

Acknowledgements

David Bainbridge would like to thank the New Zealand Vice-Chancellors' Committee for funding his work, and Tim Bell for comments on earlier drafts of this chapter. Nicholas Carter would like to acknowledge the support of the Engineering and Physical Sciences Research Council of the United Kingdom. The authors would also like to thank Cindy Grande, Ichiro Fujinaga, and Bertrand Coüasnon who contributed during the writing of this chapter.

References

[1] G. Read, *Music Notation: A Manual of Modern Practice* (Victor Gollancz Ltd., London, 1974).
[2] D. J. Grout, *A History of Western Music* (J. M. Dent and Sons, London, 1960).
[3] D. Byrd, Music notation by computer, Ph.D. Thesis, Indiana University, 1984.
[4] T. Ross, *The Art of Music Engraving and Processing: A complete manual reference and textbook on preparing music for reproduction and print* (Hansen Press, Miami, 1970).

[5] G. Heussenstamm, *The Norton Manual of Music Notation* (W. W. Norton, New York, 1987).
[6] D. Pruslin, Automatic recognition of sheet music, Sc.D. dissertation, Massachusetts Institute of Technology, 1966.
[7] M. Kassler, Optical character recognition of printed music: a review of two dissertations, *Perspectives of New Music* **11** (1972) 250–254.
[8] D. S. Prerau, Computer pattern recognition of standard engraved music notation, Ph.D. Thesis, Massachusetts Institute of Technology, 1970.
[9] D. S. Prerau, Computer pattern recognition of printed music, in *Proc. of the Fall Joint Computer Conf.*, Montvale, NJ, Nov. 1971, 153–162.
[10] D. S. Prerau, Do-Re-Mi: a program that recognizes music notation, *Computers and the Humanities* **9** (1975) 25–29.
[11] T. Matsushima, Automated high speed recognition of printed music (Wabot-2 vision system), in *Proc. of the 1985 Int. Conf. on Advanced Robotics*, Japan Industrial Robot Association (JIRA), Shiba Koen, Minato-ku, Tokyo, 1985, 477–482.
[12] T. Matsushima, T. Harada, I. Sonomoto, K. Kanamori, A. Uesugi, Y. Nimura, S. Hashimoto and S. Ohteru, Automated recognition system for musical score—the vision system of Wabot-2, *Bulletin of Science and Engineering Research Laboratory, Waseda University* **112** (1985) 25–52.
[13] C. Roads, The Tsukuba musical robot, *Computer Music J.* **10** (1986) 39–43.
[14] D. Blostein and H. S. Baird, A critical survey of music image analysis, in *Structured Document Image Analysis*, eds. H. S. Baird, H. Bunke and K. Yamamoto (Springer-Verlag, Berlin, 1992) 405–434.
[15] E. Selfridge-Field, Optical recognition of music notation: a survey of current work, *Computing in Musicology: An International Directory of Applications* **9** (1994) 109–145.
[16] W. F. McGee and P. Merkley, The optical scanning of medieval music, *Computers and the Humanities* **25** (1991) 47–53.
[17] N. P. Carter, Segmentation and preliminary recognition of madrigals notated in white mensural notation, *Machine Vision and Appl.* **5** (1992) 223–230.
[18] J. W. Roach and J. E. Tatem, Using domain knowledge in low-level visual processing to interpret handwritten music: an experiment, *Pattern Recogn.* **21** (1988) 33–44.
[19] J. Wolman, J. Choi, S. Asgharzadeh and J. Kahana, Recognition of handwritten music notation, in *Proc. of the Int. Computer Music Conf.*, San Jose, 1992, 125–127.
[20] A. Bulis, R. Almog, M. Gerner and U. Shimony, Computerized recognition of handwritten musical notes, in *Proc. of the Int. Computer Music Conf.*, San Jose, 1992, 110–112.
[21] N. P. Carter, Automatic recognition of printed music in the context of electronic publishing, Ph.D. Thesis, University of Surrey, 1989.
[22] N. P. Carter and R. A. Bacon, Automatic recognition of music notation, in *Proc. of the Int. Association for Pattern Recognition Workshop on Syntactic and Structural Pattern Recognition*, Murray Hill, NJ, June 1990, 482–487.
[23] N. P. Carter and R. A. Bacon, Automatic recognition of printed music, in *Structured Document Image Analysis*, eds. H. S. Baird, H. Bunke and K. Yamamoto (Springer-Verlag, Berlin, 1992) 456–465.
[24] N. P. Carter, A new edition of Walton's façade using automatic score recognition, in *Advances in Structural and Syntactic Pattern Recognition (Proc. of the Int. Workshop on Structural and Syntactic Pattern Recognition, Bern), Series in Machine Perception and Artificial Intelligence*, ed. H. Bunke (World Scientific, 1992) 352–362.
[25] N. P. Carter, A generalized approach to automatic recognition of music scores, Technical report STAN-M-87, Department of Music, Stanford University, 1993.

[26] N. P. Carter, Music score recognition: problems and prospects, *Computing in Musicology: An Int. Directory of Applications* **9** (1994) 152–158.

[27] N. P. Carter, Conversion of the Haydn symphonies into electronic form using automatic score recognition: a pilot study, in *Proc. of the IS&T/SPIE's Symp. on Electronic Imaging: Science and Technology; Conference 2181 - Document Recognition*, eds. L. M. Vincent and T. Pavlidis, San Jose, California, USA, Feb. 1994, 279-290.

[28] S. Glass, Optical music recognition, B.Sc. Thesis, University of Canterbury, New Zealand, 1989.

[29] J. C. Russ, *The Image Processing Handbook* (CRC Press, 1992) 96–97.

[30] A. T. Clarke, B. M. Brown and M. P. Thorne, Inexpensive optical character recognition of music notation: a new alternative for publishers, in *Proc. of the Computers in Music Research Conf.*, Lancaster, UK, 1988, 84–87.

[31] D. Bainbridge, Preliminary experiments in musical score recognition, B.Eng. Thesis, Department of Computer Science, University of Edinburgh, Edinburgh, UK, 1991.

[32] P. Martin and C. Bellissant, Low-level analysis and recognition of music drawing images, in *Proc. of the First Int. Conf. on Document Analysis*, Saint-Malo, France, 1991, 417–425.

[33] D. Bainbridge, Optical music recognition: progress report 1, Technical report tr-cosc 05.94, Department of Computer Science, University of Canterbury, New Zealand, 1994.

[34] D. Bainbridge, Optical music recognition: progress report 2, Technical report tr-cosc 02.95, Department of Computer Science, University of Canterbury, New Zealand, 1995.

[35] B. Coüasnon and J. Camillerapp, Using grammars to segment and recognize music scores, in *Int. Association for Pattern Recognition Workshop on Document Analysis Systems*, Kaiserslautern, Germany, Oct. 1994, 15–27.

[36] B. Coüasnon, P. Brisset and I. Stephan, Using logic programming languages for optical music recognition, in *The Third Int. Conf. on the Practical Application of Prolog*, Paris, France, April 1995, 115–134.

[37] I. Bratko, *Prolog: Programming for Artificial Intelligence* (Addison-Wesley, 1990) 431–458.

[38] C. H. Chen, L. F. Pau and P. S. P. Wang, eds., *Handbook of Pattern Recognition and Computer Vision* (World Scientific, 1993).

[39] H. S. Baird, H. Bunke, and K. Yamamoto, eds., *Structured Document Image Analysis* (Springer-Verlag, Berlin, 1992).

[40] H. Fahmy and D. Blostein, A graph grammar for high-level recognition of music notation, in *Proc. of the First Int. Conf. on Document Analysis and Recognition*, Saint Malo, France, 1991, 70–78.

[41] H. Fahmy and D. Blostein, Graph grammar processing of uncertain data, in *Advances in Structural and Syntactic Pattern Recognition (Proc. of the Int. Workshop on Structural and Syntactic Pattern Recognition, Bern), Series in Machine Perception and Artificial Intelligence*, ed. H. Bunke (World Scientific, 1992) 373–382.

[42] H. Kato and S. Inokuchi, The recognition system for printed piano music using musical knowledge and constraints, in *Proc. of the Int. Association for Pattern Recognition Workshop on Syntactic and Structural Pattern Recognition*, Murray Hill, NJ, June 1990, 231–248.

[43] S. Baumann and A. Dengel, Transforming printed piano music into MIDI, in *Advances in Structural and Syntactic Pattern Recognition (Proc. of the Int. Workshop on Structural and Syntactic Pattern Recognition, Bern), Series in Machine Perception and Artificial Intelligence*, ed. H. Bunke (World Scientific, 1992) 363–372.

[44] B. Alphonce, B. Pennycook, I. Fujinaga and N. Boisvert, Optical music recognition: a progress report, in *Proc. of the Small Computers in the Arts*, 1988, 8–12.

[45] I. Fujinaga, Optical music recognition using projections, M.Sc. Thesis, McGill University, Montreal, Canada, 1988.

[46] I. Fujinaga, B. Alphonce and B. Pennycook, Issues in the design of an optical music recognition system, in *Proc. of the Int. Computer Music Conf.*, Ohio State University, Nov. 1989, 113–116.

[47] I. Fujinaga, B. Alphonce, B. Pennycook and N. Boisvert, Optical recognition of music notation by computer, *Computers in Music Research* **1** (1989) 161–164.

[48] I. Fujinaga, B. Pennycook and B. Alphonce, Computer recognition of musical notation, in *Proc. of the First Int. Conf. on Music Perception and Cognition*, 1989, 87–90.

[49] I. Fujinaga, B. Pennycook and B. Alphonce, The optical music recognition project, *Computers in Music Research* **3** (1991) 139–142.

[50] I. Fujinaga, B. Alphonce, B. Pennycook and K. Hogan, Optical music recognition: progress report, in *Proc. of the Int. Computer Music Conf.*, Montreal, 1991, 66–73.

[51] I. Fujinaga, B. Alphonce, B. Pennycook and G. Diener, Interactive optical music recognition, in *Proc. of the Int. Computer Music Conf.*, San Jose, 1992, 117–120.

[52] I. Fujinaga, An optical music recognition system that learns, in *Enabling Technologies for High-Bandwidth Applications*, ed. J. Maitan (SPIE 1785, 1992) 210–217.

[53] T. M. Cover and P. E. Hart, Nearest neighbor classification, *IEEE Trans. Information Theory* **13** (1967) 21–27.

[54] I. Witten, A. Moffat and T. Bell, *Managing Gigabytes: Compressing and Indexing Documents and Images* (Van Nostrand Reinhold, 1994).

[55] R. Lippman, An introduction to computing with neural nets, *IEEE ASSP Mag.* (1987).

[56] O. Yadid-Pecht, E. Brutman, L. Dvir, M. Gerner and U. Shimony, RAMIT: Neural network for recognition of musical notes, in *Proc. of the Int. Computer Music Conf.*, San Jose, 1992, 128–131.

[57] R. Boyle and R. Thomas, *Computer Vision: A First Course* (Blackwell Scientific Publications, 1988).

[58] D. Bainbridge, A complete optical music recognition system: looking to the future, Technical report tr-cosc 06.94, Department of Computer Science, University of Canterbury, New Zealand, 1994.

[59] D. Bainbridge and T. Bell, Playing musical computers, *New Zealand Science Monthly* (1994) 10–11.

[60] D. Bainbridge, Optical music recognition: a generalised approach, in *Proc. of the Second New Zealand Computer Science Research Students' Conf.*, Waikato, New Zealand, April 1995, 23–30.

[61] J. D. Foley, A. van Dam, S. K. Feiner and J. F. Hughes, *Computer Graphics: Principles and Practice, Second Edition* (Addison-Wesley, 1990) 471–515.

[62] J. T. Tou and R. C. Gonzalez, *Pattern Recognition Principles* (Addison-Wesley, 1974).

[63] K. S. Fu, *Syntactic Pattern Recognition and Applications* (Prentice-Hall, 1982).

[64] H. Bunke and A. Sanfeliu, eds., *Syntactic and Structural Pattern Recognition: Theory and Application* (World Scientific, 1990).

[65] D. Dori, Self-structural syntax-directed pattern recognition of dimensioning components in engineering drawings, in *Structured Document Image Analysis*, eds. H. S. Baird, H. Bunke and K. Yamamoto (Springer-Verlag, Berlin, 1992) 359–384.

[66] D. Antoine, S. Collin and K. Tombre, Analysis of technical documents: the REDRAW system, in *Structured Document Image Analysis*, eds. H. S. Baird, H. Bunke and K. Yamamoto (Springer-Verlag, Berlin, 1992) 385–402.

[67] S.-W. Lee, Recognizing hand-drawn electrical circuit symbols with attributed graph matching, in *Structured Document Image Analysis*, eds. H. S. Baird, H. Bunke and K.

Yamamoto (Springer-Verlag, Berlin, 1992) 340–358.
[68] H. Bunke and B. Haller, Syntactic analysis of context-free plex languages for pattern recognition, in *Structured Document Image Analysis*, eds. H. S. Baird, H. Bunke and K. Yamamoto (Springer-Verlag, Berlin, 1992) 500–519.
[69] H. Ehrig, Introduction to the theory of graph grammars, in *Graph Grammars and Their Application to Computer Science*, eds. H. Ehrig, H.-J. Kreowski and G. Rozenberg (Springer-Verlag, Berlin, 1978) 1–69.
[70] H. Bunke, Attributed programmed graph grammars and their application to schematic diagram interpretation, *IEEE Pattern Analysis and Machine Intell.* **4**, 6 (1982) 574–582.

SPECIAL APPLICATIONS AND SYSTEMS

Handbook of Character Recognition and Document Image Analysis, pp. 605–621
Eds. H. Bunke and P. S. P. Wang

CHAPTER 23

ALGORITHMS FOR AUTOMATIC SIGNATURE VERIFICATION

G. DIMAURO, S. IMPEDOVO, G. PIRLO
Dipartimento di Informatica degli Studi di Bari,
via Amendola 173, 70126 Bari, Italy

In this paper an introduction to automatic signature verification is presented. The concepts of dynamic and static signature verification are addressed and the fundamental phases in the process of signature verification are discussed. Throughout the paper, the activities of the main research teams working in the field are presented and steps towards the development of advanced systems are described.

Keywords: Dynamic signature; Dynamic time warping; Personal verification; Signature verification system; Static signature.

1. Introduction

The use of electronic computers in gathering and processing information on local and geographical communication networks makes the problem of high-security access basically important in many applications. For this purpose, several systems for automatic personal verification are used at present [1]: physical mechanism belonging to the individual (i.e., key or badge for terminal access control), information based systems (i.e., numeric combination or password) and personal characteristics (i.e., fingerprint or signature).

In order to evaluate the effectiveness of such systems, several factors must be considered. Among others, the fundamental characteristics are the accuracy of the system, the verification speed, the simplicity of the enrolment process, the capability to work on distributed systems in remote access mode and the consensus of the users in utilizing specific devices for personal verification. In this sense, the fundamental advantage of signature is that it is a common form in legal attestation and also the customary way of identifying an individual in daily operations such as banking transactions and fund transfers. On the basis of these considerations, the systems for signature verification are of great interest to commercial companies which realized the future commercial profits involved to the development of such systems.

Unfortunately, signature verification is not a trivial pattern recognition task and a wide community of scientists is currently involved in the improvement of both acquisition devices and algorithms for signature verification [2,3]. In the field of acquisition technology for handwritten data, graphic tablets and integrated tablet-display devices are generally used for on-line signature acquisition, while cameras or scanners are used for off-line data acquisition. Two basic contributions to on-line and off-line acquisition technologies and devices are reported in Refs. [4] and [5], respectively. In the field of verification algorithms, traditional approaches are being re-addressed according to emerging strategies [3,6].

The aim of this contribution is to present the basic processing phases concerning signature verification, to illustrate the fundamental algorithms, to highlight the advances currently in progress, and also to focus on the most interesting issues in the field.

2. Signature Verification System

A Signature Verification System (SVS) consists of several processing phases as shown in Fig. 1.

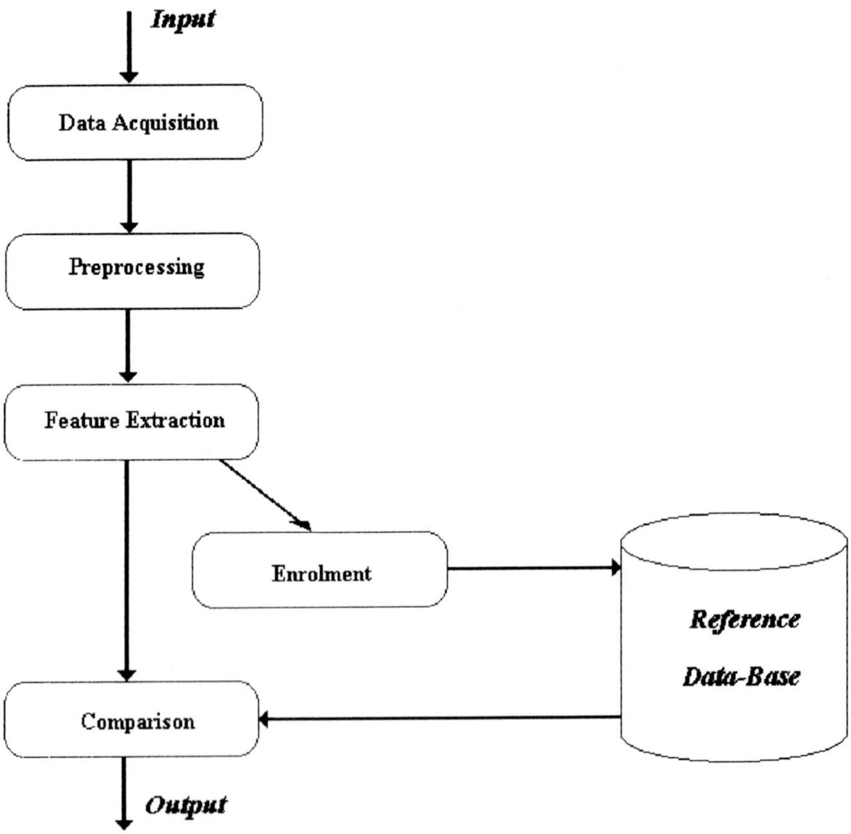

Fig. 1. The signature verification process.

After data acquisition, the preprocessing phase removes spurious noise from the input signature to obtain significant information from the raw data. In the feature extraction phase, the discriminant features are extracted. Such features will be used to discriminate genuine signatures from forgeries. During the training process, the features extracted from the set of reference signatures are enrolled into the personal database. In the comparison phase, they are matched against those belonging to the input (test) signature. The result is used to judge the authenticity of the input signature. Two types of errors can occur in the signature verification process: false-rejection errors (called type I errors) caused by the rejection of genuine signatures, and false-acceptance errors (called type II errors) caused by the acceptance of forgeries [2].

2.1. *Data Acquisition*

Two different kinds of acquisition devices can be used for data acquisition: on-line and off-line.

On-line acquisition devices [4], like graphic tablets, generate electronic signals representative of the signature trace during the writing process. The signals generated can be coordinate signals, velocity signals, acceleration signals, pressure and force signals, pen-down and pen-up signals (a pen-down movement is the operation of pulling down the tip of the pen toward the writing plane, while a pen-up movement is the operation of lifting the tip of the pen away from the writing plane) and so on. Figure 2 shows an example of a dynamic signature described by means of its coordinates $(x(i),y(i))$, $i=1,2,\ldots,N$, where N is the number of samples acquired during the signing process. In Fig. 2 the pen-down and pen-up signals are also reported. They are marked with "$*$" and "o", respectively.

Fig. 2. An example of dynamic signature.

Off-line acquisition devices [5], like scanners or cameras, perform data acquisition after the writing process has been completed. In this case the signature is represented as a grey level image $I = \{ g(x,y) \mid 1 \le x \le n_x, 1 \le y \le n_y \}$, where $n_x \times n_y$ is the size of the image, and $g(x,y)$ is the grey level of the signature image at pixel (x,y). In Fig. 3, a static signature is shown.

Fig. 3. An example of static signature

Two classes of systems have been designed for signature verification depending on the kind of input device: on-line or dynamic SVSs and off-line or static SVSs. Generally, dynamic SVSs are more effective and they offer some advantages such as high man-computer interactivity and the possibility for the user to adapt himself to the machine and vice versa.

2.2. *Preprocessing*

In the preprocessing phase, a dynamic signature is generally filtered to remove spurious signals from the image. Subsequently, a specific procedure to normalize the signatures in the time-duration domain is usually required reducing the representation of the signature to a standard number of samples. This normalization procedure can be easily performed by different approaches. Among others, a very simple approach is based on a b-spline interpolation and resampling process [7]. Of course, size normalization in the horizontal and vertical direction can also be necessary depending of the requirements of the feature extraction phase [6,7].

A critical preprocessing task is the segmentation of the signature into strokes. This task is specifically necessary when signature verification is carried out by the analysis of parts of the signature. In the approach proposed by Castellano, Dimauro, Impedovo and Pirlo [8] a segment is defined as a piece of the written trace between a pen-down and a pen-up movement. This approach is based on the consideration that the signature can be regarded as a sequence of writing units delimited by abrupt interruptions. Writing units are the regular parts of the signature, while interruptions are the singularities of the signature [8,9]. Experimental exidence has shown that singularities occur in definite positions in the signature of an individual. Therefore only a finite set of writing units, called *components or fundamental strokes*, can be generated by each writer [9]. Figure 4 shows the components of the signature in Fig. 2. Furthermore, it has been shown that although signatures of the same writer may be composed of a different number of components, and also signatures with the same number of components may be very different from each other because of the different pen-lift distribution, the components that each writer uses to produce his signature are well defined and belong to a finite set of fundamental components

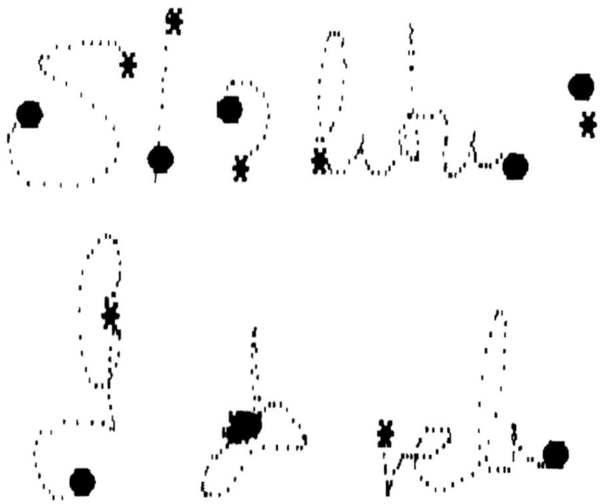

Fig. 4. Components of an dynamic signature.

The segmentation approach proposed by Plamondon, Yergeau and Brault [10] is based on the curvilinear and angular velocity signals of the pen movements during signing. In this approach it is assumed that the basic elements of the handwritten signature are the *components* and the *strings*. To locate these segments, four threshold values must be defined: the upper bound for the angular velocity, the lower bound for the curvilinear velocity, the minimal period of time for a handwritten string and the minimal period of time for lifting the pen tip. Each handwritten component is delimited by two successive significant liftings the pen along the trajectory of the signature, and each handwritten string is delimited by two successive portions of the trajectory which verify specific characteristics in terms of time-duration and angular/curvilinear velocity.

A more recent segmentation technique is based on a dynamic splitting procedure [11]. The basic idea is to perform the splitting by using the information from both the reference signatures and the input signature, in fact, the reference signatures are segmented into basic strokes taking into consideration the characteristics of the input signature. This technique first detects the Candidate Splitting Points of each reference signature and of the input signature as the local maxima and minima in the vertical direction of each specimen. Then, an elastic matching procedure is applied and the best coupling between the Candidate Splitting Points of the input signature and of the reference signatures leads to the identification of the set of coupled strokes.

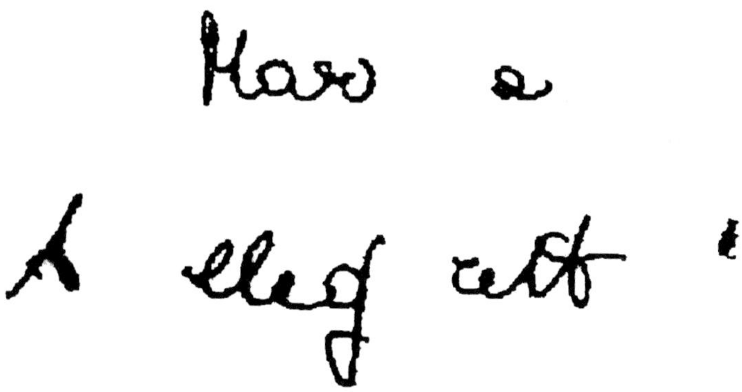

Fig. 5. Connected components of a static signature.

A static system for signature verification must be able to process a signature already generated. Under real conditions signatures are generally produced on specific paper forms. For instance, in the case of a bankcheck, the signature is written on a surface with a background pattern, which makes the signature image very noisy. In this kind of systems the preprocessing phase is more complex and time consuming. Typical preprocessing steps are the localization of the signature in the image, the extraction of the signature from the background, the smoothing of the signature and so on [7]. Also the segmentation of static signatures is very complex since no dynamic information is available. The approach of Dimauro, Impedovo and Pirlo [12] is based on the analysis of the connected components of the static signature detected by a contour following algorithm. Figure 5 shows the connected components of the signature in Fig. 3. Ammar, Yoshida and Fukumura [13] proposed the segmentation of a signature by a tree structure which identifies fundamental segments in the static image. In another approach, Brault and Plamondon [14] obtained the segmentation points with the help of a two steps procedure. The first step weights up the perceptual importance of each point of the signature, while the second step identifies the segmentation points as those which are locally more significant from a perceptual point of view.

2.3. *Feature Extraction*

Two types of features can be used for signature verification: parameters or functions [2]. In dynamic signature verification, typical parameters [15] used as personal features are the total duration of the signing process, the pen-down time ratio which is the ratio of pen-down time to total time, the average and the root-mean-square of the velocity and acceleration, the correlation between the magnitude of V_x and V_y, the number of pen lifts and specific coefficients derived from mathematical transforms [16,17]. Specifically, based on the consideration that the signature generation process of each writer is constrained by the particular characteristics of the handwriting system (nerves and muscles of the harm, hand, fingers and so on), suitable Fourier descriptors have been used as discriminant features for dynamic signature verification [16]:

Let $(x(i),y(i))$, i=0,1,...,N-1, be the sequence of coordinates representing a normalized signature, i.e., a signature with a number of samples $N=2^\beta$, where β is an integer. If it is assumed that the signature has been generated on a two-dimensional complex space, its description becomes $z(i) = x(i) + j \bullet y(i)$, $i = 0,1,...,N-1$, This complex sequence can be transformed by a Fast Fourier Transform algorithm, obtaining the Fourier coefficients Z_i, i=0,1,...,N-1. The Fourier descriptors defined in [16] are the complex values:

$$d_{1,i} = \frac{Z_{i+1}}{Z_1} \quad , \quad i = 0,1,\ldots,N-2$$

It is easy to verify that these features are related to signature dynamics. Naturally, other features can also be considered which are related to the shape of the signature, like the length-to-width ratio, the direction histogram and the histogram of the direction changes [15].

In static signature verification, some studies identify two classes of characteristics [18]. The first class consists of characteristics which are very difficult to imitate: the rhithmic line of the signature, which consists of the positional variation of the maximum coordinate of each character composing the signature, the local variation in the width of the signature line, which also depends on the personal dynamics in signing; the variation in aspect ratio of the whole signature followed by the local feature variation, like aspect ratio, orientation, relative position etc., measured between pairs of characters. The characteristics of the second class are easily perceived and therefore they are easy to imitate. They are the general design of the letter shape, the overall orientation of the signature and its position on the document, the number of local extrema and also some coefficients derived from the Hadamard transform [19].

It must be pointed out that in both dynamic and static signature verification, the selection of the optimal set of parameters is very arduous due to the difficulty in obtaining useful statistical models for genuine and forged signatures [20].

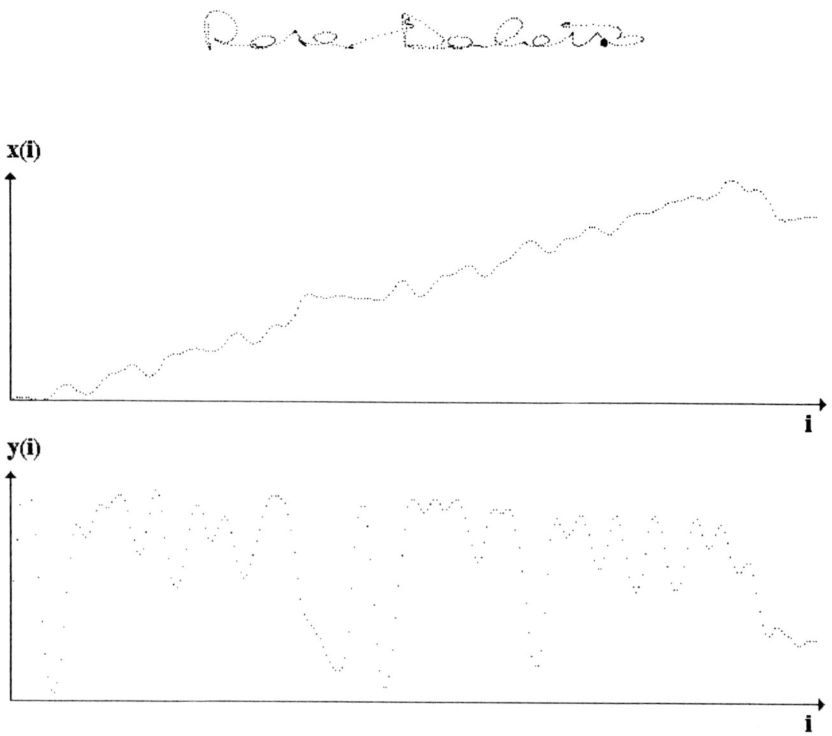

Fig. 6. A dynamic signature and its position functions.

When functions are used as features, the signature is characterized in terms of a time-function $F(i)$, i=1,2,...,N, whose values constitute the feature set. Feature functions are, for instance, the coordinates identifying the position of the tip of the pen during the writing process [21], its velocity [22], or acceleration [23,24], and also the pressure or force of the tip of the pen on the writing tablet [25]. Figure 6 shows a dynamic signature and its position functions $F(i)=(x(i),y(i))$, i=1,2,...,N. Generally speaking, function features allow better performances than the parameters in signature verification, but they usually require time-consuming matching procedures.

2.4. *Comparison Process*

In the comparison process the authenticity of the test specimen is evaluated by matching its features against those of the reference database. This process produces a single response given in the form of a Boolean value R which states the authenticity of the test signature:

$$R = \begin{cases} 0 & \textit{iff} \quad \textit{the test signature is a forgery} \\ 1 & \textit{iff} \quad \textit{the test signature is genuine.} \end{cases}$$

For this purpose, several matching strategies can be adopted to compare the test signature S^t against the N^r reference signatures S^r, r=1,2,...,N^r which are available in the reference database [6]:

- wholistic matching,
- regional matching
- multiple regional matching.

The simplest strategy is based on the *wholistic approach* and it consists of matching the test signature S^t against each one of the N^r reference signatures S^1, S^2,...,S^{N^r}, considered as a whole. Of course this approach does not allow any regional evaluation of the signature. In fact, each matching of S^t with S^r produces the response R^r:

$$R^r = \begin{cases} 0 & \textit{iff} \quad S^t \textit{ results a forgery when compared with } S^r \\ 1 & \textit{iff} \quad S^t \textit{ results genuine when compared with } S^r. \end{cases}$$

Then, the final response R can be easily defined as:

$$R = \begin{cases} 0 & \textit{iff} \quad \forall r = 1,2,\ldots,N^r : R^r = 0 \\ 1 & \textit{otherwise.} \end{cases}$$

Another strategy is based on a *regional matching* approach. In this case the test signature S^t is split into n segments:

$$(S^t_1, S^t_2, \ldots, S^t_k, \ldots, S^t_n)$$

as well as S^r which is split into n segments:

$$(S^r_1, S^r_2, \ldots, S^r_k, \ldots, S^r_n).$$

The matching between S^t and S^r is performed by evaluating the local responces R^r_k obtained by matching S^t_k against S^r_k, for k=1,2,...,n:

$$R^r_k = \begin{cases} 0 & \textit{iff} \quad S^t_k \textit{ results a forgery when compared with } S^r_k \\ 1 & \textit{iff} \quad S^t_k \textit{ results genuine when compared with } S^r_k. \end{cases}$$

This approach allows a regional analysis of the signature, but it is carried out in a one-by-one comparison process: i.e. the test signature is judged to be a genuine specimen if and only if a reference signature exists for which, in the comparison process, a suitable number of segments of the test signature are found to be genuine.

The most powerful matching approach is the *multiple regional matching*. In this case each segment S^t_k of the test signature is matched against the entire set of the corresponding segments

$$\{S_k^1, S_k^2, \ldots, S_k^{N^r}\}$$

of the N^r reference signatures $S^1, S^2, \ldots, S^{N^r}$. Therefore for each segment S_k^t of the test signature, a local verification response R^t_k is obtained as:

$$R^r_k = \begin{cases} 0 \;\; iff\; \forall r = 1,2,\ldots,N^r : S^t_k \; results\; a\; forgery\; when\; compared\; with\; S^r_k \\ 1 \;\; otherwise. \end{cases}$$

Successively, the test signature is judged to be a genuine specimen if a suitable number of segments are found to be genuine. This approach allows a regional evaluation of the signature without requiring a large set of reference signatures.

Furthermore, each comparison technique is based on a suitable similarity (or dissimilarity) measure. Let S^t be the test signature and S^r a reference signature. Two kinds of comparison techniques can be distinguished according to the type of features that are used (parameters or functions).

When parameters are considered to be features, the signatures are described by vectors of parameters in a multidimensional feature space. Let m be the number of features used for the test signature S^t and for the reference signature S^r, r=1,2,...,N^r, the feature vector of the signature S^t is:

$$(p^t_1, p^t_2, \ldots, p^t_m)$$

and the feature vector of S^r is:

$$(p^r_1, p^r_2, \ldots, p^r_m).$$

In this case, a simple dissimilarity measure $D(S^t,S^r)$ can be considered between the specimens S^t and S^r. For example, the Euclidean distance of the multidimensional feature space can be used:

$$D(S^t, S^r) = \sqrt{\sum_{i=1}^{m} (p^t_i - p^r_i)^2}\,.$$

When functions are considered to be features, the signatures are described as time functions which represent complete dynamic signals acquisited directly from the acquisition device or derived from the input signals. In this case the matching techniques must take into account the variations of signal duration from one signature to another (even if the specimens are produced by the same writer). Furthermore, random variations, due to the writer's pauses or hesitations, can create portions of signals, such as deletions, additions and gaps, which complicate the problem of matching. Dynamic non-linear time warping is one of the most common matching methods used for this purpose. Dynamic non-linear time warping, which has been firstly used for speech recognition [26], allows the compression or expansion of the time axis of two signatures in order to match two specimens in such way as to obtain the minimum of a given distance value. A simple formulation of dynamic non-linear time warping is given in the following. Let the function feature for S^t be

$$F^t(1), F^t(2), \ldots, F^t(M^t),$$

and the function feature for S^r be

$$F^r(1), F^r(2), \ldots, F^r(M^r),$$

A dynamic non-linear time warping procedure permits the detection of the coupling sequence

$$W(t,r) = c_1, c_2, \ldots, c_K,$$

(where $c_k=(i_k,j_k)$, and k, i_k, j_k are integers such that $1 \le k \le K$, $1 \le i_k \le M^t$, $1 \le j_k \le M^r$) which minimizes the dissimilarity measure:

$$D(S^t, S^r) = min \sum_{k=1}^{K} d(c_k)$$

where $d(c_k) = d(i_k, j_k)$ is a suitable distance measure between the samples $F^t(i_k)$ and $F^r(j_k)$, and where the following conditions are satisfied:

- monotonicity (i.e. $i_{k-1} \le i_k$, $j_{k-1} \le j_k$ for $k=2,..,K$)
- continuity (i.e. $i_k - i_{k-1} \le 1$, $j_k - j_{k-1} \le 1$ for $k=2,..,K$)
- boundary condition (i.e. $i_1=1$, $j_1=1$, $i_K=M^t$, $j_K=M^r$).

The optimization problem can be solved using the following equations:

$$warp(1,1) = 2 \bullet d(1,1)$$

$$warp(i,j) = \min \begin{cases} warp(i, j-1) + d(i,j) \\ warp(i-1, j-1) + 2 \bullet d(i,j) \\ warp(i-1, j) + d(i,j) \end{cases}$$

where

$$D(S^t, S^r) = \frac{1}{(M^t + M^r)} \bullet warp(M^t, M^r).$$

The pseudocode for this procedure follows a recursive strategy:

```
Main
...
...
compute  D(S^t,S^r) = Warp(M^t,M^r) / (M^t + M^r)
...
...
end

Function warp(M^t,M^r)
  do case:
   - t=1 and r=1 then
      warp=2•d(1,1)
   - t=1 and r>1 then
      warp=warp(1,r-1) + d(t,r)
   - t>1 and r=1 then
      warp=warp(t-1,1)+d(t,r)
   - t>1 and r>1 then
      warp = Min [Warp(t-1,r)+d(t,r),Warp(t-1,r-1)+2•d(t,r),Warp(t,r-1)+d(t,r)]
  end case
return.
```

An example of the application of such procedure for the detection of the best coupling sequence between S^t and S^r is shown in Fig. 7.

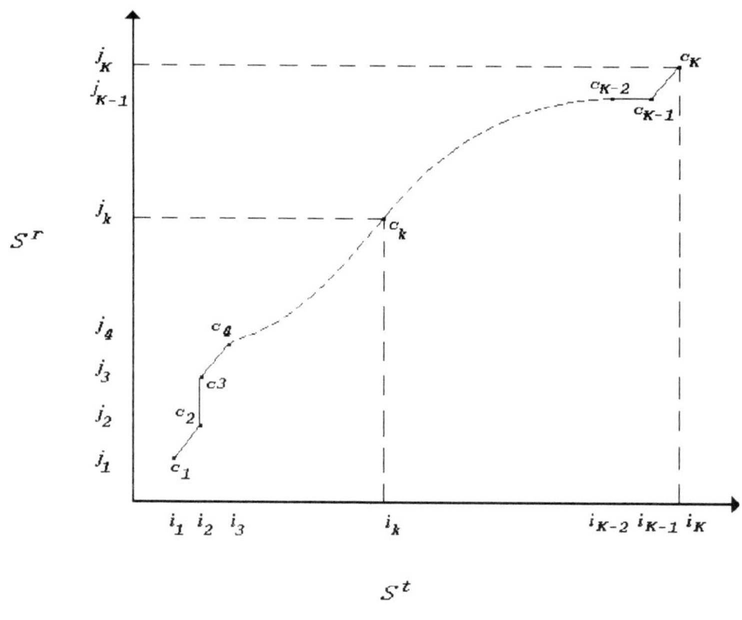

Fig. 7. Best coupling sequence.

The similarity (or dissimilarity) measure is used in the decision rule to judge the test signature according to a suitable threshold value T_0. Of course, the selection of T_0 permits the tuning of the system obtaining different Type I and Type II errors. For instance, in Fig. 8 the value of T_0 is suitably defined in order to obtain an equal number of Type I and Type II errors.

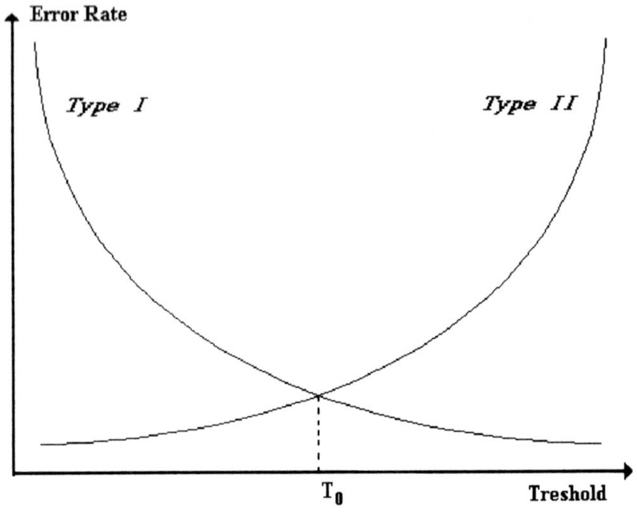

Fig. 8. Threshold selection.

More recently, the signature comparison process has been dealt with by neural networks. Both for static and dynamic signature verification, neural networks seems to be more efficient than traditional approaches in discovering personal characteristics in signatures. They are also able to modify automatically the knowledge of a writer with the analysis of fresh specimens. Several models of neural networks have been considered for this purpose: propagation classifiers [27], multilayer perceptrons [28], ART1 neural network [29] and others. The main problems related to the use of neural networks for automatic signature verification is the need for large sets of genuine samples and also for forgeries for the training phase. Unfortunately, in many real applications is not always possible to have a large number of genuine specimens for each signer as well as a significant set of forgeries [6]. However, other techniques can also be considered for signature comparison as, for instance, the techniques based on hidden Markov models, as recently proposed by Yang, Widjaja and Prasad [30].

3. Systems Implementation

Many systems for signature verification have been developed in research laboratories all over the world and some of them have already been commercially distributed [31]. All of them have been designed by choosing suitable solutions in terms of devices and algorithms for each processing phase. In the following, some of the most recent systems for dynamic and static signature verification, which focus on the major trends in current scientific research, are introduced.

For dynamic signature verification systems, research is mainly dedicated to the detection of original features to be used for high-performance signature verification and to suitable matching techniques also with respect to the local analysis of signatures.

In the field of the selection of the best features for signature verification, it is generally accepted that velocity is more informative than position and acceleration for dynamic signature verification. However, Yoshimura and Yoshimura [32] proposed a new function feature. In their work, they acquire the signature as coordinate and pressure signals and perform the comparison by an elastic pattern matching technique. In this approach the formulation of the dissimilarity function is different from the others proposed in the past. In fact, it incorporates neither the velocity of pen movements nor the acceleration of the pen, but it uses the direction of pen movements [33].

Another research trend has been recently investigated by Plamondon, Yergeau and Brault [10] which tries to combine both a parameter and a function approach. They developed a multilevel signature verification system that uses one representation of the signature based on global parameters and two representations based on functions.

In order to emphasize the role of local characteristics for signature verification two new systems have been proposed by Dimauro, Impedovo and Pirlo [11,34]. The first system performs signature verification by a local verification strategy based on the spectral analysis of *components* (also called *fundamental strokes*). It is based on a segmentation of signatures in components. In the learning phase the components of the training specimens are classified by an improved *k*-means clustering technique using suitable topological features. For each cluster, some statistical parameters are computed from the set of Fourier descriptors derived from each component, then a structured knowledge-base for each individual is obtained. This knowledge base consists of the *Component Reference Table* and the *Structural Description Graph*. The Component Reference Table contains the features representative of the components of the writer. The Structural Description Graph reports the acceptable sequences of components in the genuine signatures. The verification process is then accomplished in two successive steps. In the first step the structural organization of the unknown signature is verified. In the second step each fundamental component of the signature is verified to obtain the conclusive verification response [34]. The second system [11] uses a stroke-oriented verification strategy based on a new dynamic segmentation procedure. From the sets of Candidate Splitting Points, detected on each reference signature and on the test signature, a suitable set of coupled strokes is obtained. After the segmentation, regional thresholds are automatically computed for each set of coupled strokes of the reference signatures. The verification process follows a two-level strategy. At the first level, the segmentation results are used to perform a fast rejection of poor forgeries. This decision is accomplished by evaluating the number of strokes coupled between the test signature and the set of reference signatures and matching this result against the number of expected coupled strokes. At the second level, each stroke of the test signature is matched against each corresponding stroke of the reference signatures by an elastic matching procedure and the overall result is used to judge the whole signature [11].

In the field of static signature verification, the definition of methodologies for the description of the signature is very important. A structural approach has been presented by Sabourin, Plamondon and Beaumier [18]. In this work the signature is analysed according to a local interpretation process, in which some basic primitives detected from the signatures are evaluated. A global interpretation process is also defined which permits the evaluation of a similarity measure between two structural graphs representing the signatures. In the approach proposed by Ammar, Yoshida and Fukumura [13], the signature is segmented horizontally into zones and vertically into elements. Information from the signature and its parts is organized into features and relations which are extracted quantitatively. In the *global description*, the signature is represented by a character string formed by concatenating the symbols corresponding to the value of the following five Global Constituents (features): the dominant slant in the signature, the presence of the lower zone parts, the number of elements in the signature, the ratio between the signature length and its width, and the ratio between middle zone width and signature width. In the *local description*, the signature is represented by a tree structure and is considered as composed of a set of ordered elements which are its vertically separable parts. Each element is described by two groups of constituents. *Permanent Constituents* are the global slant of the element, high density factor, comparison relation between the length of each element and the preceding one, and the position relation between the Element Baseline and the Global Baseline. The *Nonpermanent Constituents* are the upper and lower zone parts, the ratio between middle zone width and vertical extension of the element's upper zone parts, the ratio between middle zone width and the element width, and the ratio between middle zone width and the vertical extension of the element's lower zone parts.

More recent systems for signature verification are based on neural network models. Sabourin and Drouhard [27] propose a new approach which uses the Directional Probability Density Function as a global feature vector. The comparison step is based on a completely connected feed-forward neural network classifier with a back-propagation learning algorithm. In the work of Cardot, Revenu, Revillet and Victorri [35], the static information is used in three representation: geometrical parameters, outline and image. Several neural networks cooperate together to perform training and testing.

4. Improvement of the Algorithms for Signature Verification

Notwitstanding the work carried out by scientists and technological research teams for the development of systems for automatic signature verification, the improvement of such systems depends on the solutions that will be provided to certain challenging issues.

One of the most important issue is the development of effective strategies for personalizing systems for signature verification, for instance, by the automated selection of the optimal set of reference signatures and thresholds for each individual. In fact, in some cases the procedure for the selection of reference signatures considers both genuine specimens and forgeries. This approach is weak since in many practical cases fraudulent specimens are not available or they are of poor quality. Then it is much more significant to select the reference signatures directly from the set of genuine specimens, without considering any forgeries. Following this trend, Yoshimura, Kato, Matsuda and Yoshimura [33] verified that three specimens are sufficient to be representative. Then they proposed a system which selects three signatures out of the reference writings. The selection is achieved by clustering the reference writings into three clusters and selecting one specimen with minimax property from each cluster. More recently, Congedo,

Dimauro, Impedovo and Pirlo [35] proposed a measure of the personal variability in signing and used such a measure to automatically detect the near-optimal set of reference signatures for each writer. The same tool can also be used to select a suitable representation space for the verification of dynamic signatures. Personal thresholds are generally obtained by comparing the reference signatures with each other, taken two-by-two and storing the worst result of verification [9]. The proper selection of the threshold value for the acceptance/rejection rule strongly affects the performance of the system and trains it for working in practical environments, which generally require different system characteristics.

When parameters are considered as features, automatic feature selection is very important. In a recent paper, Fairhurst and Brittan [37] concentrate on possible strategies for a parallel implementation of a feature selection algorithm, particularly suited for automatic signature verification. They demonstrate how the inherent parallelism, which exists within a generic model for signature verification, can be exploited to provide an optimized general-purpose framework for verification processing.

Another fundamental problem concerns the realization of a common database for testing a system. At present, systems developers cannot compare their results due to the lack of a widely accepted protocol for experimental tests, as well as the absence of public signature databases. In fact, experimental protocols can be very different from one research laboratory to another. Laboratory tests are completely different from field-tests, in which there is no control on the writer. Furthermore, genuine signatures are not easy to forge and laboratory tests generally take into consideration poor forgeries, obtaining, of course, better experimental results. The traditional attempt to handle this problem consists in considering a suitable set of classes of forgeries [2]: the class of *random* forgeries, in which the forger uses his own signature instead of the signature to be tested; the class of *simple* forgeries, in which the forger makes no attempt to simulate or trace a genuine signature, and the class of *freehand* or *skilled* forgeries, in which the forger tries and practices, imitating as closely as possible the static and dynamic information of a genuine signature. Naturally, this solution is not sufficient and the development of a standard database of true signatures and forgeries, collected over an extended period of time, is crucial for making reliable predictions on the performance of various signature verification algorithms.

5. Conclusion

In recent years, there has been much progress in the field of automatic signature verification. Results of the scientific research and the interest of many industrial companies in this field demonstrate the feasibility and the extreme usefulness of automatic signature verification systems in everyday applications. In this paper the basic problems involved in the development of each phase of the signature verification process have been presented and some traditional solutions discussed. Furthermore, the main problems in the development of advanced systems for signature verification have been highlighted and a useful bibliography given.

Acknowledgements

This work was supported by MURST 60%, Italy.

References

[1] *Proc. of SECURTECH*, Washington D.C., April 1992.
[2] R.Plamondon and G.Lorette, "Automatic signature verification and writer identification: The state of the art", *Pattern Recog.* **22**, 2 (1989) 107-131.
[3] F. Leclerc and R. Plamondon, "Automatic signature verification: The state of the art – 1989-1993", *Int. J. Pattern Recog. Artif. Intell.* **8**, 3 (1994) 643-660.
[4] C.A.Higgins and D.M.Ford, "Stylus driven interfaces-The electronic paper concept", in *Proc. of ICDAR 91: Int.Conf. on Document Analysis and Recognition*, Sept.-Oct.1991, pp.853-862.
[5] S.Impedovo, L.Ottaviano and S.Occhinegro, "Optical character recognition-A survey", *Int. J. Pattern Recog. Artif. Intell.* **5**, 1-2 (1991) 1-24.
[6] G. Pirlo, "Algorithms for Signature Verification", in *Fundamentals in Handwriting Recognition*, ed. S. Impedovo, Springer Verlag, Berlin, 1994, pp. 433-454.
[7] T.Pavlidis, *Algorithms for Graphics and Image Processing*, Springer Verlag, Berlin, 1982.
[8] M.Castellano, G.Dimauro, S.Impedovo G.Pirlo, "On-line signature verification system through stroke analysis", in *Proc. AFCET Int. Conf. on New Concepts in Computer Science*, 1990, pp. 47-53.
[9] G.Dimauro, S.Impedovo, G.Pirlo, "A stroke-oriented approach to signature verification", in *From Pixels to Features III - Frontiers in Handwriting Recognition*, S. Impedovo and J.C.Simon eds., Elsevier Publ., 1992, pp. 371-384.
[10] R.Plamondon, P.Yergeau and J.J.Brault, "A multi-level signature verification system", in *From Pixels to Features III - Frontiers in Handwriting Recognition*, S.Impedovo and J.C.Simon eds., Elsevier Publ., pp. 363-370, 1992.
[11] G.Dimauro, S.Impedovo and G.Pirlo, "On-line Signature Verification by a Dynamic Segmentation Technique", in *Proc. of IWFHR 3th Int. Workshop on Frontiers in Handwriting Recognition*, Buffalo, May 1993, pp. 262-271.
[12] G. Dimauro, S. Impedovo, G. Pirlo, "Off-line Signature Verification through Fundamental Strokes Analysis", in *Progress in Image Analysis and Processing III*, ed. S.Impedovo, World Scientific Publ., 1994, pp.331-337.
[13] M.Ammar, Y.Yoshida and T.Fukumura, "Structural Description and Classification of Signature Images", *Pattern Recog.* **23**, 7 (1990) 697-710.
[14] J.J.Brault and R.Plamondon, "Segmentation des Signatures Manuscrites", in *Proc. of Vision Interface '89*, London, Ontario, 1989, pp. 110-116.
[15] W. Nelson, W. Turin and T. Hastie, "Statistical methods for on-line signature verification", *Int. J. Pattern Recog. Artif. Intell.* **8**, 3 (1994) 749-770.
[16] M.Castellano, S.Impedovo, A.Mingolla, G.Pirlo, "A spectral analysis based signature verification system", in *Lecture Notes in Computer Science:Recent Issues in Pattern Analysis and Recognition*, eds. G. Goos and J.Hartmanis, Springer Verlag, Berlin, 1988, pp. 316-323.
[17] C. F. Lam and D. Kamins, "Signature recognition through spectral analysis", *Pattern Recog.* **22**, 1 (1989) 39-44.
[18] R. Sabourin, R. Plamondon, L. Beaumier, "Structural interpretation of handwritten signature images", *Int. J. Pattern Recog. Artif. Intell.* **8**, 3 (1994) 709-748.
[19] W.F.Nemcek and W.C.Lin, "Experimental investigation of automatic signature verification", *IEEE Trans. on Sys. Man and Cybern.* **4**, (1974) 121-126.
[20] M.Castellano, G.Dimauro, S.Impedovo and G.Pirlo, "Decision making process in a signature verification system", in *Progress in Image Analysis and Processing*, ed. V. Cantoni, World Scientific Publ., 1989, pp. 610-614.

[21] Y.Sato and K.Kogure, "On-line signature verification based on shape, motion and handwriting pressure", in *Proc. 6th Int. Conf. on Pattern Recognition*, Munich, 1982, vol. 2, pp. 823-826.
[22] G.Lorette and R.Plamondon, "On-line handwritten signature recognition based on data analysis and clustering", in *Proc.* 7th *Int. Conf. on Pattern Recognition*, Montreal, 1984, vol. 2, pp. 1284-1287.
[23] N.M.Herbst and C.N.Liu, "Automatic signature verification based on accelerometry", in *IBM Journal Res. and Develop.* pp.245-253, 1977.
[24] J.S.Lew, "Optimal accelerometer layouts for Data Recovery in Signature Verification", *IBM J. Res. Dev.* **24** (1980) 496-511.
[25] H.D.Crane and J.S.Ostrem, "Automatic Signature Verification Using a Three-Axis Force-Sensitive Pen", *IEEE Trans. on Sys. Man and Cybern.* **13**, 3 (1983), 329-337.
[26] L.R.Rabiner, S.E.Levinson, "Isolated and connected word recognition-Theory and selected applications", *IEEE Trans. on Communications* **29**, 5 (1981) 621-659.
[27] R. Sabourin and J.P. Drouhard, "Off-line signature verification using directional PDF and neural networks", in *Proc. of 11th Int. Conf. on Pattern Recognition*, 1992, pp.321-325.
[28] S. Barua, "Neural Networks and their applications to computer security", in *Proc. SPIE-Int. Society for Optical Engineering*, 1992, pp. 735-742.
[29] L.Y. Tseng and T.H. Huang, "An on-line Chinese signature verification scheme based on the ART1 neural network", in *Proc. of Int. Joint Conf. on Neural Networks*, Maryland, 1992, pp. 624-630.
[30] L.Yang, B.K.Widjaja, R.Prasad, "On-line signature verification applying hidden Markov models", in *Proc. of 8th Scandinavian Conf. Image Analysis*, Tromso, 1993, pp. 1311-1316.
[31] R. Plamondon (ed.), *Progress in Automatic Signature Verification*, World Scientific Publ., Singapore, 1994.
[32] I.Yoshimura and M.Yoshimura, "On-line signature verification incorporating the direction of pen movement-An experimental examination of the effectiveness", in *From Pixels to Features III - Frontiers in Handwriting Recognition*, S.Impedovo and J.C.Simon eds., Elsevier Publ., pp. 353-362, 1992.
[33] M.Yoshimura, Y.Kato, S.Matsuda and I.Yoshimura, "On-line Signature Verification Incorporating the Direction of Pen Movement", *IEICE Transactions.* **74**, 7 (1991) 2083-2092.
[34] G. Dimauro, S. Impedovo, G. Pirlo, "Component-oriented algorithms for signature verification", *Int. J. Pattern Recog. Artif. Intell.* **8**, 3 (1994) 771-794.
[35] H. Cardot, M. Revenu, B. Victorri, M.-J. Revillet, "A Static Signature Verification System based on a cooperative Neural Networks Architecture", *Int. J. Pattern Recog. Artif. Intell.* **8**, 3 (1994) 679-692.
[36] G. Congedo, G. Dimauro, S. Impedovo, G. Pirlo, "A New Methodology for the Measurement of Local Stability in Dynamical Signatures", in *Proc. of IWFHR 4th Int. Workshop on Frontiers in Handwriting Recognition*, Taiwan, 1994, pp. 135-144.
[37] M.C. Fairhurst and P. Brittan, "An evaluation of parallel strategies for feature vector construction in automatic signature verification systems", *Int. J. Pattern Recog. Artif. Intell.* **8**, 3 (1994) 661-678.

Handbook of Character Recognition and Document Image Analysis, pp. 623–651
Eds. H. Bunke and P. S. P. Wang

CHAPTER 24

BANK CHECK ANALYSIS AND RECOGNITION BY COMPUTERS

A. AGARWAL, A. GUPTA, K. HUSSEIN*
Sloan School of Management, Massachusetts Institute of Technology
Cambridge, MA 02139, USA

and

P.S.P. WANG†
College of Computer Science, Northeastern University
Boston, MA 02115, USA

This chapter discusses recent developments of bank check analysis and recognition by computers. A technique for locating the courtesy amount block on bank checks and a deterministic finite autmaton for the segmentation and recognition of *courtesy amount* will be presented. In the analysis and recognition process, connected components in the image are identified first. Then, strings are constructed on the basis of proximity and horizontal alignment of characters. Next, a set of rules and heuristics are applied to these strings to choose the correct one. The chosen string is only accepted if it passes a verification test, which includes an attempt to recognize the currency sign. A deterministic finite automaton system is then used for segmenting the handprinted courtesy amount. Finally, the separated components are passed on to a neural network based recognition system.

Keywords: Image processing; Pattern recognition; Hough transform; Optical character recognition; Check analysis and processing; Block detection; Courtesy amount recognition; Segmentation; Heuristic rules.

1. Introduction

In modern days, millions of dollars change hands everyday in the form of handwritten or machine printed bank checks. Image processing and pattern recognition techniques offer the potential for reducing costs involved in automated processing of bank checks. When a check, such as the one shown in Fig. 1, is deposited for credit into one's account via an automated teller machine (ATM), the depositor's bank is interested in two numerical fields: the account number, which is already written in MICR ink and can be handled using automated techniques; and the amount of the check, which is currently read and keyed in by a human operator. The name of the recipient, the date, the signature, and other fields on the check are normally ignored in routine check processing, unless the transaction is contested or the check

*agupta@mit.edu
†pwang@ccs.neu.edu

is presented for immediate encashment to a bank employee.

Currently, several steps involved in processing the check are done by computers. The number of the account from which a check has been drawn and the bank code are encoded in magnetic ink at the bottom of a check to facilitate this process. One step in the process that is, for the most part, still done by people is the identification of the dollar (or other unit of currency) amount for which the check has been written. Bank employees read the amount, usually from the courtesy dollar amount box, of which a sample set of typical writings is shown in Fig. 3, enter these data into a computer system, which then prints the amount in magnetic ink at the bottom of the check. From this point on, processing can be automated.

For the past decade, Optical Character Recognition (OCR) technology has evolved to a stage that would allow automation of the process of converting the written or printed image of the numerals in the courtesy dollar amount box to a machine understandable form [1, 2, 3]. The implementation of OCR systems tailored to the job of recognizing the courtesy amount on checks would greatly facilitate the design of a completely automated check processing system.

A main reason why the use of such systems has not become widespread in the U.S. is that checks do not conform to a single standard. Checks come in a variety of sizes, and the courtesy amount is not located in the same place on all checks. Some courtesy amount recognition systems under development, including the one being developed by MIT's OCR group [4, 5, 6, 7], will achieve reasonably accurate results when the input presented is either the image of the courtesy amount only, or an image of the entire check along with explicit information as to the exact location of the courtesy amount block [5], with various styles of writing shown in Fig. 3.

Another key problem with the recognition of courtesy amounts is the segmentation of the amount into the individual characters. Due to the variability in the size of these characters as well as the random connection points associated with handwritten numerals segmentation is quite difficult. Furthermore, very little context is available to determine the number of characters in the courtesy amount string unlike the case with postal zipcode readers where the number of characters is fixed at five. These reasons inhibit the preprocessing of the bank checks on a commercial basis without human intervention.

The processing of bank checks is one application of the broader technology of document image understanding. Section 2 describes the overall architecture of the check processing system. Section 3 describes methods currently used in Document Understanding Systems (DUS's). Section 4 provides a description of the CABL subsystem that has been designed for use in MIT's OCR group's check processing system. Section 5 discusses some of the heuristics that can be applied to the specific problem of locating the courtesy amount block within a check image. Section 6 outlines the performance of the CABL on actual check data, in terms of both speed and accuracy. Section 7 describes the segmentation system, followed by discussion and summary in Sec. 8.

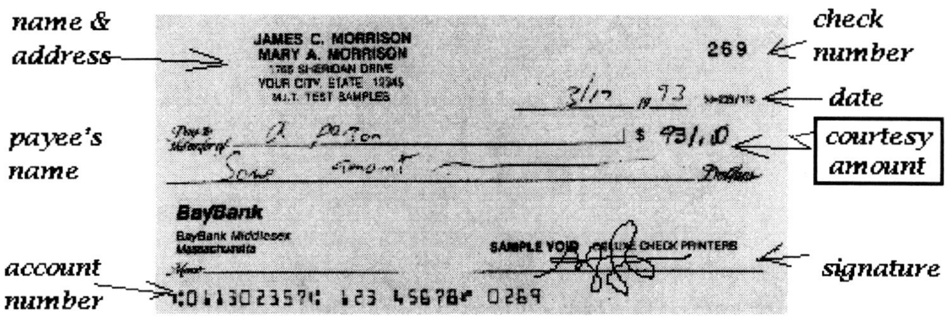

Fig. 1. An example of a typical bank check.

2. Check Recognition System Architecture

The architecture of the prototype system for reading handwritten numerals on checks, shown in Fig. 2, consists of six modules and is discussed below.

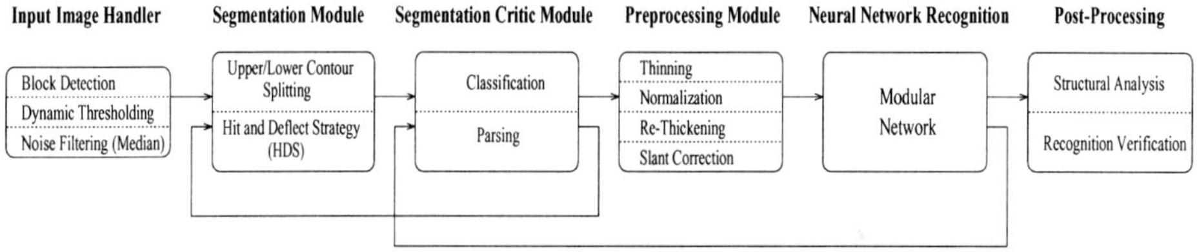

Fig. 2. A block diagram of an analysis and recognition system for bank checks.

Image Handler: The position of the courtesy dollar amount on the check varies greatly; hence, a block detection algorithm has been developed to locate and extract the handwritten amount for further processing. The courtesy amount detector is described in detail in Secs. 3 and 4. Checks in the U.S. are also characterized by wide differences in background textures, writing devices, and colors and shades of the written material. A dynamic thresholding algorithm utilizing the histogram of the input image is used to remove noise and to enhance the clarity of characters. A median filter is then employed to remove any spurious noise left in the image.

Segmentation Module: The binary bitmap of the courtesy amount is passed to the segmentation module which breaks the bitmap into distinct and meaningful pieces in three stages: extraction of connected components, splitting of upper and lower contours, and utilization of the hit and deflect strategy (HDS) for drawing segment cut lines. This segmentation procedure is detailed fully in Sparks [5] and Bunke *et al.* [15].

Segmentation Critic Module: The segmented components are crudely classified into a list of probable primitives, such as digits and commas. The order of the primitives is analyzed using a parser. Failures in parsing cause the segmentation module to reevaluate its segmentation strategy. This module is discussed in detail in Sec. 7.

Preprocessing: This stage reduces the variability in the slant and thickness of the various characters. Detailed description can be found in Lam *et al.* [8], Suen and Wang [9], Nagendraprasad *et al.* [10].

Neural Network Based Recognizer: The recognition stage consists of an adaptive modular neural network which is trained using a modified back propagation algorithm [7, 11].

Postprocessing: When the neural network classifies a character without adequate confidence, the particular digit is passed to a postprocessor, which utilizes traditional pattern recognition techniques to complement the connectionist approach. These techniques include structural analysis described in Wang and Gupta [12].

3. Document Image Analysis

3.1. *Background*

There are various document image forms that may constitute the input to a Document Understanding System (DUS). The DUS may be designed to utilize certain known constraints on the document format. Many forms restrict the writer to using black pens. Often there are guidelines to help align written text with the margins of the page, and to fix the location where a particular piece of information is to be recorded. Sometimes there are boxes to constrain the length of a piece of data, or the height of individual characters. More constraining than this, there may be individual boxes for characters, with instructions to print one character per box. A guide to forming letters in print may appear, restricting the general shape of a letter or number, but not removing variation due to a particular person's penmanship. If the DUS were designed to understand the information a writer is trying to convey in a totally unconstrained document, there would be no restriction on the information that is presented in the document. The system would have to account for many different stylus types and colors, variations in the size of print, texture of the surface, location of important data, as well as other variables. For more details of recent developments in PR and OCR in document analysis, please refer to [9, 10, 11, 13, 14, 15, 16, 17, 18, 19, 20, 12], and the various other chapters in this book.

Most forms that are subject to machine analysis fall between these two extremes. Before such documents can be analyzed and understood, several preprocessing steps, namely binarization [21], noise reduction [22] and form line removal [23] stages have to be applied to reduce the amount of redundant information. An overall analysis and recognition system is shown in Fig. 2 (see [24]).

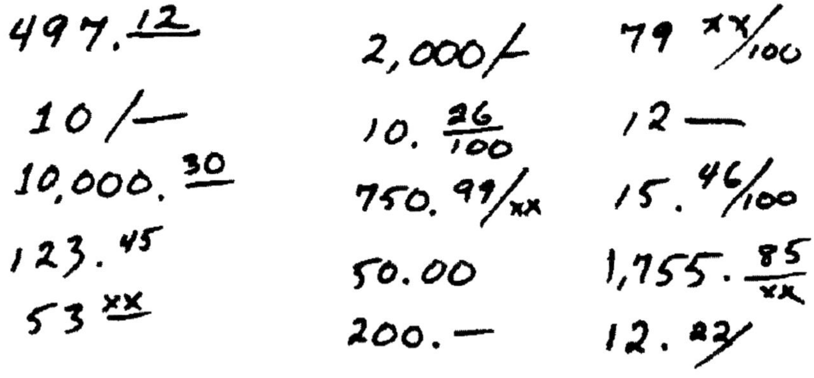

Fig. 3. Various ways of writing the courtesy dollar amount.

3.2. *Extraction of Data*

Since the purpose of this chapter is to describe a system that recognizes the courtesy amount on checks, attention will be focused on existing methods that extract just text. Two approaches are now discussed that are available to locate bits of relevant text in an image.

3.2.1. *Bottom-up processing*

In a bottom-up method [25, 21, 26], characters are first located. These characters are then grouped together to form words and lines. Bottom-up approaches, however, may be more robust than top-down approaches when faced with noise, and document skew [27]. The character location process and the character grouping process are discussed in the following subsections.

Character Location: There are two typical methods for locating characters: a stroke method and a connected component method. The rest of this subsection describes these two methods in more details.

Stroke Method: An algorithm suggested by Srihari [25] first computes run-lengths. These runs represent a horizontal slice of a vertical or slanted handwritten stroke (assuming that most of the data on the page is handwritten or printed text). If one now examines the length of all of the horizontal runs which have lengths between 3 and 20 pixels, one can determine the average thickness of a handwritten stroke on that particular document. Next, sets of 3 or more runs that approximate this length, and are vertically connected to each other, are grouped together to form complete strokes. These primitive strokes are then grown by adding adjacent runs that satisfy a more relaxed criterion. The length of these additional runs need not be as close to the average stroke width as the runs that make up the original backbone of the stroke. This process is repeated until no stroke can be grown any further. Strokes are then grouped together to form characters, words and blocks on the basis of proximity and size.

Connected Component Method: Other common bottom-up segmenters start by identifying the connected components in the image [21, 26]. This can be accomplished using a depth first search of black pixels, using 8-connectivity or 4-connectivity as an adjacency criterion. These connected components can be represented by their minimum bounding rectangles (MBR's). An MBR is the smallest rectangle whose sides are parallel to the boundaries of the image that encloses the component.

Depending on the particular application, some components may be rejected based on their size or shape. For instance, a component with an aspect ratio (height / width) of .02 cannot possibly be a character. In addition, if most of the components in the image have an area of about 500 pixels, a component that has an area of 10,000 pixels can be discarded as a graphic or other non-character piece of data. Fletcher and Kasturi [26] accept or reject each component as a character on the basis of its size, black pixel density, aspect ratio, area, and position in the image.

Character Grouping: The next step is the grouping of characters to form words and lines. One simple way is to just group together characters that are adjacent to each other horizontally, and are of similar shape and size [21]. If there is skew in the image, however, or if the document is sloppily handwritten so that neighboring characters are not necessarily directly horizontally adjacent, a different approach is necessary.

The Hough transform [28, 22] can be applied to the centroids of all of the characters. Through this method, one can determine which characters in the image are nearly collinear. Characters which are closer than some threshold distance to each other, and are also nearly collinear can be assumed to belong to the same string, or line of text [26].

3.2.2. *Top-down approaches*

In a top-down method, the system tries to determine the overall structure of a page, then breaks this down into regions, lines, words, and possibly characters. Top-down approaches tend to be less computationally intensive, since they deal mainly with image blocks that are larger, but fewer in number.

Two techniques that are useful in this process are smearing and smoothing. Downton and Leedham [29, 30] have developed a system to locate the address block on a piece of mail. As a first step, the image is smeared horizontally. The effect here is that adjacent characters and words become smeared into one another. Now, when connected components are extracted, they will be few and large. Fisher *et al.* [27] suggest choosing the smearing factor dynamically, based on the average interline and intercharacter spacings in the image, to be sure that image items are grouped together in an optimal manner.

Smoothing is similar to smearing. If the resolution of the image is lowered, and the values assigned to the new larger pixels are determined as a function of the values of the high resolution pixels they encompass, the connected components in the low resolution image will represent the large elements of the image, such as words or blocks.

Many of these common document understanding methods discussed above assume that the input document consists mostly of dense lines of text, in a small number of type sizes. But a check does not fit this description due to the presence of many fonts and text sizes, as well as the short length of many text strings. Therefore, modification of these methods was necessary. The next section will describe particular methods chosen to implement a Courtesy Amount Block Locator (CABL). Actually, the CABL is a small part of a much larger system designed to determine the amount for which a check has been written, given only the image of the check as input.

4. A Courtesy Amount Block Locator

As illustrated in Fig. 1, the job of locating the courtesy amount block can be divided into several steps, designed to systematically decrease the amount of information required to represent a check, so that computation becomes manageable. A check is initially represented by its bitmap. If the resolution of scanning is 300 dots per inch (DPI), a typical check involves 1.5 million bytes of information. A set of connected components in the image is generated, so that there are typically only about 300 chunks to deal with. Finally, the components are grouped into strings, of which there may be about 30. Once the check is characterized by a manageable amount of data (the list of about 30 strings), the heuristics described in Sec. 5 can be applied.

4.1. *Extraction of Components*

Assuming good thresholding, most of the black pixels in a check image will correspond to text characters. Text characters will, in general, be individual connected components, islands of black pixels in a sea of white. Therefore, identifying connected components in the image is a good way to start the process of developing a high level representation of the information on the check.

The method chosen to extract connected components from the image was a depth first search. A Depth First Search (DFS) for blocks of black pixels can be computationally intensive, so an intermediate representation of the image was used. The bitmap was converted to a set of horizontal runs of black pixels, before the DFS was applied. Section 3.2 described the use of run length representation for a slightly different purpose by Srihari. This is a common way of reducing the amount of information used to represent the image [31]. Since black pixels on a check are relatively sparse, there might only be 10,000 runs, as opposed to the 1.5 million pixel intensities. Moreover, as DFS is linear in the number of pieces of information through which to search, computational complexity is decreased markedly through the use of this representation.

The DFS implemented is a modification of the generic Depth First Search algorithm described in [32, p.478]. Its result is a set of connected components in the check image. Most of these components correspond to individual characters, either handwritten or machine printed. Some will actually be pieces of a character that were somehow not connected to the rest of the character, others will be comprised of several characters that touch each other, and still others will consist of graphics, lines, or other non-text items. It is also possible that some characters may be grouped together with blacks lines which they touch.

An attempt is made to weed out non-text components, by eliminating components with an aspect ratio (width/height) larger than 15. This should exclude the long lines on a check from consideration. This pruning value was not made too small, in order to reduce the likelihood of incorrectly removing any piece of the courtesy amount from consideration. If any digits touch a line or box around the amount, the aspect ratio of the associated component may be quite high. However, it is unlikely to be greater than 15.

It may seem as though components whose aspect ratios are too *small* should also be eliminated. This is not done, for fear of removing the digit "1" from the courtesy amount. We also do not try to remove very small components which could represent noise in the input, for fear of eliminating a decimal point from the courtesy amount, the presence of which may later be a strong indication that a particular string is the courtesy amount.

4.2. *Generation of Strings*

Many of the character grouping methods described in Sec. 3.2 may not be well suited to the check understanding problem. On a document such as a page of text, there are usually just one or two fonts or type sizes. There are also many lines of text, words and characters. Therefore one can use the average intercharacter gap and interword gap to classify characters as part of distinct strings. On a check, there may be ten or more fonts, in many different sizes. In addition, handwriting size may vary depending on the amount of space given for a particular piece of handwritten data. For these reasons, the intercharacter gap and interword gaps may not be well defined entities. This makes character grouping a very difficult endeavor.

Lam *et al.* [33] recognized a similar problem in the domain of automated reading of newspaper text. Like check images, newspaper pages contain many different type sizes. Lam *et al.* take a local average of the intercharacter and interword gaps in different areas of the page, and use these data to set dynamic proximity thresholds for character grouping. Even this method is not well suited to the check problem. The check understanding problem has the added difficulties that there may be some fonts which are represented only by three or four characters, and that text is sparse, as compared to text on a newspaper page.

The use of the Hough transform is suspect as well, due to the fact that many of the strings on a check are just two or three characters long. *Any* two characters will be collinear, not just the ones that together comprise a string. Moreover, a handwritten string of just three or four characters, such as the components of the date, or the courtesy amount, may not exhibit enough collinearity to be selected as part of the same string by a Hough transform based method. Mathematical methods like this are better suited to problems that involve many data points, such as characterizing a fuzzy edge, filling in the gaps in a broken line [22], or grouping together many characters of the same size and font that make up an entire line of text on a page.

The available approaches for grouping characters into strings make decisions on the basis of horizontal alignment of characters and proximity of adjacent characters. They differ only in their methods of determining these characteristics. The approach chosen for this system takes advantage of several pieces of information. First, it is not necessarily important that individual words are discovered. If a person's name appears on a check, it does not matter much whether first, middle and last names are grouped together as one string or not. The only important thing is that the digits and symbols in the courtesy amount be grouped together into one string, and that the string does not contain any characters that are *not* part of the courtesy amount.

For these reasons, the following methods were chosen. As a measure of horizontal collinearity of two characters, that is, whether or not they are part of the same line of text, the only requirement is that the centroid of one of them horizontally projects onto some part of the other. An alternative test that was considered was, whether any part of one character horizontally projects onto any part of the other. This approach was not chosen due to the high likelihood that it will erroneously group together two strings that do not belong together.

The other variable that needed to be measured was proximity. As mentioned above, in order to be considered part of the same string, the proximity that two characters must have to each other depends on the font, and type size of the characters. Since there are so many different sizes and fonts on a check, a global proximity threshold was not used. Instead, a proximity threshold that depends on the size of the characters involved was considered. Characters in smaller typefaces need to be closer together in order to be considered part of the same string, as compared to characters in a larger font.

Heuristically it was found that the height of a character was a more accurate representation of its size than its width, for several reasons: first, certain letters or numbers, e.g. *i, l* and *1* have widths that do not accurately reflects the font size. Second, it is possible that some of the connected components are not really characters, but rather two or more characters that overlap or touch each other. These components will have a height, but not a width that accurately reflect the size of the font. As a result, a possible proximity criterion would be that a connected component must be within a distance equal to its height from its closest neighbor,

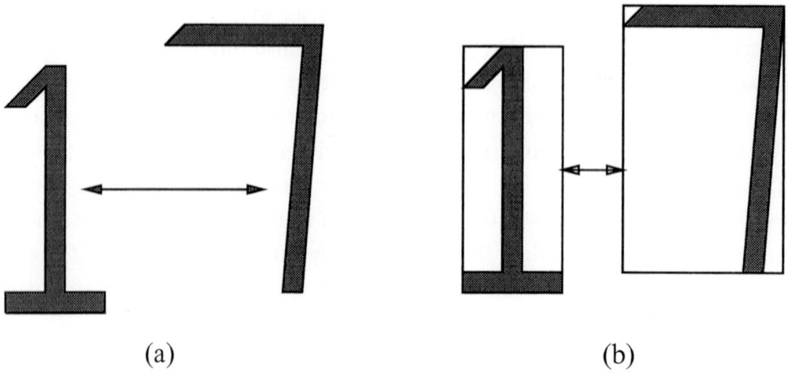

Fig. 4. (a) Distance between digits. (b) Distance between MBR's.

to be considered part of the same string as that of its neighbor.

At first glance, this appears to be an appropriate proximity threshold; however, a problem exists when a period or decimal point is encountered. These components have a very small height, so it is unlikely that there will be any neighboring characters within this distance from it. Given the high likelihood that a decimal point occurs within the courtesy amount, and the importance of correctly grouping the connected components that comprise the courtesy amount, this problem was treated directly.

A global proximity threshold is not used, rather, a threshold is computed for each character, as the weighted average of the character's height, and the average height of all of the characters in the image. A character C is put into a string with its neighbor if the neighboring character is horizontally aligned with it, and the distance between them is less than the proximity threshold associated with C. Therefore, characters in a larger font can be farther apart than characters in a smaller font, and still be grouped together.

Once a definition of the proximity threshold was achieved, the method of *measuring* distance between two characters was considered. It was decided that proximity would mean horizontal distance between the MBR's of the characters, as opposed to the minimum horizontal distance between pixels actually on the digits. The reason for this is shown in Fig. 4. In the number *17*, if the crown of the 7 is higher than the top of the 1, the minimum horizontal distance between pixels on the digits will be much larger than the real intercharacter gap of the typeface.

The proximity and horizontal collinearity criteria described above were implemented as follows. An image was created containing only rectangular boxes representing the MBR's of the original image. This image will be referred to as the highlight image. Before the highlight image is created, connected components that have a width to height ratio greater than 15 are discarded.

For each connected component in the smeared highlight image, a string is constructed by grouping connected components from the original image together, if their MBR's are contained within the connected component.

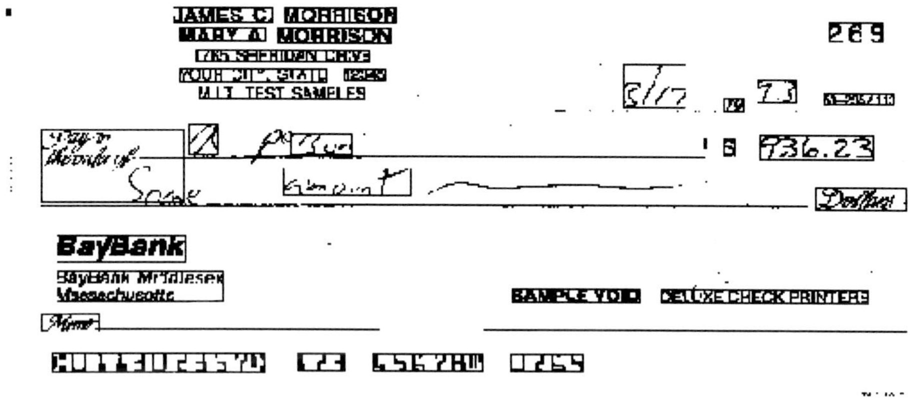

Fig. 5. Boxes around strings of components.

Before the strings are built up, any connected components in the smeared highlight image that have too small a height to be anything but noise are removed. We no longer have the danger of inadvertently throwing away periods or decimal points, since those meaningful marks must occur within strings, and will be part of a connected component that is much larger.

Once the grouping process is complete, the check is represented by a list of strings. The grouping of characters into strings can be seen in Fig. 5.

5. Bank Check Image Understanding

Usually one must decide what assumptions can be made about the input, what problems one must account ofr in the process, and around what constraints one must work, before implementing a system to recognize the courtesy amount block on a bank check. Unfortunately, the format of checks is not so strictly regulated as to make the amount recognition process, or even the job of locating the courtesy amount block, trivial. Most checks fall within certain size restrictions, although there is no specific size that a check must be. In the U.S., there are at least three accepted sizes for bank checks; however if one is considering checks from around the world, many other sizes are possible. This makes it impossible to just look at one particular location in the input image to obtain the courtesy amount block.

Although variable from one state to another, and from one bank to another, a check is valid in the United States only if it contains the date, the amount of the check in words and numerals, a signature, and an indication as to whom the check is made out to. Most checks also contain the account number of the writer, machine printed in magnetic ink by the bank who issued the checks. In other nations, these constraints may not be valid, or alternatively, may be too weak. In France,

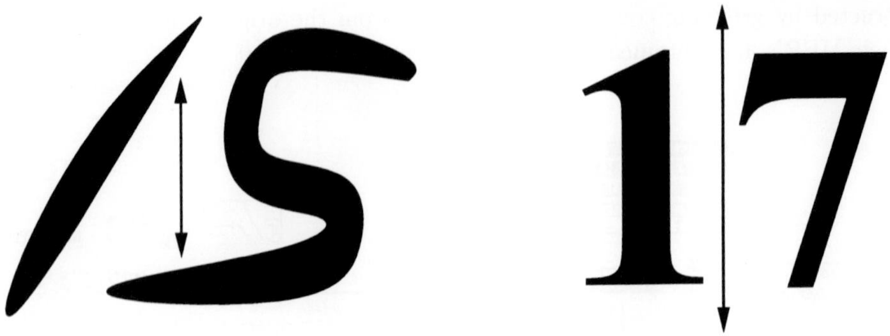

(a) (b)

Fig. 6. (a) Vertical overlap in handwritten characters. (b) A vertical line of white space between printed characters.

for example, the amount for which the check is being drawn must be written in numerals in two separate places. Many countries require that the courtesy amount block appear on the right half of the check. There are exceptions to this rule as well.

5.1. *Heuristic Rules*

In order to simplify the implementation of a courtesy amount block recognition system, as well as to boost the accuracy of the system, it is necessary to provide a context which will guide the system's search. This context will take the form of heuristic rules.

The heuristics have been divided into two categories. The first set is for distinguishing between handwritten and machine printed text. The second set is for identifying the courtesy amount block, irrespective of whether the amount has been machine printed or handwritten. Problems foreseen in using each heuristic are also described.

5.1.1. *Locating handwritten strings*

1. *If adjacent connected components overlap, then the block is handwritten*

   Machine printed text consists of characters separated by vertical lines of white pixels. Therefore, the MBR's surrounding adjacent characters will be non-overlapping. In unconstrained handwritten numerical strings, however, the right end of a character may appear in the same vertical column as the left end of the next character (See Fig. 6). In this case, the MBR's surrounding the two characters will overlap (Fig. 7). This overlap is a strong indication that a particular string is handwritten.

(a) (b)

Fig. 7. (a) Intersecting MBR's. (b) Intersecting MBR's of printed characters.

An exception to this occurs when there is a large skew in the input. Then, MBR's of adjacent machine printed characters may overlap (See Fig. 7). To prevent this, skew correction [34] can be done prior to the application of this rule.

2. *There is less regularity in handwritten text than in machine printed text*

   This regularity is manifested in many ways. The bottom-most points of all of the characters in a machine printed string will lie on a straight line. Methods for determining whether there is enough regularity in a string to say that it is indeed machine printed include the use of the Hough transform [26] and histograms [21]. Some measure of regularity may also be used with respect to the width of characters, the height of characters, and the average pixel density.

3. *Handwritten text is in a different color than machine printed text*

   Any check contains material preprinted on it by machine. This includes the guidelines on which to write, the number of the check, the century indicator on the date line, and the bank account information at the bottom. In a gray scale image, one can obtain a histogram to determine the most common intensity level of text on the check. If one assumes that the most common intensity level will correspond to the color of the machine printed text, text that has a significantly different intensity level can be assumed to be handwritten.

4. *Some characters are connected to each other in handwritten text*

   Unconstrained handwriting can be very sloppy. When writing quickly, adjacent characters can be inadvertently connected to each other. If this occurs, the two will be identified together as one character by a connected component extraction algorithm. This *character* will have a width that is much larger than the widths of the other characters in the string.

This heuristic is risky as well. Due to imperfections in the scanning, smoothing and other preprocessing of a check image, it is possible for two adjacent machine printed characters to appear connected when they are in fact merely very close to each other.

5.1.2. *Identifying the correct block*

In this section, heuristics are outlined for labeling one of the strings on the check as the courtesy amount block. If the check is handwritten, the heuristics above can be used to locate all of the handwritten strings. Then the following heuristics can be applied to only those strings that are handwritten. If the check is machine printed, they may be applied to all of the strings, since the handwriting locators above would have failed to eliminate any string from consideration.

1. *The courtesy amount should have at least two characters*

   In the U.S., the courtesy amount usually includes the dollar and cent values for which the check has been written. Therefore the amount contains at least two characters, usually at least three, since checks are seldom written for less than a dollar. Even when other units of currency are used, there is usually an integer part and a fractional part.

2. *A box may surround the courtesy amount*

   On many check formats, a rectangle denotes the area on the check where the courtesy amount should be written. If this rectangle is a black box that is dark enough to be identified as a connected component, the courtesy amount block can be chosen as the one string that appears completely within another connected component. The difficulty with this heuristic arises when the written amount overlaps, departs from, or makes contact with the box. Then, the offending character(s) will be grouped together with the box as one connected component. In this case, the form lines must be removed [34].

   Usually, the rectangle will be light enough so that it will be considered to be a part of the background by the thresholding algorithm. In this case, the rectangle can still be identified if the original gray scale image is used.

3. *A decimal point may appear in a courtesy amount*

   A decimal point is fairly easy to distinguish within a string. It may be a connected component within a string that has smaller dimensions than the others. It must also appear in the bottom portion of the string. Unfortunately, small stray marks or noise may be identified as decimal points as well.

4. *There are two digits after the decimal point in a courtesy amount*

This heuristic combats, to some extent, the problem described with the last heuristic. The decimal point in a courtesy amount is not in a random location, as a stray mark may be, but rather to the left of the rightmost two characters in the string.

5. *Some digits may be smaller than others*

   Very often, a person will write the numerals denoting the fraction amount in a smaller size than the whole amount. For example, the cents amount (or fractions of other units of currency) may also occur above other characters, e.g., $\frac{32}{100}$. By analyzing the positions of the various characters in the string, and their relative sizes, one can determine whether a fraction appears in a string.

6. *Other blocks may be in cursive*

   Handwritten portions of the check in areas other than the courtesy amount block may be written in cursive style. Examples are the signature, the payee, and the courtesy amount in words. Cursive strings will be extracted as single connected components, rather than as strings of adjacent components. Since the courtesy amount will be a string of adjacent characters, these longer cursive components can be rejected.

7. *The date will appear above the courtesy amount*

   On most checks, the date appears above the courtesy amount block. Even if it is not directly above the amount, the date is usually closer to the top of the check. Therefore if two strings are both likely to be numeric strings about six characters long, the one further down from the top of the check is more likely to be the courtesy amount.

8. *The date has distinguishing characteristics*

   The date on a check has two characteristics that distinguish it from the courtesy amount. First, the date may contain several short handwritten strings of one to four digits, separated by a machine printed string. (Many checks print the first two digits of the year on the check.) In addition, the date may be written in a smaller size than the courtesy amount, and generally contains shorter strings.

9. *The courtesy amount is in a particular general location*

   This heuristic will only work for checks of certain countries. Many countries, including the U.S., have the courtesy amount appearing on the right half of the check, and near the center of the check in the vertical direction.

5.2. *Using Redundant Information*

Further verification will take place once a choice has been made. Again, this is warranted by the fact that these heuristic methods are subject to a great degree of variability, headed by the problems described above with each rule. The following rule can be used to verify the choice of string: **A dollar sign, or other special characters will appear to the left of the courtesy amount**

Most checks will have a machine printed dollar sign (or pound, mark, etc.) on the check. In addition, a number of individuals write their own dollar sign out of *force of habit.*

When a string has been chosen, the system can find the string to the left of the chosen string, and determine if that string contains only one character. If only structural information is to be used, the verification routine can use the knowledge that there is indeed a lone symbol to the left of the chosen block. The following algorithm may be used to determine which string is the one to the left of the chosen string.

```
GetLeft(String, Stringlist)
{
  LeftString = NULL;
  VertCenter = (String->top + String->bottom) / 2;
  for (each string S in Stringlist)
    {
      if (VertCenter is Between S->top and S->bottom &&
  S->right < String->left)
if (LeftString == NULL ||
    S->right > LeftString->right)
  LeftString = S;
    }
  return(LeftString);
}
```

Alternatively, a simple symbol recognizer can be used on the `LeftString` to determine if it is indeed a currency notation symbol such as a dollar sign. A neural network can be trained to recognize a dollar sign, as well as currency symbols from other nations. The neural network can be of the same variety that will eventually be used to recognize individual digits of the courtesy amount.

6. Courtesy Amount Classification and Segmentation

Once the courtesy amount block of a check is detected, the strings within the block can be segmented and analyzed by a finite state automaton with optimum (minimized) states as shown in Fig. 11 (see [24]).

As described in Sec. 2 the bitmap image of the courtesy amount is split into probable primitives by the segmentation module. The segmentation critic module takes the list of connected components provided by the segmentation module, validates and orders the courtesy components, and transforms the list into a coarsely

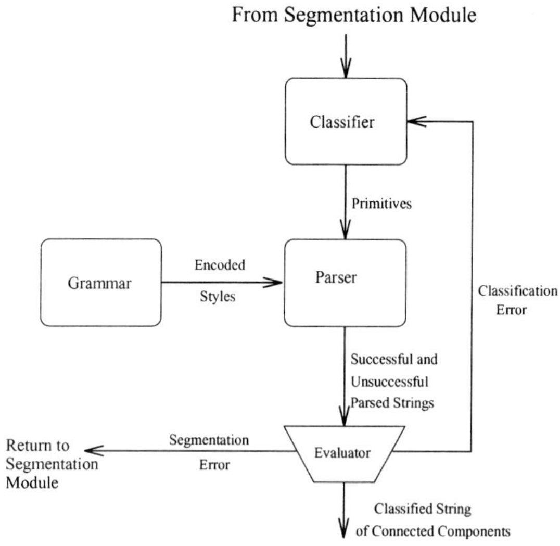

Fig. 8. A block diagram of the segmentation critic module.

classified string of components which permits the best recognition by the neural network based recognizer. The segmentation critic module is comprised of three parts: a classifier, a parser, and an evaluator (see Fig. 8). The classifier unit matches the input component with one of seven predefined primitives. These primitives include delimiters as well as some special characters. The parser unit encodes all the context knowledge associated with the fifteen styles shown in Fig. 3. The evaluator unit locates the error, if any, and instructs the classifier unit or segmentation module on the appropriate action to be taken.

6.1. *Classifier Unit*

As described earlier in this paper, the segmentation module decomposes the input string into connected components using three algorithms geared towards different cases. Each component represents a character, presently of unknown identity, and is encoded as a minimum bounding rectangle (MBR) expressed in terms of each component's end positions in two dimensions denoted by a set of four values: x_{min}, x_{max}, y_{min} and y_{max}. The scale of the coordinate system used to determine the positional information is arbitrary and most syntactic relationships are based on the position of components relative to each other. The permitted component primitives and their respective syntactic categories are shown in Fig. 9. A general classification is also given and the position of each character delineates its class.

The following heuristic features are used to assist in the preliminary classification process that is performed by the decision tree shown in Fig. 10:

Standard Height (*SH*): The standard height is the mean average of the heights

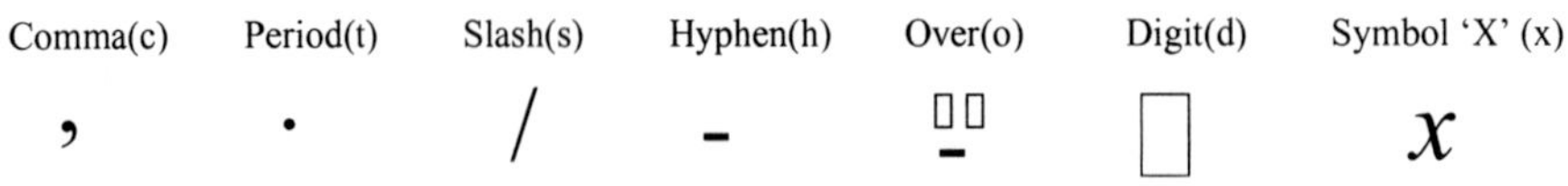

Fig. 9. Permitted component primitives and their respective classes.

of the segmented components. $SH = \frac{1}{n}\sum_{i=1}^{n}(y_{i_{max}} - y_{i_{min}})$

Midline (*M*)**:** The standard height (*SH*) divided by two. $M = \frac{SH}{2}$

Midline Offset (*MO*)**:** The distance of the top left corner of the MBR of a component in relation to the midline of the string. This distance is normalized to the standard height (*SH*). $MO = \frac{y_{min} - M}{SH}$

Aspect Ratio (*AR*)**:** This is the ratio of length to width of the component. $AR = \frac{y_{max} - y_{min}}{x_{max} - x_{min}}$

Relative Height (*RH*)**:** This is the height of a component normalized to the standard height (*SH*). $RH = \frac{y_{max} - y_{min}}{SH}$

Principal Axis Width (*PA*)**:** The principal axis is calculated by determining the axis of least inertia as described in Horn [35]. The principal axis is an approximation of the slant of each component. The width of the component along the principal axis is then determined and is normalized to the standard height(*SH*).

Above (*AB*)**:** The portion of a previous component that overlaps with the current component is normalized to the standard height. $AB = \frac{x_{(i-1)_{max}} - x_{i_{min}}}{SH}$

Span (*SP*)**:** This is the height of the current component divided by the height of the last digit encountered. $SP = \frac{y_{i_{max}} - y_{i_{min}}}{y_{(i-1)_{max}} - y_{(i-1)_{min}}}$

The classification tree, shown in Fig. 10, tests against some basic heuristics in an attempt to coarsely classify the components into the primitives shown in Fig. 9.

Associated with the above list of heuristics is a constant value that indicates the decision boundary in the classification tree. These are indicated by a letter T with a subscript indicating the heuristic on which the threshold is based. The "X" primitive is not directly determined by this simple classifier. Once a primitive is classified as a digit, the bitmap is sent to a neural network tuned to recognize the special character "X". The confidence value returned by the neural net must also pass a predetermined threshold T_X.

For each threshold in the classification tree there are two permissible values: a constrained value and a relaxed value. The classifier initially uses the constrained values until it is instructed otherwise by the evaluator described in Sec. 6.3.

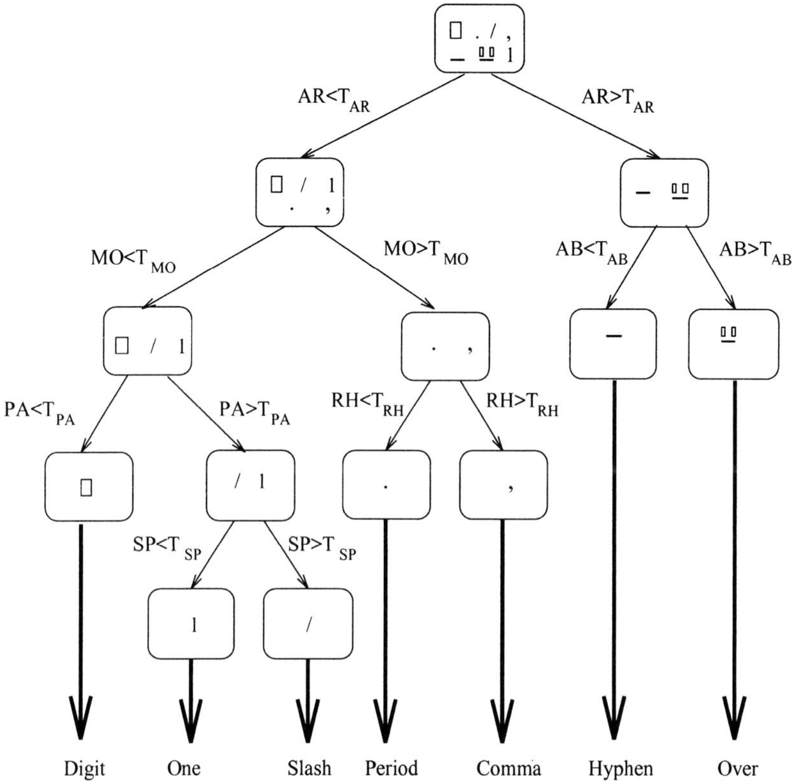

Fig. 10. Primitive classification tree.

6.2. *Parser Unit*

A survey paper [36] discusses several techniques used to store the context for a given recognition application. The efficiency in storage and the speed of retrieval and matching are essential for effective contextual checking of recognized primitives. It would be highly inefficient to store every possible primitive combination that is legal within a certain application. Therefore we have opted for a reasonably compact model for our contextual dictionary. Since our dictionary is not large, we have concentrated on a quick retrieval and matching structure. The finite state automaton described below parses the input sequence according to a grammar that contains the contextual information related to courtesy dollar amounts.

The parser unit validates the connected component sequence generated by earlier segmentation stages. For the discussion of the parser, let us call the input set of labeled connected components simply a **string**. The set of primitives from which the strings are built is called the **alphabet** of the language denoted by Σ. Languages of interest are not comprised of arbitrary sets of strings but rather of strings with specified forms, defining the syntax of the language. For the courtesy dollar amount we have defined a *deterministic finite automata* (DFA), whose operation is determined by the input string as descibed below.

A DFA is a quintuple $M = (V, \Sigma, \delta, s, F)$ where V is a finite set of states, Σ (the alphabet) is a finite set of terminal symbols, $s \in V$ is the initial state, and $F \subseteq V$ is the set of final states, and δ, the transition function, is a function from $V \times \Sigma$ to V.

$$
\begin{aligned}
\mathbf{V} &= \{\mathbf{q}_0, \mathbf{q}_1, \mathbf{q}_2, \mathbf{q}_3, \mathbf{q}_4, \mathbf{q}_5, \mathbf{q}_6, \mathbf{q}_7, \mathbf{q}_8, \mathbf{q}_9, \mathbf{q}_{10}, \mathbf{q}_{11}, \mathbf{q}_{12}, \mathbf{q}_{13}, \mathbf{q}_{14}, \\
&\quad \mathbf{q}_{15}, \mathbf{q}_{16}, \mathbf{q}_{17}, \mathbf{q}_{18}, \mathbf{q}_{19}, \mathbf{q}_F\} \\
\boldsymbol{\Sigma} &= \{d, c, t, s, o, h, x\} \\
\mathbf{s} &= \mathbf{q}_0 \\
\mathbf{F} &= \{\mathbf{q}_F\}
\end{aligned}
$$

The rules according to which the automaton M picks its next state are encoded into the transition function. Thus if M is in state $q \in V$ and the symbol read from the input string is $\sigma \in \Sigma$, then $\delta(q, \sigma) \in V$ is the uniquely determined state to which M passes. The configuration of DFA M is therefore any element of $V \times \Sigma^*$. The transition functions are shown in tabular form in Table 1. The number of states in the DFA were minimized using the algorithm in Aho *et al.* [37].

Successful and failed parses are passed on to the evaluator unit. In case $\delta(q, \sigma)$ is not defined, the parser reacts with certain errors. These errors may stipulate either errors in the classification or errors due to incorrect segmentation. Upon a parsing failure, the state and position of the error are transmitted to the evaluator unit described below to determine the appropriate feedback process.

The DFA encodes two major context constraints, the length of the dollar segment and the length and style of the cents section. States **Q0-Q8** encode the constraints on the dollar section. These states enforce the fact that a comma delimiter must be followed by three digits. States **Q9-Q19** encode the constraints imposed by the 15 styles of writing the cents portion. The transitions from states **Q0-Q9** to states **Q9-Q14** occur upon the detection of a period, slash, over or hyphen delimeter (these are typical delimeters that separate the dollar portion from the cents portion). The remaining states ensure that the cents portion has the appropriate length and type for the particular style that is used. It is clear from the above description that the available context for the cents portion is significantly richer than the dollar section. Therefore errors in segmentation of the cents portion can be detected more often than errors in segmentation of the dollar section. However, this limitation of our system is only due to the lack of contextual constraints associated with the dollar section.

6.3. *Evaluator Unit*

The evaluator unit performs the final critique of segmentation and dispatches the errors to the appropriate stages for correction. There are essentially two feedback paths provided by the evaluator unit:

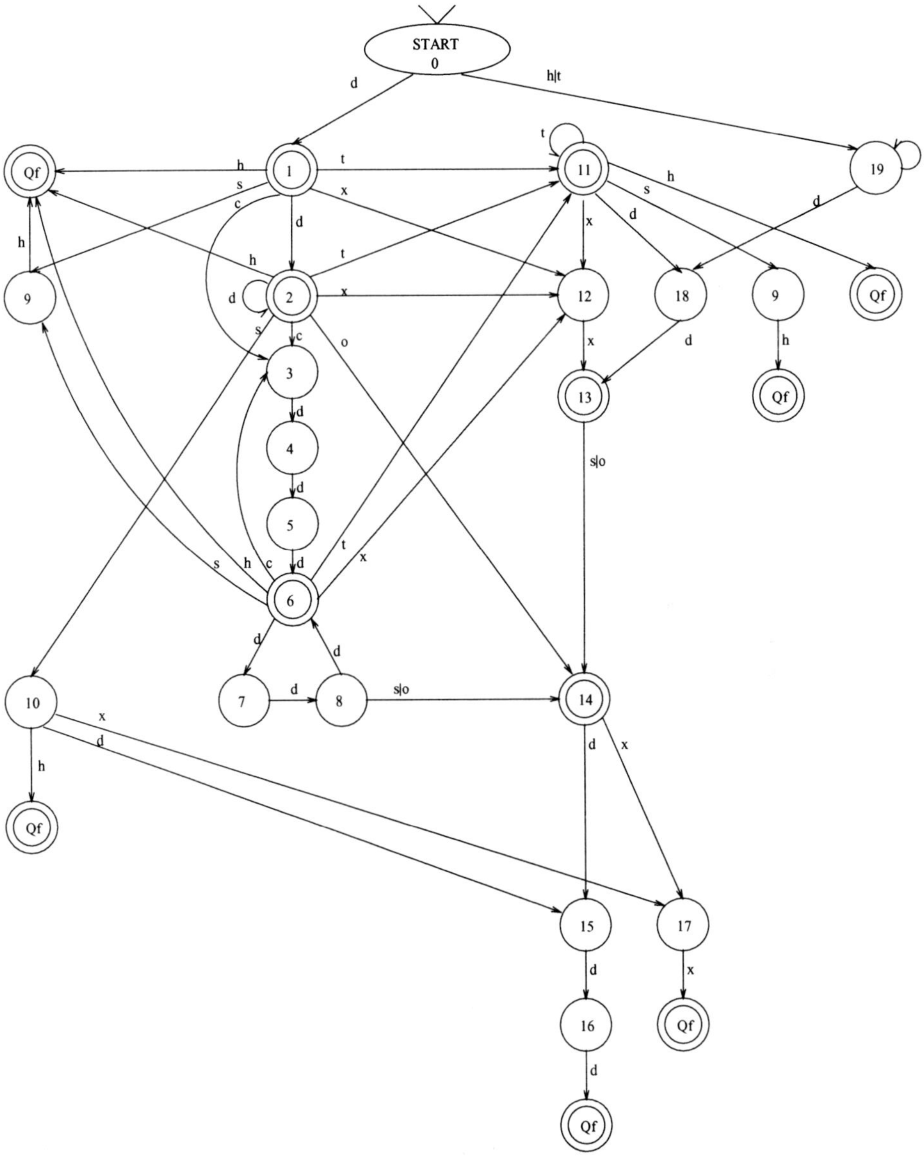

Fig. 11. State diagram of DFA for check analysis.

Feedback to Classifier Unit: Due to the inherent similarity between some of the primitives, namely the comma and period and the one and slash, the classification tree must be dynamically tuned. Failures in parsing that can be attributed to either of these substitution errors, prompt the evaluator to tune to the appropriate thresholds in the classification process.

Feedback to Segmentation Module: The ultimate goal of the segmentation critic is to improve the segmentation process. If parsing failures are not due to the substitution errors described above, the evaluator verifies the segmentation of the connected components generated by the segmentation module and identifies the most likely candidate for further segmentation.

This unit is encoded with knowledge pertaining to the probable errors associated with failed parses. It is a rule based system that attempts several feedback paths until an appropriate parse is produced. The rules are prioritized to provide a deterministic solution to the process. However, these rules do not encompass all the errors that could have occured. After consideration of a larger set of rules, the ones that follow were chosen based on their effectiveness and relative simplicity in implementation. The basic form of all the rules is as follows:

$$
\begin{array}{rcl}
Given & Failed & Parse\ State\ \mathcal{R} | \mathcal{R} \in \mathcal{V}: \\
if\ < condition > & then & < action > \\
< condition > & \Rightarrow & < condition > AND < subcondition > | \\
 & & < condition > OR < subcondition > | \\
 & & < subcondition > \\
< subcondition > & \Rightarrow & String(n) = P\ \ where\ \ (0 < n < len(String) and (P \in \Sigma). \\
< action > & \Rightarrow & Threshold\ Adjustment | \\
 & & Concatenate\ last\ two\ characters | \\
 & & Adjust\ segmentation\ criteria.
\end{array}
$$

Threshold Adjustment: This indicates replacement of any of the heuristic thresholds in the classifier unit with constrained or relaxed value.

Concatenation: This process involves the "glueing" of two primitives. It is geared to the special case when the horizontal bar above the number "5" is detached from the body of the numeral. Before concatenation, a hyphen is tested for its vertical position and proximity to the previous primitive to verify that it is part of the body of the previous primitive.

Segmentation Adjustment: There are various thresholds that direct the use of the three segmentation algorithms described earlier. These include thresholds on the aspect ratio, the primitive weight and the number of contours that are

Table 1. Parser reaction table (in part).

State	Condition	Action
Q0	c	relax T_{RH}
	s	relax T_{SP}
	x	constrain T_X
	o	relax T_{AB}
	o AND T_{AB_R}	Concatenate
Q1	o	relax T_{AB}
	o AND T_{AB_R}	Concatenate
Q2	NULL	NULL
Q3	s	relax T_{SP}
	x	constrain T_X
	h,o	Concatenate
Q4	c,t	Check Segmentation
	End	check c at Q2 relax T_{RH}
	s	relax T_{SP}
	x	constrain T_X
	h,o	Concatenate
.	.	.
.	.	.
Q18	c,t	Check Segmentation
	End	Check Segmentation
	x	constrain T_X
	s	relax T_{SP}
	h,o	Concatenate (if error Check Segmentation)

split. Relaxed values are used for these thresholds to allow for greater segmentation flexibility upon the detection of a parsing error due to segmentation by the evaluator unit.

The left column of the parser reaction table shown in Table 1 indicates the if condition while the right column indicates appropriate action to be taken. The name of the threshold indicates the module to which the feedback is directed.

6.4. *Segmentation Example*

In order to illustrate the effectiveness of the system, consider the initial scanned input image shown in Fig. 1. After the courtesy amount block is detected, the image is dynamically thresholded to separate the ink object points from the background. The result is shown in Fig. 12. The segmentation process splits the image as shown in Fig. 13. The components are then passed to the classifier unit of the segmentation critic module resulting in the string *ddscd*. Next the parser generates the following output describing the failure stage:

```
Char => d, State => Q0
Char => d, State => Q1
Char => s, State => Q2
Char => c, State => Q10
Failed Parse at State Q10 with Input 'c'
```

As described in Table 1, the evaluator relaxes the spanning threshold T_{SP} resulting in the string *dddcd*. This again fails the parser at state **Q4**.

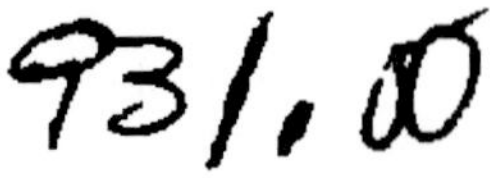

Fig. 12. Thresholded courtesy amount.

Fig. 13. Initial segmentation.

Fig. 14. Final segmenatation.

The left column of the parser reaction table shown in Table 2 indicates the if condition while the right column indicates appropriate action to be taken. The name of the threshold indicates the module to which the feedback is directed.

The T_{RH} threshold is relaxed and the comma becomes a period. The string *dddtd* again fails the parsing process at state **Q18**. The encoded reaction table then prompts the segmentor to relax its aspect ratio threshold resulting in the segmentation shown in Fig. 14. Finally, the correct string of components is generated and passed to the later stages of recognition, yielding the required answer of *931.00*.

7. Experimental Analysis, Results, and Discussion

7.1. *Choosing a Block*

Once the set of strings is determined (Sec. 4), application of the heuristics described in Sec. 5 can begin. Associated with each string is a likelihood value, representing the likelihood that that particular string is the courtesy amount block. Initially, all likelihood values are set to zero. As the heuristics are applied to each string, the string's likelihood value is incremented for each test passed by the string.

Heuristics 1 and 2 from the handwritten string location set have been implemented from the data structures involved. For Heuristic 2, the test for regularity involves determining whether the bottom of any character in the string deviates from the bottom of any other by more than 1/4 of the average character height. Heuristic 3 could not be applied, since color or gray scale images were never used. Heuristic 4 was determined not to be reliable enough for use in the system. Machine printed characters too often appeared to be connected due to low image resolution.

From the correct block identification set, the result of Heuristic 6 is achieved implicitly through the implementation of Heuristic 1. Strings written cursively should contain only one connected component. This string will fail the test of having at least two or three components. Most of these heuristics are based on the structure of the string, or of the connected components contained within it, and therefore involve simple computation of sizes, positions, and numbers of characters. The heuristics involving distinguishing characteristics of the date were not implemented due to the variations in the appearance of the date. Depending on how it is written, it may appear as one, two or three separate strings.

7.2. *Verification of Block Choice*

After all of the heuristics above are applied, the string with the highest likelihood value is extracted. This string S becomes the subject of the verification process described here. If the choice is verified, S is returned by the CABL as the correct courtesy amount string. If it is not, NULL is returned.

First, the string closest to the left of S is determined. If this string contains only one character, and that character has an aspect ratio that makes it likely to be a dollar sign or any other special currency symbol, the likelihood that string S is the courtesy amount is incremented.

Now, one would like to verify whether or not this single character is indeed a dollar sign. A neural network recognizer was trained to recognize the dollar sign and other currency signs. If the results suggest that it is indeed a correct sign, the choice is verified.

If the recognizer can find no currency sign, another decision must be made. Is the CABL confident enough in its answer even without this verification, to report that string S is the courtesy amount? If the difference between the likelihood of string S and any other string is less than three, the confidence level is not very high. String S is not a *clear* choice so the string is rejected. However, if no other string has a likelihood within three of the likelihood of string S, S is verified.

7.3. *Implementation and Testing*

Preliminary testing of the CABL described in Sec. 4 was conducted in the following manner. Test images of U.S. and British Checks were input to the system. Key features of the image were varied among the tests. For example, some of the courtesy amounts included decimal points, while others included fractions. The dates were written in several different fashions. The dollar amount in words sometimes included a fraction, and sometimes contained a line at the end of it. Often, words or digits extended below formed lines on the check. The payee was written in either cursive style or in print.

Results of these preliminary tests were very promising. Out of 54 U.S. checks, the dollar amount was correctly located 51 times. On the other three checks, the string generator did not correctly group together all digits in the dollar amount. The four digits of the integer dollar amount were grouped together, and that string was reported as the dollar amount. The cents part, which was written in the form of a fraction, was grouped as a separate string.

The CABL correctly located the courtesy amount block on nine of the ten British checks. The decimal point followed by two digits was a strong cue in that check, however only an integer amount appeared on the check which was not correctly analyzed. The CABL reported that a choice could not be made from this image. In no case was a string reported as the courtesy amount that had been incorrectly identified.

7.4. *Analysis and Performance*

The images were originally scanned at 300 DPI (Dots Per Inch). A typical check is approximately 2.75 inches by 6 inches, so that at least 1.4 million bits are needed to store the image. If gray scale is used, that number increases to about 1.5 million bytes. The CABL was implemented in C on UNIX workstations. The time necessary just to read an image file 1.5 megabytes long from disk was approximately 30 seconds. The application of all of the algorithms and heuristics took several minutes to complete. Since it takes a human an average of 4-8 seconds to locate *and* recognize the courtesy amount [1], these times are unacceptable.

In order to decrease both computational intensity and storage required for the image files, the images were converted to 100 DPI. The CABL required a more reasonable several seconds to correctly locate the courtesy amount in an image of this size.

7.5. *Future Work*

7.5.1. *Generation of strings*

Improvements can be made on many of the methods used in the implementation of the CABL. Currently, the only way these lines are eliminated from consideration is by discarding connected components that have a width/height ratio greater than a threshold value. However, if some text character touches a line, or passes through it, it will be part of a connected component with the line, and may be discarded as well. Therefore, line removal as a preprocessing step would add to the robustness of the system.

One place where improvement may be necessary is in the grouping of connected components to form strings. An optimal solution to this problem must involve feedback from a character recognizer. This can be verified by looking at the highlighted image of a check, that is, an image containing only MBR's of characters that appeared in the original check image. When one looks at such an image, which is devoid of semantic information, it is often impossible for a human to decide which MBR's should be grouped together into a string.

The likelihood value associated with each string depends on which tests the string passes. A weighting factor was given to each heuristic, based on an informal survey of types of checks and types of writing that are likely to be encountered. A more rigorous method of assigning weighting factors to the heuristics would make the heuristics more reliable.

There is also room for improvement in the heuristics themselves. More complicated heuristics can be developed to try to classify each string on the check as part of some structural piece of the image, e.g. the date, courtesy amount, signature. In this way, the CABL can be less likely to return a partial string, incorrect string, or string that includes parts of more than one structural element of a check.

The neural net that was trained to recognize dollar and pound signs could be further trained to recognize currency symbols from other nations, or other special symbols. In this way, verification would be possible for more checks.

8. Summary

Bottom-up information from connected component extraction was combined with the top-down method of smearing, in order to segment the check image into text strings accurately, while minimizing computational complexity. Dynamic information was used in the grouping of characters into strings. Top level information about the likely format of the check and the handwritten courtesy amount was encoded in the heuristics that were applied to the strings.

The methods used by the CABL were chosen to be general, wherever possible, so that changes in the problem domain do not necessitate implementing an entirely new system; however, any constraints specific to the check understanding problem were accounted for at each step, so as to make the system accurate as well as robust.

The segmentation critic suggested in this paper aids the segmentation process significantly by encoding the context knowledge associated with handwritten courtesy dollar amounts.

References

[1] J.V. Moreau, A new system for automatic reading of postal checks, in *From Pixels to Features III*, eds. S. Impevido and J.C. Simon (Elsevier, Amsterdam, 1992) 171–184.

[2] Y. Le Cun, O. Matan, B. Boser, J.S. Denker, D. Henderson, R.E. Howard, W. Hubbard, L.D. Jackel, and H.S. Baird, Handwritten ZIP code recognition with multilayer networks, in *Proc. of the 10th Int. Conf. on Pattern Recognition*, vol. 2, Atlantic City, NJ, Jun. 1990, 35–40.

[3] I. Guyon, Applications of neural networks to character recognition, in *Character and Handwriting Recognition: Expanding Frontiers*, ed. P.S.P. Wang (World Scientific, Singapore, 1991) 353–382.

[4] A. Gupta, A collection of papers on handwritten numeral recognition, Technical Report IFSRC 219-92, MIT Sloan School of Management, 1992.

[5] P. Sparks, M.V. Nagendraprasad, and A. Gupta, An algorithm for segmenting handwritten numerical strings, Technical Report IFSRC 214-92, MIT Sloan School of Management, 1992.

[6] M.V. Nagendraprasad, A. Liu, and A. Gupta, A system for automatic recognition of totally unconstrained handwritten numerals, Technical Report IFSRC 218-92, MIT Sloan School of Management, 1992.

[7] L. Mui, A. Agarwal, A. Gupta, and P.S.P. Wang, An adaptive modular neural network with application to unconstrained character recognition, *Int. J. of Pattern Recognition and Artificial Intelligence* **8** (1994), 1189–1204.

[8] L. Lam, S. W. Lee, and C. Y. Suen, Thinning methodologies-a comprehensive survery, *IEEE Trans. on Pattern Analysis and Machine Intelligence* **14** (1992) 869–885.

[9] C.Y. Suen and P.S.P. Wang eds., *Thinning Methodologies for Pattern Recognition* (World Scientific, Singapore, 1994).

[10] M.V. Nagendraprasad, P.S.P. Wang, and A. Gupta, Algorithms for thinning and rethickening digital patterns, *J. of Digital Signal Processing* **3** (1993) 97–102.

[11] I. Guyon and P.S.P. Wang eds., *Advances in Pattern Recognition Systems using Neural Network Technologies* (World Scientific, Singapore, 1994).

[12] P.S.P. Wang and A. Gupta, An improved structured approach for automated recognition of handprinted characters, *Int. J. of Pattern Recognition and Artificial Intelligence* **5** (1991) 97–121.

[13] A. Gupta, M.V. Nagendraprasad, A. Liu, P.S.P. Wang, and Ayyadurai, An integrated architecture for recognition of totally unconstrained handwritten numerals, *Int. J. of Pattern Recognition and Artificial Intelligence* **7** (1993) 757–773.

[14] H. Bunke, H. Baird, and K. Yamamoto eds., *Structured Document Image Analysis* (Springer-Verlag, Berlin, 1992).

[15] H. Bunke, P.S.P. Wang, and H. Baird eds., *Document Image Analysis and Recognition* (World Scientific, Singapore, 1994).

[16] T.M. Ha and H. Bunke, Model-based analysis and understanding of check forms, *Int. J. of Pattern Recognition and Artificial Intelligence* **8** (1994) 1053–1080.

[17] L. O'Gorman and R. Kasturi eds., *Document Image Analysis* (IEEE-CS Press, Los Alamitos, 1994).

[18] P.S.P. Wang ed., *Intelligent Chinese Language Pattern and Speech Processing* (World Scientific, Singapore, 1988).

[19] P.S.P. Wang, Learning, representation, understanding and recognition of words-an intelligent approach, in *Fundamentals in Handwriting Recognition*, ed. S. Impevido (Springer-Verlag, Berlin, 1994) 81–112.

[20] P.S.P. Wang ed., *Character and Handwriting Recognition: Expanding Frontiers* (World Scientific, Singapore, 1991).
[21] S. Srihari, C. Wang, P. Palumbo, and J. Hull, Recognizing address blocks on mail pieces. *The AI Mag.* **8** (1987) 25–40.
[22] R. Gonzales and P. Wintz, *Digital Image Processing*, (Addison-Wesley, Reading, 1987).
[23] E. Cohen, J. Hull, and S. Srihari, Understanding handwritten text in a structured environment: Determining zip codes from addresses, in *Character and Handwriting Recognition: Expanding Frontiers*, ed. P.S.P. Wang (World Scientific, Singapore, 1991) 221–264.
[24] K. Hussein, A. Agarwal, A. Gupta, and P.S.P. Wang, A knowledge based segmentation algorithm for enhanced recognition of handwritten courtesy amounts, in *Proc. of the Fifth Workshop on Structural and Syntactic Pattern Recognition SSPR '94*, Nahariya, Israel, Oct. 1994.
[25] S. Srihari, Feature extraction for locating address blocks on mail pieces, in *From Pixels to Features*, ed. J.C. Simon (North-Holland, Amsterdam, 1989) 261–275.
[26] L. Fletcher and R. Kasturi, A robust algorithm for text string separation from mixed text/graphics images, *IEEE Trans. on Pattern Analysis and Machine Intelligence*, **10** (1988) 910–918.
[27] J. Fisher, S. Hinds, and D. D'Amato. A rule based system for document image segmentation. in *Proc. of the 10th Int. Conf. on Pattern Recognition*, vol. 2, Atlantic City, NJ, Jun. 1990, 567–572.
[28] R. Duda and P. Hart, *Pattern Classification and Scene Analysis* (Wiley, New York, 1973).
[29] A. Downton and C. Leedham, Preprocessing and presorting of envelope images for automatic sorting using OCR, *Pattern Recognition* **23** (1990) 347–362.
[30] A. Downton, R. Tresidso, and C. Leedham, Recognition of handwritten British postal addresses, in *From Pixels to Features III* (Elsevier, Amseterdam, 1992) 129–144.
[31] S. Hinds, J. Fisher, and D. D'Amato, A document skew detection method using run length encoding and the Hough transform, in *Proc. of the 10th Int. Conf. on Pattern Recognition*, vol. 2, Atlantic City, NJ, Jun. 1990, 464–468.
[32] T. Cormen, C. Leiserson, and R. Rivest, *Introduction to Algorithms* (MIT Press, Cambridge, 1991).
[33] S. Lam, D. Wang, and S. Srihari, Reading newspaper text, in *Proc. of the 10th Int. Conf. on Pattern Recognition*, vol. 2, Atlantic City, NJ, Jun. 1990, 703–708.
[34] R. Casey, D. Fergueson, K. Mohiuddin, and E. Walach, Intelligent forms processing system, *Machine Vision and Appl.* **5** (1992) 143–155.
[35] B. K. P. Horn, *Robot Vision* (MIT Press, Cambridge, 1986).
[36] D. Elliman and I. Lancaster, A review of segmentation and contextual analysis techniques for text recognition, *Pattern Recogn.* **23** (1990) 337–346.
[37] A. Aho, R. Sethi, and J. Ullman, *Compilers, Principles, Techniques and Tools* (Addison Wesley, Reading, 1986).

Handbook of Character Recognition and Document Image Analysis, pp. 653–677
Eds. H. Bunke and P. S. P. Wang

CHAPTER 25

INFORMATION EXTRACTION FROM PAPER DOCUMENTS

THOMAS BAYER, ULRICH BOHNACKER, INGRID RENZ
Daimler–Benz AG, Research and Technology
P.O. Box 2360, 89013 Ulm, Germany

Extracting information from paper documents opens a variety of innovative applications by supporting people in their daily processing of documents. In this chapter, a system that interprets text on paper documents given the restricted domain of a certain application is presented . The system consists of four components. The *Document Image Analysis* component transforms the text of the scanned document image into an electronic format represented by a sequence of word hypotheses. Based on this sequence, three components extract the information necessary for automatic processing of documents. First, the information being enclosed in structured text is extracted, such as the sender and recipient of business letters, or title and author of scientific papers. Second, the text body of a message is mapped to a certain pre-defined category. In the final step, this text is analyzed and the information which is relevant for the current application is extracted. It is shown that for a real-world application the paper documents can be completely interpreted, resulting in an automatically generated answering letter. The system is fast, fault tolerant with respect to misspelling or recognition errors, and readily adaptable to new applications.

Keywords: Document understanding; Document processing; OCR; Text categorization; Information extraction; Linguistic analysis.

1. Introduction

In recent years, the importance of document image processing (DIP) and optical character recognition has strongly increased since a paper document is still the most dominant medium for exchanging information, while the computer is the most appropriate device for processing this information. Although existing OCR systems help to reduce the manual work for processing paper documents, they do not yet sufficiently cover the needs of end users because they just convert the binary document data into an equivalent representation on the character level.

On the other hand, a more intelligent reading system would be able to extract more information from the document image. It supports the information processing significantly, more and more taking over the role of a computerized assistant. For instance, if we take a closer look at the daily in-house mail processing in large companies, tasks like distributing documents, classifying documents according to certain topics, establishing relations of documents to existing ones, etc., are done manually in most cases. Many of these tasks, however, can be accomplished automatically, particularly those that have a certain routine character and are encountered in large numbers. In these cases, the automation process could cover the whole document processing from receiving the document to the generation of the answered letter, thus comprising all steps of the document work flow.

Text understanding systems, as described above and as any other cognitive system, suffer from several limitations in successfully interpreting text messages printed on paper. First, the document image must be transformed into its electronic equivalent with reasonable accuracy. Thus, little information can be automatically extracted if, e.g., the document is of very poor image quality or the text is written in cursive script. Second, the domain which the text is concerned with must be restricted to a certain message type. In this case, the essence of the message can be described by a template which is filled by the text understanding algorithms. Thus, the literacy of humans, who deal with any kind of textual contents without any restrictions, is far from being reached by reading systems.

However, in many applications the ability of human literacy is not necessary in order to automate certain tasks in document processing. Consider, for instance, the processing of bills (where certain amounts have to be detected), of inquiries about certain items (regarding the item, number of items, and other item's attributes), or of orders (items, number of items, price), etc. Common to all the examples is that they deal with a certain message type or text category, the essence of which can be expressed by templates—some items of these templates are listed in the parentheses. Hence, the task of understanding text and thus automatically performing the appropriate actions, like automatic indexing or automatically preparing data for the subsequent work flow by accessing available databases, is mapped into the task of extracting a pre-defined set of concepts from the text. Taking this into account, the applications mentioned above can be approached with the techniques described in the sequel promising significant support in many daily processing tasks.

Several aspects of this system—covering image analysis, layout generation, character classification, principles of document modeling and system aspects—have been described in detail in [1, 2, 3, 4, 5]. The focus of this chapter is to give an overview and to demonstrate what functions are necessary to accomplish this challenging text understanding task and what automation rate can be expected for real-world applications. The system has not been developed to extract the information from one distinct document class; rather, it is an open system which can be easily adapted to different applications of the kind described above. However, the system's features can be best described by selecting a certain application; thus, the requirements for each processing step and their properties are exemplified by the processing of business letters concerned with inquiries about certain items.

First, an overview of the understanding system is given and the application and the underlying data set are briefly described. The organization of the chapter then follows the functional decomposition of the system; it covers the four components *Document Image Analysis*, *Information Extraction from Structured Text*, *Text Categorization*, and *Information Extraction from Unstructured Text*. For each component the objectives and the techniques applied are briefly described, followed by a short discussion of recognition results and restrictions of the approach. The chapter concludes with an assessment of the overall system covering essential issues such as recognition accuracy, speed and portability.

2. System Overview and Application

The system's outline is sketched in Fig. 1. The task of understanding text on documents is divided into four steps:

Document Image Analysis: It transforms the text components of the document image into an electronic format. The result is a sequence of word hypotheses with their geometric properties. [a]

Information Extraction from Structured Text: The structured text—such as sender and recipient of a business letter—is extracted and filled into a pre-defined template.

Text Categorization: The text is classified according to pre-defined categories, such as order, offer, inquiry, invitation, etc.

Information Extraction from Unstructured Text: The text is analyzed with respect to its category resulting in the filling of a pre-defined template with the semantic structures.

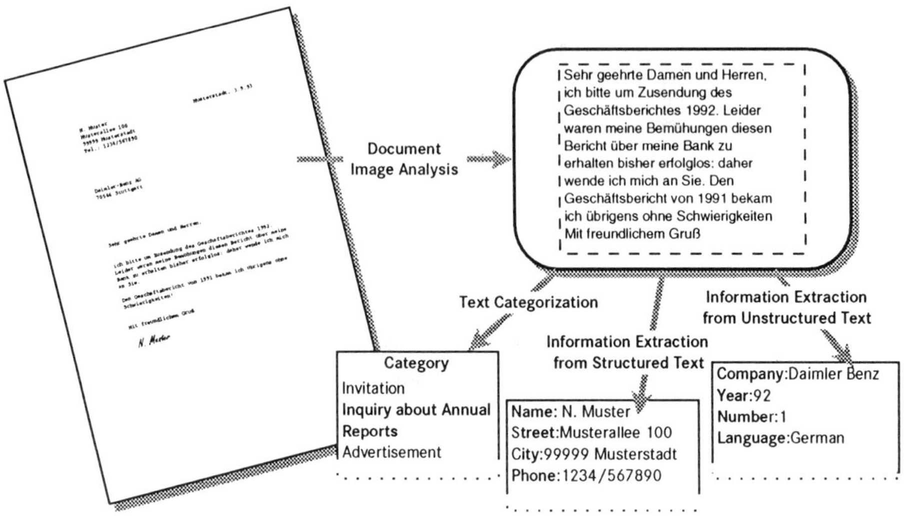

Fig. 1. System overview.

In the first step, the Daimler-Benz Recognition Engine performs the task of *Document Image Analysis*, thus generating the electronic text and the layout structure of the text. For understanding text documents, structured and unstructured text must be distinguished and processed separately (see Fig. 1). Structured text

[a]If electronic text is to be analyzed, the OCR step is omitted. In this case, the list of word hypotheses considers potential misspellings rather than OCR errors.

is defined to have a well-defined layout, a rather formal structure (syntax) and a limited set of lexical entities. The semantics of the remaining text—which is called unstructured because it is just a sequence of words—is analyzed by two modules. The Text Categorizer restricts the domain of the text to a certain message type, e.g., to an order, by calculating features from the text and mapping this representation into the decision space of text categories. Having determined the category, the domain to be modeled is restricted and the subsequent text analysis fills the template defined by this category with the concepts found in the text.

All intermediate results of the analysis are stored in an object oriented database. This database serves as a blackboard enabling the processing steps to access all available data. After *Document Image Analysis*, for instance, the subsequent three steps have access to all geometric attributes of text objects, to recognition alternatives of the character classifier, and to hypotheses of character segmentation. This data management is one prerequisite to adapt the system to different applications with little effort: the blackboard architecture allows the control of all necessary parameters and to best adapt all analysis steps to each new application.

The list of word hypotheses resulting from document image analysis is the basis for the following three steps. Since the results contain errors (recognition errors and misspelled words), these steps must be fault tolerant to a certain degree, but remain efficient. The subsequent sections zshow that the two requirements—fault tolerance and processing speed—can be met.

The functionalities of the system are exemplified by the understanding of certain business letters which reveal the potentials of document understanding techniques too. Think of a company which receives an immense number of mail pieces per day. In several thousands of letters, investors ask for annual reports of this company. These letters are currently processed manually which takes a considerable amount of time: identifying the department which is responsible for such requests, identifying the sender, the essence of his request and so forth. The goal of the system is to extract all information necessary for automating the above mentioned tasks. This information comprises the following items: sender, recipient, date, name of the company, number of copies, year(s), and the desired language. Having extracted these items, the proper actions can be performed automatically: an answered letter can be automatically formulated, a robot can be controlled to collect the desired reports, put them into an envelope along with the answered letter, and ship it to the sender of the inquiry.

In this example, the data set consists of approximately 1000 documents of which all are concerned with requests for annual reports of certain companies. The data set contains typical business letters, faxes and postcards, either from private persons or companies. Almost all the text on business letters consists of machine printed text, whereas the requests on postcards and faxes are mixed, machine printed, handprinted or in cursive script. Most of the documents have been written in German, with only a fraction of them in English, French or Spanish. The system described here focuses on the material which is machine printed and in German.

3. Document Image Analysis

The input of our document image analysis system is a binary image received either by scanning or by a fax machine. The system's goal is to provide as complete as possible an electronic representation of the text typed on a document. The following description of the Daimler-Benz Recognition Engine shows what functions are necessary to construct a powerful OCR system and how the system architecture should be designed, enabling an easy adaptation to new document types in new applications.

The objectives of the recognition engine in detail are:

- gaining a high recognition rate on single characters and on words.
- providing ranked alternatives on single characters and on words in case of ambiguities.
- providing the page's layout structure, i.e. the position of paragraphs, text lines, words and characters.

The first objective reflects the fact that the better the OCR results are, the better is the final understanding result. The second one enables the subsequent steps the access to all recognition alternatives which have been generated; this access will improve the final understanding result. And finally, the access to the layout data is necessary because many text portions on documents get their semantics from both their position and their word meaning.

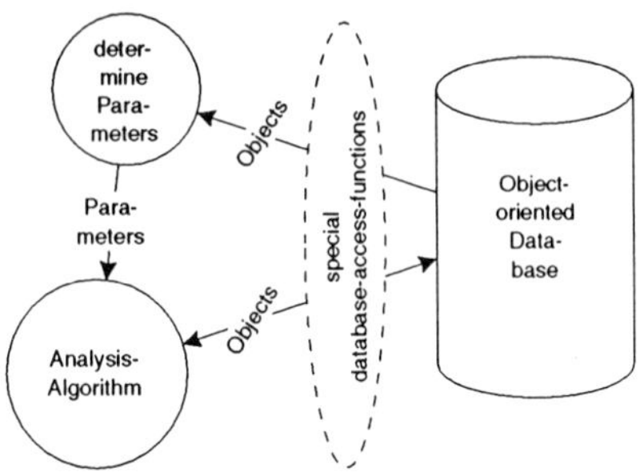

Fig. 2. Each analysis step consists of two parts: first, determination of the parameters based on examining the analysis results already available, and second, execution of the analysis algorithm.

Further objectives of the recognition engine concern the system architecture. Since the system does not focus on one specific application, it has to be built up very modularly, because different applications may require different analysis algorithms or parameters. The process of document image analysis is composed of a set of loosely coupled algorithms, thus enabling easy adaption of the system's behavior to the needs of a new application. All algorithms are collected in a toolbox and each algorithm consists of two parts (see Fig. 2), the first part determining the parameters (like the window size for smearing), and second, running the algorithm using these parameters.

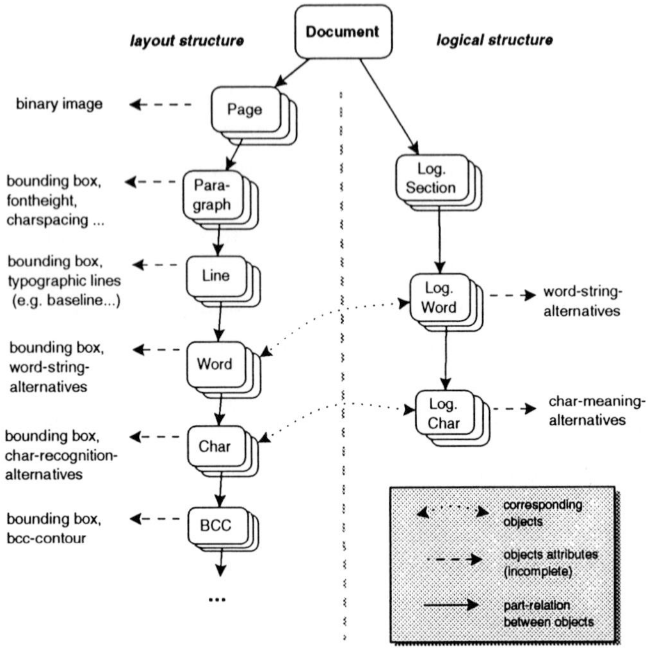

Fig. 3. The object hierarchy after full document image analysis.

All results are stored in an object-oriented database enabling access to all intermediate analysis results. This data management has the advantage that the processing sequence need not be defined in advance. The possible combinations of analysis steps result from the input and output conditions of each procedure in the toolbox. It depends on the needs of the application which processing sequence is finally chosen (for details see [5]).

The analysis results are organized in a hierarchical way, representing the document's layout structure (Fig. 3, left side). In parallel to the layout structure, a logical structure is created on the word level (Fig. 3, right side). This structure is generated at the end of the document image analysis step.

Document image analysis can conceptually be divided into the three major tasks of *layout analysis*, *character classification* and *contextual post-processing*. The theoretical and practical aspects of character classification can be found in other chapters of this handbook. In the following, we sketch the algorithms of the layout analysis and post-processing used in this system, focussing on their input/output behavior:

- *BCC*: Computing Binary Connected Components ([6, 7]). *Input*: binary image. *Output*: hierarchy of BCC objects.
- *Deskew*: Calculating the skew angle ([8]) and deskewing the text objects. *Input*: set of BCC objects. *Output*: deskewed BCC objects.
- *Page Segmentation*: Detecting text paragraphs and fixing their logical order ([9, 10]). *Input*: set of BCC objects. *Output*: ordered paragraph objects,

containing sets of text objects of homogenous properties; font size of the paragraph.

- *TextParagraphs*: Deciding, whether a paragraph object contains text or not. *Input*: set of paragraph objects. *Output*: decision for text/non-text paragraph.

Each subsequent step is executed for each text paragraph detected in the procedure *Page Segmentation*.

- *Line Segmentation*: Finding text lines
  - by profiling, i.e., the projection of the centers of text objects to the vertical axis. *Input*: set of text objects, font height. *Output*: line objects, containing sets of text objects
  - or alternatively by smearing, a closing operation on the binary image. *Input*: set of text objects, binary image, font height. *Output*: line objects, containing sets of text objects.
- *Character Creation*: Combining text objects belonging to one character, e.g. the i-stroke and the i-dot. *Input*: line objects, containing text objects. *Output*: line objects, containing character objects.
- *Character Spacing*: Deciding, whether the text is in fixed or proportional pitch ([5]). *Input*: line objects, containing character objects. *Output*: decision for fixed pitch or proportional font.
- *Character Pitch*: Estimating the pitch of fixed spaced characters by evaluating a histogram of the distances between the center columns of adjacent characters. *Input*: line objects, containing character objects. *Output*: character pitch.
- *Merged Characters—Fixed Pitch*: Detecting merged character hypotheses and re-segmenting them. *Input*: fixed pitch character objects, character pitch. *Output*: new character objects, replacing the merged ones.
- *Merge Broken Fixed Pitch Characters*: Character hypotheses are merged if they have a small distance. *Input*: fixed pitch character objects, character pitch. *Output*: new character objects, replacing the broken ones.
- *Word Segmentation*: clustering character objects to layout words. *Input*: line objects, containing character objects. *Output*: layout word objects, containing character objects.

Contextual post-processing is to improve the system's recognition performance and to provide additional information like the document's logical structure and word hypotheses to the user. The logical structure is important for dictionary look-up—a dictionary is a list of words without any annotations, in contrast to a lexicon—and for processing unstructured text as shown in the following example: In the layout structure, one text line may end with the layout word "document-ana-" and the next line may start with the layout word "lysis.". A dictionary look-up could never deal with this text. It needs exactly what is represented by the corresponding sequence of logical entities, namely, ,,document", ,,analysis" and ,,.".

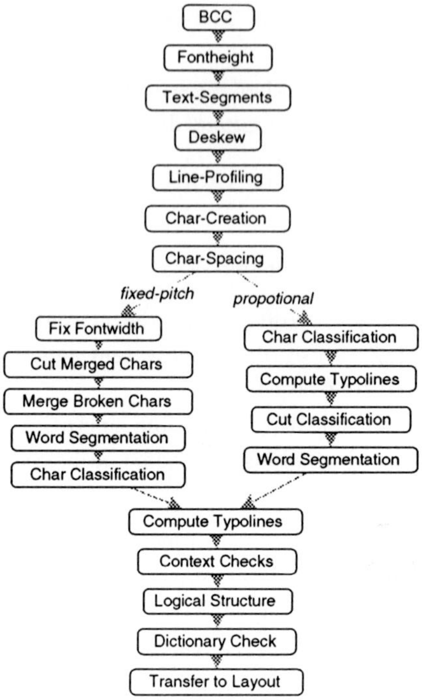

Fig. 4. One possible processing sequence.

- *Merged Characters—Variable Pitch*: Detecting and segmenting merged characters with proportional font, using a polynomial cut classifier ([11]). *Input*: character objects. *Output*: new character objects, replacing the merged ones.
- *Compute Typographic Lines*: Computing typographic lines (e.g. baseline, top-x-line, top-H-line, ...) from the positions, boxes, and shapes of characters within a line object ([3]). *Input*: line objects, containing character objects. *Output*: typographical lines for each line object.
- *Context Checks*: Using all available information about a character's box, font height, typographic class, word objects, number of segments per character, etc. to reduce or re-sort the list of hypotheses for each character. *Input*: line objects, containing word objects. *Output*: reduced lists of the characters' shape hypotheses, replacing the former lists.
- *Logical Structure*: Building up the logical structure of the document shown in Fig. 3, right side, with a set of rules. *Input*: complete layout structure. *Output*: logical structure.
- *Dictionary Check*: A fast and robust dictionary look-up to reduce errors from OCR and misspellings ([3]). *Input*: logical words, dictionary (i.e. a sorted list of strings). *Output*: ranked word alternatives.
- *Transfer to Layout*: transferring the word alternatives from logical words to the corresponding layout words. *Input*: logical words and corresponding layout words. *Output*: ranked word alternatives for layout words.

Figure 4 displays one of the possible processing sequences, containing alternative paths for segmentation of characters with fixed pitch and proportional font. Using this sequence, we tested the system on an extract of the binary scanned documents *UW English Document Image Database 1*, made publicly available by the Intelligent Systems Laboratory at the University of Washington. This set is divided into four subsets representing varying image quality. The extract consists of text zones from about 10% randomly chosen documents from each of the four categories of document images. On this subset, we gained the recognition rates shown in Table 1.

Table 1. Recognition rates for the OCR system on a subset of text zones from the *UW English Document Image Database 1* using the matching algorithm attached.

	matched	changed	deleted	inserted
Lines	58.8	41.2	0.0	1.1
Characters	97.9	1.6	0.5	0.7

The substitution rate of 1.6% on single characters (recognition errors) indicates that the performance of single character recognition still needs improvements. However, since the subsequent processes of information extraction are able to deal with alternatives of characters and words ranked after the first hypothesis and, furthermore, are able to compensate errors to a certain degree, we get a powerful overall system when using the described components for document image analysis.

4. Information Extraction from Structured Text

As described in the system overview, structured text portions are characterized by a well-defined layout, a rather formal structure (syntax) and a limited set of lexical entities within such text. The objective of the sub-task described in this section is the extraction and interpretation of such structures. The results are entries which are filled into a pre-defined template. In Fig. 5, two examples are shown, the first for tax return documents and the second for business letters. In the first example, the tax identifier, the recipient name, and the tax values have to be filled in. In the second example, the objective is to fill in the names of the sender and the recipient, the address of the sender, the opening and the closing formula.

In order to extract the information captured by such structures, the system has to have an explicit model (a kind of grammar) about the entities to be interpreted along with an interpreter. Thus, typical techniques of Artificial Intelligence (AI) may be applied, but also well known compiler techniques.

An example which is closely related to compilers is the interpretation system presented by Lorie ([12]). It can be divided into the three analysis steps of *scanning*, *syntactic* and *semantic* analysis. As a scanner, an OCR system is applied by returning the terminal symbols which are parsed in the following two steps; the predefined context free grammar describes certain entries on forms, such as the names, telephone numbers, etc. Since character alternatives are returned by the OCR system, the parser uses backtracking if the analysis comes to a dead end. This system is restricted to character meanings and does not consider any geometric or layout properties during analysis.

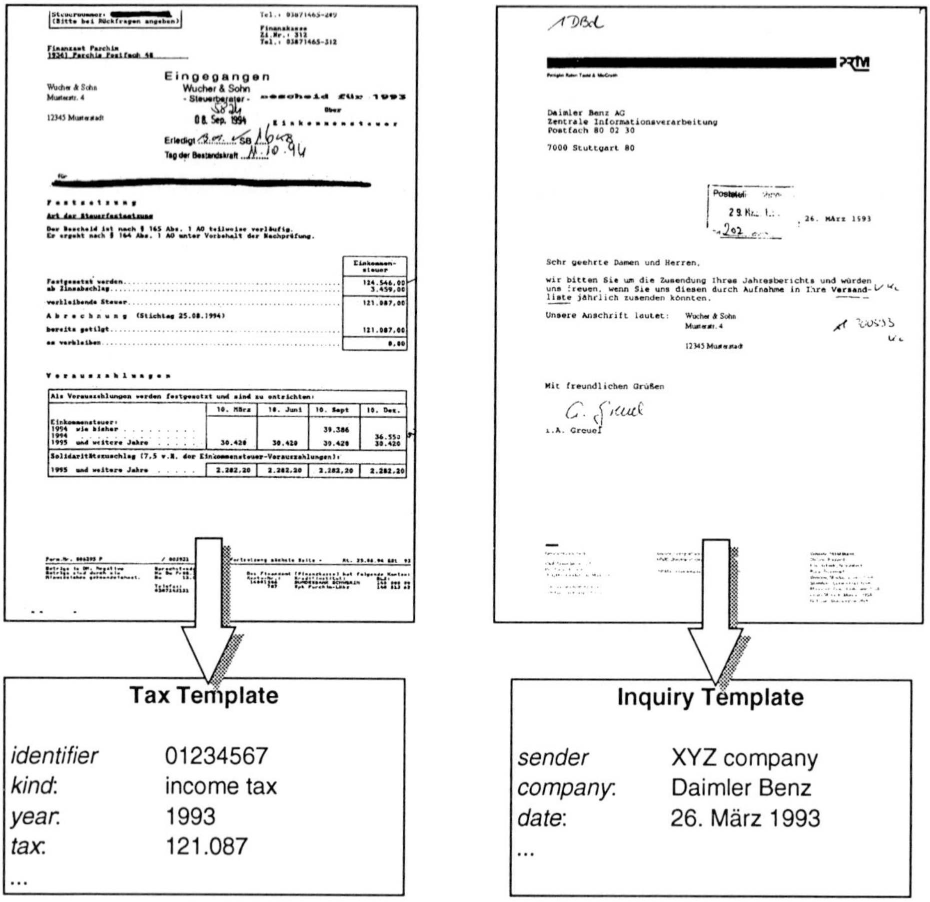

Fig. 5. Left, a subset of the information which is to be extracted from tax return forms, and, right, from inquiries about annual reports.

Other systems use representation schemes common in AI, such as semantic nets or frames. In [13], a system that extracts logic entities from a structured document without using any content knowledge, like dictionaries, etc., is presented. The entities are therefore not further analysed. In contrast to this system, the document models described in [14], [15] take into account both the layout and the content properties. The inference algorithms first employ pre-processing steps up to a certain object level; from this level, these objects are matched against the model knowledge. A different approach for understanding and classification of structured documents is presented by Lam [16] who uses a blackboard architecture and different experts in a toolbox which also use layout and content knowledge.

The system presented here differs from the first approach in that it incorporates layout knowledge within the document model which helps to better identify the correct choice from a set of alternatives. The most similar approach to the system presented is that described by [15], since the goal is actually to extract the content of the document structures, such as the name of the recipient. The other three approaches do use similar representation techniques, but differ in their recognition power; they all classify structural entities, but do not interpret their parts in full detail.

The representation language for modeling entities of structured documents is a semantic net language, called Fresco, which is described in detail in [4]. The task is to find and interpret four structural entities on business letters: recipient and sender (more precisely, the text blocks containing the address of both), date, and the text body. The last entity is not explicitly defined, since it can be inferred as the text region lying between the opening and closing formula; having verified the opening and closing, the text body is verified too. Thus, five structural entities are extracted.

In Fresco two relations are supported, the part relationship, defining the structure of the entity to be modeled, and the *is-a* relationship, enabling the user to construct a taxonomy of concepts and hence, inheriting common properties. The transitive closure of the part relationship of a concept specifies all concepts that have to be verified during analysis. In the case of the concept *BusinessLetter*, the first part-hierarchy level looks like this (in an intuitive meta-notation[b]):

BusinessLetter :: **set_of** *(sender-Role Sender) (rec-Role Recipient) (date-Role Date) (open-Role Opening) (closing-Role Closing)*

It defines that a *BusinessLetter* consists of five concepts which are defined as the second identifier in each expression within the braces. Each *„xy-Role"* defines the part-relationship more precisely: the location of each part-concept in the context of the father-concept, the cardinality and the concept itself. For example, *rec-Role* contains the following entries: **location**: (**and** *top left*), **min/max cardinality**: *1 / 1*, **partConcept**: *Recipient.*

The expression (**and** *top left*) defines a rectangular region which is calculated at the beginning from the binary image dimensions, the *Recipient* exists only once on the document (**cardinality**) and the name of the part-concept is *Recipient.* The part structure defines all concepts to be verified in a compact syntax and defines in what image regions matching objects are to be identified.

At this point, the knowledge representation seems to be equivalent to a context-free grammar. However, each concept is extended by its properties (attributes) and relations to other concepts. In this application, two attributes are sufficient in extracting objects from the pre-processed object set: box and content. They describe layout and lexical features, respectively. An example of two attributes is:

[b]Keywords of the Fresco Language are typed in boldface, identifiers in italics.

HalfLineBox:		*CityNameContent*:	
Relevance:	*required*	**Relevance**:	*required*
Range:	*Box*	**Range**:	*String*
Test:	*HalfLineBox_CF*	**Test**:	*string_cmp*
ValFunc:	*union_boxes*	**ValFunc**:	*cat_strings*
Val:	**undef**	**Val**:	*city_dict*

The **Relevance** denotes that these attributes must be calculated, that their values are of type *Box* or *String*, that the values are tested by a function given in **Test** and that values are computed by the functions *union_boxes* and *cat_strings*, respectively. Additionally, only strings are accepted which are contained in the dictionary *city_dict*.

The important entry for content-attributes is the **Val** value which contains a dictionary. The test function *string_cmp* matches the attribute's string value against the specified dictionary. Particularly, the string-match function is efficient even with large dictionaries, character alternatives, and segmentation errors, like mergings and splits.

All test functions return a certainty score from the interval $[0,1]$ which can be interpreted as a probability measure. Thus, the test function *HalfLineBox_CF* is a kind of classifier responsible for box values. Since the size of the training data is very small, the classification principle is often a heuristic rule—based on fuzzy membership functions—rather than an elaborated classifier.

An important issue for knowledge bases is the completeness of the knowledge. In this domain, the knowledge for the structures *Recipient, Date* and *Opening/Closing* formula can be specified completely. For example, the dictionaries (attached to content attributes) for the concepts *Opening/Closing* or *Month* are complete. However, the knowledge about the sender is incomplete, since neither the dictionary about the sender names will be complete nor can the location or structure of the sender be modeled sufficiently.

The model about business letters has been generated with a graphical interface. 20 documents have been used in order to adapt the fuzzy parameters of the test functions for the attributes and constraints.

Time complexity plays an important role for this system since it allows backtracking during the search process which may lead to exponential growth of the search space S. Depending on the number of concepts C to be verified and primitive objects o to be matched, S is of order o^C, i.e. $S = O(o^C)$. Table 2 shows the results on the training data.

For the first four concepts, lines have been matched; for the last concept *Date* characters were assessed. Considering these primitives and the above formula, the figures for S are very low in comparison to the worst case exponential complexity; even a number of roughly 500 nodes (for *Date*) is low, since the number of primitive objects o (characters) is 500 to 1000 in this case. Theoretically, it can be shown that $S = O(o^2)$ in the worst case, if all its parts are ordered totally rather than partially.

Table 2. Complexity for extracting structural entities on business letters. The search space size S is described by the mean value of search nodes computed from a sample set of 20 business letters.

Concept	Search space S	No. of concepts
Opening/Closing	30.4	3
Recipient	20.8	8
Sender	75.9	15
Date	448.2	19

The recognition accuracy of this step depends on the accuracy gained in the first step and on the precision of the properties of the concepts to be interpreted. If the text of the documents can be recognized with 100% accuracy and the properties of concepts can be modeled completely, the recognition accuracy for these concepts must also be 100%.

Since the knowledge about the first four concepts in this application can be completely formalized, the interpretation result only depends on the OCR quality. However, this does not hold for the sender since the lexical knowledge about this concept must be incomplete in principle; for instance, one cannot foresee all potential senders of incoming mail. Thus, the recognition rate is high, if the sender can be detected more or less by layout features. This is the case if it is a so-called *OneLineSender*—one line right above the recipient—or a *BlockSender* which has the same formal structure as a recipient and is always used by private persons.

5. Text Categorization

The Text Categorization has the task of assigning a pre-defined category (message type) to a text consisting of a sequence of words. From the application point of view, the information about the message type can be used to forward automatically the text to the proper person for further processing. From the system point of view, text categorization is an indispensable prerequisite for natural language processing which is performed in the subsequent step and which can only be successful if the domain is restricted with respect to lexical units, as well as syntactic and semantic structures. Only in this case can the knowledge be represented with reasonable effort.

The task of categorizing text is composed of two steps; in the first step, features, in this context also called *descriptors*, are extracted from the text. Such features are typically certain words from which, in the second step, the message type is inferred.

When transforming a text into a list of features (with or without further information, e.g., frequency) for the task of text categorization, the features should meet the following conditions:

- They have to be typical for the category which the text belongs to, and not to be specific for the text itself.
- They should cope with errors (OCR errors and misspellings).

- Even in the case of short texts (e.g., the text body of business letters or abstracts), the features have to be statistically significant.
- It should be easy to adapt the feature extraction stage to a new application.

The computational effort which must be spent to extract features may vary significantly depending on the type of features. Features which are easy to compute are n-grams (n typically three or four) and words (as they occur in the text). In morphologically rich languages like German, a word may be reduced to a normalized stem in order to reduce the number of different features. Moreover, full phrases consisting of several words can serve as features. However, the more complex the features get, the more computational effort must be spent. For instance, if morphological word stems are used, a lexicon (including proper names, abbreviations etc.) is necessary which is, in principle, never complete and time-consuming to generate.

In previous works, several approaches to derive suitable sets of features are described. For example, in [17] which parts of the text are suited as features is discussed. Parts of words can result from a linguistic lexicon-based analysis (normalized forms are morphemes, see [18]) or from a statistical computation (normalized forms are n-grams, see [19]). Whereas n-gram based features can be easily computed, the OCR errors drastically increase their number. On the other hand, the number of linguistically analyzed morphemes is well defined, but the respective lexicon must be made available. An analysis which statistically reduces German word forms without a lexicon is described in [20]. Reference [21] discusses whether features are dependent or independent of the specific domain. In a recent work of corpus-based morphology, [22] reduces complex morphemes using only simpler morphemes detected in the training corpus. Also [23] generates a domain specific lexicon using a corpus of training texts.

Our approach to transform a text into features is divided into two steps: first, in a training phase, a list of features is generated by reducing the text of a training corpus to meaningful parts of word. The list consists of part of words resembling the stems of words. In the second step, this list of features is used to transform the text to be categorized. Every word of this text is replaced by one or more features if these features are part of the word; otherwise it is deleted. Therefore, a text is transformed into a certain number of features on which the classification can proceed.

Whereas the second step is a simple matching procedure, the first step (reducing text of a training corpus to features) needs some elaboration. In a pre-processing step, (domain specific) stop words are defined according to their frequency in the training corpus and eliminated. The resulting word forms are used to statistically compute the typical affixes of the training corpus. The next step is the most expensive one: Extensive pattern matching of the word forms in the corpus break the complex word forms into smaller ones (under consideration of linguistically motivated restrictions like the minimum length or structure of a meaningful part of the word). After the termination of this recursive procedure, the affixes are removed from the remaining meaningful parts of the word which are then used as features.

In the categorization step a classifier is fed with these features. Depending on the nature of the features described above, different approaches are suitable for classification. If the features describing the properties of a text are meaningful entities (like word forms or phrases), rule based classifiers, like decision trees generated by algorithms of the class ID3 (see [24]), tend to be more appropriate (for example [25]). If the features consist of more or less meaningless elements and can be represented in a vector of fixed length (like n-grams), statistical classification techniques, like nearest neighbor, back-propagation neural networks etc., are employed (see e.g. [26], [27]). However, this distinction should be considered more as a general recommendation rather than as a fixed rule.

Since we are using features which are close to n-grams, a statistical classification technique is employed. The dimension L of the vector space typically ranges from 1000 to 10.000 or even higher. Before constructing the classifier, the vector space dimension is reduced for two reasons. First, such high number of dimensions would cause too much computational effort to construct and to employ a classifier. Second, a specific text is represented by a sparse vector of this size since most of the entries are zero; hence, the important feature elements should be extracted.

In order to reduce the vector space, the principal component analysis is employed which is basically a transformation of the coordinate system. This coordinate system has the property that the reconstruction error of vectors which are represented by the first l eigenvectors ($l < L$) is minimal. In the Information Retrieval community, this transformation is also called *Latent semantic indexing* (see [28]). After reduction of dimensionality, l is typically in the range of 50–500 with a reconstruction error of approximately 10 to 20%.

The classification principle is functional approximation based on polynomials. It is described in detail in two chapters of this handbook, in Franke for the recognition of handprinted characters and in Kressel and Schürmann from the theoretical point of view and . Depending on the feature vector size and on the training set size N a linear or higher order classifier can be constructed. A higher order polynomial classifier—which is more powerful than a linear one—is only possible when the feature vector size is small and N is high, since the number of parameters P to be adapted by the training samples grows with $P = \binom{l}{k}$, k being the order of the polynomial. Hence in most cases, a linear classifier is better suited than a higher order one, since the training set only comprises a few thousand text examples and the dimension of the feature space is around 250.

This approach to text categorization has been applied to a text set of roughly 1000 documents consisting of 6 classes being nearly equally distributed in this set. The length of the text ranges from five lines to a full page and the language is German. The sample set was divided into a training set of 900 texts and a test set of 150, the six classes being equally distributed in both sets. The dimension of the feature space which has been generated as described above was 2518. Approximately only 1% of the elements of a feature vector for one text was non-null. The principal component analysis was calculated from the training set. The reconstruction error (algorithmic values) is illustrated in Fig. 6 for different numbers of eigenvectors used. For example, using the first 50 eigenvectors the reconstruction error is approximately 70%, using 500 eigenvectors the error is reduced to roughly 10%.

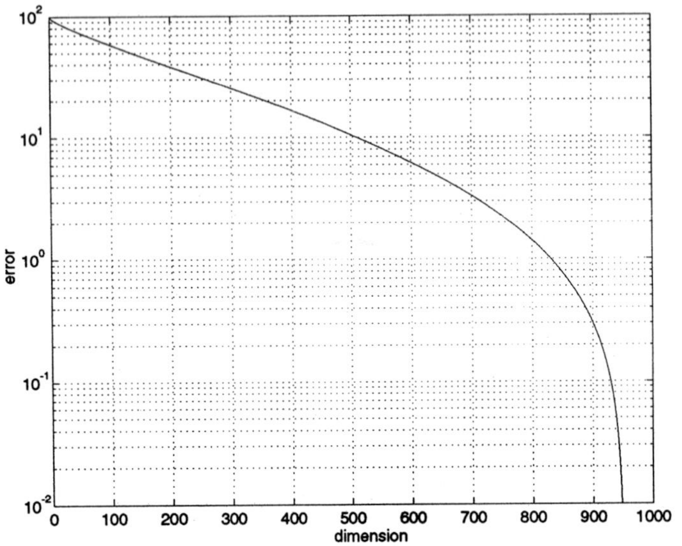

Fig. 6. Reconstruction error for a different number of eigenvectors.

In general, the recognition accuracy can be expected to lie between 60% and 90%, considering the first choice and forced recognition (compare for example [19] and [26]). This seems to be poor in comparison with the accuracy gained in other applications, like in character recognition, where 97% and higher accuracy is common. Two explanations exist for the current behavior. First, the training set is often limited to several hundred examples, making it difficult to train powerful classifiers. Second, the classification task is in fact hard in many cases. For instance, some texts may be assigned to more than one class, but finally, they had to be assigned to one class. Additionally, based on several examples, experiments with humans have demonstrated that they also have difficulties selecting the proper class. In our system we gained recognition scores of approximately 80% using the feature extraction and transformation described above.

The most significant benefits of this approach of text categorization is the way the system is constructed: it can be built automatically within a very short period (few hours) once the training and test data are labeled. It is dictionary independent and no morphological rules are necessary. Hence, time consuming manual construction of background knowledge and classification rules are avoided.

6. Information Extraction from Unstructured Text

Information extraction from unstructured text has to analyze and interpret complete text consisting of whole sentences. The basis of this step is the text category derived by the previous step since it determines the kind of information that has to be interpreted and restricts the required knowledge bases for the linguistic analysis (grammar, lexicon, which contains word forms and further annotations, like category and meaning). Information extraction has to be as efficient and robust as key word spotting, but since the correct interpretation of the information depends on the linguistic context (for example the interpretation of negation or anaphora), key word spotting would be inappropriate. On the other hand, text understanding techniques, which try to cover all the syntax of a specific language and the complete semantics of every word form and phrase, are time and space consuming (for the use of a complex text understanding system for the task of information extraction, see [29]).

At this point, we have to stress the fact that the analysis of text can only be successful if the domain is suitably restricted (for successful examples of information extraction in a restricted domain, see the *Proceedings of the Message Understanding Conference*, for example [30]). The research in natural language processing in recent years has revealed the difficulty of modeling general world knowledge and processing natural language of unrestricted domains.

The output of the information extraction is a template consisting of different slots which are filled with concepts of the text being under investigation. If the information extraction fails, the whole analysis is rejected.

The extraction component has to meet four important conditions.

- It has to analyze the text *efficiently*, according to computing time and space.
- The interpretation must be *correct*. In order to recognize the writer's intention, not only words have to be considered, but the linguistic context must be analyzed to some extent.
- The processing has to be *fault tolerant* to a certain degree. In order to cope with OCR errors and misspellings, a robust lexical look-up is often part of a pre-processing step. Ungrammatical phrases must be handled by the whole processing component.
- A *separation into processing and declarative specification* is desirable in order to maintain the component and to transfer it to other domains.

Whereas the first extraction systems like FRUMP ([31]) skimmed the texts filling out sketchy scripts without further linguistic processing, most actual systems for information extraction use several, sequentially ordered components for natural language processing according to different linguistic levels (syntax, semantics, pragmatics and discourse analysis) and/or structures (lexical items, simple phrases, complex phrases). Systems like ULINK ([32]), SCISOR ([33]), CIRCUS ([34]) analyze complete sentences. Others like FASTUS ([35]) combine the analysis of smaller phrases in order to analyze more complex ones. For further details of these and other information extraction systems see the *Proceedings of the Message Understanding Conferences* ([30]).

Our processing component for information extraction consists of two parts. The first one is a general and completely domain independent syntactic parser, or, more precisely, an analyzer for specifications of finite-state automata. It transforms the sentences of the input text into a list of semantic items according to the declarative linguistic knowledge bases (lexicon, specifications of finite-state automata as grammar). In order to handle adequately combinatorial variations, the analyzer parses the input text according to sequences of simple automata specifications.

In computational linguistics, finite-state automata are primarily applied in phonology and morphology. In the last years, though, approaches based on them were applied to more areas like lexicon, syntax, and semantics (see, e.g., [36, 37, 38]). In the field of message understanding and information extraction, the FASTUS system (see [35]) uses cascaded finite-state automata.

The second part of the processing component is a domain independent template filler which transforms domain specific semantic items into templates. This template filler uses knowledge about the possible content and structure of the templates and constitutes the major part of the semantic text model of the domain.

In contrast to analyzer and filler, the grammar, i.e. the specifications of automata, and the lexicon, which are used by the analyzer to parse the input, are the declarative linguistic knowledge bases. They are modeled on the most common linguistic expressions used in the training text. If we tried to analyze all possible expressions, the knowledge bases would be more complex, probably more faulty, and surely less efficient. Thus, we concentrate on modeling the relevant, typical, and common expressions in a correct and efficient way.

The relevant expressions in business letters can be divided into trigger expressions which set the linguistic context (e.g., order, advertisement, invitation) and description expressions which contain the specific content (what is ordered or advertised, who is invited to which event). The distinction into trigger and description is modeled in trigger automata which only have syntactic function, and description automata which furthermore describe the content of the expression. For expressions which are neither trigger nor description and which are therefore considered as irrelevant, no specification is available and the analyzer ignores them. These specifications of finite-state automata for triggers and descriptions constitute the grammar which is, together with the lexicon, the declarative component of the extraction system. Text is successfully analyzed, provided that it contains these typical expressions. On the other hand, text is rejected when triggers or descriptions cannot be properly analyzed.

As mentioned before, the problem of information extraction from unstructured text is solvable if the necessary linguistics can be restricted by further knowledge about the domain, the application, or, in our system, the text category. In the following, we consider the example of inquiries about annual reports as mentioned in Sec. 2. Most texts of this domain are short and contain between 10 and 50 words in the text body. The vocabulary with which the relevant information is expressed consists of about 1000 German word forms. The structure of the relevant parts of these documents is fairly homogeneous: the sender expresses the request and then describes the annual reports. Usually, verbal phrases are used as the requests. Examples of these expressions (with an English transcription in square brackets) are *ich bitte Sie um die Zusendung*—[I ask you for sending], *bitte senden Sie mir*—[Please send me], *wenn Sie uns zusenden könnten*—[if you could send us].

For the description of the reports, we find extensive, but simply structured noun phrases with attributes (adjective phrases, prepositional phrases etc.) like *zwei Ihrer deutschsprachigen Berichte von 1990 der Mercedes Benz AG*—[two reports of 1990 of Mercedes Benz in German]. One problem with these texts is the combinatorial complexity of the noun phrases: their attributes may occur in a different order which does not affect the meaning.

In the following, we present some examples of text structures of these inquiries which are to be handled adequately by the information extraction component (note that all of these examples are taken from original documents).

The first example illustrates the ideal case. In a sentence like *Wir bitten Sie um zwei englischsprachige Berichte der Mercedes Benz AG von 1990*—[we ask for two reports of Mercedes Benz of 1990 in English], the request is expressed explicitly (*wir bitten Sie um*) and none but the relevant facts are mentioned (number: *zwei* [two], language: *englischsprachige* [English], company: *Mercedes Benz*, year: *1990*).

Every application of information extraction has to cope with irrelevant information. This means that the documents contain more text than just the relevant parts: everywhere in the text, further linguistic material that is irrelevant may occur for which an analysis is neither possible nor necessary. This irrelevant text may express the motivation of or facts about the sender. The same information as in the first example augmented with such—for this domain—irrelevant text is in *Als Aktionär möchte ich mich genauer mit der Struktur des Unternehmens beschäftigen. Deshalb bitte ich um zwei englischsprachige Berichte der Mercedes Benz AG von 1990*—[As a shareholder I would like to know more about the structure of the company. Therefore, I ask for two reports of Mercedes Benz of 1990 in English].

Whereas irrelevant text has to be ignored, the information that is missing from the text has to be inferred. If the reports requested are insufficiently specified, default values must be inserted. In the extreme case of the example *Bitte Bericht*—[please report] the default values are Daimler Benz for the company, one copy for the number, the current year, and German for the language.

The next example is an argument against key-word spotting, since senders sometimes mention reports which they do not require. In a sentence like *Der Bericht von 1991 liegt mir dagegen bereits vor*—[I have already received the report of 1991], no request is expressed and therefore no report is ordered.

The last example illustrates more complex text structures: *Könnten Sie mir bitte 50 Exemplare des Geschäftsberichts 1991 sowie je ein Exemplar der Berichte der AEG für 1987 bis 1989 zusenden*—[could you send me, please, 50 copies of the report of 1991 as well as one copy of the reports of the AEG company from 1987 to 1989].

These examples of expressions of the request followed by a description of the report are typical for inquiries and therefore, the extraction component can concentrate on them. Other linguistic structures occur so rarely that they can be simply neglected (augmenting the rejection rate).

Whereas these inquiries about annual reports belong to a fairly specific domain, the considered text structures are common for most kinds of business letters. In this example, the specific trigger structure is the request and the specific description is the report. Similar structures can be identified elsewhere: all kinds of orders, offers, invitations, and advertisements use such trigger and description structures and an extraction component that manages one such text structure can easily be adapted to another, given that the extraction task consists of the recognition of the trigger and the interpretation of the description.

It is not intended to enlarge the linguistic resources needed for the text of this text category in order to manage the text of other text categories since such an augmentation would lead to a less efficient and more complex (therefore less tractable) system. Instead of enlarging the existing system, our objective is to adapt the concept of trigger and description expressions to other text categories. Then, the domain independent components can be used together with special declarative linguistic resources for different text categories.

For the example presented, all linguistic resources used by the information extraction were developed manually. A set of 200 documents was taken and after a conventional linguistic investigation, the result of this investigation was transformed into the linguistic knowledge bases. Since the manual construction of the required linguistic resources—even if they are restricted—is expensive, research efforts begin in order to support this task automatically.

In order to examine the performance of our information extraction component, we evaluated the implementation. The results are promising: since we modeled carefully the most typical linguistic expressions, the error rate is below 5% (accepting a rejection rate of about 20%). As to the efficiency of the components, all results are immediately returned, independently of the length of the input. These evaluations indicate that restricted linguistic resources are sufficient for a correct and efficient information extraction in the selected domain.

7. Conclusion

In the previous sections, we described the components necessary for document understanding: on the basis of the *Document Image Analysis*, the text is processed in order to extract its information by three components, *Extraction from Structured Text*, *Text Categorization*, and *Extraction from Unstructured Text*. The general approach was exemplified by analyzing the application of business letters inquiring annual reports. The essential issues of such an understanding system—recognition accuracy, speed, and portability—are summarized and discussed in the following.

The accuracy of the analyses is the essential criterion for the quality of the processing system. Like any artificial cognitive system, a 100% accuracy will never be reached. Therefore, the system is conceived as an assistance system supporting employees in handling routine tasks. While the system manages the anticipated cases, an employee can deal with the rejected documents, which are the unforeseen, complicated, or troublesome cases.

Moreover, it is even difficult to determine precisely the performance of such a complex system in general since it depends on three issues, namely image quality, complexity of the language of the domain, and the prerequisites of the application.

The effect of the quality of document images is obvious; a nearly perfect OCR result in the case of good image quality is the best basis for a successful interpretation result. However, the inverse is also true. The spectrum of the quality of document images is so poor in some applications that the document processing is not amenable to automation. Second, the complexity of the phrases occurring in the domain of a certain application also influences the system's recognition accuracy. The simpler the phrases the more successful can the information be extracted and filled into predefined templates. However, up to now no application relevant for automating the information extraction could be identified which cannot be solved with the techniques of this system. In all examples, the relevant information of these domains can be transformed to a template and the relevant phrases can be modeled. Finally, the prerequisites of each application affect the behavior of the system. For instance, the result of the anlysis of documents that are concerned with legal or financial issues, like contracts or bills, must not contain any error. On the other hand, higher error rates can be tolerated in the case of information inquiries. Hence, the requirements of the application define the ratio between errors and rejects.

This discussion reveals that the issue of recognition accuracy of document understanding systems must be assessed under these conditions. In the exemplary application of inquiries for annual reports, the image quality ranges from poor to good and is good on average; the domain is well restricted and the phrases can be modeled rather precisely; an error rate of less than 5% should be acceptable. Since every component is designed to be as fault-tolerant as possible, the non-correct analyses of preceding components will be compensated to a certain degree. Under these conditions, we expect 50 to 70% correct analyses.

For practical reasons, processing speed plays an important role since there are indeed applications which have to process several hundreds or thousands of documents (pages) per day leading to a processing speed of at most one minute for each document. Hence, particularly the text interpretation tasks, the *Extraction from Structured Text* and *Extraction from Unstructured Text*, must be carefully designed in order not to exceed certain time limits. For example, the contributions to the MUC-3 ([39]) conference have shown that several hours may be spent for extracting the information desired from one text. In the system presented the temporal requirements are met by using restricted formalisms (e.g. finite-state automata for

Extraction from Unstructured Text) or by restricting more expensive formalisms with heuristics (e.g., *Extraction from Structured Text*), thus pruning the search tree significantly.

The third essential issue is the portability of the system to different applications. The portability of this system is supported by the software architecture and restricted only by the knowledge needed. The software architecture is blackboard oriented: all analysis results are stored in the blackboard, enabling the access of all further analysis steps to this data. The processing steps are loosely coupled to allow any processing sequence which the current application requires (particularly during image analysis). The second and more crucial factor influencing the portability is the knowledge having to be modeled. Concerning *Document Image Analysis* and *Text Categorization*, this knowledge is adapted to a new application by automatic processing of large training sets. This property strongly facilitates the use of these components. On the other hand, *Extraction from Structured Text* and, especially, *Extraction from Unstructured Text* still need manual efforts to adapt the knowledge bases to a new application because here it is more difficult to automatically learn the knowledge needed.

Automatic learning needs first a large amount of training data and, second, labels being attached to each object. Both aspects are more or less trivial in the case of simple entities, like characters. It is just a matter of (computation) time to select a large sample set of characters (either live material or generating them synthetically by degradation methods, see, e.g., [40]) and attaching each character image with its label (a shape code). But learning and labeling is non-trivial in the case of complex data. If, e.g., someone tries to label geometric, syntactic or content properties of logical objects on business letters or the semantic content of linguistic phrases in a special application, tools are neither available to support the process of generating labeled sample sets nor to support the evaluation of recognition results (see e.g. [41]). Furthermore, the more complex the data is, the more difficult it becomes to collect a sufficient number of samples. In case of simple data, such as characters, up to 2000 examples can be extracted from a document; hence, a few hundreds of documents are sufficient in order to compile a sufficient number of training examples. However, if the data to be analysed are certain phrases occurring in business letters, it needs more than a million of such documents to reach an equivalent number of training samples. Collecting, managing, labeling and evaluating even thousands of documents is a difficult problem.

Hence, learning of complex knowledge bases—particularly from examples—is a key issue in building systems for text understanding which can be adapted to a new application with reasonable effort.

It has been shown that the algorithms are feasible for a wide range of real-world applications. In practice, the system presented in this chapter will help to reduce the manual effort for processing documents significantly, thus taking over more and more the role of a computerized assistant.

References

[1] T. A. Bayer, J. Franke, U. Kressel, E. Mandler, M. F. Oberländer and J. Schürmann, Towards the understanding of printed documents, in *Structured Document Image Analysis*, eds. H. S. Baird, H. Bunke and K. Yamamoto (Springer, Berlin, 1992) 3-35.

[2] T. A. Bayer, Representing and utilizing knowledge for understanding structured documents, *Proc. Conf. for Machine Visions Applications (MVA)*, Tokyo, 1992, 75–78.

[3] J. Schürmann, N. Bartneck, T. Bayer, J. Franke, E. Mandler and M. Oberländer, From pixels to contents, *Proc. IEEE* **80** (1992) 1101–1119.

[4] T. A. Bayer, Understanding structured text documents by a model based document analysis system, *Proc. 2nd Int. Conf. on Document Analysis and Recognition*, Tokyo, 1993, 448–453.

[5] T. A. Bayer, U. Bohnacker and H. Mogg-Schneider, InfoPortLab—An experimental document analysis system, *Proc. 1st Workshop on Document Analysis Systems*, Kaiserslautern, 1994, 297–312.

[6] N. Bartneck, A general data structure for image analysis based on a description of connected components, *Computing* **42** (1989) 17–34.

[7] E. Mandler and M. F. Oberländer, One-pass encoding of connected components in multi-valued images, *Proc. 10th Int. Conf. on Pattern Recognition (ICPR)*, Atlantic City, 1990, 64–69.

[8] H. S. Baird, The skew angle of printed documents, *Proc. 40th Conf. Symp. Hybrid Imaging Systems*, Rochester, NY, 1987, 21–24.

[9] F. M. Wahl, K. Y. Wong and R. G. Casey, Block segmentation and text extraction in mixed text/image documents, *Comput. Graphics and Image Processing* **20** (1982) 375–390.

[10] G. Nagy, S. C. Sethm and S. D. Stoddard, Document analysis with an expert system, *Pattern Recognition in Practice II*, eds. E.S. Gelsema and L.N. Kanal, 1986, 149–159.

[11] T. A. Bayer, U. Kressel and M. Hammelsbeck, Segmenting merged characters, *Proc. 11th Int. Conf. on Pattern Recognition (ICPR)*, The Hague, 1992, 346–350.

[12] R. A. Lorie, A system for exploiting syntactic and semantic knowledge in automatic recognition, *Proc. 1st Workshop on Document Analysis Systems*, Kaiserslautern, 1994, 277–294.

[13] A. Dengel, About logical partitioning of document images, *Proc. 3rd Annual Symp. on Document Analysis and Information Retrieval*, Las Vegas, 1994, 209–218.

[14] H. Yashiro, T. Murakami, Y. Shima, Y. Nakano and H. Fujisawa, A new method of document structure extraction using generic layout knowledge, *Proc. Int. Workshop of Industrial Applications of Machine Intelligence and Vision (MIV-89)*, Tokyo, 1989, 282–287.

[15] J. Kreich, Robust recognition of documents, *Proc. 2nd Int. Conf. on Document Analysis and Recognition*, Tsukuba, 1993, 444–447.

[16] S. W. Lam, An adaptive approach to document classification and understanding, *Proc. 1st Workshop on Document Analysis Systems*, Kaiserslautern, 1994, 231–251.

[17] D. Lewis, Feature selection and feature extraction for text categorization, *Proc. of the Speech and Natural Language Workshop*, Harriman, NY, 1992, 212–217.

[18] R. Hoch, Using IR techniques for text classification in document analysis, *Proc. 17th Int. ACM-SIGIR Conf. on Research and Development in Information Retrieval*, Dublin, 1994, 31–40.

[19] W. Cavnar and J. Trenkle, N-gram-based text categorization, *Proc. 3rd Annual Symp. on Document Analysis and Information Retrieval*, Las Vegas, 1994, 161–176.

[20] R. Kuhlen, *Experimentelle Morphologie in der Informationswissenschaft* (Verlag Dokumentation, München, 1977).

[21] C. Apté, F. Damerau and S. Weiss, Towards language independent automated learning of text categorization models, *Proc. 17th Int. ACM-SIGIR Conf. on Research and Development in Information Retrieval*, Dublin, 1994, 23–30.

[22] A. Black, J. Plassche and B. Williams, Analysis of unknown words through morphological decomposition, *Proc. Meeting of the European Chapter of the Association for Computational Linguistics (EACL)*, Berlin, 1991, 101–106.

[23] E. Riloff, Automatically constructing a dictionary for information extraction tasks, *Proc. 11th National Conf. on Artificial Intelligence*, Cambridge, MA, 1993, 811–815.

[24] R. S. Michalski, J. G. Carbonell and T. M. Mitchell, *Machine Learning* (Springer, Berlin, 1984).

[25] P. J. Hayes, P. M. Andersen, I. B. Nirenburg and L. M. Schmandt, TCS: A shell for content-based text categorization, *Proc. 6th Conf. on AI Applications*, Santa Barbara, CA, 1990, 320–326.

[26] J. Karlgren and D. Cutting: Recognising text genres with simple metrics using disriminant analysis, *Proc. Int. Conf. on Computational Linguistics (Coling'94)*, Kyoto, 1994, 1071–1075.

[27] D. Hull, Improving text retrieval for the routing problem using latent semantic indexing, *Proc. 17th Int. ACM-SIGIR Conf. on Research and Development in Information Retrieval*, Dublin, 1994, 282–290.

[28] S. Deerwester, S. Dumais, G. Furnas, T. Landauer and R. Harsman, Indexing by latent semantic analysis, *J. American Society for Information Science* **41** (1990) 391–407.

[29] J. Hobbs, D. Appelt, J. Bear, M. Tyson and D. Magerman, Robust processing of real-world natural-language texts, in *Text-Based Intelligent Systems*, ed. P. Jacobs (Lawrence Erlbaum, New Jersey, 1992) 13–33.

[30] B. Sundheim ed., *Proc. 4th Message Understanding Conf. (MUC-4)* McLean, VA, 1992.

[31] G. DeJong, Prediction and substantiation: A new approach to natural language processing, *Cognitive Sci.* **3** (1982) 251–273.

[32] J. Kirtner and S. Lytinen, ULINK: A semantics-driven approach to understanding ungrammatical input, *Proc. 9th Conf. of Artificial Intelligence*, Menlo Park, CA, 1991, 137–142.

[33] L. Rau and P. Jacobs, SCISOR: Extracting information from on-line news, *Commun. of the ACM* **33** (1990) 88–97.

[34] C. Cardie and W. Lehnert, A cognitively plausible approach to undertanding complex syntax, *Proc. 9th National Conf. on Artificial Intelligence*, Anaheim, CA, 1991, 117–124.

[35] D. Appelt, J. Hobbs, J. Bear, D. Israel and M. Tyson, FASTUS: A finite state processor for information extraction from real world text, *Proc. Int. Joint Conf. on Artificial Intelligence*, Chambery, 1993, 1172–1178.

[36] M. Gross, The use of finite automata in the lexical representation of natural language, in *Electronic Dictionaries and Automata in Computational Linguistics* eds. M. Gross and D. Perrin (Springer, Berlin, 1989) 34–50.

[37] F. Pereira and R. Wright, Finite-state approximation of phrase structure grammars, *Proc. 29th Annual Meeting of the Association for Computational Linguistics (ACL)*, Berkeley, CA, 1991, 246–255.

[38] E. Roche, Text disambiguation by finite state automata, an algorithm and experiments on eorpora, *Proc. Int. Conf. on Computational Linguistics (Coling'92)*, Nantes, 1992, 993–997.

[39] B. Sundheim ed., *Proc. 3rd Message Understanding Conf. (MUC-3)*, San Diego, CA, 1991.
[40] H. S. Baird, Document image defect models, in *Structured Document Image Analysis*, eds. H. S. Baird, H. Bunke and K. Yamamoto (Springer, Berlin, 1992) 546–558.
[41] J.J. Hull and T.A. Bayer, Document analysis and learning, in *Document Analysis Systems*, eds. A.L. Spitz and A. Dengel (World Scientific Publishing, New Jersey, 1995) 450–451

Handbook of Character Recognition and Document Image Analysis, pp. 679–702
Eds. H. Bunke and P. S. P. Wang

CHAPTER 26

AUTOMATIC INTERPRETATION AND EXECUTION OF MANUAL CORRECTIONS ON TEXT DOCUMENTS

D. MÖRI
Institut für Informatik und angewandte Mathematik
Universität Bern, Neubrückstr. 10
CH-3012 Bern, Switzerland

and

H. BUNKE
Institut für Informatik und angewandte Mathematik
Universität Bern, Neubrückstr. 10
CH-3012 Bern, Switzerland

A system for the off-line recognition and execution of correction instructions on text documents is described. It is assumed that a corrector manually marks corrections on a paper document using a color pen. Then the document is scanned. Our system automatically detects the correction marks, infers their meaning, and updates the underlying text file according to the desired corrections. A prototype of the system has been implemented and successfully tested on a number of documents.

Keywords: Document image analysis; Text processing; Text documents; Color image processing; Symbol recognition; Document correction.

1. Introduction

Document image analysis is an area where rapid scientific progress has taken place over the past few years. For example, many commercial systems became available for the automatic conversion of printed text into an ASCII representation. Other applications of document image analysis include the automatic processing and interpretation of maps and technical drawings, page layout analysis, and the understanding of music notation. Some of these subjects are described in greater detail in other chapters of this book. In the present chapter, we are concerned with another, novel application in the area of document image analysis. It is the automatic interpretation and execution of correction instructions manually drawn on a printout of a text document.

Today, a text document is typically generated on a computer under some text processing system. If a WYSIWYG (what you see is what you get) system is used, the text file is usually corrected directly on the screen. But some of the well-known text processing systems applied in science and engineering are of a non-

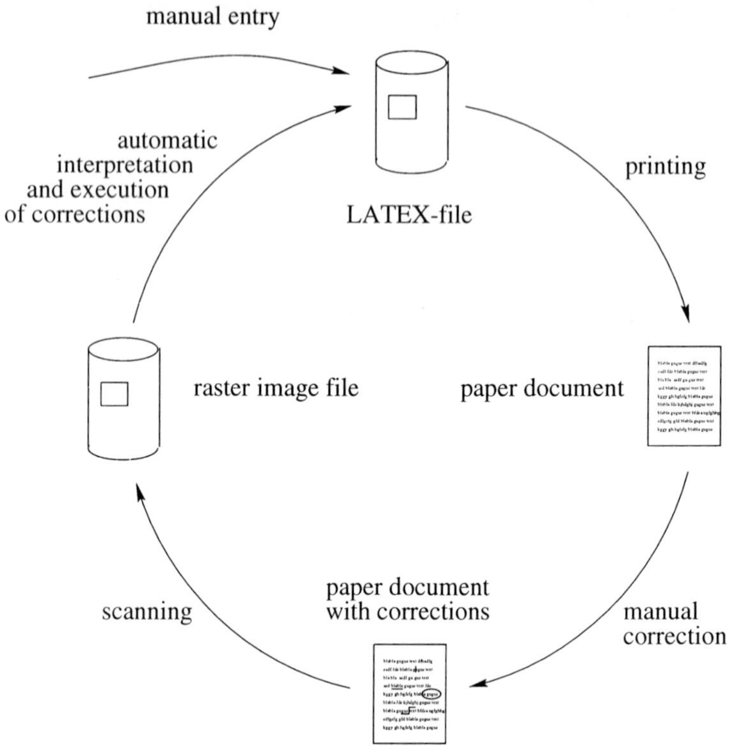

Fig. 1. The process of creating a text document using the proposed system.

WYSIWYG type. An example is LaTeX [1]. In case of a non-WYSIWYG system, the text file is often printed out on paper and proofread. Any corrections of the document are marked on the printout. After proofreading, the text file is manually updated according to the corrections marked on the printout, and a new printout is generated. Experience shows that the generation of a complex document may require several such correction cycles. So the total process of generating a text document in final form may be rather time consuming. Clearly, proofreading is a process that is hard to automate. But the interpretation and execution of the correction instructions on a page of text are susceptible to automation.

In this chapter, we describe a prototype system for the automatic interpretation and execution of correction instructions on a paper document. Input to the system is a digitized printed page — or part of a page — of text with manually marked corrections on it, and the corresponding text file. The output generated by the system is an updated version of the text file with all corrections that are marked on the document done. We assume LaTeX as the underlying text processing system [1]. The whole process of document generation using the proposed system is shown in Fig. 1. In the present chapter we describe the steps that lead from the raster image to the corrected LaTeXfile. An example of a text document with corrections on it is shown in Fig. 2. (For a more detailed description of the meaning of the correction

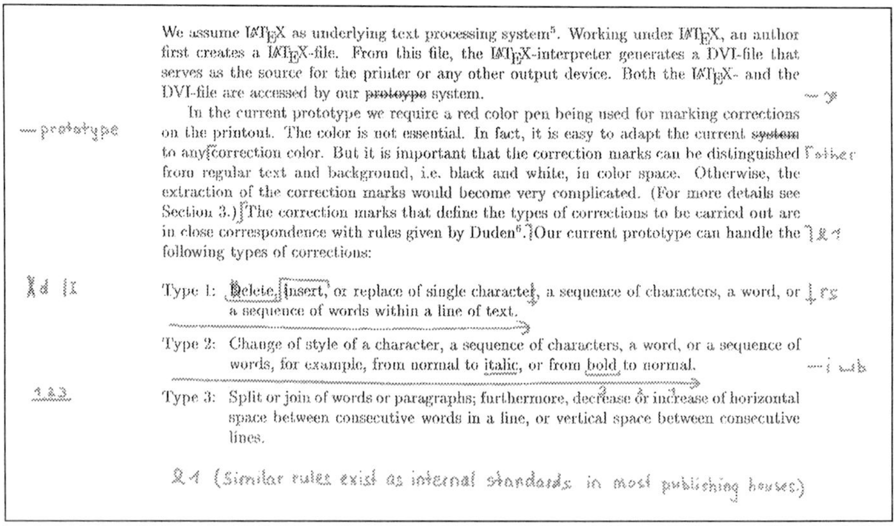

We assume LaTeX as underlying text processing system[5]. Working under LaTeX, an author first creates a LaTeX-file. From this file, the LaTeX-interpreter generates a DVI-file that serves as the source for the printer or any other output device. Both the LaTeX- and the DVI-file are accessed by our protoype system.

In the current prototype we require a red color pen being used for marking corrections on the printout. The color is not essential. In fact, it is easy to adapt the current system to any correction color. But it is important that the correction marks can be distinguished from regular text and background, i.e. black and white, in color space. Otherwise, the extraction of the correction marks would become very complicated. (For more details see Section 3.) The correction marks that define the types of corrections to be carried out are in close correspondence with rules given by Duden[6]. Our current prototype can handle the following types of corrections:

Type 1: Delete, insert, or replace of single character, a sequence of characters, a word, or a sequence of words within a line of text.

Type 2: Change of style of a character, a sequence of characters, a word, or a sequence of words, for example, from normal to italic, or from bold to normal.

Type 3: Split or join of words or paragraphs; furthermore, decrease or increase of horizontal space between consecutive words in a line, or vertical space between consecutive lines.

Fig. 2. Text document with corrections.

We assume LaTeX as underlying text processing system[5]. Working under LaTeX, an author first creates a LaTeX-file. From this file, the LaTeX-interpreter generates a DVI-file that serves as the source for the printer or any other output device. Both the LaTeX- and the DVI-file are accessed by our system.

In the current prototype we require a red color pen being used for marking corrections on the printout. The color is not essential. In fact, it is easy to adapt the current prototype to any other correction color. But it is important that the correction marks can be distinguished from regular text and background, i.e. black and white, in color space. Otherwise, the extraction of the correction marks would become very complicated. (For more details see Section 3.)

The correction marks that define the types of corrections to be carried out are in close correspondence with rules given by Duden[6]. (Similar rules exist as internal standards in most publishing houses.) Our current prototype can handle the following types of corrections:

Type 1: Insert, delete, or replace of single characters, a sequence of characters, a word, or a sequence of words within a line of text.

Type 2: Change of style of a character, a sequence of characters, a word, or a sequence of words, for example, from normal to *italic*, or from **bold** to normal.

Type 3: Split or join of words or paragraphs; furthermore, increase or decrease of horizontal space between consecutive words in a line, or vertical space between consecutive lines.

Fig. 3. Text document with applied corrections

symbols see Sec. 2.) The output generated by our system is given in Fig. 3.

The prototype system described in the following operates in the off-line mode. This has several advantages over on-line methods as only a printout of the document and a pen are required in order to correct a document. Consequently, a document can be corrected without the need of having a computer available. On the other hand, for somebody permanently working at a workstation, it might be more desirable to avoid the generation of a printout and correct the document directly on the screen. Gesture-based operations on documents are proposed in [2, 3, 4]. Therefore, the on-line and off-line correction modes are complementary to each other and both are useful, depending on the particular conditions under which documents are being prepared. As a matter of fact, we are presently extending our prototype by a module for the on-line correction mode. The design and implementation of this on-line module is definitely easier than that of the off-line system because the location and the type of each correction can be unambiguously input by the author via the keyboard or a mouse. Eventually, we are working towards a system that provides both the on-line and off-line correction modes.

Independent of the actual application, we expect some new scientific insights as a result of our present research efforts. The most important one perhaps addresses the question of how to handle multiple document representations. Clearly, in our system we are dealing with two different representations of each document, namely, the raster image and the LATEXfile. In order to modify the LATEXfile after the meaning of a correction symbol in the raster image has been recognized, we have to create links between these two representations. Similar issues concerning the representation of documents as raster images and ASCII files occur, for example, in modern information retrieval systems. This topic is addressed in greater detail in the chapters by G. Nagy *et al.* and K. Taghva *et al.* in this book.

An earlier version of our system has been described in [5]. The version reported in the present paper includes a number of new features, for example, an extended set of correction symbols, a more robust symbol matching procedure, and a more flexible method for color image processing.

2. Basic Assumptions and Definition of Correction Symbols

As mentioned before, we assume LATEX as the underlying text processing system. Working with LATEX, an author first creates a so-called LATEXfile which is an ASCII file consisting of the raw text mixed with LATEXcommands. From this file, the LATEXinterpreter generates a DVI file that serves, after it has been run through a DVI-to-Postscript converter, as the primary source for the printer or any other output device. Both the LATEXand the DVI files are accessed by our system.

In the current prototype we require the corrector using a color pen for marking corrections on the printout. The actual color is not essential. But for reasons that will become apparent in Sec. 3, the correction color should have a high degree of saturation. Otherwise, the correction pixels can't be automatically distinguished well from the black text and the white background pixels. Examples of "good" correction colors are bright red or green, while correction colors to be avoided are brown or dark blue.

The correction marks that define the types of corrections to be carried out are in close correspondence with rules given by Duden [6]. Similar rules exist as internal standards in most publishing houses. Our current prototype can handle the following types of corrections:

Type 1: Insert, delete, or replace a single character, a sequence of characters, a word, or a sequence of words within a line of text.

Type 2: Change the style of a character, a sequence of characters, a word, or a sequence of words, for example, from normal to italic, or from bold to normal.

Type 3: Split or join words or paragraphs; furthermore, increase or decrease the horizontal space between consecutive words in a line, or the vertical space between consecutive lines; swap adjacent characters or words.

Type 4: Change the order of a sequence of $n > 2$ words.

As an example, the corrections specified in Fig. 2, from top to bottom, are:

1. Delete the word "protoype".
2. Replace the word "system" by "protoype".
3. Insert "other" before "correction".
4. Start a new paragraph before "The".
5. Insert the text "(Similar rules exist as internal standards in most publishing houses.)" before "The".
6. Replace the character "D" by "d" in "Delete,".
7. Replace the character "i" by "I" in "insert,".
8. Swap the words "Delete," and "insert," after the corrections 6 and 7 of this list have been performed.
9. Replace "r" by "rs" in "character," (which is equivalent to the insertion of character "s").
10. Decrement the vertical space between the paragraph ending with "text." and the following paragraph.
11. Change the style of the word "italic" from normal to italic.
12. Change the style of the word "bold" from normal to bold.
13. Decrement the vertical space between the paragraph ending with "normal." and the following paragraph.
14. Change the order of the words "decrease or increase" into "increase or decrease".

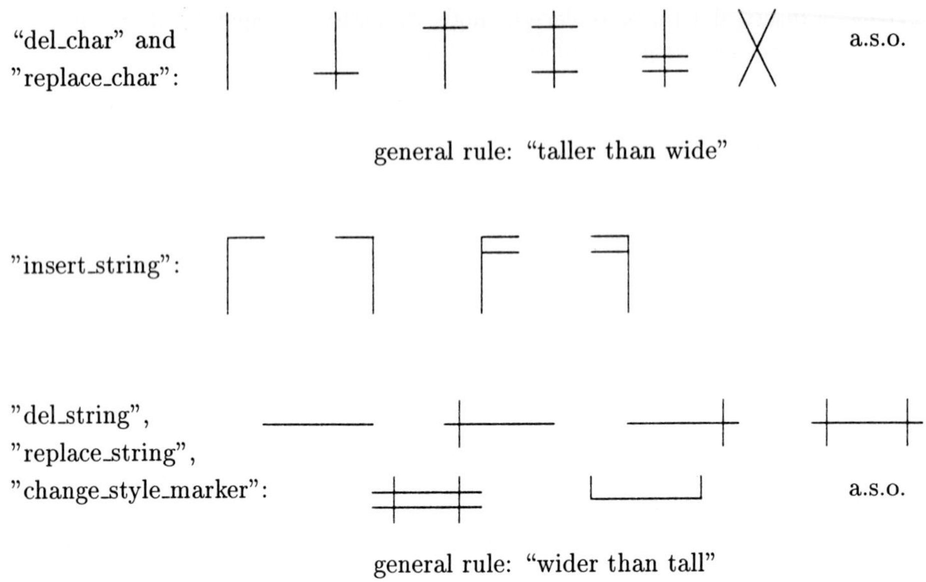

Fig. 4. Position markers for type 1 corrections.

In the following, the correction syntax, i.e. the rules the correction marks have to follow, are explained in greater detail. For the purpose of illustration, the reader may refer, throughout the following section, to Figs. 2 and 3.

Type 1 corrections need a so-called position marker in the actual text line that indicates the location of the correction. A summary of the position markers for type 1 corrections admissible in the actual prototype is shown in Fig. 4. A position marker consists of a vertical stroke — the main stroke — with some smaller horizontal strokes optionally attached to it if a single character has to be replaced or deleted. If more than one consecutive character are involved, a position marker is usually represented by a horizontal stroke — the main stroke — with some smaller vertical strokes optionally attached to it. As can be seen in Fig. 4, any other marker, for example, an X-shaped pair of lines, is possible. But a position marker must not be identical to any predefined special correction symbol (see corrections of type 3 in Fig. 5). An important rule is that each position marker that occurs in a text line must be repeated on the left or right margin of the actual text line. Immediately to the right of the repeated instance of a position marker on the margin is a sequence of characters that specify the particular correction in greater detail. That is, if a symbol or a text string is to be replaced by a new text string, or if a new text string is to be inserted, the new text has to be written by the corrector immediately to the right of the instance of the position marker on the margin. If text is to be deleted, a special deletion symbol (deleatur) has to appear to the right of the position marker. For examples, see the corrections 1–3, 5–7 and 9 from the above list. The symbol appearing in Fig. 2 in the first correction to the right of the position marker, on the margin is the deletion symbol.

Similar to type 1 corrections, position markers are used for type 2 corrections. For corrections of this type, the position markers are underlinings that indicate the characters or words the style of which has to be changed. In order to indicate the new style, each position marker for a type 2 correction has to be repeated on the left or right margin with at least one identifier following it that denotes the new style. For example, style italic is denoted by "i", style bold is denoted by "b", and so on. The corrections 11 and 12 from the above list are examples of such corrections.

A summary of the position markers for type 3 corrections and their meaning is given in Fig. 5. These predefined special markers need not be repeated on the left or right margin. For examples look at the corrections 4, 8, 10 and 13 from the above list.

The type 3 corrections "exchange-chars" and "exchange-words" allow the swapping of adjacent strings (consisting of at least one character each) within the same line of text. A more flexible way to arrange the order of several words over multiple lines of text is available with the "transpose" correction, which is of type 4 (see Fig. 6). To indicate a transpose-correction one has to draw a horizontal stroke on the margin. Directly above this stroke there are markers which have a corresponding position marker over a single word in or below the actual line of text. The sequence of the markers indicates the desired sequence of the marked words. To use numbers as markers is convenient but not mandatory. For an example, we refer to the last correction in the above list.

If several corrections are to be made on the same text line, several position markers will usually appear on this line. Each of them has to have a unique shape. Each position marker of type 1 or type 2 correction has to have a position marker of the same shape on the margin left or right to the actual line of text. To reach a higher degree of comfort, it is allowed that a position marker in the text area refers to a symbol on the margins that was earlier drawn. Thus it is possible, for example, to delete several lines of text by indicating the deletion just once. Generally, this feature is useful if we want to apply the same correction multiple times.

No attempt has been made yet to integrate a module for handprinted character recognition in the actual prototype. This means that all text strings written by the corrector on the margin (i.e., text strings to be inserted or replacing other text) are presently interactively classified. Only the delete symbol is automatically recognized. In order to provide a fully automatic system it is planned to include one of the available methods for handprinted character recognition in a future version of the system.

3. Extraction of Correction Symbols

The system described in this paper consists of three modules that are concerned with the extraction of correction symbols, their interpretation, and the execution of edit commands, respectively. In the present section, the correction symbol extraction module will be described in greater detail.

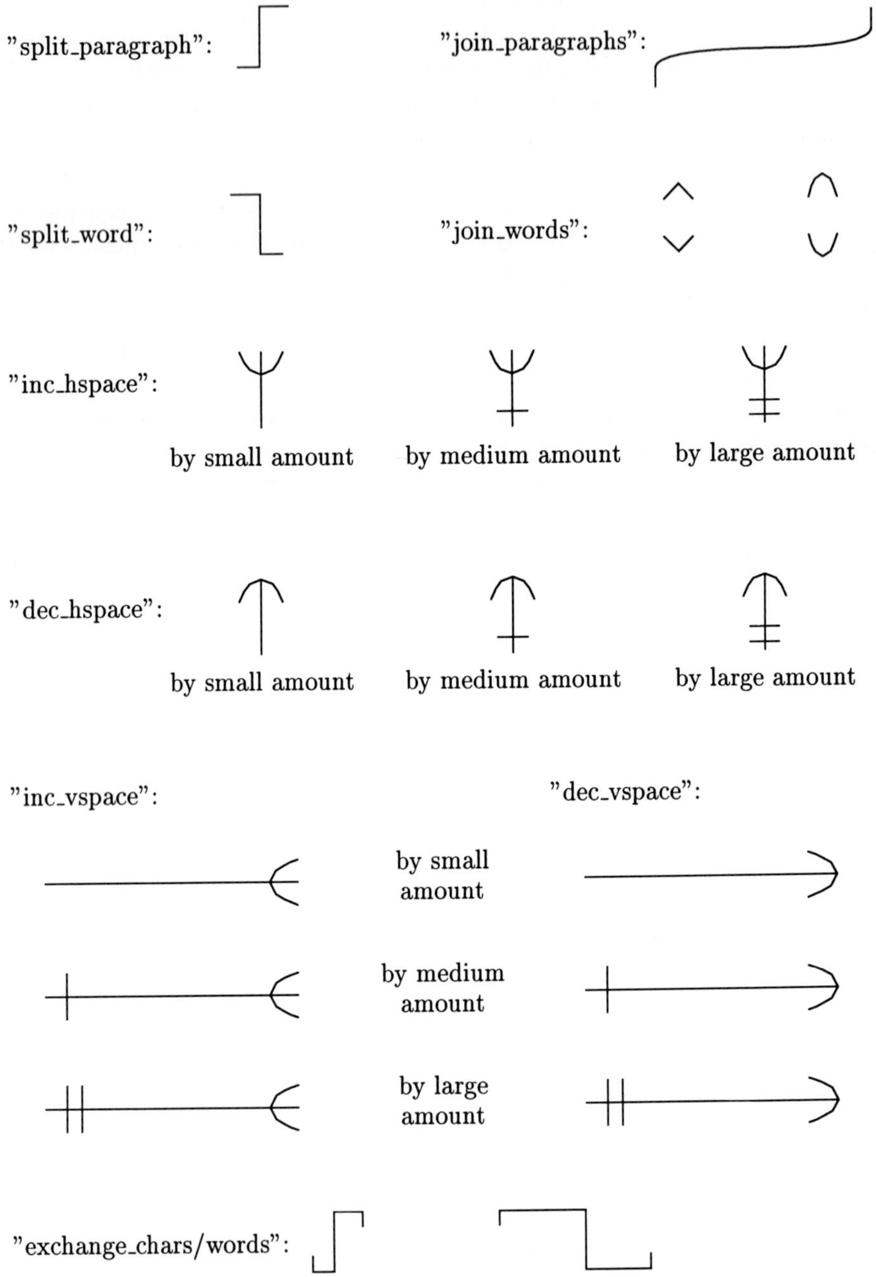

Fig. 5. Position markers for type 3 corrections.

"transpose": 3 2 4 1 1 2 3 4

Fig. 6. Position markers for type 4 corrections.

In our experiments we are digitizing the documents by means of a color scanner at an intensity resolution of at least 3 bits per channel, and a spatial resolution of at least 300 dpi. After image acquisition, an array with RGB values (red-green-blue) is given. The goal of the correction symbol extraction procedure is to separate all correction marks, which were manually entered on the document by the corrector, from the original text. More precisely, we aim at a classification of each pixel in the document image into one of the classes `TEXT`, `BACKGROUND`, and `CORRECTION`.

Our classification procedure consists of two steps. In the first step, only color features of the pixels are used. As some of the pixels cannot be assigned to a class unambiguously, taking only the color features into account, an additional class `UNKNOWN` is provided. In the second step, all `UNKNOWN` pixels will be reclassified as either `TEXT`, `BACKGROUND`, or `CORRECTION` using contextual, i.e. spatial, information.

In order to solve the classification problem, a transformation from the RGB color space into the HSI space (hue-saturation-intensity) is done first [7, 8]. This transformation is based on the following equations,

$$\begin{pmatrix} I \\ V_1 \\ V_2 \end{pmatrix} = \begin{pmatrix} \frac{1}{3} & \frac{1}{3} & \frac{1}{3} \\ -\frac{1}{\sqrt{6}} & -\frac{1}{\sqrt{6}} & \frac{2}{\sqrt{6}} \\ \frac{1}{\sqrt{6}} & -\frac{2}{\sqrt{6}} & 0 \end{pmatrix} \begin{pmatrix} R \\ G \\ B \end{pmatrix}$$

$$H = \tan^{-1}\left(\frac{V_2}{V_1}\right)$$

$$S = \sqrt{(V_1)^2 + (V_2)^2}$$

Let $R(p)$, $G(p)$, and $B(p)$ be the intensity in the R, G, and B channel, and $H(p)$, $S(p)$, and $I(p)$ denote the hue, saturation, and intensity of a pixel p, respectively. The first step of the classification is performed according to the following rules:

1. **if** $\max(R(p), G(p), B(p)) \leq minValue$

   then $p \in$ `TEXT`

2. **else if** $I(p) > maxIntensity$

   then $p \in$ `BACKGROUND`

3. **else if** $H(p) \in [H_{min}, H_{max}] \ \wedge \ S(p) \geq S_{min} \ \wedge \ I(p) \in [I_{min}, I_{max}]$

   then $p \in$ `CORRECTION`

4. **else if** $H(p) \in [H_{min2}, H_{max2}] \ \wedge \ S(p) \geq S_{min2} \ \wedge \ I(p) \in [I_{min2}, I_{max2}]$

   then $p \in$ `UNKNOWN`

5. **else**

   if $I(p) \leq maxTextIntensity$

   then $p \in$ `TEXT`

   else $p \in$ `BACKGROUND`

Intuitively, the above classification rules have the following meaning:

1. For a "real" color or the white background, at least one of the RGB values is larger than *minValue*. Therefore, if none of the values is greater than *minValue*, the pixel must be "dark", i.e., it belongs to `TEXT`.

2. Colors with a very high intensity are assigned to `BACKGROUND`.

3. If hue, saturation and intensity of p are within predefined ranges, it is considered a correction pixel.

4. The ranges for hue, saturation and intensity are extended and a check similar to step 3 is performed. If the conditions are fulfilled now, we classify p as `UNKNOWN`.

5. If not yet classified, the pixel must be either `TEXT` or `BACKGROUND`.

In the classification procedure several thresholds are used. The thresholds *minValue, maxIntensity* and *maxTextIntensity* can be experimentally determined and are independent of the correction color if we assume that the correction color is not too dark or bright, and has a certain degree of saturation. (This means that not only all greyish colors, but also pastel colors are forbidden.) Therefore, these thresholds are predefined. If the color of the correction pen is assumed to be always the same, then the remaining thresholds can be set in advance. But this assumption would be quite restrictive. Therefore, we have developed a method to determine these thresholds dynamically based on the actual correction color. First, we use *minValue* and *maxIntensity* to exclude all very dark or very bright pixels from being possible representatives of the actual correction color. Experiments have shown that after this elimination step many different colors still appear in the color table. In case of a red correction pen, for example, we also detected blue and green pixels. But these colors appear only at the border of the text in the image of the scanned document. Therefore, for all remaining colors surviving the elimination based on *minValue* and *maxIntensity*, we count the pixels which are not in the 4-neighborhood of an excluded color. As a result of this procedure we get quite high total counts of all colors that are produced by the correction pen. As mentioned before, the other colors in the image are due to pixels at the border of the printed text. Because most of these pixels are adjacent to white or black pixels, they will not be added to the total count. Therefore, this count is a good measure to distinguish the correction color from the other colors. Having determined the color of the correction pen, the remaining thresholds are automatically determined from the color with the greatest total count.

After all pixels of the document image have been classified, all UNKNOWN pixels are reconsidered taking their local neighborhood into account. UNKNOWN pixels typically belong to the border between text and background, or between a correction mark and the background. Consequently, we reclassify an UNKNOWN pixel as TEXT if at least one of its 4-neighbors belongs to TEXT. If this rule doesn't apply, we check if at least one of its 4-neighbors belongs to CORRECTION and reclassify it as CORRECTION in this case. If neither of the two rules is applicable, the pixel is reclassified as BACKGROUND. After all UNKNOWN pixels have been reclassified, all pixels in the document image belong to one of the classes TEXT, BACKGROUND, or CORRECTION. Next, two binary images are generated, the text and the correction image. The first consists of all TEXT and the second of all CORRECTION pixels.

An example of the text and correction image generated from Fig. 2 is shown in Fig. 7 and 8, respectively. From Fig. 8, we notice that the correction color remains invisible on text. Therefore, any text pixel that is marked by the correction pen is not properly classified as CORRECTION, but as TEXT. Consequently, large gaps will occur within position markers in the correction image. An example is shown in Figs. 9, 10 and 11. In order to fill these gaps, any background pixel in the correction image will be changed into a correction pixel if there is a correction pixel p_1 to its left and another correction pixel p_2 to its right such that all pixels between p_1 and p_2 are text pixels. This rule will close gaps in the horizontal direction. Similar rules are used in order to close vertical and diagonal gaps. The effect of the gap filling procedure is shown in Fig. 12. The correction image of Fig. 8 after gap filling is shown in Fig. 13.

4. Interpretation of Correction Symbols

After the processing steps described in Sec. 3 have been carried out, the correction marks and the original text have been separated from each other. All correction marks are contained in the binary correction image. In the present section it will be described how the meaning of these correction marks is determined.

First, the correction image is thinned using Hilditch's algorithm [9]. Each connected component in the thinned image is transformed into a graph structure. In this graph structure, the nodes correspond to end points, junctions, or crossing points of lines. These points are defined by having one, or more than two neighbors, respectively. The edges of the graph consist of the thinned lines in the correction image, i.e. of chains of pixels with two neighbors each. Edges are approximated by sequences of straight line segments. These sequences are stored as edge attributes. The bounding box of each connected component of the thinned correction image, i.e. graph, is determined and each node gets two attributes assigned. These attributes are the degree of a node (i.e. the number of incident edges), and its relative position inside the bounding box. Let $x_{min}(g)$, $x_{max}(g)$, $y_{min}(g)$, and $y_{max}(g)$ denote the extremal x- and y-coordinates defining the bounding box of a graph g. Then the

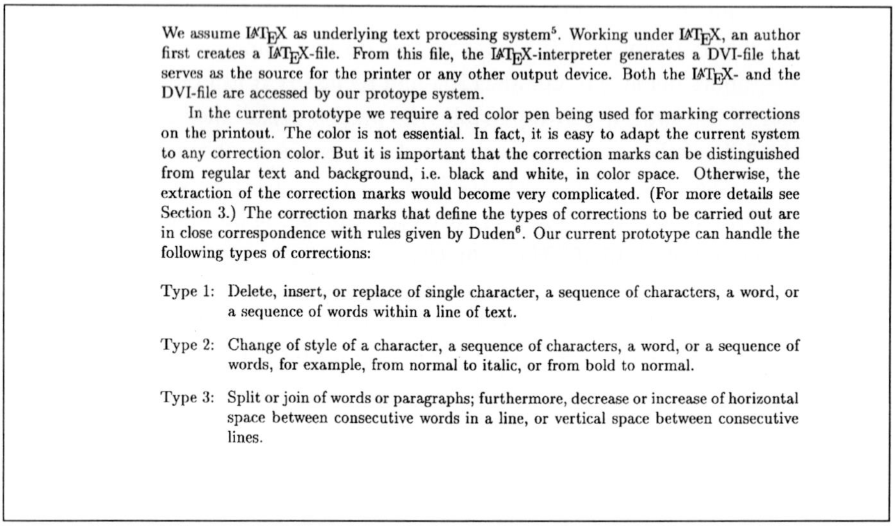

We assume LaTeX as underlying text processing system[5]. Working under LaTeX, an author first creates a LaTeX-file. From this file, the LaTeX-interpreter generates a DVI-file that serves as the source for the printer or any other output device. Both the LaTeX- and the DVI-file are accessed by our protoype system.

In the current prototype we require a red color pen being used for marking corrections on the printout. The color is not essential. In fact, it is easy to adapt the current system to any correction color. But it is important that the correction marks can be distinguished from regular text and background, i.e. black and white, in color space. Otherwise, the extraction of the correction marks would become very complicated. (For more details see Section 3.) The correction marks that define the types of corrections to be carried out are in close correspondence with rules given by Duden[6]. Our current prototype can handle the following types of corrections:

Type 1: Delete, insert, or replace of single character, a sequence of characters, a word, or a sequence of words within a line of text.

Type 2: Change of style of a character, a sequence of characters, a word, or a sequence of words, for example, from normal to italic, or from bold to normal.

Type 3: Split or join of words or paragraphs; furthermore, decrease or increase of horizontal space between consecutive words in a line, or vertical space between consecutive lines.

Fig. 7. Text image.

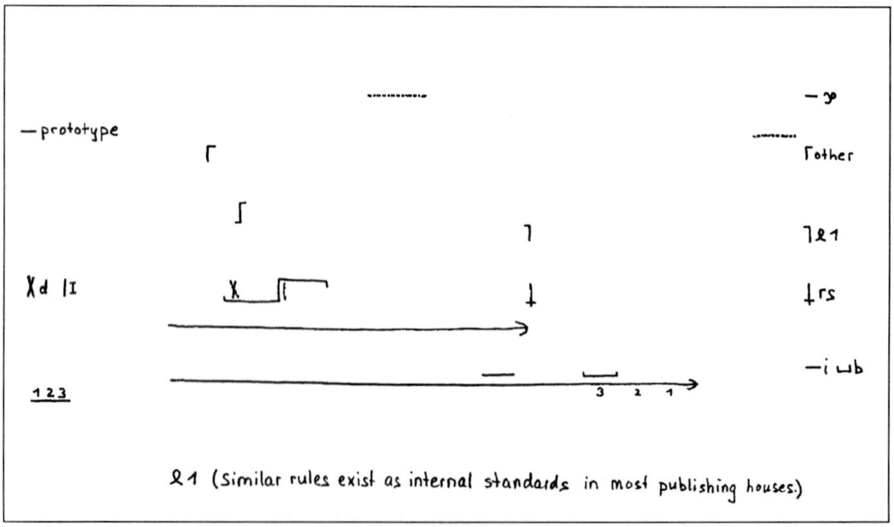

Fig. 8. Correction image.

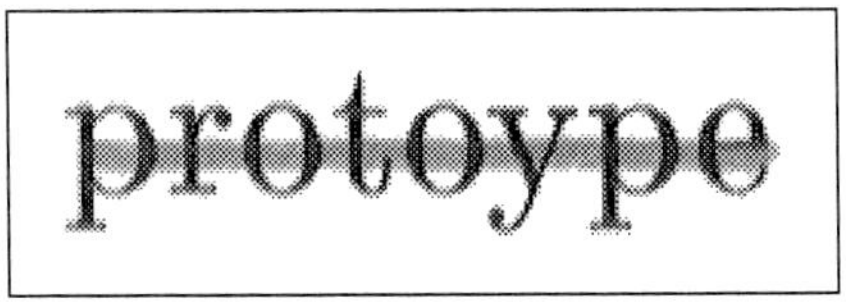

Fig. 9. Part of the original image shown in Fig. 2.

Fig. 10. The part of the correction image that corresponds to Fig. 9.

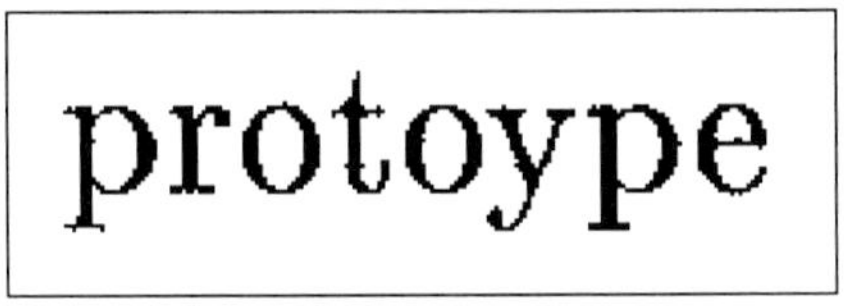

Fig. 11. The part of the text image that corresponds to Fig. 9.

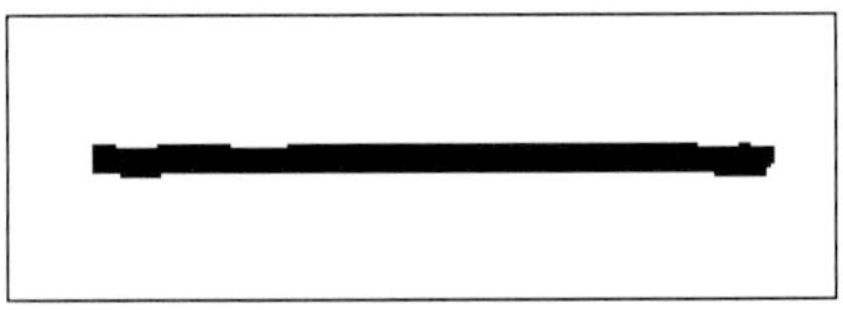

Fig. 12. The correction image of Fig. 10 after gap filling.

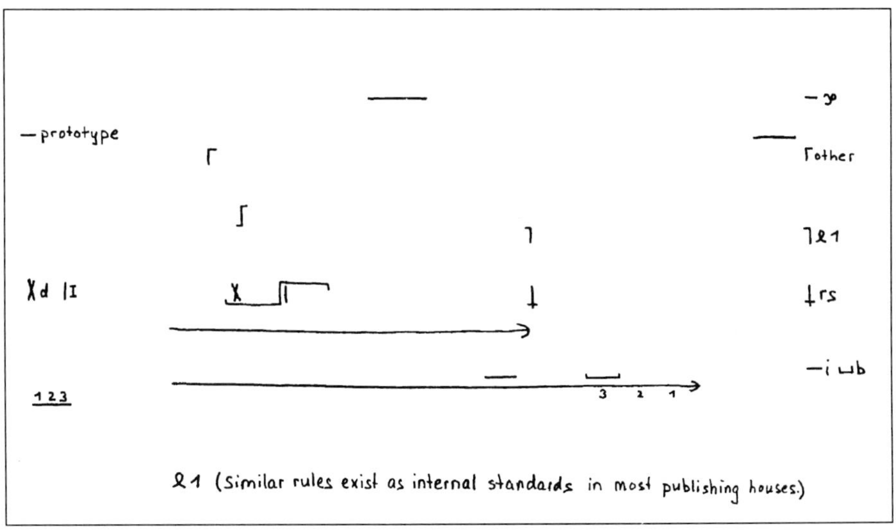

Fig. 13. The correction image of Fig. 8 after gap filling.

relative position coordinates of a node n of g with absolute coordinates $(x(n), y(n))$ are given by

$$x_{rel}(n) = \frac{x(n) - x_{min}(g)}{x_{max}(g) - x_{min}(g)}$$

$$y_{rel}(n) = \frac{y(n) - y_{min}(g)}{y_{max}(g) - y_{min}(g)}$$

Notice that after a graph has been generated, it is not yet known if it represents a position marker or some alphanumeric handprinted character that serves as a specification to some position marker. That is, any graph structure generated by the procedure described until now represents either

- a position marker of type 1, 2, 3 or 4 correction, or
- an alphanumeric handprinted character following some position marker as a specification.

Any position marker of a type 1, 2 or 4 correction must appear both inside the text area and on the left or right margin, while type 3 correction position markers appear only inside the text area.

The crucial observation for inferring the meaning of a correction symbol is that a position marker of a type 1, 2 or 4 correction inside the text area doesn't have meaning in itself. Rather, its meaning is defined by the handprinted text string that follows the repeated instance of this position marker on the left or right margin. Therefore, in the interpretation procedure we first look at the margins for horizontal strokes of type 4 corrections (transpose). If such a horizontal stroke is found, we have to detect the position markers directly above it and check for corresponding position markers in the text area at or below the actual line of text. After having handled all transpose corrections, we consider one position marker after the other in a line of text and try to find a corresponding position marker of identical shape on either the left or right margin. The regions corresponding to the text, the left, and the right margin, respectively, can be easily found by generating and analyzing horizontal and vertical projection profiles in the text image. The text lines and margins extracted from Fig. 2 are shown in Fig. 14. Clearly, the method only works if the projections of different text lines don't overlap, i.e., the skew of the document must not be large. But a skew correction algorithm could be easily integrated in the current prototype, if necessary. (For more details about skew correction, see the chapter by T.M. Ha and H. Bunke in this book.)

Once all position markers on a line of text have been matched with position markers on the margin at or above the actual line, the remaining connected components, i.e. graphs, on the margins are considered. They correspond to alphanumeric characters that define the meaning of the intended corrections. As an example, look at the first correction in Fig. 2. There is one horizontal stroke, i.e. a position

Fig. 14. Extracted text lines and margins

marker, inside the text area. It has a matching instance with an identical shape on the right margin. The remaining connected component on the right margin is the delete symbol. Thus, the position marker inside the text area is identified as representing the deletion of a string. In the second correction, the word "system" has to be replaced by "prototype". This can be inferred from matching the position marker in the text area with the corresponding position marker on the left margin. As there is no other position marker in the same line of text, the components on the margin different from the matched position marker are identified as text that replaces the text that is marked by the position marker in the text area.

The procedure that matches a position marker inside the text area with a position marker of identical shape on the right margin is guided by structural features. Let g_t and g_m denote two graphs representing a position marker inside the text area, and a position marker or an alphanumeric symbol on the margin, respectively. In the matching procedure it is checked if for each node n of g_t there exists a node n' of g_m with a relative position that differs in x- (y-)direction at most by a certain percentage ε of the width (height) of the circumscribed rectangle from the relative position of n. Furthermore, g_t and g_m must be isomorphic to each other, i.e., they must have the same topological structure.[a] For a graphical illustration see Fig. 15. In order to avoid mismatches, we start with a small value of ε. If no match for g_t is found, ε is gradually increased up to a maximum value. If still no match is found, the matching procedure allows some structural errors, like additional nodes and edges in one of the two graphs. The errors that are admissible have been heuristically derived from a sample set. In Fig. 15, for example, we need to match a single edge e to a node n with two incident edges e_1 and e_2. Such a match is admissible

[a]For large graphs, the isomorphism test can be computationally quite costly. In the present application, however, the graphs are rather small.

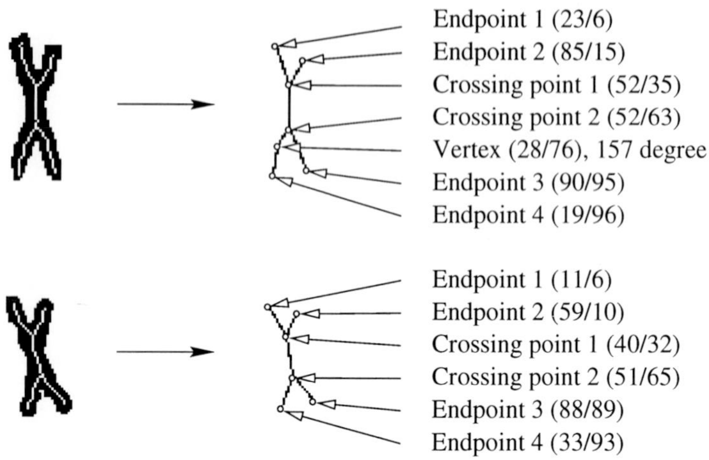

Fig. 15. Graphical illustration of the matching procedure. The numbers in brackets represent the relative x- and y-coordinates of a node, where the origin of the coordinate system is in the left upper corner of the bounding box of the correction symbol.

if the angle between e_1 and e_2 is close to 180°.

Type 3 corrections have to be handled in a different way, as they occur only within the text area and are not repeated on the margin. These symbols are directly classified by means of structural features. For example, in order to interpret a position marker inside the text area as the correction instruction "new-paragraph" (see Fig. 5), its graph must satisfy the following conditions:

- It consists of two nodes, both of degree 1; one of the nodes is at the left lower and the other at the right upper corner of the circumscribed rectangle.
- It contains exactly one edge connecting the two nodes; this edge consists of three dominant straight line segments, two being approximately horizontal — those incident to the nodes —, and one being approximately vertical.

The rules for the other position markers of type 3 corrections are similar.

In summary, in order to infer the meaning of the correction marks, we first look for all type 4 corrections and determine the corresponding position markers on the margin and in the text area. All parts of a type 4 correction are excluded from further processing. Then we try to interpret each position marker in the text area as representing a correction of type 3. Any position marker that doesn't match a type 3 correction symbol is interpreted as representing a type 1 or 2 correction, and a corresponding position marker of identical shape on the right margin is looked for. All remaining connected components on the margins are interpreted as alphanumeric characters that serve as specifications to type 1 or 2 corrections.

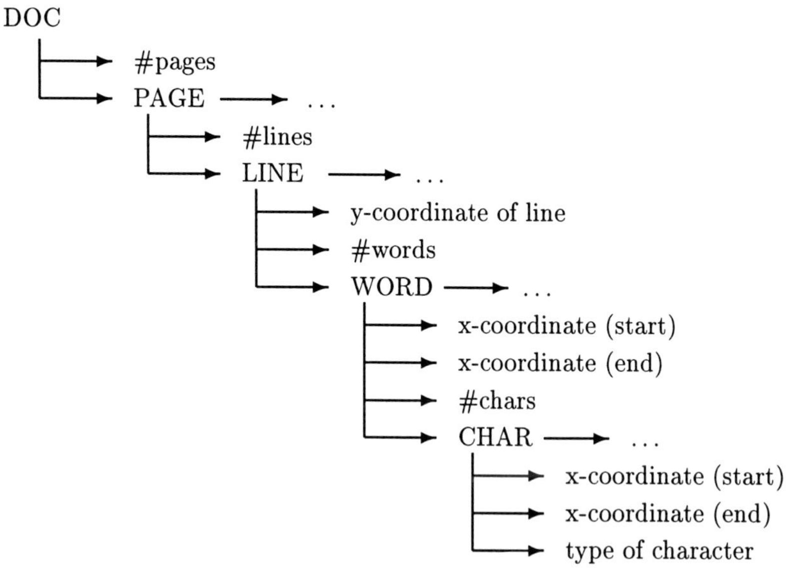

Fig. 16. Hierarchical data structure (HDS) inferred from the DVI file

5. Execution of Edit Commands

The procedures described in Sec. 4 generate the complete list of corrections for the scanned document. For each individual correction, the following information is available after the procedures have been run:

- location of the bounding box of the position marker in the text area
- type of correction; in case of an insert or replace operation, the new text string is known.[b]

In order to determine the edit operations that actually have to be carried out, the characters affected by the corrections and their positions in the LaTeXfile must be found. First we have to determine the text. For this reason we use the DVI file that is generated by the LaTeXinterpreter [10]. The DVI file contains specifications about the precise position, size, and font of every character in the document. By means of a special procedure, we convert the DVI file into a hierarchical data structure (HDS). This data structure is shown in Fig. 16. It contains all information from the DVI file that are required for our application in an easily accessible way.

By definition of the DVI file format, the standard origin of any document printed from a DVI file is always one inch to the right and one inch below the upper left corner of the page. Therefore, if we know the resolution of the scanner used for image acquisition, we can apply a linear transformation in order to map the image coordinate system to the DVI file, i.e. HDS, coordinate system. However, this is not sufficient in order to find a character in the HDS, given its location in the image,

[b]The characters in this text string are interactively classified, as mentioned before.

because it is admissible to scan only part of a page. Therefore, in order to find the definite mapping from the image to the HDS coordinate system, the actual offsets in x- and y-directions have to be determined. Given these offsets, for any location (x, y) in the image coordinate system the corresponding location (x', y') in the HDS coordinate system can be computed.

To determine the offsets in the x- and y-directions, we first calculate the bounding box of each word in the first line of text in the text image. This computation is based on the horizontal and vertical projection profiles and a threshold which is used in order to decide whether two consecutive characters belong to the same or to different words. The threshold is calculated as a predefined percentage of the height of the actual line of text. Having determined the bounding boxes from the text image, we scan the HDS line by line and check if the sequence of bounding boxes in a line matches that in the text image. If a match is found, the offsets in the x- and y-directions can be determined. If more than one matching lines in the HDS are found, we additionally calculate the bounding boxes of the words in the second line of the text image and check the HDS for a pair of matching lines. If there are multiple pairs of matching lines, we include more lines from the text image until the match is unique.

Note that the matching procedure described in the last paragraph allows us to cope very easily with documents consisting of more than one page. Clearly, given the bounding boxes of the words in the first line of the scanned document, we check the first page of the document in the HDS for matching lines first. If no match is found, we continue with the second page, a.s.o. Consequently, it doesn't matter which page of the document we are actually dealing with.

The procedure described until now allows us to transform coordinates of the scanned image, i.e. text and correction image, into coordinates in the HDS. Now given the location of a position marker in the correction image in terms of its bounding box, we first calculate the corresponding coordinates of the bounding box in the HDS. Assume that the actual correction is the replacement of an "old" character x by a string $y_1 \dots y_n$ of "new" characters. By intersecting the (transformed) bounding box of the position marker with the bounding boxes of the characters in the HDS, we can determine the occurrence of the character in the HDS that is actually affected by the intended correction. As character x probably has many occurrences in the LaTeXfile we need to determine part of its context in order to uniquely locate it in the LaTeXfile. Determining part of the context can be easily done by retrieving the next k characters $l_k, l_{k-1}, \dots, l_1$ to its left, and the next k characters $r_1, r_2, \dots, r_k$ to its right. Then, given the string $u = l_k l_{k-1} \dots l_1 x r_1 r_2 \dots r_k$ of length $2k+1$, we scan the LaTeXfile and retrieve all occurrences of u. If there are multiple occurrences of u, we have to go back to the HDS and enlarge the context. That is, we determine the $k+s$ next characters to the left and right of x, respectively, and retrieve all occurrences of $u' = l_{k+s} l_{k+s-1} \dots l_1 x r_1 r_2 \dots r_{k+s}$. Extending the context is repeated until a unique occurrence of the search string in the LaTeXfile has been found. Once such a unique occurrence has been found, we replace the "old" character x in the LaTeXfile by the "new" string $y_1 y_2 \ldots y_n$.

In the last paragraph, the replacement of a single character by a text string was described. All other correction instructions (delete, insert, and type 2, 3 and 4 corrections) can be handled in a similar way. It must be remarked that the search for a matching string in the LaTeXfile is not trivial. The searched string may contain LaTeXcommands which have to be taken into account. For example, if we are looking for the string "Möri", the string to be matched may be `M\"ori` or `M\"{o}ri`. The editing operation is also dependent on the actual LaTeXenvironment (and therefore on LaTeXcommands at previous positions in the LaTeXfile). For example, if "missing" is to be inserted as text of normal font this often results in a trivial insertion. But it might be that the location of the insertion is inside a previously opened math environment. Therefore it must be protected against being interpreted as mathematical text, e.g. the text to insert is `\mbox{`missing`}`.

Finally, we want to mention that the method that was described in this section for locating characters or words in the LaTeXfile is just one possibility to solve the problem. In an earlier version of the system, we didn't make use of the DVI file. Instead, we applied a (commercial) OCR software package for automatic machine printed character recognition [11] and determined the identity and context of the affected characters as well as the offsets in x- and y-directions directly from the text image. But even with the most sophisticated OCR technology, a recognition rate of 100% cannot be expected [12]. Therefore, we conclude that it is more reliable and elegant to infer the context from the DVI file, and to use the text image only for extracting shape information, i.e. bounding boxes of words, but not character classes.

6. Experimental Results and Examples

All procedures described in this paper have been implemented in C. They run on Sun SPARC workstations. The current version of the system comprises about 30000 lines of code, including a comfortable graphic interface and several tools for system development. As mentioned earlier, no OCR routines have been integrated yet. All alphanumeric handprinted characters are automatically located, based on position and context information (see Sec. 4). Then they are interactively classified.

The system has been tested on many documents with hundreds of correction symbols. From these tests it can be concluded that the system performs 100% correctly if the position markers are carefully drawn. Correct performance means that the system did exactly the type of corrections that were desired.

The execution time for the extraction of the correction marks is proportional to the size of the given image. In the following, we give the time that was needed for processing the example shown in Fig. 2 on a Sun SPARC 10 workstation. The resolution of the image is about 2500 × 1400 pixels. Extraction of the correction marks, i.e. the generation of the correction and the text image, took 6.5s. Generally, thinning depends on the number of correction marks and the width of the strokes of the correction pen. In our example, the average width was 6 pixels. This led to

2.4s needed for the thinning step. The time required for graph extraction depends on the number of correction marks. In our example, 0.2s were needed. Because the graphs that are to be matched are small, the complete matching and interpretation step was done in 0.01s. Finally, editing took 2.4s. The time needed by the editing procedure depends on the size of the text file and the number of corrections. In total, about 11.5s were needed for completely processing the example shown in Fig. 2. Our programs are not optimized with respect to execution speed. Moreover, on a multiprocessor platform, the computation time could be greatly reduced because most of the time is spent on operations on the image array, which could be easily distributed to several processors.

For the purpose of illustration, some more examples are shown in Figs. 17, 18.

7. Conclusions

The generation of a text document under a text processing system can be quite time consuming. Particularly if a non-WYSIWYG system is used, several iterations of preparing a draft and correcting it may be necessary. In the present paper, a prototypical system for the off-line recognition and execution of correction instructions on text documents is described. The underlying correction syntax is intuitively straightforward and easy to learn. Potential applications of a system like the one described in this paper are in the personal domain, or in a publishing house where galley proofs are sent to authors and existing text files have to be updated according to the authors' corrections.

The prototype described in this paper has been tested on many documents and shown very good performance. Nevertheless, to develop it into a product that meets commercial standards, a number of improvements are required. First of all, we assume a "cooperative" corrector in all our experiments. This means that all correction marks on the document must be rather carefully drawn. It is not unrealistic to assume a cooperative user in a real application scenario, but it would be necessary to test the system with a larger number of correctors, in order to get a more representative assessment of its robustness. Moreover, some additional classes of corrections should be incorporated. Examples are cut and paste of (long) passages of text on the same page or across different pages. Also, the system should be extended to handle documents with more than one column. Another useful extension of the current prototype would be a front end to text processing systems other than LaTeX. If the system is to be adapted to another text processsing system, at least three major changes are required. First, the parsing of the original text file has to consider all formatting commands and all special features of the underlying text processing system. Second, the generated editing commands must be properly adapted. A further change is required because of the missing DVI file. Therefore, the creation of the HDS must be based on the input file for the printer, for example, on the corresponding Postscript file. In any case, there will exist such a file submitted to the printer which contains not only the pure text but also the complete layout information for each page of the document.

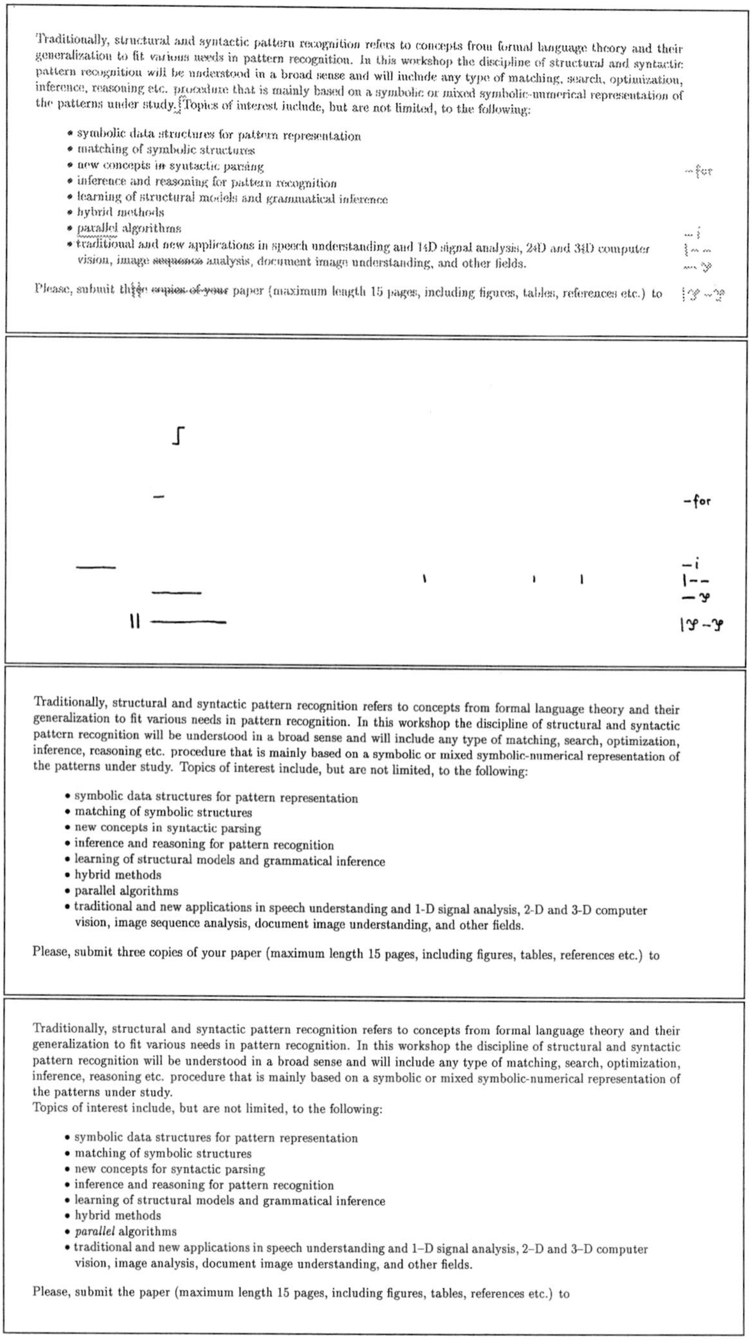

Fig. 17. Another example (from top to bottom): image with correction marks; correction image; text image; final result.

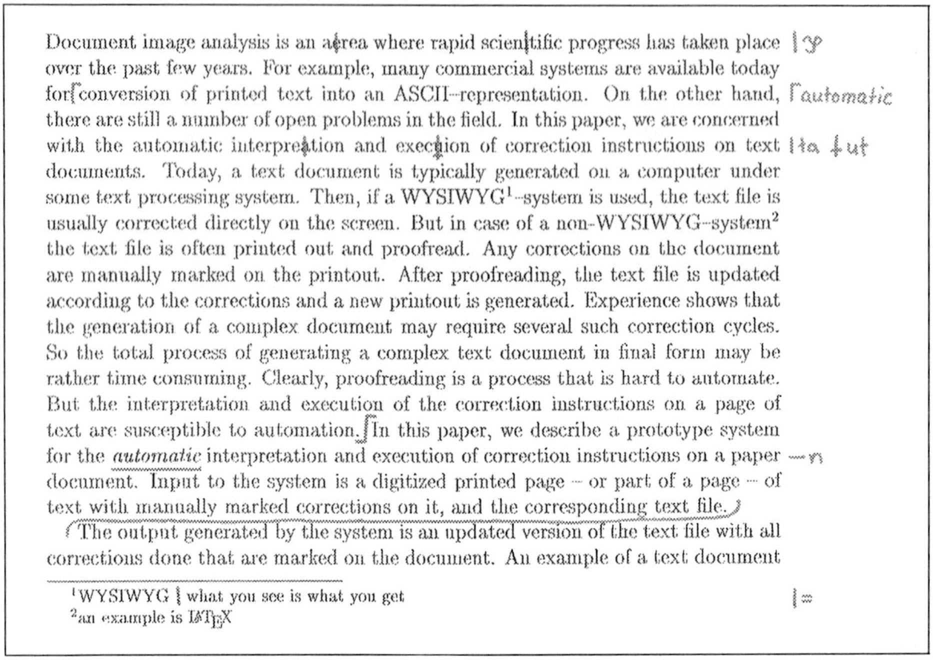

Document image analysis is an aerea where rapid scien tific progress has taken place over the past few years. For example, many commercial systems are available today for conversion of printed text into an ASCII–representation. On the other hand, there are still a number of open problems in the field. In this paper, we are concerned with the automatic interpretation and execution of correction instructions on text documents. Today, a text document is typically generated on a computer under some text processing system. Then, if a WYSIWYG[1]–system is used, the text file is usually corrected directly on the screen. But in case of a non-WYSIWYG–system[2] the text file is often printed out and proofread. Any corrections on the document are manually marked on the printout. After proofreading, the text file is updated according to the corrections and a new printout is generated. Experience shows that the generation of a complex document may require several such correction cycles. So the total process of generating a complex text document in final form may be rather time consuming. Clearly, proofreading is a process that is hard to automate. But the interpretation and execution of the correction instructions on a page of text are susceptible to automation. In this paper, we describe a prototype system for the *automatic* interpretation and execution of correction instructions on a paper document. Input to the system is a digitized printed page – or part of a page – of text with manually marked corrections on it, and the corresponding text file.
The output generated by the system is an updated version of the text file with all corrections done that are marked on the document. An example of a text document

[1]WYSIWYG what you see is what you get
[2]an example is LaTeX

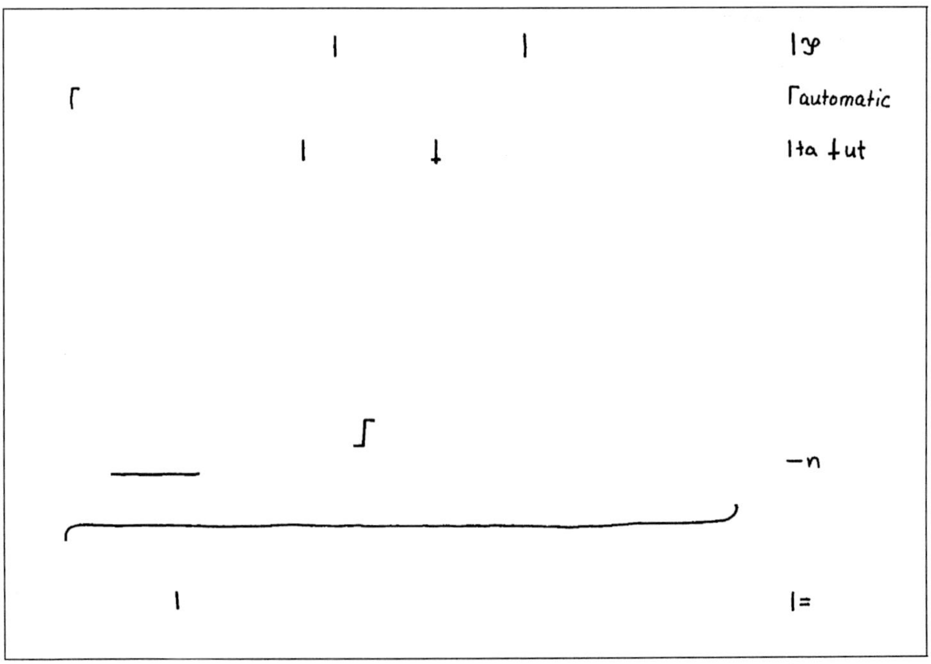

Fig. 18. Another example (from top to bottom): image with correction marks; correction image; text image; final result.

Document image analysis is an aecrea where rapid scienitific progress has taken place over the past few years. For example, many commercial systems are available today for conversion of printed text into an ASCII–representation. On the other hand, there are still a number of open problems in the field. In this paper, we are concerned with the automatic interpreation and execuion of correction instructions on text documents. Today, a text document is typically generated on a computer under some text processing system. Then, if a WYSIWYG[1]–system is used, the text file is usually corrected directly on the screen. But in case of a non-WYSIWYG–system[2] the text file is often printed out and proofread. Any corrections on the document are manually marked on the printout. After proofreading, the text file is updated according to the corrections and a new printout is generated. Experience shows that the generation of a complex document may require several such correction cycles. So the total process of generating a complex text document in final form may be rather time consuming. Clearly, proofreading is a process that is hard to automate. But the interpretation and execution of the correction instructions on a page of text are susceptible to automation. In this paper, we describe a prototype system for the *automatic* interpretation and execution of correction instructions on a paper document. Input to the system is a digitized printed page – or part of a page – of text with manually marked corrections on it, and the corresponding text file.

The output generated by the system is an updated version of the text file with all corrections done that are marked on the document. An example of a text document

[1]WYSIWYG : what you see is what you get
[2]an example is LaTeX

Document image analysis is an area where rapid scientific progress has taken place over the past few years. For example, many commercial systems are available today for automatic conversion of printed text into an ASCII–representation. On the other hand, there are still a number of open problems in the field. In this paper, we are concerned with the automatic interpretation and execution of correction instructions on text documents. Today, a text document is typically generated on a computer under some text processing system. Then, if a WYSIWYG[1]–system is used, the text file is usually corrected directly on the screen. But in case of a non-WYSIWYG–system[2] the text file is often printed out and proofread. Any corrections on the document are manually marked on the printout. After proofreading, the text file is updated according to the corrections and a new printout is generated. Experience shows that the generation of a complex document may require several such correction cycles. So the total process of generating a complex text document in final form may be rather time consuming. Clearly, proofreading is a process that is hard to automate. But the interpretation and execution of the correction instructions on a page of text are susceptible to automation.

In this paper, we describe a prototype system for the automatic interpretation and execution of correction instructions on a paper document. Input to the system is a digitized printed page – or part of a page – of text with manually marked corrections on it, and the corresponding text file. The output generated by the

[1]WYSIWYG = what you see is what you get
[2]an example is LaTeX

Fig. 18. (continued.)

Despite some limitations, we believe that the current prototype is a very promising step toward document image analysis systems that support users in the creation of text documents. It is evident that the system described in this chapter is useful for people correcting documents in an off-line manner. But also the on-line correction mode may be desirable in certain situations. Therefore, we are developing an on-line graphical user interface with the possibility to preview text documents. With this system, the user may indicate the desired changes directly by choosing the type of correction and position by means of a cursor, a mouse, or a light pen, and typing in the correction text via the keyboard. Under this procedure, the extraction and interpretation of correction symbols is no longer needed, while the editing phase will remain the same. Integrating this on-line facility into our system will increase its flexibility and usability, and we expect it to become more attractive for daily routine use.

References

[1] L. Lamport, LaTeX, *A Document Preparation System,* fifth printing (Addison-Wesley, 1986).

[2] C.A. Higgins and R.J. Duckworth, The PAD (Pen and Display) – A Demonstrator for the Electronic Paper Project, in *Computer Processing of Handwriting,* eds. R. Plamondon and C.G. Leedham (World Scientific, 1990) 111–131.

[3] L.K. Welbourn and R.J. Whitrow, A Gesture Based Text and Diagram Editor, in *Computer Processing of Handwriting,* eds. R. Plamondon and C.G. Leedham (World Scientific, 1990) 221–234.

[4] R. Zhao, *Handsketch–Based Diagram Editing* (B.G. Teubner Verlagsgesellschaft, Stuttgart, 1993).

[5] D. Möri and H. Bunke, Off–line interpretation and execution of corrections on text documents, in *Preproceedings of the Workshop on Document Analysis Systems,* eds. L. Spitz and A. Dengel, Kaiserslautern, 1994, 401–416.

[6] Duden, Band 1: Die deutsche Rechtschreibung, Meyers Lexikonverlag, Mannheim, 1991, 77–80.

[7] A.K. Jain, *Fundamentals of Digital Image Processing* (Prentice Hall,1989).

[8] W.K. Pratt, *Digital Image Processing,* second edition (John-Wiley, New York, 1991).

[9] C.J. Hilditch, Linear skeleton from square cupboards, in *Machine Intelligence* vol. 6 (Edinburgh Univ. Press, 1969), 403–420.

[10] D.E. Knuth, *TeX: The Program, volume B of Computer and Typesetting,* (Addison Wesley, 1986).

[11] ScanWorX, Xerox Imaging Systems Inc., API Ver. 1.0, 1992.

[12] S.V. Rice, J. Kanai, and T.A. Nartker, *The Third Annual Test of OCR Accuracy, Annual Report,* Information Science Research Institute, University of Nevada, Las Vegas, 1994, 11–38.

Handbook of Character Recognition and Document Image Analysis, pp. 703–728
Eds. H. Bunke and P. S. P. Wang

CHAPTER 27

AUTOMATIC READING OF BRAILLE DOCUMENTS

APOSTOLOS ANTONACOPOULOS*
Systems Engineering Group, Department of Computation,
University of Manchester Institute of Science and Technology (UMIST),
P.O. Box 88, Manchester, M60 1QD, United Kingdom

Braille is the most widely used system for written communication using tactile means. For visually handicapped people, Braille documents have proved to be an invaluable source of information, essential for education and for leisure. Although more Braille documents are produced today, there are still very few copies of the small number of books, for instance, available. Automation of the Braille reading process will effectively enable the transcription and duplication of existing documents as well as their preservation. By converting a Braille document into electronic form, all the benefits of developments for the electronic form of printed documents can be enjoyed: easier transmission, efficient storage and manipulation etc. The possibility of sighted people also using an automatic Braille reading system to communicate in writing with blind people is a further motivation. In this chapter, the aspects of the automatic reading problem are analysed and solutions to each of them are examined. Firstly, the characteristics and production of Braille and Braille documents are described and the issues involved in the problem are discussed. Secondly, each of the stages towards the automatic reading of Braille documents is analysed and the approaches that have been so far designed to carry it out are critically presented. Finally, the most important factors affecting the success of the whole process are discussed.

Keywords: Braille; Braille documents; Braille document image acquisition; Braille character segmentation; Braille character recognition; Verification of recognition results; Automatic reading system; Computing for visually handicapped people.

1. Introduction

Visually impaired people, just as many other people with disabilities, experience difficulties in integrating in society. One of the main reasons is the inaccessibility of information necessary for education as well as leisurely and cultural interests. Although some information has been available on audio, the advantages of reading and writing as methods for communication have made necessary the practice of encoding information in symbolic form. The best known writing convention is Braille, according to which,

*Present address: Department of Computer Science, University of Liverpool, P.O. Box 147, Liverpool, L69 3BX, United Kingdom, e-mail: aa@csc.liv.ac.uk.

symbols are represented by arrangements of dots recognisable by touch. Since its introduction by Louis Braille in 1829, it has been proven to be an efficient means of written communication for blind people, and hence, has been widely adopted. Braille has been further developed over the years to express different types of information and to render a more efficient way of representing text.

A significant amount of development has taken place for the production of Braille and Braille media. There now exist interfaces for computers, such as a tactile display for Braille representation of the computer screen data and a Braille mouse, which are dynamic. However, at present, Braille documents dominate the scene of written information dissemination in the majority of application areas. Considerable effort has been devoted over the years for the transcription of printed material into Braille. In fairly recent years the use of computers has considerably facilitated this. A document can be scanned and an optical character recognition (OCR) process can be applied to the image. The resulting ASCII character codes can then be translated into Braille representation and either embossing plates can be made for volume production or a hard copy can be obtained by a Braille printer.

Although the production of Braille documents is relatively easy now, there is a missing link in the written communication process: Braille documents cannot be readily encoded into electronic form, e.g. ASCII. The process of identifying and recognising Braille characters is to some extent of a different nature than that of reading printed characters yet there are many interesting parallels that can be drawn.

This chapter explores the problem of automatic reading of Braille documents and the issues related to document analysis and OCR. Firstly, the rationale behind reading Braille documents automatically is examined in Sec. 2. Secondly, the nature of Braille and the characteristics of the reading problem are discussed in Secs. 3 and 4, respectively. Each of the stages involved in the reading process are considered separately in Sec. 5 and existing approaches designed to deal with them are described along with possibilities for further work. Finally, before concluding, the most important factors that affect the success of the reading process are discussed.

2. The Need

As it was pointed out in the previous section, the automatic reading of Braille documents enables the bridging of the written communication gap. This gap is primarily created by the ability to convert printed material (readable by sighted people) to Braille and the difficulty, on the other hand, in converting Braille to a representation that can be read by sighted people. Nitta *et al.* [1] point out the need for a Braille reading system to assist blind people in learning Braille as the motivation for their research. Benjelloun *et al.* [2] also refer to the fact that such a system will enable sighted people to read Braille documents, thus enhancing the communication and collaboration possibilities with blind people. For example, blind people in a distance-learning programme can submit their work to their tutors, who will probably be sighted.

Blind people may even use a Braille reading system as an interface for the use of fax machines. The Braille document could be scanned at one end and at the other end the encoded document could either be translated to ASCII or printed with a Braille printer. A similar use can address another significant problem faced by blind people and libraries: copying of Braille documents. Through the years only a small proportion of printed books have been available in Braille. The fact that there are not many copies of each available title compounds the problem. Since the original Braille document most probably has not originated in electronic form or this encoding may not be available, a system that reads Braille documents and produces the encoding for further printing is a necessity. Without such a system the only option is manual transcription which is both costly and time consuming.

Furthermore, by converting them to their electronic representation, Braille documents can benefit from the progress made in the (electronic) handling of printed documents. Braille documents occupy considerably greater volume than printed ones (40 to 60 times according to Otake *et al.* [3]) for recording the same information. This is due to the thickness of the material (page), the height of the protruding dots and the low information recording density resulting from the relatively large spatial extent of Braille characters. Therefore, by converting Braille documents into electronic form, substantial benefits in storage and transmission can be gained. Apart from enabling wider dissemination of information, the usual advantages of indexing and manipulation, associated with document management systems, can also be enjoyed.

Finally, there is a significant need to preserve old Braille documents many of which are rare and in a delicate condition. Such documents may be converted to electronic form and reproduced accordingly.

It should be noted at this point that an automatic Braille reading system must be practical and cost-effective in order to be adopted by the people who need it. Ease of use is as important as efficiency in this case. As considerable developments have taken place in the fields of pattern recognition and document image analysis, it is natural to take advantage of them and attempt to employ them in the solution of the reading problem. Although an attempt was made to use a mechanical system [4], the automatic reading by optical means has been more favourable with the rest of the research community.

3. Braille

Since the seventeenth century there have been attempts by Francesco Lana-Tarzi to devise a tactile writing system. His ideas were further pursued by Charles Barbier and were implemented by Valentin Haüy in the Paris Institute for the Blind. Although that tactile writing system was not efficient due to the large number of sound combinations used, the general idea proved successful. One of the blind students, Louis Braille adapted and simplified the system of tactile dots and set the foundations of the Braille writing of today. The system consisted of tactile representations for each letter of the Latin alphabet along with numbers and punctuation marks.

During the years that passed since then, the system has been further developed to include more representations of printed symbols as well as rules to increase the efficiency of reading and writing. Different languages are supported as well as a variety of other types of information such as mathematical expressions, computer programs and music. In the following subsections, the nature of Braille characters and documents is described in order to develop an understanding of the problem of automatic reading of Braille.

3.1. *Braille Characters*

Printed text is represented by Braille characters. Each character is constructed as a set of six points arranged in two columns of three, as it can be seen in Fig. 1. This arrangement is called a Braille *cell*. Each point position is identified by a number and can be either raised (a protrusion) or flat. A raised point is also called a Braille *dot*.

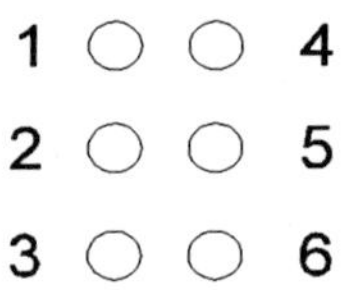

Fig. 1. The Braille cell.

The dimensions of a Braille dot have been set according to the tactile resolution of the fingertips of person. In theory, the dimensions of Braille dots must fall within certain bounds as indicated in Fig. 2. The horizontal and vertical distance between dots in a character, the distance between cells representing a word and the inter-line distance are also specified by the Library of Congress. In practice, although most instances of Braille are close to these standards, there are slight variations between Braille produced by different devices.

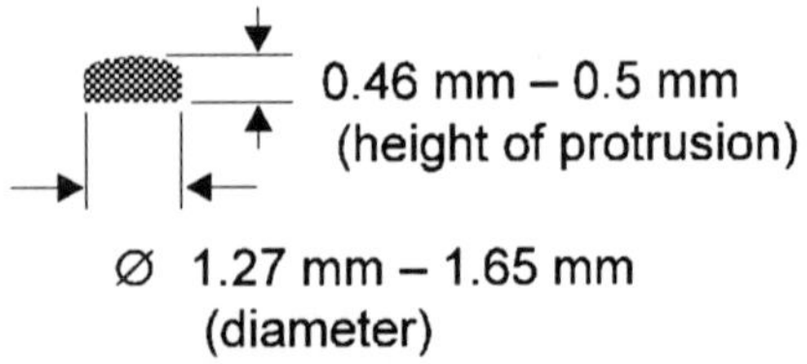

Fig. 2. Braille dot (profile) and its specified dimensions.

Each Braille character is a unique combination of dots within the cell. So far, only Braille cells with six points have been mentioned. There is, however, an eight-point version of Braille cells whose use is not very widespread compared to that of the six-point ones. The two extra points allow for the representation of a greater variety of printed symbols. It has mainly been used in computer programming where a variety of symbols is used and each must be represented unambiguously by a single Braille character (see Sec. 3.2.3.). The larger number of possible dot combinations has also been used for the representation of Japanese characters [5]. Since from the recognition point of view the principle is the same, in the rest of this chapter Braille characters will be assumed to be composed of six points, unless otherwise explicitly stated.

3.2. *Interpretation of Braille Characters*

A printed word is represented by a set of Braille characters forming a Braille word. The reason for distinguishing between the two kinds of word is that they often differ in the number of characters as it will be explained below. The characters inside a Braille word are separated by a distance greater than the horizontal one between points and less than that between words. The distance between words in a line of Braille characters consists of one or more space characters. The space character is a Braille cell with no dots (all points are flat).

Apart from the space character there are 63 other combinations of dots which can be used to represent printed symbols. According to the context and to the set of representation rules followed in a given Braille document, a particular combination of dots may correspond to more than one different printed characters (at different occasions). On the other hand, a printed symbol may require more than one Braille character for its representation. Furthermore, a Braille character may in some circumstances correspond to a concatenation of printed characters (see below). Different sets of rules exist for the production of Braille documents containing textual information. Further still, there are different representation rules for other types of documents, like those containing music or mathematical expressions. Some of the main sets of rules that are widely used are briefly described in the following.

3.2.1. Grade 1 Braille

In *grade 1* Braille, each printed letter corresponds to one Braille character as can be seen in Fig. 3. Punctuation marks are also represented by one Braille character with the exception of the opening quotes and the exclamation mark which are represented by the same Braille character (see Fig. 3). In this case, while reading, the interpretation is made according to the context. The combination of dots in each Braille character is designed in such a way that symbols are distinguished from certain others by a translation of the same dot arrangement (e.g. the letter "d" and the full stop "."). Furthermore, the symbols that are represented by only one dot are likely to be found next to others with dots in other positions which will help define the position for the reader.

It is not possible to represent the large number of existing symbols by a combination of dots in a single Braille character. Therefore, some symbols (e.g. the asterisk) are represented by two Braille characters. Furthermore, the digits "1", "2", ..., "9", "0" are represented by the Braille characters corresponding to the range of letters "a",...,"j"

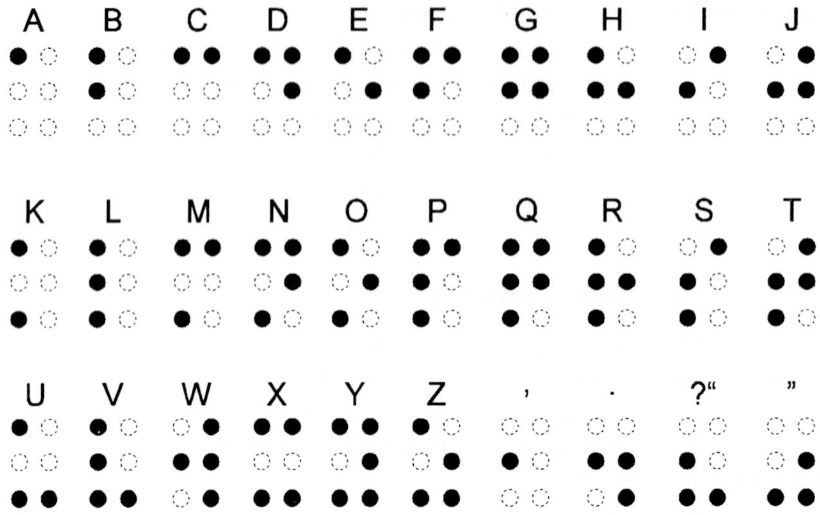

Fig. 3. Braille characters corresponding to letters of the Latin alphabet and some punctuation marks (black areas denote dots).

preceded by the *numeral* symbol as can be seen in Fig. 4. Still, grade 1 Braille is quite straightforward to interpret. The numeral symbol is an example of a *mode* symbol. Another mode symbol is that used to indicate the capitalisation of a letter (Braille character composed of dot six only). It is mainly used in North American Braille and precedes the symbol of the character to be capitalised. Similarly, mode symbols are used in many languages to denote that a particular accent should be placed on the character that follows.

3.2.2. Grade 2 Braille

The letter-by-letter representation of a word as in grade 1 Braille leads to the production of quite voluminous braille documents (a dictionary can occupy a bookcase). Furthermore, human readers have to spell each one of the characters of a word before they can recognise the meaning. This fact makes the reading process slower and obstructs fluent reading. Therefore, attempts have been made to represent frequently occurring letter strings by one or two Braille characters. These strings are usually words, word stems, prefixes, suffixes, syllables or just concatenations of letters such as "dd". The current standard which defines the representations of these contractions in Braille and the rules for using them is referred to as *grade 2* Braille (also called *contracted* or *literary* Braille).

The strings of characters that can be contracted have been selected according to the frequency that they occur and therefore, differ between languages. In English grade 2 Braille, for instance, there are over 200 contractions and short form words. The rules for replacing character strings by a Braille short form take into account the phonetic use of

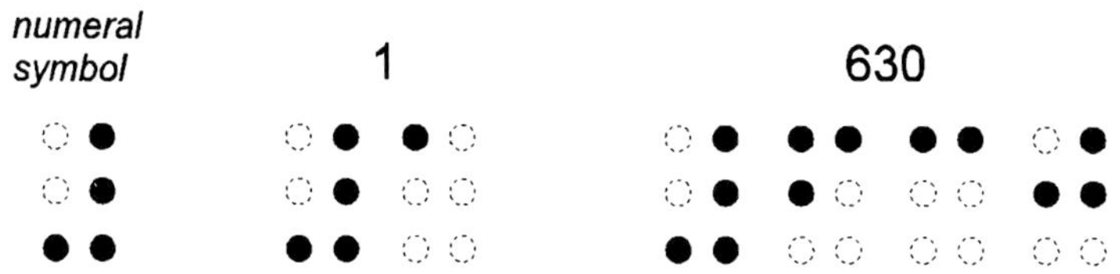

Fig. 4. Composition of numbers in Braille.

the language and are based on logical as well as linguistic and semantic rules. This fact makes the reading process more complicated and very context sensitive since the same Braille character may represent a particular contraction in some circumstances and a different one in another [6,7]. For example the Braille character representing the exclamation mark (which is the dot pattern of the letter "f" shifted downwards) takes the following meanings: (i) "!" at the end of a word, (ii) "ff" in the middle of a word and (iii) the preposition "to" if it is on its own before a word.

3.2.3. *Braille for specialised applications*

There is a number of other Braille writing conventions for the representation of symbols in more specialised applications. Computer Braille, for instance, is a one-to-one correspondence of Braille characters to ASCII ones [8] thus removing any ambiguity in the representation of computer programs. For example the digits "0",...,"9" are represented by single Braille characters in contrast with grade 1 or 2 Braille. However there are differences between the North American (six-point) and the British Braille computer codes. As it was previously mentioned, Braille cells with eight points are also used sometimes for computer access purposes. But, although this enables the representation of more symbols, there is no generally agreed standard.

Mathematical expressions are represented in Braille according to special rules which specify the encoding of the necessary symbols. Another type of information that can be expressed in Braille is music. In Braille music scores the notes are represented by different letters. The length of notes is determined by a preceding Braille character (composed of either dot 3 or dot 6). Other music symbols are also represented. Finally, a person who writes Braille (for personal use) may sometimes do so using his/her own contractions and rules.

3.3. *Braille Documents*

Most Braille documents are formed by embossing the dots on the back side of the medium sheet so that they can be read from the facing side. Most often the medium is thick paper in order to withstand the embossing. Another, less common, medium is a transparent plastic sheet (similar to an overhead projector transparency). In addition to the embossing production technique, a process of forming dots on paper from drops of hot-melt material has also been reported [3]. This chapter, however, will concentrate on the embossed-paper Braille documents because they essentially constitute the body of the information available in Braille form.

One way of embossing the sheet of paper is by hand using a Brailler, which resembles a typewriter with a key corresponding to each of the six points (plus a space key). To produce a dot, the user applies force to the appropriate key which embosses the paper. For mass production, plates are formed which are then pressed against the paper sheets to form pages of Braille. Furthermore, as it has already been mentioned, there are Braille printers, connected as peripherals to computers, which emboss the dots on paper according to the information sent in electronic form.

The size of the page varies between documents produced by different means. In addition, the colour of the paper varies, as it does not play any role in conveying the written information. The majority of Braille documents, however, are either buff-coloured or white. In addition, their surface usually has a low gloss finish.

The Braille characters are embossed in lines from the top of the document to the bottom much like printed documents. Braille documents can have the dots embossed on one side or on both. As the volume of Braille documents is quite large, it is advantageous to use both sides of the sheet. This is achieved by carefully embossing the dots of one side between the dots embossed on the other, as can be seen in Fig. 5. Because the points of Braille cells in one side are interleaved with the points of the cells of the other, these double-sided documents are also referred to as *inter-point*.

Fig. 5. Two cells on different sides of an inter-point Braille document.

4. Characteristics of the Automatic Reading Problem

In ordinary documents, the information is printed in a contrasting colour so that it is visually distinguishable from the background. As Braille documents are not designed to convey any visual information, their colour is uniform. Since the information is only intended to be read by tactile means, the first problem in automatic reading is to identify the tactile information visually. The issue in this case is the separation of the protrusions on the paper sheet from the flat background. The solution to this problem comes from the natural visual perception of three-dimensional information from a two-dimensional view. If the document is illuminated at an angle close to the horizontal then the anomalies on the surface of the paper are lit differently than the flat areas. As it will be described later, the various approaches to Braille recognition have exploited this fact in different ways.

However, the identification of Braille dots from the differences in the intensity of reflected light is not straightforward. As the paper is only used for tactile purposes, the quality of its appearance is not important. Hence, there are sometimes visual imperfections in the form of small dark fibres usually associated with cardboard. These blemishes, together with dirty marks that may also be present due to repeated use, interfere with the light intensity patterns created by the protrusions and make the identification task more difficult.

Moreover, damaged documents with creases or punched holes compound the problem. In addition, in some old documents, a few of the dots may be less prominent due to heavy use. In this case, the effect produced by the lighting is reduced.

As many of the Braille documents produced are inter-point, another problem becomes evident. Inter-point documents have both protrusions and depressions (the dots of the other side) on their surface. This complicates the dot extraction stage because shadows and bright areas are produced from both. It should be noted that the majority of existing automatic Braille reading systems are only capable of reading single-sided documents.

Once the Braille dots have been located in the image, the next task is to identify the characters. The problem is to identify which dots belong to which Braille cell. In theory, Braille cells are aligned in the horizontal and vertical directions in the form of a grid. In practice there are noticeable deviations from this due to the manufacturing process or due to distortions introduced in the image of the document.

In general, Braille characters have a more regular structure than printed ones. Their size and style is fixed throughout a document, although there are certain variations between different documents. The spacing between Braille dots is quite regular and can be used to segment characters. Nevertheless, Braille characters are not single connected entities and do not have distinguishing structural differences between them (they are all combinations of dots). These facts do not give enough information for the recognition and verification of Braille characters in the sense of providing degrees of confidence and alternatives.

Further details on the above as well as additional issues will be raised in each of the relevant subsections describing a particular stage of the reading process.

5. Stages

For a better analysis of the different aspects of the automatic Braille reading problem and for ease of comparison between current solutions, the stages of the automatic Braille reading process are examined individually. The first stage to be considered is the acquisition of the image of the tactile surface of the document. This is followed by various enhancements performed on the image prior to the extraction of any information. In the third stage, the Braille dots are identified in the image. The two stages that follow are concerned with the segmentation and recognition of Braille characters. The possibilities for verification of the recognition results are examined next and, finally, the main uses of the encoded Braille characters are described.

Each of the different stages is the subject of one of the following subsections. The characteristics of the relevant issues are described and the various approaches devised so far to deal with them are outlined.

5.1. *Image Acquisition*

The first step towards obtaining an electronic representation of a Braille document is to capture its image. As it was mentioned above, the nature of this kind of document is quite different from that of a printed one. The relief information is present in the image in the form of the bright and the shadow areas created by the differences in the intensity of the light reflected from the paper surface to the camera. The protrusions create a light area next to a dark one (shadow) while the grey-level value of the background is somewhere in-between. In inter-point Braille documents, the depressions also give rise to a dark and a light area, but in reverse order. To make sure that all the relief information is present in the image, the acquisition must be done under controlled conditions.

First, the illumination angle must be oblique. To acquire the image, most of the existing approaches have used a camera placed above the centre of the document. For example, François and Calders [9] have used a vidicon camera to capture the image of a Braille document illuminated under a very oblique angle. Their goal is to identify the bright spots corresponding to the top of Braille dots. However, the Braille documents used are assumed to have no visual defects.

Visual defects can interfere with the representation of relief information in two ways. Firstly, if there are small areas of higher reflectivity due to the materials of the paper, unwanted bright spots will appear in the image. If the reflectivity of the paper varies along wider areas then the contrast between the grey levels of the dots and those of the background will be lower in some regions. Secondly, dirty marks or other dark particles in the paper give rise to small dark regions in the image. These could be mistaken for shadows produced by dots. This is especially the case when an approach attempts to identify dots based on the assumption that each one of them creates a shadow (e.g. Dubus *et al.* [**Error! Bookmark not defined.**]). To circumvent this problem, Hentzschel [**Error! Bookmark not defined.**] adopts an interesting approach: two images of the same Braille document are captured, each with the document being lit at the same angle but from the opposite side of the camera. In this way, each dot produces a different shadow area in each image whereas dark areas on the paper appear at the same place in both images. By subtracting one image from the other the unwanted dark regions can be eliminated.

In the rarer case of transparent single-sided Braille documents, Nitta *et al.* [1] place each document on a sheet of black paper and illuminate it from a relatively higher point (30°–40° from the vertical) than the other approaches. The camera is placed directly above the document in order to capture the bright circular areas resulting from the bases of the transparent dots. In this case there are no shadows cast by the protrusions.

However, cameras introduce aberrations that must be corrected before any useful processing takes place. François and Calders [9] calibrate their camera set-up by obtaining an image of a checkerboard pattern and correcting it to achieve the original pattern. The sequence of transformations required to correct the test image is applied to every subsequent Braille document image captured for reading.

Apart from these problems associated with cameras, the arrangement of the lighting and the camera is quite complicated for use by a person with no technical knowledge. It also occupies a fair amount of space. Moreover, the cost of the equipment is prohibitive for many users.

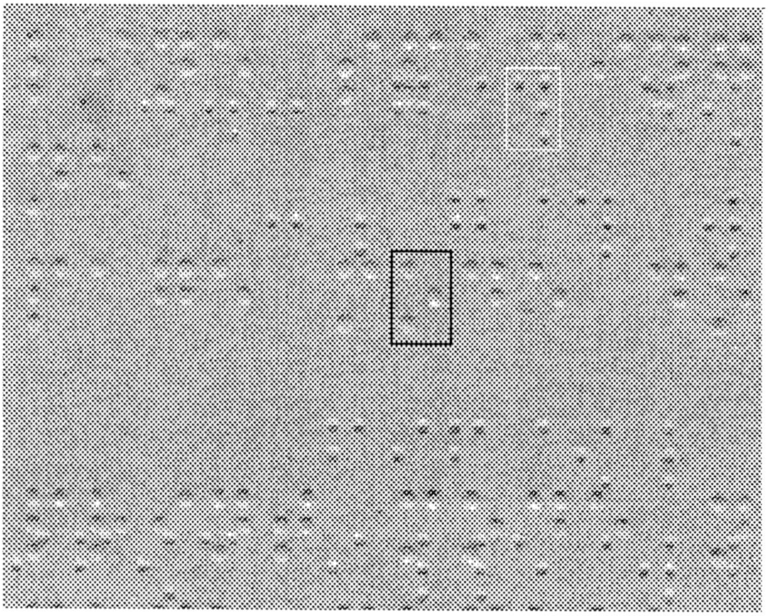

Fig. 6. Image of a part of a Braille document (cells on different sides of the document are delineated with rectangles of different colours).

The increasing popularity of flat-bed document scanners has provided a cost-effective and practical alternative. In contrast to purpose-built dedicated hardware associated with camera arrangements, a scanner is affordable and can also be used for other purposes, such as scanning of printed documents for conversion to Braille. Its compact design is also easy to use by any person. The only requirement is that the illumination of the surface of the paper is asymmetrical (i.e. the document is lit at an oblique angle). As many scanners attempt to illuminate the documents in a very symmetrical and uniform way, care must be taken to choose a scanner that produces the desired results [**Error! Bookmark not defined.**,**Error! Bookmark not defined.**]. An image of part of a double-sided Braille document captured with a scanner can be seen in Fig. 6. The protrusions are represented by a dark region above a light one and the depressions by a sequence of regions in the opposite order. Note the dirt mark at the top-left part of the image.

To distinguish the dots adequately, the spatial resolution setting should be between 80 and 200 dots per inch (dpi). Below 80 dpi there is an insufficient amount of information and above 200 dpi the extra amount of information is superfluous [**Error! Bookmark not defined.**]. The number of grey levels should also be sufficient so that the regions corresponding to dots are accurately distinguished. Although 16 grey levels (4 bits) yield acceptable results and minimise the memory space required, the use of 256 (8 bits) is recommended for better results [**Error! Bookmark not defined.**].

5.2. *Pre-Processing*

The grey-level image of a Braille document can be enhanced in order to correct problems such as skew and noise, and improve the appearance of the areas of interest for the benefit of the recognition process.

Skew may be introduced during image acquisition by not carefully aligning the document or it may arise due to a production fault. If a large skew angle is present, recognition of Braille characters may be impossible because the dots will not be aligned in rows as expected. In this case, it is preferable to recapture the image, after aligning the document, rather than attempting to correct the skew by rotating the image. Skew correction is quite time consuming and should best be avoided if possible. Although small skew angles may be tolerated in some approaches [**Error! Bookmark not defined.**], in others skew must be corrected before the recognition process. As it will become evident in later sections, this requirement is due to stricter assumptions about the position of Braille characters on the document (e.g. [**Error! Bookmark not defined.**]).

One method to detect skew is by assessing the sharpness of the peaks of the grey-level histogram of the rows of an image region [**Error! Bookmark not defined.**]. The image is first divided into rectangular subimages. The standard deviation of grey-level values inside each subimage is calculated and used as an indicator for the presence of Braille dots inside the subimage. A subimage likely to contain Braille dots is selected and a histogram of the grey-level values across its rows is computed. The standard deviation over the row sums is calculated and the image is rotated by a given angle α. The histogram is calculated again and its standard deviation is compared with the value of the previous one. The same step is repeated again until the standard deviation is maximised, in which case the Braille dots in the image should be aligned horizontally. However, the rotation of the image at every step is very time consuming. An alternative could be the calculation of the histogram in different orientations while the image is only rotated when the skew angle is established. It should be noted that this method is similar to the analysis of projection profiles in binary images of printed documents.

The standard deviation of grey-level values in each of the subimages is further used to calculate a threshold corresponding to the grey-level value of the background.

According to this threshold, pixels corresponding to light areas are assigned the value *1* while darker pixels are set to *-1*. All other pixels are set to zero.

A different technique for detecting the skew angle is that used by Hentzschel [**Error! Bookmark not defined.**]. After filtering and thresholding (see below), three of the corners of the Braille document are searched for and identified in the image. It should be noted that it is assumed that the image area is larger than the area covered by the Braille document. The skew angle is calculated from the co-ordinates of the identified corners and the image is rotated accordingly.

Another form of pre-processing is the enhancement of the image in order to improve the separability between the regions corresponding to dots and the background. Light shadows covering broad regions in the image may be introduced if the flat areas of the paper are not perfectly level. Mennens *et al.* [**Error! Bookmark not defined.**] subtract a locally averaged image from the original in order to eliminate this problem. The approach of Dubus *et al.* [**Error! Bookmark not defined.**] and Benjelloun *et al.* [2] is based on the assumption that the Braille dots give rise to bright areas in the image. According to this, the region of a Braille dot in the image will have the form of a second degree convex two-dimensional function. The image is then processed with a median filter based on a second-degree polynomial function in a 1×5 pixel neighbourhood in the horizontal direction. The regions of higher intensity corresponding to protrusions are considerably enhanced in contrast to the background. A second pass with a filter of the same structure but with different parameters is performed and the image regions whose grey-level value is constant throughout are set to zero. In the following, the image is further processed with a vertical median filter consisting of a 3×1 window. Finally, the image is binarised and a morphological closing operation is performed to connect fragmented regions corresponding to the same dot. It should be noted, however, that this approach is only applicable to images of single-sided Braille documents. Median filtering has also been employed by Hentzschel [**Error! Bookmark not defined.**] for noise removal. Binarisation is achieved by an adaptive thresholding process. An optional dilation operation is used to enlarge regions that are very small.

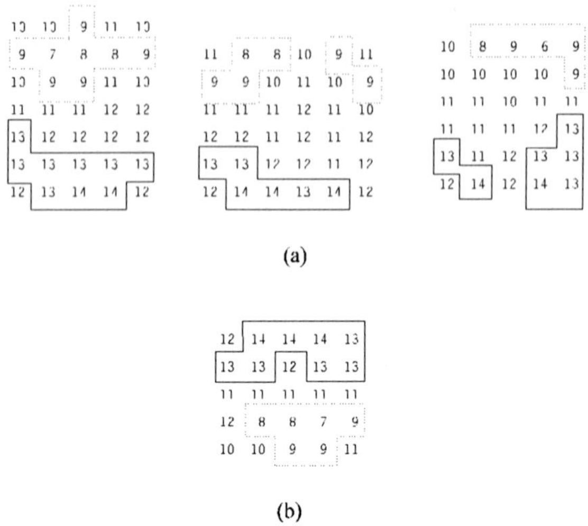

Fig. 7. Regions corresponding to (a) protrusions and (b) a depression.

5.3. *Identification of Dots*

At this stage, the regions in the image corresponding to Braille dots are identified. In a binary image, dots can be identified as small connected components. In the case of the twin shadow approach of Hentzschel **[Error! Bookmark not defined.]**, a Braille dot corresponds to two such components. It should be noted that this situation is valid only when single-sided Braille documents are encountered. In this case, the above approaches are able to identify dots from either the dark (shadow) or the light areas in the image which, after thresholding, are represented as connected components. In the special case of transparent Braille documents, Nitta *et al.* [1] identify connected components having the form of circular arcs. These are then processed so that complete circles are obtained. The centre of each of these circles is identified as the location of the Braille dot.

In a grey-level image, the identification of dot regions needs more careful consideration. It is necessary to have a specification of the image area corresponding to a dot. A dot is represented by a combination of a dark and a light region. The intermediate grey-level pixels of the background surround and, possibly, also separate the pair. Naturally, the orientation and order of the light-dark pair of regions depends on the direction of the illumination. In images of inter-point Braille documents, however, the protrusions give rise to region pairs in which the order is opposite to that of the region pairs produced by depressions. For example, in Fig. 7 some region pairs corresponding to protrusions and depressions in the document of Fig. 6 are shown. The numbers in the figure correspond to grey levels in the range 0 (darkest) to 15 (lightest).

The consistency of the appearance of grey-levels at the positions of depressions and protrusions renders the use of template matching a possible option. This is the technique adopted in the approach of Mennens *et al.* **[Error! Bookmark not defined.]**. In the images captured by their system, protrusions are represented in the image by a dark area followed by a light one and depressions are represented in the opposite order, similarly to the example of Fig. 7. In the pre-processed images (see previous section), pixels of light regions have the value *1*, background pixels have value *0* and pixels of dark

-1
0
0
0
1

regions have been set to *-1*. The mask
is then correlated with that image. The number of 0's in the middle depends on the distance between the centres of the light and dark region pairs in the document image. For instance, the above mask is constructed on the assumption that the distance is five pixels. The result is an image where groups of pixels with value *2* correspond to regions in the centre of protrusions and regions with pixels having value *-2* correspond to centres of depressions. These two kinds of regions are termed *core* regions and are used to denote the protrusions and depressions of the Braille document.

Different masks have also been tried **[Error! Bookmark not defined.]** but, in images of inter-point documents, it is often the case that the template matching technique leads to the detection of non-existent dots. For example, there are situations where a dark area belonging to a protrusion and a light area belonging to a depression may lie in such an arrangement that gives rise to the detection of a dot which is not actually present on the document. The application of the technique of Mennens *et al.* **[Error! Bookmark not defined.]** results in the detection of a number of such non-existent dots. To cope with this problem, they adopt a strict model of the structure of a Braille document (see the next section) which dictates the valid positions for dots of both sides. The identified dots which do not lie in valid positions are disregarded in subsequent stages.

A more flexible approach **[Error! Bookmark not defined.]** first identifies the dark and light regions in the grey-level image and describes them with their minimum enclosing rectangles. The relationships between light and dark components (describing rectangles) are then examined in order to identify the depressions and the protrusions. There is another problem, however. Light regions of protrusions may be merged in the image with light regions of depressions and identified as one large component instead of two distinct ones. The same situation may occur for dark regions. An example of dots belonging to different sides of the document having merged regions can be seen in Fig. 8. On the other hand, as shown in the example of Fig. 7, a dark or light region may be fragmented. In this case, only one dot should be identified.

The algorithm proceeds as follows:

for each dark region
    if a light region exists *above* within limits of the height of a dot **then**
        a *depression* is found
        check regions and mark the appropriate ones as *"used"*
    else
        if a light region exists *below* within limits of the height of a dot **then**
            a *protrusion* is found
            check regions and mark the appropriate ones as *"used"*

When a protrusion or a depression has been identified, then the dark and light regions that comprise it are marked as "used" so that they will not be considered as parts of another dot. However, only the correct part of a region or regions must be marked. Hence, there are rules that guide the algorithm to correctly identify which regions, or which part of a merged region, should be considered as the dark or the light region of a dot. For example, in Fig. 9, the dark region is under consideration and the light region *A* above it fulfils the criteria for the existence of a depression. Region *A* is marked as "used" but the dark region is wider than the maximum width of a Braille dot and so it is divided into two parts and only *B1* is marked as "used". The dark region *B2* is considered next and the two light regions *C* and *D* are found below it. Both these regions satisfy the requirements for the identification of protrusions but the distance between *C* and *D* is smaller than the minimum distance between Braille dots. Hence, it is concluded that both regions constitute the light region of one protrusion. Now, the regions *B2*, *C* and *D* are all marked as "used". The protrusions identified are written as black dots of nominal size in a new blank image. The depressions are similarly written in another blank image. Upon completion of the algorithm, the dots of each side of the Braille document have been separated from those of the other. Each side will be considered separately in the stages that follow.

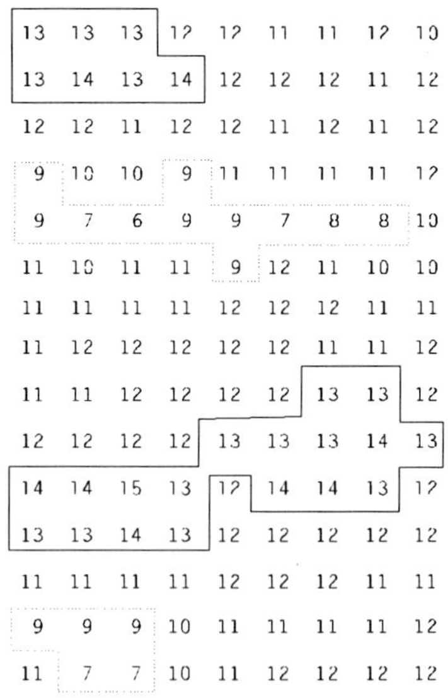

Fig. 8. Two depressions (left) and a protrusion (right) with merged regions.

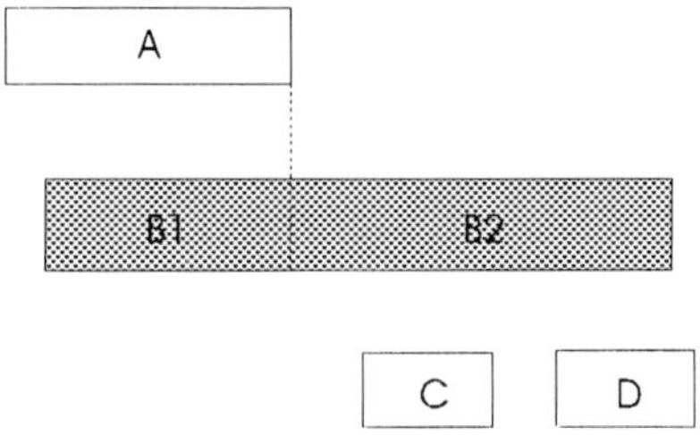

Fig. 9. Identification of Braille dots when regions are merged or fragmented.

5.4. *Braille Character Segmentation*

By this stage, the Braille dots have been detected in the image of the document. Now, groups of these dots must be identified as Braille characters. The aim of this stage is to examine the locations of the dots and deduce the positions of the characters.

One way to identify dots belonging to the same character is by examining the distance between consecutive dots in the horizontal and vertical directions. In contrast with proportionally spaced printed characters, Braille characters are printed in regular intervals (or multiples thereof) and the points in each cell are in regular positions. Therefore, it is quite straightforward to distinguish between different lengths of space between dots. It should be noted, however, that in practice there are deviations from this regularity arising from the Braille production process and the various effects of the imaging process. Nevertheless, for the purpose of spacing analysis there is no significant requirement for very high accuracy in adhering to the standard distances.

The spacing in the vertical direction falls into two categories: spacing between dots belonging to the same character and spacing between adjacent lines of Braille characters. One technique [**Error! Bookmark not defined.**] is to examine the vertical projection-profile of the binary image containing the dots. From this, the two spacing values can be obtained. According to the vertical distance between dots, lines of Braille characters can be identified as consecutive rows of dots. Across a line of Braille characters, dots are usually found in all three rows. However, there may be lines with dots in only one or two rows. For instance, there may be a line of hyphens separating text areas. A hyphen is represented by a Braille character having dots in the bottom row only (dots 3 and 6). In the image, such a character line is ambiguous with regard to the identity of the characters it contains. If the dots are in the top row then the meanings of the characters is different than what will be if the dots are in the second or in the third row. The relative position of a row of dots inside a line of Braille characters should therefore be established. For this reason, the lines having dots in all three rows are detected first. Having determined the average line height and the average inter-line spacing, the vertical positions of all the lines of Braille characters and of the blank lines are determined. From these positions it is straightforward to identify the relative positions of rows of dots inside the lines.

Having located the lines of Braille characters, the next step is to determine the positions of the characters themselves inside each line. In order to do so, the positions of character columns having at least one dot are first identified from the horizontal projection-profile. The space between columns is then examined to determine whether a column is the left or the right one of a Braille character. The spacing between adjacent columns can vary beyond the obvious intra-cell and inter-cell values. This is due to the fact that a Braille character may have dots in only one column. Hence the distances between columns may correspond to one of the following:

(i) space between the two columns of the same character (*a* in Fig. 10),
(ii) space between a *right* column and the *left* one of the next character in the word, i.e. the space between adjacent Braille cells (*b* in Fig. 10),
(iii) space between a *right* column and a *right* columned character (*c* in Fig. 10),
(iv) space between a *left* columned character and the *left* column of the next character (*d* in Fig. 10),
(v) space between a *left* columned character and a *right* columned one (*e* in Fig. 10), and
(vi) space between words (e.g. *f* in Fig. 10), which varies according to the number of space characters present.

To segment the characters in a line, the distance between consecutive columns is examined. Accordingly, each column is characterised as a left or a right one. If the distance between two columns is greater than the sum of the inter-character distance and the width of a Braille character (space character in this particular case), it is determined as inter-word distance and the number of Braille space characters is calculated. Special consideration is given to the situation where two single-columned characters are encountered. Ambiguity can arise if they are both left-columned or both right-columned (distance *c* or *d* in Fig. 10).

A similar method has been adopted by Hentzschel [**Error! Bookmark not defined.**] which, because of the twin shadow approach (see previous section), identifies two peaks for each column in the horizontal projection-profile of the line of Braille characters. The use of the height of peaks in this profile is also suggested as it can provide information on the number of dots in the column.

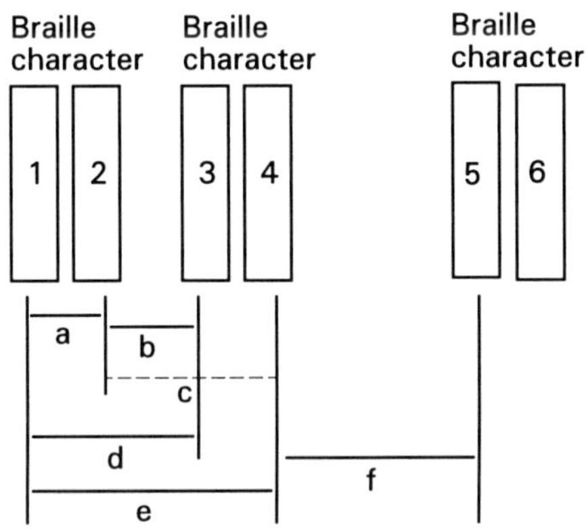

Fig. 10. Different spacing possibilities between two Braille characters.

Benjelloun *et al.* [2] start with the horizontal projection-profile. They first approximate the peak regions in the profile with Gaussians and identify their centres. The mean intra-character and inter-character distances are also calculated. With the aid of a knowledge base describing expected positions of left and right columns of dots in the image, they attempt to characterise the column positions, identified in the profile, as left or right. Identified positions that are very close to the pre-specified ones are corrected. In the next step, column positions which have not yet been characterised as left or right are considered. The distance between two of them or between one of them and an already characterised column are examined. If, for example, the distance between two consecutive column positions is equal to the inter-character one, the first is characterised as right and the second as left. Most of the column positions in the profile will have been characterised upon completion of this step. However, column positions whose distance from the closest neighbouring column positions is different from the intra- or inter-character ones will still remain uncharacterised.

From the known positions of left and right columns, an attempt is made to identify the remaining positions that columns of dots could exist. Based on the two mean distances previously computed, the positions of left and right columns in Braille cells are identified. If some of these columns actually contain dots they are characterised accordingly. At this point all positions in the image at which a column of dots may exist have been characterised as left or right.

In order to segment the Braille characters which have now been identified as columns of dots in the vertical direction, the lines of the Braille characters must be detected in the image. In doing so, the vertical projection-profile is obtained for each column axis identified in the previous step. The mean height of a Braille dot is calculated from its projection and the rows of Braille dots and lines of characters are identified accordingly.

Another approach uses a model of Braille documents [**Error! Bookmark not defined.**]. The assumption is that Braille cells are positioned in fixed rows and columns on the page. Consequently, dots of Braille characters will only be present at intersections of horizontal and vertical grid lines passing through the positions of rows and columns of Braille cells. Hence, the positions of Braille characters and their dots can be identified from the model. The most important step of this technique comprises the identification of the location of the horizontal and vertical lines of the model grid. In other words, the grid must be positioned on the current input image. The initial examination takes place in each of the subimages (see Sec. *5.2.*). The horizontal grid lines are tentatively identified from the vertical projection-profile of core regions (see previous section). The vertical grid lines are identified from the horizontal projection profile of the core regions lying on horizontal grid lines only. The positions of the horizontal grid lines are next updated and finalised by examining a new vertical projection-profile of core regions situated along the identified vertical grid lines only. The global grid for the whole image is then assembled from the pieces identified in the subimages. The dots that lie on the intersections of appropriate grid lines form the Braille characters.

Finally, in the approach of Nitta *et al.* [1] a local model is used to identify dots belonging to a Braille character. This model is actually a specification of the relative positions of the points in a Braille cell with reference to a single given point. For example, if the actual position of a dot in the image is known, its relative position inside the Braille cell could be found by searching for other dots in certain locations around it. It is these locations which are pre-determined by the model. For each identified dot in the image the specified locations in its neighbourhood are examined and, consequently, the Braille character is identified.

5.5. *Recognition*

At this stage the identity of each of the Braille characters is recognised. It should be noted that, at this point, no attempt is made to interpret the character. Since each character is distinguished from the rest by its unique combination of dots, it will suffice to identify the (relative) positions of its dots. Then, the character can be described in terms of the numbers of dot positions (see Sec. *3.1.*.

Since the positions of Braille characters in the image have already been identified, it is straightforward to check for the existence of a dot at each of the six positions of points in a Braille cell. If the Braille character is described as a region in the image, this region is divided into six equal compartments (two across, three down) in which the search for dots is performed. If the exact positions of dots in the image are known **[Error! Bookmark not defined.]**, it is easy to check if one of them is included in a compartment. Alternatively **[Error! Bookmark not defined.]**, the region in the image corresponding to a compartment is searched and the number of pixels corresponding to dot components is counted. If the number exceeds a pre-specified threshold then the compartment is considered to contain a dot. In any case, the recognised character can now be represented as a binary number with six bits. Each bit corresponds to a compartment and it is set only if there is a dot in that compartment. The principle of dividing the character region into compartments and checking for the presence of dots is illustrated in Fig. 11. Also shown is an example of encoding of a Braille character.

In other approaches the Braille characters identified in the image are described explicitly [2] or implicitly **[Error! Bookmark not defined.]** by the positions of the rows and the columns of a cell. In these cases it is necessary to check along specified directions for the existence of dots. In the approach of Mennens *et al.* **[Error! Bookmark not defined.]**, where the global grid has been constructed (see previous section), the possible positions of Braille dots are determined by the intersections of horizontal and vertical grid lines. The six possible positions of the dots are known for each individual Braille character position in the image. Hence, all the possible character positions are examined. Within each one of them, the intersections of grid lines are considered. A local correlation test is performed to determine whether a dot exists or not and the recognised Braille character is encoded as a binary number, similarly to the method previously described.

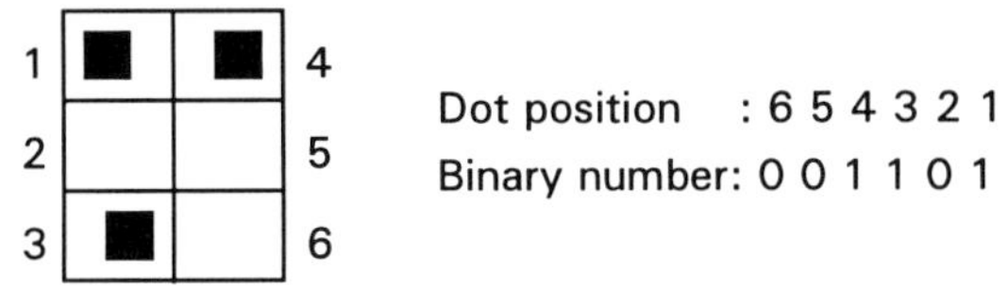

Fig. 11. Example of recognition and encoding of a Braille character.

5.6. *Verification*

Due to a number of factors, each pertinent to one of the previous stages of the reading process, some of the characters may not be correctly recognised. Due to the nature of the Braille characters as combinations of dots, the errors are mostly substitutions. The presence of an extra dot at a valid position within a character or the absence of a dot (see Sec. 6.) both lead to the recognition of a different character. Furthermore, for the same reason, verification of the recognition results is quite difficult on a character-by-character basis. Apart from the presence or absence of dots at specified positions, there are no shape or structural properties that can uniquely distinguish between Braille characters. This is in contrast with printed character recognition where structural and topological properties of characters can be used to eliminate possibilities and at worst select a ranked set of alternatives. With Braille characters, the presence or absence of dots gives rise to many possibilities, each with equal probability of being valid.

The only other option is contextual analysis. Each Braille character could be assessed and be assigned a degree of confidence according to how well it fits with the other characters in the Braille word. Similarly to the well known techniques used in OCR, different possibilities can be suggested with the use of a dictionary and so on. There is a problem here, however. The characters in a Braille word must be interpreted first in order to assess the validity of this word. This not an easy task, considering the variety of interpretations that a Braille character may assume under different circumstances. The type of the information (text, mathematics, music etc.) contained in the document must be known first. If textual information is contained, the grade must also be known. Finally, the country of origin must also be known, so that, apart from the language, the writing conventions are also taken into account. As it was mentioned before, even for the same language there may be different contractions etc. used in different countries.

However, it could still be difficult to rank alternative Braille characters. For instance, if the document contains grade 2 Braille, an isolated character will correspond to a short printed word, e.g. *for, of* or *with*. In this case, comprehension of the whole sentence is required.

Particular attention must be paid to situations where more than one Braille character corresponds to a single printed one (see Sec. *3.2.*). Example cases are the Braille representations of numbers. As it has already been mentioned, in textual Braille documents a digit is represented by the numeral mode character followed by the letter assigned to represent this particular digit. For example if the Braille character corresponding to the letter "a" follows the numeral mode character, it is interpreted as "1". Numbers are represented by the numeral mode character followed by the letters corresponding to each of the digits. Hence, if the string "cab" follows that mode character it is interpreted as "123". Failing to recognise the numeral mode character will lead to a wrong result which is, nevertheless, valid in its own right. Another example is the Braille character representing the letter "D". In grade 2 Braille, if it is on its own it is interpreted as the word "do" but if it is preceded by a Braille character having only dot 5 it is interpreted as "day". Again, failure to recognise the special character leads to a different—but valid—meaning or, if instead of the special character another one is recognised, to a different combination of characters.

Finally, it should be kept in mind that in many cases, especially in grade 2 Braille, the interpretation of each of the characters in a word depends on its position in the word. For example, in grade 2 Braille, a character is interpreted differently if it is in the beginning, at the end or anywhere else in a word. Therefore, any failure to correctly recognise other characters that effectively changes its position will also change its meaning. In some cases the wider context beyond a word will also need to be analysed. If, for instance, the Braille character representing the closing parenthesis is wrongly recognised as that representing a question mark (only one dot difference), it may even pass unnoticed. To verify its correctness, the whole paragraph may need to be examined, to check for an opening parenthesis for example.

5.7. *Further Processing*

Having obtained the codes of the Braille characters on the document, a number of possibilities exist for their use. First of all, the Braille document can be reformatted, edited and distributed by electronic means. Copies can then be sent for reproduction to a Braille printer. Alternatively, the Braille characters may be translated to ASCII in order to be read by sighted people. A further possibility is the translation from one grade of Braille to another and from the conventions system of one country to that of another.

In addition, having converted the information of Braille documents into electronic form, the benefits of the usual applications available for the electronic form of printed documents can be enjoyed.

6. Factors Affecting Automatic Reading

In the previous subsections each of the stages involved in the automatic reading of Braille documents was examined. In this section, the main factors that affect the success of the whole process are outlined.

Different approaches may be susceptible to different problems and to various degrees due to their design. However, there are some factors that affect all methods universally. The main of these are mentioned next, followed by a consideration of the reaction of the different methods to them.

The most important problem is the presence of light or dark regions which have not been created by a protrusion or a depression. These noise regions may affect the result of recognition if they fulfil the structural and spatial conditions set for the recognition of Braille dots. Extra dots may be found in a character or a non-existent character may appear instead of a space. Attempting to identify Braille dots by just identifying either dark or light regions is more susceptible to errors. The identification of dark-light region pairs leads to better results as it is more rare for two noise regions to form the same combination as a Braille dot. However, ambiguity may arise if a valid dot region-pair and a noise region are close in such an arrangement that gives rise to two possibilities for dots. In this case the problem will be in deciding whether a protrusion or a depression is present.

Another issue is that of non-detected dots. One reason for this may lie in the image acquisition and pre-processing stages during which a light or a dark area produced by a dot is not represented in the image. In later stages, the dots with missing regions will not be detected and erroneous results will be produced by the recognition stage. A protrusion or a depression may also not be detected because its light or dark region is merged with that of a dot of the other side. If one of the dots is recognised and the merged region is not sufficiently large to suggest the presence of another dot, the remaining region of that other dot may be thought of as noise and be ignored.

Finally, although they may be identified, the dots may not be in the expected positions within a Braille cell. Apart from the image being skewed, there may be defects of the embossing process which compromise accuracy. In the case of skew, it can be detected and corrected in the pre-processing stage (see Sec. *5.2.*). Alternatively, if the skew angle is relatively small, a flexible method may be able to overcome the shift in the positions of dots. A similarly flexible approach will also give better results in the case where arbitrary shifts of dot positions exist due to the Braille production process. The assumption of a fixed grid is rather limiting in that it prevents dots which are not positioned on the intersection of grid lines from being recognised as part of a Braille character. However, caution is also needed in the flexible approach in order to avoid recognising dots resulting from noise.

Perhaps, more than one method should be combined in an approach in order to draw from the areas of strength of each individual one and to complement each other in a way that the overall performance will be less affected by the problems mentioned above.

7. Concluding Remarks

The problem of automatic reading of Braille documents is an interesting one in that, despite its similarities to printed document analysis, it involves some challenges of a different nature. The aim of this chapter was to introduce the aspects of the problem and examine solutions to each of them. The essence and production of Braille and Braille documents were described and the characteristics of the problem were discussed. Following that, each of the stages towards the automatic reading of Braille documents was analysed and the approaches that have been so far designed to carry it out were presented.

Having therefore studied the whole problem, one can arrive at some conclusions. First of all, the quality of the image is important. Adequate information about the protrusions and, possibly, the depressions must be present. On the other hand, the quality of Braille documents should not be expected to be necessarily very high. Older documents may have dirt marks and be in a state of deterioration. Furthermore, defects of the production process are responsible for minor shifts of the positions of dots from the expected locations of points in a Braille cell. This calls for a flexibility in the recognition approach. Assuming that all dots lie in the intersections of fixed grid lines may lead to the omission of some protrusions or depressions. However, caution is needed in allowing an approach to be very flexible as dot regions resulting from noise should not be considered as part of characters. Finally, inter-point Braille documents should be recognised as well as single-sided ones.

In terms of the performance of the various approaches, good results have been reported for the type of images tested by each one of them. More specifically, for parts of transparent single-sided documents, Nitta *et al.* [1] report a maximum correct recognition rate of 80%. This figure, however, does not take into account the Braille characters at the edges of the image which cannot be recognised due to camera lens aberrations. Furthermore, Braille characters in skewed images cannot be identified. Dubus *et al.* **[Error! Bookmark not defined.]** do not refer to any results but in a later paper Benjelloun *et al.* [2] quote a success rate of 99% for single-sided paper sheets of Braille. François and Calders [9] also report a high average recognition rate of 99% for single-sided opaque Braille documents. The approach of Hentzschel **[Error! Bookmark not defined.]**, applied to parts only of single-sided paper documents, achieves a maximum recognition rate of 92%.

For test samples including both single-sided and double-sided paper documents, Mennens *et al.* **[Error! Bookmark not defined.]** achieve a range of results depending on the presence and nature of defects on the document. If there are no defects, they report that no errors are encountered, while some errors (0.25%) arise from marks on the surface of the paper. However, if there are shifts in the positions of dots, they mention that the results are unacceptable. This is because their approach assumes the existence of a fixed grid on which the Braille dots can be found. Finally, the more flexible approach of Ritchings *et al.* **[Error! Bookmark not defined.]** does not encounter such errors. The approach was applied to low quality images of double-sided documents only. Despite the low resolution (100 dpi compared to that of 200 dpi in **[Error! Bookmark not defined.]**) and small range of grey levels (16 compared to that of 256 levels in **[Error! Bookmark not defined.]**) the average success rate for recognising Braille dots was 98.5% for the protrusions and 97.6% for the depressions of the same image. The overall rate for simultaneous recognition of characters on both sides of a Braille document was reported as 90.8%. The authors explain that the expected success rate for higher resolution 256 grey-level images will be correspondingly higher.

A comparison of all methods on the same image data is however needed in order to obtain an objective measure of recognition performance and to progress forward. Apart from a high recognition performance, the time taken for the completion of the reading process is of significant importance. Repeated image accesses and the need for time-consuming operations can render an approach less practical than others. For example, dependence on skew correction, even for very small angles, is not desirable. The use of computationally intensive image transforms should also be kept to a minimum. Finally, it is advantageous to obtain a representation of the dot regions, in terms of their co-ordinates and size, and use this through the various stages instead of having to search and index through the volume of the image data.

References

[1] S. Nitta, J. Oshima and M. Okhawa, Recognition algorithm on transparent Braille, *Proc. 28th SICE Annual Conference*, Matsuyama, Japan, 1989, 1017–1020.

[2] M. Benjelloun, V. Devlaminck, F. Wauquier, P. Altmayer and J. P. Dubus, Application du traitement d'images à la conception d'un système de transmission de relief Braille en texte noir, *Traitement du Signal* **6** (1989) 291–300.

[3] T. Otake, T. Saito and Y. Yonezawa, Braille printer using hot-melt material, *J. Microcomputer Applications* **13** (1990) 123–131.

[4] F. Germagnoli and G. Magenes, Computerised system for helping blind people to learn Braille code, *Proc. 15th Annual Int. Conf. on Engineering in Medicine and Biology*, 1993, 1318–1319.

[5] O. Sueda, Braille translation by microcomputer and a paperless Braille dictionary, in *Computerised Braille Production Today and Tomorrow*, eds. D. W. Croisdale, H. Kamp and H. Werner (Springer-Verlag, 1979) 272–289.

[6] Royal National Institute for the Blind, *Braille Primer* (RNIB, 1992).

[7] W. Slaby, Computerised Braille translation, *Computers for Handicapped Persons, Proc. Conf. on Computers for Handicapped Persons*, eds. A. M. Tjoa, H. Reiterer and R. Wagner (R. Oldenbourg, Vienna, Austria, 1989) 26–34.

[8] P. Blenkhorn and T. Maley, Computer Braille code, *Computer Assisted Learning for the Visually Handicapped*, Information Sheet 4 (Research Centre for the Visually Handicapped, Faculty of Education, University of Birmingham, 1985).

[9] G. François and P. Calders, The reproduction of Braille originals by means of optical pattern recognition, *Proc. 5th Int. Workshop on Computer Braille Production*, Heverlee, 1985, 119–122.

[10] J. P. Dubus, M. Benjelloun, V. Devlaminck, F. Wauquier and P. Altmayer, Image processing techniques to perform an autonomous system to translate relief Braille into black-ink, called: lectobraille, *Proc. IEEE Engineering in Medicine and Biology Society 10th Annual Int. Conf.*, New Orleans, 1988, 1584–1585.

[11] T. W. Hentzschel, An optical Braille reading system, M.Sc. Dissertation, Faculty of Technology, University of Manchester, 1992.

[12] J. Mennens, L. Van Tichelen, G. François and J. Engelen, Optical recognition of Braille writing, *Proc. 2nd Int. Conf. on Document Analysis and Recognition*, Tsukuba, Japan, Oct. 1993, 428–431.

[13] R. T. Ritchings, A. Antonacopoulos and D. Drakopoulos, Analysis of scanned Braille documents, *Proc. Int. Association for Pattern Recognition Workshop on Document Analysis Systems*, Kaiserslautern, Germany, Oct. 1994, 417–424.

[14] D. Drakopoulos, Analysis of scanned Braille document, M.Sc. Dissertation, Faculty of Technology, University of Manchester, 1993.

Handbook of Character Recognition and Document Image Analysis, pp. 729–754
Eds. H. Bunke and P. S. P. Wang

CHAPTER 28

DIA, OCR, AND THE WWW

GEORGE NAGY[a]
SHARAD SETH[b]
MAHESH VISWANATHAN[c]

The rapid growth of the World-Wide Web (WWW) offers digital image analysis and optical character recognition researchers a golden opportunity to bring ideas to bear on an important and timely application. To focus our discussion, we examine the needs of the Internet community for archival technical material. Some of these needs can be satisfied by posting articles now on paper in coded text form, others by posting them in facsimile image form. Most can be satisfied by a combination of both forms, but dual representation imposes certain additional requirements on document analysis. Potential sources of archival electronic documents include existing CD-ROM databases, publishers, libraries, and individual members of the Internet community. However, the cost of manual conversion of a significant portion of the pre-1980 technical material is staggering. This material will remain unavailable on the network unless certain specific optical character recognition (OCR) and document image analysis (DIA) tasks, most of which are the subject of current research, can be automated.
Keywords: Document browsing; Document display; Document conversion; Conversion accuracy; Internet; Digital library; Automatic indexing; Document indexing; Display format; Hypertext; Information retrieval.

1. Introduction

Most of us in the DIA-OCR community were caught unprepared by the exponential growth of the World-Wide Web. Although the WWW is the culmination of decades of research and development on computer networking, for our purpose it can be considered simply a set of distributed file servers and client stations linked by digital networks. The encoded information is accessed by individuals through local soft-

[a] *Electrical, Computer, and Systems Engineering Department; Rensselaer Polytechnic Institute; Troy, NY 12180, USA*
[b] *Computer Science and Engineering Department, University of Nebraska, Lincoln, NE 68588, USA*
[c] *IBM Thomas J. Watson Research Center, Yorktown Heights, NY 10598, USA*

ware called "viewers" or "browsers" (e.g., Netscape), which include hypertext-based information retrieval facilities. The document encoding method used is HTML (HyperText Markup Language). For additional buzzwords and acronyms, please refer to the Glossary, and for an excellent overview of the WWW, to [1].

Only a negligible portion of the enormous amount of material now on the Internet is the result of automated processing that incorporates our research. In fact, there is remarkably little reviewed scholarly material to be found. Publishers, including our technical societies, have been invited to the party of the century, but have nothing to wear except baubles. This paper addresses the question of what needs to be done in order to make the WWW, and its eventual successors, the prime source of serious technical material for scientists, engineers, teachers and students.

Adding archival material to the WWW could be prohibitively expensive. If we consider only the leading 1000 technical periodicals — most research libraries stock two or three times that many titles — and assume six issues per year at 100 pages each, then twenty-five years' accumulation totals 15,000,000 pages. At a manual conversion cost of $10 per page [2], the cost would be 150 million dollars. The IEEE and the ACM, to name only two technical societies, each year publish about as many pages of conference proceedings as of periodicals. And we haven't counted yet all the theses and dissertations nestling on dusty shelves, or the daily and weekly trade publications. Furthermore, $10 may well be an underestimate for dense technical pages with formulas, tables, and illustrations.

In contrast to key entry, optical scanning is comparable in cost to microfilming — about $0.10 per page. If DIA and OCR can perform the rest of the task automatically, then the figures reach the realm of realizability. Surprisingly, the recent NSF-ARPA-NASA initiative for "Research on Digital Libraries" gives little emphasis to document conversion.

Besides the pressing technical questions, there are many economical and intellectual property issues to consider [3]. What will be the role of libraries and librarians if everything is available on the Internet? How will publishers, who now exercise essential quality control over publications, as well as assume the cost of production, recover their costs? Will articles remain substantially "as published," or will they continue to evolve as new findings take place? Will it be possible to find common reference points if every paper appears in numerous versions? Can the current growth rate be sustained, or will the WWW eventually collapse under its own weight?

Fortunately, it is possible to discuss many of the technical aspects of automated document conversion independently of these issues. In the remainder of the paper we consider the principal scholarly-material requirements of the technical community, examine options in making this material available on the Internet, and discuss the principal areas where additional progress in DIA and OCR is necessary to render archival technical material accessible through the World-Wide Web.

2. User Needs

From the user's perspective, automated document conversion must address both retrieval and presentation issues.

2.1. *Retrieval*

A modern scholar, overburdened by the publication glut and the proliferation of sources, is increasingly at the mercy of electronic search tools and incomplete databases to keep up to date. Finding out what is out there on the WWW is even more frustrating and is a major limitation to its use. With the current information retrieval resources available, a typical Web user will most likely miss a document that is available on the Net. A related problem is remembering where documents were found when attempting to access them again ("bookmarks" provide only a partial solution). These limitations of WWW are well recognized and apply to accessing *any* available resource. A satisfactory resolution of the information retrieval problem will have to go well beyond the baby steps already taken [1].

Documents on the Internet can be located in a variety of ways. Using an index on wide area information servers (*WAIS*), the user can directly locate the document. Or, by accessing selected sites for *Archie* indexes of file names available in public domain, the user can determine the anonymous FTP site for the document. However, unless the user has substantial prior knowledge of network resources, the huge number of possible paths through index/menu trees render this a very unreliable way to locate a document. W3Catalog and WWWW (World Wide Web Worm) are searchable indexes available on WWW. Similar search capabilities are available at other starting points on the Internet, e.g., Veronica word search can find resources whose titles appear in gopher menus.

Keyword searches on bibliographic databases are also possible but often lead to just pointers to documents. Examples include the collection of computer science bibliographies (about 300,000 citations) in bibTeX format that can be searched flexibly through the Glimpse server located at University of Arizona [4] and the DIA database of 1500 PLUS ENTRIES available on DIMUND, the Document Image Understanding Information Server at University of Maryland [5].

Internal and external hypertext links must be created any time a new document is converted and becomes accessible on WWW. This is particularly important for citations. Users must be able to find any cited article via the on-line database of pre-indexed articles. When a new article is brought on-line, each citation in the article must be compared against the existing database of documents to see if the material is already available on-line. If such a document exists, a link must be inserted from the citation in the new article to the document in the database. Further, each pre-existing on-line article must also be checked to see if links have to be inserted to the newly added article. Since readers often find articles by searching via keywords, subject, and author indexes, any change in the database would have to find its way into all the indexes.

Rudimentary methods of cross-referencing between text and image form have already been developed [6]. However, we are not aware of any research on introducing the necessary links automatically.

2.2. *Presentation*

To understand user needs subsequent to location of a document, consider the many ways in which a reader absorbs the information in a journal article. The process starts typically with browsing through the pages, scanning the abstract, section headers, figures, captions, and the list of references. Partial access to the document may be sufficient for this type of pre-reading. Occasionally, the pre-reading may encourage the reader to analyze the detailed contents (equations, proofs, figures, tabular data, and so forth), make careful annotations, attach book marks, and mark suspected errors. This form of intensive reading requires complete and accurate access to all information in the document. Facsimile (bitmap) images meet the requirement but pose difficulties in annotation, editing, and other forms of computer manipulations. Another alternative is to convert all the text and retain bitmap images for non-textual entities. This requires high accuracy in OCR. To avoid unpredictable and annoying delays in transmission of bitmap images on demand from a remote server, it may be necessary to download the whole document initially or use a caching scheme. The added burden of storage at the user end appears not to be a major issue, considering that even low-end PCs these days come equipped with tens of Megabytes of RAM and hundreds of Megabytes of disk space.

To summarize, even if the required document can be located, neither a coded version that cannot reproduce the original faithfully, nor a facsimile image that cannot be electronically manipulated, will satisfy all users completely.

3. Potential Sources Of Electronic Documents

As mentioned, most of the material currently on the Internet consists of contemporary text produced on computers for electronic distribution. Here, however, we consider only archival technical material originally produced for distribution as hardcopy (but possibly already available in electronic form).

3.1. *CD-ROM*

With the rapid proliferation of CD-ROM players, publishers have rushed to make existing popular reference works, such as encyclopedias, directories, and law books, available on optical disks. The distribution of disks parallels that of conventional books and involves no special copyright or marketing problems (except for competing disk formats). An increasing amount of technical material is also available on CD-ROM and, once copyright and cost-recovery problems are solved, can be readily posted on the WWW without major additional cost. In many instances, the CD material already contains links to facilitate navigating within the disk itself. External links will, of course, have to be added. Most CD-ROMs contain formatted text augmented with illustrations rather than page images, but examples of dual representation exist [6]. The IEEE is reissuing most of its recent publications on CD-ROMs in page-image form.

3.2. *Publishers*

Either the current creators of technical material — professional societies, university presses, and for-profit commercial publishers — adapt to the challenges of the WWW, or they will be displaced by a new generation that finds a way to survive in the electronic age [7]. In some sense, electronic journals [8], Prodigy, America On-Line, and Compuserve can be considered the forerunners of this new generation. Wire service stories going back more than ten years are already on line. The IEEE and the ACM have already posted thousands of pages — consisting of tables of contents, abstracts, bibliographies — though most of them serve only as pointers to the substantive material that is not yet accessible online.

For already archived material, the major problem, facing both non-profit and for-profit publishers, is the recovery of the cost of conversion of hardcopy to electronic form. Ironically, most technical publishers have made heavy use of computerized editing and typesetting for at least two decades, but because of the patchwork nature of the operations, the conversion of preserved files may be more expensive than conversion from the final printed copy.

3.3. *Large-scale Conversion*

Publishers are one type of organization that have in place many of the components necessary for massive conversion operations but service bureaus and research libraries also have the required resources.

Service bureaus have set up large-scale document scanning and editing operations, primarily to satisfy government needs. Examples of large-scale government document conversion operations include social security records, internal revenue forms, state vital records, vehicle records, unemployment insurance applications, and the 42,000,000 page (estimated) Licensing Support System for the Nuclear Regulatory Commission [9]. Perhaps closest to technical document conversion are longstanding military efforts to make service manuals available in electronic form [10, 11]. More recently, the National Institute of Standards and Technology has conducted large scale tests for the Bureau of Census [12–15].

Research libraries also have a vested interest to reduce the cost of making their holdings available to their clientele. Fax has largely displaced postal delivery in inter-library loans, and it is only a small additional step to retain frequently-faxed articles in digital form. At Rensselaer, the course handouts deposited at the library Reserve Desk are routinely scanned and posted for access through Netscape within twelve hours of receipt [16]. The research library at AT&T Bell Laboratories is promulgating the RightPages service [17, 18].

Just as research libraries have cooperated to avoid duplicate indexing and cataloging, they will probably take the initiative in large scale conversion of electronic materials. The lead organization may be OCLC (Online Computer Library Center), a consortium for resource sharing with over 17,000 member libraries [19] or ARL (Association of Research Libraries). The US Department of Education College Libraries program has recently sponsored several research projects to pave the way [20], and university libraries are full partners in the projects sponsored under the NSF-ARPA-NASA Digital Libraries initiative mentioned in the introduction.

3.4. *Distributed Contributors*

Although publishers, libraries, and service bureaus are likely to play the most significant role in making archival material available on the Internet, it is also possible that individual Web surfers will contribute. Optical scanners and OCR systems are now widely available. Researchers, teachers, and students may occasionally convert an old article from a journal or conference proceedings for their own use. They might then be willing, with or without specific incentives, to post the converted material on the Web. Already many individuals have taken it upon themselves to act as "doorkeepers" for various specialized collections of interest to a segment of the research community. A well-known example is the Los Alamos high-energy physics publications database [21].

If technical text is first posted in the form of images, then it will be even easier for individuals to convert it to (logically tagged) symbolic form (e.g., SGML) without first having to locate the hardcopy and scanning it. Their major contribution will be the post-editing necessary to correct OCR errors and adding formatting information. It is also possible that this process will be shared among many individuals, and that the accuracy and usability of posted material will gradually improve. It is, however, unlikely that a system can be devised that allows individuals to correct OCR errors, add formatting information, and still guarantee the integrity of the conversion. In contrast, with our current system of hardcopy dissemination, the publisher's good name serves as the guarantee that the authors' work is faithfully reproduced.

Even if only a small fraction of the technical community makes an occasional contribution, the aggregate effect may be significant.

3.5. *Location of the Material*

Where the material is stored may be determined by security considerations related to cost recovery and by network efficiency. Except for the on-line corrections just mentioned, most of the access to the technical material is likely to be read-only. Therefore, its availability on multiple servers would enhance service. It is possible that large organizations, such as corporations, research centers, and universities, will want to keep frequently-used material accessible on local file servers. The Uniform Resource Locator (URL) and Uniform Resource Citation (URC) schemes in the Web can be used to facilitate the economic retrieval of multiply-distributed material. Except for differences in transmission delay, the storage location will be transparent to readers.

4. Presentation Alternatives

For the purpose of this paper, "technical articles" are the contents of scholarly and technical periodicals, conference proceedings, and collections of reprints. Technical articles typically consist of single or multi-column printed text, intermixed with line art and half-tones. Each article includes a number of different functional components (such as running header, title, byline, author's affiliation, copyright notice, abstract, figures, figure captions, tables, footnotes, references, etc.) that are visually separated by well-established layout conventions. The bulk of the textual material is "bodytext" typically printed in a single 8–12 point typeface including boldface, italics, or small caps. This section discusses alternative encoding methods that, in turn, affect how the material appears on the Web. The examples provided are presented in the currently available version of HTML (HTML+, soon to be renamed as HTML3.0), which is referred to hereafter simply as HTML.

Technical articles may be presented to WWW users in several different forms that differ greatly in cost of preparation, functionality, and storage requirements. The different types are formatted and unformatted text, and images.

4.1. *Unformatted Text*

The simplest format is plain text represented simply as a character stream. Minimal formatting may be included in this form by the use of tabs and line breaks. The cost of conversion of archival articles into this form depends on the desired accuracy. If a few mistakes in each paragraph can be tolerated, then the output of most commercial OCR devices can be accepted with only minimal proof-reading to locate major mishaps. Line art, half-tones, complex tables, and formulas are omitted. Since no typeface, type size, or leading information is retained from the original article, every article is reproduced in essentially the same format — somewhat like current electronic mail.

Since the disappearance of EBCDIC, 7-bit ASCII and its various 8-bit extensions have been the dominant coding standard. ASCII is gradually being overtaken by Unicode, which encodes each symbol in the world's major languages, including Chinese [22]. Many commercial OCR systems can recognize the symbols used in major European languages (including Cyrillic typefaces for Russian), and Chinese and Japanese OCR systems have also been available for several years. [d]

An OCR-converted paragraph can be presented in WWW simply by enclosing it in the paragraph "container" in HTML. Tabs can be set by the <LIT> (literal) element and line breaks can be effected by the <L> (explicit line break) element. This would work as intended as long as the word wrap is disabled for the current paragraph. Problems can arise when the technical article is set in a single column and its line length is too long to be displayed on the screen. HTML currently uses the 8-bit Latin-1 character set (ISO 8859) for English text. Conversion to and from the user set (ASCII or EBCDIC) is likely to be handled by the network protocol used to retrieve the document. Unusual characters are included by using SGML entity definitions such as &name; as in "Poincaré" for Poincaré. It is also possible to encode subscripts and superscripts, for example, e^{log^x} would produce e^{log^x}.

4.2. *Formatted Text*

Current document interchange formats (ODA, SGML) make a clear distinction between physical and logical layouts. The former is the publisher's domain of concern — typeface, type size, leading, line, column, and page breaks, placement of floating figures and tables — while the latter is the author's domain of concern — titles, sections, paragraphs, footnotes, references. In the Web the author can specify the logical layout in HTML but, for the most part, must defer to the local browser for decisions related to the physical layout.

4.2.1. *Physical layout*

While there have been some demonstrations of combined DIA and OCR that preserve almost all formatting information [23], current commercial systems cannot achieve sufficiently accurate symbol and typeface recognition without extensive post-editing. The cost of the necessary post-editing exceeds, in general, the cost of key-entry from scratch. For an example of faithful page recomposition, see discussion of Adobe's Acrobat [24, 25].

[d] See also the chapter by Spitz in this book.

Even when the complete physical layout is known, reproducing it exactly on the Web may be nearly impossible because of the autonomy accorded to the browsers. HTML however, has certain features useful for specifying physical layout. Some of these were pointed out in the previous section. Others include the following:

• Presentation-only tags may be used to give hints to the browser as to how the text should be rendered, e.g., italic, bold, or underline. These tags may be nested for combined effect. Tags used are <B> for bold text, <I> for italic text, and <TT> for fixed-width ("teletype") text.

• Computer output and plain text files are rendered in the fixed-width font using the <PRE> (preformatted) tag. The white spaces and line breaks are preserved. Hence,

```
<PRE>
for (i = 1; i < MAXSIZE; i++)
   array[i] += 2*array[i-1];
</PRE>
```

will be displayed as is.

• Images and figures can be included as character-like or (floating) paragraph-like elements respectively. An in-line image is included using the image tag: <IMG SRC=image-URL>, where image-URL is the Uniform Resource Locator of the image file.

4.2.2. *Example 1*

We will use Figure 1 as a running example. Figure 2(a) shows an HTML encoding to preserve the ASCII content up to the end of first paragraph. The title is used primarily for document identification purpose. The *body* of code following the title is bracketed by the preformatted tag pairs <PRE> and </PRE>. The example also uses two tags that were not discussed above: <HR> produces a horizontal rule line across the page and <STRONG> displays bold type for strong emphasis. The resulting output in Figure 2(b) does preserve the lines in the original document but looks very different otherwise, primarily because of the use of a fixed-width font. The loss of right column justification is also noticeable.

4.2.3. *Logical layout*

Automatic extraction of logical entities in a document is beyond the current state-of-the-art, except when the problem is posed for a very restricted domain. More than half the cost of conversion in the current CD-ROM projects involves the cost of manual "markup" [2].

HTML provides only rudimentary capabilities for logical markup because of the need to cater to a large variety of platforms with widely differing display capabilities. Headers (six levels), paragraph styles (plain, quotes, abstracts, bylines, and notes) and lists (four types) are well supported but "support for tables and figures is currently thin on the ground" [26] and only relatively simple ways are provided for representing mathematical equations. With increasing commercialization of the Web, however, work is underway to determine how HTML can support style sheets that give a unique appearance to a particular vendor's documents.

VOL 67 No 10 PHYSICS ILLUSTRATED

Symmetry Breaking Mechanisms In Closed Laundry Loads

R BENNINK, N CHOU, H LI-TSUNG, W XIANG

Virtually everyone who has done laundry is familiar with the phenomena of finding, at the end of the whole procedure, a single unmatched sock. Exactly what has happened to the other sock is a question which physicists have grappled with for over two centuries. For indeed, the following note was found scribbled in the margin of one of Isaac Newton's mathematical textbooks:

> I have discovered a most curious and contrary phenomenon: when the chamber-maid washes the breeches and stockings, such being sloshed and spun about for some while, and thence retrieves them from the tub, there are on rare occasions discovered to be fewer stockings in the tub than were previously put into the tub. Does not every contemplation indicate stockings are by nature manifest, and therefore ought not be subject as phantoms to transience?[1]

Numerous theories have been put forth over the decades to explain the mysterious disappearance of socks. Bound by the law of mass conservation, classical physicists sought to describe the process in terms of statistical thermodynamics: In the Ideal Wash model, socks undergo elastic collisions with one another, constantly transferring kinetic energy back and forth. If socks obey the well known Hosé-Einstein statistics, there is a finite probability that a single sock will be temporarily excited to a high energy state. This energy, some proposed, would be sufficient to completely vaporize the sock. However, a number of experiments performed by Carnot and Lord Kelvin in the mid-19th century proved this theory wrong. It was found that socks have a high heat of vaporization, much higher than the energy supplied by hand scrubbing (or even modern washers and driers).

The dryer may be thought of as a macroscopic version of an ordinary atom (*Figure 1*).

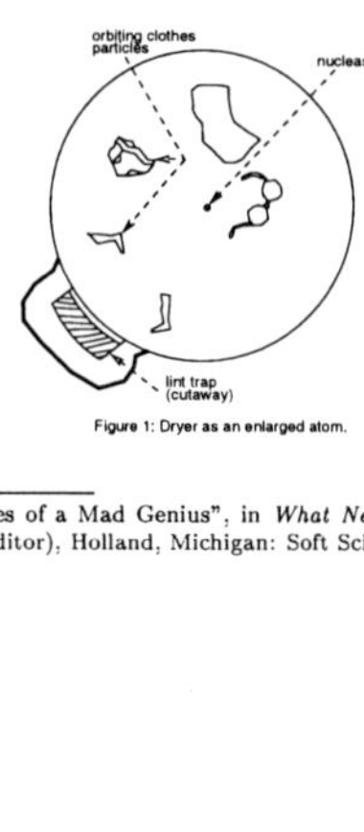

Figure 1: Dryer as an enlarged atom.

[1] L. A. Brown, "Anecdotes of a Mad Genius", in *What Newton Means to the Twentieth Century*, S. I. Hemenway (editor), Holland, Michigan: Soft Science Press, 1986

Figure 1.: Sample document used throughout to illustrate HTML encoding, browser results, and hyperlinks.

(a) HTML

```
<TITLE>Example 1: Reproducing physical layout</TITLE>
<PRE>
<L> <b>VOL 67  No 10                                    PHYSICS ILLUSTRATED </b>

<HR>
<STRONG>Symmetry Breaking Mechanisms In Closed
Laundry Loads</STRONG>

R BENNINK, N CHOU, H LI-TSUNG, W XIANG
<HR>


Virtually everyone who has done laundry is familiar with the phenomena of
finding, at the end of the whole procedure, a single unmatched sock.  Exactly
what has happened to the other sock is a question which physicists have grappled
with for over two centuries.  For indeed, the following note was found scribbled
in the margin of one of Isaac Newton's mathematical textbooks:
</PRE>
```

....

(b) Web Output

```
VOL 67  No 10                                    PHYSICS ILLUSTRATED
```

Symmetry Breaking Mechanisms In Closed
Laundry Loads

R BENNINK, N CHOU, H LI-TSUNG, W XIANG

```
Virtually everyone who has done laundry is familiar with the phenomena of
finding, at the end of the whole procedure, a single unmatched sock.  Exactly
what has happened to the other sock is a question which physicists have grappled
with for over two centuries.  For indeed, the following note was found scribbled
in the margin of one of Isaac Newton's mathematical textbooks:
```

....

Figure 2.: (Example 1). HTML code for preserving the ASCII content of Fig. 1 and the corresponding Web output. Only the first paragraph is shown for illustration.

4.2.4. *Example 2*

An HTML encoding of the previous example, intended to preserve the logical layout, is shown in Figure 3(a). The HTML code has the <PRE> tag for the preformatted document header but otherwise uses several of the logical tags mentioned above. The number following H in the header tag indicates the level. The <P> tag is used as a paragraph separator. Further formatting options can be specified within the <P> tag, as in <P align=center> to obtain a centered paragraph. The displayed output in Figure 3(b) is in a proportional font but does not preserve text lines of the original. Web users can change the font of the displayed material from a pulldown menu, thus any attempt to reproduce the original font would be futile.

4.3. *Hypertext*

HTML can link parts of text (or regions of images) to other resources (documents, images) by way of hypertext references or <HREFS>. Both the starting point and destination are called anchors. Anchors may have several attributes but they must have one or both of the NAME and HREF attibutes. The syntax of such a statement would be:

<A NAME="name" HREF="destination URL"> text-to-be-highlighted</A>

The browser highlights the region between the <A> and </A> tags and interprets clicks to retrieve the referenced document.

We separate *internal* links from *external* links because the two pose very different requirements on DIA. It is not unreasonable to expect DIA to deduce the links internal to a document (citations, such as, *see [Brown:86, p. 113]*, or *see Figure 2*) so that they may be inserted automatically in HTML. On the other hand, automatic insertion of external links belongs properly in the domain of information retrieval.

4.3.1. *Example 3*

We illustrate how an internal link at the end of the quoted paragraph can be tied to the footnote reference. In the HTML code, shown in Figure 4(a), the anchor for the target (footnote) is named "footnote_ref" using the NAME attribute and referenced in the citation by "#footnote_ref". The resulting output in Figure 4(b) has the citation "[1]" highlighted on the display which, when clicked, brings the footnote reference into the display window. Another approach to display the target material is illustrated for the figure citation in our example. When "Figure 1" after the icon (highlighted on the display) is clicked, the figure is displayed in a new window. In the HTML source (Figure 4(a)), the target gif image, "dryer.gif," is specified in the HREF attribute. The example also shows how an iconic image can be used as a "hot button" in addition to the highlighted text.

4.4. *Facsimile*

Pages of print are most easily posted on the Internet in the form of bit-mapped images. The typical work-station screen of about 800×1100 dots accommodates a full page of text at about 100 dpi, in contrast to the 300-1200 dpi used in office

(a) HTML

```
<TITLE>Example 2: Logical layout specification in HTML</TITLE>
<PRE>
<B>VOL 67  No 10                      PHYSICS ILLUSTRATED </B>
</PRE>

<HR>
<H1>Symmetry Breaking Mechanisms In Closed Laundry Loads</H1>

<H3>R BENNINK, N CHOU, H LI-TSUNG, W XIANG</H3>
<HR>
<P>
        Virtually everyone who has done laundry is familiar with
the phenomena of finding, at the end of the whole procedure, a
single unmatched sock.  Exactly what has happened to the other
sock is a question which physicists have grappled with for over
two centuries.  For indeed, the following note was found
scribbled in the margin of one of Isaac Newton's mathematical
textbooks:
```

. . . .

(b) Web Output

VOL 67 No 10 **PHYSICS ILLUSTRATED**

Symmetry Breaking Mechanisms In Closed Laundry Loads

R BENNIK, N CHOU, H LI-TSUNG, W XIANG

Virtually everyone who has done laundry is familiar with the phenomena of finding, at the end of the whole procedure, a single unmatched sock. Exactly what has happened to the other sock is a question which physicists have grappled with for over two centuries. For indeed, the following note was found scribbled in the margin of one of Isaac Newton's mathematical textbooks:

. . . .

Figure 3.: (Example 2). HTML code for preserving the logical layout of Fig. 1 and the corresponding Web output. Only the first paragraph is shown for illustration. Compare this with Fig. 2.

(a) HTML

```
<BLOCKQUOTE>
        I have discovered a most curious and contrary phenomenon:
when the chamber-maid washes the breeches and stockings, such
being sloshed and spun about for some while, and thence retrieves
them from the tub, there are on rare occasions discovered to be
fewer stockings in the tub than were previously put into the tub.
Does not every contemplation indicate stockings are by nature
manifest, and therefore ought not be subject as phantoms to
transience? <A HREF = "#footnote_ref">[1]</A>
</BLOCKQUOTE>
<P>
The dryer may be thought of as a macroscopic version of
an ordinary atom.

<A HREF = "dryer.gif"><IMG ALIGN=bottom SRC=dryer_small.gif>Figure 1</a>.

<P>
<A NAME = "footnote_ref">
[1] L. A. Brown, "Anecdotes of a Mad Genius", in
<EM> What Newton Means to the Twentieth Century</EM>, S. I. Hemenway
(editor), Holland, Michigan: Soft Science Press, 1986.
</A>
```

(b) Web Output

...

> I have discovered a most curious and contrary phenomenon: when the chamber-maid washes the breeches and stockings, such being sloshed and spun about for some while, and thence retrieves them from the tub, there are on rare occasions discovered to be fewer stockings in the tub than were previously put into the tub. Does not every contemplation indicate stockings are by nature manifest, and therefore ought not be subject as phantoms to transience? [1]

...

The dryer may be thought of as a macroscopic version of an ordinary atom.  Figure 1.

[1] L. A. Brown, "Anecdotes of a Mad Genius", in *What Newton Means to the Twentieth Century*, S. I. Hemenway (editor), Holland, Michigan: Soft Science Press, 1986.

Figure 4.: (Example 3). HTML source code and Web output illustrating citation links for a footnote and a figure. Only the relevant portions of the document in Fig. 1 are shown in this illustration.

printers and OCR scanners. Ten or eleven point type remains quite legible, but subscripts and some mathematical symbols may be lost. The effect is roughly comparable to low-resolution fax. Many line drawings, already reduced as much as possible because of the page-count limitations of journal editors, cannot be interpreted. Visual noise is also introduced when the original bitmap is resampled for a slightly different screen resolution. However, even for high-contrast material, scanning and displaying the pages at a six- or eight-bit grey-scale improves legibility by reducing aliasing (staircase effect).

Of course, if the original journal pages were sampled initially at a sufficiently high sampling rate, and all of the data was preserved, then selected sections could be magnified. This works for columnar layouts, because magnified text organized in narrow columns can be readily scrolled vertically. However, scrolling wide paragraphs horizontally impedes readability to an unacceptable degree. One solution is "image text-line wrapping" [27, 28].

At 300 dpi, a typical bitmap image of a page requires one megabyte (without compression), therefore downloading a typical 20-page article requires substantial storage. At current line speeds, transmission delays may also be unacceptable. Although much greater bandwidth channels are already being installed, we will be fortunate if the increase in bandwidth keeps up with the increase in traffic, and delays do not increase overall. Current compression methods (TIFF, CCITT Group-4), reduce packet sizes by a factor of ten to twenty, but without special hardware the decompression time is not negligible.

Facsimile images can be displayed on a Web browser by specifying the location of the image or images using the <href> tag, e.g.,

<A HREF="http://www.someschool.edu/~nvs/page1.gif"> Page one in facsimile format.</A>

By clicking on the highlighted text ("Page one in ..."), the facsimile image is retrieved from the remote site and displayed. Documents prepared for the Web often have table-of-contents hyperlinks to facilitate non-linear browsing. However, providing similar access to the page images of a thirty-page technical article would require building the table of contents manually. Instead, it may be adequate to provide only sequential access to the article. Implied "next" and "previous" buttons could be placed for this purpose in the header field of each page-image file. These buttons will be linked to subsequent and preceding pages of the same article on the remote server.

4.5. *Formatted Facsimile*

The large transmission delays, storage requirements, and display size associated with page-based facsimile can be mitigated by subdividing the pages into functional components. This gives the user the option of viewing, for instance, only the title block, abstract, consecutive columns of texts, specific figures, or the references. Each block can be separately encoded into image form, possibly with higher resolution for some components and with optimal compression based on block classification [29]. Links may be inserted between the various components of the image in order to facilitate navigation through the article, and each page may be displayed in iconic form at highly reduced resolution [27]. Small digitization errors in scanned images may be interactively corrected with a bitmap editor such as Image-EMACS, which can also be used for limited searches [30].

Fully decomposed page images may be retrieved using the following HTML incantation:

<A HREF="http://www.someschool.edu/~nvs/page1.gif"> Page one of article.</A>

Clicking on the highlighted text would retrieve a geometric representation of page one of the article with rectangular outlines of components with their functional identities inscribed [27]. Now, by clicking on different parts of this "page," the image-blocks corresponding to the selected components of the document page are retrieved and displayed.

The IMG tag element in HTML can be used to set up "image maps," a feature of HTML designed to produce just this effect. An image map is designated by including the ISMAP attribute in the IMG element as follows:

<A HREF="http://www.someschool.edu/~nvs/imagemap.spec">
<IMG SRC="page1.gif" ISMAP ></A>

The coordinates of the point where the user clicks is sent to the server. The server must be configured to handle coordinate information. The image map specification file contains definitions of the correlation between regions of the image (derived from DIA) and other documents which contain the image data corresponding to these regions. A typical image map entry looks as follows:

rect http://www.someschool.edu/~nvs/page1/title.gif 0,0 299,30

It specifies a rectangular region on the Web display by its upper-left and lower-right points (X-window tools, such as *xv*, can be used to obtain this information) and includes the information required to access the image when the user clicks in this region. In general, regions may be defined by polygons and circles.

4.5.1. *Example 4*

The HTML code in Figure 5(a) would display the geometric representation of our example page shown in Figure 5(b). We assume that these image blocks have previously been segmented using DIA and stored as GIF images. The image map for the example is shown in Figure 5(c).

4.6. *Dual Representation: Text and Facsimile*

Plain text lacks many of the visual clues, developed over centuries, that improve legibility and comprehension. Facsimile cannot be easily searched for content words and, once downloaded, cannot be edited or otherwise manipulated (but see [28]). Making technical material available in both forms avoids these shortcomings, and may still be less expensive than producing symbolic representations of articles that include all the necessary formatting conventions.

Dual representation requires simple access from one representation to the other. For instance, when the reader sees a reference in the plain text to a mathematical

(a) HTML

```
<TITLE>Example 4: Selective retrieval of segmented blocks </TITLE>
<CAPTION> Click on the desired block to view full image </CAPTION>
<A HREF = /img/~seth/e4_imagemap.spec><IMG SRC = "icon.gif" ISMAP></A>
```

(b) Iconic Display
of Page Layout

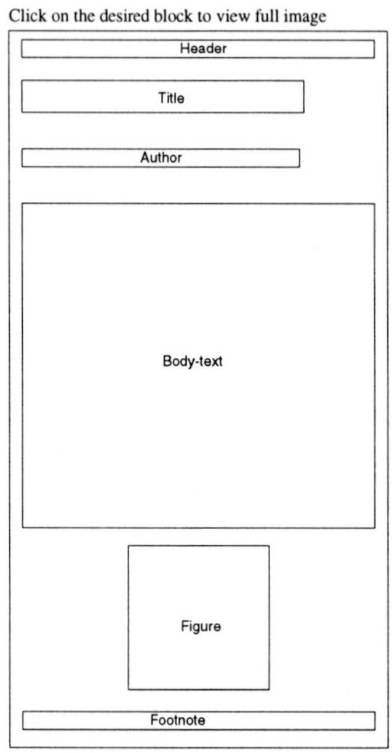

(c) Image-map
Specification File

```
rectangle (9,8) (409,28)          http://http.cs.unl.edu/~seth/header.gif
rectangle (9,54) (329,88)         http://http.cs.unl.edu/~seth/title.gif
rectangle (10,29) (324,149)       http://http.cs.unl.edu/~seth/author.gif
rectangle (10,189) (410,548)      http://http.cs.unl.edu/~seth/text.gif
rectangle (130,569) (288,730)     http://http.cs.unl.edu/~seth/figure.gif
rectangle  (10,753) (410,774)     http://http.cs.unl.edu/~seth/footnote.gif
```

Figure 5.: (Example 4). The geometric representation of the example page is shown. By clicking the mouse in the desired rectangle the corresponding document component is displayed in image form. The coordinates of these rectangles (which correspond to a functional component of the page) are stored in a separate image-map specification file.

formula or a figure, it should be possible to switch quickly to the facsimile form of that component.

In dual representation non-text material and document components, such as equations and tables whose formats must be preserved, can be retained in image form while all other components can be stored in either form. (HTML does contain rudimentary tags to format tables, horizontal rules, and math expressions; these capabilities are likely to improve because they are used so widely in technical communication.) When most of the article is presented in ASCII, the image portions can be displayed as in-line images. With new features to permit text to flow around images a rendition closer to the original is possible.

4.6.1. *Example 5*

We illustrate the dual representation in Figure 6 for the example shown in Figure 5. To accomplish this, we modify the image map specification file as shown in Figure 6(a) so that clicking the mouse on a rectangular block results in the Web display window driven by an HTML file uniquely associated with the block. An example file for the title block is shown in Figure 6(b). Along with the image of the title block, it creates a hyperlink to the dual HTML file (Figure 6(c)) that contains the ASCII text of the title block. The latter contains a reverse link that allows the user to switch between the two representations shown in Figures 6(d) and 6(e).

The formatting achieved using HTML may only be a reasonable approximation of the original. Therefore, when a document is converted, the original information may be lost not only during OCR/DIA conversion but also during HTML conversion. Even with all the enhancements in HTML3.0 it cannot match the sophistication of Latex. HTML is still evolving with major standardization efforts underway. The latest information on this topic may be obtained from the Web [31].

5. DIA and OCR Requirements

Since it is already possible to prepare *any* material for posting on the WWW in acceptable form, the first challenge for DIA and OCR is to perform the conversion less expensively than is possible by key-entry of the text and interactive reformatting of all illustrations. While cost figures are obtainable, the required level of "acceptable" quality is not yet understood. We list below the tasks necessary for the various methods of presentation discussed above.

5.1. *Low-accuracy OCR*

Commercial systems typically achieve 98–99% character accuracy on pre-zoned magazine articles [32], and similar performance can be expected on technical articles. At 98%, the typical 4000-character page of dense print will have 80 errors, but most of the material will be quite legible, and the accuracy is acceptable for full-text search [33].[e] If the material is not manually pre-zoned, then considerably higher error-rates can be expected, because of occasional failures in locating columns, paragraphs, indentations, and problems with tables [32, 34, 35]. Fine-grained graphic fragments are interpreted by OCR as nonsense text that may affect information

[e]See also the chapter by Taghva *et al.* in this book.

(a) Image Map File

```
rectangle (9,8) (409,28)        http://http.cs.unl.edu/~seth/header_image.html
rectangle (9,54) (329,88)       http://http.cs.unl.edu/~seth/title_image.html
rectangle (10,29) (324,149)     http://http.cs.unl.edu/~seth/author_image.html
rectangle (10,189) (410,548)    http://http.cs.unl.edu/~seth/text_image.html
rectangle (130,569) (288,730)   http://http.cs.unl.edu/~seth/figure_image.html
rectangle  (10,753) (410,774)   http://http.cs.unl.edu/~seth/footnote_image.html
```

(b) HTML File to Display Image

```
<TITLE> <HREF=title_image.html>
Example 5: Dual Representation - Image
</TITLE>
<IMG ALIGN = top SRC=title.gif>
<BR>
<BR>
(<A HREF = "ascii.html"> Click here for an ASCII display of
the title block </A>)
```

(c) HTML File to Display ASCII

```
<TITLE>Example 5: Dual Representation - ASCII </TITLE>
Symmetry Breaking Mechanisms In Closed Laundry Loads
<BR>
<BR>
(<A NAME = "title"><A HREF = "title_image.html"> Click here to
see the image of the title-block </A>)
```

(d) Web Display of Title-block Image

Symmetry Breaking Mechanisms In Closed
Laundry Loads

(Click here for an ASCII display of the title block)

(e) Web Display of Title-block ASCII

Symmetry Breaking Mechanisms In Closed Laundry Loads

(Click here to see the image of the title-block)

Figure 6.: (Example 5). An illustration of the dual representation in the last example. Now, by clicking the mouse in a desired rectangle, the user can alternately view the ASCII and the bitmap image of the block. The example illustrates how this can be achieved by changing the image map specification to point to an HTML file that contains the bitmap image and a hypertext reference to the an HTML file that contains the ASCII text. The latter, in turn, has a reverse link to the image file, thus allowing the user to switch between the two representations.

retrieval efficiency [33]. So the major requirement for conversion to plain text is discrimination between text and nontext areas. Research on this topic has been actively pursued for many years.

Variations in thresholding (binarization) greatly affect OCR accuracy. Automated methods cannot find the ideal threshold setting, which varies from system to system [36–38]. Benchmarking also shows that significant skew leads to higher error rate, mainly because of formatting failures. Many commercial systems do not perform any skew correction, in spite of extensive research on this topic. An impediment to skew correction and OCR is the proliferation of color and white-on-black text in many journals [32].

5.2. *High-accuracy OCR with Format Preservation*

Current OCR systems are not very reliable in identifying boldface and italics, and are marginal in font identification. (Accurate font identification is also difficult for human operators.) Some systems accept common mathematical symbols (including the Greek alphabet), while others have provisions for training the system on new symbols. Preservation of all format information has been accomplished in research settings on carefully selected material [23]. However, considerable additional work is necessary for robust large-scale technical document conversion where typeface changes may carry significant information (e.g., in computer program listings) [39].

5.3. *Layout Analysis*

As mentioned, there are two principal aspects of layout analysis. The first is the separation of text, line art and halftones, and determination of the reading order of the text. The second is locating and labeling the twenty or thirty customary functional components of a technical article. This is necessary for "value-added" display of the article in image form, and for the insertion of hyperlinks.

Functional labeling has hitherto been accomplished only with document-specific systems [40, 41, 27]. The common components of a family of documents, such as articles from a single technical journal, are described by some set of rules. There are several alternative formalisms for embedding the rules [42], but no matter which one is used, the compilation of the rule-base for each document family is time-consuming and error-prone. Therefore the major challenge here is the development of automated learning schemes for document format analysis.

Once we have the labeled functional components of a document, we can generate HTML tags automatically to format the document for browsing on the Web. The most important requirement is that there be a one-to-one correspondence between the DIA-assigned page component identity and a HTML tag element. As each page is analyzed, HTML tags for the components must be generated, the textual regions converted to ASCII using OCR, and inserted between the appropriate tags. Each textual component must be searched to insert links from the citation to the image versions of figures, tables, math expressions and anything else that may be stored in image form. Code to highlight the appropriate phrases to invoke the links must also be introduced at this point. In the case of bibliographic citations, links are needed from the abbreviated citation in the text to the full citation in the references section. To provide a link to the actual cited document, some post-processing will be required to search indexes to determine the location of the cited document and use that to insert links in the document being browsed.

In order to produce HTML versions of the document used in Examples 1–5, there are certain requirements imposed on DIA and OCR systems. Clearly, OCR systems must recognize horizontal rules which today they have difficulty with. Reliable bold and italic font recognition is a must in order to depict document components such as titles and footnotes correctly. References to footnotes, a number within square brackets in this case, must be very accurately detected since they form anchors to hyperlinks. Another OCR challenge is to distinguish between hyphenated words and words broken up for line formatting. The use of a lexicon is called for when converting text for entry into a word-processing system. DIA systems must produce accurate functional descriptions of key document components for accurate conversion to HTML tags. Examples are headers, title, horizontal rules, authors, paragraphs, and footnotes. In the examples dealt with here, of particular importance is the delineation of quotations which often appear as indented paragraphs. A special classification is required so that it may be presented as a <blockquote> when the HTML source is generated automatically from the DIA and OCR output.

5.4. *Semi-graphics*

Tables and formulas are examples of common technical material containing predominantly symbolic information whose meaning is determined by their spatial relationship to one another.

Tabular material may simply be converted into lines of text for the benefit of human readers. This is what most OCR systems attempt, with mixed success [32]. The danger is columnizing them because table columns are superficially similar to columns of text. However, for computer manipulation, tables should be converted into spreadsheet or relational database format. Some research projects have been reported [43, 44, 45], but there has been no large scale demonstration. Indeed, some tables have a complex enough structure to require substantial subject-matter expertise for their comprehension. As an exercise, the interested reader may attempt to identify the appropriate row and column headings in tables from foreign-language publications.

Mathematical formulas also require considerable expertise for their interpretation. Here the crucial question is that of a desirable target representation. Equation handlers in TeX, Troff, and WYSIWYG word processors preserve only the appearance of formulas. They do not retain sufficient structural information for interpretation by a symbolic formula-manipulation program such as Mathematica, Maple or MathCad. Although a few attempts have been reported on parsing mathematical formulas [46] for the time being it seems likely that only the surface structure can be encoded. For WWW users, this is not necessarily preferable to a bitmap representation, since about the same amount of manual work will be involved in translating the formula into a form suitable for computer interpretation.

Similar considerations apply to other semi-graphics such as structural formulas in organic chemistry, organization charts, and musical notation.

5.5. *Uniformly Formatted Material*

Technical directories, catalogs, standards, dictionaries, printed bibliographies, and component lists contain a large amount of material in a complex but highly repetitive format. The utility of such material is greatly enhanced if it can be searched rather than presented in facsimile form. Therefore it may be cost-effective to develop specific rules for the conversion even if such rules apply to only a few volumes of such material. Many of the formalisms developed for rule bases are probably adequate to the task, but few large-scale experiments have been described. A rule-base for telephone-directory yellow pages, that elegantly combines OCR and DIA, is described in [28].

5.6. *Compression*

Image and text compression is not traditionally included in DIA and OCR. Nevertheless, techniques borrowed from these disciplines can greatly enhance compression performance, which in turn is critical to the usability of WWW for information retrieval.

The currently popular compression schemes were developed for facsimile a decade ago. They are typically based on one- or two-dimensional run-length encoding and yield a compression ratio of 10:1 to 20:1 on a typical page of text scanned at 300 dpi. Of course, higher compression ratios are achievable on more densely sampled pages.

Intelligent compression, based on partial recognition of the symbols on the page, can easily double or triple the compression ratio [47, 48]. Although the performance approaches that of textual encoding only asymptotically, it offers the advantage of complete format preservation. Furthermore, encoding the functional units of each page separately allows the transmission of partial pages, serving a role analogous to progressive resolution transmission for greyscale images.

For text, the most effective general coding methods currently in use seems to be arithmetic encoding [49, 50]. A compression ratio of 3:1 is typical for technical text.

6. Conclusion

There is little doubt that the popularity of the World-Wide Web has caught a lot of people by surprise. The number of users of the Web is doubling every 50 days or so and the amount of information on the net approaches four terabytes. Almost anybody with a net address and a server can post information accessible to the rest of the user community, but there is a shortage of serious material that can be used for educational and research purposes. While businesses muse about the use of the Web for interactive shopping, the DIA and OCR community has the means to address the shortfall in technical literature.

Among major challenges facing the DIA researchers are:

1. sensible and robust text/non-text discrimination,
2. accurate recognition of special symbols, type styles, and punctuation,
3. faithful page reconstruction (matching typeface, size, and layout),

4. identification of logical document components, and
5. development of standards for the representation of tables and formulae.

While all of these have been the subject of intensive research for many years, two additional requirements specific to the WWW have surfaced. The first is the automated generation of HTML code from DIA/OCR output. The second is the automated insertion of internal and external (bidirectional) links.

Unless we find effective solutions for these problems, there is a good chance that many scientific findings from the past will have to await rediscovery by a generation reluctant to extend scholarship beyond their display screens.

Acknowledgements

George Nagy gratefully acknowledges the support of the of the Central Research Laboratory, Hitachi, Ltd., of the Educational and Research Networking of Northern Telecom – BNR, Inc., and of the New York State Center for Advanced Technology (CAT) in Automation, Robotics and Manufacturing at Rensselaer Polytechnic Institute. The CAT is partially funded by a block grant from the New York State Science and Technology Foundation. We are grateful to Mahesh Chugani (RPI), Julie Borsack (UNLV), Kazem Taghva (UNLV), David Boas (U. Pennsylvania) and Don Sylwester (Concordia College, Seward) for comments on the drafts, and to Professor Hal Berghel (U. Arkansas) for guidance on cyberspeak. We also acknowledge Mr. R. Bennink (Hope College, Holland, Michigan) for permission to include and adapt a part of his student paper in our examples and to Professor S. Hemenway for bringing this paper to our attention.

Bibliography

[1] R. J. Vetter, C. Spell, and C. Ward, Mosaic and the World-Wide Web, *IEEE Computer* 27, 10 (1994) 49–57.

[2] R. G. Fuller, Personal Communication, 1993.

[3] E. A. Fox, R. M. Akscyn, R. K. Furuta, J. J. Leggett, Special Issue on Digital Libraries, *Comm. ACM* 38, 1 (1995).

[4] U. Manber and S. Wu, Glimpse: A Tool to Search Through Entire File Systems, TR 93–94, Department of Computer Science, University of Arizona, 1993. Available at: http://glimpse.cs.arizona.edu:1994/bib/.

[5] DIMUND, http://documents.cfar.umd.edu/.

[6] K. Preas and B. Preas, About the DALibrary: Its Design and Development, The DALibrary Version 1.0 Documentation Packet, (ACM Press 1992).

[7] G. Stix, The Speed of Write, *Scientific American* 12, (1994) 106–111.

[8] A. Okerson, Directory of Electronic Journals, Newsletters, and Academic Discussion Lists, Association of Research Libraries, (Office of Scientific and Academic Publishing, 1994).

[9] R. Bradford and B. Cerny and T. Nartker, A Preliminary Report on UNLV/GT1: A Document Image Library for Ground Truth Testing in Document Analysis and Character Recognition, *Symposium on Document Analysis and Information Retrieval*, Las Vegas, NV, Apr. 1992, 300–315.

[10] J. E. Knoerdel and S. W. Watkins, Document Interchange Format, *Naval Data Automation Standards*, *NAVDAC*, PUB 17.11, Naval Data Automation Command, Washington D. C. (1984) 1–33.

[11] P. Borsook, U. S. Navy Brings Out Standard for Word Processing Interchange, *Data Communications* (1984).

[12] R. A. Wilkinson, et al., *Proc. First Census Optical Character Recognition Systems Conference*, NISTIR 4912, NIST, Advanced Systems Development Image Recognition Group, Gaithersburg, MD, August 1992.

[13] D. K. Harman, Overview of the first TREC conference, *Proc. 16th ACM SIGIR International Conference on Research and Development in Information Retrieval*, 1993, 36–47.

[14] G. Geist, et al., *Proc. Second Census Optical Character Recognition Systems Conference*, NISTIR 5452, NIST, Advanced Systems Development Image Recognition Group, Gaithersburg, MD, May 1994.

[15] D. K. Harman (Ed.), *The Second Text Retrieval Conference (TREC-2)*, Special Publication 500–215, National Institute of Standards and Technology, Gaithersburg, Maryland, 1994.

[16] Rensselaer Libraries Public Services, The Electronic Reserves System (E–Res), September 1994.

[17] G. A. Story, L. O'Gorman, D. Fox, L. Levy Schaper, and H. V. Jagadish, The Right Pages Image-Based Electronic Library for Alerting and Browsing, *IEEE Computer* 25, 9 (1992) 17–26.

[18] L. O'Gorman, Document Spectrum for Page Layout Analysis, *IEEE Trans. Patt. Anal. Mach. Intell.* 15, 11 (1993) 1162–1173.

[19] K. W. Smith, OCLC: Changing the Tasks of Librarianship, *Library Hi Tech* 11, 3 (1993) 7–17.

[20] *Library Hi Tech Journal.* A Special Issue on: College Library Technology and Cooperation Grants, 11, 3 (1993).

[21] P. Ginsparg, First Steps Towards Electronic Research Communication, *Computers in Physics* 8, 4 (1994) 390–397.

[22] *The Unicode Standard Version 1.0*, Volumes 1 and 2, (Addison-Wesley Publishing Company, Reading MA, 1991).

[23] M. Okamoto and A. Miyazawa, An Experimental Implementation of Document Recognition System for Papers Containing Mathematical Expressions, *Structured Document Image Analysis*, eds. H. S. Baird, H. Bunke and K. Yamamoto, (Springer-Verlag, Berlin, 1992) 36–53.

[24] L. Press, The Internet is not TV: Web Publishing, *Comm. ACM* 38, 3 (1995) 17–23.

[25] P. Ames, *Beyond Paper: The Official Guide to Adobe Acrobat*, (Adobe Press, Mountain View, 1995).

[26] D. Raggett, A Review of the HTML+ Document Format, *Computer Networking ISDN Systems* 27, 2 (1994), 135–145.

[27] G. Nagy, S. Seth, and M. Viswanathan, A Prototype Document Image Analysis System for Technical Journals, *IEEE Computer* 25, 7 (1992), 10–22. Reprinted: *Document Image Analysis*, eds. L. O'Gorman and R. Kasturi, (IEEE Computer Society Press, Los Alamitos, 1995), 369–381.

[28] G. E. Kopec, and P. A. Chou, Document Image Decoding Using Markov Source Models, *IEEE Trans. Patt. Anal. Mach. Intell.* 16, 6 (1994) 602–617.

[29] R. Sennhauser and K. W. Ohnesorge, Document Image Compression Using Document Analysis and Block-Class-Specific Data Compression Methods, *Proc. SPIE Image and Video Compression*, 2186, San Jose, CA, Feb. 1994, 146–155.

[30] S. C. Bagley and G. E. Kopec, Editing Images of Text, *Comm. ACM* 37, 2 (1994), 63–72.

[31] http://www.w3.org/hypertext/WWW/MarkUp/MarkUp.html.

[32] S. V. Rice, J. Kanai, T. A. Nartker, The Third Annual test of OCR Accuracy, UNLV ISRI 1994 Annual Report, Information Science Research Report, University of Nevada, Las Vegas, NV, Apr. 1994, 11–38.

[33] K. Taghva, J. Borsack, A. Condit, and S. Erva, The Effect of Noisy Data on Text Retrieval, *J. Am. Soc. Inf. Sc.* 45, 1 (1994), 50–58.

[34] R. M. Haralick, Document Image Understanding: Geometric and Logical Layout, *Proc. 1994 IEEE Computer Society Conference on Computer Vision and Pattern Recognition*, Seattle, WA, Jun. 1994, 385–390.

[35] S. Randriamasy and L. Vincent, A Region-Based System for the Automatic Evaluation of Page Segmentation Algorithms, *Proc. IAPR Workshop on Document Analysis Systems*, Kaiserslautern, Germany, Oct. 1994, 29–41.

[36] P. W. Palumbo, P. Swaminathan, S. N. Srihari, Document Image Binarization – Evaluation of Algorithms, *SPIE Applications of Digital Image Processing IX* (697) 1986, 278–285.

[37] R. Smith, C. Newton, P. Cheatle, Adaptive Thresholding for OCR: A Significant Test, TR HPL–93–22, HP Laboratories, Bristol, UK, Mar. 1993.

[38] O. D. Trier and A. K. Jain, Goal-directed Evaluation of Binarization Methods, To appear: *IEEE Trans. Patt. Anal. Mach. Intell.* (1995). Preliminary version in *Proc. of the NSF/ARPA Workshop on Performance Versus Methodology in Computer Vision*, Seattle, WA, Jun. 1994, 206–217.

[39] Nagy, G., On the Frontiers of OCR, *Proc. IEEE* 40, 8 (1992), 1093–1100.

[40] L. O'Gorman and R. Kasturi, Document Image Analysis, (IEEE Computer Society Press, Los Alamitos, 1995).

[41] H. S. Baird, H. Bunke and K. Yamamoto, eds., *Structured Document Image Analysis*, (Springer-Verlag, Berlin, 1992).

[42] M. S. Krishnamoorthy, G. Nagy, S. Seth, and M. Viswanathan, Syntactic Segmentation and Labeling of Digitized Pages from Technical Journals, *IEEE Trans. Patt. Anal. Mach. Intell.* 15, 7 (1993), 737–747.

[43] A. Laurentini and P. Viada, Identifying and Understanding Tabular Material in Compound Documents, *Proc. 11th International Conference on Pattern Recognition (ICPR-11)*, The Hague, Netherlands, 1992, 405–409.

[44] T. Watanabe, H. Naruse, Q. Luo, and N. Sugie, Structure Analysis of Table-form Documents on the Basis of the Recognition of Vertical and Horizontal Line Segments, *Proc. International Conference on Document Analysis and Recognition*, Saint-Malo, France, 1991, 638–646.

[45] E. Green and M. S. Krishnamoorthy, Recognition of Tables Using Table Grammars, *Proc. Fourth Annual Symposium on Document Analysis and Information Retrieval*, Las Vegas, NV, Apr. 1995, 261–278.

[46] P. A. Chou, Recognition of Equations Using a Two-dimensional Stochastic Context-Free Grammar, *Proc. SPIE Conference on Visual Communications and Image Processing*, Philadelphia, PA, Nov. 1989, 852–863.

[47] R. N. Ascher and G. Nagy, A Means for Achieving a High Degree of Compaction on Scan-Digitized Printed Text, *IEEE Trans. Comput.* C-23, 11 (1974), 1174–1179.

[48] W. K. Pratt, et al., Constrained Symbol Matching Facsimile Data Compression System, *Proc. IEEE* 68, 7 (1980), 786–795.

[49] V. S. Miller, Data Compression Algorithms, *Proc. Symposium on Applied Mathematics* 34, American Mathematical Society, 1986, 107–117.

[50] C. B. Jones, An Efficient Coding System for Long Source Sequences, *IEEE Trans. Information Theory* IT-27, 5 (1981) 280–291.

Handbook of Character Recognition and Document Image Analysis, pp. 755–777
Eds. H. Bunke and P. S. P. Wang

CHAPTER 29

INFORMATION RETRIEVAL AND OCR

KAZEM TAGHVA, JULIE BORSACK, and ALLEN CONDIT

Information Science Research Institute
University of Nevada, Las Vegas
Las Vegas, Nevada 89154-4021, USA

This chapter undertakes an in-depth analysis of the interaction between OCR and information retrieval. In particular, after providing an introduction to information retrieval, we report on retrieval effectiveness from OCR generated document collections. It will be shown that, in general, average precision and recall is not affected while document term assignment, weighting and document ranking may be affected. We also point out that even though OCR errors do not affect average retrieval effectiveness, there are other consequences that should be considered when OCR text is applied.

Keywords: Information retrieval; OCR errors; Weighting systems; IR models; Post processing; Retrieval performance; Recall and precision; Relevancy; Document ranking; Feedback.

1. Introduction

Although optical character recognition (OCR) and information retrieval (IR) both manipulate text, their initial objectives were very different. In fact, their objectives began at opposite ends of the linguistic spectrum. OCR devices, as their name implies, were designed to recognize characters and convert them to another format; whereas, the goal of information retrieval systems is to represent a collection of documents where a single document is usually their smallest unit of information. But as OCR technology progressed and its uses broadened, a more global view of its input became inevitable. OCR devices that recognize printed text no longer perceive an input page as a set of unrelated, isolated, character patterns. A page's continuity is used to disclose more information about the text being recognized.

IR technology's purpose is to typify document content and be able to present this information upon request. Unfortunately, semantically representing natural language is problematic, especially for a diverse document collection. Therefore, information retrieval systems used for general application have resorted to simpler methods of text representation. One of the more useful tools applied in IR is statistical analysis. This kind of analysis breaks the document into smaller segments. Sections, paragraphs, sentences, words, and even character combinations can be used to delineate document content. Examination of document constituents has

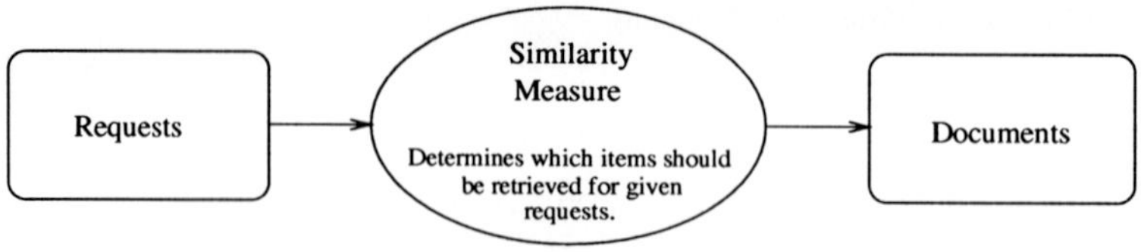

Fig. 1. A conceptual view of an information system.

contributed to our understanding of language structure and the qualities of printed text. So a relationship that may not have been initially obvious becomes more clear. From a progressive perspective, IR can be considered an extension of OCR, where the unified systems begin with an indivisible particle, the pixel, and produce an accessible collection of information.

It is the integration of these two technologies that is presented in this chapter. The first few sections introduce conceptual models as well as structures and methods applied in an IR system's implementation. Section 5 describes experiments combining these two technologies. The introduction to IR given here is by no means complete; we include only enough background for understanding the concepts presented in this chapter. Interested readers are referred to [1, 2].

2. Information Retrieval Systems

Information can be presented in many forms but in this chapter we concentrate on information in the form of typed text. Further, we will assume that the database of information consists of discrete units called documents. With this in mind, a high-level information system can be illustrated as shown in Fig. 1[1]. A set of information needs or requests is compared to a document collection to determine which documents satisfy the requests.

Figure 1 is only a conceptual depiction of an information system. A user request must be formalized and an analogous representation of the documents in the collection must be built prior to the comparison. So before the similarity measure can be applied, some resolution between the requests and the documents is performed. This resolution is defined through information retrieval models.

IR models provide a framework for the three elements of an IR system: its requests, its documents, and its similarity measure. The distinctions in their infrastructure give rise to the different outcomes and consequences related to introducing OCR text as input. Moreover, characteristics not inherent in a particular model, but nonetheless specific to certain systems, introduce other matters of concern. These underlying implementation considerations common to most IR systems will be discussed. Sections 2.2, 2.3, and 2.4 describe three well-developed models that currently influence information retrieval and have been examined with respect to OCR generated data. These models include: boolean model, vector space model, and probablistic indexing model.

document 1: OCR Devices

```
Optical character recognition devices translate printed
text to a computer readable form.
```

document 2: IR Systems

```
Information retrieval systems extract information from
computer readable text.
```

document 3: OCR and IR

```
Optical character recognition devices and information
retrieval systems can be used sequentially to translate
printed text to its computer readable form and then, with
an information retrieval system, extract this text upon
request.
```

Fig. 2. A document collection.

2.1. Implementation Considerations

The most convenient way of perceiving a database collection is as a set of documents. But in practice, the most common structure for document storage in a retreival model is an *inverted index.* An inverted index transposes the document-term relationship to a term-document relationship. For each term in the collection, the documents in which that term occurs are assigned to that term. An example collection with three short documents appears in Fig. 2, and its inverted index appears in Table 1. This implementation allows for immediate responses to user requests; the documents which correspond to terms that are found in both the query and the index can be easily identified and returned. Our example index stores all the words in our document collection, the documents containing those words, and the freqencies of the words in each document. Other information can be stored in an inverted index too, but as the amount of information increases, so does the overhead for storing this structure. In our various projects with OCR text, we use this information for analysis and correction[3, 4, 5, 6, 7]. Although not explicitly stated in the description of the models below, this is the document database representation employed.

Another common practice in most IR implementations is the removal of *stopwords.* Stopwords can be defined as those words in the text that do not add to a document's substance or meaning. There is usually a minimal set of stopwords provided by an IR system for exclusion during indexing, although this list can usually be tuned to a collection. An example stopword list might include: `the`, `and`, `to`, `a`, `in`, `that`, `through`, `but`, etc. Notice that the inverted index in Table 1 does not contain the stopwords in the document collection it represents.

Another system-specific enhancement that may influence the use of OCR data is *word stemming.* Word stemming removes suffixes (and in some systems prefixes)

Table 1. Document collection's corresponding inverted index.

Term	Docid:term frequency		
character	1:1	3:1	
computer	1:1	2:1	3:1
devices	1:2	3:1	
extract	2:1	3:1	
form	1:1	3:1	
information	2:2	3:2	
IR	2:1	3:1	
OCR	1:1	3:1	
optical	1:1	3:1	
printed	1:1	3:1	
readable	1:1	2:1	3:1
recognition	1:1	3:1	
request	3:1		
retrieval	2:1	3:2	
sequentially	3:1		
system	2:2	3:2	
text	1:1	2:1	3:2
translate	1:1	3:1	

to form *root words*. This normalization reduces many forms of the same word to a single common word stem to increase system effectiveness as well as reduce term storage overhead. The same algorithm is applied to query terms as well. There are usually three stemming choices offered in an IR system: full stemming, s-removal stemming, and no stemming. Full stemming removes an affix with the longest string of matching characters using a set of predefined rules. This kind of stemming usually produces word stems that are not actually words at all. To illustrate, `sacrifice` stems to `sacr`. S-removal replaces plural terms with their singular equivalent. In the experiments run with OCR text, only s-removal stemming was applied. Our intent was to eliminate as many variations to the text as possible without negatively affecting retrieval performance. If we had used full stemming, comparison of the index to the original OCR text would have been much more complex.

While IR systems accept free text, structure is applied for its implementation. The representations of both the documents and the requests are modified to improve efficiency and effectiveness. A modified representation of an IR system, shown in Fig. 3, illustrates these changes [8]. The modifications shown change the original input; this is true of both clean and OCR text. The impact of these changes on the recognized versions, together with the effect of OCR errors on retrieval performance, is the focal point of this chapter.

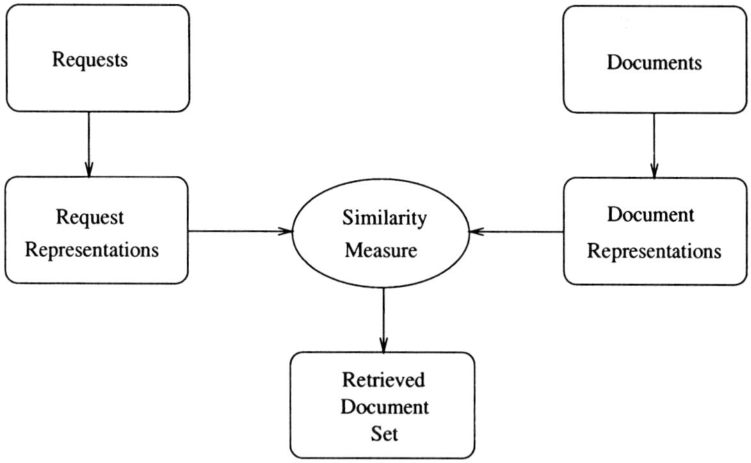

Fig. 3. Expanded information system.

2.2. Boolean Model

The boolean model is named for its method for formulating user requests. In a pure boolean model, the request or *query* is represented as a set of terms joined by the logical operators `or`, **and**, and `not`. The similarity measure becomes the evaluation of a boolean query against the document collection. The documents retrieved represent the satisfiability of the boolean expression. If a document satisfies the expression, then a *true* value results and the document is considered relevant; a *false* value indicates non-relevance. True and false in this context refer to the presence or absence of the query term in a document in the database. More formally, if we let A and B represent query terms, then:

(A `or` B) is true if and only if either A or B is true.

(A **and** B) is true if and only if both A and B are true.

(`not` A) is true if and only if A is false.

Referring again to Table 1, we can show some concrete examples:

Query 1: `recognition or retrieval`
documents returned to the user: 1 2 3

Query 2: `sequentially` **and** `readable`
documents returned to the user: 3

Query 3: `not translate`
documents returned to the user: 2

In theory, the inverted index is searched for each query term, the documents assigned to each term are returned, and the logical operator(s) are applied to the term results. This final operation produces the set returned to the user.

Two problems associated with the boolean model are:

- The complexity of query formulation and its interpretation.
- The lack of ranked document output.

The syntactic structure of a boolean query language is simple. With unambiguous parsing rules and a set of axioms, the evaluation of a query is clear. But the more simplistic the language, the more tedious it becomes for the user to express complex relationships. To further confuse the issue, the order in which operations are executed may change a query's results. If the parsing rules are not fully understood by the user, it may not be clear why a certain set of documents was returned and its complement excluded. For example, using the inverted index in Table 1 and the following query:

Query: `character and recognition or retrieval`

If the parsing starts at the left and moves right, the documents retrieved will be: 1 2 3. If the parsing is done from right to left the results will be: 1 3. Salton states, "In general, formulating boolean queries is an art that is not accessible to uninitiated users [8]."

A boolean query returns a result set by partitioning the document collection into two parts—the retrieved part and the non-retrieved part. Even if we assume that all the documents retrieved are relevant, it is still left to the user to determine which documents are *most* relevant. In a large document collection, this filtering may not be feasible. Document ranking addresses this problem but is not easily incorporated into the boolean model [8].

2.3. Vector Space Model

In the vector space model, each document is denoted by a vector of concepts (i.e. index terms). Further, this model extends the vector representation to its queries. Every term in the collection is represented in each document and query vector. If a term occurs in a document, then a 1 would be placed in the corresponding position in that document's vector, the absence of a term in a document is represented by a 0. To illustrate this, looking back at the inverted index in Table 1, the binary vector for document 1 would be:

(1 1 1 0 1 0 0 1 1 1 1 1 0 0 0 1 1)

The 1 in the first position of the vector corresponds to the word `character`—the first word in the inverted index.

An alternative approach would be to weigh the value of term i in its vector based on its importance to the document or query. For example, a term's frequency in a document could be used to give more weight to terms that occur more often. The most commonly employed term weighting algorithm is the $tf \bullet idf$ weight which uses the frequency of a term in a single document (tf or term frequency) balanced by the number of documents to which the term has been assigned (idf or inverse document frequency). In this weighting scheme, the weight of term i in document j is defined as:

Table 2. The weights for three terms in the example collection applying $tf \bullet idf$ weighting.

Term i	Document j	df_i	t_{ij}	w_{ij}
character	1	2	1	0.17609
computer	1	3	1	0.00000
retrieval	3	2	2	0.35218

$$w_{ij} = tf_{ij} \cdot \log \frac{N}{df_i} \tag{2.1}$$

where

$N \equiv$ total documents in the collection
$df_i \equiv$ number of documents containing term i
$t_{ij} \equiv$ frequency of term i in document j

Table 2 shows the weighting for three terms using the frequencies from the inverted index in Table 1 with N constant at 3. Note that the term `computer` is assigned a weight of 0. Since `computer` is in every document in our collection it has no discrimination value. We can see from the term weights that the objective of this weighting scheme is to assign a higher weight to document terms that have high occurence in a few documents. Many other term weighting methods can be used; details of these are described in [9].

Queries can be submitted to an IR system as either a list of terms or as a natural language command. In either case, the system treats the query like a document and builds a vector representation. The query vector can now be easily compared with the documents. There are a number of similarity measures that can be used to make this comparison, but the most natural measure is the cosine of the angle between two vectors.

If we let,
$D_j = (w_{1j}, w_{2j}, \cdots, w_{mj})$
denote a document vector with m term weights, and
$Q_r = (q_{1r}, q_{2r}, \cdots, q_{mr})$
denote a query vector with m term weights, then document-query similarity can be computed as:

$$cosine(D_j, Q_r) = \frac{\sum_{k=1}^{m} w_{kj} * q_{kr}}{\sqrt{\sum_{k=1}^{m} w_{kj}^2 * \sum_{k=1}^{m} q_{kr}^2}}. \tag{2.2}$$

Since the terms are considered uncorrelated, the term vectors are orthogonal and therefore linearly independent. Other similarity measures can be found in [8].

The vector space model resolves some of the problems of the boolean model:

- The query is easier for the user to formulate since it consists of a set of relevant terms, usually entered using natural language. No logical operators need be considered.
- Since ranking can easily be introduced into the retrieval system through term weighting, the user has more information about the probability of relevance in the retrieved set.

Some of its disadvantages are its assumed term independence, its arbitrary selection of a weighting technique, and its similarity function to determine relevance [10].

In the vector space system we used for our experimentation, a technique known as *relevance feedback* is implemented. Although not unique to this model, we discuss it here because feedback was incorporated into our vector space experiment with the SMART retrieval system. Relevance feedback is an automatic process that uses information given to the system from previously run queries. For instance, when a user makes a request, certain documents are returned by the system. From this retrieved set, some are relevant and some are not. The relevant documents are selected by the user. The next time a similar query is run, information from these previous query results can be used. Relevant document terms can be used to expand this new query and existing query terms can be reweighted. This method of query modification has proven effective in experimental environments [11]. The employment of relevance information is also the fundamental principle behind the probabilistic indexing model described in the next section.

2.4. Probabilistic Indexing Model

In 1976, Robertson and Spark Jones [12] introduced a probabilistic retrieval model that uses relevance information to weigh search terms. In principle, for a given query, we can assign higher weights to terms that appear in previously retrieved relevant documents. Following Robertson and Spark Jones [12], for a term t and query q, if we let

$$\begin{aligned} N &\equiv \text{number of documents in a collection} \\ R &\equiv \text{number of relevant documents to } q \\ n &\equiv \text{number of documents containing } t \\ r &\equiv \text{number of relevant documents containing } t \end{aligned}$$

then Table 3 represents the distribution of terms with respect to relevant and non-relevant documents.

This table can be used to obtain formulas reflecting the relative distribution of terms. One formula in particular,

$$\mathcal{F} = \log\left[\frac{\frac{r}{R-r}}{\frac{n-r}{(N-n)-(R-r)}}\right] \tag{2.3}$$

represents the ratio of the proportion of relevant documents to the proportion of non-relevant documents in which the term occurs. The formula $\mathcal{F}$ can be used to assign weights to query terms if the relevancy information is available.[a] Spark Jones

[a] This is not a realistic assumption.

Table 3. Term distribution.

		Document relevance		
		+	–	
Term in	+	r	n - r	n
document	–	R - r	(N-n) - (R-r)	N - n
		R	N - R	N

in 1979 [13, 14] showed that by adding the relevance weighting from even an initial query search can lead to a better reweighting of search terms, and consequently, better system performance.

Later, work by Croft and Harper [15] and Croft [16] led to the reformulation of the formula $\mathcal{F}$ without relevance information. This formula is applied in INQUERY [17], the probabilistic indexing retrieval model used in our experiments. This system uses the following formula to assign term weights:

$$k + \left[(1-k) * \frac{\log t_{ij}}{\log maxtf_j} * \frac{\log \frac{c}{f_i}}{\log c} \right] \tag{2.4}$$

where

k ≡ constant, adjusted based on collection size
c ≡ collection size
f_i ≡ number of documents in which term i occurs
t_{ij} ≡ frequency of term i in document j
$maxtf_j$ ≡ maximum term frequency in document j

3. Measuring Retrieval Effectiveness

Determining an OCR system's effectiveness is typically conducted with character accuracy evaluation. And of course, words, sentences, and documents are made of characters; so, in some sense this is the most basic measure. But for full text documents to be accessed in an IR environment, character accuracy may not be the most applicable criterion for judging the input's usefulness. The ability of an IR system to retrieve OCR generated documents would be more meaningful and would give a more appropriate assessment of the goodness of an OCR system for this application. For this reason, we examine and evaluate retrieval of OCR text within the IR environments described.

In Sections 2.2, 2.3, and 2.4, we discuss the fundamental principles of three information retrieval models. These are presented to give a more complete understanding of the interaction of IR systems with OCR text. But to fully grasp our projects' results, the methods used to evaluate IR systems must also be understood. In this section, we describe some generally accepted methods for evaluating IR system effectiveness.

Information retrieval systems can be evaluated using various criteria. If we direct our attention to the concerns of the users of an IR system, Salton and Van

Table 4. Using the contingency table below, recall and precision formulas can be calculated.

	Relevant	Not relevant
Retrieved	a	b
Not retrieved	c	d

$$a + b + c + d = \text{total document collection}$$

$$\text{Recall} = \frac{a}{a+c}$$
$$\text{Precision} = \frac{a}{a+b}$$

Rijsbergen [1, 18] both point to six criteria which are considered critical in an IR evaluation:

- **Coverage:** the extent to which the system includes relevant documents.
- **Time lag:** the average time it takes to produce an answer to a search request.
- **Presentation:** the quality of the output.
- **Effort:** the energies put forth by the user to obtain the information he seeks.
- **Recall:** the proportion of relevant material received from a query.
- **Precision:** the proportion of retrieved documents that are actually relevant.

Effort, time, and presentation are easily evaluated [1]. Coverage deals with the breadth of the collection and is not directly related to system performance. The last two criteria, *recall* and *precision*, measure a system's effectiveness. How well can a system find documents that are *relevant* to a user's request?

Relevancy is difficult to quantify because of its subjectivity. If the same query is run by different searchers, their judgement of document relevancy will differ. In experimental situations, relevancy assessments are made by experts. A set of queries are defined for which the correct responses are known. In this way, the documents retrieved by the system can be compared to this known set of correct responses. The assumption is that if a system fares well under experimental conditions, the same performance can be expected in an operational situation. Different relevancy judgements have been noted among users and experts, but in general, the differences are small and therefore do not invalidate experimental testing [18].

To quantify relevance then, two measures, recall and precision, have endured since their introduction by Kent in 1955 [19]. Put simply, recall is the ratio of the number of relevant documents retrieved to the total number of relevant documents in the collection. Precision is the ratio of the number of relevant documents that have been retrieved to the total number of retrieved documents in the query result [1]. Table 4 illustrates the partitioning of the document collection after a query has been run and shows the formulas that define recall and precision with respect to this table.

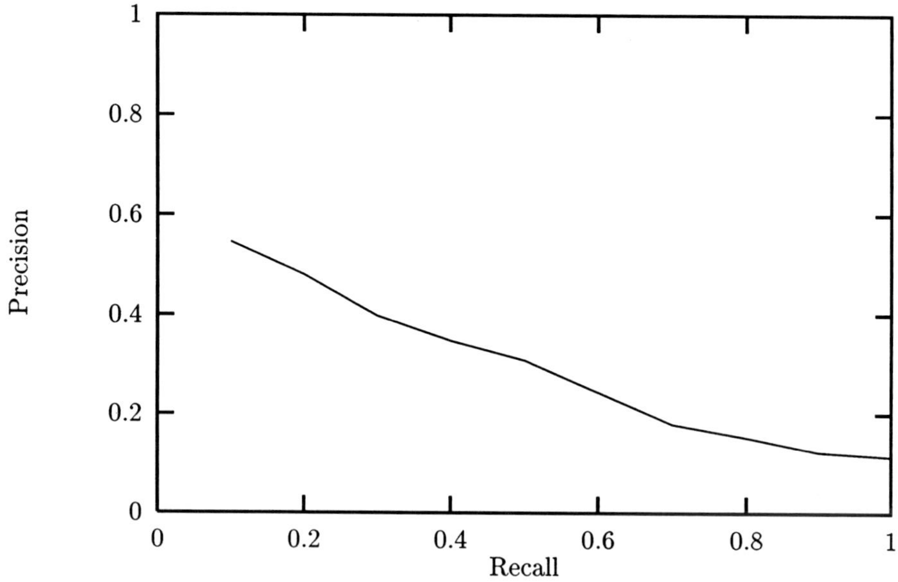

Fig. 4. IR performance can be measured using a precision and recall graph.

Recall and precision, in combination, are important values of measurement. If recall alone were high, then the user would have to filter through all the documents returned by the system to find those that were useful. If only precision were high, then a number of relevant documents may not have been retrieved by the system. So, one way of presenting their relationship is through a *recall-precision graph.* A sample graph is shown in Fig. 4.

Each query submitted to the system would produce a separate graph. So to evaluate a system with respect to many queries, an averaging technique is required. The technique used for the projects in this chapter is called *macro-evaluation* [18]. It specifies a set of standard recall values and averages single query precision values at each recall point. Since the graph is continuous, and since between any two plotted precision values no more documents are retreived, interpolation is used to join the observed values.

In Sections 2.3 and 2.4 we point out that the vector space model and the probabilistic indexing model rank their query results. Document ranking should also be considered at evaluation time. For example, if we are comparing the results of two ranking systems, where system 1 returns three relevant documents to a query with rank 2, 3 and 5 and system 2 returns three relevant documents to a query with rank 5, 7 and 9, obviously system 1 has done a better job of selecting relevant documents. This information is measured using *normalized recall and precision.* These values give an estimation of the effectiveness of the document ranking returned. Normalized recall and precision are calculated using the number of relevant documents to a given query, the ranks of those documents, and the best and worst possible ranking scenarios. A more explicit description and the formulas used can be found in [18].

In the following sections, we point out several issues related to the use of OCR text in an IR environment, but our predominate focus in our experimentation has been retrieval effectiveness.

4. Experimental Components

We do not propose that in our testing we have tried to evaluate recognition devices or retrieval systems. We simply compare retrieval results between unaltered OCR text and what is currently widely used—typed or *corrected* OCR text. From this comparison, we are able to discern differences in retrieval due to the application of recognized text in an IR environment. In the following sections we describe the collection and the procedures we used to conduct these experiments.

4.1. The Collection

IR experimentation, due to its nature, is difficult to conduct. In part this is true because of its inherent difficulties in obtaining appropriate collections, queries, and relevancy judgements; but also, considerable amounts of data are necessary to obtain worthwhile results. Parts of the problems associated with designing an IR experiment were precluded by our research environment. An online collection of Department of Energy (DOE) documents was made available to us for OCR and IR experimentation and research. This document collection is representative of the kinds of collections that many IR reasearchers, government organizations, and private businesses require for various applications:

- The documents are related to a particular topic.
- The documents have varied sources and authors.
- The documents are scientific and technical.
- The documents are rich in structure.

This collection makes up the Licensing Support System (LSS) Prototype designed to track information related to the Yucca Mountain Nuclear Waste Storage Facility. It covers a variety of topics that relate to this site and to matters that pertain to its construction. The collection includes scientific reports, journal articles, correspondence between involved parties, legal material, and government publications. Since there is such a variety of sources, the documents differ widely in format. There is a myriad of fonts and various quality levels of hard copy. Most of the scientific documents contain formulas, graphs, photos, maps, and other graphics. For each document in the collection, we recieved the typed ASCII file with approximately 99.8% accuracy and the scanned images for many[b] of the hard copy pages.

For the LSS documents to be used in our experiments, we required that they pass, what we call, *minimum document verification* (MDV)[20]. MDV ensured that

[b]Those image files that were corrupt, missing, or unusable were regenerated by us.

Test Query INJD-T3-Q1

LSS Prototype Test Question: Your office is trying to trace the evolution of NRC's position on repository sealing concepts (e.g. shaft and borehole seals). You need to produce a listing of all documents (including meeting material) discussing seals.

Boolean translation: `find document where text include phrase like 'repository' & 'seal' or text include phrase like 'shaft' & 'seal' or text include phrase like 'borehole' and 'seal' order by docid`

Natural language translation: *Find documents discussing repository sealing concepts (shaft and borehole seals).*

Fig. 5. Example query translations.

the typed ASCII and the images matched, page for page. This correspondence of course is crucial to our experimentation. Documents were excluded for the following reasons:

- Hardcopy pages were missing from a document.
- A document was primarily made up of figures with virtually no text.
- Text was written in a foreign language.
- Hardcopy document was in multicolumn format but was not decolumnized in the corrected ASCII version.
- Large amounts of main body text were missing from the ASCII file.
- Text sequence of a majority of the document did not match the original hardcopy.

Two other important pieces to any IR experimental collection are the set of queries run against the collection and the relevancy judgements used to evalute them. The queries in our collection were written by and obtained from the LSS contractors. These queries were constructed to evaluate the usability and the effectiveness of the LSS prototype design. Since this system was built with special fields, including title, author, publication source, etc., many of the queries were written to access these specific fields. Our concerns were with the text of the documents, so only 71 queries dealing with the documents' contents were used in our experimentation. An example query appears in Fig. 5.

Unfortunately, none of the queries were accompanied by relevancy judgments. We have painstakingly constructed the judgements ourselves over many months with the aid of geology graduate students. Each document is read and examined with respect to each query, and the documents are classified as either relevant or not relevant. Comparative usefulness or *ranked relevancy* was not assessed. Excluding seven of the queries with no relevant documents, the average number of relevancy assessments per query is ten.[c]

At the time of our experiments, verification and relevancy judgements were incomplete. We identified a set of 204 documents[d] for our first experiment and 674 documents for our second and third experiments.

4.2. The Recognized Versions

Section 4.1 describes the characteristics of our IR collection, and as it stands, it could be used for pure IR experimentation. But for our testing, we needed an additional component: the corresponding recognized collection. Each page in the collection was recognized by three OCR devices.[e] These devices represent different levels of character accuracy: OCR 1 with 98.14%, OCR 2 with 97.06%, and OCR 3 with 94.63% for our collection [21]. With 26,467 pages, this in itself was no small task. The steps for creating these recognized versions were:

- Rescan missing, corrupt, or bad images.
- OCR each image page with each device.
- Compare the generated text to its image ensuring usability.
- Concatenate document pages and remove any special characters output by the device that might interfere with IR loading or retrieval.

The page images we received from the DOE contractors were binary page images produced with either a Ricoh or Fujitsu scanner at 300 dpi; the threshold values are not known [20]. The pages we rescanned were created using a Fujitsu M3096G scanner with a resolution of 300 dpi and a median threshold of 127.

For those who use OCR devices regularly, there may seem to be a step missing in the enumeration: *manual zoning*. We used devices that offer *automatic zoning* and we exploited this feature for two reasons:

- We would have introduced human intervention into the automatically generated data we were hoping to test. Since the OCR text is the independent variable in our testing, it would have been much more difficult to evaluate the experimental results had we added this uncontrollable factor. What is more, manual zoning of 26,467 pages would have been an extremely labor-intensive task.
- A set of complex rules for inclusion and exclusion of text had been set up for the corrected version of this collection. There was no guarantee that we would be able to match their set of zones exactly. Rather than take the chance of removing text by manually selecting zones, we decided to use automatic zoning in our experimentation.

---

[c]For those queries used in our experimentation.

[d]Since we had no relevancy judgements at this time, recall and precision measures could not be used.

[e]Our second OCR/IR experiment used the data from all three.

OCR devices are typically interactive systems. In an experimental situation such as this, more control is required. ISRI provides an environment with a rich set of tools for conducting OCR experiments. The process of recognizing the over 26,000 document pages was automated and controlled from a UNIX workstation.

Any irregularities in the generated text were examined with respect to the text's corresponding image page. Problems in the text may indicate possible problems with the image necessitating rescanning and recognition. For example, if an OCR generated text page had blocks of meaningless characters, it could indicate a landscape page had been improperly rotated.

The next step was to concatenate pages into complete documents for IR loading. At this stage, we also inspected the document files for "problem" characters. To illustrate, some non-ASCII characters cause an IR system to truncate input files; other characters meaningful to the IR system may occur randomly in the OCR generated text. These characters were either displaced or replaced by white space. All our projects use these prepared data in their testing.

4.3. The Post-Processing System

In preparation for our experiments, we constructed an automatic OCR text post-processing system [3, 6]. This system was designed to correct OCR errors in the text without human intervention. After we loaded our three OCR databases, we ran the post-processing system on all three of these collections to see what improvements we could attain.

The system consists of numerous steps. We start with a full-text database of documents and obtain an index listing all the unique indexed words in the database. Then, using word frequency information, an English dictionary [22], and an auxiliary domain-specific dictionary, we divided the words into a list of what we call *centroids*, and a list of misspellings. The centroids are the words which are spelled correctly and have a frequency greater than 1. We then use approximation matching [23] to cluster the misspellings around the centroids. This matching is based on the edit distance between the misspellings and centroids. In general, edit distance is taken to be 1, while for longer words we increase it. The idea is that if we could find misspelled words that are close to a centroid word, then we could equate those misspelled words to its centroid. This procedure worked for the most part, but sometimes a misspelled word would be close to more than one centroid. We developed three heuristics to alleviate this problem. The first involves using document frequency information about the words (as opposed to collection frequency). The second process takes information about the kinds of mistakes the OCR device was likely to make, and eliminates clusterings that are unlikely to occur. Our final check considers certain suffixes to improve the preciseness of centroid selection. Finally, a list of misspelled words and their correct words are produced and the replacements are made.

Another problem we found with using OCR text with the weighted systems was the amount of "junk" in the text, and subsequently in the index, that caused unreliable term frequencies. One difficulty in removing these indexed strings is determining which terms are "junk" and which terms are misspellings that could be rectified by the post-processing system. After analyzing the list of indexed words we implemented the following heuristics: we set limits on the minimum and maximum lengths of words (3 and 27, respectively) and removed words with four or more identical consecutive characters. We also ignored strings that had a high ratio of non-alphabetic characters or an irregular vowel-to-consonant ratio. These words were not indexed.

5. Retrieval in the Presence of OCR Errors

The retrieval systems we used in these experiments were selected with their underlying models in mind. Everything else constant, one would expect the results from a boolean system, a probabilistic system, and a vector space system to produce results similar to what we report here. The underlying models should be reflected upon as the observations and experiences of these projects unfold in the following section.

5.1. Effects on Implementation Components

Recall in Section 2, we identified three facets of IR systems that impact or are impacted by the use of OCR text. The first of these, the inverted index, is the heart of the IR system. The words of the document, together with data about their distribution, are stored here. The inverted index is affected in several ways by the use of OCR text, but most notably, its size is increased. In our first experiment with 204 documents, the index of the OCR text was three times the size of the corrected text index. In the next two experiments, with 674 documents, the inverted index increased to 5 times its corrected version size. Some interesting statistics comparing OCR text with its corrected counterpart can be derived from an IR system's index. Table 5 reveals some of these variations. This table lists various statistics for the corrected collection and for the three recognized versions of 674 documents. These statistics indicate that:

- An OCR collection requires more overhead (database size and terms occurring only once).
- There will be an increase in per document processing time (average number of terms per document).
- There is a considerable increase in the size of the index (number of unique indexed terms).
- There is an increase in useless terms (terms occurring only once).[f]
- At least in our collection, the corrected text is missing some information that is picked up by the device (correctly spelled words).

Probably the most unanticipated set of figures is the *correctly spelled words.* There are more correctly spelled words in each of the three OCR collections than there are in the corrected collection. We attribute this anomaly to the removal of document text from the corrected collections [20].

[f] These are mostly misspellings and "graphic text" strings.

Table 5. Statistics for the corrected collection and the three OCR collections.

Statistic	Corrected	OCR 1	OCR 2	OCR 3
Database size (bytes)	15,686,772	37,080,489	40,918,148	42,247,537
Average number of terms per document	6,114	9,321	8,583	7,925
Number of unique indexed terms	78,494	320,338	387,276	414,715
Terms occurring only once	36,742	223,058	278,907	296,572
Terms occurring more than once	41,752	97,280	108,369	118,143
Correctly spelled words	22,833	25,241	24,728	23,552

Another source of problems not readily apparent are end-of-line hyphenations. Hyphenations are commonly used when documents are typeset for printing. With no special handling, the OCR outputs exactly what it "sees" and the IR system indexes what it "reads". The consequence: a percentage of words in the index are split consuming twice as much index space as they should. Further, the words are useless as they are and have a negative impact on retrieval since the *full* word is not available for querying.

Recall that some inconsequential words called stopwords do not get indexed by an IR system. Each incoming text word is compared to the list of stopwords; if the word appears in the list, it will be skipped. Sometimes, stopwords are misrecognized by the OCR device and in turn, are inadvertently indexed. In a boolean system, this recognition error probably will not make much difference.[g] But for systems that weight the importance of terms using term frequencies, this kind of transformation can disrupt term assignment and therefore document to query relevance in those documents affected. Two recognized documents exposing exactly this error can be found in Table 6. Also note that since stopwords are used frequently, these misrecognized terms became the *most* frequent "terms" for these documents.

Stemming is another common practice employed by IR systems. In all our projects, we applied only S-removal stemming to the OCR text. We disclosed no adverse effects from this application. Although it does not seem as if full stemming would produce very different results, actual experiments followed by a thorough analysis would be necessary to verify this claim.

[g] Erroneous query results would occur only if a stopword had been translated to a correctly spelled query term (possible but not probable).

Table 6. Stopwords in these documents were misrecognized consistently.

Document	OCR 2		OCR 3	
j	Term	Frequency	Term	Frequency
87			fhe	129
583	ihe	256	thc	176

Table 7. Results from querying the raw OCR 1 collection of 204 documents using a boolean system.

Total number of documents retrieved for corrected data:	632
Total number of documents retrieved for OCR data:	617
Percentage returned:	97.6%
Number of queries for which result sets were identical:	63
Number of queries for which result sets were different:	8

5.2. Effects on IR Models

A boolean system relies on the presence of query terms within documents to determine relevance. We found in our first experiment using a boolean system that redundancy in the document text could compensate for most of the problems associated with OCR errors. The results in Table 7 show most of the documents returned from querying the corrected collection were also returned from querying the OCR collection. After applying our post-processing system described in Section 4.3, seven more documents were retrieved. 98.7% of the corrected documents were returned when the OCR set was queried, as shown in Table 8.

For systems that apply term weighting and document ranking, the translation of text produced by the device causes a more significant impact. This consequence can be attributed directly to the mutations to the text that affect term frequencies, and hence cause skewed term weighting. An example of such an effect is the misrecognition of the stopword `the` discussed above. In the probabilistic indexing system, the term with the maximum frequency is used to normalize term weights (Eq. 2.4). When this value is uncharacteristically high, as it was in the stopword examples, the weights of other document terms are underestimated. This effect eventually manifests itself to the user when the documents are ranked to a query. Table 9 illustrates the ramifications of changes in term frequencies on ranking.

The maximum term count is not the only frequency that may get altered. Other normalization values suffer as well. For example, we found in the vector space model that cosine normalization negatively affects document ranking. The SMART system, the vector space implementation used in our experiments, allows the applica-

Table 8. Results from querying the post-processed OCR 1 collection of 204 documents using a boolean system.

Total number of documents retrieved for corrected data:	632
Total number of documents retrieved for OCR data:	624
Percentage returned:	98.7%
Number of queries for which result sets were identical:	65
Number of queries for which result sets were different:	6

Table 9. Relationship between *max term frequency* and ranking of *document*$_j$ using a probabilistic indexing system.

query	document	corrected		OCR 1		OCR 2		OCR 3	
	j	$maxtf_j$	rank	$maxtf_j$	rank	$maxtf_j$	rank	$maxtf_j$	rank
14	193	73	1	714	32	630	27	109	1
28	468	96	27	94	22	179	> 600	396	453

tion of several weighting techniques. One of these techniques applies vector length (i.e. square root of the sum of squares of the weights in the vector) to normalize the length of the documents. It functions as a way of equalizing the importance of terms in short documents even though they occur less frequently. This application works well for documents that are clean and manually typed. But for an OCR-generated collection where the vectors are artificially padded with misspellings and "graphic text" this normalization technique causes problems with term weighting, and in turn, document ranking. Table 10 demonstrates this variablity. We show the query number, a document determined relevant to that query, and the output ranking for that document for the corrected and the OCR collection. A complete analysis showed the variability was attributed directly to the cosine weighting component [7].

Pure document length (number of non stopwords) is another factor that is sometimes used in term weighting calculations. But as we can see from Table 5 in Section 5.1 there is a marked increase in document length for the recognized text (note: *average number of terms per document*). None of the systems with which we experimented applied this component in their weighting technique, so results are unknown. But the averages in Table 11 indicate differences in results should be expected. Even in the presence of these variations in document frequencies between the corrected text and its OCR generated counterpart, for these weighted systems, average precision was hardly affected. For our second experiment using the same data sets and applying our post-processing system, the average precision results appear in Table 12. Note that these small percentage differences are not considered significant.

Table 10. Ranking variability between corrected and OCR 1 collection using a vector space system.

		Vector length applied to normalize term weights	
Query	Document	Corrected document ranking	OCR 1 document ranking
4	13	20	204
5	13	75	263
5	24	7	128
6	264	15	47
7	22	20	72
7	53	34	110
10	253	60	215
10	573	18	47
11	504	44	178
16	403	46	141
16	618	73	42

Table 11. Average precision for the corrected document collection and the three OCR collections using a probabilistic indexing system.

Recall	Precision (% change) – 68 queries						
	Corrected	OCR 1		OCR 2		OCR 3	
Average	28.2	27.8	(−1.5)	28.2	(−0.1)	28.8	(+2.1)

Table 12. Average precision for the corrected document collection and the three OCR collections after post-processing using a probabilistic indexing system.

Recall	Precision (% change) – 68 queries						
	Corrected	OCR 1-pps		OCR 2-pps		OCR 3-pps	
Average	28.2	28.9	(+2.5)	28.7	(+1.8)	28.7	(+1.5)

Table 13. Average precision improved for the corrected collection as more terms were added to the feedback queries but remained fairly constant for the OCR 1 collection.

	Corrected		OCR 1		
Partial expansion	Feedback run	% change over original run	Feedback run	% change over original run	% of difference
20	0.2000	30.1	0.1928	36.6	-3.6
50	0.2211	43.8	0.2045	44.8	-7.5
100	0.2164	40.7	0.2089	48.0	-3.5
250	0.2240	45.6	0.2071	46.7	-7.5
500	0.2328	51.4	0.2031	43.9	-12.8
750	0.2422	57.5	0.2092	48.2	-13.6
1000	0.2536	64.9	0.2095	48.4	-17.4

Our experiment with SMART gave us the opportunity to test OCR text within a vector space framework, and examine several weighting techniques, as discussed. These techniques were applied to our OCR collection, and again, these results demonstrated that no significant difference is apparent when average precision is compared.

Relevance feedback, as we discussed in Section 2.3, is a technique designed to use information from known relevant documents to improve query effectiveness. In our experiment with SMART we were able to apply this technique to the OCR-generated text. We discovered an interesting phenomenon: as the queries were expanded with more feedback terms, the precision for the corrected collection improved while the precision for the OCR collection leveled off after the initial twenty word expansion. Table 13 shows several feedback runs for both the corrected and OCR collections. The first column lists the number of terms that were used to expand the query. The second column shows the average precision for the feedback run and the third column shows its improvement over the original query run for the corrected collection. The next two columns contain the same data for the OCR collection. The last column demonstrates the increasing difference for the corrected set over the OCR. After analysis of the expanded queries, we found that the increasing difference was not due to query degradation but was attributable to a lack of improvement in rank for a few of the relevant documents in the OCR collection. The documents that did not improve in rank for the feedback runs were the same documents with a discrepancy in rank for the initial runs. In general, these documents represent poor candidates for recognition due to the poor quality of the original images or the large amounts of graphical material contained in the documents. In any case, these results show that feedback cannot overcome some of the shortcomings found in OCR generated collections.

6. Conclusion

Optical character recognition together with information retrieval encompass the task of producing accessible information. Instead of analyzing each technology in separate domains we believe the two can and should be assessed as a singular system. This is what we established in our testing.

We have introduced three IR models that have been analyzed with respect to OCR-generated text. Further, we emphasized the features of these systems that we felt were impacted most by using these input data. The magnified size of the index was a consequential issue for all three models. This consideration becomes crucial for actual document databases that are made up of hundreds of thousands of documents.

Our experimentation elicited a few answers. For example, we demonstrated that average precision is hardly affected by the use of OCR text for any of the models we tested. In addition, for simple IR models, like the boolean system, redundancy in full text alone can overcome the errors generated by OCR. With weighted systems, we found that methods of normalization disrupt ranking for OCR-generated documents with certain characteristics. Further, some errors made by the device can be corrected automatically by knowing the kinds of errors to look for and by using information obtained at a more global level. But our experimentation also raised a number of questions:

- How correct is *corrected* text?

  We found from our research that it wasn't as clean as we had hoped.

- At what accuracy level should a document be retyped and not OCR'd?

  A few documents in our collection generated text that was unusable.

- Is there a weighting technique that is more compatible with OCR-generated text?

  The applicability of current IR weighting schemes may lose their effectiveness when OCR text is used.

- Can an IR system use the information provided by an OCR device to compensate for the errors that occur?

  We believe it can if the IR system is designed to take advantage of the information available.

Our experiments have given OCR evaluation a new perspective and so, a stimulus for future research. Continued investigation and experimentation should bring these two technologies even closer.

References

[1] G. Salton and M. J. McGill, *Introduction to Modern Information Retrieval* (McGraw Hill, New York, 1983).

[2] W. B. Frakes and R. Baeza-Yates, eds., *Information Retrieval, Data Structures & Algorithms* (Prentice Hall, Englewood Cliffs, NJ, 1992).

[3] K. Taghva, J. Borsack, B. Bullard, and A. Condit, Post-editing through approximation and global correction, Technical Report 93-05, Information Science Research Institute, University of Nevada, Las Vegas, March 1993.

[4] K. Taghva, J. Borsack, A. Condit, and S. Erva, The effects of noisy data on text retrieval, *J. American Soc. for Inf. Sci.* **45**, 1 (1994) 50–58.

[5] K. Taghva, J. Borsack, and A. Condit, Results of applying probabilistic IR to OCR text, in *Proc. 17th Intl. ACM/SIGIR Conf. on Research and Development in Information Retrieval*, Dublin, Ireland, July 1994, 202–211.

[6] K. Taghva, J. Borsack, and A. Condit, An expert system for automatically correcting OCR output, in *Proc. IS&T/SPIE 1994 Intl. Symp. on Electronic Imaging Science and Technology*, San Jose, CA, Feb. 1994, 270–278.

[7] K. Taghva, J. Borsack, and A. Condit, Effects of OCR errors on ranking and feedback using the vector space model, Technical Report 94-06, Information Science Research Institute, University of Nevada, Las Vegas, Aug. 1994.

[8] G. Salton. *Automatic Text Processing* (Addison-Wesley, Reading, MA, 1989).

[9] G. Salton and C. Buckley, Term-weighting approaches in automatic text retrieval, *Inf. Proc. and Management* **24**, 5 (1988) 513–523.

[10] W. Bruce Croft and H. R. Turtle, Text retrieval and inference, in *Text-based Intelligent Systems*, ed. P. S. Jacobs (Lawrence Erlbaum, 1992) 127–155.

[11] G. Salton and C. Buckley, Improving retrieval performance by relevance feedback, *J. American Soc. for Inf. Sci.* **41**, 4 (1990) 288–297.

[12] S. E. Robertson and K. Sparck Jones, Relevance weighting of search terms, *J. American Soc. for Inf. Sci.* **27**, 3 (1976) 129–46.

[13] K. Sparck Jones, Experiments in relevance weighting of search terms, *Inf. Proc. and Management* **15**, 3 (1979) 133–44.

[14] K. Sparck Jones, Search term relevance weighting given little relevance information, *Journal of Documentation* **35**, 1 (1979) 30–48.

[15] W. B. Croft and D. J. Harper, Using probabilistic models of document retrieval without relevance information, *Documentation* **35**, 4 (1979) 285–295.

[16] W. B. Croft, Experiments with representation in a document retrieval system, *Information Technology: Research and Development* **2**, 1 (1983) 1–21.

[17] J. P. Callan, W. B. Croft, and S. M. Harding, The INQUERY retrieval system, in *Proc. 3rd Intl. Conf. on Database and Expert Systems Applications*, 1992, 78–83.

[18] C. J. van Rijsbergen, *Information Retrieval* (Butterworth & Co. Ltd., Reading, MA, 1975).

[19] F. Wilfrid Lancaster, *Information Retrieval Systems* (John Wiley, New York, 1979).

[20] J. Borsack and B. Huey, Verification of GT1, Technical Report 94-05, Information Science Research Institute, University of Nevada, Las Vegas, July 1994.

[21] T. A. Nartker, S. V. Rice, and J. Kanai, OCR accuracy: UNLV's second annual test, *INFORM* **8**, 1 (1994) 40–45.

[22] R. E. Gorin, P. Willisson, W. Buehring, G. Kuenning *et al.*, Ispell: a free software package for spell checking files, The UNIX community, 1971, version 2.0.02.

[23] S. Wu and U. Manber, Fast text searching allowing errors, *Commun. of the ACM* **35**, 10 (1992) 83–91.

DATABASES AND BENCHMARKING

Handbook of Character Recognition and Document Image Analysis, pp. 779–799
Eds. H. Bunke and P. S. P. Wang

CHAPTER 30

DATA SETS FOR OCR AND DOCUMENT IMAGE UNDERSTANDING RESEARCH

ISABELLE GUYON
AT&T Bell Laboratories
955 Creston Road
Berkeley, CA 94708
isabelle@research.att.com

ROBERT M. HARALICK
Department of Electrical Engineering FT-10
University of Washington
Seattle, WA 98195

JONATHAN J. HULL
RICOH California Research Center
2882 Sand Hill Road, Suite 115
Menlo Park, CA 94025
hull@crc.ricoh.com

IHSIN TSAIYUN PHILLIPS
Department of Computer Science/Software Engineering
Seattle University
Seattle, WA 98125
yun@seattleu.edu

Several significant sets of labeled samples of image data are surveyed that can be used in the development of algorithms for offline and online handwriting recognition as well as for machine printed text recognition. The method used to gather each data set, the numbers of samples they contain, and the associated truth data are discussed. In the domain of offline handwriting, the CEDAR, NIST, and CENPARMI data sets are presented. These contain primarily isolated digits and alphabetic characters. The UNIPEN data set of online handwriting was collected from a number of independent sources and it contains individual characters as well as handwritten phrases. The University of Washington document image databases are also discussed. They contain a large number of English and Japanese document images that were selected from a range of publications.

Keywords: Data sets, databases, text images, online handwriting, offline handwriting, machine printed text, CEDAR, NIST, CENPARMI, UNIPEN, University of Washington.

1. Introduction

The availability of a data set that contains an appropriate number and selection of samples is a critical part of any experimental research project [8]. This is especially true in an image-based application such as optical character recognition (OCR) where it is sometimes difficult for individual researchers to gather the number and type of data they need because of the costs involved. Ideally, a data set would allow a researcher to project the performance achieved experimentally to the application domain represented by the data set. This implies that the data set used for algorithm development should reflect the application domain as closely as possible.

A significant advance in the experimental integrity of OCR research has been made possible in recent years with the availability of several image data sets. These data sets allow researchers to train and test algorithms on significant numbers of data items and to compare performance on specific images. This has improved productivity since researchers can conduct experiments without first gathering data.

There are three areas of OCR research (offline handwriting, online handwriting, and machine printed text) that require specialized data sets. Offline handwriting is produced by an individual, typically by writing with a pen or pencil on paper, and is scanned into a digital format. Online handwriting is written directly on a digitizing tablet with a stylus. The output is a sequence of x-y coordinates that express pen position as well as other information such as pressure. Machine printed text occurs commonly in daily use and is produced by offset processes, laser, inkjet, or dot matrix printing. Each area has unique characteristics that affect the design of a data set. These characteristics determine how well experimental results on that data set can be generalized to another application.

Offline handwritten text data sets typically contain isolated alphanumeric characters or words. These data sets are often produced by choosing subjects to write on sample forms that are subsequently digitized. An important consideration in the design of data sets in this area is how well the subjects that produce the data and the conditions under which the data are gathered represent the eventual application. Ideally, the subjects should be chosen from the same population and the data gathered under the same conditions (e.g., data gathered in the field, same data form used, etc.) that will be used in the final application. Otherwise, significant differences in performance can occur between the data set and the data used in practice.

Databases of online handwriting also contain isolated characters and words as well as text phrases. Research in online handwriting recognition has been conducted for many years at a large number of research labs. Each project typically developed their own databases for internal use. The *UNIPEN* project has taken advantage of this existing body of data by asking individual research groups to contribute their databases to a common set. This data set will first be used for a comparative benchmark. It will then be shared with the rest of the research community. A significant issue in UNIPEN was the large number of different formats used for online handwriting. This was solved by the development of a common data format which is described in this chapter. The background, history, and future of the UNIPEN effort are also discussed.

Machine-printed text data sets often include images of complete pages. These pages can be chosen from a large number of different classes of machine printed document (e.g., scientific papers, memorandums, newspapers, etc.) and can vary widely in format (fonts, point size, column layout, etc.). A significant consideration is the document classes and variations in format represented in the data set. Also, the ground truth information supplied with such a data set should be comprehensive so that it can be used to improve the performance of OCR algorithms. This is significant since modern OCR systems are affected by many aspects of the data format which are often inter-related. This chapter describes the CDROM data sets developed by the University of Washington which include several document classes and extensive truth data.

The rest of this paper describes several data sets of offline and online handwriting as well as machine printed text. We discuss the contents of these data sets and describe how they address various research considerations.

2. Offline Handwritten Text

There are at least two classes of databases of offline handwritten text that are of interest to most researchers. Datasets that contain isolated handprinted characters and digits have been used in the development of many pattern recognition algorithms. The recognition problem represented by these data is well defined (e.g., 10, 26, 52, or 62 classes) and the recognition task is relatively easy for ordinary people. A reliable solution to this problem could have significant economic consequences since many handwritten forms could be read automatically.

Data sets can also contain isolated handwritten words and phrases. The recognition problem represented by these data is more challenging since the number of classes is larger (e.g., there are more than 40,000 legal ZIP Codes in the U.S.) and there may be few constraints on how users prepare the samples. However, a robust solution to this problem would be even more significant than for isolated character recognition since it would provide a general purpose computer input medium.

Important considerations when evaluating the usefulness of a data set for experimental purposes include the number of samples it contains, the sampling rate at which it was captured, and the pixel depth at which the data is furnished. Commonly used sampling rates include 200 or 300 pixels per inch (ppi) and pixel depths are usually one or eight bits. Such image characteristics are also important system design parameters since they influence the ultimate cost of an implementation.

2.1. Isolated Digits and Alphabetic Characters

Isolated digit and character recognition is one of the most extensively studied application domains in pattern recognition. Some of the earliest databases of digital images, beginning at least in 1966, were composed of single characters (see [7] for a review). In recent years, several significant collections of such data have been issued.

The CEDAR data set contains nearly 28,000 examples of isolated handwritten characters and digits that were extracted from images of postal addresses [6]. Envelopes with handwritten addresses on them were sampled from the mail as it was being processed in a post office. This assured that the people that prepared the original data were not aware that their addresses would be included in the data set. The address images were scanned at 300 ppi in 8-bit gray scale and later reduced to one-bit images. Individual characters and digits were segmented from the address images by a semi-automatic process that extracted single connected components. These were displayed to a human operator who entered their truth values.

An additional group of about 21,000 digits were also extracted from ZIP Codes in the CEDAR data set by a fully automatic segmentation algorithm. These digits were manually screened for quality and placed into a *good* set if they contained no segmentation artifacts that could be considered errors such as one digit split into two. All the digits output by the segmentation algorithm are also provided in the data set. This provides some challenging examples for testing the performance of an algorithm since artifacts caused by segmentation are preserved in the individual images.

The CENPARMI data set contains 17,000 isolated digits that were extracted from images of about 3400 postal ZIP Codes [11]. The ZIP Codes were selected from dead letter mail in a working post office and thus the preparers of the data also were separated from the data collection task. The ZIP Codes were scanned at 200 ppi with a one-bit digitizer. The isolated digits were extracted from the ZIP Codes by a manual segmentation process that displayed individual images to operators who entered their identities.

The NIST data set SD3 contains one of the largest selections of isolated digits and characters that are publically available. Altogether it contains over 300,000 character images that were extracted from data collection forms filled out by 2100 individuals. Employees of the U.S. Bureau of the Census were instructed to fill the boxes on a form with a particular sequence of digits or characters and were asked to make sure that the individual digits or characters did not touch one another or the surrounding box. The forms were subsequently scanned at 300 ppi in binary mode and automatically segmented. The string of digits that should have been written in each box was used to assign a truth value to each individual image. The truth values were subsequently verified by a human operator.

An interesting issue in data set design is presented by the NIST images. The initial collection process described above was done in preparation for a competition that evaluated the performance of more than 20 different algorithms in isolated digit and character recognition. The images on SD3 were provided as a training set and were used by most of the competing organizations in developing their methods. A separate data set (called TD1) was collected to provide test data. The same collection process was used. However, the subjects were high school students. The quality of the data varied more widely in TD1 than it did in SD3 and was on the whole more sloppy. This caused a significant drop in performance on the test data because most systems had been trained on the neater images. This experience shows that a given technique may perform very well on a specific data set. However, this does not necessarily mean that the recognition problem represented by that data has been solved. It can mean that the recognition problem has been solved for that data set only. The generalization of any results to a larger population should only be made after careful consideration and comparison of the training and test data to the real-life application.

2.2. Word and Phrase Recognition

Word and phrase recognition is a less frequently studied application of pattern recognition than digit or character recognition. However, words or phrases offer contextual constraints such as dictionaries that make it possible to model interactions between image processing operations such as segmentation and the recognition of isolated symbols given that only certain sequences of symbols occur in the dictionary.

ZIP Codes, city names, and state names are examples of handwritten word images that are available in the CEDAR data set. Approximately 5,000 ZIP Codes, 5,000 city names, and 9,000 state name images are included. These data were scanned by the process described above. However, the 300 ppi 8-bit gray scale versions of the whole word images are provided and the specific address in which each image occurred is identified. Thus, experimentation can be performed with isolated word recognition on gray scale data and a comprehensive system could be developed that used partial results from recognition of the city or state name to influence the recognition of specific digits in the corresponding ZIP Code.

A significant amount of running English text is also available in the NIST SD3 and TD1 data sets. Each collection form contained a *Constitution box* in which subjects wrote the 52-word preamble to the Constitution of the United States. This data was scanned at 300 ppi in 1-bit format as part of the normal data capture process. These images also make it possible to develop algorithms that integrate early processing with contextual analysis. The domain is constrained enough that a range of contextual constraints can be investigated and at the same time the physical format is sufficiently unconstrained (the subjects could have written the preamble anywhere in a given box) that it reasonably represents the way people would like to use handwriting to communicate with a computer.

There are three other NIST data sets (SD11-SD13) that contain examples of phrases that were written by respondents to the U.S. Census to describe their jobs. Three boxes were filled in with the title of a person's job, the work they do, and the work done by the company they are employed by. SD11-SD13 differ from SD3 and

TD1 in that the people that prepared the data were from the general population and had no idea that their writing would be scanned. Altogether, 91,500 handwritten phrases were scanned at 200 ppi in binary mode. 64,500 of the phrases were scanned from microfilm and 27,000 were scanned from the original paper versions. These data are also provided with a dictionary of *legal* entries for each box that was derived from the previous census. Thus, there is no guarantee that the letter-for-letter transcription of each image appears in the dictionary. The recognition task is to identify the dictionary entry that is the *closest* match to the phrase written in the box.

3. Online Handwritten Text

In this section, we present the design of a database for On-Line Handwriting Recognition (OLHR). The database is composed of isolated characters, words, and sentences, written in a variety of styles (handprinted, cursive or mixed). The alphabet is restricted to the ASCII keyboard set. The data were donated by 40 different institutions and therefore includes a variety of writers and recording conditions. The database size approaches 5 million characters. We provide some details about the data exchange platform which could inspire other similar efforts. The database will be distributed to the public for a small fee by the Linguistic Data Consortium (LDC) in 1996.

3.1. History of UNIPEN

On-line handwriting recognition (OLHR) addresses the problem of recognizing handwriting from data collected with a sensitive pad which provides discretized pen trajectory information. OLHR has long been the poor parent of pattern recognition in terms of publicly available large corpora of data. To remedy this problem, the UNIPEN project was started in September 1992 at the initiative of the Technical Committee 11 of the International Association for Pattern Recognition (IAPR). Two IAPR delegates (Isabelle Guyon and Lambert Schomaker) were appointed to explore the possibility of creating large OLHR databases.

In the field of OLHR there exist many privately owned databases. These databases constitute a potential resource which is much richer than any single database, because of the diversity of text entered, recording conditions and writers. Therefore data exchange is a natural way to constitute a sizable database which is representative of the tasks of interest to research and development groups.

A small working group of experts from Apple, AT&T, HP, GO, IBM and NICI laid the foundations of UNIPEN in May, 1993 and proposed that a common data format would be designed to facilitate data exchange. In the summer of 1993, the UNIPEN format was designed, incorporating features of the internal formats of several institutions, including IBM, Apple (Tap), Microsoft, Slate (Jot), HP,

AT&T, NICI, GO and CIC. The format was then tested independently by members of the working group, soon followed by many other volunteers. A second iteration of the test was organized in autumn 1993 to check the changes and additions to the format [4]. In parallel, a set of tools to parse the format and browse the data were developed at NICI (with the sponsorship of HP) and at AT&T.

In January 1994, NIST and LDC committed to help UNIPEN. NIST has been supervising the data gathering and the organization of a benchmark test. LDC will publish a CD ROM and distribute the data. There is already an FTP site at LDC where data and programs can be exchanged.

In June 1994, the instructions for participation in the first UNIPEN benchmark, limited to the Latin alphabet, were released. Potential participants were requested to donate isolated characters, words or sentences containing at least 12,000 characters. Forty institutions responded to the call for data. Further negotiations between owners of large databases succeeded in gathering larger data sets. There are nearly five million individual character samples in the database.

In February 1995, the data donators met for a one day workshop to determine how the data will be split into training and test sets. A benchmark test using these data will take place in 1995. During the test the data will remain the property of the data donators. After the test, it will become publicly available and will be distributed by LDC for a nominal fee.

The activities of UNIPEN will expand in the future according to the needs and desires of the participants.

3.2. Collecting donated samples

3.2.1. A common data format

The UNIPEN format is an ASCII format designed specifically for data collected with any touch sensitive, resistive or electro-magnetic device providing discretized pen trajectory information. Users can easily convert their own format to and from the UNIPEN format or collect data directly in that format.

The minimum number of signal channels is two: X and Y, but more signals are allowed (e.g., pen angle or pressure information). In contrast with binary formats, such as Jot [2], the UNIPEN format is not optimized for data storage or real time data transmission and it is not designed to handle ink manipulation applications involving colors, image rotations, rescaling, etc. However, in the UNIPEN format, there are provisions for data annotation about recording conditions, writers, segmentation, data layout, data quality, labeling and recognition results.

Efforts were made to make the format human intelligible without documentation (keywords are explicit English words), easily machine readable (an awk parser was developed in conjunction with the development of the format itself), compact (few keywords), complete (enough keywords), and expandable.

The format is a succession of instructions consisting of a keyword followed by arguments. Keywords are reserved words starting with a dot in the first column of a line. Arguments are strings or numbers, separated by spaces, tabs or new lines. The arguments relative to a given keyword start after that keyword and end with the appearance of the next keyword or the end of file (see Fig. 1).

Almost everything is optional, so that simple data sets can be described in a simple way. All variables are global: declared variables retain their values until the next similar declaration. Databases written in the UNIPEN format may be concatenated in a single file or they may be organized in different files and directories.

The format can be thought of as a sequence of pen coordinates, annotated with various information, including segmentation and labeling. The pen trajectory is encoded as a sequence of components. A pen-down component is a trace recorded when the pen is in contact with the surface of the digitizer. A pen-up component is a trace recorded when the pen is near the digitizer without touching it. Components are not necessarily delimited by pen-lifts and may or may not coincide with strokes. *.PEN_DOWN* and *.PEN_UP* contain pen coordinates (e.g. *XY* or *XYT* as declared in *.COORD*). The instruction *.DT* specifies the elapsed time between two components. The database is divided into one or several data sets starting with *.START_SET*. Within a set, components are implicitly numbered, starting from zero.

Segmentation and labeling are provided by the *.SEGMENT* instruction. Component numbers are used by *.SEGMENT* to delineate sentences, words, and characters. A segmentation hierarchy (e.g. *SENTENCE WORD CHARACTER*) is declared with *.HIERARCHY*. Because components are referred to by a unique combination of set name and order number in that set, it is possible to separate the *.SEGMENT* from the data itself.

The format also provides a unified way of encoding recognizer outputs to be used for benchmark purposes. To obtain more information about the format, it is possible to access its full definition electronically (see next section).

3.2.2. Internet connections

Without the internet, the Unipen project would not have been possible. Electronic mail has been the primary means of communication between organizers and participants. The data and the tools were exchanged by FTP.

In March 1994, UNIPEN advertised its existence on several electronic mailing lists, resulting in nearly 200 subscriptions to the UNIPEN newsletter. People interested in UNIPEN can send a request to be added to the Scrib-L mailing list. Scrib-L is a mailing list for researchers and developers in the field of handwriting. Electronic mail to:

`SCRIB-L@NIC.SURFNET.NL`

```
.VERSION                1.0
.DATA_SOURCE            ATT
.DATA_ID                Example

.COMMENT                Documentation
                        -------------
.DATA_CONTACT           Isabelle Guyon (isabelle@research.att.com).
.DATA_INFO              Latin alphabet, isolated characters. Data cleaned manually.
.SETUP                  Volunteer staff members. People sitting at a desk.
.PAD                    WACOM HD-648A LCD Digitizer.

.COMMENT                Declarations
                        ------------
.X_DIM                  4160
.Y_DIM                  3200
.H_LINE                 450 1900 2300
.X_POINTS_PER_INCH      508
.Y_POINTS_PER_INCH      508
.POINTS_PER_SECOND      200
.COORD                  X Y
.HIERARCHY              PAGE TEXT WORD

.COMMENT                Data
                         ----
  Most of the point have been removed to shorten the example.
.INCLUDE                lexicon.lex
.DATE                   9 20 93
.WRITER_ID              08171408_14804
.STYLE                  MIXED
.START_BOX
.SEGMENT PAGE           0-58
.SEGMENT TEXT_CHUNK     0-29 ? "that nothing more happened ,"
.SEGMENT WORD           0-6 ? "that"
.PEN_DOWN
707 2417
707 2424
590 2319
.PEN_UP
.DT 151
.PEN_DOWN
588 2377
586 2377
695 2393
.PEN_UP
   etc.
.COMMENT For more examples, please ftp data samples.
```

Fig. 1. Example of UNIPEN formatted data. This example is simplified. Real data are more richly annotated.

will be forwarded to all subscribers. Please refrain from sending messages which are not in the general interest of researchers in handwriting.

Scrib-L resides on the computers of the national node of the Nijmegen University Computing Centre in The Netherlands. Scrib-L subscribers represent as many as 23 countries of the world. Messages are in ASCII, max 77 columns wide, concise, and formal.

```
------------------------- Summary ----------------------------
ASCII MESSAGES (<77 chars/line) to: Scrib-L@NIC.SURFNET.NL
COMMANDS to the boss of Scrib_L:   LISTSERV@NIC.SURFNET.NL
     Subscribe:                    SUBSCRIBE SCRIB-L Name (Fax:...)
     Get list of subscribers:  REVIEW SCRIB-L (COUNTRIES
     Get Archive of June'93:   SEND SCRIB-L log9306
-------------------------------------------------------------
```

An FTP site has been set up at the Linguistic Data Consortium for data and software exchange. Currently, most of the directories can be read only by data donators. When the database becomes public, more directories will be open.

To access the directories that are publicly available, proceed as follows:

```
ftp ftp.cis.upenn.edu
Name: anonymous
Password: [use your email address]
ftp> cd pub/UNIPEN-pub/documents
ftp> get call-for-data.ps (benchmark instructions)
ftp> cd ../definition
ftp> get unipen.def       (format definition)
ftp> quit
```

People having access to WWW through Mosaic will find images of UNIPEN example files as produced by the Upview program developed at NICI at:

`http://www.nici.kun.nl/unipen`

3.3. Organizing the database

3.3.1. Computing statistics

For each data set donated, a data sheet with relevant statistics was computed (see Fig. 2). These statistics serve as a basis to determine how to split the data into different subsets.

3.3.2. Defining tasks

Because the database is composed of many data sets of limited size, it is important to group the data sets that address similar tasks. This will also be important for defining a set of standard benchmarks.

```
========================================= SEGMENTATION ======================================
Type            Totals                                    Intersections
----            ------                                    -------------
                                          both               WORD             CHAR          neither
TEXT:        69 writers                      0                 69                0                0
          26279 segments                     0              26279                0                0
                                          both               TEXT             CHAR          neither
WORD:        94 writers                      0                 69                0               25
          69527 segments                     0              61127                0             8400
                                          both               TEXT             WORD          neither
CHAR:         0 writers                      0                  0                0                0
              0 segments                     0                  0                0                0
TOTAL:       94 writers
          95806 segments
         302843 characters
         588183 components
======================================== ALPHABET ============================================
                  s     l    u     d   sl   su    lu   sd   ld   ud  slu  sld  sud  lud  slud
TEXT: segments  417 11011 3048   594 1039   70  7991 1183    0    0  836    0   18    0    72
       writers   21    69   69    22   69   22    69   69    0    0   69    0   18    0    37
WORD: segments 4827 30028 7942 11186  382   22 14422  675    0    0   43    0    0    0     0
       writers   69    94   94    94   66   22    94   25    0    0   22    0    0    0     0
CHAR: segments    0     0    0     0    0    0     0    0    0    0    0    0    0    0     0
       writers    0     0    0     0    0    0     0    0    0    0    0    0    0    0     0
======================================== STYLE ===============================================
                        PRINTED         CURSIVE         MIXED           _unspec           _bad
TEXT: segments            9174            0             17105              0                 0
WORD: segments           26956            0             42571              0                 0
WD_LU: segments          16853            0             35539              0                 0
CHAR: segments               0            0                 0              0                 0
TOTAL: segments          36130            0             59676              0                 0
======================================== LEXICON =============================================
From:                   labels          lexicon
TEXT:                     1196            1123
WORD:                     1744               0
WD_LU:                       0               0
======================================== WRITER ==============================================
Total number of characters:                 302843
Total number of writers:                    94
Average number of characters/writer:        3221.73
Std. dev. of characters/writer:             52.63
Minimum number of characters/writer:        0
Maximum number of characters/writer:        9796
==============================================================================================
```

Fig. 2. Example of data sheet. The statistics of each data set donated to UNIPEN are computed and summarized in a data sheet. This example shows one of the HP data sheets. In the ALPHABET section, the symbols 's', 'l', 'u', and 'd' stand for symbols, lowercase letters, uppercase letters and digits. In the STYLE and LEXICON sections 'WORD_LU' means words contain only lowercase or uppercase letters (no symbols or digits).

Examples of tasks that were defined include:

1. Isolated characters, case separated
2. Isolated characters, mixed case
3. Isolated characters in the context of words (with a dictionary)
4. Isolated printed words
5. Isolated cursive words
6. Text (sentences)

The number of characters available for each task was computed. Tasks may overlap: some characters may be used in more than one task. There is a total of 4,422,863 characters. For tests 1 and 2, there are 90,305 digits, 155,494 uppercase letters, 367,214 lowercase letters and 85,952 symbols. For tests 3, 4, and 6, there are respectively 320,916, 302,251, 341,017 and 98,785 characters available.

3.3.3. Defining training and test set

The next problem is to determine what size test set will give statistically significant results for each task. The remainder of the data (if any) will serve as training data. Since training data is valuable, it is important not to be wasteful when defining the test sets.

Obviously, defining a statistically significant test set size is a chicken and egg problem: before obtaining recognizer performance, it is not possible to determine statistical significance. Nevertheless, since approximate values of the error rates of particular recognizers on given tasks are known, it is possible to estimate the size of a test set using straightforward statistical arguments.

For identically and independently distributed (i.i.d.) data, if E is the error rate of a recognizer, it is possible to compute the number n of test examples such that with probability $(1-\alpha)$ the actual error rate (on an "infinite" size test set) will not be higher than βE [3]:

$$n = \left(\frac{z_\alpha}{\beta}\right)^2 \frac{(1-E)}{E} \qquad (1)$$

where z_α is a tabulated coefficient which is a function of α. For typical values of the parameters, $\alpha = 0.05$ ($z_\alpha = 1.65$), $\beta = 0.05$ and $E = 1\%$ character error, the number of test examples obtained is less than 10,000.

In reality, data are far from being i.i.d. In particular, the data usually come from a limited number of writers which necessarily introduces correlations. The data donators, who are experts in on-line handwriting recognition, advocated a minimum of 100 different writers in every test set, each writer providing at least

1000 characters. Different writers must figure in the training data and the test data for writer independent tests.

The test set size recommended by the donators is 10 times larger than what the theory predicts. At the time of writing this paper, there were still open discussions regarding this matter.

4. Machine Printed Text

The *UW-I* and *UW-II* document image databases [10] contain the following types of document image pages: *English* technical journal articles, 1147 pages *UW-I*, 623 pages *UW-II*; *Japanese* technical journal articles, 477 pages *UW-II*; *English* memorandums, 62 pages *UW-II*. Each document image page is zoned and the text zones have line by line associated character ground truth generated by a double data entry and triple verification protocol [5]. The *UW-III* document image database, which will be issued in 1996, will have line drawings and word bounding boxes for each English page in the *UW-I* and *UW-II* document image databases.

UW-I contains a large set of isolated degraded characters generated by Baird's degradation model [1]. In addition *UW-I* contains software for determining OCR accuracy by comparing OCR generated text with the ground truth text and software for degrading a document image. *UW-II* contains a document image viewer, called Illuminator provided by RAF, Inc. The databases are distributed on a CDROM media [9].

The pages from both English and Japanese journals and reports come from the University of Washington libraries. Some of the English report pages come from the University of Nevada Information Sciences Research Institute database. The memorandums come from staff members of Seattle University. The pages contain a diverse sample of document styles found in technical journals and other documents.

The document image pages are scanned at 300 ppi on either a Fujitsu 3096 document scanner or a Ricoh IS50 scanner. They consist of: binary images scanned directly from the original document pages; binary images scanned from first generation photocopies of the original document pages; binary images scanned from the second or later generation photocopies of the original document pages; gray scale images scanned from the original document pages; and synthetic noise-free binary images from LaTeX generated documents.

In addition to these images, the document image database is annotated with information about the contents of the pages. Qualitative information about the condition of each page in terms of the nature of noise present, including the page rotation angle, are presented in its page condition file. Information about the document page such as the journal from which it is taken, its page number, specification of the dominant font on the page, specification as to whether figures etc. are present on the page are in its page attribute file.

Most OCR algorithms proceed first with a segmentation of the page into "zones" which are usually rectangular areas that are semantically homogeneous. The *UW* document image data bases provide this kind of annotation. At the coarsest level, a page is decomposed into "header," "footer" and "live-matter" areas. Standard definitions of what constitutes a header, etc., from the publishing and page description world are used. The *header* is defined to be ancillary text that appears above the main body of the page. For the world of technical journals this usually includes information such as the name of the article, the journal, the authors and the page number. A similar field may be present below the main body of the page and is referred to as the *footer*. The main body of the page is referred to as the *live matter*. Information about the pixel location and size of each of these zones on the page are provided in its associated "page bounding box" annotation file.

At the next finer level, the page is decomposed into "zones". Zones can be of various types: text, figures, tables, half-tones and mathematical equations among others. Zone delineation information for each page is provided in its "zone bounding box" annotation file.

Each zone has attributes which include things such as the semantic meaning of each zone (for example, a text zone could be a section heading, reference list item or page number), the dominant font in the zone, the font-style, etc. This information is provided in the page's "zone attribute" annotation file.

Finally, there is the "ground-truth" data files. For each "non-text" zone, the zone-type (figure, displayed math, table, line drawing etc.) is given. For each text zone, the text within the zone is specified in terms of its ASCII text, line for line.

4.1. Page Attributes

For each document page in the database, there is a set of attributes that describe the top level attributes of the page. The page attributes contain the page ID, the page contents, the page layout, the font and publication information of a document page. The group of attributes associated with publication information includes: name, volume number, issue number and publication date of the journal. It also has the corresponding page number of the document page from the publication. The page attributes also include the type of language, script, font type, and the character orientation and reading direction of the document page. The font types are defined to be of two varieties: those with and without serifs. Thus fonts such as Times are part of the Serif font type and fonts such as Helvetica are part of the Sans-Serif font type. Where more than one font type is present, the dominant font type is defined to be the one which occupies the largest fraction of the page in terms of physical area.

The various page attributes and their possible values are as follows: Document ID, *8 character string*; Document language, *English, Japanese*; Document script, *Roman, Katakana, Kanji, Hiragana*; Document type, *journal, letter, memo*; Publication Information; Multiple pages from the same article, *yes, no*; Text zone present, *yes, no*; Special symbols present in text zone, *yes, no*; Displayed Math zone present, *yes, no*; Table zone present, *yes, no*; Half-tone zone present, *yes, no*; Drawing zone present, *yes, no*; Page header present, *yes, no*; Page footer present, *yes, no*; Maximum number of text columns, *1, 2, 3, 4, non-text*; Page Column layout, *regular, irregular, non-text*; Character orientation, *up-right, rotated-right, rotated-left, non-text*; Text reading direction, *left-right, right-left, top-down, bottom-up, non-text*; Dominant font type, *serif, sans-serif, non-text*; Dominant character spacing, *proportional, fixed, non-text*; Dominant font size (pts), *4-8, 9-12, 13-18, 19-24, 25-36, 37-99, non-text*; Dominant font style, *plain, bold, italic, underline, script, non-text.*

4.2. Page Condition

The page condition of a document page describes the visual condition (or qualities) of a given document page. For example, it contains information on the presence or absence of visible salt and pepper noise, or visible vertical and horizontal streaks, or extraneous symbols from other pages. It also indicates if the document page is smeared (or blurred) because of poor focusing. It contains the measured page rotation angle and its standard deviation. It contains information about how many times the document page was successively copied. Thus, if the page is scanned from a first generation copy, the "Nth copy" attribute will have the value of `1`.

Quite often when a bound journal is scanned or photocopied, the portion of the page that is close to the spine of the journal is subject to perspective distortion. In regions of the document page that are close to the spine, the lines of text appear to curve (skew) towards either the top or bottom of the page. The page skewed left or right attribute value pairs indicate whether such distortion is present on the document page.

When a bound journal page is scanned or photocopied, there are sometimes sections of the page close to the spine that contain dark blotches which smear the text together. This is because the page appears darker (in grayscale) close to the spine and a uniform binarization threshold would then result in dark blotches. The page smeared left or right attribute values indicate whether such distortions are present on the document page. The page rotation angle is defined as the orientation of the lines of text relative to the horizontal. These orientations and their standard deviations are estimated using a triangulation scheme on multiple sets of manually entered groups of three points on each document page.

The page condition file has the following fields: Document ID, *8 character string*; Degradation type, *original, photocopy, fax*; Nth copy, *noise-free, original, 1, 2, ...*; Visible salt/pepper noise, *yes, no*; Visible vertical streaks, *yes, no*; Visible horizontal streaks, *yes, no*; Extraneous symbols on the top, *yes, no*; Extraneous symbols on the bottom, *yes, no*; Extraneous symbols on the left, *yes, no*; Extraneous symbols on the right, *yes no*; Page skewed on the left, *yes, no*; Page skewed on the right, *yes, no*; Page smeared on the left, *yes, no*; Page smeared on the right, *yes, no*; Page rotation angle (in degrees); Page rotation angle standard deviation.

4.3. Zones on a Page

A document page can be geometrically partitioned into several rectangular regions, called *zones*. In general any section of text that is clearly demarcated from adjacent areas of a page by "white space" is a *text zone*.

The rules for defining zones on a page are as follows: A zone is geometrically defined as a rectangular region on a document page. A zone is confined to a single column of text. Font type, style and size are mostly homogeneous across one zone. A zone must not be nested completely within another zone. A drawing, table, or half-tone (without its caption) is a zone. A line that demarcates two sections of text or lines that make up a box that encloses a section of text is a special kind of "drawing" and is called a *ruling zone*. Another special case of a "drawing" is called the *logo zone* and usually consists of the business logo of the company that publishes the journal. Other kinds of drawings that make up distinct zones are geographic maps (*map zones*) and advertisements (*advertisement zone*). The caption of a drawing, table or half-tone is a zone. A *list* is defined as any sequence of text zones, each associated with an alphanumeric "counter" or "index tag". The index tag could be a "•" symbol or any other string for e.g. references are often indexed by a string made up of the initials of the authors and the year of publication. Every item in a list constitutes a zone. Every paragraph of text that remains unbroken (with or without in-line mathematical equations) constitutes a zone. A *drop-cap* is in a zone by itself. Every displayed mathematical equation is a zone. *Section headings* (which are distinguished in many cases from the text body by being either bold-faced or underlined) constitute a zone. The section heading could be part of a line of text. Headings that indicate that the following text is part of the abstract of a paper, or the keywords used to index the paper or the start of a list of references constitute legitimate zones. They are referred to as *abstract*, *keyword* and *reference heading* zones respectively. The *page number* of a page constitutes a separate zone even if there is no white space separating it from nearby text.

Sections of text that represent computer algorithms in "pseudo-code" are also zones (called *pseudo-code zones*). Sections of the text that form part of the title area of a journal article have special semantic meanings and are each assigned to a separate zone. These include the title (*title zones*), the names of the authors (*author zone*) (where more than one appears on a line they are all part of the same zone), the organizational affiliation of the authors (*affiliation zone*), any diplomas or educational qualifications of the authors (*diploma zone*), and memberships in academic societies (*membership zones*). Information pertaining to the date on which an article was submitted or accepted for publication has important semantic information and constitutes an *article submission information* zone. Some articles contain a "blurb" of text that appears on the same page in order to emphasize the point the authors are trying to make. These text regions are called *highlight zones.* Some articles contain a brief summary of the contents of the page in a separate text region. Such a zone is referred to as a *synopsis zone.* The area that contains the key words used to index a paper has special semantic meaning and constitutes a *keyword zone.*

Some of the document pages contain handwritten annotations. These constitute zones which are called *handwriting zones.* No ground truth is entered for these zones. Sometimes there are extra text symbols from the opposite page that appear on a document image (this happens when photocopying from a bound journal). These symbols are not zoned.

4.3.1. Bounding box information

Bounding boxes are given relative to the page as a whole and relative to each zone on the page. Page bounding box information specifies the size and the location of the three types of special zones, i.e. page header zone, page footer zone and live matter zone. The location of the zone is specified in terms of the row and column pixel coordinates of the top left hand corner of each type of zone box. The size is specified by giving the row and column pixel coordinates of the bottom right hand corner of the zone box. These zone boxes have been delineated by hand using an interactive zone boxing tool. These zone boxes are by no means the smallest bounding rectangle but are guaranteed to contain the page area they are meant to. Where unavoidable (as a result of page rotation or too small a separation between zone boxes), there may also be an overlap area between adjacent zone boxes.

For each of the zones in the page the size and location of the zone bounding boxes is similarly specified.

4.3.2. Zone threading

The zones of each document page are grouped into several logical units. Within each logical unit, the reading order is sequential. Such a logical unit is called a *semantic thread.* Thus associated with each zone in a thread is a pointer to the next zone within the thread. Figures and their captions make up a thread. So do tables and their captions. Zones in the header or the footer areas make up threads individually. All other sets of text zones constitute the main thread of the document page. In general, the last zone within a thread has a "`nil`" pointer. In some cases, zone threading may cross pages. When consecutive document pages are presented within the database, the last zone of a given page that is in the live matter area may thread with the appropriate zone on the following page.

4.3.3. Zone attributes

Zone attributes define the properties of a zone on a document page. The zone attributes contain information on the Page ID, the Zone ID, the zone contents, the zone label, the text alignment, the font, the column number, the language and the script, the character orientation and reading direction of the zone. It also has zone threading information, which was explained in the previous section.

The text alignment attribute is defined relative to the zone bounding box for the zone. If the lines of text are aligned with the left edge of the bounding box, it is referred to as *left aligned.* There are similar definitions for right and center aligned zones. If the text is aligned with both the right and left edge of the zone bounding box it is referred to as *justified.* Text alignment within the zone, can have the values: *left, center, right, justified, justified hanging*, and *left hanging.*

Zone attributes include: Dominant font type, *serif, sans-serif, non-text*; Dominant character spacing, *proportional fixed non-text*; Dominant font size (pts), *4-8, 9-12, 13-18, 19-24, 25-36, 37-99, non-text*; Dominant font style, *plain, bold, italic, underline, script, non-text*; Character orientation, *up-right, rotated-right, rotated-left, non-text*; Text reading direction, *left-right, right-left, top-down, bottom-up, non-text*; Zone's column number, *header-area, footer-area, 1-1, 1-2, 2-2, 1-3, 2-3, 3-3, non-text*; Next Zone ID within the same thread, *___, nil*;

Other zone attributes include: Document ID, *8 character string*; Zone ID *3 character string*; Language, *English, French, German, Japanese*; Script, *Roman, Katakana, Kanji, Hiragana*; Zone content, *text, text with special symbols, text with non japanese symbols, text with special symbols non-japanese, text with rubi, text with non-japanese and rubi, text with special symbols non-japanese and rubi, math, table, halftone, drawing, ruling, bounding box, logo, map, form, advertisement, announcement, handwriting, seal, halftone with drawing, figurative text english, figurative text chinese, figurative text korean, signature, initials, new definition, block, figurative text, figurative text japanese*; Text zone label, *text body, list item, drop cap, caption, section heading, synopsis, highlight, pseudo-code, reference heading, definition, reference list, reference list item, footnote, biography, list, not clear.* Page headers and footers can have these zone labels: *page header, page footer, page number.*

Title pages of documents can have zone labels: *title, author, affiliation, diploma, membership, abstract heading, abstract body, abstract heading and body, correspondence, executive abstract heading, executive abstract body, keyword heading, keyword body, reader service, keyword heading and body, publication infomation, article submission information, not clear.*

Memos and Letter documents can also have the following zone attributes: *date, to, from, subject, cc, memo heading, complement, street address, city address, PO Box, phone number, e-mail address, fax number, telex number, laboratory, department, group, division, institution, subject matter, sender's reference, sender's name, recipient's name, secretary's initials, sender's title, sender's initials, opening salutation, closing salutation, home office information, founding date, enclosure.*

Japanese documents can have these other label values: *subject title, illustrator, author affiliation membership, title, author affiliation.*

4.4. Ground Truth Data

The database provides the ground truth for all text zones on a document page. For non-text zones, such as the displayed mathematical formulas, line-art, figures, table zones and etc., an indication of such will be given.

In *UW-II* a zone-based ground truth method is used. The zone-based ground truth consists of the symbol strings which are contained in the text zone. This includes the standard ASCII characters as well as special symbol escape sequences for non-ASCII characters. The lines in the ground-truth data are broken at the same position of the string where the physical line is broken on the page. Tabbing or indentations are ignored. Single blank characters (spaces) are used for one or more spaces within a text line. Ground truth for Japanese text is provided in unicode. The methodology used for accurate ground truthing involves double data entry and triple verification [5]. The character substitution error rate is estimated at about 60 characters per million for *UW-I*, and 40 characters per million for *UW-II*.

5. Discussion and Conclusions

Several significant data sets for OCR and document image understanding research were surveyed. The CEDAR, CENPARMI, and NIST data sets were designed to address the needs of researchers in offline handwriting recognition.

The UNIPEN data set contains a large number of samples of online handwriting. The success of the UNIPEN project demonstrates that it is possible to exchange data on a large scale. The keys to the success of UNIPEN were to develop a common data format and to keep all decisions democratic so that they would reflect the desires of the majority of the participants. Exchanging data also raises some difficulties. One of them is that it can take a long time (the project is already two years old). Another one is that transforming many different data sets into a standard format can be time consuming.

The University of Washington data sets of machine printed text images have helped many researchers in document recognition. The large number of truthed images on the UW CDROMs have seen widespread use and have been one of the factors in improving the performance of various commercial OCR packages. This project recently provided a significant sample of scanned and truthed Japanese text.

Some interesting practical problems in the design of a data set are illustrated by these efforts. For example, in the UNIPEN project, the splitting of the data set into training and test sets while retaining a statistically significant test set size was considered. Another open problem in data set design is in non-English language text. The University of Washington has begun to address this issue, but collections of truthed images from other languages besides English and Japanese (e.g., French, Russian, Chinese, etc.) are needed so that problems in non-English OCR can be investigated by many researchers.

Acknowledgments

Isabelle Guyon would like to thank all the persons who provided help to the UNIPEN project and in particular John Makhoul for discussions on statistical significance and the co-organizers of UNIPEN Lambert Schomaker, Stan Janet, Réjean Plamondon and Mark Liberman for sustaining several years of common effort.

References

[1] H.S. Baird, Document image defect models, in *Structured Document Image Analysis,* eds. H.S. Baird, H. Bunke, and K. Yamamoto, Springer-Verlag, Heidelberg, 1992, 546–556.

[2] D. Gerrety, JOT, A specification for an ink storage and interchange format, Technical Report draft version 0.99, Slate Corporation, San Mateo, California, 1993.

[3] I. Guyon, J. Makhoul, R. Schwartz, and V. Vapnik, What size test set gives good error rate estimates, AT&T Bell Laboratories Technical Memorandum BL0115540-951206-07, submitted to IEEE Transactions on Pattern Analysis and Machine Intelligence, 1995.

[4] I. Guyon, L. Schomaker, R. Plamondon, M. Liberman, and S. Janet, UNIPEN project of on-line data exchange and benchmarks, *Proc. of the 12th IAPR International Conference on Pattern Recognition*, Jerusalem, Israel, Oct. 1994, 29–33.

[5] J. Ha, R.M. Haralick, S. Chen, and I.T. Phillips, Estimating errors in document databases, *Proc. of the Third Annual Symposium on Document Analysis and Information Retrieval*, University of Nevada, Las Vegas, NV, April 1994, 435–459.

[6] J.J. Hull, A database for handwritten text recognition research, *IEEE Trans. on Pattern Analysis and Machine Intelligence*, 16, 5 (1994) 550–554.

[7] G. Nagy, At the frontiers of OCR, *Proc. of the IEEE*, 80, 7 (1992) 1093–1100.

[8] I.T. Phillips, S. Chen, J. Ha, and R.M. Haralick, English document database design and implementation methodology, *Proc. of the Second Annual Symposium on Document Analysis and Information Retrieval*, Las Vegas, NV, April 1993, 65–104.

[9] I.T. Phillips, S. Chen, and R.M. Haralick, CD-ROM Document Database Standard, *International Conference on Document Analysis and Recognition,*

Tsukuba Japan, Oct. 1993, 478–483; reprinted in *Document Image Analysis* by L. O'Gorman and R. Kasturi, Los Alamitos, CA, IEEE Computer Society Press, 1994, 198–203.

[10] I.T. Phillips, J. Ha, R.M. Haralick, and D. Dori, The implementation methodology for a CD-ROM English document database, *International Conference on Document Analysis and Recognition*, Tsukuba, Japan, October 1993, 484–487.

[11] C.Y. Suen, C. Nadal, R. Legault, T.A. Mai, and L. Lam, Computer recognition of unconstrained handwritten numerals, *Proc. of the IEEE*, 80, 7 (1992) 1162–1180.

Handbook of Character Recognition and Document Image Analysis, pp. 801–820
Eds. H. Bunke and P. S. P. Wang

CHAPTER 31

BENCHMARKING DIA SYSTEMS

THOMAS A. NARTKER
Information Science Research Institute
University of Nevada, Las Vegas
4505 Maryland Parkway
Las Vegas, NV 89154-4021, USA

Benchmarking document image analysis (DIA) systems is important for users, for researchers, and for developers of these technologies. From the user's point of view, the appropriate measure of a DIA system is the total cost of document conversion. This cost is typically dominated by the cost of correcting residual errors in the output.

The cost measure needed by a user in selecting a system, however, is not necessarily appropriate for a researcher who seeks to measure progress in developing a new algorithm. Furthermore, the task of comparing the performance of two algorithms, whose internal operation is visible, is quite different from that of comparing several large systems whose source code is a carefully guarded secret. The latter is referred to as "black-box" testing.

All DIA systems are composed of different sub-algorithms often applied in sequence. Measuring the performance of such systems when neither the algorithms nor the sequence is known is an especially difficult task.

In this chapter, we present the important issues in benchmark testing from the "black-box" point of view by reviewing a selection of performance metrics for systems that recognize text from images of machine-printed pages. We present the functional steps for automated evaluation of OCR system performance. We also discuss the relevance of these steps to benchmarking other DIA technologies.

Keywords: benchmarking, performance measurement, document image analysis, optical character recognition.

1. Introduction

1.1. Document Image Analysis Systems

The general problem of all document image analysis (DIA) systems is one of appropriate extraction of coded information from a two-dimensional array of numbers.

$$\text{“Sequence of Code Values”} \Leftarrow R(A)$$

A DIA system can be thought of as a function R that maps a numeric array A into a sequence of codes, each representing an object that has been recognized.

Each code value in the sequence is sometimes referred to as a label. In the case of simple character recognition, the sequence might contain ASCII codes. In more complicated cases, attributes could be associated with each code value. For example, a system that recognizes typeface or type size information might attach these attributes to each ASCII code.

In more advanced DIA systems, the coded objects can themselves possess structure and can thus (recursively) be made up of other coded objects with (possibly) attributes attached to each object. In all cases, the system output is an ordered (sometimes partially ordered) sequence of such codes. For notational simplicity, we will refer to this sequence as a "string" and denote it as S. Thus, a DIA system R accepts an array A and produces a string S.

$$S \Leftarrow R(A)$$

The array A is a digitized "image" of one page (or part of a page) of a document and is produced by a scanner. A is an m x n array of elements with m and n taking on values in the thousands, depending on page size and the scanning resolution. If the image is black and white, the element type is binary and one value represents black (or foreground) pixels while the other value represents white (or background) pixels. For gray images (from a gray scale scanner), the element type is integral with the range of each element value defined by the gray range of the scanner used (typically 0 to 255). For color images, the element type is multi-valued integral. In this chapter, we are mostly concerned with systems that recover information from binary or gray scale page images.

As these technologies have improved, R has more and more become a rule-based "expert system" incorporating diverse recognition techniques (such as those described in previous chapters) trained to maximize overall performance. Because most DIA systems are "rule-based," their performance is not predictable from first principles of an underlying theory.

Performance improvement is not necessarily associated with a newer, or more expensive, version of a system. Even for the same system, it is common to observe significant variation in performance due to very slight changes in a given input image. For systems that can be trained to a given type of input document, performance can vary even for the exact same input image. In general these systems will accept as input any image from the universe of machine-printed documents. Thus, although the behavior of DIA systems is technically deterministic, their performance in practice is highly variable. For most DIA systems, it is difficult to predict how they will perform for any specific class of input. Even the approximate state-of-the-art is difficult to determine.

1.2. Definition of Benchmarking

In the last ten years, benchmarking has become a standard technique for improving not only industrial products but also industry operations themselves [1]. Benchmarking is a process of identifying superior performance among competing systems through careful selection, measurement, and comparison of critical aspects of system operation. It is the standard technique for analyzing complex systems when there is no underlying theory available for more conventional analysis.

The key component of this type of analysis is measurement. It involves the selection and acquisition of appropriate types of data, devising special metrics that reflect interesting aspects of performance, and conducting large-scale tests of competing systems to identify (and understand the nature of) systems that exhibit superior performance.

1.3. Benchmarking DIA Systems

The benchmarking paradigm for DIA systems requires comparing an *observed variable* with a *reference variable* under *controlled conditions* [2]. Controlled conditions are necessary to ensure reproducible tests. They are achieved in part by careful acquisition of a representative set of test images I from a document domain of interest. The observed variable is some aspect of the output string S when one element of I (denoted I_i) is used as input. The reference variable is the corresponding aspect of the desired (or correct) output string.

Some measure of system performance p is produced by computing the "similarity" between these two strings. Benchmarking experiments are conducted by measuring values of p (the dependent variable) while varying either some attribute of the set of input images or of device operation (the independent variable).

Thus, a measure of performance p of a DIA system R is computed as some function F (i.e., some measure of the similarity) of the observed string S^o and the reference string S^t. The essential point is that the more similar S^o is to S^t, the better is the performance of the system.

For an image I_i, the string S_i^o is the string produced by R given I_i as input. The string S_i^t is the string deemed to be correct (i.e., the "Ground-Truth" string) and p_i is the value of the measure.

$$S_i^o \Leftarrow R(I_i)$$

$$p_i \Leftarrow F(S_i^o, S_i^t)$$

For an optical character recognition system, F could simply count the number of correct characters produced (i.e., the characters in S^o that are correct). Even for this simple measure of performance, the number of "correct" characters is not obvious.

To illustrate the problems, Figure 1 shows a paragraph of text extracted from a technical report [3]. There are 595 characters in this paragraph, including line-breaks and interword spaces. Figure 2 shows the output produced by a contemporary OCR device when given the image of the paragraph from Figure 1.

Note that the OCR system produced 7 tilde (˜) characters. For this OCR device, the tilde is the "reject" character, i.e. it is the symbol produced when the system is unable to recognize a character.

There are three problems. First, OCR devices routinely generate both "many for one" and "one for many" substitutions in unpredictable patterns. Second, because we have no access to the internal operation of the device and because current devices do not generate position information, we have no information about how and where such substitutions have been made. Third, we must determine not only that the correct characters have been generated but also in the correct order.

Head contours in the saturated zone underlying Yucca Mountain, Nevada, and its environs are derived on the basis of alternative interpretations of the influence of geologic structure on the flow of fluids. Numerical experiments examine the sensitivity of flow and transport to uncertainties in existing data and the response to assumed catastrophic changes in hydraulic conductivity. The calculations are intended to delineate the data that a performance assessment analyst might wish to ask for. From these experiments, it appears that faults controlling the flow are not a dominating feature.

Fig. 1. Correct text (S^t)

Flead contours in the satu˜fated zone unds,Iying Yucca MounEisin.
Nevada, and its environs are derived on the bi˜Rsis of alternstLve
interptet\˜ltions of the infLuence of geologic structuse on the flow
o f f Iu ids . Numer i ca 1 expe r iment s exami ne the sens i u ivi ty
of f low and triinspost to uncerEainties in exisEing data and the
respionse to a s sumed ciB t3 B t f ophi c chango s in hydsau L ic
conduc t ivi ty . The calculations are i˜f˜.Eended to delineate the data
that a pesfosmance assessment analyst might wish to as% for.
B'˜.som these expesiments. it appeass that faults controlling the flow
ase not a domini˜Rting f@iR LUI[@ .

Fig. 2. OCR generated text (S^o)

For these reasons, a correspondence between S^o and S^t must be established and any measure of the correctness of operation will depend upon the correspondence that is chosen. This "dependence upon correspondence" exists for any accuracy measure of any OCR system or, in general, for any DIA system in which S^t is an ordered string.

The lesson is that algorithms to identify which characters of S^o correspond with those in S^t (i.e., "matching" algorithms) are an important component of benchmarking DIA systems.

1.4. Computing Performance Measures

Thus, for a performance measure p of a DIA system R, it is necessary to determine how elements of the string S^o "match" with elements of S^t. If we denote the matching operation as a function M, the performance measure p_i can be computed for page I_i by:

$$p_i \Leftarrow F(M(R(I_i), S_i^t))$$

Figure 3 shows the output of one such "matching" algorithm given the strings from Figurel and Figure 2. Non-matching characters are identified, in order of first occurrence, by an index {enclosed in braces} identifying a specific confusion.

{1}ead contours in the satu{2}ated zone und{3}ying Yucca
Moun{4}in{5} Nevada, and its environs are derived on the b{6}sis
of altern{7}t{8}ve interp{9}et{10}tions of the inf{11}uence of
geologic structu{12}e on the flow o{13}f f{14}u{13}ids{13}.
Numer{13}i{13}ca{15} expe{13}r{13}iment{13}s exami{13}ne
the sens{13}i{16}ivi{13}ty of f{13}low and tr{17}nspo{12}t to
uncer{18}ainties in exis{18}ing data and the resp{19}onse to
a{13}s{13}sumed c{20}t{21}t{22}ophi{13}c chang{23}s in
hyd{12}au{24}ic conduc{13}t{13}ivi{13}ty{13}. The calculations
are i{25}ended to delineate the data that a pe{12}fo{12}mance
assessment analyst might wish to as{26} for. {27}om these
expe{12}iments{5} it appea{12}s that faults controlling the flow
a{12}e not a domin{6}ting f{28}.

Fig. 3. Result of matching S^o with S^t

The exact confusion associated with each index is shown in Figure 4. Note that <nil> refers to the empty (zero-length) string and <space> refers to one blank character.

Although the function F can now easily calculate, say, the number of "correct" characters in S^o (i.e., the number of characters in Figure 3 that are not enclosed in braces), this number is not necessarily a good measure of performance. (In this case, there are 495 "correct" characters.)

It is clear from this example that two different devices, R_1 and R_2, might generate two very different output strings, S_1^o and S_2^o, that have the same number of "correct" characters. Thus, the number (or even the percentage) of "correct" characters is not necessarily a good measure of "similarity." The point is that, in measuring superior performance, it is frequently more meaningful to focus on the "errors" in S^o than on the number of correct characters.

1.5. Automating Benchmarking Operations

In the last few years, the need for widely published benchmark comparisons of DIA system performance has been recognized. Large-scale tests are needed to obtain statistically significant results. Such tests are practical only when all calculations can be performed by a computer.

We note that conducting such tests is a formidable task requiring a significant amount of supporting software. Thus, the goal of this chapter is not only to review issues in benchmark testing, but to relate these issues to the design of the requisite software.

In automating benchmarking functions, the set of images I, the set of strings S^t corresponding to I, and the intermediate results would normally be implemented as files.

The functions

$$F(M(R()))$$

	Correct	Generated
1	H	Fl
2	r	~f
3	erl	s,I
4	ta	Eis
5	,	.
6	a	i~R
7	a	s
8	i	L
9	r	t
10	a	\~l
11	l	L
12	r	s
13	<nil>	<space>
14	l	<space>I
15	l	<space>1
16	t	<space>u<space>
17	a	ii
18	t	E
19	<nil>	i
20	a	iB<space>
21	as	3<space>B<space>
22	r	<space>f<space>
23	e	o<space>
24	l	<space>L<space>
25	nt	~f~.E
26	k	%
27	Fr	B'~.s
28	eature	@iR<space>LUI[@<space>

Fig. 4. Non-matching characters

would be implemented as operations (or commands) on files typically called from the operating system command language. Using this functional notation, in which R represents the command language call to the system, M represents the matching step, F the function to compute a performance measure, and $\Leftarrow$ represents a file assignment operation, we will discuss the components of a benchmark testing environment in the context of a selection of different performance measures.

1.6. Identifying Superior Performance among Competing DIA Systems

Therefore, for a given set of input images I, superior performance of a set of competing systems $R = (R_1, R_2, \ldots, R_k)$ can be determined by examining performance values $p_{j,i}$ for each image i and each device j.

$$p_{j,i} \Leftarrow F(M(R_j(I_i), S_i^t))$$

In practice, there are still other functions that need to be provided. For example, simply examining performance values on individual pages for individual devices yields little insight. Typically, it is desirable to examine values of p averaged over different subsets of I (i.e., for sets of pages {or documents}, and also for sets of documents) for each device. Also, it is necessary to investigate the significance of these results by computing confidence intervals for the values of p that are calculated.

In Section 3 we will discuss each of these issues in connection with several measures of performance of OCR systems. In general, we will focus on the important characteristics of the functions M and F for each performance measure.

2. Testing of Isolated Character Classifiers

Many of the test results reported in the DIA literature have been concerned with comparing the performance of different classification algorithms in recognizing individual printed characters (i.e., the image I_i contains only one character and tests were conducted one character at a time). Usually, there was interest in measuring the average rate of classification error for different algorithms using the set I of "character" images.

These tests reflected an emphasis on basic research in pattern recognition. In this Section, we review the considerations involved in conducting such tests.

2.1. Selection of Test Characters

When comparing isolated character classifiers, the data utilized are typically in the form of test alphabets with an equal number of samples of each class (digits, upper case, lower case, punctuation, special symbols), or of characters extracted from a number of sample documents roughly corresponding to their frequency of usage. The inputs to each algorithm are the coordinates of the box bounding the image of each individual character. Test data are chosen only to provide a meaningful comparison of different approaches to the classification of characters.

2.2. Representation of Output Characters

When tests are performed on known algorithms, issues regarding differences in output representation are seldom important. For example, if the algorithms generate different reject characters, this difference can easily be removed by a change in the source code of one of the algorithms. Non-ASCII characters would not be used in the test.

2.3. Matching Generated with Test Characters and Evaluating Accuracy

When testing is focused on evaluating classifier performance at the character level, S^o and S^t are each a single character. There is no order in the output and matching is not required.

The percentage of character errors is evaluated by counting the number of incorrectly recognized characters and the number of total characters in a set of tests and computing the ratio. Percent error is evaluated from,

$$F = \frac{|Errors(\bigcup_{i=1}^{n} S_i^o)|}{n} \times 100, \quad (n = |I|)$$

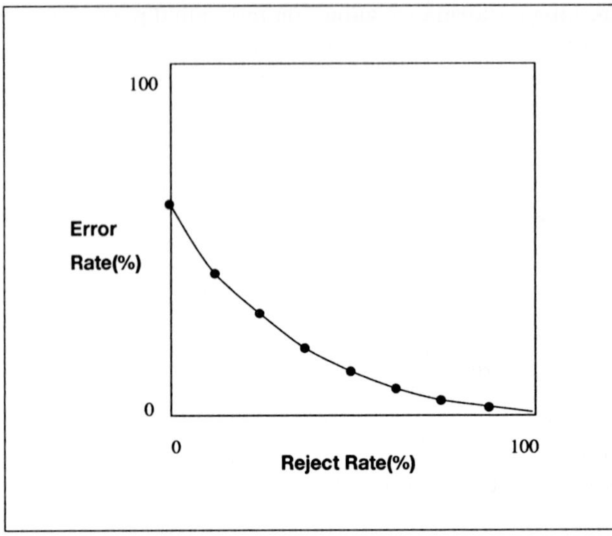

Fig. 5. Error/Reject Curve

where *Errors* is a subset operation that denotes the identification of the set of errors from the characters output from a set of tests and n is the number of elements in the set I. Superior performance is demonstrated by the classifier with the lowest percentage over the set I.

2.4. Other Measures of Isolated Recognition

In 1970, C. K. Chow proposed an especially good metric for characterizing classifier performance [4]. Chow's metric focused on measuring error as a function of an operating parameter.

As mentioned in Section 1.3, most classifiers emit a special character in the output (called a reject character) when a clear classification is not possible. In this manner a potential misrecognition is converted into a rejection that must be corrected manually. This trade-off between errors and rejects, however, is not one for one. Inevitably, some correct recognitions are also converted into rejects.

The percentage of characters that are rejected is termed the *reject rate* r. For characters that are not rejected, the percentage that are identified incorrectly is termed the *error rate* e. Chow proposed that a good overall measure of classifier performance was to plot the error rate versus various reject rates (see Figure 5). This performance measure is called an *error-reject curve.*

It is common for classifiers to accept a confidence value v that controls the threshold for reject determination. Data for plotting *error-reject curves* is obtained by comparing the output produced by a classifier with the correct characters S^t for several different confidence values ($V = v_1, v_2, \ldots, v_m$) for R. For each different confidence value chosen, an error rate e, and a reject rate r, can be computed from the output of,

$$M(R(v_k, I), S^t).$$

The function for computing values of the error rate e is,

$$F = \frac{|Errors(\bigcup_{i=1}^{n} S_i^o)|}{n - |Rejected(\bigcup_{i=1}^{n} S_i^o)|} \times 100 \quad (n = |I|)$$

where both *Errors* and *Rejected* are subset operations that denote the identification of the indicated subset of the characters output from a set of experiments and n is the number of elements in the set I. Values of e are plotted as a function of r.

Not only did Chow devise an especially good metric for comparing the performance of classifiers, he also demonstrated the value of measuring performance as a function of critical operating parameters. Thus, superior performance was determined by comparing a set of curves and not a set of points. Superior performance is shown by classifiers that produce curves that are uniformly closer to the origin.

Although error/reject curves have traditionally been used to characterize the performance of competing character classifiers, similar metrics are possible for other kinds of image classifiers.

2.5. Summary of issues in testing Isolated Classifiers

Historically, researchers have been interested in measuring and improving the performance of classification algorithms. While such algorithms are an important component of a page-reading OCR system, the performance of the system is only partially related to the performance of its classification algorithm. From the user's point of view, it is more important to measure the performance of the overall system.

Measuring system performance, however, involves problems not present when dealing with known algorithms. There are different issues involved with both the choice and preparation of test data and the representation of system output. As was shown in Section 1.3, an especially difficult issue is the need to match strings when no information concerning internal operation is available.

3. Automated Testing of Page-Reading OCR Systems

In the last few years, there has been growing recognition that large-scale benchmarking of competing systems is needed for progress in OCR. The combination of increasing OCR device complexity and the extreme diversity of input images and applications has placed emphasis on the need for large-scale automated tests.

In the field of constrained handprint recognition, large-scale comparisons of existing technologies have been conducted at the National Institute of Standards and Technology [5, 6]. For recognition of machine print, tests have been conducted using both commercial products and research prototypes [7, 8, 9, 10, 11]. Although these tests provide information useful to OCR users, their main purpose has been to make the general state-of-the-art visible to a wide audience and especially to the research and development communities.

In this section, we discuss the functions necessary to automate a series of measures of the accuracy of page-reading OCR systems.

3.1. Page-Based Testing of Character Recognition Performance

The most fundamental measure of an OCR system (or any DIA system) is the accuracy of the elements of S^o. To evaluate complete commercial systems, however, a number of interconnected problems must be solved.

3.1.1. Preparation of truth data

When automated tests of recognition from page images are to be performed, preparation of the correct string S^t presents special problems. On the surface, S^t would seem to be a simple ASCII string with one ASCII label for each character on the page.

First, large numbers of "real world" pages are desirable for tests. Unless machine readable code is available, each page must be manually zoned followed by repeated rekeying/resolution steps for each zone to achieve high accuracy. Manual zoning is required both to ensure that only valid text fields are included in the test and that correct zone information is stored as part of the input images. By including zone information as part of both I and S^t, it is possible to insure that each test device receives the exact same input image data in each test. Such test data is very expensive to prepare.

Second, how is the correct number of blank spaces to be determined? Should a device be penalized for generating one space where two were present on the page? One common solution is to compress consecutive blank spaces into one single space. A device is then charged one error only if no space is recognized whenever one or more exist. One problem with this solution is that proper registration of the columns in a table is not retained.

Third, how are non-ASCII characters to be represented. Should one device be penalized for recognizing the "degree" symbol and creating its own internal label? Should another device be penalized for substituting a blank space for this symbol? Should a device be penalized for generating a + followed by a - when the symbol ± is encountered?

Conventions must be adopted that will affect the measure computed and the implementation of the matching algorithm.

3.1.2. Document data representation / Device invocation

The output strings, S^o, produced by different systems frequently deal with recognition issues in different manners. For example, systems may use different symbols as reject characters and/or use a different symbol to mark "suspect" characters in the output.

It is also possible that systems may run on different hardware and/or are invoked by very different calls. To deal with such problems it is convenient to insert another function into the benchmarking environment that compensates for different calling procedures and produces a normalized S^o as output.

Thus, N accepts the index j of a recognition device and an input image I_i and invokes $R_j(I_i)$ in an appropriate manner. The output is "normalized" by, for example, substituting a common reject character in the output for all systems.

$$N(j, I_i) = \text{"Normalize"}(R_j(I_i))$$

N serves as a software interface between the various systems to be tested and a "device independent" test environment.

$$p_{j,i} \Leftarrow F(M(N(j, I_i), S_i^t))$$

New or modified recognition devices can be tested without modifying the routines that match strings or compute performance measures. All changes to the set of test devices can be accommodated simply by modifying the "device dependent" routines in N that ensure a standard representation for all output strings [12].

3.1.3. Algorithms for matching / Cost models

The most significant difference between measuring the accuracy of open algorithms and of closed systems is the need to provide an "arbitrary" match between S^o and S^t. Such a matching is necessarily "blind" when no information regarding internal operation is available. Of course, the matching must not really be "arbitrary." An algorithm must be provided to determine a match that is well-defined, reproducible, and in some sense is "appropriate" for the measure to be computed.

Although the first authors to employ a matching algorithm to compute the accuracy of commercial OCR systems were Handley and Hickey [13], the theoretical problem was first solved by Wagner and Fischer [14]. They presented an algorithm, based on dynamic programming, that identifies a sequence of edit operations for transforming S^o into S^t with minimum cost. Finding this "optimal" correspondence is known as the "string editing problem."

If we let $S^o = s_1, s_2, \ldots, s_n$ be a string of n characters selected from an alphabet Σ, there are three types of edit operations that can be applied to S^o.

1. *insertion:* any symbol $x \in S$ can be inserted before s_1, after s_n, or between s_i and s_{i+1} $(1 \leq i < n)$;
2. *deletion:* the symbol s_i can be deleted $(1 \leq i < n)$;
3. *substitution:* the symbol s_i can be replaced by any symbol $x \neq s_i$ $(1 \leq i \leq n)$.

The edit model $\gamma = (\gamma_I, \gamma_D, \gamma_S)$ specifies the cost of each insertion, deletion, and substitution, respectively, where γ_I, γ_D, and γ_S are non-negative real numbers. The cost of transforming S^o into S^t can be expressed as

$$\gamma_I \times \#Insertions + \gamma_D \times \#Deletions + \gamma_S \times \#Substitutions$$

and for any value of $(\gamma_I, \gamma_D, \gamma_S)$, the Wagner-Fischer algorithm finds a set of insertions, deletions, and substitutions that minimizes this cost. Note that these are the "errors" in S^o. The characters in S^o that are neither deleted nor substituted are said to match with those in S^t and are considered to be correct.

The Wagner-Fischer algorithm matches S^o with S^t in $O(n^2)$ time and $O(n^2)$ space. Since 1974, many authors have developed more efficient algorithms for solving the string editing problem. Among the algorithms available, one published by

Ukkonen [15] is an excellent choice for M. A good survey and comparison of these algorithms can be found in [16].

Thus, the use of cost models and cost minimization algorithms has become the standard means of choosing a match between S^o and S^t when benchmarking "black-box" systems.

3.1.4. Calculation of system (character) accuracy

If we assume that the human editing effort required to perform single character insertions, deletions, and substitutions is about equal, we can assign equal weight to each parameter in the cost model. If we then choose the match that results from finding the minimum cost sequence of edit operations using $\gamma = (1,1,1)$, we can compute a character accuracy value that reflects the minimum effort required of a user to correct the output (i.e., to transform S^o into S^t).

If, as $\gamma = (1,1,1)$ implies, we consider each edit operation to be one error, then

$$Errors(S_i^o) = Insertions + Deletions + Substitutions$$

and the character accuracy can be calculated by:

$$F_i = \frac{|S_i^t| - |Errors(S_i^o)|}{|S_i^t|}.$$

The subscript i is a reminder that experiments are conducted one page (or one zone) at a time.

3.1.5. Identifying Superior Character Accuracy

It is frequently the case that examining character accuracy values on individual pages for individual devices yields little insight. Typically, it is desirable to examine performance values averaged over different subsets of I for each device.

One approach is to extend the functionality provided by F. Suppose, for example, that instead of simply calculating values of character accuracy as shown above, F were to compute a 3-tuple as follows:

$$(p_i, |S_i^t|, |Errors(S_i^o)|) \Leftarrow F(M(N(j, I_i), S_i^t))$$

Essentially, F computes not only the character accuracy for a page but also the number of characters on the page and the number of errors. For convenience, we will call this 3-tuple a T^c for "character accuracy tuple."

It is now easy to provide yet another function C that accepts as input any number of 3-tuples (or T^c's) and generates another as output.

$$T_j^c \Leftarrow C(T_{j,i}^c, T_{j,i+1}^c, \ldots)$$

Thus, F produces values of character accuracy for each page. Subsequent applications of C can produce average character accuracy values at the document level, over a collection of documents or over any subset of pages desired [12].

There is one final consideration in determining superior performance. How can we be confident that

$$p_j > p_k$$

for any collection of pages or documents implies that

$$\text{Device}_j \text{ is significantly more accurate than Device}_k?$$

The answer can be determined by employing an appropriate test of statistical significance. Rice [11], was the first to employ a technique from statistics known as the *jack knife estimator* [17], to establish 95% confidence intervals for character accuracy.

Although we will not describe this calculation here, yet another function application after M, F, and C, is commonly required to determine superior performance.

3.2. Page-Based Testing of Word Recognition Performance

In many cases, users of an OCR system are not concerned as much with individual characters as with words. They may not be concerned with numeric digits or with punctuation. Thus, another interesting measure of performance is word accuracy.

3.2.1. Representation of words

As was the case with character accuracy, there are issues regarding the "correct" representation of words. For example, how are words defined in S^t? One definition is to accept any sequence of one or more letters to be a word. Whatever definition is chosen, it is possible to use the definition to parse S^t (or S^o) into an ordered string of words. We denote the ordered string of words corresponding to S^t by $S^{t'}$ and we define the function P to be the appropriate parsing operation.

$$S^{t'} \Leftarrow P(S^t)$$

It appears that word accuracy would be calculated from:

$$p_{j,i} \Leftarrow F(M(P(N(j, I_i)), S_i^{t'}))$$

3.2.2. Word matching / Cost model

For word accuracy however, the matching operation needed is different from that used for character matching.

When matching words, we wish to compute the number of correctly recognized words. Thus, the appropriate cost function to use when aligning strings $S^{o'}$ and $S^{t'}$ is $\gamma = (1, 0, 1)$. It allows us to count the number of "essential" edit operations (in this case word insertions and word substitutions) needed to restore all words from $S^{t'}$ into $S^{o'}$. The count of word insertions plus word substitutions is the number of misrecognized words in $S^{o'}$. An efficient algorithm that can be used for this matching operation (M') was developed by Myers [18].

3.2.3. Word accuracy

Having achieved the appropriate match between $S^{o'}$ and $S^{t'}$, F' can then tabulate the number of erroneous (or misrecognized) words in $S^{o'}$ and the total number of words in $S^{t'}$ for any page in I. If

$$Errors(S_i^{o'}) = \text{the set of misrecognized words from page } i$$

then

$$F_i' = \frac{|S_i^{t'}| - |Errors(S_i^{o'})|}{|S_i^{t'}|}$$

As was the case for character accuracy, F' can be extended to produce a 3-tuple,

$$(p_i', |S_i^{t'}|, |Errors(S_i^{o'})|)$$

that we call a "T^w" for "word accuracy tuple."

$$T_j^w \Leftarrow F'(M'(P(N(j, I_i)), S_i^{t'}))$$

C' can then be applied to any subset of the output of F' to calculate an average word accuracy for a document, for a collection of documents, or for any subset of pages desired [12].

$$T_j^w \Leftarrow C'(T_{j,i}^w, T_{j,i+1}^w, \ldots)$$

3.2.4. Non-stopword recognition

A frequent use of the output of OCR systems is in text retrieval databases. In text retrieval applications, common words known as *stopwords* are normally not indexed because they offer little or no retrieval value. Examples of the most common stopwords are *the*, *of*, *and*, and *to*.

Only words that are not stopwords, called *non-stopwords*, are indexed. Typically, users of text retrieval systems search for one or more non-stopwords in the database. Therefore, the accuracy of non-stopwords in the database is especially relevant in text retrieval applications.

Using an appropriate function to remove stopwords from $S^{t'}$, the set of functions used to compute word accuracy can be used to compute the accuracy of non-stopwords from any desired subset of images.

3.3. Testing of Automatic Zoning

The first step in recognizing characters on a page is to locate and order the text to be recognized. Although page-reading OCR systems accept the coordinates (or zone information) for each text block on a page, manual determination of these coordinates is expensive and time consuming. For this reason, most OCR systems will attempt to determine an ordered set of zone coordinates for each page automatically. For some applications, it is important to be able to estimate the accuracy of this segmentation operation separate from the accuracy of classification.

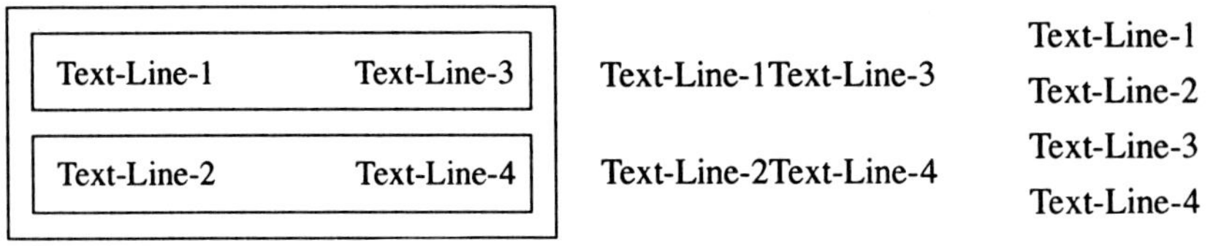

(a) Incorrectly Zoned Page (b) Generated Text (c) Correct Text

Fig. 6. Zoning error

This measure can be especially valuable when some pages to be processed contain single column text and some contain multiple column (newspaper style) text. For multi-column text, it is important to properly decolumnize text paragraphs in the output (see Figure 6).

For scientific documents that contain much tabular information, it is important that tables are not decolumnized. Thus, a metric that measures a system's ability to properly decolumnize multi-column text while not decolumnizing tabular information is needed. One such measure of zoning performance is called the "Cost of Automatic Zoning" [19]. Although the complete description of this metric requires several pages, even a brief description will reveal similarities with previous measures.

3.3.1. Zoning performance as a function of editing effort

To correct the output produced from an automatically zoned page, a human editor would utilize text insertion, deletion, and move operations. If a zoning algorithm mis-classifies a text region as a graph, or does not detect text, the editor must insert the missing characters into S^o.

If a graph is mis-classified as text, characters in the graph, or other graphic objects, could be included as text in S^o that must be deleted. When a multi-column page is incorrectly zoned, as shown in Figure 6, Text-Line-3 must be moved between Text-Line-2 and Text-Line-4. Thus, the cost of correcting the reading order in S^o is related to the performance of the zoning algorithm.

In calculating a cost of automatic zoning we can estimate the overall effort required by a human operator to correct the errors in S^o, produced from an automatically zoned image, by counting the number of insertions, deletions, and block move operations required.

Part of the effort of correcting S^o however, are edit operations that are due to mis-classification of characters and not to zoning error. We assume that OCR systems make the same classification errors when a page is manually zoned and when automatically zoned.

Thus, the "Cost of Automatic Zoning" is estimated by subtracting the effort required by a human operator to correct the errors in the S^o produced from the manually zoned image. Because we have used I_i to designate manually zoned page images, we denote their automatically zoned counterpart as I_i^Z.

3.3.2. Zone matching by matching strings / Number of moves

To estimate the overall editing effort required to correct the output from automatically zoned images, it is necessary to calculate the number of insertions, deletions, and move operations required. Because block moves may be necessary, the matching algorithm used must be able to find transposed matches. Such an algorithm (M^T) is described in [19].

After determining the matches between S^o and S^t, an algorithm derived by Latifi (denoted L) [20] can be used to estimate the minimum number of moves necessary to rearrange the matches in S^o into the proper order.

3.3.3. Cost of correcting automatically zoned pages

As was the case with previous measures, an edit model is employed. In this model, $\gamma = (\gamma_I, \gamma_D, \gamma_M)$, γ_I is the cost of an insertion, γ_D is the cost of a deletion, and γ_M is the cost of a move.

Although a detailed description of this model requires several pages, the essential parameters used are:

$$\gamma_I = 1, \gamma_D = 0, \gamma_M = min(SizeOfMove, T)$$

where T is an integer representing the maximum number if insert/delete pairs that a user would employ to achieve a block move instead of a single cut and paste. This model interprets errors in terms of the "equivalent number" of insertions needed for correction. Using this model, the number of errors in S_i^o produced from recognition of an automatically zoned image I_i^Z for device j can be calculated from:

$$|Errors(S_i^o, I_i^Z)_j| \Leftarrow F(L(M^T(N(j, I_i^Z), S_i^t)))$$

where F converts all moves into an equivalent number of insertions (see 19).

3.3.4. Cost of Automatic Zoning

To eliminate the effect of character mis-classifications, the cost of correcting errors from a manually zoned image must be subtracted from the overall cost. Note that the errors in S_i^o produced from recognition of a manually zoned image I_i for device j can be estimated from:

$$|Errors(S_i^o, I_i)_j| \Leftarrow F(M^T(N(j, I_i), S_i^t))$$

where F counts the number of insertions needed. If we denote the cost of employing automatic zoning as C^Z:

$$C_{j,i}^Z = |Errors(S_i^o, I_i^Z)_j| - |Errors(S_i^o, I_i)_j|$$

or

$$C_{j,i}^Z \Leftarrow F(L(M^T(N(j, I_i^Z), S_i^t))) - F(M^T(N(j, I_i), S_i^t))$$

In order to remove the effect of an arbitrary choice of the value for T, it is beneficial to plot C_i^Z for each device j (for a set of pages) versus a range of possible values for T.

The detailed description of this measure is given by Kanai [19]. Examples of its use are shown in [11].

3.4. The Pattern of Function Calls

A recurring pattern is the use of a matching algorithm (frequently an optimization algorithm), and an editing model. Matching provides a well-defined association between the generated and the correct output strings that might be chosen by users of the device. The editing model is chosen to reflect the human effort needed to correct the aspect of performance being measured.

For character accuracy, we used a string-matching algorithm that minimized the number of character insertions, deletions, and substitutions. In the model, we chose $\gamma = (1, 1, 1)$ and counted each such edit operation as one "character error."

For word (and non-stopword) accuracy, we used a string-matching algorithm that minimized the number of word insertions and substitutions. In this model, we chose $\gamma = (1, 0, 1)$ and counted each insertion and substitution operation as one "misrecognized word."

For zoning accuracy, we used a string-matching algorithm that found transposed matches. In this editing model, we chose a value for γ that made it possible to consider the likelihood that block moves were performed using cut and paste operations.

The use of matching algorithms and editing models in this manner makes possible the computation of well-defined (and reproducible) performance measures for complete (closed) DIA systems that reflect the users point of view.

4. Summary of Considerations in Designing Automated Performance Measures of DIA Systems

In each case above, the pattern of concerns has been the same. The issues are:

1. selection and preparation of the set of test page images I and the set of "truth" strings S^t, corresponding to I,
2. normalization of device output (N) and representation of S^o and S^t in a form suitable for algorithmic matching and comparison,
3. choice of an algorithm (M) for matching elements from S^o and S^t,
4. choice of an edit model (γ) that reflects the cost of correction, and
5. design of the function (F) for computing the measure of interest from the matching output.

Of these concerns, it is the second that differentiates the benchmarking of *systems* from the activity of benchmarking *algorithms*. When complete DIA systems (such as a page-reading OCR system) are created, they seldom are capable of achieving 98% accuracy using any measure of performance. Improvements in accuracy beyond this are very difficult to achieve.

Most importantly, as asymptotic improvements are gradually made, more and more obtuse issues can easily dominate any single test. A good example in testing OCR systems is the issue of leader dots, commonly used in a Table of Contents. Different approaches to recognizing these dots, each equally acceptable in some context, are frequently used by different OCR devices. Depending on how the ground-truth S^t is prepared for these dots, the existence of a small number of Contents pages in a test set can greatly skew the results of the test. Different approaches to recognizing non-ASCII characters can also dominate the results of a test.

All such issues must be carefully checked to insure that the accuracy results computed provide a correct comparison of the devices tested. The representation of these characters in the ground-truth string and by each test device in their output strings must be examined. The simplest solution to these problems is to provide appropriate mappings to device outputs as part of the "Normalization" step to insure that the results of tests reflect the measurement desired.

In fact, no matter how carefully a test environment is prepared, it is prudent to review test output to insure that some unanticipated new issue has not dominated the test results. An interesting approach to measuring algorithm performance that contrasts with system benchmarking is shown in Haralick [21] and Jaisimha [22].

4.1. Open Issues

The selection of test data is an especially difficult problem. In fact, both the sampling policy for choosing pages and the quantity of pages needed for testing are not well understood issues. For any significant number of pages, the cost of preparing S^t can be very high. Much current research in the field is related to developing satisfactory methods for automatically creating test images and their associated truth data.

Open issues can also be involved in the representation chosen for S^t. Except for the ASCII standard for characters, there are no widely accepted standards. For example, in testing the recognition of mathematical or musical symbols, some non-standard choice would need to be made.

A special set of issues is introduced when benchmarking DIA systems that recognize non-ordered image objects (i.e., when no reading order is involved and S^t is a non-ordered, or partially ordered, string). All of the metrics described above measure accuracy when S^t is an ordered string.

Finally, for DIA systems that recognize structured objects (or even printed text with paragraph and/or character level attributes) tree or graph matching algorithms might need to be employed. However, if accurate position information were produced by the systems being tested (and included in S^o) the matching step could be simplified. A format for representing structured documents that allows for position information as an attribute has been proposed [23]. Although this format has the potential for simplifying benchmarking operations, the cost of preparing truth data S^t (that includes position information) would be significantly higher.

4.2. Summary

In benchmarking any DIA system, choices must be made among alternatives and frequently, no one alternative stands out as the obvious choice. The choice in one category, however, usually affects issues in another category.

Overall, the alternatives selected must:

- support the computation of a metric which in some way provides a measure of the performance characteristic of interest, where
- all steps of the computation can be automated, and which
- makes completely reproducible experiments possible.

It is frequently the case that no single metric provides a complete picture of relative performance. Several different measures are necessary to gain insight into the behavior of DIA systems.

Acknowledgment

The author is indebted to S. V. Rice and Prof. J. Kanai of UNLV/ISRI, and Prof. G. Nagy from RPI for their contributions to the benchmarking methodology described herein. The assistance provided by Andy Bagdanov in preparing this manuscript is gratefully acknowledged.

References

[1] R. C. Camp, "BENCHMARKING: The Search for Industry Best Practices that Lead to Superior Performance," *ASQC Quality Press,* Milwaukee, WI, 1989.

[2] J. Kanai, T. A. Nartker, S. V. Rice, and G. Nagy, "Performance Metrics for Document Understanding Systems", *Proc. Second International Conference on Document Analysis and Recognition,* Tsukuba Science City, Japan, Oct. 1993, 424-427.

[3] G. E. Barr and W. B. Miller, "Simple Models of the Saturated Zone at Yucca Mountain," Sandia National Laboratories Technical Report, SAND87-0112, 1987.

[4] C. K. Chow, "On Optimum Recognition Error and Reject Tradeoff," *IEEE Transactions on Information Theory,* Volume IT-16, No. 1, Jan. 1970, 41-46.

[5] R. A. Wilkinson, et. al., "The First Census Optical Character Recognition Systems Conference," NISTIR 4912, National Institute of Standards and Technology, Gaithersburg, MD, May 1992.

[6] J. Geist, et. al., "The Second Census Optical Character Recognition Systems Conference," NISTIR 5452, National Institute of Standards and Technology, Gaithersburg, MD, May 1994.

[7] S. Chen, S. Subramaniam, R. M. Haralick, and I. T. Phillips, "Performance Evaluation of Two OCR Systems," *Proc. SDAIR'94,* Las Vegas, NV, April 1994, 299-317.

[8] S. V. Rice, J. Kanai, and T. A. Nartker, "A Report on the Accuracy of OCR Devices," ISRI Technical Report TR-92-02, University of Nevada, Las Vegas, March 1992.

[9] S. V. Rice, J. Kanai, and T. A. Nartker, "An Evaluation of OCR Accuracy," ISRI 1993 Annual Research Report, University of Nevada, Las Vegas, April 1993, 9-32.

[10] S. V. Rice, J. Kanai, and T. A. Nartker, "The Third Annual Test of OCR Accuracy," ISRI 1994 Annual Research Report, University of Nevada, Las Vegas, April 1994, 11-38.

[11] S. V. Rice, F. R. Jenkins, and T. A. Nartker, "The Fourth Annual Test of OCR Accuracy," ISRI 1995 Annual Research Report, University of Nevada, Las Vegas, April 1995, 11- 49.

[12] S. V. Rice, "The OCR Experimental Environment, Version 3," ISRI 1993 Annual Research Report, University of Nevada, Las Vegas, April 1993, 83-86.

[13] J. C. Handley and T. B. Hickey, "Merging Optical Character Recognition Outputs for Improved Accuracy," *Proc. RIAO'91 Conference,* Barcelona, Spain, (1991) 160-174.

[14] R. A. Wagner and M. J. Fisher, "The String-to-String Correction Problem," *Journal of the ACM,* 21(1), (1974) 168-173.

[15] E. Ukkonen, "Algorithms for Approximate String Matching," *Information and Control,* 64, (1985) 100-118.

[16] S. V. Rice, "Measuring the Accuracy of Page-Reading Systems," PhD Dissertation, Department of Computer Science, Univ. of Nevada, Las Vegas, Las Vegas, NV, May 1996.

[17] E. J. Dudewicz and S. N. Mishra, Modern Mathematical Statistics, (John Wiley & Sons, 1988) 743-748.

[18] E.W. Myers, An O(ND) Difference Algorithm and its Variations, Algorithmica, 1, (1986) 251-266.

[19] J. Kanai, S. V. Rice, and T. A. Nartker, "Automated Evaluation of OCR Zoning," *IEEE Trans. on Pattern Analysis and Machine Intelligence,* Vol. 17, (Jan. 1995) 86-90.

[20] S. Latifi, "Correcting OCR-Generated Page Images Using Permutations," *Proc. ICEE'93,* Amir Kabir University, Tehran, Iran, May 1993.

[21] R. M. Haralick, "Performance Characterization in Image Analysis: Thinning, a case in point," *Pattern Recognition Letters,* 13, (1992) 5-12.

[22] M.Y. Jaisimha, R.M. Haralick, and D.Dori, "A Methodology for Characterization of the Performance of Thinning Algorithms," *Proc. ICDAR'93,* Tsukuba Science City, Japan, Oct. 1993, 282-286.

[23] RAF Technology, Inc., "DAFS: Document Attribute Format Specification," Redmond, WA (1995).

INDEX

ERRATA

Handbook of Character Recognition and Document Image Analysis, pp. 703–728
Eds. H. Bunke and P. S. P. Wang

CHAPTER 27

AUTOMATIC READING OF BRAILLE DOCUMENTS

APOSTOLOS ANTONACOPOULOS*
Systems Engineering Group, Department of Computation,
University of Manchester Institute of Science and Technology (UMIST),
P.O. Box 88, Manchester, M60 1QD, United Kingdom

Braille is the most widely used system for written communication using tactile means. For visually handicapped people, Braille documents have proved to be an invaluable source of information, essential for education and for leisure. Although more Braille documents are produced today, there are still very few copies of the small number of books, for instance, available. Automation of the Braille reading process will effectively enable the transcription and duplication of existing documents as well as their preservation. By converting a Braille document into electronic form, all the benefits of developments for the electronic form of printed documents can be enjoyed: easier transmission, efficient storage and manipulation etc. The possibility of sighted people also using an automatic Braille reading system to communicate in writing with blind people is a further motivation. In this chapter, the aspects of the automatic reading problem are analysed and solutions to each of them are examined. Firstly, the characteristics and production of Braille and Braille documents are described and the issues involved in the problem are discussed. Secondly, each of the stages towards the automatic reading of Braille documents is analysed and the approaches that have been so far designed to carry it out are critically presented. Finally, the most important factors affecting the success of the whole process are discussed.

Keywords: Braille; Braille documents; Braille document image acquisition; Braille character segmentation; Braille character recognition; Verification of recognition results; Automatic reading system; Computing for visually handicapped people.

1. Introduction

Visually impaired people, just as many other people with disabilities, experience difficulties in integrating in society. One of the main reasons is the inaccessibility of information necessary for education as well as leisurely and cultural interests. Although some information has been available on audio, the advantages of reading and writing as methods for communication have made necessary the practice of encoding information in symbolic form. The best known writing convention is Braille, according to which,

*Present address: Department of Computer Science, University of Liverpool, P.O. Box 147, Liverpool, L69 3BX, United Kingdom, e-mail: aa@csc.liv.ac.uk.

symbols are represented by arrangements of dots recognisable by touch. Since its introduction by Louis Braille in 1829, it has been proven to be an efficient means of written communication for blind people, and hence, has been widely adopted. Braille has been further developed over the years to express different types of information and to render a more efficient way of representing text.

A significant amount of development has taken place for the production of Braille and Braille media. There now exist interfaces for computers, such as a tactile display for Braille representation of the computer screen data and a Braille mouse, which are dynamic. However, at present, Braille documents dominate the scene of written information dissemination in the majority of application areas. Considerable effort has been devoted over the years for the transcription of printed material into Braille. In fairly recent years the use of computers has considerably facilitated this. A document can be scanned and an optical character recognition (OCR) process can be applied to the image. The resulting ASCII character codes can then be translated into Braille representation and either embossing plates can be made for volume production or a hard copy can be obtained by a Braille printer.

Although the production of Braille documents is relatively easy now, there is a missing link in the written communication process: Braille documents cannot be readily encoded into electronic form, e.g. ASCII. The process of identifying and recognising Braille characters is to some extent of a different nature than that of reading printed characters yet there are many interesting parallels that can be drawn.

This chapter explores the problem of automatic reading of Braille documents and the issues related to document analysis and OCR. Firstly, the rationale behind reading Braille documents automatically is examined in Sec. 2. Secondly, the nature of Braille and the characteristics of the reading problem are discussed in Secs. 3 and 4, respectively. Each of the stages involved in the reading process are considered separately in Sec. 5 and existing approaches designed to deal with them are described along with possibilities for further work. Finally, before concluding, the most important factors that affect the success of the reading process are discussed.

2. The Need

As it was pointed out in the previous section, the automatic reading of Braille documents enables the bridging of the written communication gap. This gap is primarily created by the ability to convert printed material (readable by sighted people) to Braille and the difficulty, on the other hand, in converting Braille to a representation that can be read by sighted people. Nitta *et al.* [1] point out the need for a Braille reading system to assist blind people in learning Braille as the motivation for their research. Benjelloun *et al.* [2] also refer to the fact that such a system will enable sighted people to read Braille documents, thus enhancing the communication and collaboration possibilities with blind people. For example, blind people in a distance-learning programme can submit their work to their tutors, who will probably be sighted.

Blind people may even use a Braille reading system as an interface for the use of fax machines. The Braille document could be scanned at one end and at the other end the encoded document could either be translated to ASCII or printed with a Braille printer. A similar use can address another significant problem faced by blind people and libraries: copying of Braille documents. Through the years only a small proportion of printed books have been available in Braille. The fact that there are not many copies of each available title compounds the problem. Since the original Braille document most probably has not originated in electronic form or this encoding may not be available, a system that reads Braille documents and produces the encoding for further printing is a necessity. Without such a system the only option is manual transcription which is both costly and time consuming.

Furthermore, by converting them to their electronic representation, Braille documents can benefit from the progress made in the (electronic) handling of printed documents. Braille documents occupy considerably greater volume than printed ones (40 to 60 times according to Otake *et al.* [3]) for recording the same information. This is due to the thickness of the material (page), the height of the protruding dots and the low information recording density resulting from the relatively large spatial extent of Braille characters. Therefore, by converting Braille documents into electronic form, substantial benefits in storage and transmission can be gained. Apart from enabling wider dissemination of information, the usual advantages of indexing and manipulation, associated with document management systems, can also be enjoyed.

Finally, there is a significant need to preserve old Braille documents many of which are rare and in a delicate condition. Such documents may be converted to electronic form and reproduced accordingly.

It should be noted at this point that an automatic Braille reading system must be practical and cost-effective in order to be adopted by the people who need it. Ease of use is as important as efficiency in this case. As considerable developments have taken place in the fields of pattern recognition and document image analysis, it is natural to take advantage of them and attempt to employ them in the solution of the reading problem. Although an attempt was made to use a mechanical system [4], the automatic reading by optical means has been more favourable with the rest of the research community.

3. Braille

Since the seventeenth century there have been attempts by Francesco Lana-Tarzi to devise a tactile writing system. His ideas were further pursued by Charles Barbier and were implemented by Valentin Haüy in the Paris Institute for the Blind. Although that tactile writing system was not efficient due to the large number of sound combinations used, the general idea proved successful. One of the blind students, Louis Braille adapted and simplified the system of tactile dots and set the foundations of the Braille writing of today. The system consisted of tactile representations for each letter of the Latin alphabet along with numbers and punctuation marks.

During the years that passed since then, the system has been further developed to include more representations of printed symbols as well as rules to increase the efficiency of reading and writing. Different languages are supported as well as a variety of other types of information such as mathematical expressions, computer programs and music. In the following subsections, the nature of Braille characters and documents is described in order to develop an understanding of the problem of automatic reading of Braille.

3.1. *Braille Characters*

Printed text is represented by Braille characters. Each character is constructed as a set of six points arranged in two columns of three, as it can be seen in Fig. 1. This arrangement is called a Braille *cell*. Each point position is identified by a number and can be either raised (a protrusion) or flat. A raised point is also called a Braille *dot*.

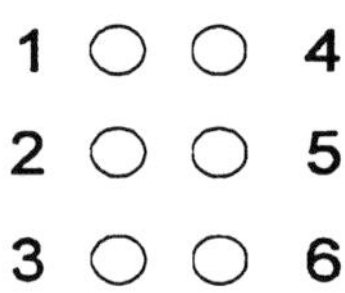

Fig. 1. The Braille cell.

The dimensions of a Braille dot have been set according to the tactile resolution of the fingertips of person. In theory, the dimensions of Braille dots must fall within certain bounds as indicated in Fig. 2. The horizontal and vertical distance between dots in a character, the distance between cells representing a word and the inter-line distance are also specified by the Library of Congress. In practice, although most instances of Braille are close to these standards, there are slight variations between Braille produced by different devices.

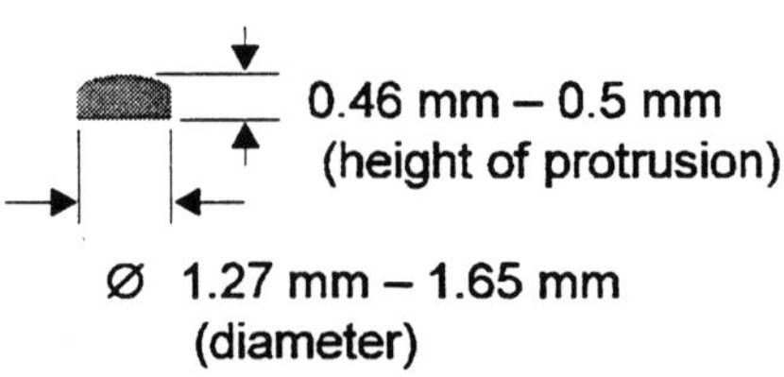

Fig. 2. Braille dot (profile) and its specified dimensions.

Each Braille character is a unique combination of dots within the cell. So far, only Braille cells with six points have been mentioned. There is, however, an eight-point version of Braille cells whose use is not very widespread compared to that of the six-point ones. The two extra points allow for the representation of a greater variety of printed symbols. It has mainly been used in computer programming where a variety of symbols is used and each must be represented unambiguously by a single Braille character (see Sec. 3.2.3). The larger number of possible dot combinations has also been used for the representation of Japanese characters [5]. Since from the recognition point of view the principle is the same, in the rest of this chapter Braille characters will be assumed to be composed of six points, unless otherwise explicitly stated.

3.2. *Interpretation of Braille Characters*

A printed word is represented by a set of Braille characters forming a Braille word. The reason for distinguishing between the two kinds of word is that they often differ in the number of characters as it will be explained below. The characters inside a Braille word are separated by a distance greater than the horizontal one between points and less than that between words. The distance between words in a line of Braille characters consists of one or more space characters. The space character is a Braille cell with no dots (all points are flat).

Apart from the space character there are 63 other combinations of dots which can be used to represent printed symbols. According to the context and to the set of representation rules followed in a given Braille document, a particular combination of dots may correspond to more than one different printed characters (at different occasions). On the other hand, a printed symbol may require more than one Braille character for its representation. Furthermore, a Braille character may in some circumstances correspond to a concatenation of printed characters (see below). Different sets of rules exist for the production of Braille documents containing textual information. Further still, there are different representation rules for other types of documents, like those containing music or mathematical expressions. Some of the main sets of rules that are widely used are briefly described in the following.

3.2.1. Grade 1 Braille

In *grade 1* Braille, each printed letter corresponds to one Braille character as can be seen in Fig. 3. Punctuation marks are also represented by one Braille character with the exception of the opening quotes and the exclamation mark which are represented by the same Braille character (see Fig. 3). In this case, while reading, the interpretation is made according to the context. The combination of dots in each Braille character is designed in such a way that symbols are distinguished from certain others by a translation of the same dot arrangement (e.g. the letter "d" and the full stop "."). Furthermore, the symbols that are represented by only one dot are likely to be found next to others with dots in other positions which will help define the position for the reader.

It is not possible to represent the large number of existing symbols by a combination of dots in a single Braille character. Therefore, some symbols (e.g. the asterisk) are represented by two Braille characters. Furthermore, the digits "1", "2", ..., "9", "0" are represented by the Braille characters corresponding to the range of letters "a",...,"j"

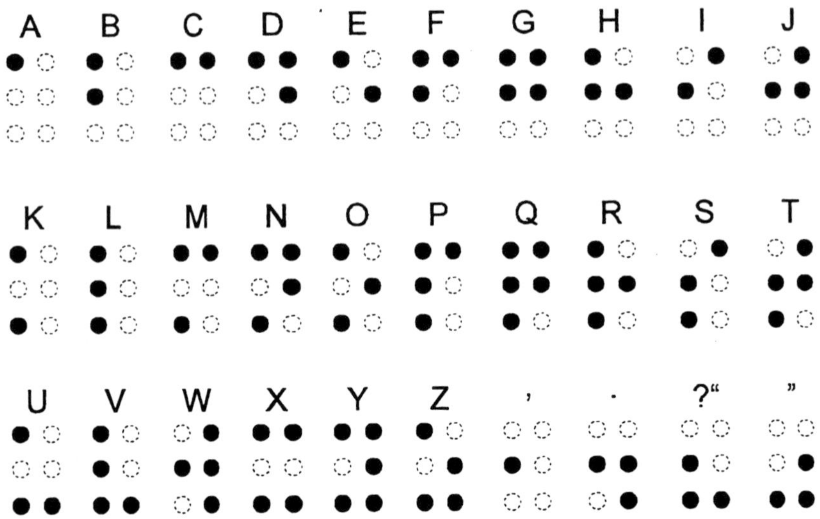

Fig. 3. Braille characters corresponding to letters of the Latin alphabet and some punctuation marks (black areas denote dots).

preceded by the *numeral* symbol as can be seen in Fig. 4. Still, grade 1 Braille is quite straightforward to interpret. The numeral symbol is an example of a *mode* symbol. Another mode symbol is that used to indicate the capitalisation of a letter (Braille character composed of dot six only). It is mainly used in North American Braille and precedes the symbol of the character to be capitalised. Similarly, mode symbols are used in many languages to denote that a particular accent should be placed on the character that follows.

3.2.2. Grade 2 Braille

The letter-by-letter representation of a word as in grade 1 Braille leads to the production of quite voluminous braille documents (a dictionary can occupy a bookcase). Furthermore, human readers have to spell each one of the characters of a word before they can recognise the meaning. This fact makes the reading process slower and obstructs fluent reading. Therefore, attempts have been made to represent frequently occurring letter strings by one or two Braille characters. These strings are usually words, word stems, prefixes, suffixes, syllables or just concatenations of letters such as “dd”. The current standard which defines the representations of these contractions in Braille and the rules for using them is referred to as *grade 2* Braille (also called *contracted* or *literary* Braille).

The strings of characters that can be contracted have been selected according to the frequency that they occur and therefore, differ between languages. In English grade 2 Braille, for instance, there are over 200 contractions and short form words. The rules for replacing character strings by a Braille short form take into account the phonetic use of

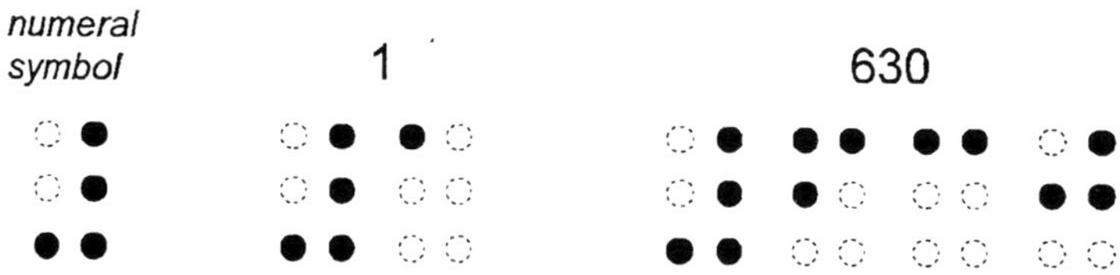

Fig. 4. Composition of numbers in Braille.

the language and are based on logical as well as linguistic and semantic rules. This fact makes the reading process more complicated and very context sensitive since the same Braille character may represent a particular contraction in some circumstances and a different one in another [6,7]. For example the Braille character representing the exclamation mark (which is the dot pattern of the letter "f" shifted downwards) takes the following meanings: (i) "!" at the end of a word, (ii) "ff" in the middle of a word and (iii) the preposition "to" if it is on its own before a word.

3.2.3. Braille for specialised applications

There is a number of other Braille writing conventions for the representation of symbols in more specialised applications. Computer Braille, for instance, is a one-to-one correspondence of Braille characters to ASCII ones [8] thus removing any ambiguity in the representation of computer programs. For example the digits "0",...,"9" are represented by single Braille characters in contrast with grade 1 or 2 Braille. However there are differences between the North American (six-point) and the British Braille computer codes. As it was previously mentioned, Braille cells with eight points are also used sometimes for computer access purposes. But, although this enables the representation of more symbols, there is no generally agreed standard.

Mathematical expressions are represented in Braille according to special rules which specify the encoding of the necessary symbols. Another type of information that can be expressed in Braille is music. In Braille music scores the notes are represented by different letters. The length of notes is determined by a preceding Braille character (composed of either dot 3 or dot 6). Other music symbols are also represented. Finally, a person who writes Braille (for personal use) may sometimes do so using his/her own contractions and rules.

3.3. *Braille Documents*

Most Braille documents are formed by embossing the dots on the back side of the medium sheet so that they can be read from the facing side. Most often the medium is thick paper in order to withstand the embossing. Another, less common, medium is a transparent plastic sheet (similar to an overhead projector transparency). In addition to the embossing production technique, a process of forming dots on paper from drops of hot-melt material has also been reported [3]. This chapter, however, will concentrate on the embossed-paper Braille documents because they essentially constitute the body of the information available in Braille form.

One way of embossing the sheet of paper is by hand using a Brailler, which resembles a typewriter with a key corresponding to each of the six points (plus a space key). To produce a dot, the user applies force to the appropriate key which embosses the paper. For mass production, plates are formed which are then pressed against the paper sheets to form pages of Braille. Furthermore, as it has already been mentioned, there are Braille printers, connected as peripherals to computers, which emboss the dots on paper according to the information sent in electronic form.

The size of the page varies between documents produced by different means. In addition, the colour of the paper varies, as it does not play any role in conveying the written information. The majority of Braille documents, however, are either buff-coloured or white. In addition, their surface usually has a low gloss finish.

The Braille characters are embossed in lines from the top of the document to the bottom much like printed documents. Braille documents can have the dots embossed on one side or on both. As the volume of Braille documents is quite large, it is advantageous to use both sides of the sheet. This is achieved by carefully embossing the dots of one side between the dots embossed on the other, as can be seen in Fig. 5. Because the points of Braille cells in one side are interleaved with the points of the cells of the other, these double-sided documents are also referred to as *inter-point*.

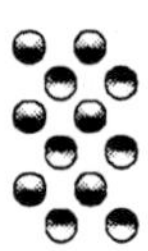

Fig. 5. Two cells on different sides of an inter-point Braille document.

4. Characteristics of the Automatic Reading Problem

In ordinary documents, the information is printed in a contrasting colour so that it is visually distinguishable from the background. As Braille documents are not designed to convey any visual information, their colour is uniform. Since the information is only intended to be read by tactile means, the first problem in automatic reading is to identify the tactile information visually. The issue in this case is the separation of the protrusions on the paper sheet from the flat background. The solution to this problem comes from the natural visual perception of three-dimensional information from a two-dimensional view. If the document is illuminated at an angle close to the horizontal then the anomalies on the surface of the paper are lit differently than the flat areas. As it will be described later, the various approaches to Braille recognition have exploited this fact in different ways.

However, the identification of Braille dots from the differences in the intensity of reflected light is not straightforward. As the paper is only used for tactile purposes, the quality of its appearance is not important. Hence, there are sometimes visual imperfections in the form of small dark fibres usually associated with cardboard. These blemishes, together with dirty marks that may also be present due to repeated use, interfere with the light intensity patterns created by the protrusions and make the identification task more difficult.

Moreover, damaged documents with creases or punched holes compound the problem. In addition, in some old documents, a few of the dots may be less prominent due to heavy use. In this case, the effect produced by the lighting is reduced.

As many of the Braille documents produced are inter-point, another problem becomes evident. Inter-point documents have both protrusions and depressions (the dots of the other side) on their surface. This complicates the dot extraction stage because shadows and bright areas are produced from both. It should be noted that the majority of existing automatic Braille reading systems are only capable of reading single-sided documents.

Once the Braille dots have been located in the image, the next task is to identify the characters. The problem is to identify which dots belong to which Braille cell. In theory, Braille cells are aligned in the horizontal and vertical directions in the form of a grid. In practice there are noticeable deviations from this due to the manufacturing process or due to distortions introduced in the image of the document.

In general, Braille characters have a more regular structure than printed ones. Their size and style is fixed throughout a document, although there are certain variations between different documents. The spacing between Braille dots is quite regular and can be used to segment characters. Nevertheless, Braille characters are not single connected entities and do not have distinguishing structural differences between them (they are all combinations of dots). These facts do not give enough information for the recognition and verification of Braille characters in the sense of providing degrees of confidence and alternatives.

Further details on the above as well as additional issues will be raised in each of the relevant subsections describing a particular stage of the reading process.

5. Stages

For a better analysis of the different aspects of the automatic Braille reading problem and for ease of comparison between current solutions, the stages of the automatic Braille reading process are examined individually. The first stage to be considered is the acquisition of the image of the tactile surface of the document. This is followed by various enhancements performed on the image prior to the extraction of any information. In the third stage, the Braille dots are identified in the image. The two stages that follow are concerned with the segmentation and recognition of Braille characters. The possibilities for verification of the recognition results are examined next and, finally, the main uses of the encoded Braille characters are described.

Each of the different stages is the subject of one of the following subsections. The characteristics of the relevant issues are described and the various approaches devised so far to deal with them are outlined.

5.1. *Image Acquisition*

The first step towards obtaining an electronic representation of a Braille document is to capture its image. As it was mentioned above, the nature of this kind of document is quite different from that of a printed one. The relief information is present in the image in the form of the bright and the shadow areas created by the differences in the intensity of the light reflected from the paper surface to the camera. The protrusions create a light area next to a dark one (shadow) while the grey-level value of the background is somewhere in-between. In inter-point Braille documents, the depressions also give rise to a dark and a light area, but in reverse order. To make sure that all the relief information is present in the image, the acquisition must be done under controlled conditions.

First, the illumination angle must be oblique. To acquire the image, most of the existing approaches have used a camera placed above the centre of the document. For example, François and Calders [9] have used a vidicon camera to capture the image of a Braille document illuminated under a very oblique angle. Their goal is to identify the bright spots corresponding to the top of Braille dots. However, the Braille documents used are assumed to have no visual defects.

Visual defects can interfere with the representation of relief information in two ways. Firstly, if there are small areas of higher reflectivity due to the materials of the paper, unwanted bright spots will appear in the image. If the reflectivity of the paper varies along wider areas then the contrast between the grey levels of the dots and those of the background will be lower in some regions. Secondly, dirty marks or other dark particles in the paper give rise to small dark regions in the image. These could be mistaken for shadows produced by dots. This is especially the case when an approach attempts to identify dots based on the assumption that each one of them creates a shadow (e.g. Dubus *et al.* [10]). To circumvent this problem, Hentzschel [11] adopts an interesting approach: two images of the same Braille document are captured, each with the document being lit at the same angle but from the opposite side of the camera. In this way, each dot produces a different shadow area in each image whereas dark areas on the paper appear at the same place in both images. By subtracting one image from the other the unwanted dark regions can be eliminated.

In the rarer case of transparent single-sided Braille documents, Nitta *et al.* [1] place each document on a sheet of black paper and illuminate it from a relatively higher point (30°–40° from the vertical) than the other approaches. The camera is placed directly above the document in order to capture the bright circular areas resulting from the bases of the transparent dots. In this case there are no shadows cast by the protrusions.

However, cameras introduce aberrations that must be corrected before any useful processing takes place. François and Calders [9] calibrate their camera set-up by obtaining an image of a checkerboard pattern and correcting it to achieve the original pattern. The sequence of transformations required to correct the test image is applied to every subsequent Braille document image captured for reading.

Apart from these problems associated with cameras, the arrangement of the lighting and the camera is quite complicated for use by a person with no technical knowledge. It also occupies a fair amount of space. Moreover, the cost of the equipment is prohibitive for many users.

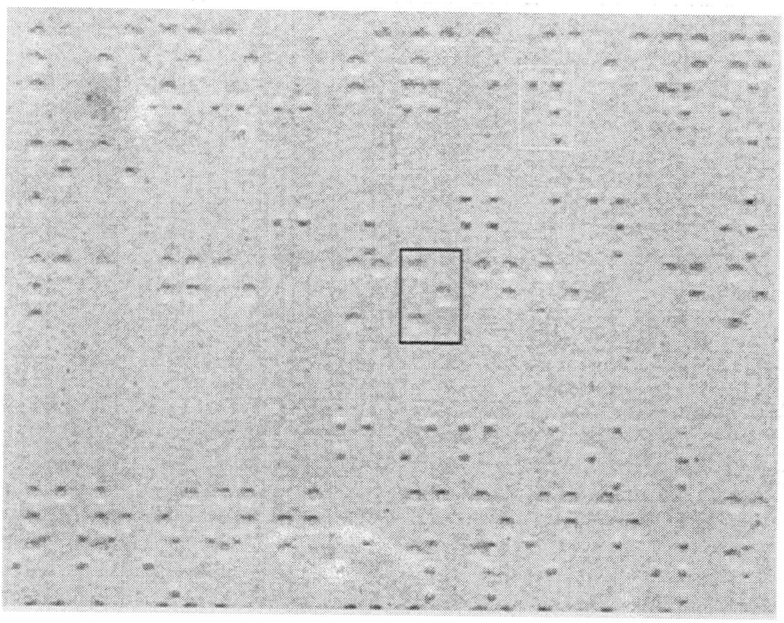

Fig. 6. Image of a part of a Braille document (cells on different sides of the document are delineated with rectangles of different colours).

The increasing popularity of flat-bed document scanners has provided a cost-effective and practical alternative. In contrast to purpose-built dedicated hardware associated with camera arrangements, a scanner is affordable and can also be used for other purposes, such as scanning of printed documents for conversion to Braille. Its compact design is also easy to use by any person. The only requirement is that the illumination of the surface of the paper is asymmetrical (i.e. the document is lit at an oblique angle). As many scanners attempt to illuminate the documents in a very symmetrical and uniform way, care must be taken to choose a scanner that produces the desired results [12,13]. An image of part of a double-sided Braille document captured with a scanner can be seen in Fig. 6. The protrusions are represented by a dark region above a light one and the depressions by a sequence of regions in the opposite order. Note the dirt mark at the top-left part of the image.

To distinguish the dots adequately, the spatial resolution setting should be between 80 and 200 dots per inch (dpi). Below 80 dpi there is an insufficient amount of information and above 200 dpi the extra amount of information is superfluous [12]. The number of grey levels should also be sufficient so that the regions corresponding to dots are accurately distinguished. Although 16 grey levels (4 bits) yield acceptable results and minimise the memory space required, the use of 256 (8 bits) is recommended for better results [13].

5.2. *Pre-Processing*

The grey-level image of a Braille document can be enhanced in order to correct problems such as skew and noise, and improve the appearance of the areas of interest for the benefit of the recognition process.

Skew may be introduced during image acquisition by not carefully aligning the document or it may arise due to a production fault. If a large skew angle is present, recognition of Braille characters may be impossible because the dots will not be aligned in rows as expected. In this case, it is preferable to recapture the image, after aligning the document, rather than attempting to correct the skew by rotating the image. Skew correction is quite time consuming and should best be avoided if possible. Although small skew angles may be tolerated in some approaches [13], in others skew must be corrected before the recognition process. As it will become evident in later sections, this requirement is due to stricter assumptions about the position of Braille characters on the document (e.g. [12]).

One method to detect skew is by assessing the sharpness of the peaks of the grey-level histogram of the rows of an image region [12]. The image is first divided into rectangular subimages. The standard deviation of grey-level values inside each subimage is calculated and used as an indicator for the presence of Braille dots inside the subimage. A subimage likely to contain Braille dots is selected and a histogram of the grey-level values across its rows is computed. The standard deviation over the row sums is calculated and the image is rotated by a given angle α. The histogram is calculated again and its standard deviation is compared with the value of the previous one. The same step is repeated again until the standard deviation is maximised, in which case the Braille dots in the image should be aligned horizontally. However, the rotation of the image at every step is very time consuming. An alternative could be the calculation of the histogram in different orientations while the image is only rotated when the skew angle is established. It should be noted that this method is similar to the analysis of projection profiles in binary images of printed documents.

The standard deviation of grey-level values in each of the subimages is further used to calculate a threshold corresponding to the grey-level value of the background. According to this threshold, pixels corresponding to light areas are assigned the value *1* while darker pixels are set to *-1*. All other pixels are set to zero.

A different technique for detecting the skew angle is that used by Hentzschel [11]. After filtering and thresholding (see below), three of the corners of the Braille document are searched for and identified in the image. It should be noted that it is assumed that the image area is larger than the area covered by the Braille document. The skew angle is calculated from the co-ordinates of the identified corners and the image is rotated accordingly.

Another form of pre-processing is the enhancement of the image in order to improve the separability between the regions corresponding to dots and the background. Light shadows covering broad regions in the image may be introduced if the flat areas of the paper are not perfectly level. Mennens *et al.* [12] subtract a locally averaged image from the original in order to eliminate this problem. The approach of Dubus *et al.* [10] and Benjelloun *et al.* [2] is based on the assumption that the Braille dots give rise to bright areas in the image. According to this, the region of a Braille dot in the image will have the form of a second degree convex two-dimensional function. The image is then processed with a median filter based on a second-degree polynomial function in a 1×5 pixel neighbourhood in the horizontal direction. The regions of higher intensity corresponding to protrusions are considerably enhanced in contrast to the background. A second pass with a filter of the same structure but with different parameters is performed and the image regions whose grey-level value is constant throughout are set to zero. In the following, the image is further processed with a vertical median filter consisting of a 3×1 window. Finally, the image is binarised and a morphological closing operation is performed to connect fragmented regions corresponding to the same dot. It should be noted, however, that this approach is only applicable to images of single-sided Braille documents. Median filtering has also been employed by Hentzschel [11] for noise removal. Binarisation is achieved by an adaptive thresholding process. An optional dilation operation is used to enlarge regions that are very small.

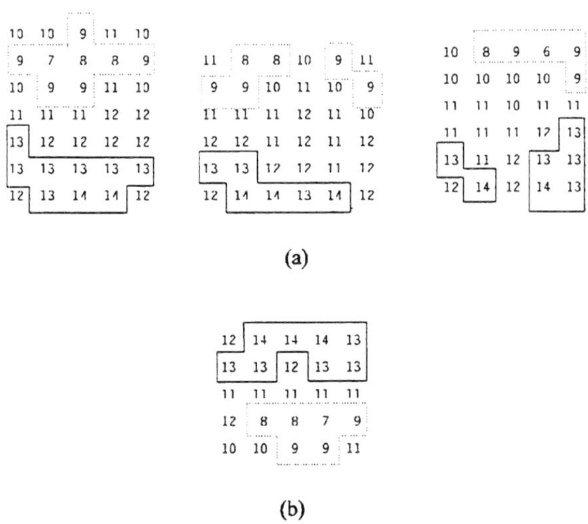

Fig. 7. Regions corresponding to (a) protrusions and (b) a depression.

5.3. *Identification of Dots*

At this stage, the regions in the image corresponding to Braille dots are identified. In a binary image, dots can be identified as small connected components. In the case of the twin shadow approach of Hentzschel [11], a Braille dot corresponds to two such components. It should be noted that this situation is valid only when single-sided Braille documents are encountered. In this case, the above approaches are able to identify dots from either the dark (shadow) or the light areas in the image which, after thresholding, are represented as connected components. In the special case of transparent Braille documents, Nitta *et al.* [1] identify connected components having the form of circular arcs. These are then processed so that complete circles are obtained. The centre of each of these circles is identified as the location of the Braille dot.

In a grey-level image, the identification of dot regions needs more careful consideration. It is necessary to have a specification of the image area corresponding to a dot. A dot is represented by a combination of a dark and a light region. The intermediate grey-level pixels of the background surround and, possibly, also separate the pair. Naturally, the orientation and order of the light-dark pair of regions depends on the direction of the illumination. In images of inter-point Braille documents, however, the protrusions give rise to region pairs in which the order is opposite to that of the region pairs produced by depressions. For example, in Fig. 7 some region pairs corresponding to protrusions and depressions in the document of Fig. 6 are shown. The numbers in the figure correspond to grey levels in the range 0 (darkest) to 15 (lightest).

The consistency of the appearance of grey-levels at the positions of depressions and protrusions renders the use of template matching a possible option. This is the technique adopted in the approach of Mennens *et al.* [12]. In the images captured by their system, protrusions are represented in the image by a dark area followed by a light one and depressions are represented in the opposite order, similarly to the example of Fig. 7. In the pre-processed images (see previous section), pixels of light regions have the value *1*, background pixels have value *0* and pixels of dark regions have been set to *-1*. The mask

$$\begin{matrix} -1 \\ 0 \\ 0 \\ 0 \\ 1 \end{matrix}$$

is then correlated with that image. The number of 0's in the middle depends on the distance between the centres of the light and dark region pairs in the document image. For instance, the above mask is constructed on the assumption that the distance is five pixels. The result is an image where groups of pixels with value *2* correspond to regions in the centre of protrusions and regions with pixels having value *-2* correspond to centres of depressions. These two kinds of regions are termed *core* regions and are used to denote the protrusions and depressions of the Braille document.

Different masks have also been tried [14] but, in images of inter-point documents, it is often the case that the template matching technique leads to the detection of non-existent dots. For example, there are situations where a dark area belonging to a protrusion and a light area belonging to a depression may lie in such an arrangement that gives rise to the detection of a dot which is not actually present on the document. The application of the technique of Mennens *et al.* [12] results in the detection of a number of such non-existent dots. To cope with this problem, they adopt a strict model of the structure of a Braille document (see the next section) which dictates the valid positions for dots of both sides. The identified dots which do not lie in valid positions are disregarded in subsequent stages.

A more flexible approach [13] first identifies the dark and light regions in the grey-level image and describes them with their minimum enclosing rectangles. The relationships between light and dark components (describing rectangles) are then examined in order to identify the depressions and the protrusions. There is another problem, however. Light regions of protrusions may be merged in the image with light regions of depressions and identified as one large component instead of two distinct ones. The same situation may occur for dark regions. An example of dots belonging to different sides of the document having merged regions can be seen in Fig. 8. On the other hand, as shown in the example of Fig. 7, a dark or light region may be fragmented. In this case, only one dot should be identified.

The algorithm proceeds as follows:

> **for** each dark region
>   **if** a light region exists *above* within limits of the height of a dot **then**
>   a *depression* is found
>   check regions and mark the appropriate ones as *"used"*
>   **else**
>   **if** a light region exists *below* within limits of the height of a dot **then**
>   a *protrusion* is found
>   check regions and mark the appropriate ones as *"used"*

When a protrusion or a depression has been identified, then the dark and light regions that comprise it are marked as "used" so that they will not be considered as parts of another dot. However, only the correct part of a region or regions must be marked. Hence, there are rules that guide the algorithm to correctly identify which regions, or which part of a merged region, should be considered as the dark or the light region of a dot. For example, in Fig. 9, the dark region is under consideration and the light region *A* above it fulfils the criteria for the existence of a depression. Region *A* is marked as "used" but the dark region is wider than the maximum width of a Braille dot and so it is divided into two parts and only *B1* is marked as "used". The dark region *B2* is considered next and the two light regions *C* and *D* are found below it. Both these regions satisfy the requirements for the identification of protrusions but the distance between *C* and *D* is smaller than the minimum distance between Braille dots. Hence, it is concluded that both regions constitute the light region of one protrusion. Now, the regions *B2*, *C* and *D* are all marked as "used". The protrusions identified are written as black dots of nominal size in a new blank image. The depressions are similarly written in another blank image. Upon completion of the algorithm, the dots of each side of the Braille document have been separated from those of the other. Each side will be considered separately in the stages that follow.

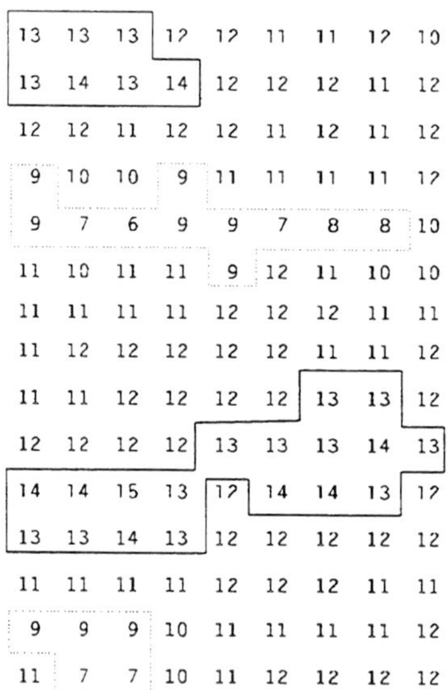

Fig. 8. Two depressions (left) and a protrusion (right) with merged regions.

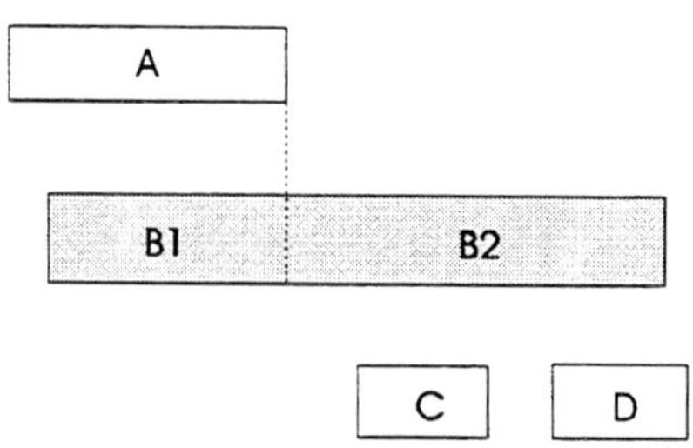

Fig. 9. Identification of Braille dots when regions are merged or fragmented.

5.4. *Braille Character Segmentation*

By this stage, the Braille dots have been detected in the image of the document. Now, groups of these dots must be identified as Braille characters. The aim of this stage is to examine the locations of the dots and deduce the positions of the characters.

One way to identify dots belonging to the same character is by examining the distance between consecutive dots in the horizontal and vertical directions. In contrast with proportionally spaced printed characters, Braille characters are printed in regular intervals (or multiples thereof) and the points in each cell are in regular positions. Therefore, it is quite straightforward to distinguish between different lengths of space between dots. It should be noted, however, that in practice there are deviations from this regularity arising from the Braille production process and the various effects of the imaging process. Nevertheless, for the purpose of spacing analysis there is no significant requirement for very high accuracy in adhering to the standard distances.

The spacing in the vertical direction falls into two categories: spacing between dots belonging to the same character and spacing between adjacent lines of Braille characters. One technique [13] is to examine the vertical projection-profile of the binary image containing the dots. From this, the two spacing values can be obtained. According to the vertical distance between dots, lines of Braille characters can be identified as consecutive rows of dots. Across a line of Braille characters, dots are usually found in all three rows. However, there may be lines with dots in only one or two rows. For instance, there may be a line of hyphens separating text areas. A hyphen is represented by a Braille character having dots in the bottom row only (dots 3 and 6). In the image, such a character line is ambiguous with regard to the identity of the characters it contains. If the dots are in the top row then the meanings of the characters is different than what will be if the dots are in the second or in the third row. The relative position of a row of dots inside a line of Braille characters should therefore be established. For this reason, the lines having dots in all three rows are detected first. Having determined the average line height and the average inter-line spacing, the vertical positions of all the lines of Braille characters and of the blank lines are determined. From these positions it is straightforward to identify the relative positions of rows of dots inside the lines.

Having located the lines of Braille characters, the next step is to determine the positions of the characters themselves inside each line. In order to do so, the positions of character columns having at least one dot are first identified from the horizontal projection-profile. The space between columns is then examined to determine whether a column is the left or the right one of a Braille character. The spacing between adjacent columns can vary beyond the obvious intra-cell and inter-cell values. This is due to the fact that a Braille character may have dots in only one column. Hence the distances between columns may correspond to one of the following:

(i) space between the two columns of the same character (*a* in Fig. 10),

(ii) space between a *right* column and the *left* one of the next character in the word, i.e. the space between adjacent Braille cells (*b* in Fig. 10),

(iii) space between a *right* column and a *right* columned character (*c* in Fig. 10),

(iv) space between a *left* columned character and the *left* column of the next character (*d* in Fig. 10),

(v) space between a *left* columned character and a *right* columned one (*e* in Fig. 10), and

(vi) space between words (e.g. *f* in Fig. 10), which varies according to the number of space characters present.

To segment the characters in a line, the distance between consecutive columns is examined. Accordingly, each column is characterised as a left or a right one. If the distance between two columns is greater than the sum of the inter-character distance and the width of a Braille character (space character in this particular case), it is determined as inter-word distance and the number of Braille space characters is calculated. Special consideration is given to the situation where two single-columned characters are encountered. Ambiguity can arise if they are both left-columned or both right-columned (distance *c* or *d* in Fig. 10).

A similar method has been adopted by Hentzschel [11] which, because of the twin shadow approach (see previous section), identifies two peaks for each column in the horizontal projection-profile of the line of Braille characters. The use of the height of peaks in this profile is also suggested as it can provide information on the number of dots in the column.

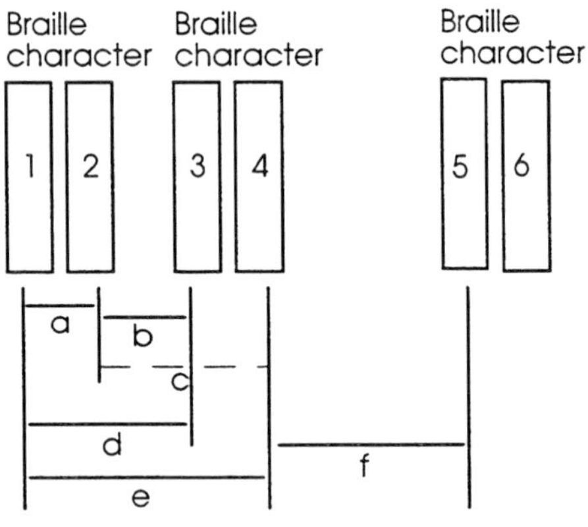

Fig. 10. Different spacing possibilities between two Braille characters.

Benjelloun *et al.* [2] start with the horizontal projection-profile. They first approximate the peak regions in the profile with Gaussians and identify their centres. The mean intra-character and inter-character distances are also calculated. With the aid of a knowledge base describing expected positions of left and right columns of dots in the image, they attempt to characterise the column positions, identified in the profile, as left or right. Identified positions that are very close to the pre-specified ones are corrected. In the next step, column positions which have not yet been characterised as left or right are considered. The distance between two of them or between one of them and an already characterised column are examined. If, for example, the distance between two consecutive column positions is equal to the inter-character one, the first is characterised as right and the second as left. Most of the column positions in the profile will have been characterised upon completion of this step. However, column positions whose distance from the closest neighbouring column positions is different from the intra- or inter-character ones will still remain uncharacterised.

From the known positions of left and right columns, an attempt is made to identify the remaining positions that columns of dots could exist. Based on the two mean distances previously computed, the positions of left and right columns in Braille cells are identified. If some of these columns actually contain dots they are characterised accordingly. At this point all positions in the image at which a column of dots may exist have been characterised as left or right.

In order to segment the Braille characters which have now been identified as columns of dots in the vertical direction, the lines of the Braille characters must be detected in the image. In doing so, the vertical projection-profile is obtained for each column axis identified in the previous step. The mean height of a Braille dot is calculated from its projection and the rows of Braille dots and lines of characters are identified accordingly.

Another approach uses a model of Braille documents [12]. The assumption is that Braille cells are positioned in fixed rows and columns on the page. Consequently, dots of Braille characters will only be present at intersections of horizontal and vertical grid lines passing through the positions of rows and columns of Braille cells. Hence, the positions of Braille characters and their dots can be identified from the model. The most important step of this technique comprises the identification of the location of the horizontal and vertical lines of the model grid. In other words, the grid must be positioned on the current input image. The initial examination takes place in each of the subimages (see Sec. 5.2). The horizontal grid lines are tentatively identified from the vertical projection-profile of core regions (see previous section). The vertical grid lines are identified from the horizontal projection profile of the core regions lying on horizontal grid lines only. The positions of the horizontal grid lines are next updated and finalised by examining a new vertical projection-profile of core regions situated along the identified vertical grid lines only. The global grid for the whole image is then assembled from the pieces identified in the subimages. The dots that lie on the intersections of appropriate grid lines form the Braille characters.

Finally, in the approach of Nitta *et al.* [1] a local model is used to identify dots belonging to a Braille character. This model is actually a specification of the relative positions of the points in a Braille cell with reference to a single given point. For example, if the actual position of a dot in the image is known, its relative position inside the Braille cell could be found by searching for other dots in certain locations around it. It is these locations which are pre-determined by the model. For each identified dot in the image the specified locations in its neighbourhood are examined and, consequently, the Braille character is identified.

5.5. *Recognition*

At this stage the identity of each of the Braille characters is recognised. It should be noted that, at this point, no attempt is made to interpret the character. Since each character is distinguished from the rest by its unique combination of dots, it will suffice to identify the (relative) positions of its dots. Then, the character can be described in terms of the numbers of dot positions (see Sec. 3.1).

Since the positions of Braille characters in the image have already been identified, it is straightforward to check for the existence of a dot at each of the six positions of points in a Braille cell. If the Braille character is described as a region in the image, this region is divided into six equal compartments (two across, three down) in which the search for dots is performed. If the exact positions of dots in the image are known [13], it is easy to check if one of them is included in a compartment. Alternatively [11], the region in the image corresponding to a compartment is searched and the number of pixels corresponding to dot components is counted. If the number exceeds a pre-specified threshold then the compartment is considered to contain a dot. In any case, the recognised character can now be represented as a binary number with six bits. Each bit corresponds to a compartment and it is set only if there is a dot in that compartment. The principle of dividing the character region into compartments and checking for the presence of dots is illustrated in Fig. 11. Also shown is an example of encoding of a Braille character.

In other approaches the Braille characters identified in the image are described explicitly [2] or implicitly [12] by the positions of the rows and the columns of a cell. In these cases it is necessary to check along specified directions for the existence of dots. In the approach of Mennens *et al.* [12], where the global grid has been constructed (see previous section), the possible positions of Braille dots are determined by the intersections of horizontal and vertical grid lines. The six possible positions of the dots are known for each individual Braille character position in the image. Hence, all the possible character positions are examined. Within each one of them, the intersections of grid lines are considered. A local correlation test is performed to determine whether a dot exists or not and the recognised Braille character is encoded as a binary number, similarly to the method previously described.

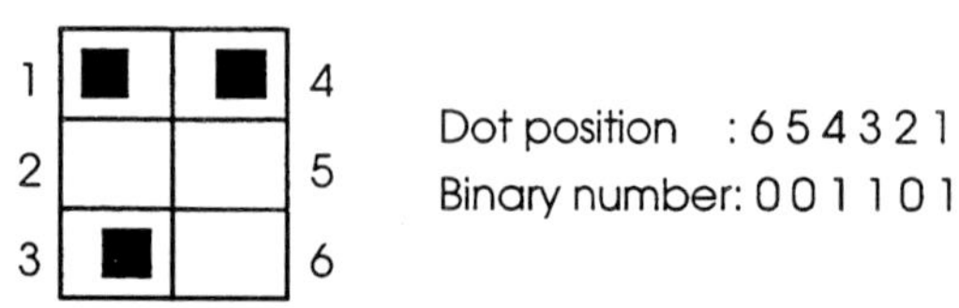

Fig. 11. Example of recognition and encoding of a Braille character.

5.6. *Verification*

Due to a number of factors, each pertinent to one of the previous stages of the reading process, some of the characters may not be correctly recognised. Due to the nature of the Braille characters as combinations of dots, the errors are mostly substitutions. The presence of an extra dot at a valid position within a character or the absence of a dot (see Sec. 6) both lead to the recognition of a different character. Furthermore, for the same reason, verification of the recognition results is quite difficult on a character-by-character basis. Apart from the presence or absence of dots at specified positions, there are no shape or structural properties that can uniquely distinguish between Braille characters. This is in contrast with printed character recognition where structural and topological properties of characters can be used to eliminate possibilities and at worst select a ranked set of alternatives. With Braille characters, the presence or absence of dots gives rise to many possibilities, each with equal probability of being valid.

The only other option is contextual analysis. Each Braille character could be assessed and be assigned a degree of confidence according to how well it fits with the other characters in the Braille word. Similarly to the well known techniques used in OCR, different possibilities can be suggested with the use of a dictionary and so on. There is a problem here, however. The characters in a Braille word must be interpreted first in order to assess the validity of this word. This not an easy task, considering the variety of interpretations that a Braille character may assume under different circumstances. The type of the information (text, mathematics, music etc.) contained in the document must be known first. If textual information is contained, the grade must also be known. Finally, the country of origin must also be known, so that, apart from the language, the writing conventions are also taken into account. As it was mentioned before, even for the same language there may be different contractions etc. used in different countries.

However, it could still be difficult to rank alternative Braille characters. For instance, if the document contains grade 2 Braille, an isolated character will correspond to a short printed word, e.g. *for*, *of* or *with*. In this case, comprehension of the whole sentence is required.

Particular attention must be paid to situations where more than one Braille character corresponds to a single printed one (see Sec. 3.2). Example cases are the Braille representations of numbers. As it has already been mentioned, in textual Braille documents a digit is represented by the numeral mode character followed by the letter assigned to represent this particular digit. For example if the Braille character corresponding to the letter "a" follows the numeral mode character, it is interpreted as "1". Numbers are represented by the numeral mode character followed by the letters corresponding to each of the digits. Hence, if the string "cab" follows that mode character it is interpreted as "123". Failing to recognise the numeral mode character will lead to a wrong result which is, nevertheless, valid in its own right. Another example is the Braille character representing the letter "D". In grade 2 Braille, if it is on its own it is interpreted as the word "do" but if it is preceded by a Braille character having only dot

5 it is interpreted as "day". Again, failure to recognise the special character leads to a different—but valid—meaning or, if instead of the special character another one is recognised, to a different combination of characters.

Finally, it should be kept in mind that in many cases, especially in grade 2 Braille, the interpretation of each of the characters in a word depends on its position in the word. For example, in grade 2 Braille, a character is interpreted differently if it is in the beginning, at the end or anywhere else in a word. Therefore, any failure to correctly recognise other characters that effectively changes its position will also change its meaning. In some cases the wider context beyond a word will also need to be analysed. If, for instance, the Braille character representing the closing parenthesis is wrongly recognised as that representing a question mark (only one dot difference), it may even pass unnoticed. To verify its correctness, the whole paragraph may need to be examined, to check for an opening parenthesis for example.

5.7. *Further Processing*

Having obtained the codes of the Braille characters on the document, a number of possibilities exist for their use. First of all, the Braille document can be reformatted, edited and distributed by electronic means. Copies can then be sent for reproduction to a Braille printer. Alternatively, the Braille characters may be translated to ASCII in order to be read by sighted people. A further possibility is the translation from one grade of Braille to another and from the conventions system of one country to that of another.

In addition, having converted the information of Braille documents into electronic form, the benefits of the usual applications available for the electronic form of printed documents can be enjoyed.

6. Factors Affecting Automatic Reading

In the previous subsections each of the stages involved in the automatic reading of Braille documents was examined. In this section, the main factors that affect the success of the whole process are outlined.

Different approaches may be susceptible to different problems and to various degrees due to their design. However, there are some factors that affect all methods universally. The main of these are mentioned next, followed by a consideration of the reaction of the different methods to them.

The most important problem is the presence of light or dark regions which have not been created by a protrusion or a depression. These noise regions may affect the result of recognition if they fulfil the structural and spatial conditions set for the recognition of Braille dots. Extra dots may be found in a character or a non-existent character may appear instead of a space. Attempting to identify Braille dots by just identifying either dark or light regions is more susceptible to errors. The identification of dark-light region pairs leads to better results as it is more rare for two noise regions to form the same combination as a Braille dot. However, ambiguity may arise if a valid dot region-pair and a noise region are close in such an arrangement that gives rise to two possibilities for dots. In this case the problem will be in deciding whether a protrusion or a depression is present.

Another issue is that of non-detected dots. One reason for this may lie in the image acquisition and pre-processing stages during which a light or a dark area produced by a dot is not represented in the image. In later stages, the dots with missing regions will not be detected and erroneous results will be produced by the recognition stage. A protrusion or a depression may also not be detected because its light or dark region is merged with that of a dot of the other side. If one of the dots is recognised and the merged region is not sufficiently large to suggest the presence of another dot, the remaining region of that other dot may be thought of as noise and be ignored.

Finally, although they may be identified, the dots may not be in the expected positions within a Braille cell. Apart from the image being skewed, there may be defects of the embossing process which compromise accuracy. In the case of skew, it can be detected and corrected in the pre-processing stage (see Sec. 5.2). Alternatively, if the skew angle is relatively small, a flexible method may be able to overcome the shift in the positions of dots. A similarly flexible approach will also give better results in the case where arbitrary shifts of dot positions exist due to the Braille production process. The assumption of a fixed grid is rather limiting in that it prevents dots which are not positioned on the intersection of grid lines from being recognised as part of a Braille character. However, caution is also needed in the flexible approach in order to avoid recognising dots resulting from noise.

Perhaps, more than one method should be combined in an approach in order to draw from the areas of strength of each individual one and to complement each other in a way that the overall performance will be less affected by the problems mentioned above.

7. Concluding Remarks

The problem of automatic reading of Braille documents is an interesting one in that, despite its similarities to printed document analysis, it involves some challenges of a different nature. The aim of this chapter was to introduce the aspects of the problem and examine solutions to each of them. The essence and production of Braille and Braille documents were described and the characteristics of the problem were discussed. Following that, each of the stages towards the automatic reading of Braille documents was analysed and the approaches that have been so far designed to carry it out were presented.

Having therefore studied the whole problem, one can arrive at some conclusions. First of all, the quality of the image is important. Adequate information about the protrusions and, possibly, the depressions must be present. On the other hand, the quality of Braille documents should not be expected to be necessarily very high. Older documents may have dirt marks and be in a state of deterioration. Furthermore, defects of the production process are responsible for minor shifts of the positions of dots from the expected locations of points in a Braille cell. This calls for a flexibility in the recognition approach. Assuming that all dots lie in the intersections of fixed grid lines may lead to the omission of some protrusions or depressions. However, caution is needed in allowing an approach to be very flexible as dot regions resulting from noise should not be considered as part of characters. Finally, inter-point Braille documents should be recognised as well as single-sided ones.

In terms of the performance of the various approaches, good results have been reported for the type of images tested by each one of them. More specifically, for parts of transparent single-sided documents, Nitta *et al.* [1] report a maximum correct recognition rate of 80%. This figure, however, does not take into account the Braille characters at the edges of the image which cannot be recognised due to camera lens aberrations. Furthermore, Braille characters in skewed images cannot be identified. Dubus *et al.* [10] do not refer to any results but in a later paper Benjelloun *et al.* [2] quote a success rate of 99% for single-sided paper sheets of Braille. François and Calders [9] also report a high average recognition rate of 99% for single-sided opaque Braille documents. The approach of Hentzschel [11], applied to parts only of single-sided paper documents, achieves a maximum recognition rate of 92%.

For test samples including both single-sided and double-sided paper documents, Mennens *et al.* [12] achieve a range of results depending on the presence and nature of defects on the document. If there are no defects, they report that no errors are encountered, while some errors (0.25%) arise from marks on the surface of the paper. However, if there are shifts in the positions of dots, they mention that the results are unacceptable. This is because their approach assumes the existence of a fixed grid on which the Braille dots can be found. Finally, the more flexible approach of Ritchings *et al.* [13] does not encounter such errors. The approach was applied to low quality images of double-sided documents only. Despite the low resolution (100 dpi compared to that of 200 dpi in [12]) and small range of grey levels (16 compared to that of 256 levels in [12]) the average success rate for recognising Braille dots was 98.5% for the protrusions and 97.6% for the depressions of the same image. The overall rate for simultaneous recognition of characters on both sides of a Braille document was reported as 90.8%. The authors explain that the expected success rate for higher resolution 256 grey-level images will be correspondingly higher.

A comparison of all methods on the same image data is however needed in order to obtain an objective measure of recognition performance and to progress forward. Apart from a high recognition performance, the time taken for the completion of the reading process is of significant importance. Repeated image accesses and the need for time-consuming operations can render an approach less practical than others. For example, dependence on skew correction, even for very small angles, is not desirable. The use of computationally intensive image transforms should also be kept to a minimum. Finally, it is advantageous to obtain a representation of the dot regions, in terms of their co-ordinates and size, and use this through the various stages instead of having to search and index through the volume of the image data.

References

[1] S. Nitta, J. Oshima and M. Okhawa, Recognition algorithm on transparent Braille, *Proc. 28th SICE Annual Conference*, Matsuyama, Japan, 1989, 1017–1020.

[2] M. Benjelloun, V. Devlaminck, F. Wauquier, P. Altmayer and J. P. Dubus, Application du traitement d'images à la conception d'un système de transmission de relief Braille en texte noir, *Traitement du Signal* **6** (1989) 291–300.

[3] T. Otake, T. Saito and Y. Yonezawa, Braille printer using hot-melt material, *J. Microcomputer Applications* **13** (1990) 123–131.

[4] F. Germagnoli and G. Magenes, Computerised system for helping blind people to learn Braille code, *Proc. 15th Annual Int. Conf. on Engineering in Medicine and Biology*, 1993, 1318–1319.

[5] O. Sueda, Braille translation by microcomputer and a paperless Braille dictionary, in *Computerised Braille Production Today and Tomorrow*, eds. D. W. Croisdale, H. Kamp and H. Werner (Springer-Verlag, 1979) 272–289.

[6] Royal National Institute for the Blind, *Braille Primer* (RNIB, 1992).

[7] W. Slaby, Computerised Braille translation, *Computers for Handicapped Persons, Proc. Conf. on Computers for Handicapped Persons*, eds. A. M. Tjoa, H. Reiterer and R. Wagner (R. Oldenbourg, Vienna, Austria, 1989) 26–34.

[8] P. Blenkhorn and T. Maley, Computer Braille code, *Computer Assisted Learning for the Visually Handicapped*, Information Sheet 4 (Research Centre for the Visually Handicapped, Faculty of Education, University of Birmingham, 1985).

[9] G. François and P. Calders, The reproduction of Braille originals by means of optical pattern recognition, *Proc. 5th Int. Workshop on Computer Braille Production*, Heverlee, 1985, 119–122.

[10] J. P. Dubus, M. Benjelloun, V. Devlaminck, F. Wauquier and P. Altmayer, Image processing techniques to perform an autonomous system to translate relief Braille into black-ink, called: lectobraille, *Proc. IEEE Engineering in Medicine and Biology Society 10th Annual Int. Conf.*, New Orleans, 1988, 1584–1585.

[11] T. W. Hentzschel, An optical Braille reading system, M.Sc. Dissertation, Faculty of Technology, University of Manchester, 1992.

[12] J. Mennens, L. Van Tichelen, G. François and J. Engelen, Optical recognition of Braille writing, *Proc. 2nd Int. Conf. on Document Analysis and Recognition*, Tsukuba, Japan, Oct. 1993, 428–431.

[13] R. T. Ritchings, A. Antonacopoulos and D. Drakopoulos, Analysis of scanned Braille documents, *Proc. Int. Association for Pattern Recognition Workshop on Document Analysis Systems*, Kaiserslautern, Germany, Oct. 1994, 417–424.

[14] D. Drakopoulos, Analysis of scanned Braille document, M.Sc. Dissertation, Faculty of Technology, University of Manchester, 1993.